ESE 2024

UPSC ENGINEERING SERVICES EXAMINATION

Main Examination

Electrical Engineering

Topicwise *Conventional* Solved Papers

Paper-I

23 Years SOLVED PAPERS

- ✓ Thoroughly revised and updated
- ✓ Fully solved with explanations
- ✓ Also useful for State Engineering Services Exams

MADE EASY PUBLICATIONS

www.madeeasypublications.org

MADE EASY Publications Pvt. Ltd.

Corporate Office: 44-A/4, Kalu Sarai (Near Hauz Khas Metro Station), New Delhi-110016
E-mail: infomep@madeeasy.in
Contact: 9021300500

Visit us at: www.madeeasypublications.org

ESE-2024 : Main Examination
Electrical Engineering : Paper-I | Conventional Solved Questions : (2001-2023)

1st Edition : 2008
2nd Edition : 2009
3rd Edition : 2010
4th Edition : 2011
5th Edition : 2012
6th Edition : 2013
7th Edition : 2014
8th Edition : 2015
9th Edition : 2016
10th Edition : 2017
11th Edition : 2018
12th Edition: 2019
13th Edition: 2020
14th Edition (Feb) : 2022
15th Edition (July) : 2022

16th Edition: 2023

B. Singh (Ex. IES)

Director's Message

In past few years ESE Main exam has evolved as an examination designed to evaluate a candidate's subject knowledge. Studying engineering is one aspect but studying to crack prestigious ESE exam requires altogether different strategy, crystal clear concepts and rigorous practice of previous years' questions. ESE mains being conventional exam has subjective nature of questions, where an aspirant has to write elaborately - step by step with proper and well labeled diagrams and figures. This characteristic of the main exam gave me the aim and purpose to write this book. This book is an effort to cater all the difficulties being faced by students during their preparation right from conceptual clarity to answer writing approach.

MADE EASY Team has put sincere efforts in solving and preparing accurate and detailed explanation for all the previous years' questions in a coherent manner. Due emphasis is made to illustrate the ideal method and procedure of writing subjective answers. All the previous years' questions are segregated subject wise and further they have been categorised topic-wise for easy learning and helping aspirants to solve all previous years' questions of particular area at one place. This feature of the book will also help aspirants to develop understanding of important and frequently asked areas in the exam.

I would like to acknowledge the efforts of entire MADE EASY team who worked hard to solve previous years' questions with accuracy. I hope this book will stand upto the expectations of aspirants and my desire to serve the student community by providing best study material will get accomplished.

B. Singh (Ex. IES)
CMD, MADE EASY Group

ESE 2024 Main Examination

Electrical Engineering *Conventional* Solved Questions

Paper-I

Contents

■■■■

1 Electrical Materials

Revised Syllabus of ESE: *Electrical engineering materials, crystal structures and defects, ceramic materials, insulating materials, magnetic materials-basics, properties and applications, ferrities, ferro-magentic materials and components, basics of solid state physics, conductors, photo-conductivity, basics of nano materials and superconductors.*

1. Introduction to Engineering Materials

1.1 **What is the basis for classifying a material as a conductor, semi-conductor or a dielectric? What is the conductivity of a perfect dielectric?**

[12 marks : 2002]

Solution:

Classification of insulators, metals and semiconductors are mainly by its conductivity and its temperature coefficient of resistivity.

Conductor:

In which conduction band and valence band are overlapping and there is no band gap, between them ($\Delta E_g = 0$). Due to this large number of electrons available for electrical conduction and therefore its resistivity is low and temperature coefficient of resistivity is positive ($\rho = 10^{-2}\ \Omega m$ to $10^{-8}\ \Omega m$).

Insulator:

In which energy gap is large ($E_g > 3eV$). So in conduction band there are no electrons and so no electrical conduction is possible. Here energy gap is so large that electrons cannot be easily excited from the valence band to conduction band by any external energy.

$$\rho = 10^8\ \Omega\text{-m}$$

Temperature coefficient of resistivity is negative

Semiconductor:

In which a finite but small band gap ($E_g < 3eV$) exists. Due to this small band gap some electrons can be thermally excited to "conduction band". These thermally excited electrons can move in conduction band and can conduct current, their resistivity is in medium range. Its $\rho = 10^5 - 10^{10}\ \Omega m$. temperature coefficient of resistance is negative. Conductivity of a perfect dielectric is zero.

1.2 **Write the notable shortcoming of the free electron model. How it is overcome by band-theory of solids? Sketch the energy bands of diamond. With the help of energy band diagram, explain an insulator.**

[12 marks : 2003]

Solution:

In free electron theory developed by Arnold Sommerfeld we assume that the electron to be completely free. But it fails mainly because most electrons are not completely free. They interact with

(i) lattice (ii) with each other (iii) impurities

Drude incorporated into his model in the form of a hard sphere interaction but completely neglected the other interactions Sommerfeld made some corrections to this purely classical model by ensuring that the electron gas obey fermi statistics. This faced some problems (like the overestimate of the electronic heat capacity but still didn't cover for the missing interaction in the model. Due to its limitations, the free electron model was unable to account for things like magnetoresistance, thermal dependence of conductivity, optical properties and the host of many body phenomenon that result from electron-electron interaction being of importance.

In band theory of e^- s, A solid is assumed to contain many bands in which electrons in it are packed. The most important are valence band and conduction band. The energy of electrons in these bands will be different. The difference in valence band and conduction determines whether solid is a conductor, semi-conductor, or insulator.

So it clearly explains the dependency of conductivity on temperature which can't be explained by free e^- theory. When temperature increases, the energy of e^- s increases, so e^- s started to jump from valence band to conduction so conduction of e^{-s} changes with temperature in CB so as conductivity changes with temperature.

It also explains many other properties like optical, magneto resistance etc. easily so shortcomings of free e^- theory is overcome by band theory in which we assume that there is an interaction between e^- s only when energy state of different e^- s is different. The interaction of e^- with lattice and impurities are taken into account.

Diamond is the crystalline form of carbon and is of interest because it lies in the same period and has the same type of bonding as silicon and germanium. The shape of the energy-band structure of diamond as a function of atomic spacing is shown in Fig. (a), it is shown that there is a crossover point at which two of the 2 P states form a band with the two 2S state, thus making available a band with four available states. The remaining four of the total of six 2 P states off and form a higher band of their own. At normal atomic spacing, the forbidden gap between the higher and lower bands is wide. Since there is a total of four electrons in the 2 S and 2 P states of carbon, all these electrons are used up in filling the lower band. The upper band is left empty, and hence diamond is a very good insulator. Fig. (b) shows the stable energy level of diamond.

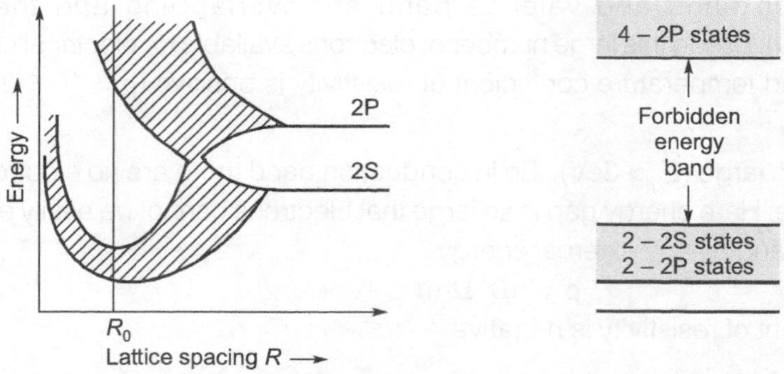

(a) The energy levels in the tetrahedral of diamond as a function of lattice spacing.

(b) The stable energy level of the diamond crystal.

1.3 **Distinguish between metals, insulators and semiconductors.** **[12 marks : 2007]**

Solution:

According to Energy Band Theory:

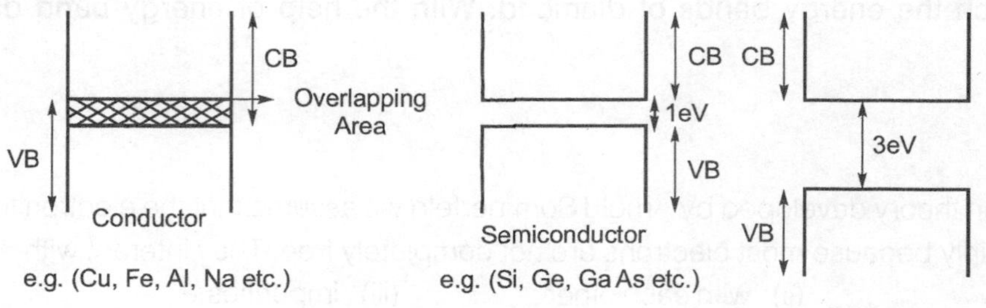

Classification of insulators, metals and semiconductors mainly by its conductivity and its temperature coefficient of resistivity.

Conductor:

In which conduction band and valence band are overlapping and there is no band gap, between them it $\Delta E_g = 0$. Due to this large number of electrons available for electrical conduction and therefore its resistivity is low and temperature coefficient of resistivity is positive (ρ = 10^{-2} to 10^{-8} Ω-m).

Insulator:

In which energy gap is large ($E_g > 3$ eV). So in conduction band there are no electrons and so no electrical conduction is possible. Here energy gap is so large that electrons cannot be easily excited from the valence band to conduction band by any external energy.

$$\rho = 10^8\ \Omega\text{-m}$$

Temperature coefficient of resistivity is negative.

Semi-conductor:

In which a finite but small band gap ($E_g < 3$eV) exists. Due to this small band gap some electrons can be thermally excited to "conduction band". These thermally excited electrons can move in conduction band and can conduct current, their resistivity is in medium range. Its ρ = (10^5 to 10^{10}) Ω-m temperature coefficient of resistance is negative. Conductivity of a perfect dielectric is zero.

1.4 Explain with the diagram band structure of metals, insulators and semiconductors.

[10 marks : 2013]

Solution:

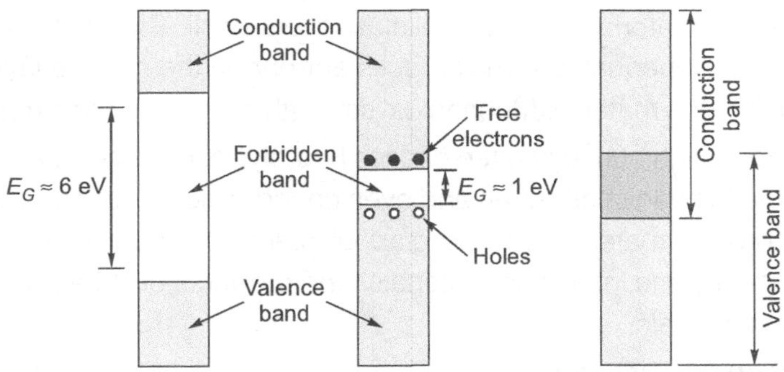

Fig. (a) **Fig. (b)** **Fig. (c)**

Insulators:

- An insulating material has an energy band diagram as shown in figure (a).
- It has a very wide forbidden-energy gap ($\simeq$ 6 eV) separating the filled valence region from the vacant conduction band. Because of this, it is practically impossible for an electron in the valence band to jump the gap, reach the conduction band.
- At room temperature, an insulator does not conduct. However it may conduct if its temperature is very high or if a high voltage is applied across it. This is termed as the **breakdown of the insulator.**
- **Example:** diamond.

Semiconductors:

- A semiconductor has an energy-band gap as shown in figure (b).
- At 0°K semiconductor materials have the same structure as insulators except the difference in the size of the band gap E_G, which is much smaller in semiconductors ($E_G \simeq 1$ eV) than in insulators.
- The relatively small band gaps of semiconductors allow for excitation of electrons from the lower (valence) band to the upper (conduction) band by reasonable amount of thermal or optical energy.
- The difference between semiconductors and insulators is that the conductivity of semiconductors can increase greatly by thermal or optical energy.
- **Example:** Ge and Si.

Metals:

- There is no forbidden energy gap between the valence and conduction bands. The two bands actually overlap as shown in figure (c).
- Without supplying any additional energy such as heat or light, a metal already contains a large number of free electrons and that is why it works as a good conductor.
- **Example:** Al, Cu etc.

1.5 State three effects of moisture on insulation and explain any two methods adopted to protect them against moisture along with salient features.

[12 marks : 2017]

Solution:

The effect of moisture on insulation:

(i) **Change in electrical properties:** Moisture absorbed by the insulation causes a decrease in the volume resistivity, and especially surface resistivity, an increase in the dissipation factor and a certain increase in dielectric constant, a reduction in the dielectric strength due to change in field distribution within the insulating material. Under high humidity and electric tension current, conducting bridge may appear across the surface of the insulating material. In some cases the thin films formed by the moisture on insulating material dries up when the equipment is working. Such places get a carbonized spot. Such carbonized spot may join together with time and build up a conducting bridge, thus a short-circuit may result.

(ii) **Physical and mechanical changes:** Due to high humidity for long time some insulating materials like plastics polymers and material filled with cellulose fillers swell. The swelling of inner and outer layers of insulating material are unequal, so small cracks appear on the surface layer, through which moisture easily penetrates into the material. Mechanical strength of the insulating material looses with moisture.

(iii) **Chemical changes:** High humidity often causes hydrolysis. For example under the effect of moisture, linseed base insulating vanishes soften and even change into liquid. This evolves organic acids which deteriorate other organic materials and bring about intensive corrosion of metals. High humidity favours the growth of fungi in some insulating materials. The products of life activity formed by fungi degrade organic insulating materials.

Protection of insulation against moisture:

(i) **Impregnation of winding:** The windings of all low voltages pieces of equipment are impregnated with baking varnishes and some time with compounds. This treatment solidifies the windings, increases their thermal conductivity, improves their electrical and mechanical strength and heat resistance. Impregnating varnishes and compounds raise the moisture resistance of windings.

(ii) **Rendering insulation hydrophobic:** Sometimes the insulating materials assemblies are rendered hydrophobic (or waterproofed) so as to protect them against moisture. This treatment is particularly effective for polymer containing hydrolysis and for cellulose-base insulating materials. It is a good practice to render some inorganic materials hydrophobic so as to decrease their wettability and to raise the surface resistivity of these materials. In addition to old widely used techniques employing asphalts, bitumens, waxes, waterproofing by means of some hydrophobic silicon compositions free of hydroxyls and carboxyls is finding ever increasing favour. Paper, cotton fabrics are rendered hydrophobic by dipping them in the solution of methyl butoxidiamine silane in carbon terrachloride or methyl triethoxisilane in absolute alcohol.

1.6 Show that the surface area to volume ratio increases with the reduction in radius of nanoparticle. Plot a graph of surface to volume ratio as a function of radius.

[8 marks : 2017]

Solution:

Surface to volume ratio of nanoparticle relation:

Surface to volume ratio in nanoparticles have a significant effect on the nanoparticles properties firstly, nanoparticles have a relative large surface area when compared to the same volume of the material.

Let us consider a sphere of radius '*r*'.

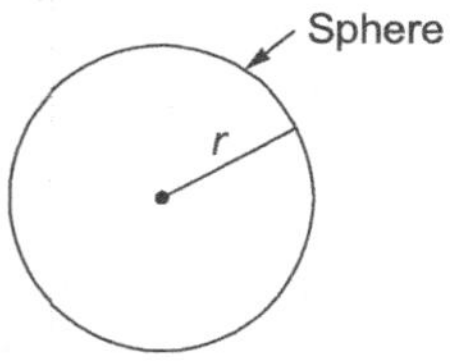

The surface area of the sphere, $S = 4\pi r^2$

and the volume of sphere, $V = \frac{4}{3}\pi r^3$

Therefore the surface area to volume ratio,

$$\frac{S}{V} = \frac{4\pi r^2}{\left(\frac{4}{3}\pi r^3\right)} = \frac{3}{r} \Rightarrow \frac{S}{V} \propto \frac{1}{r}$$

It means that the surface area to volume ratio increases with the decreases in radius of the sphere and vice-versa. It can also be concluded that when given volume is divided into smaller piece, the surface area increases. Therefore as particle size decreases, a greater portion of the atoms are found at the surface compared to those inside.

For example, a particle of size 3 nm has 50% of its atom on its surface, at 10 nm 20% of its atoms and at 30 nm has 5% of its atoms on its surface.

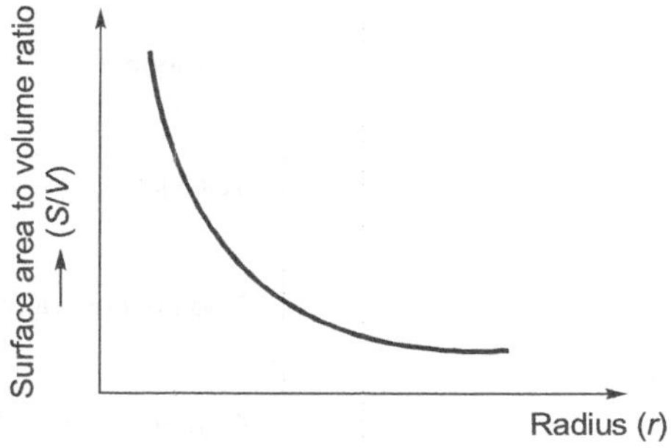

Therefore nanoparticles have a much greater surface area per unit volume compared with the large particles. It leads to nanoparticles more chemically reactive. As growth and catalytic chemical reaction occur at surfaces, therefore a given mass of nanomaterial will be much more reactive than the same mass of nano material made up of large particles. It is also found that material which are insert in their bulk form are reactive when produced in their nanoscale form. It can improve their properties. Surface area to volume ratio vs radius plots.

$$\frac{S}{V} = \frac{3}{r}; \qquad \frac{S}{V} \propto \frac{1}{r}$$

1.7 **Determine the diameter of the lead wire to pass maximum super current of 0.0021 mA at 5.0 K. The critical temperature and critical magnetic field for lead are respectively 7.18 K and 6.5 × 10^{-4} A/m.**

[8 marks : 2017]

Solution:

Given that:

Maximum supercurrent, I_m = 0.0021 mA at 5.0 K

The critical temperature = 7.18 K

The critical magnetic field = 6.5 × 10^{-4} A/m

We know that,
$$H_C = H_0\left[1-\left(\frac{T}{T_C}\right)^2\right]$$

$$H_{C(5K)} = H_{0(7.18\,K)}\left[1-\left(\frac{5}{7.18}\right)^2\right]$$

$$= 6.5 \times 10^{-4}\,[1 - 0.48494] = 6.5 \times 10^{-4}\,(0.51506)$$

$$= 3.3479 \times 10^{-4} \text{ A/m}$$

Now, maximum current, $I_m = 2\pi r \cdot H_C$

$$0.0021 \times 10^{-3} = 2\pi r \times 3.3479 \times 10^{-4}$$

Radius of lead wire, $$r = \frac{0.0021 \times 10^{-3}}{2\pi \times 3.3479 \times 10^{-4}} = 9.9831 \times 10^{-4} \text{ m}$$

$$= 0.99831 \times 10^{-3} \text{ m} = 0.99831 \text{ mm}$$

So, diameter of the lead wire,

$$d = 2r = 2 \times 0.99831 = 1.99662 \text{ mm} \simeq 2.0 \text{ mm}$$

1.8 Mention the type of Bravais space lattice, relationship of crystal axial lengths (*x*, *y*, *z*) and relationship of interaxial angles (a, b, g) in the following order of the crystal system.
(i) Triclinic (ii) Monoclinic
(iii) Orthorhombic and (iv) Trigonal

[12 marks : 2018]

Solution:

Crystal System	Axial Relationships	Interaxial Angles
Cubic	$a = b = c$	$\alpha = \beta = \gamma = 90°$
Hexagonal	$a = b \neq c$	$\alpha = \beta = 90°, \gamma = 120°$
Tetragonal	$a = b \neq c$	$\alpha = \beta = \gamma = 90°$
Rhombohedral(Trigonal)	$a = b = c$	$\alpha = \beta = \gamma \neq 90°$
Orthorhombic	$a \neq b \neq c$	$\alpha = \beta = \gamma = 90°$
Monoclinic	$a \neq b \neq c$	$\alpha = \gamma = 90° \neq \beta$
Triclinic	$a \neq b \neq c$	$\alpha \neq \beta \neq \gamma \neq 90°$

1.9 Compute the packing efficiency and density of diamond. Consider diamond as diamond cubic (DC) structure and lattice parameter, *a* = 3.57 Å and mass of atom *m* = 1.66 × 10^{-27} kg.

[12 marks : 2022]

Solution:

Packing Efficiency of Diamond :

Atoms present at CCP = 4

Atoms present at alternate $T_oV_o = 4$

So, total number of atoms in diamond,

$$Z = 8$$

So, $$\text{pf} = \frac{Z \times \frac{4}{3} \times \pi r^3}{a^3} \quad \text{...(i)}$$

Now to find relation between *a* and *r*.

We know atoms touch corner with TV.

So,
$$\frac{\sqrt{3}a}{4} = 2r$$

$$a = \frac{8r}{\sqrt{3}} \quad ...(ii)$$

Using (ii) in (i), we get,

$$pf = \frac{\frac{4}{3}\times\pi\times 8\times r^3}{\left(\frac{8r}{\sqrt{3}}\right)^3} = 0.34 \text{ or } 34\%$$

To find density :

$$\rho = \frac{Z\times m}{a^3}$$

Using the values given, $m = 1.66 \times 10^{-27}$ kg

$$a = 3.57 \times 10^{-10} \text{ m}$$

We get,
$$\rho = \frac{8\times 1.66\times 10^{-27}}{(3.57\times 10^{-10})^3} = 0.291 \times 10^3$$

$$\rho = 291 \text{ kg/m}^3$$

1.10 Zinc has HCP structure. The height of the unit cell is 0.494 Å. The nearest neighbour distance is 0.27 nm. The atomic weight of zinc is 65.37. Evaluate,

(i) The volume of the unit cell.

(ii) The density of zinc

Take NA = 6.0238 × 10^{26} Molecules/Kg Mole.

[12 marks : 2022]

Solution:

Given, Nearest neighbour distance,

$$a = 2r = 0.27 \text{ nm} = 0.27 \times 10^{-9} \text{ m}$$

Atomic weight of Z_n = 65.37

The height of the unit cell, h = 0.494 Å

(i) Volume of unit cell,

The volume of the HCP unit cell is given by

$$V = \frac{3\sqrt{3}\,a^2h}{2} = \frac{3\sqrt{3}\,a^2h}{2}$$

Substituting the values we get

$$V = \frac{3\sqrt{3}(0.27\times 10^{-9})^2\times 0.494\times 10^{-10}}{2} = 9.35 \times 10^{-30} \text{ m}^3$$

(ii) The density of zinc,

Effective number of atom's in unit cell of HCP,

$$n = 6$$

Using the relation, $\rho = \frac{nm}{N_0 \times V}$ we have,

[Where $n = 6$, $m = 65.37$ (given) $N_A = 6.0238 \times 10^{26}$]

$$\rho = \frac{6\times 65.37}{6.0238\times 10^{26}\times 9.35\times 10^{-30}} \approx 69647 \text{ kg/m}^3$$

1.11 Calculate the value of contact potential and the total width of the depletion region under unbiased condition of an abrupt silicon P-N junction at 300° K. The intrinsic carrier concentration of silicon is 1.7×10^{16} atoms/m^3 and its relative permittivity is 11.8. The junction has acceptor density $N_A = 10^{22}$ atoms/m^3 on *P*-side and donor density of $N_D = 10^{20}$ atoms/m^3 on *n*-side.
Derive the equation used for calculation.
Boltzmann's constant = 1.381×10^{-23} J/°K
Electron charge = 1.602×10^{-19} C

[20 marks : 2022]

Solution:

The contact potential of the P-N junction is

$$V_{contact} = \frac{KT}{q}\ln\left(\frac{N_A N_D}{nq^2}\right)$$

$$V_{contact} = \frac{1.381\times10^{-23}\times300}{1.602\times10^{-19}}\ln\left(\frac{10^{22}\times10^{20}}{(1.7\times10^{16})^2}\right)$$

$$V_{contact} = \frac{1.381\times10^{-23}\times300}{1.602\times10^{-19}}\ln\left(\frac{10^{42}}{(1.7)^2 10^{32}}\right)$$

$$= 258.6\times10^{-4}\ln\left(\frac{10^{10}}{(1.7)^2}\right)$$

$$= 21.96\times258.6\times10^{-4} = 0.568 \text{ V}$$

Width of deplection region, $$W = \sqrt{\frac{2\,\epsilon_r\epsilon_0\,V_{contact}}{q}\left(\frac{1}{N_A}+\frac{1}{N_D}\right)}$$

$$= \sqrt{\frac{2\times11\times8.85\times10^{-12}\times0.568}{1.6\times10^{-19}}\left(\frac{1}{10^{22}}+\frac{1}{10^{20}}\right)}$$

$$= \sqrt{69.1185\times10^{7}\left(\frac{1.01}{10^{20}}\right)} = \sqrt{69.8096\times10^{-13}}$$

$$W = 2.642\times10^{-6}\text{ m}$$

1.12 Evaluate the linear density in atoms per mm in the following directions in BCC iron, which has lattice constant of 2.89 Å :
(i) [1 0 0]; (ii) [1 1 0]; (iii) [1 1 1]

[12 marks : 2023]

Solution:

(i) BCC Unit Cell :

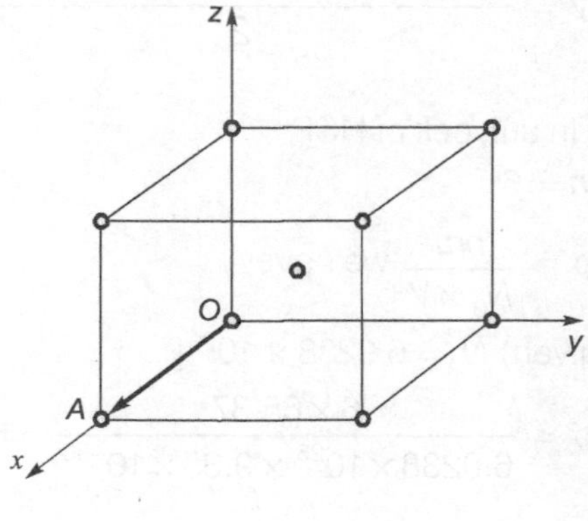

Vector OA is along [1 0 0]

$$\text{Length of vector } OA = \text{Edge length of cube} = a = 2.89 \text{ Å}$$

$$\text{No. of atoms centred on } OA = \frac{1}{2}+\frac{1}{2} = 1 \text{ atm}$$

So, linear density along OA

i.e.,
$$LD_{[100]} = \frac{1\text{ atom}}{\text{Length of } OA} = \frac{1\text{ atom}}{2.89\times10^{-7}\text{ mm}}$$

$$\Rightarrow \quad LD_{[100]} = 3.46 \times 10^6 \text{ atoms/mm}$$

(ii) Vector OB is along [1 1 0]

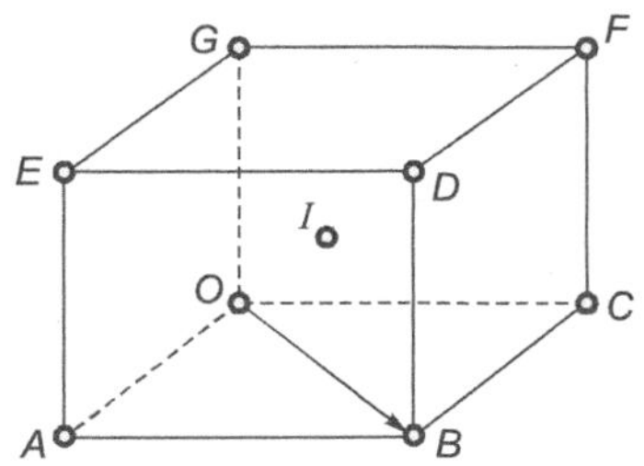

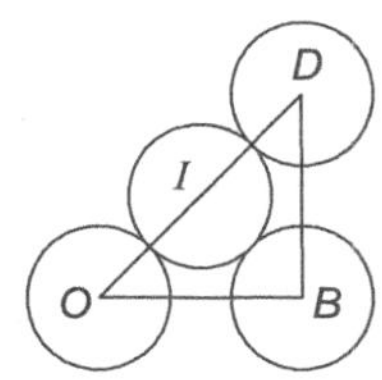

$$OD = (R + 2R + R) = 4R$$
$$BD = a$$

For BCC,
$$a = \frac{4R}{\sqrt{3}}$$

$$\therefore \quad OB = \sqrt{OD^2 - BD^2} = \sqrt{16R^2 - \frac{16R^2}{3}} = 4R\sqrt{\frac{2}{3}}$$

$$LD_{[110]} = \frac{\text{No. of atoms centred on [110] direction}}{\text{Length of [110] direction vector}}$$

$$= \frac{\left(\frac{1}{2}+\frac{1}{2}\right)\text{atom}}{OB} = \frac{1\text{ atom}}{4R\times\sqrt{\frac{2}{3}}}$$

But
$$R = \frac{\sqrt{3}a}{4}$$

$$\therefore \quad LD_{[110]} = \frac{1\text{ atom}}{4\times\frac{\sqrt{3}a}{4}\times\frac{\sqrt{2}}{\sqrt{3}}} = \frac{1\text{ atom}}{\sqrt{2}a}$$

$$\Rightarrow \quad LD_{[110]} = \frac{1\text{ atom}}{\sqrt{2}\times2.89\times10^{-7}\text{mm}} = 2.446 \times 10^6 \text{ atoms/mm}$$

$$\Rightarrow \quad LD_{[110]} = 2.446 \times 10^6 \text{ atoms/mm}$$

(iii) Vector OD is [111] along.

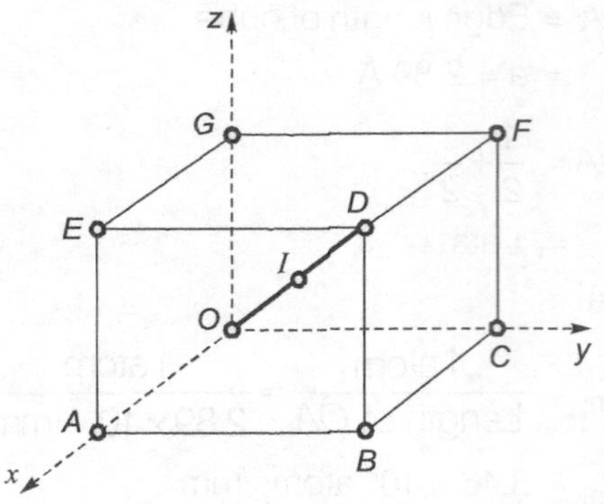

$$OD = 4R$$

$$\Rightarrow \quad OD = 4\times\frac{\sqrt{3}a}{4}$$

$$OD = \sqrt{3}a$$

Number of atoms entered on [1 1 1]

$$= \frac{1}{2}+1+\frac{1}{2} = 2$$

$$\therefore \quad LD_{[111]} = \frac{2 \text{ atoms}}{\sqrt{3}a} = \frac{2 \text{ atoms}}{\sqrt{3}\times 2.89\times 10^{-7}\text{mm}}$$

$$= 3.99 \times 10^{6} \text{ atoms/mm}$$

$$\Rightarrow \quad LD_{[111]} \simeq 4 \times 10^{6} \text{ atoms/mm}$$

1.13 **(i) Determine the volume of an HCP unit cell in terms of its *a* and *c* lattice parameters.**

(ii) Copper has an atomic radius of 0.13 nm, an FCC crystal structure and an atomic weight of 63.5 g/mol. Evaluate its theoretical density and compare the answer with its measured density. (Take Avogadro number, $N_A = 6.022 \times 10^{23}$ atoms/mol).

[10+10 = 20 marks : 2023]

Solution:

(i) HCP unit cell is shown in adjacent figure :

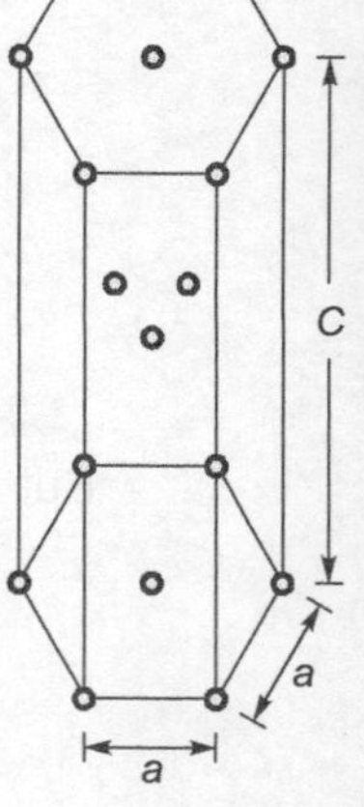

$$\frac{c}{a} = 1.633 \text{ for HCP}$$

Volume (V_c) of HCP unit cell = Base area × Cell height (c)

⇒ To calculate base area consider following figure :

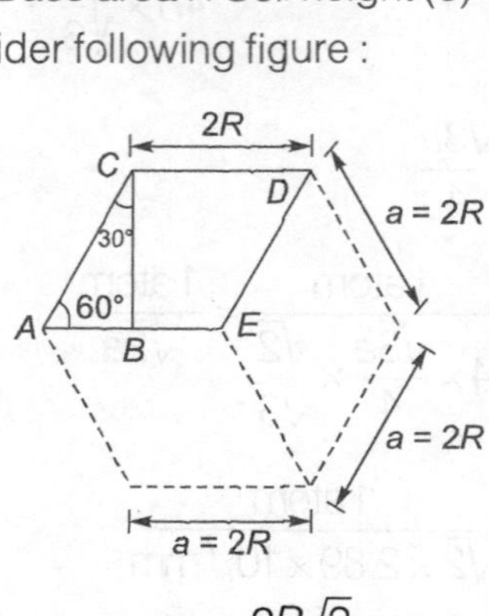

$$BC = 2R\cos 30° = \frac{2R\sqrt{3}}{2}$$

$$\Rightarrow \qquad BC = \sqrt{3}R$$

$$a = 2R \Rightarrow R = \frac{a}{2}$$

$$\text{The area of } ACDE = CD \times BC = 2R \times \sqrt{3}R$$

The base area is just three times the area of the parallelpied *ACDE* shown above.

$$\therefore \qquad \text{Area of base} = 3 \times \text{Area of } ACDE = 3 \times 2R \times \sqrt{3}R$$

$$= 6\sqrt{3}R^2 = 6\sqrt{3} \times \left(\frac{a}{2}\right)^2 = \frac{6\sqrt{3}}{4}a^2$$

Hence, volume of unit cell

$$V_c = \text{Area} \times c = \frac{6\sqrt{3}}{4}a^2 \times c$$

$$V_c = 2.598a^2c$$

(ii) Theoretical density,

$$\rho = \frac{nA_{Cu}}{V_c N_A}$$

where, n = Number of atoms per unit cell

A_{Cu} = Atomic weight of Cu

V_c = Volume of unit cell

$$V_c = a^3$$

$$= \left(2\sqrt{2}R\right)^3$$

$$\Rightarrow \qquad V_c = 16\sqrt{2}R^3$$

$$\therefore \qquad \rho = \frac{4 \times 63.5}{16\sqrt{2}(0.13 \times 10^{-7})^3 \times 6.023 \times 10^{23}} \text{ g/cm}^3$$

$$\rho = 8.48 \text{ g/cm}^3 \text{ (Theoretical density of Cu)}$$

The literature value (measured value) for the density of Cu is 8.94 g/cm^3, which is in very close agreement with above calculated value.

1.14 Define line defects in materials. Explain different types of line defects and compare them. Also, explain their cause of creation.

[12 marks : 2023]

Solution:

Line defect is the defect confined to more number of atoms in a lattice.

Linear defect: A dislocation is a linear or one dimensional defect around which some of the atoms are misaligned. There are following types of linear dislocation as given below:

1. **Edge dislocation:** In an edge dislocation, an extra portion of a plane of atoms, or half-plane appears and the edge of which terminates within the crystal. It is a linear defect that centers around the line that is defined along the end of the extra half plane of atoms. This is sometimes termed as dislocation line, which for the edge dislocation as shown in figure, is perpendicular to the plane of the page. Within the region around the dislocation line there is some localized lattice distortion. The atoms above the dislocation line are squeezed together and those below are pulled apart. This is reflected in the slight curvature for the vertical planes of atoms as they bend around this extra half plane. The magnitude of this dislocation decreases with distance away from the dislocation line. An edge dislocation may also be formed by an extra half plane of atoms that is included in the bottom portion of the crystal.

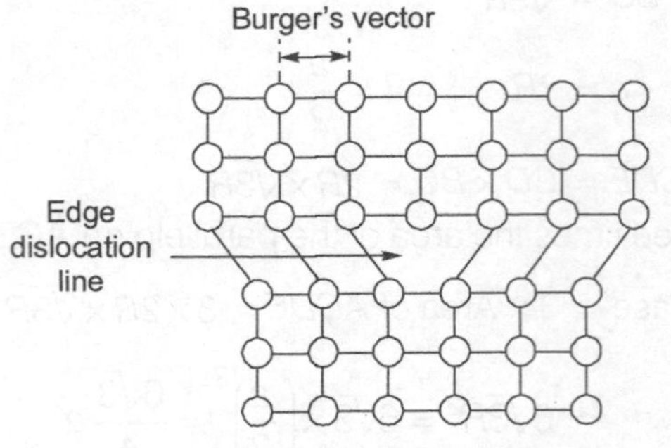

Edge dislocation

2. **Screw dislocation:** Screw dislocation is formed by shear stress that is applied to produce the distortion. The upper front region of the crystal is shifted one atomic distance to the right relative to the bottom portion. The atomic distortion associated with a screw dislocation is also linear and along a dislocation line. The screw dislocation derivers its name from the spiral or helical path or ramp that is traced around the dislocation line by the atomic planes of atoms.

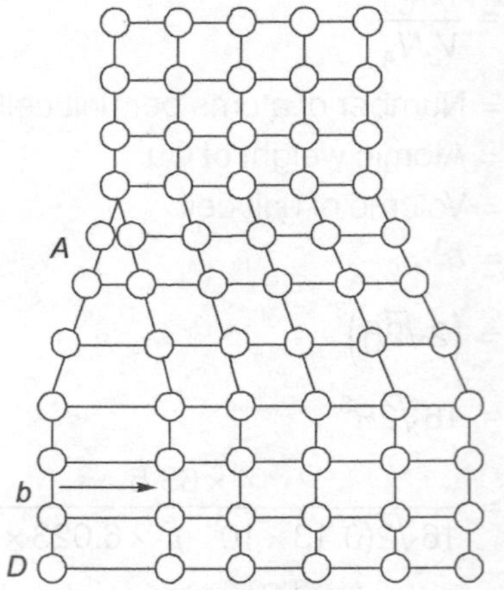

Screw dislocation

3. **Mixed dislocation:** Most dislocations found in crystalline materials are probably neither pure edge nor pure screw, but exhibit components of both types, these are termed as mixed dislocations.
   The nature of a dislocation is defined by the relative orientations of dislocation line and Burger's vector. For an edge, they are perpendicular, whereas for a screw, they are parallel. They are neither parallel nor perpendicular for a mixed dislocation. Also, even though a dislocation changes direction and nature within a crystal (e.g. from edge to mixed to screw), the Burger's vector will be the same at all points along its line.

2. Dielectric Properties of Materials

2.1 What are ceramic materials? Differentiate between two types of ceramics having their permittivity less than 12 and greater than 12. Name two ceramic materials in each of above two types and mention one application of each type.

[12 marks : 2001]

Solution:

Ceramics is a compound formed by the combination of a metallic and non-metallic elements. Unlike metals, they do not have large number of free electrons, the electrons being shared covalently as in ionic bonds. Crystal structure of ceramics contain atoms of different sizes.

Due to the presence of ionic and covalent bond, ceramics possess high hardness, brittleness, high melting, point, chemical inertness and electrical insulation.

The main characteristics are low water absorption, low thermal expansion and good electrical properties.

Ceramics material may be divided in to two groups, depending on whether the permittivity is less or greater than 12. Material with high permittivity below 12 are more advantageously used in bulk as insulators, bushing, housing.

1. **Example: Porcelain, Alumina etc. ($\in_r < 12$)**
   Porcelain is used as insulators in transmission and distribution, fuse links, plugs and sockets.
2. Material with high permittivity ($\in_r > 12$), is used in capacitor in which permittivity should be as large as possible, for reducing the size of the capacitor barium titanate ($B_aT_iO_3$), which has permittivity of 2000, is used.

2.2 Calculate the polarization induced in Al_2O_3 with dielectric constant $\in_r = 8$, when it is placed between two plates of a parallel plate capacitor separated by 1 mm. Potential difference between the two plates is 10 V.

[12 marks : 2002]

Solution:

$d = 1 \text{ mm} = 10^{-3}$ m, $V = 10$ V, $\in_r = 8$

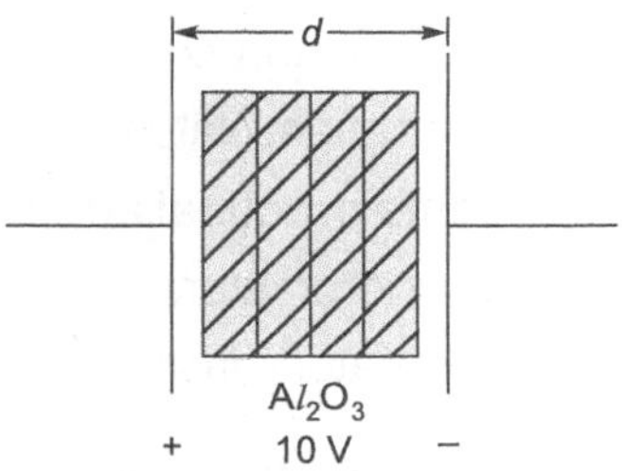

$$\text{Electric field} = \frac{\text{potential difference}}{\text{distance}} = V/d$$

$$= \frac{10}{1\text{mm}} = 10 \times 10^3 \text{ V/m}$$

$$\text{Polarization} = \bar{p} = \in_0 \chi_e E = \in_0(\in_r - 1)E$$

$$\bar{p} = 8.86 \times 10^{-12} \times (8-1) \times 10 \times 10^3$$

$$= 6.202 \times 10^{-7} \text{ c/m}^2$$

2.3 Discuss the trends of insulating materials used in modern electrical and electronic devices.

[12 marks : 2003]

Solution:

Insulating material are invariably used in the electrical and electronic fields. For the effective use of an insulating material the following factors should be considered while selecting an insulating material for a particular application.

(a) Properties of the material
(b) Ease of fabrication
(c) Availability of material
(d) Cost

Selection of an insulating material to a specific application requires thorough knowledge of the following properties.

(i) Electrical
(ii) Mechanical
(iii) Thermal
(iv) Chemical
(v) Visual

(i) Electrical properties:
- (i) Insulation resistance
- (ii) Permittivity
- (iii) Dielectric strength
- (iv) Dielectric loss.

(ii) Mechanical properties:
- (i) Tensile strength
- (ii) Compressive strength
- (iii) Flexural strength

(iii) Thermal properties: (i) Heat resistance
(ii) Frost resistance
(iii) Thermal resistance

(iv) Chemical properties: (i) Chemical resistance
(ii) Hygroscopicity
(iii) Moisture permeability

2.4 What is piezoelectricity? What are the different applications in which piezoelectricity is used? Describe materials that show piezoelectricity.

[12 marks : 2005]

Solution:

In some of the dielectric crystals it is found that mechanical strain may produce an electrostatic charge on the faces of a crystal, this is known as piezoelectricity. Piezo electric effect is the production of electricity by pressure. This property used in transducers. When an electric field is applied to a dielectric material, ionic cores and electron distribution are destroyed. This distribution appears as a strain and is proportional to square of electric field. This effect is small and is called electrostriction. However inverse effect does not exist on other hand, when a stress is applied to a crystal, which does not possess a centre of symmetry, such as quartz, it develops a polarization, which is directly proportional to the magnitude of applied stress.

Examples: Natural quartz, Rochelle salt, lithium sulphate, many group II-IV compound semi conductors etc.

Some practical applications of commercial piezoelectric are in: Gramophone pick-ups, air transducers, accelerometers. Ultrasonic flow detectors, under water sonar transducers, filters, resonators, high voltage generator (gas liters), delay lines etc.

2.5 Explain the basic properties of dielectrics. What do you understand by the loss tangent and susceptibility?

[12 marks : 2005]

Solution:

Dielectric is a non-conducting material which can be polarized (exhibit an electric dipole) by an electric field. The dielectric constant, loss tangent and dielectric strength are the three parameters which characterize dielectric. Heat resistance, is however, such an important property that standards of classification.

A dielectric is chiefly characterized by its dielectric constant $\in_r$, which is also called relative permittivity.

The dielectric constant, $$\in_r = \frac{\in}{\in_0}$$

$$\vec{D} = \in \vec{E}$$

$$\in \vec{E} = \in_0 \vec{E} + \vec{P}$$

$$\vec{P} = \in(\in_r - 1)\,\vec{E}$$

where $(\in_r - 1)$ is called susceptibility.

The absorption of electrical energy by a dielectric material subjected to an alternating electric field is known as dielectric loss. The dielectric loss caused by an ac field also result in dissipation of the electrical energy as heat in the material. An ideal dielectric does not absorb electrical energy. However a real dielectric always causes some loss of electrical energy.

Complex part of dielectric

$$\in_r^* = \in'_r - j\in''_r$$

$$\vec{D} = \in^* \vec{E} = \in_0 \times \in_r^* \vec{E}$$

$$= (\in'_r - j\in''_r) \in_0 \vec{E} \exp(j\omega t)$$

$$\vec{J} = \frac{\partial D}{\partial t} = (\in''_r + j\in'_r)\,\omega \in_0 E \exp(j\omega t)$$

$$= \in''_r \omega \in_0 E \exp(j\omega t) + j[\in'_r \omega \in_0 E \exp(j\omega t)]$$

$$J = \vec{J}_C + \vec{J}_L$$

$$\text{Loss tangent} = \frac{\vec{J}_L}{\vec{J}_C} = \frac{\in''_r}{\in'_r}$$

2.6 **What are the factors that will increase the dielectric loss in dielectric materials? What properties should the insulating materials have to avoid breakdown? What are the important applications of insulating materials?**

[12 marks : 2007]

Solution:

When an insulating material is subjected to alternating voltage, some of the electric energy is absorbed by the insulation and is dissipated as heat. Energy absorbed by material in unit time is called dielectric loss. The factors affecting the dielectric loss are:

(*i*) The loss increases proportionately with the frequency of applied voltage.
(*ii*) Humidity increase the loss
(*iii*) Temperature rise normally increase the loss.
(*iv*) Voltage increase causes increased dielectric loss. (loss $\propto V^2$)

Dielectric strength is defined as the maximum potential gradient that the insulating material can withstand without rupture. Dielectric strength of a material depends upon temperature and humidity in the case of liquid insulators the effect varies with the type of oil and its viscosity.

The dielectric strength depends upon the thickness of the insulating material and on the duration of time for which the dielectric is subjected to electric field. Moisture, contamination, elevated temperature ageing and mechanical stress usually tends to decrease the dielectric strength.

So thickness of the insultating material should be large to avoid breakdown, operation of temperature should also be optimum not very high.

Applications: Winding and cable insulation, primary dielectric in capacitance, used in transistors, used in coil formers at high frequencies, storage battery housing, coating materials it is used in base material for audio and video tapes. Transformer oil use for insulation and moisture absorbent.

2.7 **What are the characteristics of ceramics? What are the three categories of ceramic tool materials? Give the uses of ceramics.**

[10 marks : 2008]

Solution:

Characteristics of ceramics:

1. high brittleness
2. high hardness
3. high melting point
4. chemical inertness
5. electrical insulation
6. low water absorption
7. low thermal expansion
8. high temperature withstand capacity

Main ceramics material divided into two group depending whether permittivity ($\in_r$) is less than or greater than 12.

Material with ($\in_r < 12$) are more used in insulators, bushing, housing and as like. Example porcelain, alumina, etc. Porcelain is used as insulators in transmission and distributions, fuse links, plugs and socket.

Material with permittivity ($\in_r$) > 12 is mainly used as dielectric in capacitor because size of dielectric should be as small as possible one example is (Ba T_i O_3).

2.8 **A solid dielectric specimen has a dielectric constant of 4.8, and tan δ = 0.001 at a frequency of 100 Hz. If it is subjected to an alternating field of 60 kV/cm, calculate the dielectric loss of the specimen.**

[12 marks : 2009]

Solution:

$$\tan\delta = \frac{\epsilon_r''}{\epsilon_r'}$$

Dielectric constant, $\epsilon_r^* = 4.8$

$$E = 60\text{ kV/cm} = 60 \times 10^5\text{ V/m}$$

$$\omega = 2\pi \times 100 \text{ and } \epsilon_r'' = 4.8 \times \tan\delta = 4.8 \times 10^{-3}$$

$$\omega = 200\pi\text{ rad/sec}$$

$$\text{Dielectric losses/unit area} = \frac{1}{2}\,\omega\,\epsilon_0\,\epsilon_r''\,E^2$$

$$= \frac{1}{2} \times 2\pi \times 100 \times \epsilon_0 \times 4.8 \times 10^{-3} \times (60 \times 10^5)^2$$

$$= 480.66\text{ W}$$

2.9 **Write down the properties and the types of insulating materials used in the construction of power and distribution transformers.**

[12 marks : 2009]

Solution:

Properties of insultating material used in designing of power and distribution transformer:

1. High fire point (for insulating fluid)
2. A low net calorific value (< 32 MJ/kg)
3. Ecologically compatible
4. Low accute and chronic toricity
5. Good thermal properties
6. Excellent dielectric properties
7. Excellent fire resistance
8. It is not subject to the regulations for land, sea, and air transport
9. It should be easily disposed.

Normally insulating fluid are used in transformer used for preventing arcs.

It is basically used for insulating high voltage systems.

Another material used is mineral oil used in distribution transformer.

Transformer oil:

Transformer cores and chokes are diped immersed in mineral oil, known as transformer oil which acts as an insulator. It also act as a cooling medium.

Properties of transformer oil:

1. Its specific gravity 0.885.
2. Its flash point 135° C.
3. Its Resistivity is $(1 - 10) \times 10^{12}\ \Omega$ cm.
4. Its permittivity is 2.2 at room temperature.
5. Its specific heat is 0.425.
6. Its dielectric strength is 30 kV/mm.
7. Its viscosity is low.

Impregnating varnishes:

These are utilized with porons and fibrous insulation like paper, card board, fabrics, glass etc., in order to increase dielectric strength heat conductivity, resistance to moisture absorption and mechanical properties. This is known as minor insulation in transformer.

Enamels:

Enamel is a insulated coating of organic base material which is applied on conducting surface. The main application of this in the transformer is coating for wires and winding.

2.10 Explain the following:

(i) Complex dielectric constant

(ii) Ceramics as insulating materials [12 marks : 2011]

Solution:

(i)
- The absorption of electric energy by a dielectric material subjected to an alternating E-field is known as dielectric losses. The result in dissipation of the electric energy as heat in the material and in this dielectric constant in such case in a complex quantity where imaginary part corresponds the loss. So, complex dielectric constant is expressed as

$$\epsilon_r^* = \epsilon_r' - j\epsilon_r''$$

- The complex dielectric constant incorporates all the contributions of polarization.
- The imaginary part of the equation is so that rise to absorption of energy by the material from the alternating field.

(ii)
- Ceramics are dielectric material having large band gap E_g. So in working temperature range they behave as a insulator.
- Some materials (ceramic) such as glass, paper, Teflon are very good electric insulators.
- These ceramics are used for high voltage power transmission because of their good insulating properties. These are normally glass, porcelain, or composite polymer materials.
- Insulators made from porcelain rich in alumina are used where high mechanical strength is criterion.
- Porcelain has a dielectric strength of 4 – 10 kV/mm.
- Recently some electric utilities have begun converting to polymer composite materials for some types of insulators.
- Ceramic insulator are also used at railway.

2.11 **(i) Name the basic polarization mechanisms which occur in a dielectric. A dielectric material contains 2×10^{19} polar molecules/m^3, each of dipole moment 1.8×10^{-27} C-m. Assuming that all the dipoles are aligned in the direction of electric field $\vec{E} = 10^5 \vec{a}_x$ V/m, find $\vec{P}$ and ε_r.**

(ii) Describe in detail the thermal break-down of solid dielectrics under application of alternating as well as direct voltages.

[10 marks : 2011]

Solution:

(i) Following polarization mechanisms occurs in dielectric:

1. Electric or induced polarization
2. Ionic polarization
3. Orientational polarization
4. Interfacial or space charge polarization

Given that:

Polar molecules/m^3 = Number of dipoles/m^3, $N = 2 \times 10^{19}$/m^3

Element dipole moment, $P = 1.8 \times 10^{-27}$ cm

$$\text{Polarization} = \text{dipole moment/m}^3$$

$$|P| = \text{Number of dipoles/m}^3 \times \text{elemental dipole moment}$$

$$|P| = \text{N.P.}$$

$$|P| = 2 \times 10^{19} \times 1.8 \times 10^{-27}$$

$$|P| = 3.6 \times 10^{-8} \text{ c/m}^2 \quad \text{...(i)}$$

We also know that, $|P| = \epsilon_0 X_e |E|$, ϵ_0 — dielectric constant for free space.

where, $X_e = (\epsilon_r - 1)$ — Electric susceptibility,

$\therefore$ $$P = \epsilon_0(\epsilon_r - 1)E \quad \text{...(ii)}$$

where, $|E| = 10^5$ V/m

$\therefore$ From equation (*i*) and (*ii*),

We have $$3.6 \times 10^{-8} = 8.85 \times 10^{-12}(\epsilon_r - 1) \times 10^5$$

$$(\epsilon_r - 1) = \frac{3.6 \times 10^{-8}}{8.85 \times 10^{-7}}$$

$$\Rightarrow \quad \epsilon_r = 0.04067 = 1.04067$$

(ii) Solid dielectric suffers from the phenomenon of thermal breakdown.

- When the electric field applied across an insulating substance exceeds the threshold breakdown field for that substance which is proportional to band-gap energy, the dielectric suddenly twins into a resistor sometimes with catastrophic results.
- During this breakdown any free charge carrier accelerated by strong electric field will have enough velocity to knock electrons from any atom it strikes. These feed electrons and ions are in turn accelerated and strike other atoms creating more charge carriers in a chain reaction.
- Rapidly the dielectric become filled with mobile carriers and its resistance drops to a low level.
- Even a vacuum (which is dielectric) can suffer from for breakdown, but is this case the breakdown or vacuum arc involves charges ejected from the surface of metal electrodes rather than produced by vacuum itself.

Breakdown due to excessive voltage can occure in one of two ways.

1. Puncture voltage is voltage across the insulator which causes a breakdown and induction through the interior of the insulator. The heat resulting from the puncture arc usually damages the insulator irreparably.
2. Flash, over voltage is the voltage which causes the air around or along the surface of dielectric to breakdown and conduct causing a flashover arc along the outside of insulator. They are usually design to withstand this without damage.

2.12 Give and explain the temperature classification of solid insulating materials. Name at least two materials in each class.

[16 marks : 2012]

Solution:

The insulating materials are subjected to several types of stresses. Most important of these stresses is temperature in service as the chemical structure of any organic material is affected by its temperature. Hence, higher the temperature of insulation and lower will be the useful life of the insulation. Hence, for the satisfactory operation of an electrical equipment, it is necessary to determine the temperature limits for the various types of insulating materials before the degradation occurs. The degradation can be experienced by one or more of the following symptoms: hardening, cracking, distortion, weight lost and ultimately at high temperature charging. The causes for degradation may be evaporation of volatiles, oxidation, formation of gases, water within insulation and polymerisation. Degradation of insulating materials cause electrical failures. Realising the above characteristics, national and international standards have been formulated classifying the various types of insulating materials in seven major-classes according to their thermal stability in service.

Class	Limiting Temperature - °C	Materials
Y	90	Cotton, silk, paper and similar organic materials neither impregnated nor immersed in any oil, rubber, polyvinyl chloride.
A	105	Impregnated paper, silk cotton (fibrous materials), polymide resins.
E	120	Enammelled wire insulations on bases of polyvinyl formal, polyurethane and epoxy resins, moulding powder plastics etc.
B	130	Inorganic materials (mica, fibreglass, asbestos) impregnated with varnishes or other compounds.
F	155	Polyester epoxide varnishes and other varnishes with a high heat resistance.
H	180	Composite materials on mica, fibre glass and asbestos bases impregnated with silicons, silicon rubber.
C	225	Mica, ceramics, glass, teflon

2.13 A dielectric material contains 2×10^{19} polar molecules/m^3, each of dipole moment 1.8×10^{-27} C/m. Assuming that all the dipoles are aligned in the direction of the electric field $\vec{E} = 10^5\, \vec{a}_x$ V/m, find $\vec{P}$ and ε_r.

[10 marks : 2012]

Solution:

Polarization,
$$\vec{P} = N\vec{p} = 2\times10^{19}\times1.8\times10^{-27}\vec{a}_x = 3.8\times10^{-8}\vec{a}_x \text{ c/m}^2$$

$$\vec{P} = \epsilon_o(\epsilon_r - 1)\vec{E}$$
$$3.8\times10^{-8} = 8.85\times10^{-12}\times10^5\times(\epsilon_r - 1)$$
$$\epsilon_r - 1 = 0.043$$
$$\therefore \quad \epsilon_r = 1.043$$

2.14 What is polarization? In a dielectric material $E_x = 5$ V/m and $P = \frac{1}{10\pi}(3\hat{a}_x - \hat{a}_y + 4\hat{a}_z)$ nC/m^2.

Calculate:

(i) Electric susceptibility (χ_e)

(ii) Electric field intensity (E)

(iii) Electric flux density (D)

[5 marks : 2013]

Solution:

Polarization: In the dielectric materials most of the electrons are bound to the nucleus. When an external electric field is applied then the bound electrons of an atom are displaced such that the centroid of the electronic cloud is separated from the centroid of the nucleus. Hence an electric dipole is created and the atom is said to be polarized. This phenomenon is known as **Polarization.** On a macroscopic scale in field theory we define vector $\vec{P}$, called the polarization as the electric dipole moment per unit volume. Thus if N-denotes the number of molecules per unit volume of a material, then there are $N\Delta v$ molecules in a volume Δv and

$$\boxed{\vec{P} = \frac{1}{\Delta v}\sum_{j=1}^{N\Delta v}\vec{P}_j = N\vec{p}}$$

i.e. average dipole moment per unit volume. Its unit is C/m^2.

(i) Now, $E_x = 5$ V/m

$$P = \frac{1}{10\pi}(3\hat{a}_x - \hat{a}_y + 4\hat{a}_z)\ \text{nC/m}^2$$

Now as given field is in x-direction, then the polarization induced will also be in x-direction.

So using, $P = \epsilon_o \chi_e E$

$$\chi_e = \frac{P}{\epsilon_o E_x} = \frac{\left(\frac{3}{10\pi}\right)\times 10^{-9}}{(8.85\times 10^{-12})\times 5} = \frac{\left(\frac{3}{10\pi}\right)\times 10^{-9}}{(8.85\times 10^{-12})\times (5)} = 2.158$$

(ii) $$E_i = E_x + \frac{\gamma P}{\epsilon_o}$$

Taking, $\gamma = \frac{1}{3}$

$$E_i = 5 + \left(\frac{1}{3}\right)\left(\frac{[3/10\pi]\times 10^{-9}}{8.85\times 10^{-12}}\right) = 8.5967\ \text{V/m}$$

(iii) $D = \epsilon_o \epsilon_r E_i = \epsilon_o(1+\chi_e)E_i$

$D = (8.85\times 10^{-12})(1 + 2.158)(8.5967) = 0.24\ \text{C/m}^2$

2.15 Differentiate between linear, homogeneous and isotropic dielectic material. At the center of a hollow dielectric sphere ($\varepsilon = \varepsilon_0\varepsilon_r$) is placed a point charge Q. If the sphere has inner radius a and outer radius b, calculate $\vec{D}, \vec{E}$ and $\vec{P}$.

[12 marks : 2014]

Solution:

Homogenous: A medium is homogenous if its physical characteristics (mass, density, molecular structure etc.) do not vary from point to point.

Linear: A medium is linear with respect to an electrostatic field. If the flux density $\overline{D}$ is proportional to the electric field intensity $(\overline{E})$. This is the case in free space where $\overline{D} = \epsilon_0 \overline{E}$. Here the factor ϵ_0 or permittivity, is constant. In material media the permittivity ϵ may not always be constant. If it is not constant, the material is said to be non-linear.

Isotropic: An isotropic material is one whose properties are independent of direction. Generally, materials whose molecular structure is randomly oriented will be isotropic.

2.16 List at least two essential properties of each material that are necessary for construction of medium rating power transformer.

[10 marks, 2015]

Solution:

Essential properties of material for power transformer.

1. Insulating Oil:

Electrical Properties

(*i*) ***Electric strength (breakdown voltage).*** BDV is the voltage at which breakdown occurs between two electrodes when oil is subjected to an electric field under prescribed conditions. Electric strength is the basic parameter for insulation system design of a transformer. It serves to indicate the presence of contaminating agents like moisture, fibrous materials, carbon particles, precipitable sludge and sediment.

(*ii*) ***Dielectric dissipation factor (DDF).*** DDF is numerically equal to sine of the loss angle (approximately equal to tangent of loss angle for dielectrics) and is a good tool to indicate the quality of an insulation. A high value of DDF is an indication of the presence of contaminants or deterioration products such as water, oxidation products, metal soaps, soluble varnishes and resins.

2. Insulating Paper:

Characteristic	Requirement		
	Kraft paper of medium air permeability	**Kraft paper of high air permeability**	**Press paper**
Electrical Properties			
Electric strength (BDV), kV/mm, (min.)			
(i) In air at 90°C	7.5	7.0	10
(ii) In oil at 90°C	—	—	65
Chemical Properties			
Mineral ash, %, (max.)	1.0	1.0	2.0
Conductivity of 5% aqueous extract mS/m, (max.)	10	10	13.5

3. Pressboard:

Characteristic	Requirement			
	Precompressed solid	**Soft calendered solid**	**Precompressed laminated**	**Soft calendered laminated**
Composition	**100% wood pulp**	**Cotton, jute and wood pulp**	**100% wood pulp**	**Cotton, jute and wood pulp**
Mechanical Properties				
Tensile strength, MPa (min.)				
(i) Machine direction	100	47	—	—
(ii) Cross direction	75	22	—	—
Electrical Properties				
Electric strength in oil 90°C kV/mm (min.)				
(i) 1.5 mm thick	35 BDV	15.3 for one min.	70 kV for one min. on 25 mm width (edgewise)	
(ii) 3 mm thick	30 BDV	11.3 for one min.		

4. Continuously Transposed Copper Conductor (CTC):

CTC comprises an odd number of high conductivity annealed copper strips insulated with a polyvinyl-acetal based enamel, arranged in two side-by-side stacks with the individual enamelled conductors continuously transposed throughout their length.

Radial insulation thickness (mm)	Test voltage breakdown voltage in oil at 90°C (kV rms)
0.5	20
1.0	35
1.5	48
2.0	61
2.5	72

5. **Cold Rolled Grain-Oriented Electrical Steel Sheet (CRGO):**
CRGO electrical steel with an experiment silicon content of 3% is used for magnetic circuits of a transformer.

Characteristic	Grade CGO	Grade Hi–B
Magnetic Properties		
1. Specific core-loss, W/kg (max.) at 50 Hz.		
(i) At 1.5 T	0069 – 1.11	—
(ii) At 1.7 T	1.17 to 1.57	0.95–1.17
2. Magnetic induction at magnetizing force of 800 A/m (B_8), T(min.) at 50 Hz.	1.75	1.85
Electrical Properties		
Resistance of surface coating, Ω (min.)	2 for 80% of the readings 5 for 50% of the readings	

2.17 **Specific gravity of a ceramic is 3.2 g/cm³. Calculate the percentage apparent porosity and percentage true porosity with the following data:**
Ceramic when weighted dry = 360g
Ceramic when weighted after soaking in water = 385g
Ceramic weighted while suspended in water = 224g

[6 marks : 2018]

Solution:

Specific gravity of a ceramic, $\rho = 3.2\text{ g/cm}^3$
Ceramic weight when dry, $W_d = 360\text{ gm}$
Ceramic weight when soaking in water $W_{soak} = 385\text{ gm}$
Ceramic weight when suspended in water $W_{sus} = 224\text{ g}$

- $$\text{Percentage true porosity} = \left(\frac{\text{Soaked weight} - \text{Dry weight}}{\text{Soaked weight} - \text{suspended wt.}}\right) \times 100\%$$
$$= \left(\frac{385-360}{385-224}\right) \times 100\% = 15.527\%$$
- $$\text{True porosity} = \left[1 - \left(\frac{\text{apparent SP gravity}}{\text{True SP gravity}}\right)\right] \times 100$$

Apparent specific gravity (Bulk density)
$$= \frac{W_{dry}}{W_{soaked} - W_{suspended}}$$
$$= \frac{360}{385-224} = 2.236\text{ g/cm}^3$$
$$\therefore \quad \text{True porosity} = \left(1 - \frac{2.236}{3.2}\right) \times 100\% = 30.12\%$$

2.18 **List properties of ceramic materials and write their applications in technology.**

[20 marks : 2019]

Solution:

Ceramics: Ceramics are hard, strong, dense and brittle

- High temperature stability
- Generally inorganic materials
- They are non-metallic oxides, nitrides, carbides
- Generally crystalline materials (only exception is amorphous glass)

  Ex.: CdS, MgO, ZnO, garnets, $BaTiO_3$, cement, ferrites, SiC, TiO_2 (Rutile), ZnS, Quartz.

Types of Ceramics:

(a) **Porcelain ($\in_r$ < 12):** It is used for low and high voltage applications. It is used as insulator in transmission and distribution system, plugs and sockets.

(b) **Steatite ($\in_r$ < 12):** It is used in high frequency application.

(c) **Alumina ($\in_r$ < 12):** Al_2O_3

- It is used in high temperature application.
- It is used in circuit breakers and resistance cores.

(d) **Titanate ($\in_r$ > 12):** High dielectric constant.

- It is used in capacitors.

Another Classification of Ceramics:

(a) **$\in_r$ < 12:** Ceramics having relative permittivity ($\hat{I}_r$ < 12) used as insulators.

Ex.: Porcelain, steatite, alumina

(b) **$\in_r$ > 12:** This ceramics used in capacitor applications due to high dielectric constant.

Ex.: TiO_2 (Kutile)

2.19 What is piezoelectricity? Explain the construction and working of a piezoelectric device. State any two applications of piezoelectric devices.

[10 marks : 2021]

Solution:

In certain dielectric materials when an uniaxial mechanical stress is applied, they get polarized. Such materials are known as piezoelectric materials and this property of material is known as piezoelectricity. In such materials if electric field is applied then they get strained and produced strain is proportional to applied electric field.

There are two effects in piezoelectricity :

Direct Effect : Generation of charge due to applied mechanical stress.

Inverse Effect : Generation of strain due to applied electric field.

Both above effects are linear in nature. Consider the specimen shown below for piezoeletric device with area of cross-section A, thickness t with force F applied as shown :

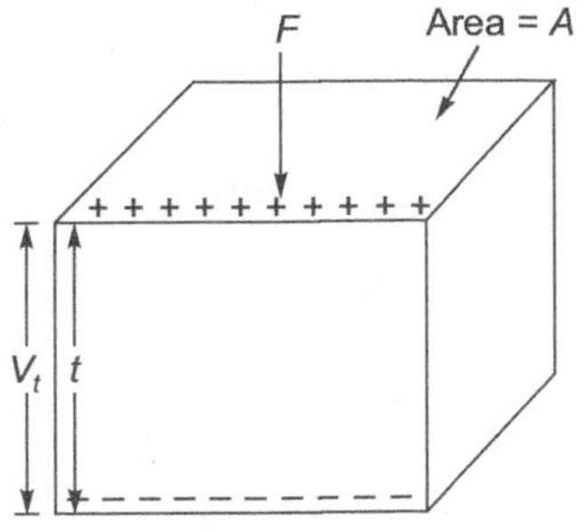

Here developed charge $Q \propto F$ (linear effect)

$Q = dF$ where d is charge sensitivity in C/N

Now the positive and negative charge developed across capacitor with separation t between plate.

$$Q = CV \text{ or } V = \frac{Q}{C} \text{ where } C = \frac{\varepsilon_o A}{t}$$

So,
$$V = \frac{dF}{\frac{\varepsilon_o A}{t}}$$

$$V = \left(\frac{F}{A}\right) \times \left(\frac{d}{\varepsilon}\right) \times t$$

∴
$$V = P \times g \times t$$

where
$$g = \frac{d}{\varepsilon} = \text{Voltage sensitivity (V-m/N)}$$

Applications of Piezoelectric devices :

1. As a crystal oscillators in clocks circuits.
2. Air transducers.

2.20 A capacitor of 0.025 μF has the loss angle, δ = 0.0286°. Calculate the dielectric loss of the capacitor when it carries a current of 100 A at a frequency of 25 kHz.

[10 marks : 2021]

Solution:

Given :
$$C = 0.025\ \mu\text{F},\ \delta = 0.0286°,\ I_c = 100\ \text{A},\ f = 25\ \text{kHz}$$

As we know,
$$|I_c| = V\omega C$$

$$V = \frac{I_C}{\omega C} = \frac{100}{2\pi \times 25 \times 10^3 \times 0.025 \times 10^{-6}}$$

Applied voltage,
$$V = 25464.80\ \text{Volt}$$

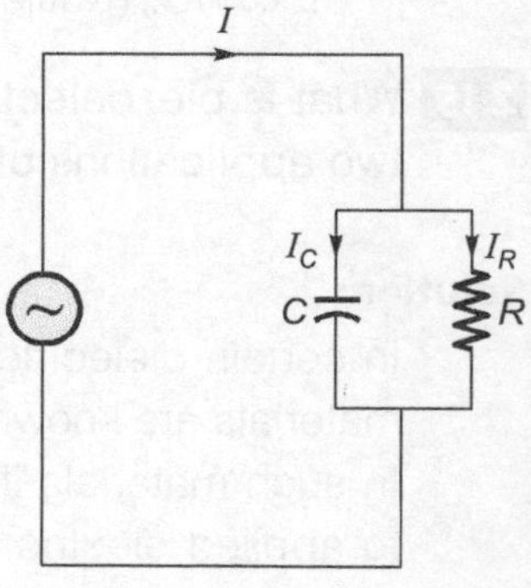

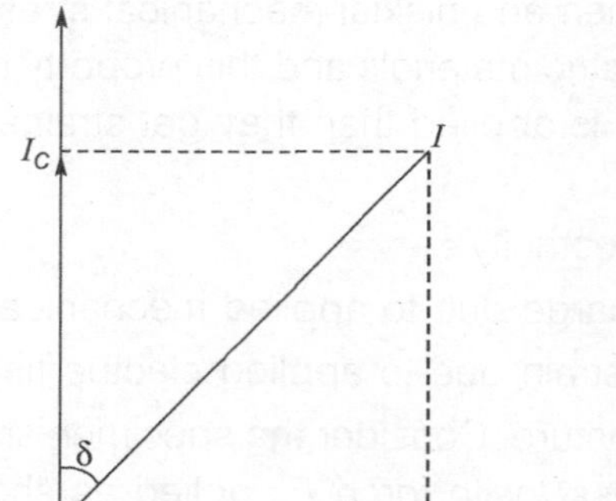

$$I_R = I_C \tan\delta$$
$$= 100 \times \tan(0.0286°)$$
$$I_R = 0.05\ \text{A}$$

Dielectric loss can be given as

$$P_L = VI_R = 25464.80 \times 0.05$$
$$P_L = 1273.24\ \text{Watt}$$

3. Magnetic Properties of Materials

3.1 Define spontaneous magnetization. How does it behave in ferromagnetic materials below Curies point? Match List-I of (Materials) with List-II of their Curie point in K.

	List-I (Material)		List-II (Curie point K)
A.	Cobalt	1.	340
B.	Iron	2.	1400
C.	Nickel	3.	630
D.	30% Permalloy	4.	1040

[12 marks : 2001]

Solution:

Starting with a virgin specimen, even in the absence of external field, the neighbouring dipoles are alligned as the allignment force between adjacent dipole is higher because of permanent dipole moment (P_m) is strong. So there are groups of large number of dipoles allign in particular direction this type of group is called magnetic domain. If the external field is applied, dipoles are going to allign with $(\bar{H})$, hence size of domain in the direction of $(\bar{H})$ increase. For a particular value of $(\bar{H})$, almost all the dipoles are allign $(\bar{H})$. This is called magnetic saturation.

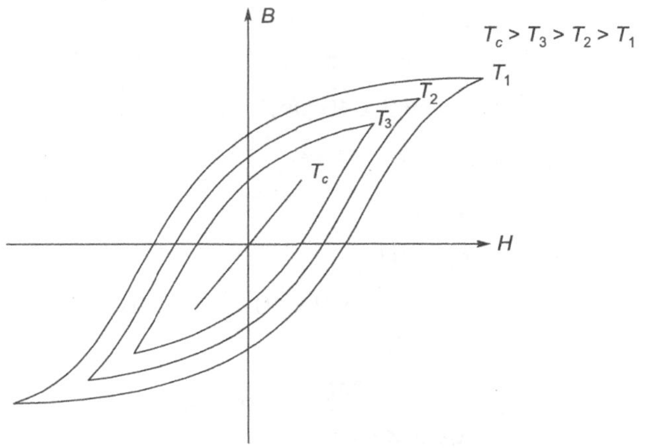

Hysteresis property of ferromagnetic material with temperature

After maximum value of $(\bar{H}_m)$, if $\bar{H}$ is decrease, dipoles are going to be dealligned from $\bar{H}$, due to thermal agitation, hence size of main domain decreases, when 'H' is reduced to zero, the size of the main domain is left some what larger than that of initial value. So spontaneous magnetization is back and hence residual density is left. If temperature is increases randomness in orientation of dipole increases. So size of the domain decreases while number of domains increases. So magnetization decreases. At a particular value of temperature (T_c) (Curie temperature), the domains are vanish as all the dipoles are randomly allign, ferromagnetic material becomes paramagnetic.

Materials		Curie point K
(a) Cobalt	–	1400 K
(b) Iron	–	1040 K
(c) Nickel	–	630 K
(d) 30% permalloy	–	340 K

Curie weiss-law

$$\chi_m = \frac{C}{T-\theta}$$

where, C – Curie constant,

θ – Paramagnetic curie temperature.

3.2 Find the magnitude of the magnetic flux density in a material for which
(i) the magnetization is 2.8 A/m and magnetic susceptibility is 0.0025
(ii) the magnetic field intensity is 1300 A/m and the relative permeability is 1.006.
(iii) there are 8.2×10^{28} atoms/m^3, each having a dipole moment of 3×10^{-30} A-m^2 in the same direction and the magnetic susceptibility is 2×10^{-4}.

[12 marks : 2002]

Solution:

(i) Magnetization = 2.8 A/m, magnetic susceptibility = 0.0025

$$\bar{M} = 2.8 \text{ A/m}, \quad \chi_m = 0.0025$$

$$\bar{M} = \chi_m H$$

$$\therefore \quad \bar{H} = \frac{\bar{M}}{\chi_m} = \frac{2.8}{0.0025} = 1120 \text{ A/m}$$

$$\text{Flux density } ((B) = \mu_0(\bar{H} + \bar{M})$$

$$= 4\pi \times 10^{-7} [1120 + 2.8] = 1.41 \times 10^{-3} \text{ wb/m}^2$$

(ii) Magnetic field intensity = 1300 A/m

$$\mu_r = 1.006$$

$$B = \mu_0 \mu_r H = 4\pi \times 10^{-7} \times 1.006 \times 1300 = 1.64 \times 10^{-3} \text{ wb/m}^2$$

(iii) Number of dipoles/unit volume $N = 8.2 \times 10^{28}$/m^3

Each dipole moment, $\bar{P}_m = 3 \times 10^{-30}$ A–m^2

Since all dipoles are in same direction

$$\bar{m} \text{ (magnetization)} = \bar{M} = \text{N} \bar{P}_m$$

$$= 8.2 \times 10^{28} \times 3 \times 10^{-30} = 0.246 \text{ A/m.}$$

Magnetic susceptibility = $\chi_m = 2 \times 10^{-4}$

$$\therefore \quad H = \frac{M}{\chi_m} = \frac{0.246}{2 \times 10^{-4}} = 1.23 \times 10^3 \text{ A/m}$$

$$\therefore \quad B = \mu_0 (H + M) = 4\pi \times 10^{-7} (1.23 \times 10^3 + 0.246)$$

$$= 1.545 \times 10^{-3} \text{ wb/m}^2$$

3.3 Discuss the $B-H$ characteristics of a ferromagnetic material below the critical temperature. Give its significance. Match the values of Curie temperatures for ferromagnetic elements Co, Fe, Gd and Ni with the values of Curie temperatures 289 K, 631 K, 1043 K and 1393 K.

[12 marks : 2003]

Solution:

Starting with virgin specimen, even in the absence of external field, in ferromagnetic material, the neighbouring are alligned as the allignment force between adjacent dipole is higher because of strong permanent dipole moment (P_m). So there is groups of large number of dipoles allign in particular direction, this type of group is called magnetic domain.

If the external field is applied, dipoles are going to allign with $(\bar{H})$ hence size of domain in the direction of $(\bar{H})$ increases. For a particular value of $(\bar{H})$, almost all the dipole are allign in $(\bar{H})$. This is called magnetic saturation.

After maximum value of $(\bar{H}_m)$ if $\bar{H}$ is decrease, dipoles are going to be dealligned from $\bar{H}$, due to thermo agitation, hence size of main domain decreases when 'H' is reduced to zero, the size of the main domain is left some what larger than that of initial value. So spontaneous magnetization is back and hence residual flux density is left. If $\bar{H}$ is increased in opposite direction, the size of spontaneous magnetization decreases at particular value of ($H = H_{coercive}$), the spontaneous magnetization become zero. Residual flux density becomes zero.

Magnetization curve of Ferromagnetic material

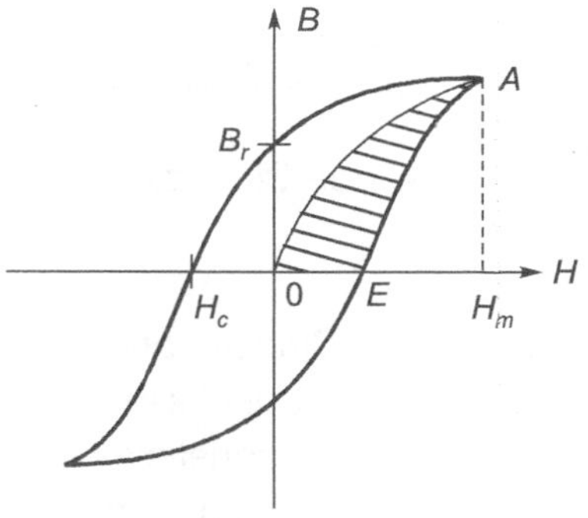

where B_r – Spontaneous magnetizations

H_c – Coercive force.

Hysteresis loop is called hysteresis loss w_h. In a nonlinear magnetic field varies cyclically in subsequent 2nd, 3rd .... nth cycles, it causes eddy current loss w_e.

The sum of the hysteresis loss and eddy current loss is known as the total core loss.

Thus the total power loss,

$$w_c = w_h + w_e$$

$$\therefore \quad w_c = \eta\, B_{max}^{1.6}\, fV + \lambda B_{max}^{2} f^2 t^2 V$$

Where η – steinmetz coefficient, λ is eddy current constant of material, $B_{max} = B_s$ is the saturation flux density, t is thickness of the material, f-frequency of magnetic reversal and V is volume of material.

Material	– Curie temperature
Co	– 1393 K
Fe	– 1043 K
Ni	– 631K
Gd	– 289 K

3.4 Give the general electrical and magnetic characteristics of ferrites. List its applications.

[12 marks : 2003]

Solution:

Ferrimagnetic materials are of great technical importance. They exhibit spontaneous magnetic moment below a curie temperature, just as iron, cobalt or nickel. Ferrimagnetic materials possess antiparallel spin coupling and incomplete magnetic moment cancellation. In ferrimagnetic materials, the moments of the two sublattices are unequal and there is a resultant magnetic moment. The material is called a ferrimagnet. This material disappeared above a curie temperature. Thermal energy randomizes the individual magnetic moments and the material becomes paramagnetic.

The main difference to ferromagnetics is, however, that ferrimagnetics are ceramic materials (oxides) and that they are poor electrical conductors. The other application of ferrimagnetic are:

(a) Magnetically hard ferrites for permanent magnets.

(b) Magnetically soft ferrites for transformers and inductors.

(c) Rectangular loop ferrites for data stores.

Ferrites is a material whose formula is MO. Fe_2O_3 where M is a bivalent metal e.g., Fe, Cu, Ni etc. The crystallography of ferrites is rather complex. Let us discuss an example nickel ferrite, NiO, Fe_2O_3. The Fe^{3+} ions are equally distributed between A and B sites and since ions on A and B sites exhibit spontaneous magnetization in opposite directions.

- Ferrite is material whose formula is MO Fe_2O_3, where M is bivalent metals e.g. Fe, Cu Mn etc. whose structure is of the spinal family.

  A spinal (Mg Al_2O_4) is a compound whose Al (trivalent) atoms and Mg atoms occupy tetrahedral sites in an FCC oxygen in an FCC oxygen lattice.
- The ferrimagnetic oxides, ferrites and garnets, are very suitable in the high frequency range.

- Being electrical insulators, they have a much higher resistivity than alloys. This reduces the eddy current losses to a negligible value.
- Ni-Zn ferrites are used for audio and TV transformers.
- Mg-Mn ferrites with high resistivity are used as microwave isolators and gyrators in the kHz and MHz range

A recent development in soft magnetic materials is the magnetic bubbles.

3.5 Describe the classification of magnetic materials. Give the examples for each class.

[12 marks : 2004]

Solution:

Classification	Examples
(i) Dia magnetic material	– Diamond, silicon
(ii) Para magnetic material	– Phosphorous, O_2
(iii) Ferro magnetic material	– Soft iron, carbon steel
(iv) Ferri magnetic material	– F_3O_4, Mn Fe_2O_4
(v) Anti ferro magnetic material	– Mn F_2

On the basis of magnetic behaviour material may be classified as:

1. Diamagnetic
2. Paramagnetic
3. Ferromagnetic
4. Ferrimagnetic
5. Antiferromagnetic

1. **Diamagnetic Materials:** These materials when placed in a magnetic field acquire weak magnetization in a direction opposite to that of applied magnetic field called as diamagnetic materials.
   The value of magnetic susceptiability for these materials is –ve and very small e.g. diamond, silicon, Ge.
2. **Paramagnetic Materials:** These materials when placed in a magnetic field acquire a weak magnetization in the same direction as the applied field are called paramagnetic materials.
   The value of magnetic susceptibility for these materials is +ve and very small e.g. phosphorus, O_2.
3. **Ferromagnetic Materials:** Ferromegnetic materials are spontaneously magnetized even in the absence of external magnetic field. All ferromagnetic materials becomes paramagnetic above certain temperature called as curie temperature.
   The value of magnetic susceptibility for these materials is +ve and very high. e.g. Soft Iron, Carbon, Steel.
4. **Ferrimagnetic Materials:** In ferrimagnetic substances the magnetic moment of adjacent atoms are alligned in opposite direction but the moments are not equal so that there is a net magnetic moment i.e. not zero. However it is less than that of ferromagnetic materials.
   They loses these properties above certain temperature called as curie temp.
   These materials have high value of magnetic susceptibility but less than that of ferromagnetic materials. e.g. Fe_3O_4, $MnFe_2O_4$.
5. **Antiferromagnetic Materials:** In antiferromagnetic materials the magnetic moments of adjacent atoms allign in opposite directions so that net magnetic moment of a specimen is nill in absence of magnetic field. But on the application of magnetic field dipoles are alligned in direction of magnetic field.
   The value of magnetic susceptibility is +ve and very small e.g MnF_2.

3.6 What is meant by magnetic bubbles? Mention the applications of magnetic bubbles.

[12 marks : 2004]

Solution:

Magnetic bubble memory consists of magnetic bubbles used to store data bits. These bubbles move in thin films of magnetic material. The bubbles are cylindrical magnetic domains having polarization opposite to that of the thin film in which they are embedded.

A thin film of magnetic garnet contains wawy strips of magnetic domains when an external magnetic field is applied perpendicular to the film and increased slowly, the wavy magnetization strips having magnetization opposite to the external field begin to narrow. At a certain value of field strength all these domains contract and look like small circular bubbles. These bubbles are tiny magnetic floating in an external field. These bubbles can be moved by applying a magnetic field parallel to the film. To control the direction of movement, special magnetic paths are created by depositing magnetically conductive material in specific pattern on the film.

The main components of magnetic bubble memory are thin film memory chip, drive coils, permanent magnetized control electronics. Logical 1s and 0s are represented by presence or absence of bubbles respectively.

It can store large number of data bits. The storage is non-variable and is not erased if power is disconnected. The only disadvantage of this memory is that access time is high. This is because of the reason that data has to be accessed serially. Random access is not possible.

3.7 **Describe the phenomenon of magnetostriction for Fe, Ni and Co. Magnetostriction is a reversible process. Explain why?**

[12 marks : 2004]

Solution:

When a magnetic material is magnetized, there is change in dimension of material is observed. This is called magnetostriction. The main reason for the magnetostriction is, change inter atomic distance.

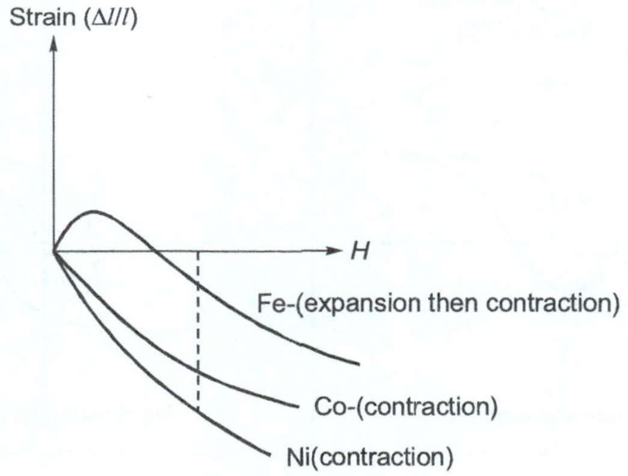

Due to the change in dimension (as a result of applied mechanical stress) there is change in the magnetization or permeability of material, because inter atomic distance is changed. This effect is called villari effect. So reverse effect is also there, so it is a reversible process.

3.8 **What do you understand by magnetic hysteresis? Differentiate between hard and soft magnetic materials?**

[12 marks : 2005]

Solution:

Below the ferromagnetic curie temperature, ferromagnetic materials show a increase in susceptibility and exhibit *B-H* characteristics. Starting with a virgin specimen ($B = H = 0$) if H is increased, B at first increases, until at the point of saturation. This point, further increase $\bar{H}$ causes negligible change in $\vec{B}$. At saturation, if now H is decreased the curve will go through another path which is shown in the figure. When $\bar{H} = 0$, some residual flux will remain in the specimen is called residual or remnant flux density.

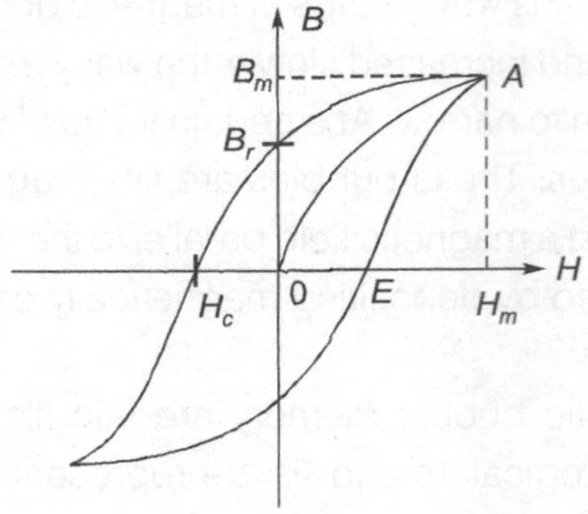

Further increase of $\bar{H}$ in negative direction, reduce the flux density to zero, is called the coercive force. Further increase of $\bar{H}$ leads to saturation.

Depending upon the coercive force and remnant flux density magnetic material are divided into two which are soft and hard magnetic materials.

Ferro magnetic material mainly we divided into two,

Ferromagnetic

Soft magnetic material	Hard magnetic material
High permeability, low retentivity, low coercivity and small hysteresis loss	High permeability, high retentivity, high coercivity, large hysteresis loss and high curie temperature.
Suitable for making electromagnets, core of transformers etc. Example: Soft iron, supermalloy (used in magnetic shielding)	Suitable for permanent magnet Example: Alnico
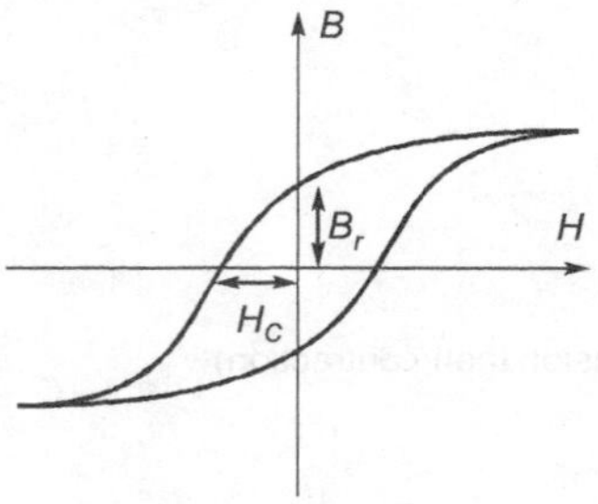 Soft magnetic material	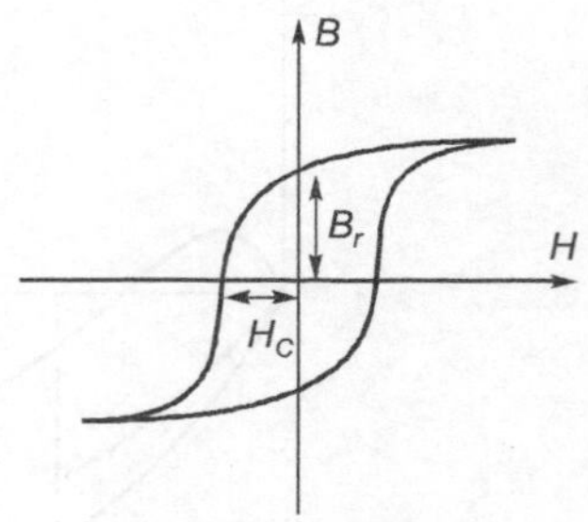 Hard magnetic material

Example for hard magnetic material:

(i) **Alnico:** It is an alloy of aluminium, nickel and cobalt. Mainly used in PMMC.

(ii) **Tungsten steel:** In which tungsten content is more than (50%).

(iii) **Carbon steel:** Used in compass needle.

3.9 What is the domain theory of magnetization? What is Curie temperature? Can Curie temperature of a material be changed?

[12 marks : 2005]

Solution:

Starting with virgin specimen, even in the absence of external field, the neighbouring dipoles are alligned as the allignment force between adjacent dipole is higher because of permanent dipole moment (P_m) is strong. So there are groups of large number of dipoles allign in particular direction this type of group is called magnetic domain.

If the external field is applied, dipoles are going to allign with ($\bar{H}$), hence size of domain in the direction of ($\bar{H}$) increase. For a particular value of ($\bar{H}$), almost all the dipoles are align ($\bar{H}$). This is called magnetic saturation.

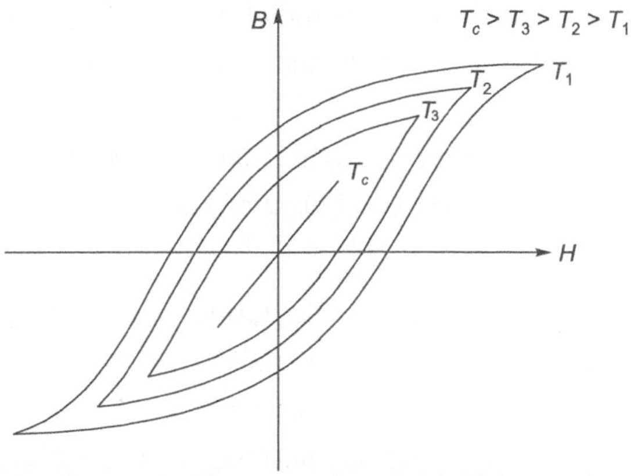

Hysteresis property of ferromagnetic material with temperature

After maximum value of $(\bar{H}_m)$, if $\bar{H}$ is decrease, dipoles are going to be dealligned from $\bar{H}$, due to thermal agitation, hence size of main domain decreases, when 'H' is reduced to zero, the size of the main domain is left some what larger than that of initial value. So spontaneous magnetization is back and hence residual density is left. If temperature increases randomness in orientation of dipole increases. So size of the domain decreases while number of domains increases. So magnetization decreases. At a particular value of temperature (T_c) (Curie temperature), the domains are vanished as all the dipoles are randomly allign, ferromagnetic material becomes paramagnetic.

	Materials		Curie point K
(a)	Cobalt	–	1400 K
(b)	Iron	–	1040 K
(c)	Nickel	–	630 K
(d)	30% permalloy	–	340 K

Curie weiss-law:

$$\chi_m = \frac{C}{T-\theta}$$

where C – Curie constant,

θ – Para magnetic curie temperature.

where, $\theta = \dfrac{\gamma N P_p^2}{3K \in_0}$

γ – Internal field constant

N – Number of dipoles/m^3

$\in_0$ – Constant permittivity.

So curie temp. is constant for a given material.

3.10 Distinguish between diamagnetic, paramagnetic and ferromagnetic materials.

[10 marks : 2007]

Solution:

Diamagnetic Materials:

1. These are the materials when placed in a magnetic field acquire weak magnetization in a direction opposite to that of applied field are called diamagnetic materials.

2. The value magnetic susceptibility for these material is –ve and small.
3. Due to –ve value of magnetic susceptibility they repeal the magnetic field (i.e. magnetize in opposite direction)
4. Diamagnetism has no ordinary practical use. Some of the material which only exhibit diamagnetic behavious are Cu, Au, Ge, Si, diamond, Al_2O_3, NaCl etc.

Paramagnetic Materials:

1. The materials when placed in a magnetic field acquire a weak magnetization in the same direction as the applied field are called paramagnetic materials.
2. The value of magnetic susceptibility is +ve and small.
3. The properties of paramagnetic materials are not much of practical importance, but it helps to understand the theory of ferro and ferrimagnetic materials.
4. It is basically used in designing of transducers.
5. e.g. Fe_2O_3, $MnSO_4$, $FeSO_4$, $FeCl_2$, $NiSO_4$ etc.

Ferromagnetic Materials:

1. Ferromagnetic materials are spontaneously magnetized even in the absence of external magnetic field.
2. All ferromagnetic materials becomes paramagnetic above certain temperature called as curie temperature
3. The value of magnetic susceptibility is +ve and very large.
4. Ferromagnetic materials are characterized by parallel allignment of dipole moment even in the absence of external field.
5. Most important ferromagnetic materials are Fe (Iron), Cobalt (CO), Nickel (Ni), Gadolinium (Gd).

3.11 **In a magnetic material, the field strength is found to be 10^6 A/m. If the magnetic susceptibility of the material is 0.5×10^{-5}, calculate the intensity of the magnetization and flux density in the material.**

[8 marks : 2008]

Solution:

Given magnetic field strength = 10^6 A/m

magnetic susceptibility = 0.5×10^{-5}

Intensity of magnetization = $M = \chi H = 0.5 \times 10^{-5} \times 10^6 = 5$ A/m

$$\mu_r = \chi + 1 = 0.5 \times 10^{-5} + 1 = 1.000005$$

Then flux density, $B_0 = \mu_0 \mu_r H = 4\pi \times 10^{-7}(1 + 0.5 \times 10^{-5})10^6$

$= 1.256$ weber/m^2

3.12 **What are the different types of magnetic materials? Distinguish between hard and soft magnetic materials.**

[16 marks : 2008]

Solution:

On the basis of magnetic properties of the materials, faraday divide materials in three classes:

(*i*) Dia magnetic (*ii*) Para magnetic

(*iii*) Ferro magnetic (*iv*) Ferri magnetic material

(*v*) Anti ferromagnetic material

Ferromagnetic

Soft magnetic material	Hard magnetic material
High permeability, low retentivity, low coercivity and small hysteresis loss Suitable for making electromagnets, core of transformers etc. Example: Soft iron, supermalloy (used in magnetic shielding) 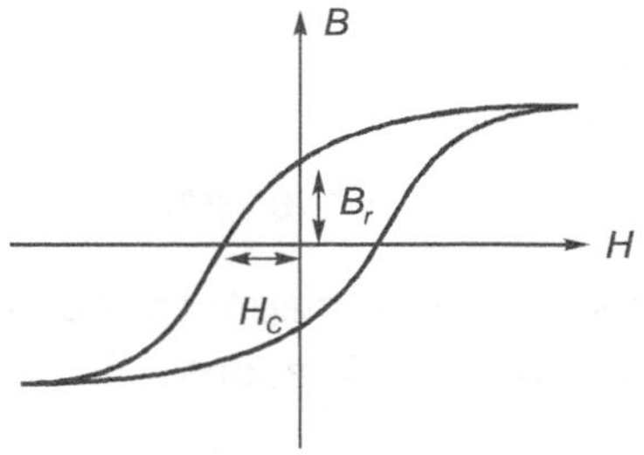 Soft magnetic material	High permeability, high retentivity, high coercivity, large hysteresis loss and high curie temperature. Suitable for permanent magnet Example: Alnico 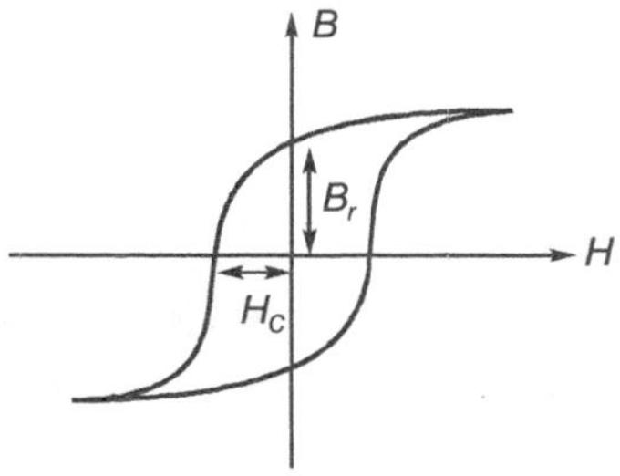 Hard magnetic material

Example for hard magnetic material:

(*i*) **Alnico:** It is an alloy of aluminium, nickel and cobalt. Mainly used in PMMC.

(*ii*) **Tungsten steel:** In which tungsten content is more than (50%).

(*iii*) **Carbon steel:** Used in compass needle.

3.13 **Distinguish between 'hard' and 'soft' magnetic materials. List the properties and applications of 'hard' magnetic materials, giving 2 examples of such materials.**

[12 marks : 2009]

Solution:

Ferro magnetic material mainly we divided into two,

Ferro Magnetic

Soft magnetic material	Hard magnetic material
High permeability, low retentivity, low coercivity and small hysteresis loss Suitable for making electromagnets, core of transformers etc. Example: Soft iron, supermalloy (used in magnetic shielding) 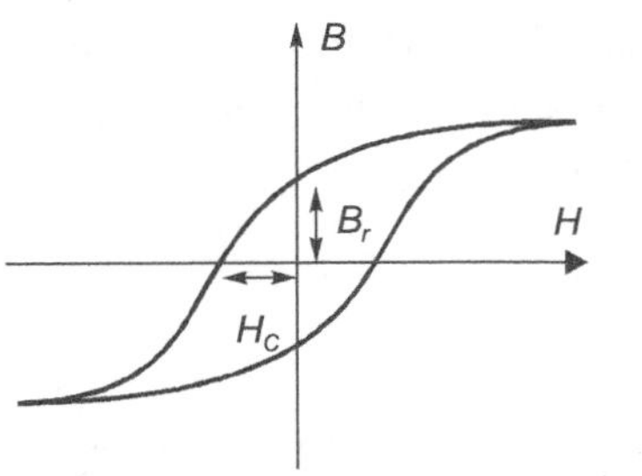 Soft magnetic material	High permeability, high retentivity, high coercivity, large hysteresis loss and high curie temperature. Suitable for permanent magnet Example: Alnico 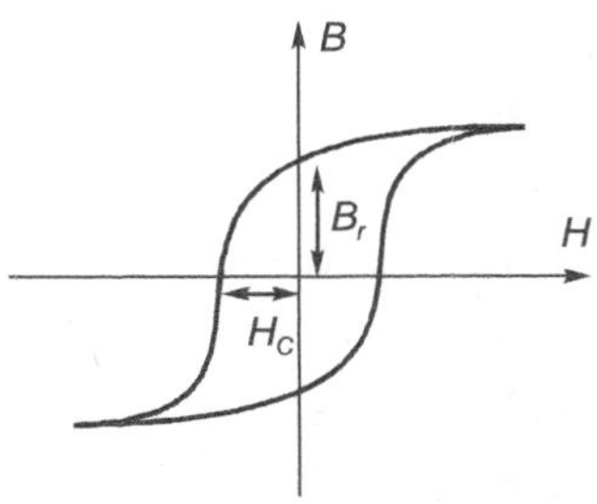 Hard magnetic material

Example for hard magnetic material:

(*i*) **Alnico:** It is an alloy of aluminium, nickel and cobalt. Mainly used in PMMC.

(*ii*) **Tungsten steel:** In which tungsten content is more than (50%).

(*iii*) **Carbon steel:** Used in compass needle.

3.14 **The magnetic susceptibility of Fe_2O_3 is 1.4×10^{-3}. Calculate the flux density and magnetization when it is subjected to a magnetic field of 10^6 A/m.**

[12 marks : 2009]

Solution:

Magnetic susceptibility = 1.4×10^{-3}

Magnetic field = $\overline{H} = 10^6$ A/m

Magnetic flux density = $\overline{B} = \mu_0(1+\chi_m)\overline{H} = (\mu_0 H + \mu_0 \overline{m})$

where, $\overline{m}$ = magnetization = $\chi_m \overline{H}$

$= 1.4 \times 10^{-3} \times 10^6 = 1400$ A/m.

$\therefore$ $B = \mu_0(1 + 1.4 \times 10^{-3}) \times 10^6 = 1.258$ weber/m^2

3.15 **With respect to magnetic behaviour, give the classification of magnetic materials. Discuss general electric and magnetic characteristics of ferrites and their applications.**

[16 marks : 2011]

Solution:

On the basis of magnetic behaviour may be classified as:

1. Diamagnetic
2. Paramagnetic
3. Ferromagnetic
4. Antiferromagnetic
5. Ferrimagnetic

Diamagnetic Materials:

- The materials when placed in a magnetic field acquire weak magnetization in a direction opposite to that of the applied field are called diamagnetic materials.
- For diamagnetic material magnetic susceptibility is small and –ve.

  e.g. Cu, Au, Ge, Si, diamond, Al_2O_3.

Paramagnetic Materials:

- The materials when placed in a magnetic field acquire a weak magnetization in the same direction as the applied field are called paramagnetic. The permanent magnetic moment of the atoms or ions are acted upon individually with no mutual interaction among them (randomly distributed) this effect is called paramagnetism.
- The magnetic susceptibility X_m is +ve and very small for paramagnetic materials.

  e.g. Fe_2O_3, $MnSO_4$, $FeSO_4$, $FeCl_2$, $NiSO_4$ etc.

Ferromagnetic Materials:

- Ferromagnetic materials are spontaneously magnetized even in the absence of external magnetic field. All ferromagnetic materials becomes paramagnetic above a temperature called curie temperature T_c.
- Magnetic susceptibility X_m of ferromagnetic material ($T < T_c$) increases with increase in temperature where it reduces with increase in temperature ($T > T_c$)
- The value X_m is large and +ve.

  e.g. Fe, Cobalt (CO) Nickel (Ni), Gadolinium (Gd)

Antiferromagnetic Materials:

- In antiferromagnetic materials the magnetic moments of adjacent atoms align in opposite directions so that the net magnetic moment of specimen is nil even in presence of applied field.
- Magnetic susceptibility is positive and of the order of 10^{-3} to 10^{-5}.
- X_m increases with temp and reaches at maximum at a certain temperature called nil temperature T_N.
- Above T_N the material becomes paramagnetic.

  e.g. MnO, MnS, Cr_2O_3, NiCr, MnF_2 etc.

Ferrimagnetic Materials:

- In ferrimagnetic substances the magnetic moment of adjacent atoms are also aligned in opposite moment i.e. not zero. However it is less than ferromagnetic materials.
- This moment disappears above curie tempered material becomes paramagnetic.
  e.g. $MO.Fe_2O_3$, where M is a bivalent metal e.g. Fe, Cu, Ni etc.

General electric and magnetic characteristics of ferrites:

- Ferrites (Ferrimagnetics) are ceramic materials (oxides) and that they are poor electrical conductors.
- They exhibit characteristic of hysteresis loop below curie temperature.
- Magnetic susceptibility X_m is +ve and very large.
- They exhibit spontaneous magnetization.
- Ferrite is a material whose formula is $MO.Fe_2O_3$ where M is a bivalent metal e.g. Fe, Cu, Ni etc.

Application:

- Magnetically hard ferrites for permanent magnets.
- Magnetically soft ferrites for transformer and inductors.
- Rectangular loop ferrites for data stores.
- Microwave application.

3.16 Discuss classification of Magnetic materials List properties of ferromagnetic materials. In a certain region, $\mu = 4.6\,\mu_0$ and $B = 10\,e^{-y}\,\vec{a}_z$. Calculate the values (i) c_m, (ii) H and (iii) M.

[12 marks : 2012]

Solution:

With respect to their magnetic behaviour, materials may be classified as diamagnetic, paramagnetic, ferromagnetic, antiferromagnetic and ferrimagnetic. Briefly the manifestation of magnetic behaviour may be attributed to electron orbital motion, to electron spin and to nuclear spin.

Diamagnetism has its origin in the circulating charges in the orbits and hence, all the materials exhibit diamagnetism. Only those materials which posses permanent magnetic dipoles can be paramagnetic, ferromagnetic, antiferromagnetic or ferrimagnetic.

When the individual dipoles do not interact with each other, and are randomly distributed in absence of the field, paramagnetism arises. In other three magnetic behaviours, there is sufficient interaction between neighbouring dipoles.

If interaction leads to parallel alignment of the neighbouring magnetic moments ferromagnetic behaviour is exhibited.

In case the result is antiparallel spin arrangement, antiferromagnetic (where the compensation of magnetic moments is complete and the net magnetism is zero) or ferrimagnetism (where the compensation is incomplete arises).

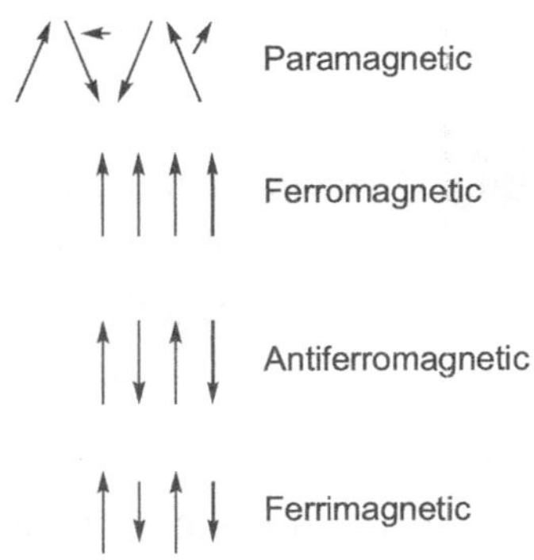

Ferromagnetic materials have following properties:

- In ferromagnetic materials, the magnetic lines of forces due to the applied magnetic field are strongly attracted towards the material.
- All ferromagnetic materials become paramagnetic above a temperature called Curie temperature T_c.
- Permeability is greater than 1.
- Magnetic susceptibility is large and positive.
- Magnetic susceptibility decreases with the rise in temperature according to Curie-Weiss law.
- Ferromagnetism is the property of a material to be strongly attracted to a magnetic field and to become a powerful magnet.

- The source of ferromagnetism is the spin of the electrons.
- Ferromagnetic materials like Fe, Co, Ni, have incomplete inner shells. These shells can be completed by using Hund's rule.
- When the specimen of a ferromagnetic material is magnetized by gradually increasing the magnetising fields, then the change of magnetic flux through the material is not continuous but in small discrete steps. Along the steep portion of the M-H magnetization curve, the discontinuous rotation of the magnetic domains give rise to Barkhauszen effect.

(i) Given, $\mu = 4.6\,\mu_o \;\Rightarrow\; \mu_r = 4.6$

$$\chi_m = \mu_r - 1 = 4.6 - 1 = 3.6$$

(ii) $$\vec{B} = \mu\vec{H}$$

$$\Rightarrow \quad \vec{H} = \frac{1}{\mu}\vec{B} = \frac{1}{4.6\,\mu_o}10e^{-y}\hat{a}_z$$

$$= 1.73\times10^6\, e^{-y}\hat{a}_z \text{ A/m}$$

(iii) $$\vec{M} = \chi_m\vec{H} = 3.6\times1.73\times10^6\, e^{-y}\hat{a}_z$$

$$= 6.23\times10^6 e^{-y}\hat{a}_z \text{ A/m}$$

3.17 In a certain region ($\mu = 4.6\,\mu_0$), $\vec{B} = 10\,e^{-y}\,\vec{a}_z$ mWb/m^2. Find:

(i) **χ_m** ***(ii)*** **$\vec{H}$** ***(iii)*** **$\vec{M}$** **[8 marks : 2014]**

Solution:

($\mu = 4.6\,\mu_0$), $$\vec{B} = 10e^{-y}\,\vec{a}_z \text{ mWb/m}^2$$

(i) $$\chi_m = (\mu_r - 1) = (4.6 - 1) = 3.6$$

(ii) $$\vec{H} = \frac{\vec{B}}{\mu} = \frac{10e^{-y}\,\hat{a}_z \text{ mWb/m}^2}{4.6\times4\pi\times10^{-7}} = \left(\frac{10\times e^{-y}\times10^{-3+7}}{4.6\times4\pi}\right)\hat{a}_z$$

$$\vec{H} = 1729.94\, e^{-y}\,\hat{a}_z$$

(iii) $$\vec{M} = \vec{H}(\mu_r - 1) = 1729.94\,(3.6)e^{-y}\,\hat{a}_z$$

$$= 6227.80e^{-y}\,\hat{a}_z$$

3.18 Give classification of magnetic materials and explain the significance of μ_r. Which of the material out of nickel, silver, tungsten and sodium chloride will require least value of magnetic strength to magnetize it. Explain briefly.

[10 marks : 2014]

Solution:

The magnetic materials are classified into:

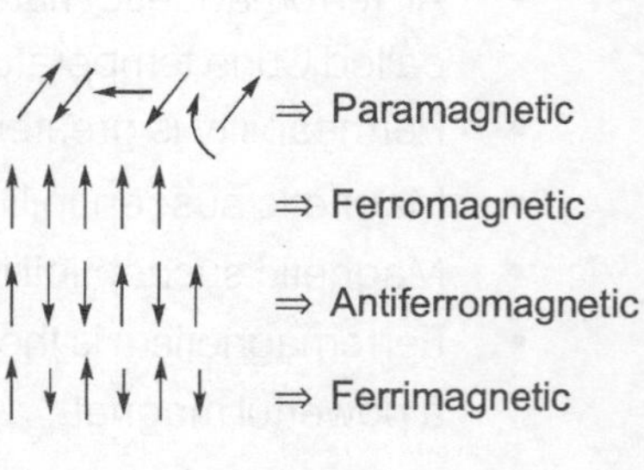

Paramagnetic: In this material, the interaction between adjacent dipole moments of atoms is negligible or zero.

Ferromagnetic: The dipole moment of neighbouring atoms are aligned in a particular direction (parallel) to each other and there is a strong magnetic field.

Antiferromagnetic: The permanent dipole moment of neighbouring atoms are aligned opposite to each other (antiparallel). So that there is no magnetization.

Ferrimagnetic: The dipole moments of neighbouring atoms are aligned antiparallel like antiferromagnetic. But these magnitudes are unequal so that there is large net magnetization like ferromagnetic materials.

Nickel is a ferromagnetic material, silver and sodium chloride is a diamagnetic material. Tungsten is a paramagnetic material.

As nickel comes under ferromagnetic which has permanent dipole moment and easy to magnetize with least value of magnetic strength to magnetize it.

3.19 The magnetic field intensity is H = 1200 A/m in a material when B = 2 Wb/m^2. When $\vec{H}$ is reduced to 400 A/m, B = 1.4 Wb/m^2. Calculate the change in magnetization M.

[10 marks : 2014]

Solution:

$B = \mu H,\ B_1 = 2,\ H_1 = 1200$

$$2 = \mu(1200)$$

$$\Rightarrow \quad \mu = \left(\frac{2}{1200}\right)$$

$$\mu_0\mu_r = \left(\frac{2}{1200}\right)$$

$$4\pi \times 10^{-7}\,\mu r_1 = \frac{2}{1200}$$

$$\mu r_1 = \frac{2\times 10^7}{1200\times 4\pi} = 1326.29$$

$$B = \mu H$$

$$1.4 = \mu\, 400$$

$$\mu = \frac{14}{400}$$

$$4\pi \times 10^{-7} \times \mu r_2 = \frac{1.4\times 10^7}{400\times 4\pi}$$

$$\mu r_2 = 2785.21$$

$$\chi_{m1} = \mu_{r1} - 1$$

$$= 1326.29 - 1 = 1325.29$$

$$M_1 = \chi_{m1} H_1 = 1325.29 \times 1200$$

$$M_1 = 15{,}90{,}348$$

$$\chi_{m2} = \mu_{r2} - 1$$

$$= 24785.21 = 2784.21$$

$$M_2 = \chi_{m2} H_2$$

$$= (2784.21)\,5\,400 = 11{,}13{,}684$$

$$\text{Change in magnetization } (M) = M_1 - M_2 = 1590348 - 1113684 = 4{,}76{,}664$$

$$\text{Percentage change} = \left(\frac{M_1 - M_2}{M_1}\right)\times 100\%$$

$$= \left(\frac{4{,}76{,}774}{1590348}\right)\times 100 = 30\%$$

3.20 **Explain the behavior of ferromagnetic materials above and below the Curie temperature.**

[10 marks, 2015]

Solution:

Curie temperature:

- Curie temperature is a certain temperature above which the spontaneous polarization of a ferroelectric material vanishes and the hysteresis loop merges into a straight line passing through the origin.
- Above the Curie temperature, the dielectric constant varies with temperature in accordance with the Curie-Weiss law as,

$$\chi = \epsilon_r - 1 = \frac{C}{T-\theta}$$

where C is a constant and θ is a characteristic temperature which is usually a few degree smaller than the ferroelectric Curie temperature θ_f.

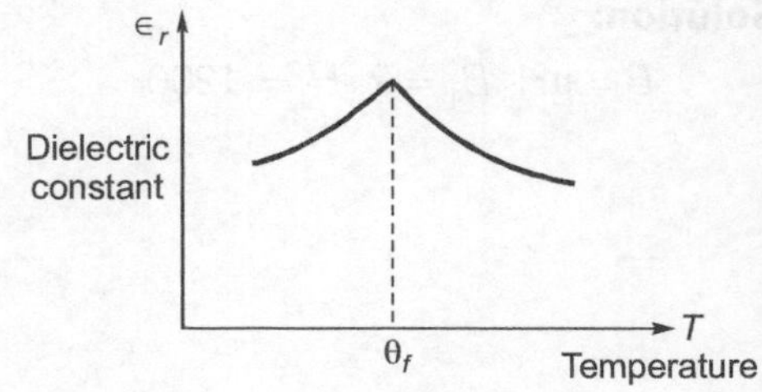

- Thus, it can be seen that for the temperature below Curie temperature θ_f, the relative dielectric constant increases with increase in temperature and material is ferroelectric. However, for the temperature above θ_f, the relative dielectric constant decreases with increase in temperature and material behaves like piezoelectric material.

3.21 **Write a note on Hysteresis with typical hysteresis loop for ferromagnetic materials.**

[5 marks : 2016]

Solution:

The ferromagnetic material show a unique B-H characteristics and this characteristics is called a hysteresis loop. Below the ferromagnetic curie temperature, ferromagnetic materials show a increase in susceptibility and exhibit *B-H* characteristics. Starting with a virgin specimen ($B = H = 0$) if H is increased, B at first increases, until at the point of saturation. This point, further increase $\bar{H}$ causes negligible change in $\vec{B}$. At saturation, if now H is decreased the curve will go through another path which shows in the figure. When $\bar{H}$ = 0, some residual flux will remain in the specimen is called residual or remnant flux density.

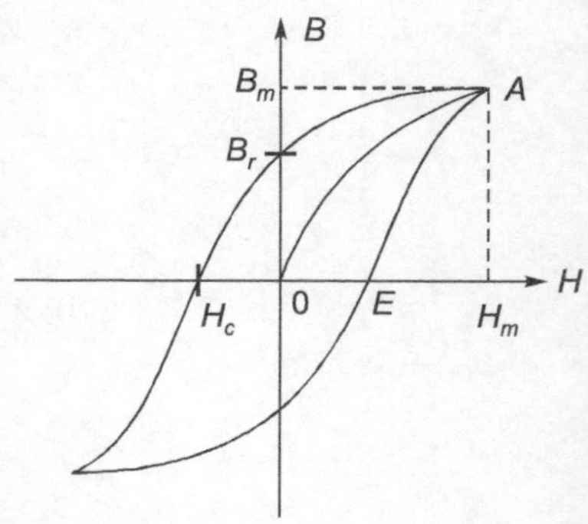

Further increase of $\bar{H}$ in negative direction, reduce the flux density to zero, is called the coercive force.

Further increase of $\bar{H}$ leads to saturation.

Depending upon the coercive force and remnant flux density magnetic material are divided into two which are soft and hard magnetic materials. Ferro magnetic material mainly we divided into two,

Ferromagnetic

Soft magnetic material	Hard magnetic material
High permeability, low retentivity, low coercivity and small hysteresis loss Suitable for making electromagnets, core of transformers etc. Example: Soft iron, supermalloy (used in magnetic shielding) 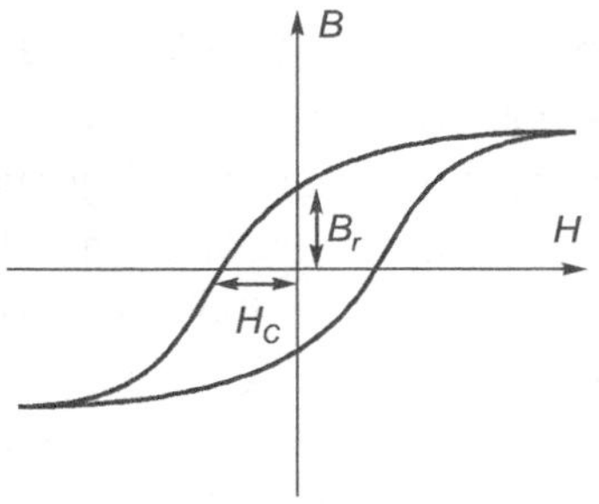 Soft magnetic material	High permeability, high retentivity, high coercivity, large hysteresis loss and high curie temperature. Suitable for permanent magnet Example: Alnico 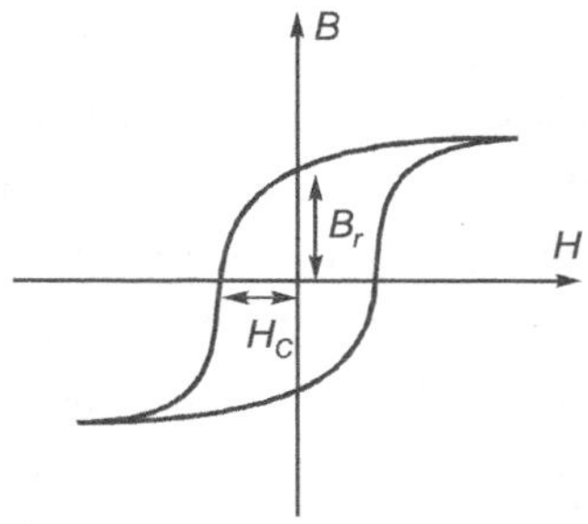 Hard magnetic material

Example for hard magnetic material

(i) **Alnico:** It is an alloy of aluminium, nickel and cobalt. Mainly used in PMMC.

(ii) **Tungsten steel:** In which tungsten content is more than (50%).

(iii) **Carbon steel:** Used in compass needle.

3.22 Compare the properties of diamagnetism, paramagnetism and ferromagnetism.

[10 marks : 2016]

Solution:

Diamagnetism:

1. These are the materials when placed in a magnetic field acquire weak magnetization in a direction opposite to that of applied field are called diamagnetic materials.
2. The value magnetic susceptibility for these material is –ve and small.
3. Due to –ve value of magnetic susceptibility they repeal the magnetic field (i.e. magnelize in opposite direction)
4. Diamagnetism has no ordinary practical use. Some of the material which only exhibit diamagnetic behavious are Cu, Au, Ge, Si, diamond, Al_2O_3, NaCl etc.

Paramagnetism:

1. The materials when placed in a magnetic field acquire a weak magnetization in the same direction as the applied field are called paramagnetic materials.
2. The value of magnetic susceptibility is +ve and small.
3. The properties of paramagnetic materials are not much of practical importance, but it helps to understand the theory of ferro and ferrimagnetic materials.
4. It is basically used in designing of transducers.
5. e.g. Fe_2O_3, $MnSO_4$, $FeSO_4$, $FeCl_2$, $NiSO_4$ etc.

Ferromagnetism:

1. Ferromagnetic materials are spontaneously magnetized even in the absence of external magnetic field.
2. All ferromagnetic materials becomes paramagnetic above certain temperature called as curie temperature
3. The value of magnetic susceptibility is +ve and very large.

4. Ferromagnetic materials are characterized by parallel allignment of eipole moment even in the absence of external field.
5. Most important ferromagnetic materials are Fe (Iron), Cobalt (CO), Nickel (Ni), Gadolinium (Gd).

3.23 What are ferrites? How they are derived? Enumerate magnetic, electrical and mechanical properties of ferrites (two each).

[ESE-2017 : 12 Marks]

Solution:

Ferrite: A ferrite is a type of ceramic compound composed of iron (III) oxide (Fe_2O_3) combined chemically with one or more additional metallic elements. They are both electrically non-conductive and ferrimagnetic i.e. they can be magnetized or attracted to a magnet. In these materials the magnetic moment (or dipole) of adjacent atoms are aligned in opposite, but the moment are not equal so that there is a net magnetic moment i.e. if the net magnetization of magnetic sublattice is not zero, the material exhibits a net magnetic moment. However, it is less than ferromagnetic material.

This magnetic moment disappears above a curie temperature T_C, at which thermal energy randomizes the individual magnetic moments and the material becomes paramagnetic.

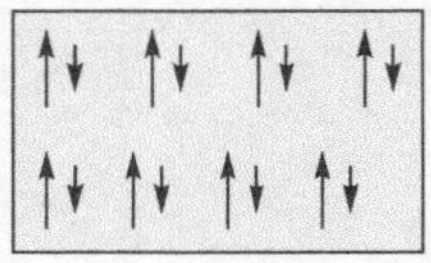
$H = 0, M \neq 0$

Properties of ferrites:

Magnetic properties:

1. High permeability
2. Low coercive force

Electrical properties:

1. Very high resistivity (> 10^5 Ω-cm)
2. Extremely low dielectric loss
3. High dielectric constant

Mechanical properties:

1. Mechanically hard and brittle
2. They have poor machinability

3.24 Explain how some ferrite is suitable for use in a magnetic memory core in computers? Explanation should be based on the hysteresis loop of the ferrite.

[8 marks : 2017]

Solution:

Some ferrite posses rectangular hysteresis loop which have been used as a magnetic memory core in computers. These ferrite are also called rectangular ferrite.

e.g.: Mn-Mg ferrite, Mn-Cu ferrite, Ni-Li ferrite etc.

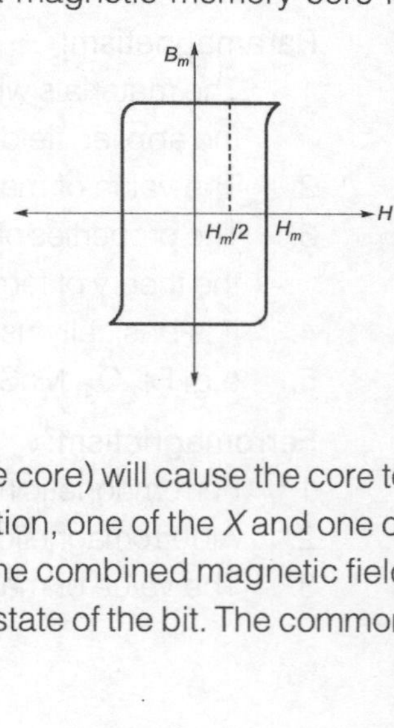

Ferrite is used for its properties of high magnetic permeability coupled with low electrical conductivity (which helps prevent eddy currents). Because of this properties, comparatively low losses at high frequencies. The memory consists of an array of ferrite cores, that are about 11 mil in diameter. A core is required for each bit of memory and has three wires passing through it, providing the means to select and detect the contents of each bit.

Only a magnetic field over a certain intensity (generated by the wires through the core) will cause the core to change its magnetic polarity (or state from '0' to '1'). To select the memory location, one of the X and one of the Y lines are driven with half the current required to cause this change. Only the combined magnetic field generated at the intersection the driven X and Y lines is sufficient to change the state of the bit. The common inhibit line for each bit, is used to alter or detect the contents of a memory location. Every memory access involves a read/write cycle. Core memory is said to be non-volatile and will not loose it contents when the power is removed.

3.25 **(i) Explain the electrochemical breakdown in insulators and discuss any two factors that accelerates the breakdown.**

(ii) A magnetic material having almost a square hysteresis loop has a coercivity of 50 A/m and a remanence of 0.5 T. If this material is used in a toroidal inductor of mean diameter 1.6 cm, with a cross-sectional area of 0.25×10^{-4} m^2, calculate the power loss at a frequency of 50 Hz when the material is driven around one complete hysteresis cycle.

[8 + 4 = 12 marks : 2018]

Solution:

(i) Electrochemical breakdown in insulators :

- In presence of air and other gases some dielectric materials undergo chemical changes when subjected to continuous stresses.
- In presence of oxygen, some materials such as rubber and polythene undergo oxidation giving rise to surface cracks.
- When moisture is present on the surface of a solid dielectric, hydrolisis occurs and the material loses their electrical and mechanical properties.
- Even in the absence of electric field, progressive chemical degradation of insulating materials can occur due to variety of processes such as chemical instability at high temperature.

Factors affecting electrochemical breakdown:

- Temperature : Electrochemcial deterioration increases very rapidly with temperature.
- Atmospheric exposure: Moist and damp condition cause fast degradation of materials.

(ii)

$$H_c = \text{coercivity} = 50 \text{ A/m}$$

$$B_r = \text{remanence} = 0.5 \text{ T}$$

$$\text{mean diameter of inductor} = 1.6 \text{ cm}$$

$$\text{cross sectional area} = 0.25 \times 10^{-4} \text{ m}^2$$

$$\text{frequency} = f = 50 \text{ Hz}$$

$$\text{energy lost per unit volume} = B \times H \text{ J/m}^3 = 50 \times 0.5 \text{ J/m}^3$$

$$E = 25 \text{ J/m}^3$$

$$\text{Volume} = \pi D \times A = 1.256 \times 10^{-6} \text{ m}^3$$

$$\text{power lost} = \frac{\text{Energy}}{\text{time}} = \text{Energy} \times \text{frequency}$$

$$\therefore \quad \text{power lost} = 25 \times 1.25 \times 10^{-6} \times 50 = 1.57 \text{ mW}$$

3.26 **(i) Explain electrical resistivity of metals in terms of thermal and residual components. Also draw the schematic variation of them with respect to temperature.**

(ii) Write the relation between magnetic susceptibility and temperature according to Curie Law, Curie-Weiss Law and Neel Law. Sketch the variation of reciprocal of susceptibility with temperature as per the above laws.

(iii) What are ferrites? Mention 3 disadvantages. State the reason, why ferrites are suitable for high frequency operation.

[8 + 6 + 6 = 20 marks : 2018]

Solution:

(i) In metals the electrons scatter where there are disruptions in the periodic lattice potential. Thus, chance encounters with impurity atoms, point defects (Vacancies, interstitials), line defects (dis locations of all types and configurations), and surface defects (grain boundaries, stacking faults) would result in scattering. it has been observed that average distance λ between scattering of each type of scatter, differ. When we have number of freely occurring scattering events, the average overall., mean free path in system in governed by mechanism with smallest λ, or

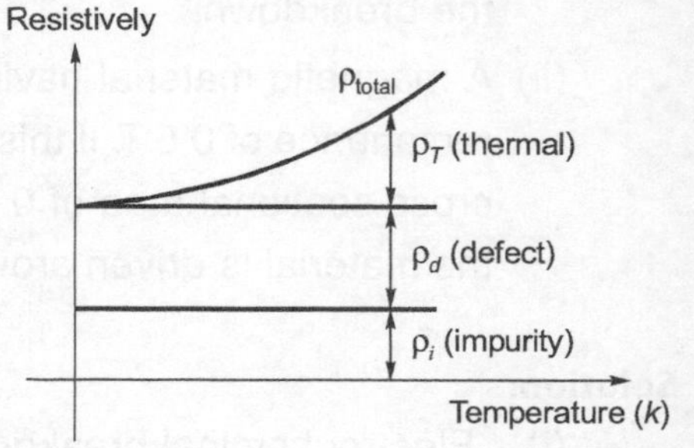

$$\lambda^{-1} = \lambda_T^{-1} + \lambda_i^{-1} + \lambda_d^{-1}$$

Since the electrical conductivity is proportional to λ, and resistivity (ρ) to λ^{-1}

$$\therefore \quad \rho_{total} = \rho_T + \rho_i + \rho_d$$

ρ_T is contribution which varies with temperature.

(ii) **Curie Law:**

It is given by;

$$\chi_m = \frac{C}{T}$$

where, $C = \frac{Np_B^2\mu_0}{kT}$ is called as curie constant.

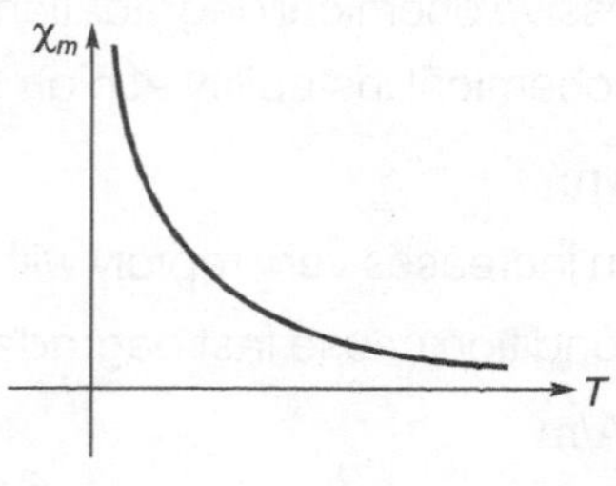

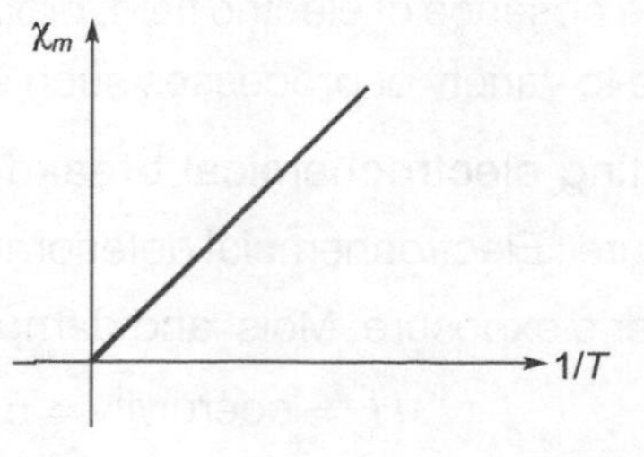

Curie-Weiss Law:

Many paramagnetic substances don't follow curie law, they follow curie weiss law.

$$\chi_m = \frac{C}{T-\theta}$$

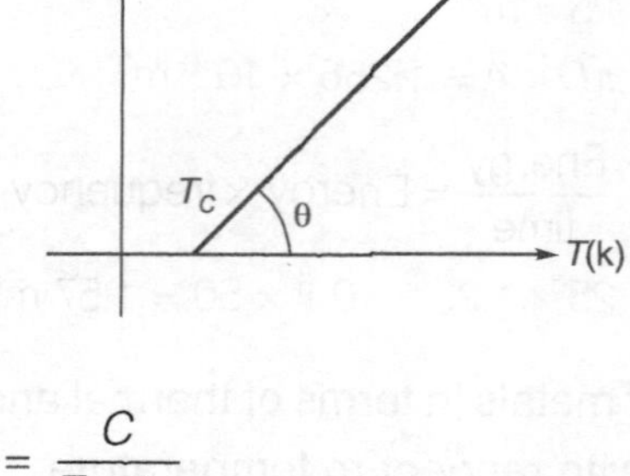

Neel Law:

$$\chi_m = \frac{C}{T+\theta_n}$$

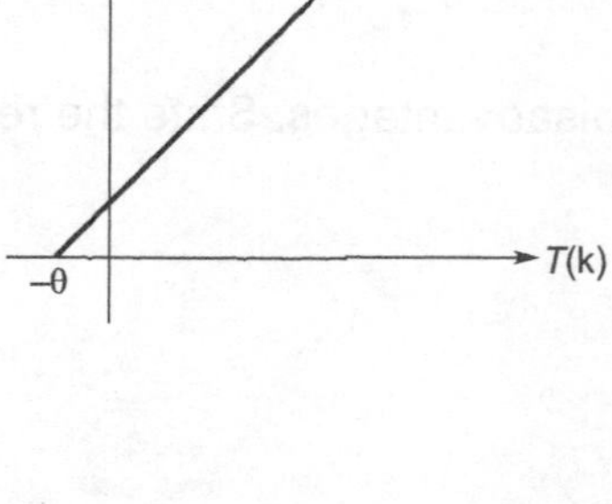

(iii) **Ferrites:** Ferrites are ceramic compounds of a mixed oxides of iron and other metals which have ferrimagnetic properties.

- Disadvantages of Ferrites:
  (i) Mechanical strength of ferrites is low, so they are easy to crack.
  (ii) Ferrites are brittle so they are difficult to mold.
  (iii) Unlike permanent magnets, ferrites require continuous supply of electric current to maintain the magnetic field.
- Ferrites are suitable for high frequency operation because they have high resistance to electric cement which reduces eddy current losses and hystesists losses at high frequency resulting into economical operation.

3.27 Supporting with suitable figures, define initial permeability and maximum permeability. Mention the approximate range of values for iron. Mention one importance each of initial and maximum permeabilities.

[8 marks : 2018]

Solution:

Initial and Maximum permeability:

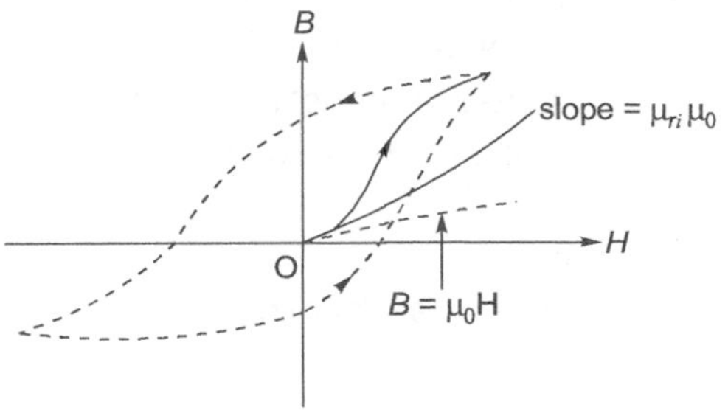

Fig. (a): Definition of intial permeability

- Initial relative permeability (μ_{ri}) represents the initial slope of the initial B versus H curve as the material is first magnetized from unmagnetized state as shown in figure (a).
- This definition is useful for soft magnetic materials that are employed at very low magnetic fields. In practice, weak applied magnetic fields where μ_{ri} is useful are typically less than $10^{-4}T$.
- μ_{ri} for iron is 150.

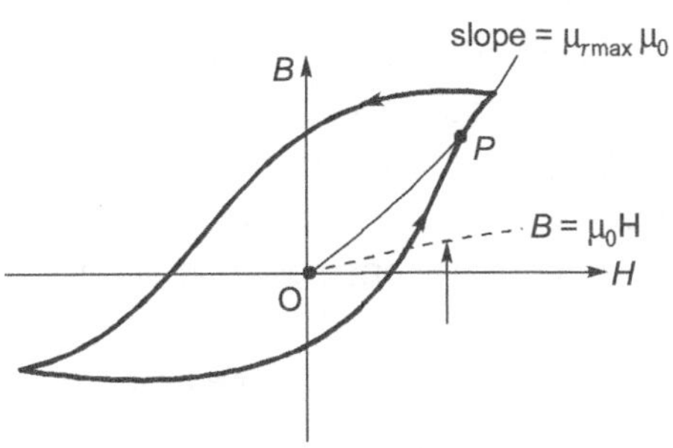

Fig. (b): Definition of maximum permeability

- $\mu_r = \dfrac{B}{\mu_0 H}$ represents the slope of the straight line from the origin O to the point P, as shown in figure (b).

  This is maximum when the line becomes a tangent to the $B-H$ curve at P, as in the figure (b). Anyother line from O to the $B-H$ curve that is not a tangent does not yield a maximum permeability.

- The maximum relative permeability, (μ_{rmax}) serves as a useful magnetic parameter. The point P in figure (b) that defines the maximum permeability corresponds to what is called the "knee" of the $B-H$ curve.
- Many transformers are designed to operate with the maximum magnetic field in the care reaching this knee point. μ_{rmax} is useful when the magnet field is not far removed from saturation.
- For pure iron, μ_{rmax} is less than 10^4 but for certain soft magnetic materials such as superalloys μ_{rmax} can be as high as 10^6.

3.28 What are magnetic materials? Give classification of magnetic materials and name some materials in each class.

[20 marks : 2019]

Solution:

Magnetic Materials: The materials which can be magnetized are called magnetic materials. All materials show some magnetic effect. In many substances the effects are so weak that the materials are often considered to be nonmagnetic. However, a vacuum is the only truly nonmagnetic medium. The response of a material at electronic, atomic, molecular and microscopic level to a magnetic field constitutes magnetic properties. Many characteristics of the magnetic materials are similar to the characteristics of the dielectric materials. Atoms and molecules give magnetic dipole moments similar to electric dipole moments. Some magnetic materials exhibit spontaneous magnetization just like spontaneous polarization in dielectrics. The study of magnetic materials can be done parallel to the study of dielectric materials. The difference is that individual electric charges of one sign do exist, whereas, magnetic monopole does not occur. While the electric field is due to fundamental charges, the magnetic field is always associated with an electric current flowing in a loop.

Classification of Magnetic Materials:

A material possesses magnetic properties on account of

1. motion of charges and
2. permanent magnetic dipoles or moments of the atom or electrons.

Hence those materials which lack permanent magnetic dipoles are called **diamagnetic.**

Those atoms which possess permanent magnetic dipoles may be **paramagnetic, ferromagnetic, antiferromagnetic or ferrimagnetic** depending on the interaction between individual dipoles.

Paramagnetic: In this material, the interaction between adjacent dipole moments of atoms is negligible or zero.

Ferromagnetic: The dipole moment of neighbouring atoms are aligned in a particular direction (parallel) to each other and there is a strong magnetic field.

Antiferromagnetic: The permanent dipole moment of neighbouring atoms are aligned opposite to each other (antiparallel) so that there is no magnetization.

Ferrimagnetic: he dipole moments of neighbouring atoms are aligned antiparallel like antiferromagnetic. But these magnitudes are unequal so that there is large net magnetization like ferromagnetic materials. Figure indicates dipole arrangements for different magnetic behaviours.

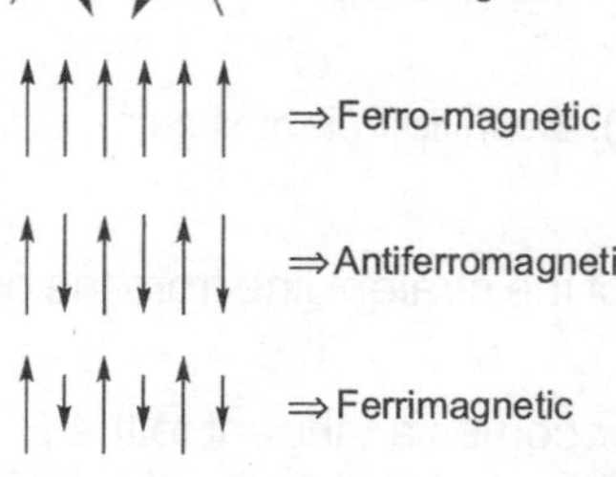

Different magnetic behaviour of materials (arrangement of dipole moments of spins)

Table summarizes features of different magnetic behaviour of materials.

Type	Susceptibility x_m	x_m Vs T relation	Examples
1. Diamagnetic	$\sim -10^{-6}$ (negative)	Independent	Atoms of solids having closed shells and some metal Au, Ge, etc.
2. Paramagnetic	$\sim -10^{-5}$ (positive)	$x_m = \frac{C}{T}$ Curie-law or $x_m = \frac{C}{T-\theta}$ Curie-Weiss law	Atoms possessing odd number of electrons, ionic crystals, etc. $MnSO_4$, Fe_2O_3 Cr_2O_3, $FeCl_2$ etc.
3. Ferromagnetic	Very large and positive	$(x_m \rightarrow \infty)$	Iron, cobalt, nickel, godolinium
4. Antiferromagnetic	Small and positive	x_m decreases with temperature	Salts and oxides of transition metals, e.g. NiO, MnF_2
5. Ferrimagnetic	Large and positive	$x_m \rightarrow \infty$	Ferrites, e.g. FeO_4

3.29 **Discuss the phenomena of Hysteresis in ferromagnetic materials. On the B-H curve show the retentivity, coercivity and saturation points. What is coercive force and the energy dissipated per unit volume of the ferromagnetic substance during the hysteresis cycle?**

[12 marks : 2020]

Solution:

When a fresh ferromagnetic material is subjected to an external field, dipoles are going to be aligned in the direction of applied H. Size of corresponding domain (D) increases. If the field is further increased, size further increases and for a particular value of $H = H_m$ no further expansion will be possible, that means saturation has been achieved

$$M_{sat} = N\beta$$

If direction of H is reversed, dipoles start dealigning and magnetization start decreasing when H is removed, there is no complete dealignment of the dipoles and have some residual magnetization remains. The magnetic flux density corresponding to this is known as **residual** or **remanent flux density** or **retentivity.**

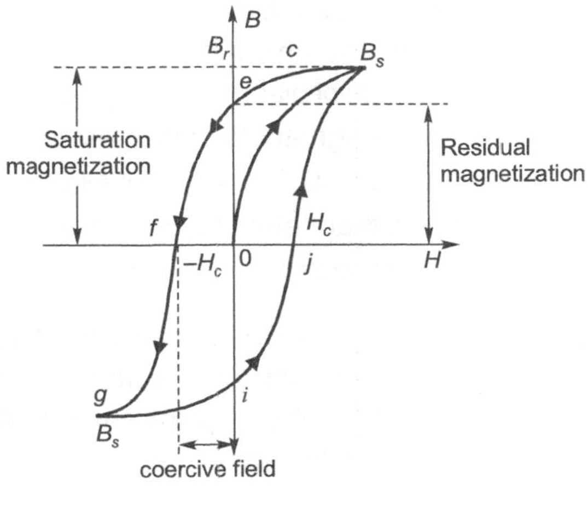

The amount of magnetic field required to be applied in reverse direction to reduce the residual magnetization to zero is known as coercive field. Thus characteristic never return to origin if once ferromagnetic material is magnetized below ferromagnetic Curie temperature. In this region the susceptibility increases with increase in temperature.

Consider small change in flux (dB),

induced emf, $$e = N\frac{d\phi}{dt} = NA\frac{dB}{dt}$$

$$dE = N\frac{d\phi}{dt} \times i \times dt$$

$$dE = NAi\frac{dB}{dt}dt$$

$$dE = NAi\,dB$$

and
$$H = N\frac{i}{l}$$

$$i = \frac{Hl}{N}$$

$\therefore$
$$dE = NA\frac{Hl}{N}dB$$

$$\frac{dE}{Al} = H \cdot dB$$

Energy loss per unit volume per cycle

$$E = \oint H \cdot dB$$

The integral gives the area enclosed by the curve.

Thus, power loss per unit volume in a hysteresis cycle equals area under the hysteresis loop.

3.30 Define magnetostriction. Explain different types of magnetostriction.

[12 marks : 2021]

Solution:

Magnetostriction is the property in which when a ferromagnetic material is magnetized, its dimension changes on application of external electric field.

There are three types of magnetostriction :

1. **Longitudinal Magnetostriction** : In this, the length of the specimen changes in the direction of magnetisation changes in dimension are observed parallel to applied magnetic field.
2. **Transverse Magnetostriction** : In this, the dimension of the specimen changes in the direction perpendicular to the direction of applied magnetic field.
3. **Volume Magnetostriction** : In this, changes in dimensions are observed in both parallel and perpendicular to the applied magnetic field (hence volume of material changes and therefore called volume magnetostriction).

Humming noise in transformer is example of magnetostriction phenomena.

Another way of classifying magnetostriction is as follows :

Positive Magnetostriction : Materials expands with increase in applied magnetic field show positive magnetostrication. e.g., Fe–Ni Alloy.

Negative Magnetostriction : Material which contrat with increase in applied magnetic field show negative magnetostriction. e.g., Co, Ni, etc.

3.31 An iron piece of 50 kg mass, 7500 kg/m^3 density is subjected to an AC supply of 50 Hz frequency; its hysteresis loop area is found to be 160 cm^2. The scale factors on ordinate and abscissa are 1 cm = 0.008 Wb/m^2 and 1 cm = 20 AT/m, respectively. Calculate the loss of energy per hour due to hysteresis in the specimen.

[10 marks : 2021]

Solution:

$$\text{Volume of 1 kg iron} = \frac{1}{\text{Density}} = \frac{1}{7500}\ \text{m}^3/\text{kg}$$

$$\text{Energy density} = \text{B.H.} = 0.008 \times 20 = 0.16$$

$$\text{Hystersis energy loss per cycle} = (\text{Area of loop}) \times \text{Energy density}$$

$$= 0.16 \times 160 = 25.6 \text{ J-kg/m}^3\text{-cycle}$$

Hystersis energy loss in 50 kg iron per cycle

$$= 25.6 \times \frac{1}{7500} \times 50 = 0.1706 \text{ J/cycle}$$

$$\text{Energy loss per hour} = 0.1706 \times 50 \times 60 \times 60$$

$$= 30720 \text{ J/hour}$$

3.32 **Classify the ways in which materials respond to magnetic fields and also specify their value of susceptibility.**

[12 marks : 2022]

Solution:

Magnetic materials are classified into following categories on the basis of their response of magnetic field applied externally.

- Diamagnetic materials
- Paramagnetic materials
- Ferromagetic materials
- Antiferromagnetic materials
- Ferrimagnetic materials

Diamagnetic Materials : They get feebly magnetised opposite to externally applied magnetic field as shown below. The value of magnetic susceptibility, χ_m for diamagnetic material is small negative of order of 10^{-8} and it is independent of temerature.

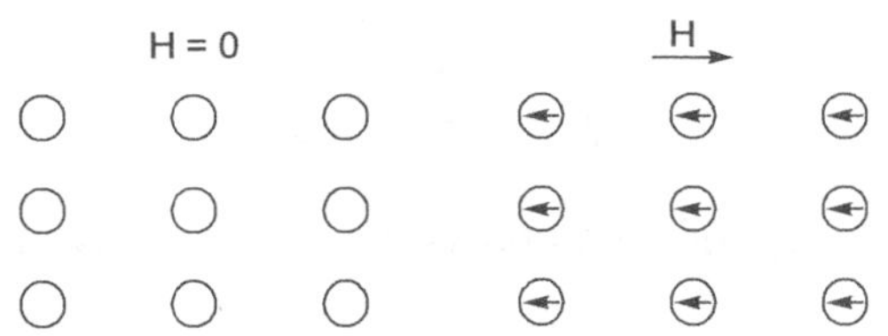

Paramagnetic Materials : These materials get feebly magnetised in the same direction of applied external magnetic field as shown below. The value of magnetic susceptibility is small positive of order of 10^{-5} to 10^{-2}. Magnetic susceptibility is inversely proportional to temperature.

Ferromagnetic Materials : These materials get strongly magnetised in the applied external field direction as shown below. The value of magnetic susceptibility is positive and very high.

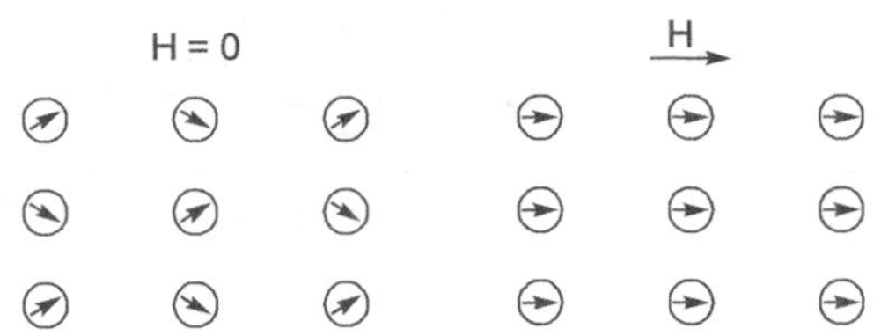

The variation of magnetic susceptibility with temperature for these material is represented in below :

$$\chi_m = \frac{C}{T - T_C}$$

where T_C is critical temperature.

Above T_C ferromagnetic materials become paramagnetic.

Antiferromagnetic Materials : In these materials magnetic moments are aligned antiparallel to each other resulting in a net zero magnetic dipole as shown below.

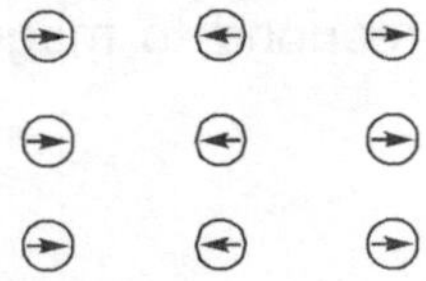

Magnetic susceptibility of antiferromagnetic material varies according to below relation

$$\chi = \frac{C}{T + T_N}$$

where T_N is Neel temperature.

Above Neel temperature antiferromagnetic material become paramagnetic.

Ferrimagnetic Materials : These materials consists of ferromagnetic parallel spins combined with some antiparallel spins such that net magnetic dipole is not cancelled to zero as in case of antiferromagnetic material which is shown below.

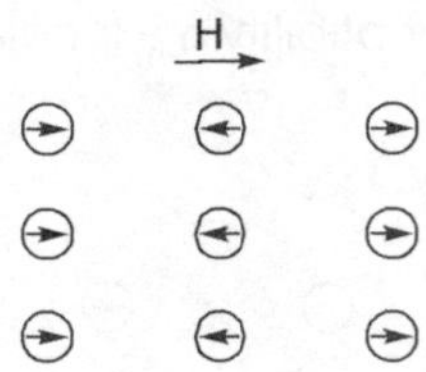

Magnetic susceptibility of ferrimagnetic material is positive, but its magnitude varies with kind of ordering and temperature.

3.33 Compare the phenomenon of Piezoelectricity and Ferroelectricity.

[8 marks : 2022]

Solution:

The phenomenon of Piezoelectricity and Ferroelectricity are compared as follows :

Piezoelectricity	Ferroelectricity
(i) Material generates potential when mechanical stress is applied across it.	(i) Material exhibits spontaneous polarisation in presence of elastic field.
(ii) Larger class of material.	(i) Ferroelectronics are subclass of piezoelectrics.
(iii) All piezoelectric materials are not ferroelectirc.	(iii) All ferroelectric materials are piezoelectrics.
(iv) Used in applications like pressure-transducer.	(iv) Used in application like transformer core.
(v) e.g., Rochelle salt quartz	(v) e.g., Lead titanate $PbTiO_3$

4. Conductive Materials

4.1 A conducting wire has a resistivity of 1.5×10^{-8} W m at 300 K. Its Fermi energy is 5 eV. Its volume of 1 m^3 contains 6×10^{28} conducting electrons. Calculate the relaxation time of conducting electrons. Hence or otherwise obtain the value of drift velocity of electrons, when the conductor is subjected to an electric field of 100 V/m. Write the significance of relaxation time and drift velocity.

Given: Mass of electron = 9.107×10^{-31} kg

Electron charge = 1.601×10^{-19} coulomb [12 marks : 2001]

Solution:

$$\text{Resistivity} = \rho = 1.5 \times 10^{-8}\ \Omega\text{-m}$$

$$W_F = 5\text{ eV}$$

$$n = 6 \times 10^{28}/\text{m}^3$$

$$\rho = \frac{m}{ne^2\tau}$$

$$\therefore \quad 1.5 \times 10^{-8} = \frac{9.11\times10^{-31}}{6\times10^{28}\times(1.6\times10^{-19})^2\times\tau}$$

$$\therefore \quad \text{Relaxation time } (\tau) = \frac{7.9\times10^{-14}}{2} = 3.95 \times 10^{-14}\text{ sec}$$

(for $\cos\theta = 0$ relaxation time = τ_c avg. collision time.)

Drift velocity,

$$V_d = \mu_n E$$

$$\therefore \quad \mu = \frac{e\tau}{m}$$

$$V_d = \frac{e\tau E}{m}$$

$$\therefore \quad V_d = \frac{1.6\times10^{-19}\times7.9\times10^{-14}\times100}{2\times9.11\times10^{-31}} = 1.38/2\text{ m/s} = 0.69\text{ m/sec}$$

Velocity of electrons corresponding to Fermi energy

$$W_F = \frac{1}{2}mv_F^2$$

$$\therefore \quad v_F^2 = \frac{5\times1.6\times10^{-19}}{1/2\times9.11\times10^{-31}}$$

$$\therefore \quad v_F = 1.4 \times 10^6\text{ m/s}$$

Effective or mean free path,

$$\lambda = V_f \times \tau$$

$$= \frac{1.4\times10^6\times7.9\times10^{-14}}{2} = \frac{1.1\times10^{-7}}{2}\text{ m} = 0.55 \times 10^{-7}\text{ m}$$

4.2 Write the properties and applications of any two noble metals used in electrical contacts. [12 marks : 2001]

Solution:

The most used noble metals and their alloys for electrical contacts are platinum and palladium.

1. **Platinum and its alloys:** Platinum has a weakness to form oxides, sulphides and chlorides etc. easily, it has high melting point and provides good resistance to electrical erosion, pure platinum is soft as electrical contact material, so is alloyed in various upto 35% common alloy elements being iridium, ruthenium and osmium. They are used for lightly loaded contacts, used upto 500 volts (a.c.) or d.c. upto 25 amps.

Example: Magneto ignition system, thermostatic relays, sensitive relays, galvanometer contacts, recording instrument, railway signal equipment, vibrator contacts etc.

2. **Palladium and its alloys:** They have similar mechanical and electrical properties to those of platinum, it is less costly. It is used for voltage upto 500 volts and upto 15 amps.

   Typical application: Telephone relays, thermostatic controls, sensitive relays and relays for chemical industry.

4.3 A copper wire of 2 mm diameter with conductivity of 5.8×10^7 S/m and electron mobility of 0.0032 m²/Vs is subjected to an electric field of 20 mV/m. Find:
(i) the charge density of free electrons
(ii) the current density
(iii) the current flowing in the wire
(iv) the electron drift velocity

[12 marks : 2002]

Solution:

$$\text{Diameter} = 2\text{ mm} \qquad \therefore r = 1\text{ mm} = 10^{-3}\text{ m}$$
$$\text{Conductivity} = 5.8 \times 10^7\text{ S/m}$$
$$\mu_e = 0.0032\text{ m}^2/\text{V-s}$$
$$E_{applied} = 20\text{ mV/m} = 0.02\text{ V/m}$$
$$\text{Cross-sectional area} = \pi \times (1 \times 10^{-3})^2 = 3.14 \times 10^{-6}\text{ m}^2 \qquad (\because A = \pi r^2)$$

(i)
$$\text{Conductivity} = ne\mu_e$$
$$\therefore \quad \text{Charge density } (ne) = \frac{5.8\times10^7}{0.0032} = 1.8 \times 10^{10}\text{ c/m}^3$$

(ii)
$$\text{Current density} = J = \sigma E$$
$$= 5.8 \times 10^7 \times 0.02 = 1.16 \times 10^6\text{ A/m}^2$$

(iii) Current flowing in the wire,
$$I = J \times A$$
$$= 1.16 \times 10^6 \times 3.14 \times 10^{-6} = 3.64\text{ A}$$

(iv)
$$\text{Electron drift velocity} = \mu_e E = 0.0032 \times 0.02 = 6.4 \times 10^{-5}\text{ m/s}$$

4.4 Explain the silver contacts. The Fermi energy for silver is determined from the relation

$$W_F = \frac{h^2}{8m}\left(\frac{3n}{\pi}\right)^{2/3}$$

Calculate the value of Fermi velocity of the electrons in silver. Given Plank's constant = 6.624×10^{-34} Js, free electrons per meter cube = 5.86×10^{28}, mass of each electron = 9.108×10^{-31} kg.

[12 marks : 2003]

Solution:

Silver contacts are alloys, which are pure silver, cadmium silver, palladium etc. Silver contacts are used upto 600 V, 50 amp. d.c. and 200 amp. a.c. They are used in relays, automatic voltage regulators, high switches etc.

The velocity of the electrons with the fermi energy is expressed as

$$v_F = \sqrt{\frac{2W_F}{m}}$$

Here,
$$W_F = \frac{h^2}{8m} \times \left(\frac{3n}{\pi}\right)^{2/3} \qquad n = \text{number of } e^{-}\text{s/m}^3$$
$$h = \text{planks constant}$$
$$m = \text{mass of free } e^{-}$$

$$= \frac{(6.624\times10^{-34})^2}{8\times9.1\times10^{-31}}\times\frac{(3\times5.86\times10^{28})^{2/3}}{(3.14)^{2/3}} = 8.82 \times 10^{-19} \text{ J}$$

Then, $$v_F = \sqrt{\frac{2\times8.82\times10^{-19}}{9.1\times10^{-31}}} = 1.39 \times 10^6 \text{ m/s}$$

4.5 Describe the factor on which the conductivity of a material depends. Give your answer discussing scattering phenomenon.

[12 marks : 2004]

Solution:

The conductivity of materials depends on

$$\sigma = qn\mu_m + qp\mu_p$$

for semiconductors and, $\sigma = qn\mu_n$ for metals

So conductivity is proportional to no. of charge carries present in material per unit volume.

Conduvtivity is proportional to mobility of charge carries which also depend on temperature (decrease with increase in temperature) so conductivity also depend on temperature.

Normally in metal conductivity decreases with temperature while in pure semiconductors conductivity increases with temperature. In extrinsic semiconductors also conductivity decreases with increases in temperature above room temperature.

Electrical resistance of solid is not the collision of free electrons with the lattice atoms but their scattering by the lattice defects which distorts the periodic lattice potential. This specific resistance independent of temperature and does not vanish at absolute zero. It is termed residual resistivity ($\rho_{residual}$). At temperature other than zero, a resistivity ρ_T due to electron scattering by the lattice vibrations is added to the residual resistivity.

$$\rho = \rho_{residual} + \rho_T = \frac{m}{ne^2}\left(\frac{1}{\tau_i}+\frac{1}{\tau_T}\right) \text{ (reciprocal of } \rho \text{ is conductivity)}$$

Now as, T increases, the amplitude of the atomic vibrations increases, hence the deviations from periodicity become more severe and the scattering by lattice vibration increases. Consequently, as T increases, τ_T decreases. Thus resistivity increases as the temperature increases. So conductivity decreases with increase in temperature.

4.6 The resistivity of copper rod as measured by Kelvin double bridge is 1.74 micro-ohm cm at 20° C. If the resistivity of the copper is 1.73 micro-ohm-cm at 20° C, calculate the impurity (%) contents in the rod.

[12 marks : 2004]

Solution:

The resistivity ρ(Scattering phenomenon)

$$= \frac{m}{ne^2}\left[\frac{1}{\tau_i}+\frac{1}{\tau_T}\right]$$

∴ Percentage of impurity $$= \frac{\sigma_{actual} - \sigma_{meas.}}{\sigma_{actual}}\times100\%$$

$$= \frac{\frac{1}{\rho_{actual}} - \frac{1}{\rho_{meas.}}}{\frac{1}{\rho_{actual}}} = \frac{\frac{1}{1.73}-\frac{1}{1.74}}{\frac{1}{1.73}}\times100\% = 0.574\%$$

4.7 **Explain electron and hole mobilities and drift velocities.** **[12 marks : 2005]**

Solution:

Conduction in conductor

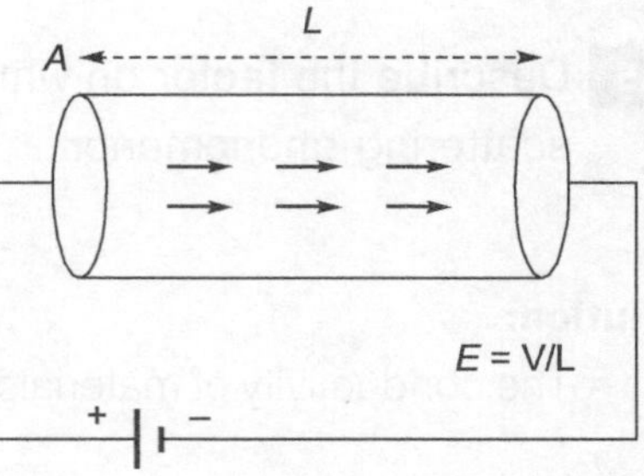

Relation between current (I) and drift velocity (v_d)

$I = neAv_d$

n = no. of electron in unit volume

$J = nev_d$

A = Cross-sectional area

$J = ne\mu E$

v_d = Drift velocity of electrons = μE

$J = \sigma E$ where, $J = I/A$ Amp/m^2 (current density)

$\sigma = ne\mu$, σ = conductivity, ρ = resistivity

Mobility, $\mu = v_d/E$, μ = mobility

Conduction in semi conductor,

Intrinsic semiconductor	*P*-type	*N*-type
$n_e = n_h$ $J = ne[v_e + v_h]$ $\sigma = \frac{1}{\rho} = en[\mu_e + \mu_h]$	$n_h >> n_e$ $J \cong en_h v_h$ $\sigma \cong \frac{1}{\rho} \cong en_h \mu_n$	$n_e >> n_h$ $J \cong en_e v_e$ $\sigma = \frac{1}{\rho} \cong en_e \mu_e$

When an electric field 'E is applied across the semi-conductor, the electrons and holes starts moving in opposite and along the direction of applied magnetic field. The steady average velocity of electrons and holes in the applied field E is called drift velocity of the carrier in the field.

$$\text{Drift velocity} = v_d = \left(\frac{e\tau}{m}\right)E$$

Here drift velocity is proportional to the field and the absolute magnitude of the proportionality is factor given by:

$$\mu_e = \frac{e\tau_e}{m_e} \text{ for electron}$$

$$\mu_h = \frac{e\tau_h}{m_h} \text{ for holes}$$

Here τ_e, and τ_h are the relaxation time of electron and holes respectively, m_e and m_h are the effective masses of electrons and holes.

4.8 **Derive an expression for the thermal conductivity of a metal. What are the desirable properties of low resistivity materials? Explain with examples and write their uses.**

[16 marks : 2008]

Solution:

Thermal conduction is the phenomenon by which the thermal energy is transferred from high temperature region to low temperature regions when both regions are brought into contact. The property that characterizes the ability of a material to transfer thermal energy (or heat) is known as thermal conductivity.

Mathematically it is defined in terms of the equation,

$$J_q = -K\frac{dT}{dx} \quad \text{unit} = \text{J/m}^2 \text{ sec.} \qquad ...(i)$$

where J_q is the heat flux (heat flow per unit time per unit area), K is the proportionality constant, called as thermal conductivity and dT/dx is the temperature gradient through the conducting medium.

The transportation of heat in solid materials is due to phonons (lattice vibration waves) and free electrons. Hence the total thermal conductivity is the sum of the two contributions;

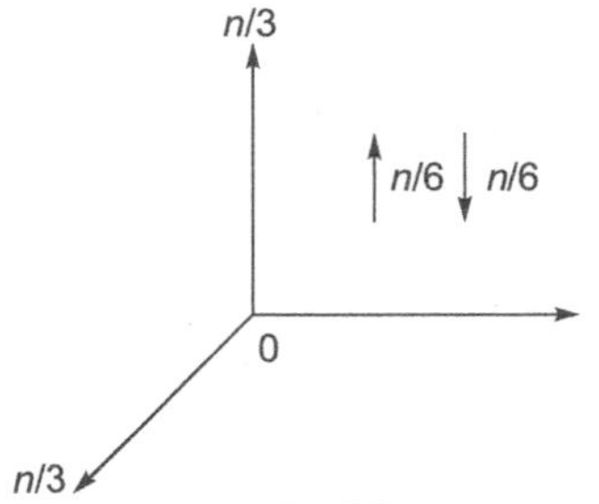

Fig. (a)
Particle moving in three directions equally

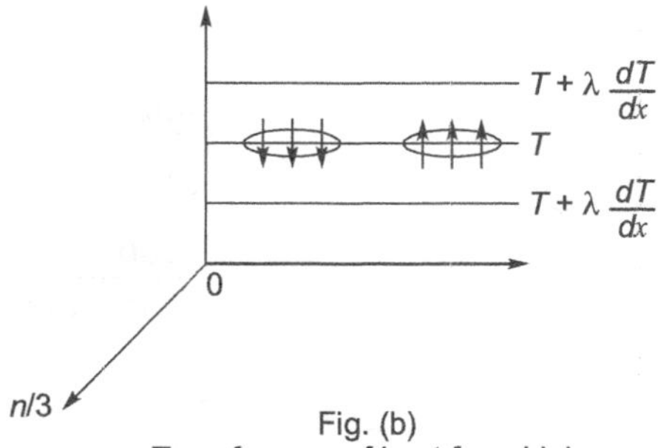

Fig. (b)
Transference of heat from high temp. to low temp. region (free path)

$$K = K_{ph} + K_e \qquad ...(ii)$$

where K_{ph} and K_e are the phonon and electron thermal conductivities, respectively.

Let us consider that metal is free electron ideal gas and electrons are moving randomly in all three possible directions [Fig. (a)]. The transference of heat from high temperature to low temperature region can be found as:

Heat transference per unit area per second upward to downward

$$= \frac{mnv}{6} C_v \left(T + \frac{\lambda dT}{dx}\right) \qquad ...(i)$$

where C_v is specific heat.

Heat transference per unit area per second downward to upward

$$= \frac{mnv}{6} C_v \left(T - \frac{\lambda dT}{dx}\right) \qquad ...(ii)$$

Net heat transfer of heat per unit area per sec is subtracting equation (*i*) and (*ii*).

$$J_q = \frac{mnv}{3} C_v \cdot \frac{\lambda dT}{dx}$$

also heat flux $\quad J_q = K\frac{dT}{dx}$

$$\therefore \quad K\frac{dT}{dx} = \frac{mnv}{3} \cdot C_v \cdot \frac{\lambda dT}{dx} \quad \text{or} \quad K = \frac{mnv}{3} \cdot C_v \cdot \lambda$$

Energy of free electron is given by:

$$mC_v T = \frac{3}{2} K_B T \qquad \text{(where } K_B \text{ is the Boltzmann's constant)}$$

$$\therefore \quad mC_v = \frac{3}{2} K_B$$

$\therefore$ Thermal conductivity, $\quad K = \frac{3}{2} K_B \frac{nv}{3} \lambda$

or

$$\boxed{K = \left(\frac{nv\lambda}{2}\right) K_B} \qquad ...(iii)$$

Specific heat of an ideal gas is given by:

$$C_v = \frac{3}{2} n K_B$$

$$\therefore \quad nK_B = \frac{2}{3} C_v$$

$$\therefore \quad \boxed{K = \frac{1}{3} C_v \lambda v} \quad \text{...(iv)}$$

As we know that electrical conductivity

$$\sigma = \frac{ne^2\tau}{m}$$

Relaxation time, $\tau = \dfrac{\text{Mean free path}}{\text{Fermi velocity}} = \dfrac{\lambda}{v_F}$

and $\dfrac{1}{2} m v_F^2 = \dfrac{3}{2} K_B T$

$$m v_F = \frac{3K_B T}{v_F}$$

$$\therefore \quad \sigma = \frac{ne^2\lambda}{m v_F} = \frac{ne^2 \lambda v_F}{3K_B T} \quad \text{...(v)}$$

Hence the ratio to thermal to the electrical conductivity [from (*iii*) and (*v*)] is

$$\frac{K}{\sigma} = \frac{n v \lambda}{2} \cdot K_B \times \frac{3K_B T}{ne^2 \lambda v} = \frac{3}{2} \frac{K_B^2}{e^2} \cdot T \quad \text{(assuming that } v = v_F\text{)}$$

$$\frac{K}{\sigma T} = \frac{3}{2}\left(\frac{K_B}{e}\right)^2 = \text{Lorentz number} \quad \text{...(vi)}$$

$$= 5.838 \times 10^{-9}\ \Omega\text{-cal/K}^2\text{-sec}$$

$$= 2.44 \times 10^{-8}\ \Omega\text{-w/K}^2$$

Hence the ratio between thermal and electrical conductivities (K/σ) is the same for all metals and is function of temperature only. This empirical law is known as Wieldeman-Franz-Lorentz relation.

Low resistivity conducting materials are used in house wiring, as conductors for power transmission and distribution, in the winding of transformer and machines like motors and generators. These types of material are in fact used in all such applications where power loss and voltage drop should be low. Copper and aluminium are example of commercially acceptable low resistivity conducting materials. Though silver has lower resistivity than copper but its very high cost restricts it to be used commercially as a conductor. Low resistivity conducting materials, besides possessing resistivity, should also possess the following properties:

1. Low temperature coefficient
2. Sufficient mechanical strength
3. Ductility
4. Solderability
5. Resistance to corrosion

e.g.: Copper: Copper possesses following properties:

1. Excellent resistance to corrosion.
2. Easy to work, it is highly ductile and malleable.
3. Good throwing power during electroplating.
4. It can be easily soldered, brazed or welded.
5. High thermal and electrical conductivity.
6. Very good machinability.

4.9 **Prove with the help of electrons flow that $W = VI$, where W is the power in watts, V voltage in volts and I current in amperes, in case of conductor materials.**

[6 marks : 2012]

Solution:

Consider a system of free electrons in a conductor which is subjected to an electric field, E V/m.

The charge flowing the section of conductor having cross-section A, with n number of electron per unit volume in a volume $A\,v_x dt$.

$$dq = en\,Av_x\,dt$$

Force on charge dq, $\quad dF = dq \cdot E$

Work done by field, $\quad dW = dqE \cdot dx$

Power, $\quad dP = \frac{dW}{dt} = \frac{dq}{dt} E \cdot dx = IEdx$

Total power consumed in a conductor,

$$P = \int IEdx = IEl \qquad \text{(where } l \text{ is the length of the conductor)}$$

$$= J \cdot AEl = \sigma E \times AEl \qquad \text{(from ohm's law)}$$

$$= \sigma \frac{V}{l} \cdot A \cdot \frac{V}{l} \cdot l = \frac{\sigma A}{l} \cdot V^2 = \frac{1}{R} \cdot V^2 = \frac{V}{R} \cdot V = VI \qquad \text{(from ohm's law)}$$

4.10 **Find the mean velocity of electron flow in a conductor having a cross-sectional area of 2.1×10^{-6} m^2 when a current of 20 amperes flows through it. Assume that there are 8.5×10^{28} electrons/m^3 of the material. Charge on an electron is 1.6×10^{-19} coulombs.**

[6 marks : 2012]

Solution:

$$A = 2.1 \times 10^{-6}\text{ m}^2, \quad I = 20\text{ A}$$

$$n = 8.5 \times 10^{28}\text{ electrons/m}^3, \; e = 1.6 \times 10^{-19}\text{ C}$$

Current, $\quad I = neAv_d$

$\Rightarrow$

$$v_d = \frac{I}{neA} = \frac{20}{8.5 \times 10^{28} \times 1.6 \times 10^{-19} \times 2.1 \times 10^{-6}}$$

$$= 0.7 \times 10^{-3}\text{ m/sec.} = 0.7\text{ mm/sec.}$$

4.11 **Explain basic the factors affecting the resistivity of electrical materials.**

[5 marks : 2013]

Solution:

Factors affecting resistivity of materials:

1. **Charge carrier density (n):** Resistivity of an electric material is inversely proportional of an electric carriers present in the material $\boxed{\rho \propto \frac{1}{n}}$. As number of charge carriers increase, conductivity increases and hence resistivity decreases.

2. **Average collision time (τ):** $\boxed{\rho \propto \frac{1}{\tau}}$ Resistivity is inversely proportional to average collision time.

   Also, $\tau \propto \frac{1}{\sqrt{\text{Temp.}}}$ so if n is constant $\rho \propto \sqrt{\text{Temp.}}$

   So, resistivity increases with increase in temperature if n is constants (in case of metals).

4.12 **Derive an equation to correlate the conductivity σ with number of charge carriers *n* and their mobilitiy, μ.**

[12 marks : 2014]

Solution:

The '*n*' number of charges are flowing through a cross section '*A*' with a velocity 'v' then the current (I)

$$I = n\,q\,v\,A$$

Velocity of charge carriers is related to electric field as

$$v = \mu E$$

$$I = n\,q\,\mu E\,A$$

$$\text{Current density } (J) = \frac{I}{A}$$

$$J = n\,q\,\mu E = \sigma E$$

$$\boxed{\sigma = nq\mu}$$

where, σ = conductivity

μ = mobility

4.13 **Discuss in detail various factors affecting the resistivity of electrical materials.**

[5 marks : 2016]

Solution:

Factors effecting the resistivity of electrical materials are listed below:

1. Temperature
2. Alloying
3. Mechanical stressing
4. Age Hardening
5. Cold Working

Temperature: The resistivity of materials changes with temperature. Resistivity of most of the metals increase with temperature. The resistivity of metals increase with increase of temperature. Means the metals are having positive temperature coefficient of resistance. Several metals exhibit the zero resistivity at temperature near to absolute zero. This phenomenon is "called the superconductivity". The resistivity of semiconductors and insulators decrease with increase in temperature. Means the semiconductors and insulators are having negative temperature coefficient of resistance.

Alloying: Alloying is a solid solution of two or more metals. Alloying of metals is used to achieve some mechanical and electrical properties. The atomic structure of a solid solution is irregular as compare to pure metals. Due to which the electrical resistivity of solid solution increases more rapidly with increase of alloy content. A small content of impurity may increase the resistivity metal considerably. Even the impurity of low resistivity increase the resistivity of base metal considerably. For example the impurity of silver (having lowest resistivity among all metals) in copper increase the resistivity of copper.

Mechanical stressing: Mechanical stressing of crystal structure of material develops the localized strains in material crystal structure. These localized stains disturb the movement of free electrons through material. Which results in increase in resistivity of material. Subsequently annealing, of metal reduces the resistivity of metal. Annealing of metal, relive the mechanical stressing of material due to which the localized stains got removed from the crystal structure of metal. Due to which the resistivity of metal decrease. For example, the resistivity of hard drawn copper is more as compare to annealed copper. Age Hardening Age hardening is a heat treatment process used to increase the yield strength and to develop the ability in alloys to resist the permanent deformation by external forces. Age hardening is also called "Precipitation Hardening". This process increases the strength of alloys by creating solid impurities or precipitate. These created solid impurities or precipitate, disturb the crystal structure of metal which interrupts the flow of free electrons through metal/Due to which the resistivity of metal increases.

Cold Working: Cold working is a manufacturing process used to increase the strength of metals. Cold working is also known as "Work hardening" or "Strain hardening". Cold working is used to increase the mechanical strength of metal. Cold working disturbs the crystal structure of metals which interfere with movement of electrons in metal, due to which the resistivity of metal increases.

4.14 Write a note on emission of electrons from metals. **[10 marks : 2016]**

Solution:

Electron Emission Definition: Generally, emission refers to a process by which any object becomes free. In the similar way, the process by which the free electrons becomes free from the metal and enters into the vacuum is called electron emission.

What is electron emission?

When a small amount of external energy is applied to the valence electrons, they gain enough energy and break the bonding with the parent atom. The electron, which breaks the bonding with the parent atom, moves freely from one place to another place. These electrons, which move freely within the metal, are called free electrons. However, these free electrons cannot escape from the surface of a metal. It is because the free electrons in the metals do not have enough energy to escape from metal. The free electrons, which try to escape from the metal, are pulled by positive electric force of atomic nucleus towards the metal. Hence, free electrons cannot escape from metal without sufficient energy. Thus, the positive atomic nucleus of the metal opposes the free electrons, which try to escape from the metal. In order to make the free electrons to escape from the metal, an additional external energy is required. This additional external energy is applied in the form of heat, electric field, or light. When the external energy in the form of heat, light, or electric field is applied to the metal, the free electrons gain enough energy and break the bonding with the metal. Free electrons, which break the bonding with metal, will jumps into the vacuum. The free electrons in the vacuum are not associated with the metal. The process by which, these free electrons break the bonding with metals and jumps into the vacuum is called electron emission. The amount of external energy required to remove an electron from the surface of a metal is called work function. Work function is measured in joules.

Different methods of Electron Emission:

The different methods to emit free electrons from the metal surface include:

1. Thermionic emission
2. Electric field electron emission
3. Photoelectric emission
4. Secondary emission
5. ionic emission

Thermionic Emission Definition: The process by which free electrons are emitted from the surface of a metal when external heat energy is applied is called thermionic emission. Thermionic emission occurs in metals that are heated to a very high temperature. In other words, thermionic emission occurs, when large amount of external energy in the form of heat is supplied to the free electrons in the metals. When a small amount of heat energy is applied to the metal, the valence electrons gain enough energy and break the bonding with the paren tatom. The valence electron, which breaks the bonding with the parent atom, becomes free. This electron, which breaks the bonding with the parent atom, is called as the free electron. The free electrons in the metal have some kinetic energy. However, they do not have enough energy to escape from the metal. The attractive force of the atomic nuclei opposes the free electrons, which try to escape from the metal. Free electrons in the metal have less energy compared to the free electrons in vacuum. Hence, free electrons require extra energy from the outside source in order to jump into the vacuum.

When heat energy applied to the metal is increased to a higher value, the free electrons gain enough energy and overcome the attractive force of the atomic nucleus, which holds the free electrons in the metal. The free electrons, which overcome the attractive force of the nuclei, break the bonding with the metal and jumps into the vacuum. The free electrons, which are escaped from the surface of a metal when heat energy is supplied, are called thermions. Thermionic emission process plays a major role in the operation of electronic devices.

Applications of Thermionic Emission: The components, which are made by the process of thermionic emission are used in the electronic devices such as cathode ray tube, radio etc.

Electric Field Electron Emission: Electric field electron emission is the process by which free electrons are emitted from the metal surface when strong electric field is applied. Electric field electron emission is also called as electron field emission, field electron emission and field emission. Electric field electron emission occurs not only in metals, but also in liquids. Field electron emission occurs in metals that are placed at very strong electric field. In other words, field electron emission occurs when large amount of energy in the form of electric field is applied to the free electrons in the metals. If the strong electric field applied to the metal is great enough, the free electrons in the metal gains enough energy and break the bonding with metal or overcome the attractive force from the atomic nuclei. The free electrons which breaks the bonding with the metal, will jumps into the vacuum.

Photoelectric Emission: The process by which free electrons are emitted from the metal surface by the application of light is called photoelectric emission. It is also defined as, the process by which free electrons are released from the metal when it absorbs light energy. Photoelectric emission is also called as photoemission or photoelectron emission or photoelectric effect. In this method, light or photons is used to remove the free electrons from the solid metal. Hence, the free electrons emitted from the solid metal are called photoelectrons and current produced due to this process is called photoelectric current. When the light energy is not applied to the metals, the free electrons cannot escape from the metals. However, some of the valence electrons become free from the atoms. When the light energy is applied to the metal, the free electrons gains energy. In other words, when the light particles (photons) hits free electrons in the metal, they transfer their energy to the free electrons.The free electrons, which gains extra energy from the light, will try to the overcome the attractive force of the nuclei. If the light energy applied to the metal is further increased to a higher value, the free electrons in the metals gain sufficient energy and overcome the strong attractive force of the nuclei. The free electrons, which overcome the attractive force of the nuclei, will jumps into the vacuum.

Secondary Electron Emission: Secondary electron emission is the emission of free electrons from the metal surface, which occurs when the high-speed electrons or primary electrons hit the free electrons or secondary electrons in the metal. Under normal temperature, some of the valence electrons gain sufficient energy from the heat source and breaks the bonding with the parent atom. The valence electrons that break the bonding with the parent atom moves freely from one point to another point within the metal are called free electrons. Free electrons have some kinetic energy. But, they do not have sufficient energy to escape from the metal. The attractive force of the nuclei resists the free electrons that try to escape from the metal. The free electrons in the metal need sufficient energy from the outside source to escape from the metal. When the high-speed electrons strike the metal surface, they transfer their kinetic energy to the free electrons in the metals. The free electrons that gains sufficient energy will overcomes the strong attractive force of the nuclei and escapes from the metal. The high-speed electrons that strike the free electrons in the metal are called primary electrons whereas the free electrons emitted from the metal surface are called secondary electrons. We are emitting the secondary electrons from the metal surface. Hence, the process of emitting the secondary electrons from the metal surface is called secondary emission. Secondary electron emission depends on the number of primary electrons and energy of primary electrons. Commonly used electron emissive materials include alkali antimonide, lead oxide, magnesium oxide, and gallium phosphide.

4.15 **Calculate the number of vacancies per cubic meter in iron at 850° C with the following data:**

Energy vacancy formation = 1.08 eV/atom
One mole = 6.022×10^{23} atoms
Density at 850° C = 7.65 g/cm^3
Atomic weight of iron = 55.85 g/mol
Boltzman's constant = 1.38×10^{-23} J/atom-K

[12 marks : 2017]

Solution:

The number of vacancies per cubic meter in iron at 850°C (1123 K) is

$$N_v = N\exp\left(-\frac{Q_v}{kT}\right) = \frac{N_A \rho_{Fe}}{A_{Fe}}\exp\left(-\frac{Q_v}{kT}\right)$$

Incorporation of values of the parameters provided in the problem statement into the above equation leads to

$$N_v = \frac{(6.022\times10^{23}\,\text{atoms/mol})\,(7.65\,\text{g/cm}^3)}{55.85\,\text{g/mol}}\exp\left[-\frac{1.08\,\text{eV/atom}}{(8.62\times10^{-5}\,\text{eV/atom-K})\,(850°\text{C}+273\,\text{K})}\right]$$

$$= 1.18\times10^{18}\,\text{cm}^{-3}$$

4.16 **Addition of 0.25 atomic percent nickel and 0.4 atomic percent silver into copper at 298 K increases the resistivity by 0.012 mW cm and 0.016 mΩm respectively. If the resistivity of copper is 0.015 mΩm at 298 K, determine the conductivity of the resulting alloy.**

[4 marks : 2018]

Solution:

$$\rho_{alloy} = \rho_{\text{pure metal}} + \rho_{\text{impurity}}$$

Given,

$$\rho_{\text{pure cu}} = 0.015\times10^{-6}\,\Omega\text{.m}$$

$$\rho_{\text{nickel impurity}} = 0.012\times10^{-6}\times10^{-2}\,\Omega\text{.m}$$

$$\rho_{\text{silver impurity}} = 0.016\times10^{-6}\,\Omega\text{.m}$$

$$\rho_{alloy} = (0.015+0.012\times10^{-2}+0.016)10^{-6}$$

$$= 0.03112\times10^{-6}\,\Omega\text{.m}$$

$$\text{Conductivity of alloy} = \sigma_{alloy}$$

$$= \frac{1}{\rho_{alloy}} = \frac{1}{0.03112\times10^{-6}}$$

$$\sigma_{alloy} = 3.2133\times10^{7}\,\Omega^{-1}\,\text{m}^{-1}$$

4.17 **Discuss the factors affecting electrical resistance of conductors.**

[10 marks : 2020]

Solution:

Factors that affect electrical resistance of conductors are shown below :

- **Temperature** : Average collision time of electron in conducting material reduces with increase in temperature. Since the conductivity of the material is directly proportional to average collision time. Therefore conductivity reduces with increasing temperature. Since resistivity is the reciprocal of the conductivity, so resistivity of conducting material increases with increase in temperature. The resistivity of conducting material as a function of temperature is given below:

$$\rho_{T_2} = \rho_{T_1}(1+\alpha\,\Delta T)$$

where, $\Delta T = T_2 - T_1$ = difference of temperature

α = temperature coefficient

Since the electrical conductivity is proportional to l, and resistivity (ρ) to l^{-1}.

Hence, $$\rho_{Total} = \rho_T + \rho_i + \rho_d = \rho_T + \rho_r$$

Equation is statement of **"Matthiessen's Rule"** and it indicates the various contribution to the resistivity of metals are independently additive.

Also, $\rho_T \rightarrow$ ideal resistivity varies with temperature.

$\rho_r \rightarrow$ residual resistivity (depends on impurity).

NOTE: The temperature coefficient of conducting material is positive.

- **Alloying :** The resistivity of alloy has two components called (a) thermal component and (b) residual component. **Thermal component** of conducting material is due to lattice vibration caused by thermal energy. This component reduces with decrease in temperature and becomes zero at absolute zero temperature. The **residual component** of resistivity arise due to impurity of material and lattice imperfection. This component is independent of temperature and will be existing even at absolute zero temperature. Therefore, alloying can induce the resistivity. Alloys generally have a less regular structure than pure metals. Consequently, the electrical conductivity of a solid solution alloy, drops off rapidly with the increase alloy content.

Resistivity of alloy (ρ_{alloy}) = $\rho_{puremetal} + s\rho_i$

where, s = atomic percentage of added impurity

ρ_i = increase in resistivity per atomic%

Unit of "ρ_{alloy}" is μ-ohm-cm.

When a copper is added to silver (Ag), the resistivity of alloy is more than resistivity of copper (Cu), and conductivity of alloy is less than conductivity of copper (Cu).

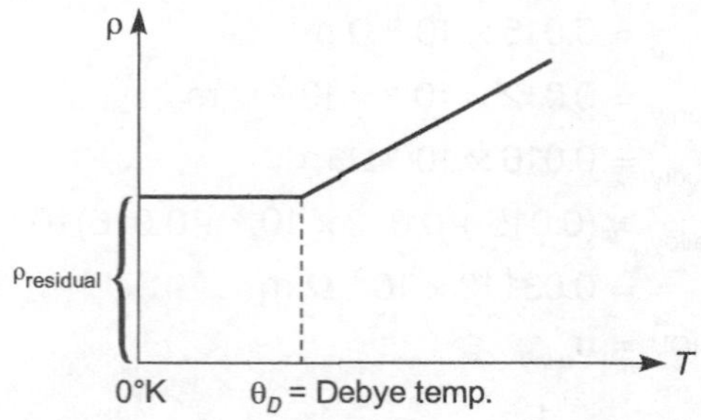

When a small amount of Ni, i.e., 1% is a added to Cu, the resistivity of Cu goes up by 1.3 μ-ohm-cm.

- **Cold Working :** The mechanical treatments such as cold working produces localized strain in the material which results in the increase in resistivity of material. **A hard drawn copper wire has lower conductivity than annealed copper.**
- **Age Hardening :** If material gets hardened with age then its resistivity increases and hence conductivity decreases.

4.18 Discuss photoelectricity and photoemissive effect along with laws of photoemissive effect.

[20 marks : 2020]

Solution:

Photo electric effect is phenomena in which electrically charged particles are released from or within a material when it absorbs electromagnetic radiation. A material that can exhibit the photo electric effect is said to be photo emissive.

The photons must have sufficient energy to overcome the work function of the material. If this condition is not met, increased irradiance and longer exposure time will not induce emission of free electrons from the surface.

Following observations were noted during photoemissive effect :

- The electrons are emitted immediately without any time lag.
- On increasing the intensity of light, there is an increase in the number of photo electrons, but not to their maximum kinetic energy.
- Red light never causes the emission of electrons, irrespective of any value of intensity of red light.
- A weak violet light ejected only a few electrons, but their maximum kinetic energies are greater than those for intense light of longer wavelengths.
- The energy of the emitted electrons is proportional to the frequency of illuminating light.
- The emission energy is independent of the total energy of illumination.

Laws of photo electricity:

1. Photo electric effect is an instantaneous process.
2. Photo electric current is directly proportional to the intensity of incident light and is independent of frequency.
3. The stopping potential and hence the maximum velocity of the electrons depends upon the frequency of incident light and is independent of its intensity.
4. The emission of electrons stops below a certain minimum frequency known as threshold frequency.

4.19 (i) **Define Hall effect. With a sketch, explain the concept of Hall effect and arrive at an equation for Hall voltage V_H.**

(ii) **The superconducting state of a lead specimen has critical temperature of 7.26 K at zero magnetic field and the critical field is 8 × 10^5 A/m at 0 K. Estimate the critical field at 5 K for this specimen.**

[10+10 = 20 marks : 2021]

Solution:

(i) **Hall Effect :** Hall effect is production of a voltage difference also known as Hall voltage across an electrical conductor, that is transverse to an electric current in the conductor and to an applied magnetic field perpendicular to the current. This effect also occurs across a void or hole in a semiconductor. When current is injected via contact on the boundary or edge and the charge flows outside the void or hole in metal or semiconductor.

Consider the specimen with current and magnetic field as shown below :

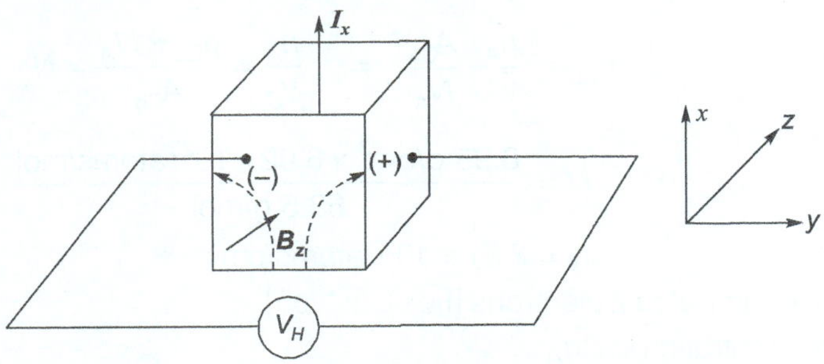

On applying magnetic field in positive z-direction (B_z), +ve charges are deflected towards positive y-direction.

–ve charges are deflected towards negative y-direction.

Magnitude of Hall voltage,

$$V_H = \frac{R_H I_x B_z}{d}$$

V_H depends on I_x, B_z and thickness d, where R_H is termed as Hall coefficient.

For metals, where conduction is in electrons,

$$R_H = \frac{1}{n|e|}$$

Electron mobility,
$$\mu_e = \frac{\sigma}{n|e|}$$

$$\mu_e = |R_H|\sigma$$

The magnitude of μ_e may be determined if the conductivity σ has also been measured.

(ii) Critical temperature at zero magnetic field : 7.26 K.

Critical field at zero kelvin, $H_C(0) = 8 \times 10^5$ A/m

Critical field at 5 K $H_C(T)$

$$H_C(T) = H_C(0)\left[1-\left(\frac{T}{T_C}\right)^2\right]$$

$$= 800{,}000\left[1-\left(\frac{5}{7.26}\right)^2\right]$$

$$= 800000 \times 0.525 = 4.205 \times 10^5 \text{ A/m}$$

4.20 **Compute the drift mobility and mean scattering time of conduction electrons in copper at room temperature, given that the density of copper is 8.98 g/cm³, the conductivity of copper is 5.95 × 10⁵ Ω⁻¹ cm⁻¹ and the atomic mass of copper is 63.5 g/mol. Take Avogadro number, N_A = 6.02 × 10²³ and charge on electron (e) = 1.6 × 10⁻¹⁹ coulomb, mass of electron (m_e) = 9.1 × 10⁻³¹ kg.**

[8 marks : 2023]

Solution:

Given :
$$\rho_{Cu} = 8.98 \text{ g/cm}^3$$
$$\sigma = 5.95 \times 10^5 \text{ cm}^{-1}$$
$$A_{Cu} = 63.5 \text{ g/mol}$$
$$N_A = 6.02 \times 10^{23} \text{ atoms/mol}$$

Atomic concentration in Cu

$$\rho_{Cu} = \frac{n_A}{V_C}\cdot\frac{A_{Cu}}{N_A} \Rightarrow \frac{n_A}{V_C} = \frac{\rho_{Cu} \times N_A}{A_{Cu}} = N$$

$$N = \frac{8.98 \text{ g/cm}^3 \times 6.02 \times 10^{23} \text{ atoms/mol}}{63.5 \text{ g/mol}}$$

$$N = 8.51 \times 10^{22} \text{ atoms/cm}^3$$

Each copper atom contributes 2 electrons [i.e., Cu^{2+} ion].

Hence, electron concentration in Cu,

$$n = 2N = 17.02 \times 10^{22} \text{ electrons/cm}^3$$

$$\Rightarrow \quad n = \frac{17.02 \times 10^{22}}{(10^{-2})^3 \text{m}^3} = 1.702 \times 10^{29} \text{ electrons/m}^3$$

Now, conductivity
$$\sigma = nq\mu_n \Rightarrow \mu_n = \frac{\sigma}{nq}$$

$$\Rightarrow \quad \mu_n = \frac{5.95\times10^5}{1.702\times10^{29}\times1.6\times10^{-19}}$$

$$= 2.185 \times 10^{-5}\ m^2/V$$

$$\Rightarrow \quad \mu_n = 2.185 \times 10^{-5}\ m^2/V\text{-}s = \text{mobility}$$

$$\mu_n = \frac{q\tau_C}{m_n} \Rightarrow \tau_C = \frac{\mu_n m_n}{q} = \text{mean scattering time}$$

$$\Rightarrow \quad \tau_c = \frac{2.185\times10^{-5}\times9.1\times10^{-31}}{1.6\times10^{-19}}$$

$$\Rightarrow \quad \tau_c = 1.25 \times 10^{-16}\ s$$

5. Semiconductor Materials

5.1 Explain Hall effect in *P*-type semiconductors. Hence or otherwise define Hall coefficient. Resistivity of a sample semiconductors is 9 milli ohm meter. Its holes have mobility of 0.03 m²/v.s. Calculate the value of Hall coefficient of the sample. State assumptions made.

[12 marks : 2001]

Solution:

When a metal or a semi-conductor carrying a current I is placed in a transverse magnetic field B, a potential difference is produced in the direction normal to both the current and magnetic field directions. This phenomenon is called Hall effect.

Suppose p-type semi-conductor:

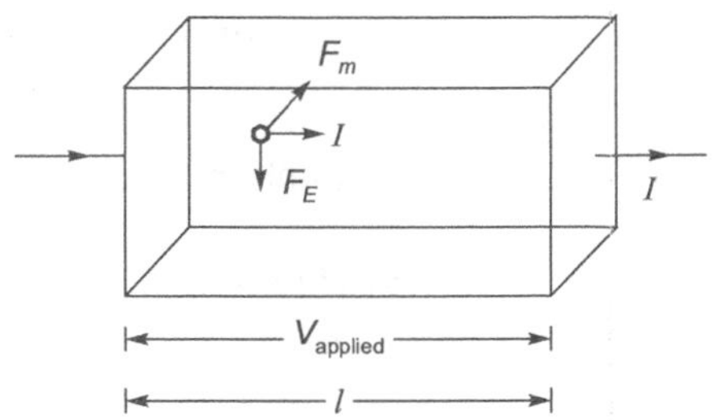

Applied longitudinal electric field $[E_l] = \frac{V_{applied}}{l}$

drift velocity of charge carriers

$$V_d = \mu_p E_l$$

$\mu_p \rightarrow$ mobility of holes

If magnetic field is applied in perpendicular direction

then, magnetic force = $F_m = eV_d B$

Induced voltage (Hall effect voltage) = V_H

$\therefore$ Induced electric field $(E_H) = \frac{V_H}{d}$

E_H is perpendicular to both E_l and B

At equilibrium, $F_e = F_m$

$$e E_H = e V_d B$$

$$\therefore \quad V_d = \frac{E_H}{B}$$

Hall coefficient, $$R_H = \frac{1}{\text{Concentration of free charge}}$$

$$= \frac{1}{\rho\times e} \quad \text{[For } p\text{-type semi-conductor]}$$

$$\text{Conductivity} = \rho \times e \times \mu_p = \frac{\mu_p}{R_H}$$

Assumption: all carriers travel with a mean speed 'V_d' however this does not happen

Here, Resistivity $= 9 \times 10^{-3}\ \Omega\text{-m} = \dfrac{R_H}{\mu_p}$

$\therefore$ Resistivity $= \dfrac{1}{\text{conductivity}}$

$\therefore$ $R_H = 9 \times 10^{-3} \times 0.03 = 0.27 \times 10^{-3}\ \text{m}^3/\text{c}$

5.2 **Find the conductivity of *n*-type Ge at room temperature, assuming one donor atom in each 10^8 atoms. The density of Ge is 5.32×10^3 kg/m³ and the atomic weight is 72.6 kg/k-mol. Comment on the result. ($e = 1.6 \times 10^{-19}$ C, $\mu_e = 0.38$ m²/Vs, $\mu_h = 0.18$ m²/Vs)**

[12 marks : 2002]

Solution:

Avogadro number $= N_A = 6.024 \times 10^{23}$ /mole

Atomic-weight = 72.6 gm/mole

$\because$ Atomic weight = 72.6 kg/k-mole

$= 72.6 \text{ gm} \times 10^3/10^3 \text{ mole} = 72.6$ gm/mole

$$\text{Number of atoms in 1 kg material} = \frac{6.024\times10^{23}}{72.6\times10^{-3}} = \frac{6.024\times10^{26}}{72.6}/\text{kg}$$

$$\text{Volume corresponding to1 kg} = \frac{1}{5.32\times10^{3}} = 1.87 \times 10^{-4}\ \text{m}^3/\text{kg} = (1/\text{density})$$

$$\text{No. of atoms in 1 m}^3 \text{ of material} = \frac{6.024\times10^{26}}{72.6\times1.87\times10^{-4}} = 4.43 \times 10^{28}$$

$$\therefore \text{Number of donor atoms/m}^3 = \frac{4.43\times10^{28}}{10^{8}} = 4.43 \times 10^{20}/\text{m}^3 \qquad (\because \text{doping is } 1 : 10^8)$$

$\therefore$ Concentration of free electrons $= n = 4.43 \times 10^{20}/\text{m}^3$

$$\text{Conductivity} = \sigma = \sigma_e + \sigma_p = ne\,\mu_e + pe\,\mu_p$$

$$\simeq ne\,\mu_e \simeq 4.43 \times 10^{20} \times 1.6 \times 10^{-19} \times 0.38$$

$$= 26.93 \text{ S/m} \qquad (\because \text{holes} << e^- s)$$

So conductivity of *n*-type Ge mainly depends e^- s concentration and it is much-much greater than that of intrinsic Ge.

5.3 **The Hall coefficient of a specimen of doped silicon is found to be 3.66×10^{-4} m³/C the resistivity of the specimen is 8.93×10^{-3} Ωm. Find the mobility and the density of the charge carriers, assuming single carrier conduction.**

[12 marks : 2002]

Solution:

$$\text{Hall coefficient} = R_H = \frac{1}{ne}$$

$$\text{Resistivity} = \frac{1}{ne\mu}$$

$$\therefore \text{Mobility} = \mu = \frac{1}{ne}\times\frac{1}{\rho} = \frac{R_H}{\rho} = \frac{3.66\times10^{-4}}{8.93\times10^{-3}}\frac{\text{m}^3/\text{c}}{\Omega\text{-m}} = 0.04\ \text{m}^2/\text{V-s}$$

and
$$\text{Density of charge carriers} = n = \frac{1}{R_H\times e} = \frac{1}{3.66\times10^{-4}\times1.6\times10^{-19}} = 1.707\times10^{22}/\text{m}^3$$

5.4 "Hall effect has played a decisive role in revealing the mechanism of conduction in semiconductors". Explain the statement. The single carrier holes in a doped silicon sample is $2.05 \times 10^{22}\ m^{-3}$. Calculate its Hall coefficient. Electron charge = 1.602×10^{-19} Coulombs.

[12 marks : 2003]

Solution:

The Hall effect has played a decisive role in revealing the mechanism of conduction in semiconductors. The conductivity of semiconductors does not determine the mobility and the density of charge carriers separately (determines the product of the two). However, in case there is only one type of charge carriers (either electron or holes); the density of charge carriers can be found from the measurement of the, so called Hall coefficient R_H of the material. Hall coefficient for holes,

$$R_H = \frac{1}{en_h} = \frac{1}{2.05\times10^{22}\times1.602\times10^{-19}} = 3.045 \times 10^{-4}\ m^3/C$$

5.5 An *n*-type germanium crystal has a current density of 100 A/m^2. The crystal has a resistivity of 0.5 ohm-m and electron mobility of 0.4 m^2/V-s. Calculate the drift velocity and the time taken by electron to travel 10 micro-m in the crystal (Assume $e = 1.6 \times 10^{-19}$ C).

[12 marks : 2004]

Solution:

Current density, $J = 100\ A/m^2$

$\rho = 0.5\ \Omega\text{-m}$

$$\sigma = \frac{1}{\rho} = 2\ \Omega^{-1}\ m^{-1}$$

mobility = $0.4\ m^2/V\text{-s}$

$J = \sigma E$

where, $\sigma = ne\mu_e$

$2 = n \times 1.6 \times 10^{-19} \times 0.4$

$\therefore$ $n = 3.125 \times 10^{19}$

and $J = nev_d$

$100 = 3.125 \times 10^{19} \times 1.6 \times 10^{-19} \times v_d$

$\therefore$ drift velocity $v_d = 20$ m/s

To travel distance (10 μm) time taken is

$t = l/v_d$

$$t = \frac{10\times10^{-6}}{20} = 0.5\ \mu s$$

5.6 Explain the current conduction mechanism in a semiconductor. Explain how doping changes the conductivity of a semiconductor? Show that the Fermi level lies halfway between the valence and conduction band of an intrinsic semiconductor.

[12 marks : 2005]

Solution:

The current flowing through a pure semiconductor is carried by two kinds of carriers, namely electrons and holes Conductivity of semiconductor is given by.

$$\sigma = \frac{n_e e^2 \tau_e}{m_e} + \frac{n_h e^2 \tau_h}{m_h}$$

Where n_e and n_h are the electrons and holes concentration per m^3 respectively τ_e and τ_h are relaxation time of electrons and holes respectively.

All of the carriers then arise from the electron-hole pairs that are generated when thermal energy breaks the covalent bonds. Electrons and holes are thus always present in equal concentration in an intrinsic semiconductor.

In intrinsic semiconductor, the conduction is due to electrons and holes from Si or Ge atoms and not from any foreign atoms. The electron and holes are collectively termed charge carriers. With doping of certain impurity, the conductivity of semiconductor can be increased to 100 times.

In n-type extrinsic semiconductor, Ge or Si is doped with an element from fifth column, for example Ge or Si doped with phosphorus, antimony and arsenic. The extra electrons increase the conductivity many times to that of intrinsic level.

In energy band diagram the energy level of this electron will close to conduction band as indicated in the following diagram.

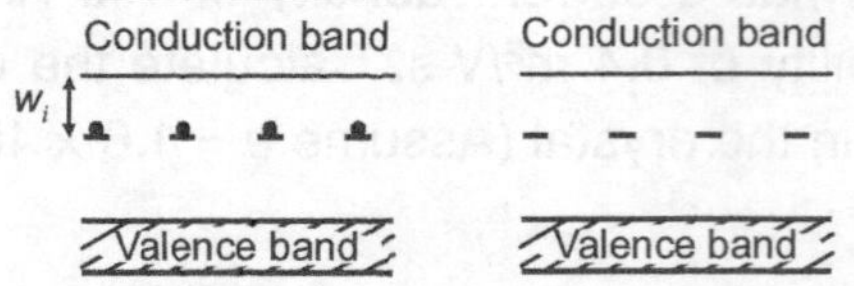

Impurity atoms like arsenic, that contribute free electrons as described above called donors, because they donate electrons.

Now for p-type intrinsic semiconductor, Ge or Si is doped with boron, gallium or aluminium. Due to doping with trivalent atoms the hole is generated. The holes are the main contributors of the conductors of the conductivity of p-type semiconductor. Energy band diagram for acceptor impurity can be shown below.

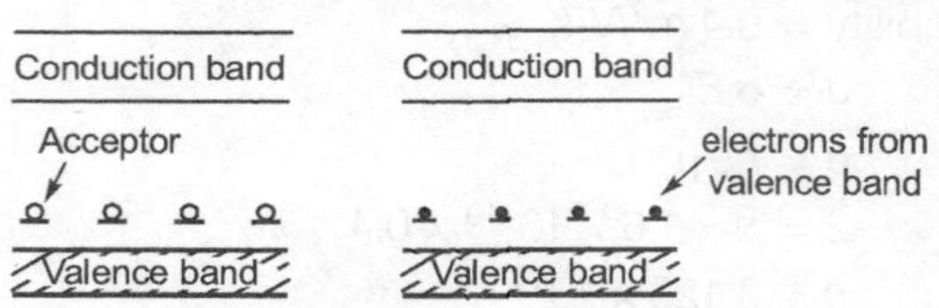

Fermi level for semiconductor lies halfway between the valence band and conduction band of an intrinsic semiconductor. This can be shown from the expression of the probability of a state corresponding to an energy w_0 being occupied by an electron which is given by

$$f(W) = \frac{1}{\exp[(W - W_F)/KT] + 1}$$

at fermi level $W = W_F$

$$f(W) = \frac{1}{\exp[0] + 1} = \frac{1}{2}$$

Hence, fermi level for an intrinsic semiconductor lies midway between the valence band and conduction band.

5.7 Describe Hall effect and find out the value of Hall coefficient, based on classical approach with specific charge carrier.

[12 marks : 2006]

Solution:

When a metal or a semiconductor carrying a current I is placed in a transverse magnetic field B, a potential difference is produced in the direction normal to both the current and magnetic field directions. This phenomenon is called Hall effect.

Hall effect measurements showed that it is the negative charge carriers namely electrons that are responsible for electrical conduction in metals. It also showed that there exist types of charge carriers in semiconductors.

Importance: The importance of Hall effect is that it helps to determine the:

(*i*) sign of charge carriers,

(*ii*) charge carriers' concentration and

(*iii*) mobility of charge carriers if conductivity of the material is known.

Let us consider an *n*-type semiconductor in which the conduction is predominated by electrons. Suppose an electric current j flows in the positive x-direction and a magnitude field B is applied normal to this electric field z-direction. A force, called the Lorentz force is exerted on each electron which causes the electron paths to bend. As a result of this, the electrons accumulate on one side of the slab and are deficient on the other side. Thus, as electric field is created in the y-direction which is called the Hall field.

$$\text{Hall force} = \text{Lorentz force}$$

$$F_H = F_L$$

$$-eE_H = -v_x B_z e$$

where v_x is the velocity of the electrons, and e the electronic charge.

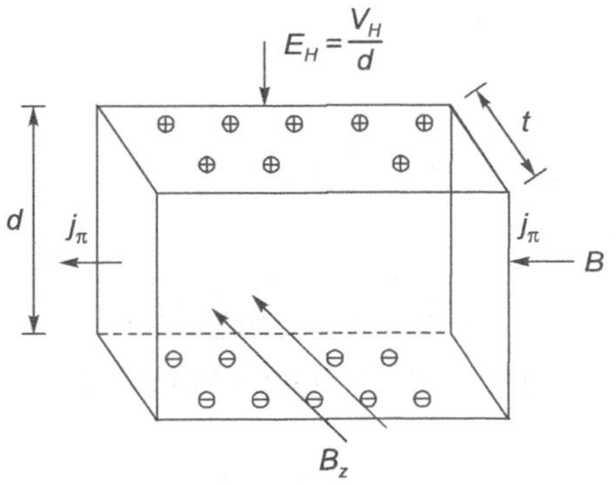

$$\therefore \quad E_H = v_x B_z \qquad \text{...}(i)$$

As current density, $\quad j_x = nv_x e \qquad \left[\text{Putting } v_x = \frac{E_H}{B_z}\right]$

$$j_x = \frac{nE_He}{B_z} = -\frac{nV_He}{B_zd}$$

$$V_H = -\frac{j_x B_z d}{ne} = -\frac{IB_z d}{neA} \qquad [A = \text{Area of cross-section of end face}]$$

If t is the thickness of the semiconductor specimen, $A = td$ and above equation reduces to

$$V_H = \frac{B_z I}{net} \qquad \text{...}(ii)$$

Hall field per unit current density per unit magnetic induction is called Hall coefficient R_H.

Thus, $$R_H = \frac{E_H}{j_x B_z} = \frac{\frac{V_H}{d}}{j_x B_z} = -\frac{B_z I}{j_x B_z \cdot net}$$

$$\boxed{R_H = -\frac{1}{ne}} \qquad \text{...}(iii)$$

In terms of Hall coefficient, Hall voltage is given by:

$$\boxed{V_H = R_H \frac{BI}{t}} \qquad \text{...}(iv)$$

The sign of the Hall coefficient R_H indicates whether electrons or holes predominate in the conduction process.
If, R_H = negative, then electrons are the predominant charge carriers
Thus, the magnitude of the μ_e may be determine if the conductivity — also been measured.
The net electric field E in the semiconductor is a vector sum of E_x (electric field component in x-direction) and E_H (Hall field). It acts at an angle θ_H to the x-axis. θ_H is called the Hall angle.

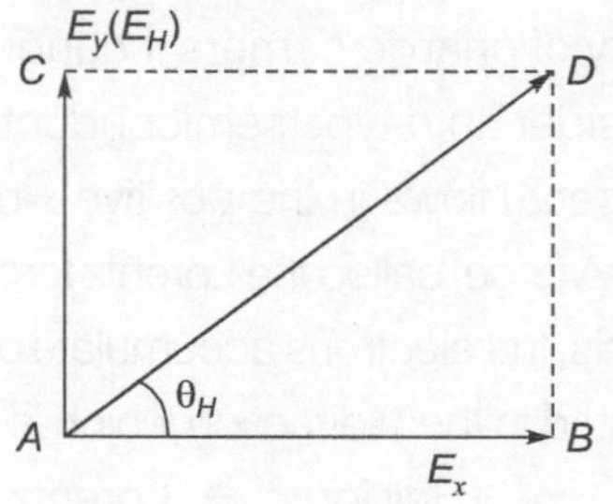

$$\tan\theta_H = \frac{BD}{AB}$$

$$\tan\theta_H = \frac{B_H}{E_x}$$

$$\tan\theta_H = \frac{E_H}{E_x}$$

As, $$E_H = \frac{V_H}{d} = \frac{B_z j_x}{n|e|}$$

Also $$E_x = \rho j_x \quad \text{...(v)}$$

$\therefore$ $$\tan\theta_H = \frac{B_z}{n|e|\rho} \quad \left[\because \sigma = \frac{1}{\rho}\right]$$

$\therefore$ $$\tan\theta_H = \sigma R_H B_z$$

The product σR_H is designated as μ_e the mobility of electrons.

$\therefore$ $$\tan\theta_H = \mu_e B_z$$

$\therefore$ $$\boxed{\theta_H = \tan^{-1}(\mu_e B_z)}$$

In the above discussions it is assumed that all carriers travel with a mean speed v_x, however this does not happen.

5.8 Draw the energy band diagram of an n type semiconductor. State the factors on which the position of Fermi level depends. A sample of *n*-type semiconductor has a resistivity of 0.1 ohm cm and a Hall coefficient of 100 cm³/coulomb. Assuming only electrons as carriers, determine the electron density and mobility.

[12 marks : 2006]

Solution:

n-type semiconductors are those in which Ge or Si is doped with an element from the fifth column, for example, Ge or Si doped with phosphorus, antimony or arsenic. These impurities from the fifth column have five electrons in its outer shell available for interaction with other atoms. Assume, that a small amount of arsenic (As), as impurity is added to silicon, such as one arsenic impurity atom per million atoms of silicon. Such impurity atoms are comparable in size to silicon atoms; they can, therefore take positions normally occupied by silicon atom as shown in figure because the impurity atoms (arsenic) are proportionally so few in number, each will ordinarily be entirely surrounded by silicon atoms.

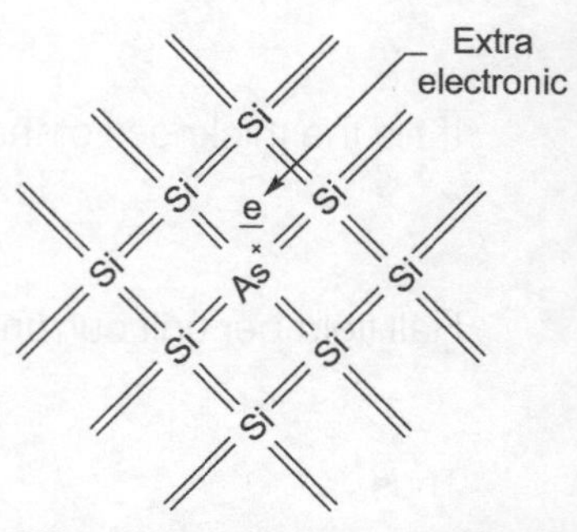

Silicon lattice with donor subsitituent

The four nearest silicon atoms then form covalent bonds with four of the five valence electrons of the arsenic atom. This ties these four electrons into the crystal structure with forces that require a considerable amount of energy to disrupt. However, no such binding force holds the fifth valence electron. In the energy band diagram, the energy level of this electron will be close to the conduction band as indicated.

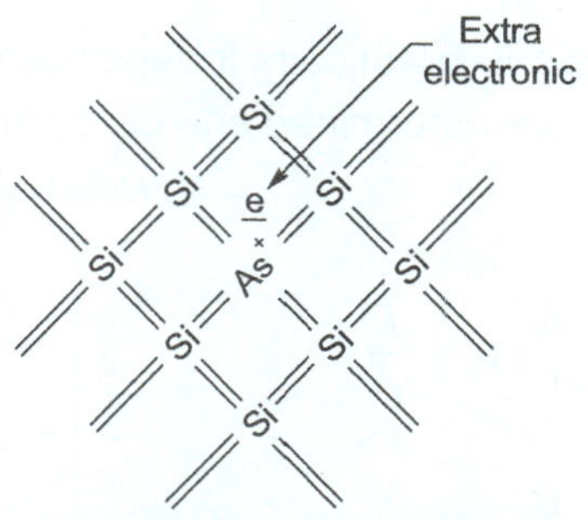

Silicon lattice with donor subsitituent

In fig. Being almost free, it can be removed from the parent impurity atom by the expenditure of very little energy W_i, only 0.01 electron volt for germanium. When free from its parent impurity atom, this fifth valence electron moves randomly through crystal in the same manner as do the freeelectrons present in an intrinsic semiconductor. When an electric field is applied, there is superimposed upon this random motion a steady drift towards the positive electrode that represents a flow of current.

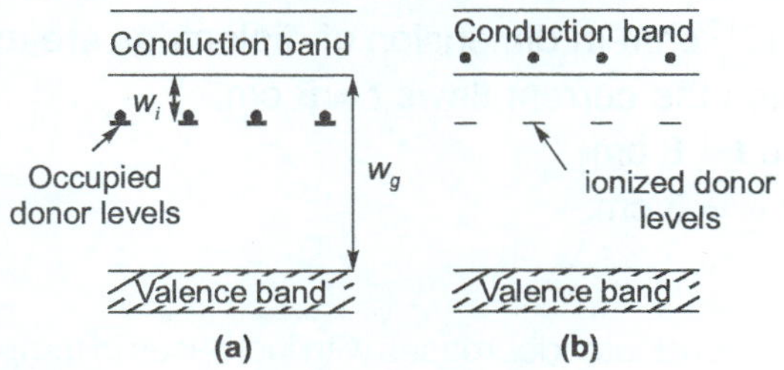

(a) Energy band diagram (b) Donor atoms are ionized by thermal excitation, leading to conduction by electron

Fermi level lies just below donor ionization level and depend on temp and doping project.

When the impurity atom loses an electron, it becomes a positively charged ion. The ion is immobile, being bound by four covalent bonds, cannot contribute to the conduction of electricity.

Impurity atoms like arsenic, that contribute free electrons as described above, are called donors, because they donate electrons.

$$\text{Resistivity} = 0.1\ \Omega\ \text{cm}$$

$$R_H = 100 = \frac{1}{ne}$$

$$\therefore \quad n = \frac{1}{100 \times 1.6 \times 10^{-19}} = 6.25 \times 10^{16}\ \text{cm}^{-3}$$

$$\text{Conductivity} = \sigma = ne\mu$$

$$\therefore \quad \frac{1}{\rho} = ne\mu$$

$$\therefore \quad \mu = \frac{1}{\rho \times n \times e} = \frac{R_H}{\rho} = \frac{100}{0.1} = 10^3\ \text{cm}^2\ \text{volt}^{-1}\ \text{sec}^{-1}$$

5.9 Explain with graphical representation the variation of mobility with temperature.

[12 marks : 2006]

Solution:

Electrical conductivity of pure metals is due to the drift of free charge carriers of one sign. In the absolute majority of pure metals, the charge carriers are free electrons. The conductivity or specific conductance of electron metals is given by

$$\sigma = ne\mu$$

The electron concentration in metals is practically independent of temperature for them, temperature dependence of specific conductance is determined entirely by the temperature dependence of mobility of the electrons.

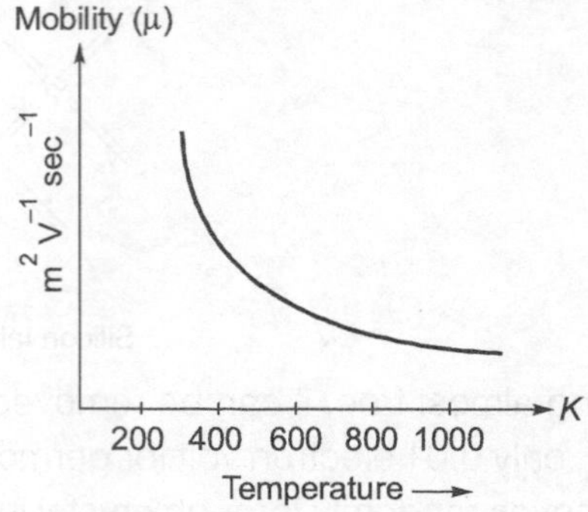

In the high temperature,

$$\sigma = \frac{A}{T},\ \rho = \frac{T}{A}$$

In the low temperature range,

$$\sigma = \frac{B}{T^5}$$

∴ In high temperature,

$$\mu \propto \frac{1}{T}$$

5.10 **Draw the variation of resistance with temperature in an intrinsic semiconductor. Find the value of insulation resistance of a dielectric whose volume resistivity is 5×10^{16} ohm-m, whose sheet resistivity is 10^{18} ohm-m dimension of dielectrics are as follows:**
The thickness through which the current flows t = 5 cm
The length of the dielectric l = 8 cm
The width of the dielectric ω = 3 cm. **[10 marks : 2007]**

Solution:

The resistance of intrinsic semiconductor decreases with increases in temperature as it conductivity increases with temperature.

R is inversely varying with temperature.

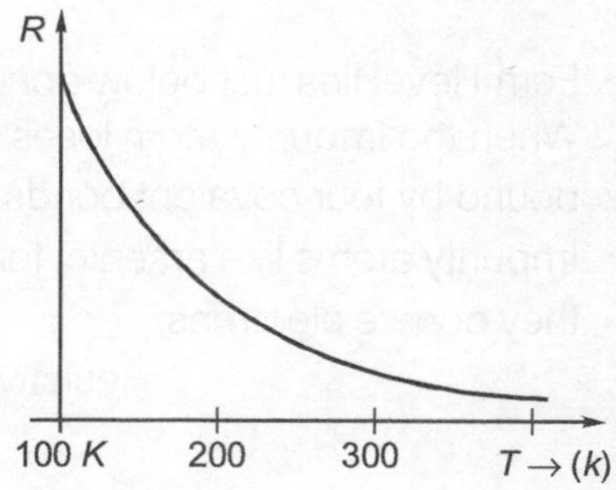

∴ $\rho \propto T^{-3/2} e^{E_G/KT}$

For pure semiconductor,

∴ $R \propto \rho$

So, $R \propto T^{-3/2} e^{E_G/KT}$

Volume resistivity (or electrical resistivity) or specific electrical resistance

$$\rho = R_1 A/l = 5 \times 10^{16} \text{ ohm-m}$$

$$R_1 = \rho l/A$$

where, $A = 3 \times 5 = 15 \times 10^{-4}\ \text{m}^2$

$l = 8 \text{ cm} = 8 \times 10^{-2}$ m

∴ $$R_1 = 5\times10^{16} \times \frac{8\times10^{-2}}{15\times10^{-4}} = \frac{8}{3}\times10^{18} \text{ ohm}$$

and surface resistivity = surface resistance

$$R_2 = 10^{18} \text{ ohm}$$

∴ net resistance is $$R = R_1 + R_2 = \frac{8}{3}\times10^{18} + 10^{18} = \frac{11}{3}\times10^{18} \text{ ohm} = 3.67 \times 10^{18} \text{ ohm}$$

5.11 **Explain Hall effect, Hall coefficient and its value in different semiconductor.** **[10 marks : 2008]**

Solution:

Hall effect:

When a transverse magnetic field is applied to a current carrying conductor bar, an electric field is induced in perpendicular direction to both magnetic field and current. This effect is called Hall effect. The voltage

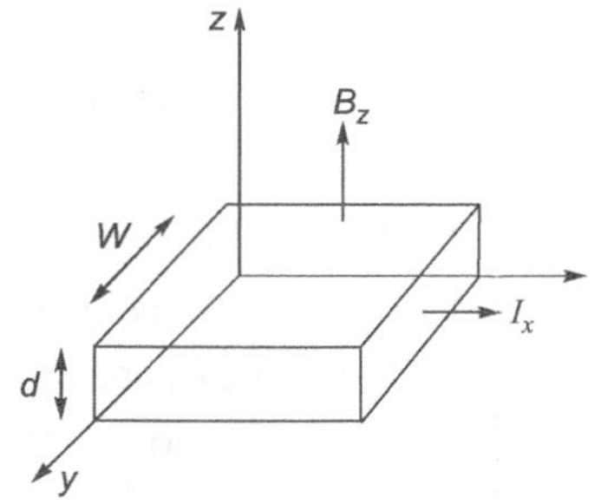

induced between two faces of conducting bar is called Hall voltage.
Consider an '*n*' type conducting material carrying current is *x*-direction.
Let B_z(magnetic field) is applied in *z*-direction if material is N-type then the electrons experienced a force in +*y*-direction. The upper part (and lower part) of the bar becomes positive, and negative respectively.
Force experienced by an electron is given by,

$$|F_e| = e E_y$$

Force due to B_z is

$$F_M = eVB_z \text{ and at equilibrium}$$

$$e E_y = eV B_z$$

$$\therefore \quad V = \frac{E_y}{B_z}$$

where *V* is velocity of charge particles.
The potential difference between upper and lower part of the bar

$$V_H = W \times E_y = WV B_z \quad \ldots(i)$$

$$V = \frac{J}{ne} = \frac{J}{\rho} = \frac{I_x}{Wd\rho} = RH\frac{I_x}{Wd} \quad [\rho = ne = \text{charge density}] \quad \ldots(ii)$$

Hall coefficient,

$$R_H = \frac{1}{\rho} = \frac{+1}{ne} \text{ for holes}$$

$$= -\frac{1}{ne} \text{ for electrons}$$

From equation (*i*) and (*ii*),

$$\therefore \quad R_H = \frac{V_H \times d}{B_z I_x} = \frac{E_y}{B_z I_x}$$

When two types of charge carriers are present at the same time, the hall effect is more complicated and measurement of the Hall coefficient becomes

$$R_H = \frac{E_y}{B_z I_x} = \frac{1}{e}\frac{(n_h \mu_h^2 - n_e \mu_e^2)}{(n_h \mu_h - n_e \mu_e)}$$

Where n_e and n_h are the electron and hole densities and μ_e and μ_h the carrier mobilities.
For p-type semi-conductor R_H = positive
and for n-type semi-conductor R_H = negative

5.12 Comment on the statement that "Hall effect" is observed in all metals but is very much prominent in semiconductor materials." Explain the Hall effect with a neat sketch.

[12 marks : 2009]

Solution:

When a metal or a semiconductor carrying a current *I* is placed in a transverse magnetic field *B*, a potential difference is produced in the direction normal to both the current and magnetic field directions. This phenomenon is called Hall effect. Hall effect measurements showed that it is the negative charge carriers namely electrons that are responsible for electrical conduction in metals. It also showed that there exist types of charge carriers in semiconductors.

Importance: The importance of Hall effect is that it helps to determine the:

(*i*) sign of charge carriers,
(*ii*) charge carriers' concentration, and
(*iii*) mobility of charge carriers if conductivity of the material is known.

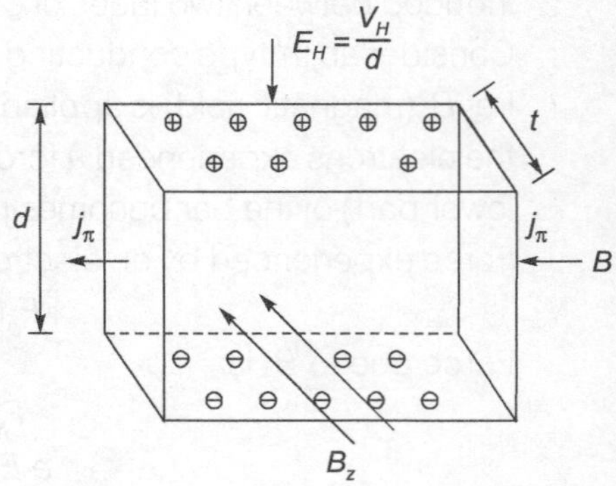

Let us consider an n-type semiconductor in which the conduction is predominated by electrons. Suppose an electric current *i* flows in the positive *x* direction and a magnitude field *B* is applied normal to this electric field z-direction. A force, called the Lorentz force is exerted on each electron which causes the electron paths to bend. As a result of this, the electrons accumulate on one side of the slab and are deficient on the other side. Thus, as electric field is created in the y-direction which is called the Hall field.

$$\text{Hall force} = \text{Lorentz force}$$

$$F_H = F_L$$

$$-e E_H = -v_x B_z e$$

where v_x is the velocity of the electrons, and e the electronic charge.

$$\therefore \quad E_H = v_x B_z \quad \text{...(i)}$$

As current density, $J_x = n v_x e$ $\left[\text{Putting } v_x = \frac{E_H}{B_z}\right]$

$$j_x = \frac{nE_H e}{B_z} = -\frac{nV_H e}{B_z d}$$

$$V_H = -\frac{J_x B_z d}{ne} = -\frac{IB_z d}{neA} \quad [A = \text{Area of cross-section of end face}]$$

If *t* is the thickness of the semiconductor specimen, $A = td$ and above equation reduces to

$$V_H = \frac{B_z I}{net} \quad \text{...(ii)}$$

Hall field per unit current density per unit magnetic induction is called Hall coefficient R_H Thus,

$$R_H = \frac{E_H}{J_x B_z} = \frac{\frac{V_H}{d}}{J_x B_z} = -\frac{B_z I}{J_x B_z \cdot net}$$

$$\boxed{R_H = -\frac{1}{ne}} \quad \text{...(iii)}$$

In terms of Hall coefficient, Hall voltage is given by:

$$\boxed{V_H = R_H \frac{BI}{t}} \quad \text{...(iv)}$$

$R_H \propto 1/n$ (number of e^{-s}/m^3)

$\because$ in metal *n* is large compare to semiconductors.

$\therefore$ V_H will be low for metals compare to semiconductors i.e. prominent in SC compare to metal.

The sign of the Hall coefficient R_H indicates whether electrons or holes predominate in the conduction process.

If R_H = negative, then electrons are the predominant charge carriers.

Thus, the magnitude of the μ_e may be determine if the conductivity — also been measured.

The net electric field *E* in the semiconductor is a vector sum of E_x (electric field component in *x*-direction) and E_H (Hall field). It acts at an angle θ_H to the *x*-axis. θ_H is called the Hall angle.

$$\tan\theta_H = \frac{BD}{AB}$$

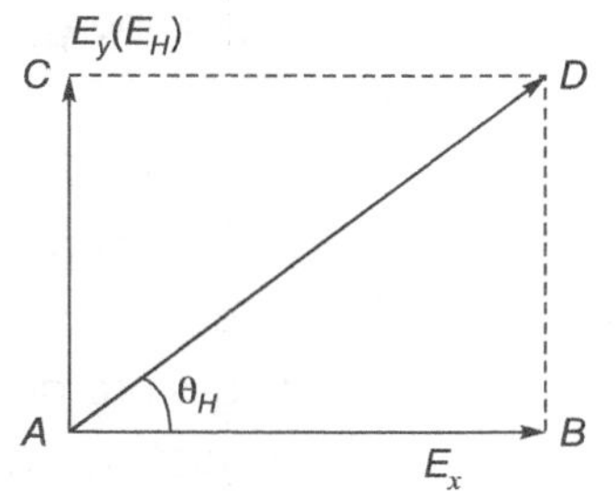

$$\tan\theta_H = \frac{B_H}{E_x}$$

$$\tan\theta_H = \frac{E_H}{E_x}$$

As $$E_H = \frac{V_H}{d} = \frac{B_z\, j_x}{n|e|}$$

Also, $$E_x = \rho j_x \qquad \left[\because \sigma = \frac{1}{\rho}\right] \qquad ...(v)$$

$\therefore$ $$\tan\theta_H = \sigma R_H B_z$$

The product σR_H is designated as μ_e the mobility of electrons.

$\therefore$ $$\tan\theta_H = \mu_e B_z$$

$\therefore$ $$\boxed{\theta_H = \tan^{-1}(\mu_e B_z)}$$

In the above discussions it is assumed that all carriers travel with a mean speed v_x, however this does not happen.

5.13 List the different groups into which semiconducting materials can be classified. Name at least one material from each group.

[12 marks : 2009]

Solution:

Classification of Semiconducting Materials:

1. Elemental semiconductivity material (ESM)
2. Compound semiconducting material (CSM)

ESMs in the periodic table are Boron, Gallium, Carbon, Silicon, Germanium, Phosphorous, Arsenic, Antimony, Selenium, Tellurium and Iodine.

As conductivity (σ) and band gap E_g are limited for ESM hence their usefulness is limited. So two groups elements (IV - IV), (III - V) or (II - VI) semi conductors known as (CSMs) are used to give the better properties.

I. **Elemental Semiconductors:**

1. **Antimony and Arsenic:** This pentavalent semiconducting element is used as a dopant to provide donor impurity in the fabrication of *n*-type intrinsic semiconductor made of germanium or Silicon.
2. **Gallium:** This is trivalent semiconducting element is used as an acceptor impurity in the fabrication of p-type extrinsic semiconductors made of germanium or silicon.
3. **Silicon:** This is tetravalent element and it is used in nearly all active solid state devices. Silicon posses diamond cubic (D-c) crystal structure. Silicon permits high temperature operation and reduces leakage. This is very useful for power transistors.

II. **Compound Semiconductors:**

III-V. Semi conductor

In these semi conductors one element is of group III and other is group V and average charge remain four. Bonding is covalent dominating. Semiconductor properties are similar to those of group-IV elements. Band gap (E_g) is different because of the presence of ionicity. Group III element electropositive and group V element are electro negative this difference in electro negativity gives rise to ionicity.

This provide wider range of energy band gaps (E_g).

Example: (GaAs) $\rightarrow$ Gallium Arsenide

- Band gap energy (E_g) → 1.34 eV (high)
- Electron mobility (μ_e) → high
- It is direct band-gap material.

II-VI. Semi conductor

- Average valence electron remains 4
- Bonding is pre-dominantly covalent
- Because of electro negativity, E_g(band gap energy) is larger than III-V semiconductor.

Example: cds, cdse, cdTe can be used for the detection and measurement of visible light.

IV-IV. Semiconductor

This class of semiconductor include Pbs, Pbse, PbTe. In this semiconductors excess Pb gives rise to n-type conduction and less Pb gives rise to p-type conduction.

III. Amorphous Semiconductor

Structure of amorphous semiconductor is similar to that of super cooled liquid. Atoms upto first nearest neighbours are arranged periodically but the atoms which are away from the first nearest neighbours are found arranged randomly.

Properties

- Switching characteristics
- They are easy to prepare.

Classified as

- Elemental amorphous-example: Ge, Si, Se, Te
- Covalent amorphous-example: GeTe, $As_2 Se_3$
- Ionic amorphous-$Al_2 O_3$, V_2O_5 etc.

5.14 Explain the Hall effect in semiconductors and define Hall constant. What do you mean by negative Hall constant?

[12 marks : 2011]

Solution:

Hall effect states that "If a specimen (metal or semiconductor) carrying the current *I* is placed in transversed magnetic field *B* then electric field intensity (*E*) is induced in a direction perpendicular *b* to both *I* and *B*".

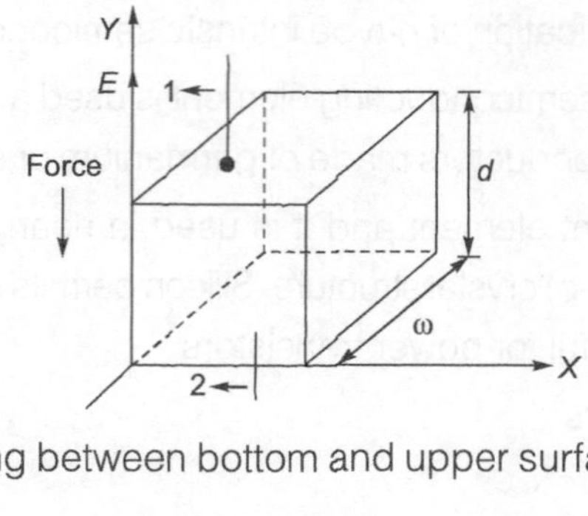

d — distance or height or spacing between bottom and upper surface of spacing.

ω — width of specimen

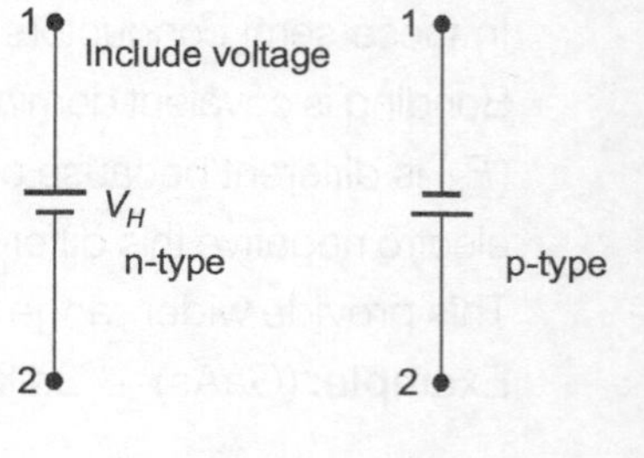

- The polarity and magnitude hall voltage V_H will indicate whether the given specimen is a metal or SC.

  From diagram

  Electric field intensity,

$$E = \frac{V_H}{d} V/m$$

Hall voltage $V_H = |E| \cdot d$ volts

or $V_H = \dfrac{BI}{\rho\omega}$ Volts

where, ρ – charge density is c/m^3

B – applied magnetic flux density

I – current flowing in specimen

ω – width of specimen

where, $\dfrac{1}{\rho} = R_H$

where, R_H is n known as Hall constant or Hall coefficient in m^3/c

$$\therefore \quad V_H = R_H \frac{BI}{\omega}$$

$$V_H \propto R_H$$

- R_H will be –ve for N-type SC and for metal as charge carriers (majority in N-type) are electrons, charge on which –ve so charge density ρ will be –ve and therefore Hall constant will be –ve.
- So –ve hall constant means specimen will be either metal or N-type.

5.15 Differentiate between intrinsic and extrinsic semiconductors. Explain doping.

[12 marks : 2012]

Solution:

An intrinsic semiconductor, also called an undoped semiconductor, is pure semiconductor, in which the number of free electrons must equal to the number of mobile holes.

Thus for intrinsic semiconductor

$$n_e = n_h = n_i \text{ (called intrinsic concentration)}$$

The conductivity of an intrinsic semiconductor is called the intrinsic conductivity.

When some impurities are added to the intrinsic semiconductor, the conductivity of semiconductor changes drastically and the resulting material called extrinsic semiconductor. The process of addition of impurities is called ***doping***.

The impurity atoms are atoms of different element than the atoms of the intrinsic semiconductor. Impurity atoms can act as either donors or acceptors to the intrinsic semiconductor. In case of Si and Ge, the doping element are either from third or fifth column of periodic table.

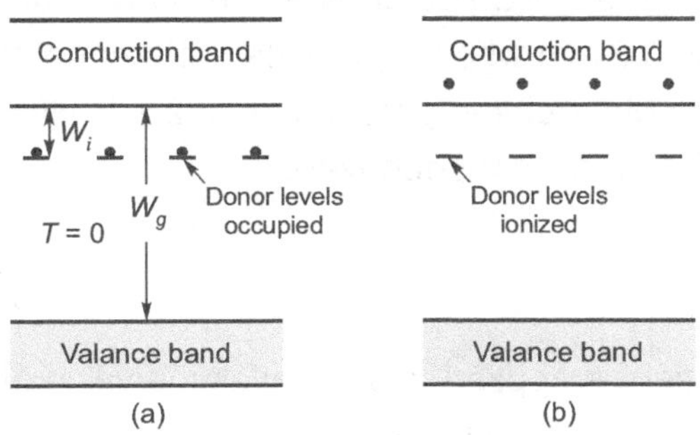

When Si or Ge is doped with an element from fifth column results in n-type semiconductor and if doped with an element of third column results in p-type semiconductor. The fifth column elements are called doner impurity and third column elements are called acceptor impurity.

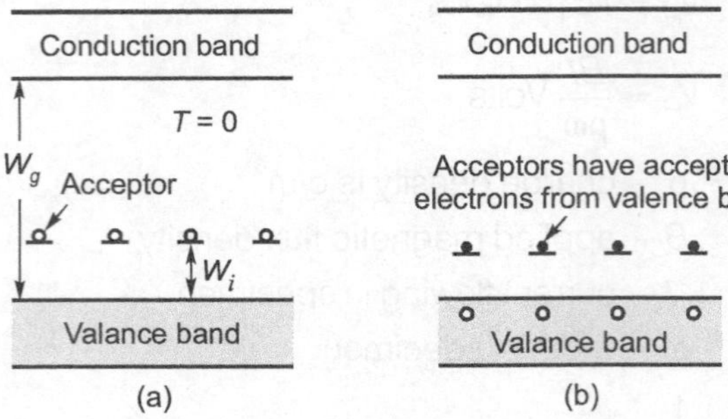

Donor impurity atoms have more valence electrons than the atoms they replace in the intrinsic semiconductor lattice. Donor impurities donate their extra balance electrons to a semiconductor's conduction band, providing excess electrons to the intrinsic semiconductor.

Acceptor impurity atoms have fewer electrons than the atoms they replace in the intrinsic semiconductor lattice. They accept electrons from the semiconductor valence band. This provides excess holes to the intrinsic semiconductor.

Group IV semiconductors	Silicon, Germanium
Donor atoms	Phosphorus, Arsenic
Acceptor atoms	Boron, Aluminium

5.16 The resistivity of pure germanium at room temperature is 0.47 ohm-m. Find out the carrier density of germanium at room temperature for the electron mobility of 0.42 m²/volt-sec and hole mobility of 0.20 m²/volt-sec. (electron charge $e = 1.6 \times 10^{-19}$ C).

[8 marks : 2012]

Solution:

Resistivity of pure germanium ; $\rho_i = 0.47\ \Omega$-m

Electron mobility, $\mu_e = 0.42\ \text{m}^2/\text{V-sec}$

Hole mobility, $\mu_n = 0.20\ \text{m}^2/\text{V-sec}$

Carrier density of germanium,

$$n_i = \frac{1}{\rho_i(\mu_e + \mu_n)\cdot e} = \frac{1}{0.47(0.42+0.20)\times 1.6\times 10^{-19}}$$

$$= 2.145 \times 10^{19}\ \text{m}^{-3}$$

5.17 What is "magnetoresistance effect"? Calculate the current produced in a small germanium plate of area 1 cm² and of thickness 0.3 mm when a potential difference of 2 V is applied across the faces. Given: Concentration of free electrons in germanium is $2\times 10^{19}/\text{m}^3$ and mobilities of electrons and holes are 0.36 m²/V-s and 0.17 m²/V-s respectively.

[12 marks : 2012]

Solution:

When an external magnetic field is applied to a material, the value of electric resistance changes. This property of material is called magnetoresistance.

Given, Plate area = $1\ \text{cm}^2 = 10^{-4}\text{m}^2$

Thickness = $0.3\ \text{mm} = 3 \times 10^{-4}\text{m}$

Concentration of free electrons in germanium

$$n = 2\times 10^{19}/\text{m}^3$$

Mobility of electrons, $\mu_e = 0.36\ m^2/V\text{-}s$

Mobility of holes, $\mu_h = 0.17\ m^2/V\text{-}s$

Resistivity of germanium, $\rho = \dfrac{1}{ne(\mu_e + \mu_n)}$

$$\Rightarrow \qquad \rho = \frac{1}{2\times10^{19}\times1.6\times10^{-19}\times(0.36+0.17)} = 0.59\ \Omega\text{-m}$$

Resistance of germanium, $R = \rho\dfrac{l}{A} = 0.59\times\dfrac{3\times10^{-4}}{1\times10^{-4}} = 1.77\ \Omega$

$$= 0.59\times\frac{3\times10^{-4}}{1\times10^{-4}} = 1.77\ \Omega$$

Current produced, $I = \dfrac{V}{R} = \dfrac{2}{1.77} = 1.13\ A$

5.18 Explain the recombination and trapping processes in a semiconductor. **[10 marks : 2013]**

Solution:

Recombination and Trapping: Recombination is the process where an electron moves from the conduction band into the valence band so that a mobile electron-hole pair disappear. Classical mechanics requires that momentum be conserved in an encounter of two particles. Since the momentum is zero after recombination, this conservation law requires that the "colliding" electron and hole must have equal magnitudes of momentum and they must be travelling in opposite direction. This requirement is very stringent, and hence the probability of recombination by such a direct encounter is very small. The most important mechanism in silicon or germanium through which holes and electrons recombine is that involving traps, or recombination centers, which contribute electronic states in the energy gap of the semiconductor. Such a location acts effectively as a third body which can satisfy the conservation-of-momentum requirement. These new states are associated with imperfection in the crystal. Specifically, metallic impurities in the semiconductor are capable of introducing energy states in the forbidden gap. Recombination is affected not only by volume impurities but also by surface imperfections in the crystal. Gold is extensively used as a recombination agent by semiconductor device manufacturers. Thus the device designer can obtain desired carrier lifetimes by introducing gold into silicon under controlled conditions. Carrier lifetimes range from nanoseconds 1 ns = 10^{-9} s to hundreds of microseconds (μs).

5.19 What is photo-conductivity? Discuss the factors that are to be considered for the selection of photo-conduction material.

[6 marks : 2018]

Solution:

Photoconductivity:

- In case of a semiconductor, when excess of electrons and holes are produced. But when the excess carriers (electrons and holes) are produced by optical luminescence, the increase in conductivity is called as photoconductivity. The photoconductive effect may be explained as follows:
  When radiant energy (greater than or equal to bandgap E_G) is supplied to a semiconductor material the covalent bands are broken. As a result, electron-hole in excess of those generated thermally are produced. Now, the increased current carriers reduce the resistance of material, thus the conductivity is increased.

- In general, semiconductors are most sensitive to photons with energies equal to band gap or slightly more than bandgap. For example *CdS* (E_G = 2.42 eV) is commonly used as a photoconductor in visible range. Further Ge(0.67 eV) and InSb(0.18 ev) are useful in infrared portion of the spectrum.
- The performance of a photoconductor depends upon the following factors:
  1. Quantum efficiency
  2. Response time
  3. Sensitivity of detection
- The quantum efficiency is defined as the number of free carriers generated per photon. The response time of a photoconductor is determined by the carrier transit time along the length of the semiconductor. The ratio of charge produced per unit time to the rate of charge flow (i.e. current) between the electrodes is defined as current gain.
- The optical sensitivity of a photoconductor is evaluated by examining the steady state excess carrier concentrations generated by optical generation rate g_{op}.

  We have, $$\delta n = \tau_n g_{op} \text{ and } \delta p = \tau_p g_{op} \quad ...(i)$$

  where τ_n and τ_p are the mean time which each carrier spends in its respective band before capture.
- The photoconductivity change is given by

  $$\Delta\sigma = q\, g_{op}[\tau_n\mu_n + \tau_p\mu_p] \quad ...(ii)$$

  where q is the charge on carrier and μ_n and μ_p are their respective mobilities.
- It is clear from equation (ii) that for maximum photoconductive response, the carriers should have long lifetimes and high mobilities.

5.20 **(i) State Hall effect and discuss the applications of Hall effect.**

(ii) A flat silver strip of width 1.5 cm and thickness 1.5 mm carries a current of 150 A. A magnetic field of 2.0 Tesla is applied perpendicular to the flat face of the strip. The emf developed across the width of the strip is measured to be 17.9 μV (Hall effect). Estimated the number density of the free electrons in the metal.

[12 + 8 = 20 marks : 2019]

Solution:

(i) **Hall Effect:** The **Hall effect** has played a decisive role in revealing the mechanism of conduction in semiconductors. "If a specimen (metal or semiconductor) carrying a current 'I' is placed in a **transverse magnetic field 'B'**, an electric field 'E' is induced in the direction perpendicular to both I and B. This phenomenon is known as the **"Hall effect"**. In figure I is in (+) ve X-direction and B is in (+) ve Z-direction; a force will be exerted in the (–)ve Y-direction on the charge carriers.

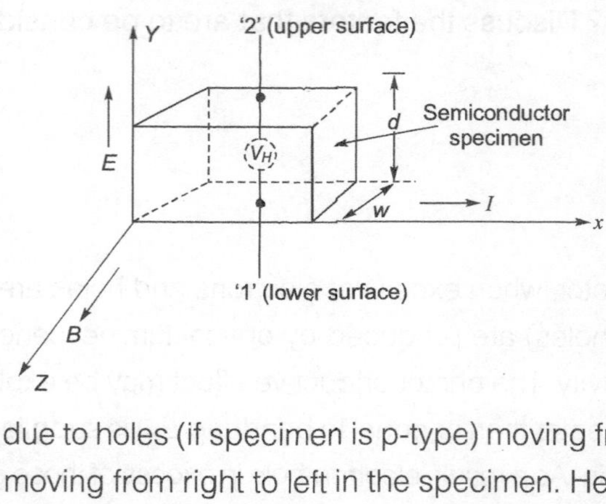

The current (I) may be due to holes (if specimen is p-type) moving from left to right or to free electrons (if specimen is *n*-type) moving from right to left in the specimen. Hence independently of whether the carriers are holes or electrons, they will be forced downward **toward side '1'**.

If the semiconductor is **n-type material**, then all free electrons will accumulate on **side '1'** and so the lower surface becomes (–)vely charged with respect to **side '2'**. Hence a potential, called the **Hall voltage** (V_H) appears between surfaces '1' and '2'. If the polarity of "V_H" is (+)ve at terminal '2', then as explained above, the carrier must be electrons and so the specimen is **n-type.**

On the other hand, if terminal '1' becomes charged positively with respect to terminal '2', the semiconductor bar must be **p-type.**

Applications of Hall Effect:

It is used to determine whether a semiconductor is **n-type** or **p-type.**

It is also used to determine:

- the carrier concentration.
- the mobility.
- the drift velocity.
- the conductivity of specimen.
- the field 'B'.

It is also used in manufacturing of **magnetic-flux meter** and **Hall-effect multiplier.**

(ii)
$$V_H = \frac{IB}{nte}$$

I = current across the conductor length in ampere

V_H = Hall voltage in volts

B = Magnetic field in Tesla

n = charge carrier density of the carrier electrons per m^3

t = Thickness of the conductor in metre

$e = 1.6 \times 10^{-19}$ C

$$V_H = \frac{IB}{nte}$$

$$17.9 \times 10^{-6} = \frac{150 \times 2}{n \times 1.5 \times 10^{-3} \times 1.6 \times 10^{-19}}$$

$$n = 6.983 \times 10^{28}/m^3$$

5.21 On the basis of specific resistance ρ, discuss the difference between conductors, semi-conductors and insulators.

[8 marks : 2020]

Solution:

Electric specific resistance also known as electrical resistivity is fundamental property of material that quantifies how strongly it resists or conducts electric current.

It is commonly represented by $\rho(\Omega - m)$

$$\rho = \frac{RA}{l}$$

where, R = Electrical resistance of uniform specimen

l = Length of specimen

A = Area of cross-section

Metals: A metal consists of a lattice of atoms, each with an outer shall of electrons that freely dissociate from their parent atoms and travel through the lattice and hence value of specific resistance of metal is comparatively lowest.

Semiconductors: The resistivity of semiconductor is generally intermediate but varies widely under different condition such as exposure of material to electric field and specific frequencies of light, degree of semiconductor doping etc.

Insulators: The insulators have generally very high value of resistivity due to non availability of free electron like in conductors.

The below table mentions the order of values of specific resistance for different electrical materials.

Material	Resistivity (specific resistance)
Super conductors	0
Metals	10^{-8}
Semiconductor	Variable
Electrolytes	Variable
Insulators	10^{16}
Super insulators	∞

5.22 **Find the diffusion coefficients of electrons and holes of a single silicon crystal at 27° C, if the mobilities of electrons and holes are 0.17 and 0.025 m²/volt-sec respectively at 27° C. (Boltzmann's constant k = 1.38 × 10^{-23} Joule/degree)**

[10 marks : 2020]

Solution:

Given :

$$T = 27°\text{C, i.e., } T = 300 \text{ K}$$

$$\text{Electron mobility, } \mu_n = 0.17 \text{ m}^2/\text{V-sec}$$

$$\text{Hole mobility, } \mu_P = 0.025 \text{ m}^2/\text{V-sec}$$

$$\text{Boltzmann's constant, } K = 1.38 \times 10^{-23} \text{ J/degree}$$

From Einstein relation,

$$\frac{D_n}{\mu_n} = V_T = \frac{kT}{q} = \frac{D_P}{\mu_P}$$

For electrons

$$\text{Diffusion coefficient, } D_n = \mu_n V_T$$

$$\Rightarrow \quad D_n = \mu_n \frac{kT}{q} = \frac{0.17 \times 1.38 \times 10^{-23} \times 300}{1.6 \times 10^{-19}}$$

$$\Rightarrow \quad D_n = 4.39 \times 10^{-3} \text{ m}^2/\text{sec} = 43.9 \text{ cm}^2/\text{sec}$$

For holes,

$$\text{Diffusion coefficient, } D_p = \mu_P V_T = \mu_P \frac{kT}{q}$$

$$\Rightarrow \quad D_p = \frac{0.025 \times 1.38 \times 10^{-23} \times 300}{1.6 \times 10^{-19}} \text{ m}^2/\text{sec} = 6.47 \text{ cm}^2/\text{sec}$$

6. Superconductors

6.1 **What is critical field in a superconductor? How is it affected by temperature? Write the application of a superconductor as an electrical switching element. The critical temperature (T_C) of Nb-60%. T_i is 8.7 K. The value of T_C for Nb_3 Al in K will be (pick up correct value) (i) 1.8 (ii) 4.8 (iii) 8.7 or (iv) 18.**

[12 marks : 2001]

Solution:

"The minimum applied magnetic field necessary to destroy superconductivity and restore the normal resistivity is called the critical field H_c". When the applied magnetic field exceeds the critical value H_c, the superconducting state is destroyed and the material goes into the normal state. H_c depends on the temperature, figure shows the critical field H_c as a function of temperature. A specimen is superconducting below the curve and normal above the curve.

For a given sustenance, value of H_c decreases as temperature increases from $T = 0$ K to T_c (critical temperature).

The curve is nearly parabolic and can be represented as:

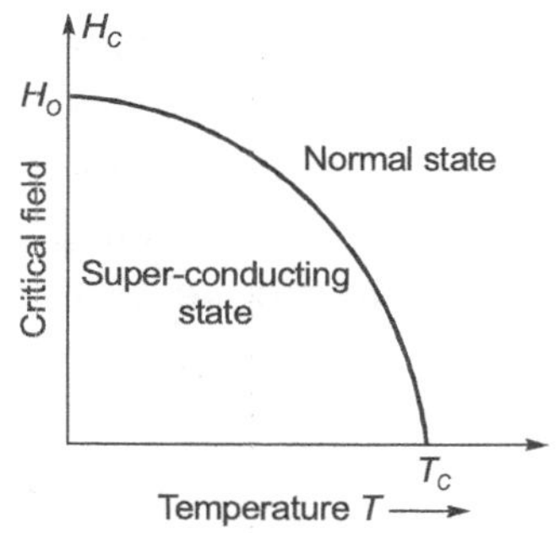

Scheme representation of the critical field H_c as a function of temperature

$$H_c = H_0\left(1-\frac{T^2}{T_c^2}\right) \quad \ldots(i)$$

where H_0 is the critical field at 0 K.

Thus the field has its maximum value H_0 at $T = 0$ K

At $T = 0$ K $\quad H_c = H_0\left(1-\frac{0}{T_c^2}\right) = H_0$

At $T = T_c$, $\quad H_c = H_0\left[1-\frac{T_c^2}{T_c^2}\right] = 0$

Equation (*i*) is the phase boundary between the normal and-super conducting state.

The super conductors can be used in the electrical switching elements (cryotrons), the switch used can be

(a) temperature controlled

(b) magnetically operated current switch

The transition temperature (T_c) of Nb_3Al is 18.

$\therefore$ *K* critical temperature is a large value at which required critical field will be zero, so in question.

$T_i = 8.7$ (at which it is in super conducting phase)

so, $\quad T_c > T_i \rightarrow T_c > 8.7$

So, option (iv) is correct.

6.2 Describe the phenomena of superconductivity. **[12 marks : 2006]**

Solution:

In many of the metals and alloys, *dc* resistivity vanishes (super conductivity) entirely below a certain temperature. The temperature at which the normal metal passes into superconducting state is called the transition or critical temperature (T_c).

According to free electron model of metals, the resistivity of the metal may be written as,

$$\rho = \frac{m}{ne^2\tau}$$

m – mass of the electron

'*n*' – the number of electrons per unit volume

'τ' – average collision time

From equation, it is clear that resistivity decreases as temperature is lowered because as temperature decreases the lattice vibrations begin to 'freeze' and hence the scattering of electron diminishes. This result in a larger 'τ' and hence a smaller resistivity (ρ). If average time of collision (τ) becomes infinite at sufficiently low temperature, then the resistivity (ρ) vanishes entirely is observed in case of super conducting materials.

Properties of superconductors:

(*i*) At room temperature, the resistivity (ρ) of super conducting materials are greater than other elements.

(*ii*) All thermo electric effects disappear in superconducting state.

(*iii*) When a sufficient strong magnetic field is applied to super-conductor below critical temperature (T_c), its super conducting property is destroyed.

(*iv*) When current is passed through the super conducting materials, the heating loss (I^2R) is zero.

As, resistivity $\rho = RA/l$

$\rho \to$ very small (zero at T_c)

$R = 0$, Hence no heating loss

(*v*) According to meissners effect, the magnetic flux density in conductor at super conductivity state is zero.

6.3 Explain the dependence of transition temperature of superconductor on external magnetic field with graphical representation.

[12 marks : 2006]

Solution:

"The minimum applied magnetic field necessary to destroy superconductivity and restore the normal resistivity is called the critical field H_c". When the applied magnetic field exceeds the critical value H_c, the superconducting state is destroyed and the material goes into the normal state.

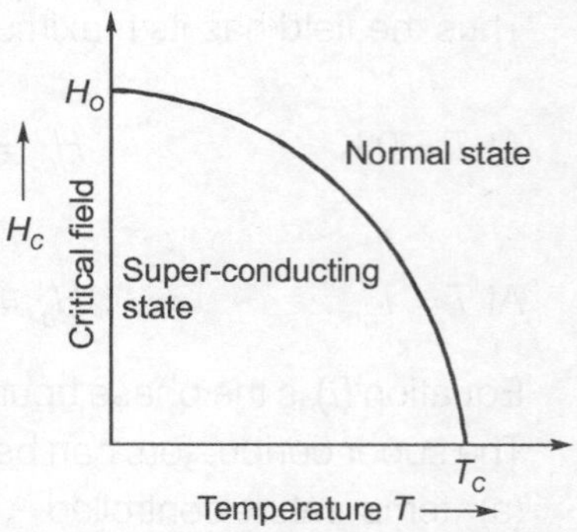

Scheme representation of the critical field H_c as a function of temperature

H_c depends on the temperature, figure shows the critical field H_c as a function of temperature. A specimen is superconducting below the curve and normal above the curve.

For a given substance, value of H_c decreases as temperature increases from $T = 0$ K to T_c (critical temperature).

The curve is nearly parabolic and can be represented as:

$$\boxed{H_c = H_0\left(1-\frac{T^2}{T_c^2}\right)} \quad ...(i)$$

where H_0 is the critical field at 0 K.

Thus the field has its maximum value H_0 at $T = 0$ K

At $T = 0$ K, $$H_c = H_0\left(1-\frac{0}{T_c^2}\right) = H_0$$

At $T = T_c$, $$H_c = H_0\left[1-\frac{T_c^2}{T_c^2}\right] = 0$$

Equation (*i*) is the phase boundary between the normal and-super conducting state.

6.4 Describe the salient features of type-I and type-II superconductors.

[12 marks : 2006]

Solution:

Type-I and Type-II Superconductors:

Superconducting materials can be divided into two types based on their magnetic response. These two types are designated as type I and Type-II.

Type-I or the ideal superconductor, while in the superconducting state are completely diamagnetic, that is when these superconductors are placed in a magnetic field, then all the lines of induction are pushed out from the specimen. (These superconductors show Meissner effect). As magnetic field is increased, the material remains diamagnetic untill the critical value H_c is reached. At this point conduction becomes normal and complete magnetic flux penetration takes place.

Type-II or hard superconductors are those in which the ideal behaviour is seen up to a lower critical field H_{c1} beyond which the magnetization gradually changes and attains zero at an upper critical field designated as H_{c2}.

The Meissner effect is incomplete in this region between H_{c1} and H_{c2}, this region is known as the vortex region as shown as figure. The normal beahaviour is observed only beyond H_{c2}. The line of induction penetrate gradually from the specimen, as the field is increased beyond H_{c1} and the penetration completes at H_{c2}. Figure shows the behaviour of Type-I and Type-II superconductors as a function of $-M$ and H.

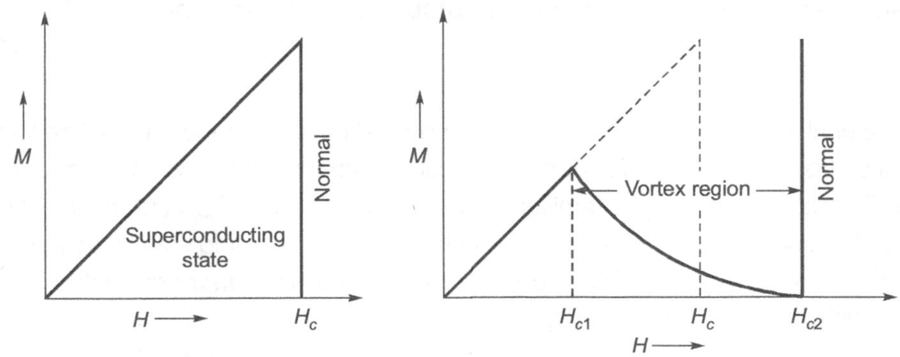

(a) Type-I Superconductor (b) Type-II Superconductor

Figure Magnetization versus magnetic field for (a) Type-I and (b) Type-II superconductors.

From figure it is clear that for Type-I superconductor, upto H_c, the magnetization of the material grows proportion to the external magnetic field and then abruptly drops to zero at the transition to the normally conducting state.

Table 1 lists some of the Type-I superconductors and the critical temperature T_c along with the critical fields H_c.

Superconductor		T_c(K)	H_c(Tesla)
Aluminium	(Al)	1.18	0.0105
Mercury	(Hg)	4.15	0.0411
Lead	(Pb)	7.19	0.0803
Zinc	(Zn)	0.85	0.0054

Table 2 lists some of the Type-II superconductor and their critical temperature T_c and critical field H_c.

Superconductor	T_c(K)	H_c(Tesla)
Nb Ti	10.0	15.0
Nb_3 Sn	18.0	24.5
V_3 Ge	14.8	2.08
V_3 Si	16.9	2.35

Variation of resistivity of a Type-I superconductor and Type-II super-conductor as a function of applied magnetic field is shown in figure. In Type-I superconductor, the transition from superconducting state to normal state in the presence of a magnetic field occurs sharply at the critical field H_c, while in Type-II superconductor the transition from superconducting state to normal state occurs gradually as the magnetic field is increased from H_{c1} to H_{c2} as illustrated in Figure (b).

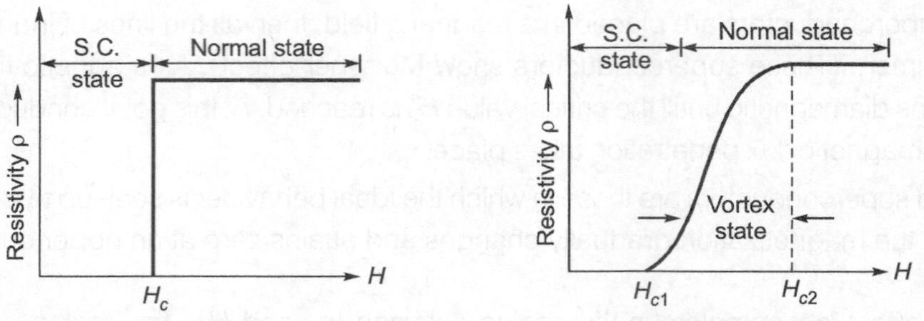

(a) Magnetic field (b) Magnetic field

ρ versus H for (a) Type-I and (b) Type-II superconductors

6.5 What are the physical important superconductors? What are the physical changes observed at the transition temperature? What are the general properties of superconductors?

[12 marks : 2007]

Solution:

The state of material at which resistivity reduce to zero is called super conductivity. The temperature at which there is transition from normal state to super conducting state is called transition temperature. Above critical temperature (T_c) the material is in the familiar normal state but below (T_c) it enters an entirely different super conducting state, resistance of these materials in the super conducting state is at least 10^{16} times smaller than their room temperature values. Two independent conditions of super-conductivity:

(*i*) the zero resistivity (*ii*) perfect diamagnetism.

Properties:

(*i*) At room temperature, the resistivity 'ρ' of super conducting materials are greater than other elements.

(*ii*) All thermo electric effects disappear in super conducting state.

(*iii*) When a sufficient strong magnetic field is applied to super-conductor below critical temperature T_c, its super conducting property is destroyed.

(*iv*) When current is passed through the super conducting material the heat loss ($I^2 R$) is zero.

(*v*) The magnetic flux density in super conductor is zero.

6.6 Prove that "superconductors are perfectly diamagnetic". Photoelectric threshold of metallic copper is at λ = 3000 A.U. Find out (*i*) the work function of the metal (*ii*) maximum kinetic energy of the photoelectrons ejected, when ultraviolet light of λ = 2536 A.U. falls on the metal surface. Given: 1 A.U. = 10^{-10} meter; 1 eV = 1.6 × 10^{-19} Joule.

[12 marks : 2007]

Solution:

Meissner Effect: Meissner and Ochsenfeld found that if a superconductor is cooled in a magnetic field to below the critical temperature T_c (transition temperature), then at the transition, the lines of induction are pushed out. "The expulsion of magnetic flux from the interior of piece of super-conducting material as the material undergoes the transition to the super-conducting phase is known as Meissner effect." Figure shows the normal sphere at $T > T_c$ and superconducting sphere at $T < T_c$ showing the expulsion of magnetic lines of induction.

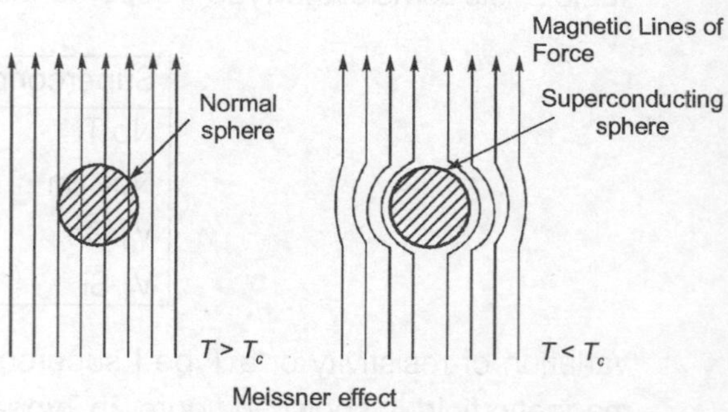

Meissner effect

Meissner effect is reversible, when the temperature is raised from below T_c the flux suddenly penetrates the specimen after it reaches T_c and the substance is in the normal state.

As, $\vec{B} = \mu_0(\vec{H}+\vec{M}) = \mu_0(1+\chi)\,\vec{H}$

where $\chi = \dfrac{\vec{M}}{\vec{H}}$

where, H = applied magnetic field

M = magnetization

χ = magnetic susceptibility

Since, $B = 0$ in superconducting state

$\therefore$ $M = H$

$\therefore$ $\boxed{\chi = -1}$ and also $\chi = \mu - 1$ hence $\boxed{\mu_r = 0}$

i.e. a superconductor exhibits perfect diamagnetism. Because diamagnetic nature, superconducting materials strongly repels external magnets. It leads to a levitation effect.

(*i*) Minimum kinetic energy ($K:E$) required to detach electrons from surfaces

$$W_s = h\nu = h\frac{c}{\lambda}$$

$\therefore$ Threshold wave length $(\lambda_T) = 3000$ Å

$\therefore$ Work function of the metal $= \dfrac{6.64\times10^{-34}\times10^8\times3}{3000\times10^{-10}}$

$= 6.64\times10^{-19} = 4.15$ eV

(*ii*) If the light of 2536 A.U. wavelength falls on the surfaces, the K : E of the ejected electrons is

$$\text{K : E} = h\gamma - W_s = h\frac{c}{\lambda} - W_s$$

$$= \frac{6.64\times10^{-34}\times3\times10^8}{2536\times10^{-10}} - 6.64\times10^{-19}$$

$$= 1.2148\times10^{-19}\text{ J} = 0.759\text{ eV}$$

6.7 List out the applications of super-conductors. The superconducting transition temperature of lead is 7.26 K. The initial field at 0 k is 64 × 10³ A/m. Calculate the critical field at 5 K.

[12 marks : 2008]

Solution:

Super conductors can be used to produce strong magnetic field.

- Super conductors are used for winding of exciting field of generators of low heating.
- Super conductors can be used for transmission purpose.
- Super conductors can be used in the electrical switching elements (cryotrons).
- Super conductors are also used in high levitated trains.

$$H_c = H_0\left[1-\left(\frac{T}{T_c}\right)^2\right]$$

Here, $H_0 = 64\times10^3$ A/m

$T_c = 7.26$ K

$$H_c(5\text{ K}) = 64\times10^3\left[1-\left(\frac{5}{7.26}\right)^2\right] = 33.6438\times10^3\text{ A/m}$$

6.8 **Give the properties of superconductors. What are type-I and type-II superconductors? Indicate their behaviour with respect to applied magnetic field.**

[12 marks : 2011]

Solution:

A state of material in which it has zero resistivity is called superconductivity.

- At the state of superconductivity material has zero resistivity as well as diamagnetism.
- The temperature at which superconductivity appears is called the critical temperature or transition temperature.
- Superconductors are materials which show super-conductivity under certain conditions of temperature and magnetic field.
- It shows meissner effect.
- Superconductor exhibits perfect diagnometism ($m_r = 0$) because of diamagnetic nature, superconductivity material strongly repel external magnets.

Type-I super-conductor:

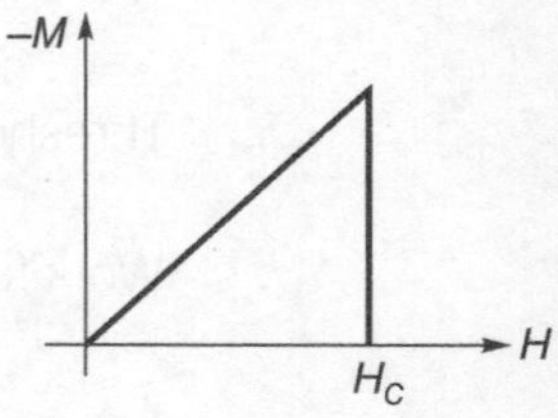

- Type-I or ideal superconductors, while in superconducting state are completely diamagnetic upto the critical field H_c and to normal state above critical field as.
- Shoft metals such as lead or indium belong to this group. The critical field of these superconductors are low.
- They have low melting point.
- These materials obey Silsbee's rule and show Meissner effect.
- The transition from normal to superconducting state is sharp. They have low value of H_C and T_C.

Type-II super-conductor:

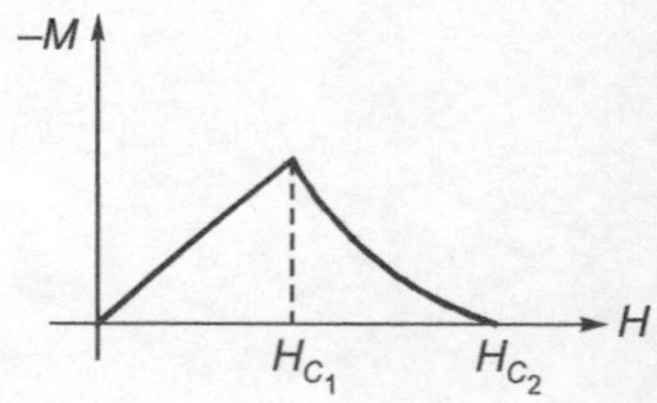

- Hard metals and alloy have different magnetization characteristic as shown figure are called type-II superconductors.
- They are characterized by high transition temperature, high critical field incomplete Meissner effect, break down of Silsbee's rule and or broad transition region.
- Type-II are those in which the ideal behaviour is seen up to a lower critical field H_{C_1} beyond which the magnetization gradually changes and attains zero at an upper critical field designated as H_{C_2}.

6.9 **Explain the phenomenon of superconductivity. Estimate the critical magnetic field for superconductivity of niobium at 5°K.**

[10 marks : 2013]

Solution:

Superconductivity:

- A state of material in which it has zero resistivity, is called super conductivity.
- At the state of super conductivity material shows **diamagnetism property.**
- **Mercury (Hg)** was first discovered super conductor material and observed that the resistivity of mercury vanished completely below **4.2 K**, the transition from normal to super conductivity occurring over a very narrow range of temperature of the order of **0.05 K**.
- The temperature at which superconductivity appears is called the **critical temperature** or **transition temperature** (T_C).

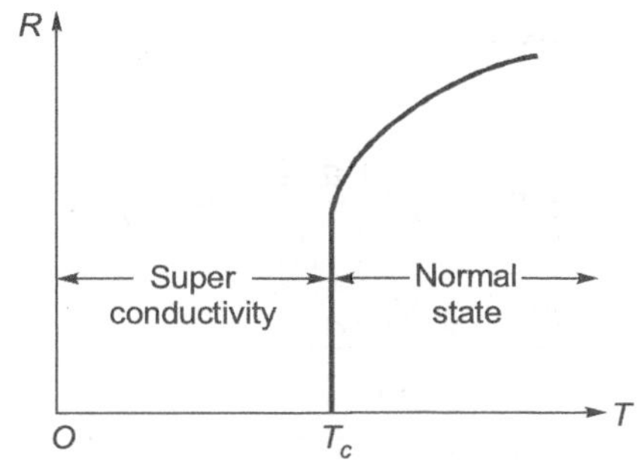

Variation of electrical resistance versus temperature

- Superconductors are materials which show super conductivity under certain conditions of temperature and magnetic field. The state in which the superconductor does not show superconductivity is called the **normal state.**
- Some superconducting materials:

Element	T_C(K)	Alloy or Compound	T_C(K)
Al	1.19	Nb-Ti	8.0
La	5.9	Nb-Zr	11.0
Nb	9.2	Nb_3 Sn	18.3
Pb	7.18	Nb_3 Ge	22.5
Sn	3.72	V_3 Si	17.0
Hg	4.2	Nb N	17.3
Ta	4.48		
Tc	8.22		
V	5.13		

6.10 For a face centered cubic metal state with reasons the number of atoms assigned per unit cell and coordination number. Also deduce a value for atomic packing factor.

[12 marks : 2017]

Solution:

Face Centered Cubic (FCC):

In FCC structure, one atom lies at each corner of the cube and one atom at the center of each cube face. For any corner atom, there are four atoms at each corner that surround it. Each face center has an atom. There are four atoms nearest to the atom on four atomic plane below it and four above it in the atomic plane. The co-ordination number of face centered cubic structure is thus 4 + 4 + 4 =12.

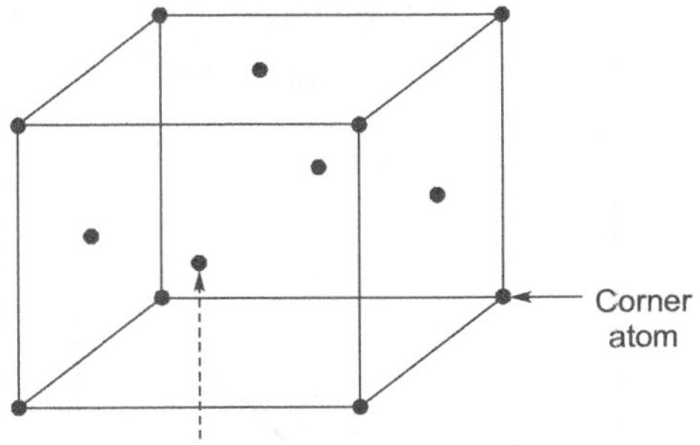

Fig. (i)

$$\text{Number of atom per unit cell} = 8\times\frac{1}{8}+\left(6\times\frac{1}{2}\right) = 4$$

From figure (ii),

$$AC^2 = AB^2 + BC^2$$
$$(r + 2r + r)^2 = a^2 + a^2$$
$$16r^2 = 2a^2$$

Radius of spherical unit cell (r) $= \dfrac{a\sqrt{2}}{4}$

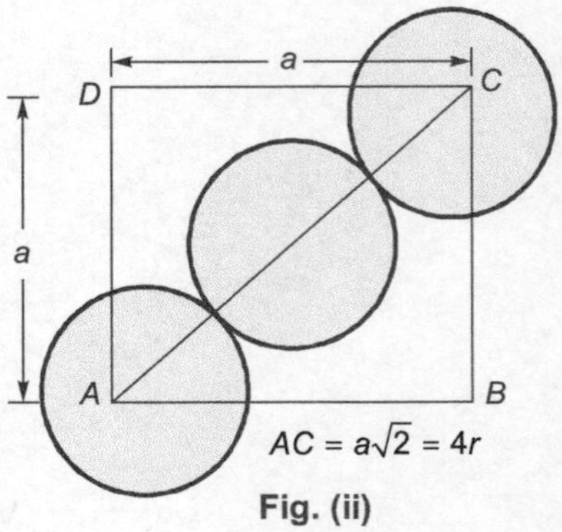

Fig. (ii)

$$\text{Volume of each sphere} = \frac{4}{3}\pi r^3 = \frac{4}{3}\pi\left(\frac{a\sqrt{2}}{4}\right)^3 = \frac{\sqrt{2}\,\pi}{24}a^3$$

Atomic packing factor, $$APF = \frac{N_{particle}\ V_{particle}}{V_{unit\ cell}}$$

where $N_{particle}$ is the number of particle in the unit cell, $V_{particle}$ is the volume of each spherical particle and $V_{unit\ cell}$ is the volume occupied by the unit cell.

Here, $N_{particle} = 4$

$$APF = \frac{4\times\left(\dfrac{\sqrt{2}\pi}{24}a^3\right)}{a^3} = \frac{\sqrt{2}\pi}{6} = 0.74$$

$$\%\ APF = 74\%\ \text{packed}$$

6.11 **Name 5 properties that are exhibited by superconductors. Draw the magnetising curve for Type-I and Type-II superconductors specifying regions of conducting states.**

[12 marks : 2017]

Solution:

Properties exhibited by superconductor:

1. Zero electrical resistance
2. Meissner effect
3. Critical magnetic field (H_C)
4. Critical current density (J_C)
5. Persistent current lead to constant magnetic flux

Type-I (or ideal) superconductor: The superconductor which show perfect diamagnetism up to the critical field (H_C) and go to normal state beyond critical field are known as type-1 or ideal superconductors. The critical field of these super conductors are low. Hence these materials are also called soft superconductors (or ideal). They have low melting point. These materials obey Silsbee's rule and show Meissner effect. The transition from superconducting state to normal state is abrupt. **e.g.:** Pb, In, Pd, Hg, etc.

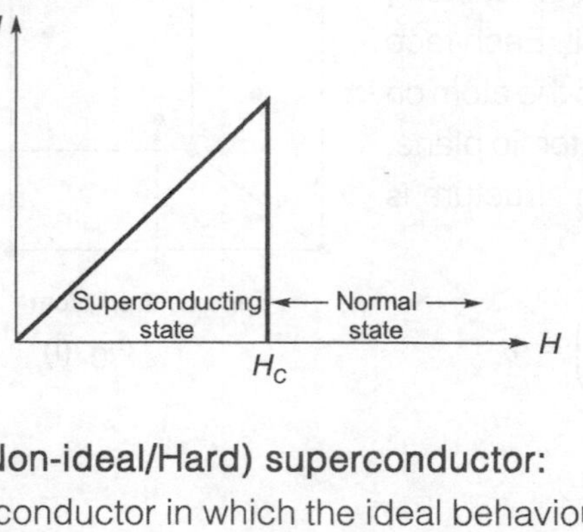

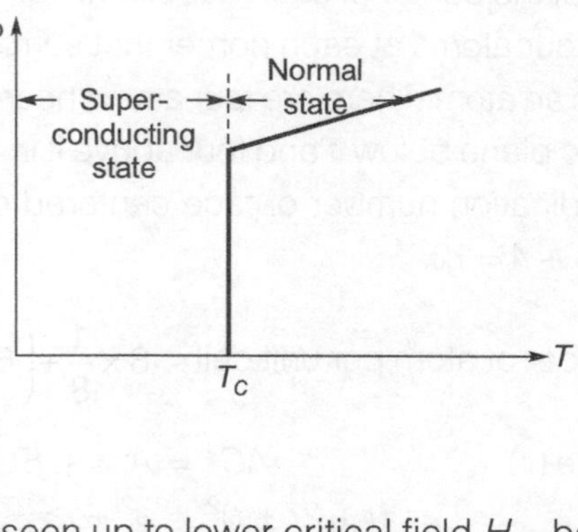

Type-II (or Non-ideal/Hard) superconductor:

Those superconductor in which the ideal behaviour is seen up to lower critical field H_{C1} beyond which the magnetization gradually changes and attains zero at an upper critical field H_{C2}. They have high critical field and critical temperature. These material incompletely follow Silsbee's rule and Meissner effect. The transition from superconducting state to normal state is gradual. **e.g.:** NbTi, Nb_3Al, Nb_3Sn etc.

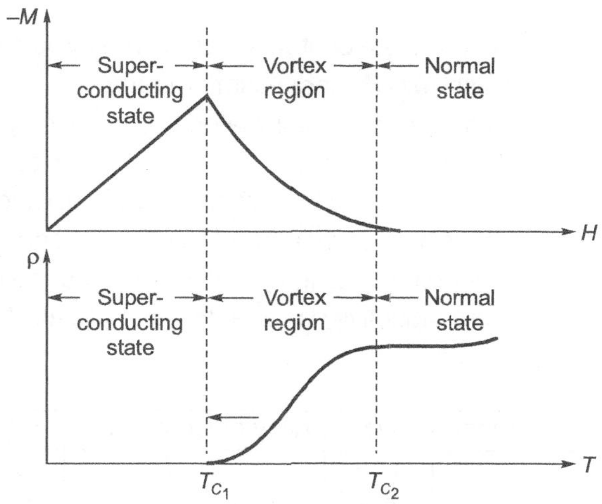

6.12 The current measured in a superconducting ring by 0.01% accuracy meter after one year shows no decay of current. If there are 10^{28} electrons/m³, calculate the conductivity. How many times larger is this conductivity than that of copper of resistivity 1.724×10^{-8} Wm.

[8 marks : 2018]

Solution:

For a superconductor, the current decays with time as,

$$I(t) = I_0 e^{-t/\tau}$$

where, τ = relaxation time of the material

The meter used for the current measurement has an error of 0.01%.

So, at t = 1 year = 365 × 24 × 3600 seconds

$$I(t) \geq 0.9999\, I_0$$

$$I_0 e^{-t/\tau} \geq 0.9999 I_0$$

$$\frac{t}{\tau} \leq -ln(0.9999)$$

$$\tau \geq \frac{t}{-ln(0.9999)} = \frac{365 \times 24 \times 3600}{-ln(0.9999)} = 3.15 \times 10^{11} \text{ seconds}$$

Taking the lower limit of the relaxation time for calculating the worst case conductivity of the superconductor.

So, $\tau = 3.15 \times 10^{11}$ sec.

Conductivity, $$\sigma_{super} = \frac{ne^2\tau}{m_e} = \frac{10^{28} \times (1.6 \times 10^{-19})^2 \times (3.15 \times 10^{11})}{(9.1 \times 10^{-31})} = 8.86 \times 10^{31} \text{ ℧/m}$$

$$\sigma_{copper} = \frac{1}{\rho_{copper}} = \frac{1}{1.724 \times 10^{-28}} = 58 \times 10^{16} \text{ ℧/m}$$

$$\frac{\sigma_{super}}{\sigma_{copper}} = \frac{8.86 \times 10^{31}}{58 \times 10^{6}} = 1.53 \times 10^{24}$$

So, the conductivity of the superconductor is 1.53×10^{24} times the conducitivity of the copper.

6.13 **Discuss superconductivity, superconducting materials and their applications.**

[12 marks : 2019]

Solution:

Superconductivity: A state of material in which it has zero resistivity, is called super conductivity. At the state of super conductivity material shows diamagnetism property.

Mercury (Hg) was first discovered super conductor material and observed that the resistivity of mercury vanished completely below 4.2 K, the transition from normal to super conductivity occurring over a very narrow range of temperature of the order of 0.05 K. The temperature at which superconductivity appears is called the critical temperature or transition temperature (TC). Superconductors are materials which show super conductivity under certain conditions of temperature and magnetic field. The state in which the superconductor does not show superconductivity is called the normal state. Some superconducting materials are shown in table here:

Element	T_C(K)	Alloy or Compound	T_C(K)
Al	1.19	*Nb – Ti*	8.0
La	5.9	*Nb – Zr*	11.0
Nb	9.2	$Nb_3\,Sn$	18.3
Pb	7.18	$Nb_3\,Ge$	22.5
Sn	3.72	$V_3\,Si$	17.0
Hg	4.2	*Nb N*	17.3
Ta	4.48		
Tc	8.22		
V	5.13		

Applications of Superconductor: Superconductor can be used for generation of strong magnetic field. It may also be used for the design of modulators, rectifiers, used for detection of modulated high frequency waves. These materials are used in electronic switching devices called "Cryotrons". Superconducting magnets find applications in the following areas:

⇒ Magnets for nuclear fusion.

⇒ Generator and motors.

⇒ Magnetically Leviated Transportation.

⇒ Magnets for high-energy physics.

⇒ Superconducting magnets for energy storage.

⇒ Magnetic resonance imaging (MRI) and other applications (in medicine).

6.14 **(i) Explain the variation of transition temperature of a superconductor with its isotopic mass *M*.**

(ii) The transition temperature of Mercury with an average atomic mass of 200.6 is 4.2 K. Evaluate the transition temperature of one of its isotopes, $_{80}Hg^{204}$.

[8 + 12 marks : 2022]

Solution:

(i) According to isotopic effect, "The critical temperature T_c of a superconductor varies inversely with the square root of the isotpic mass, i.e., *M*. The relation can be given by

$$T_c M^{1/2} = \text{Constant}$$

$$\Rightarrow \quad T_c \propto \frac{1}{M^{1/2}}$$

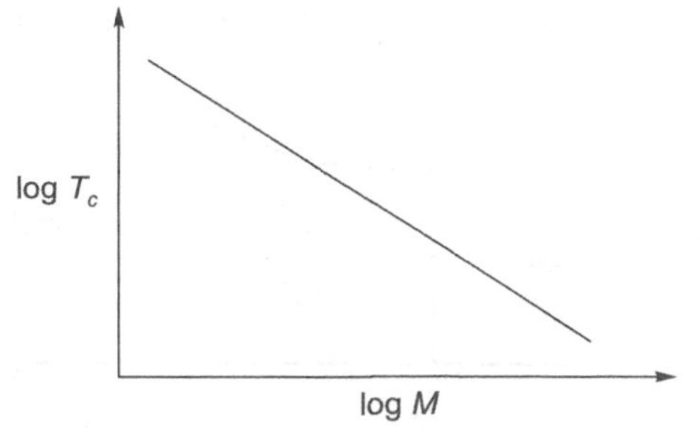

This isotopic effect was discovered in 1950 by Maxwell and Reynolds while they were obtaining critical temperature of different isotoes of Mercury (Hg).

(ii) Given : Average atomic mass,

$$M_1 = 200.6$$

Critical temperature, $T_{c1} = 4.2$ K

Isotopic mass of mercury, $M_2 = 204$

Let the critical temperature for $_{80}Hg^{204}$ be T_{c_2} in K, then

$$T_c \propto \frac{1}{M^{1/2}}$$

where T_c is critical temperlature, M is isotopic mass.

$$\Rightarrow \quad \frac{T_{c_1}}{T_{c_2}} = \sqrt{\frac{M_2}{M_1}} \quad \Rightarrow \quad T_{c_2} = T_{c_1}\sqrt{\frac{M_1}{M_2}}$$

$$\Rightarrow \quad T_{c_2} = 4.2\sqrt{\frac{200.6}{204}} \quad \Rightarrow \quad T_{c_2} = 4.165 \text{ K}$$

6.15 Give the variation of resistivity of purified mercury with temperature. Also, represent the resistivity of normal metal as a function of temperature (T) along with pure and impure superconductors.

[12 marks : 2023]

Solution:

Below the critical temperature (T_c = 4.2 K), mercury enters into superconducting state and its resistivity becomes zero.

For temperature $T > T_c$, it behaves like normal conductor.

Following figure shows the resistivity (ρ) of mercury with temperature (T).

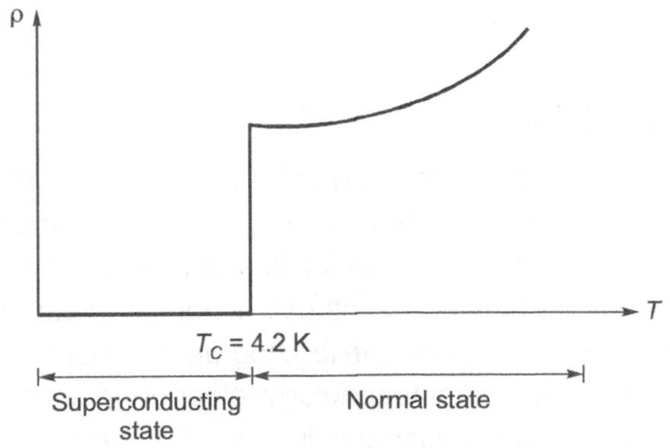

There is an abrupt change in resistivity of mercury at critical temperature (T_c).

Pure superconductors are type-I superconductors and following figure represents comparison of resistivity of pure superconductor and normal metal.

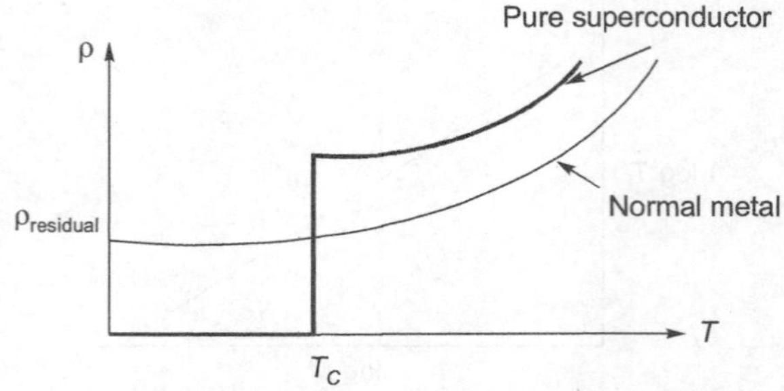

Impure superconductors are type-II superconductors and following figure represents comparison with normal metal.

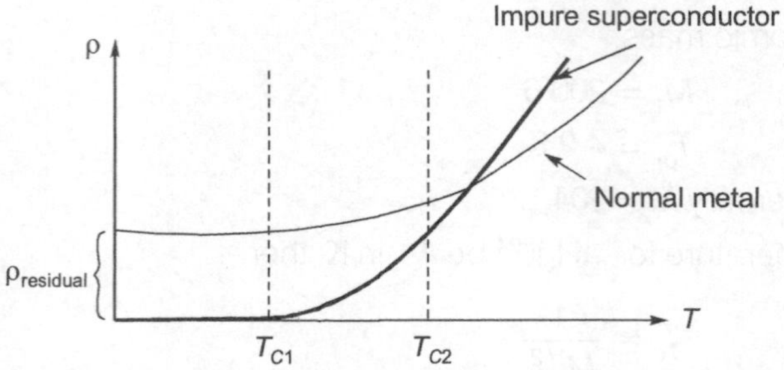

7. Nano Materials

7.1 Explain the top-bottom and bottom-up approach to produce the nanostructure. Name the methods used in each case to produce nanomaterial.

[8 marks : 2018]

Solution:

Top-down Vs Bottom-Up Technique: There are two different approaches of material fabrication. One is the top-down approach and the other is the bottom-up approach. In top-down technique, generally a bulk material is tab and machined it to modify into the desired shape and product. Examples of this type of technique are the manufacturing of integrated circuits using a sequence of steps such as crystal growth, lithography, etching, ion implantation, etc. For nanomaterial synthesis, ball-milling is an important top-down approach, where microcrystalline structures are broken down to nanocrystalline structures, but original integrity of the material is retained. Sometimes this method is used to prepare nanostructured metal oxides by chemical reaction between two constituents during crushing. The crystallites are allowed to react with each other by the supply of kinetic energy during milling process to form the required nanostructured oxide. The methods used are:

- Photolithography
- Electron beam lithography
- Focussed ion beam (FIB) machining

On the other hand, bottom-up technique is used to build something from basic materials, for example, assembling materials from the atoms/molecules up and therefore very important for nano-fabrication. Unlike lithographic technique of top-down approach, which is extensively used in silicon industry, this bottom-up nano lithograhphic approach of nanomaterial synthesis is not completely proven in manufacturing yet, but has great potential to become important alternative to lithographic process. Examples of bottom-up technique are self-assembly of nanomaterials, solgel technology, electrodeposition, physical and chemical vapour deposition (PVD. CVD), epitaxial growth, laser ablation, etc. The methods are:

- Chemical vapor deposition (CVD)
- Gas phase methods
- Sol-Gel method
- plasma based synthesis
- Microwave synthesis

2 Electric Circuits

Revised Syllabus of ESE: *Circuit elements, network graph, KCL, KVL, Node and Mesh analysis, ideal current and voltage sources, Thevenin's, Norton's, Superposition and Maximum power transfer theorems, transient response of DC and AC networks, Sinusoidal steady state analysis, Basic filter concepts, Two-port networks, Three-phase circuits, Magnetically coupled circuits.*

1. Circuit Elements, Nodal and Mesh Analysis

1.1 Find V_{AB} for the circuit shown below using source transformation.

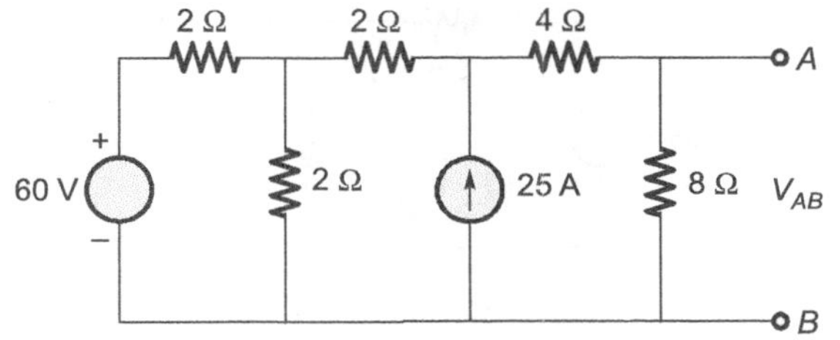

[10 marks : 2002]

Solution:

By(60 V) voltage source changes to respective current source.

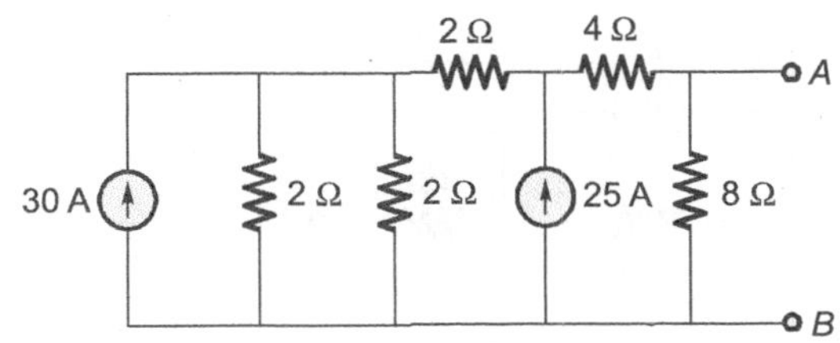

Now we combine the parallel resistance

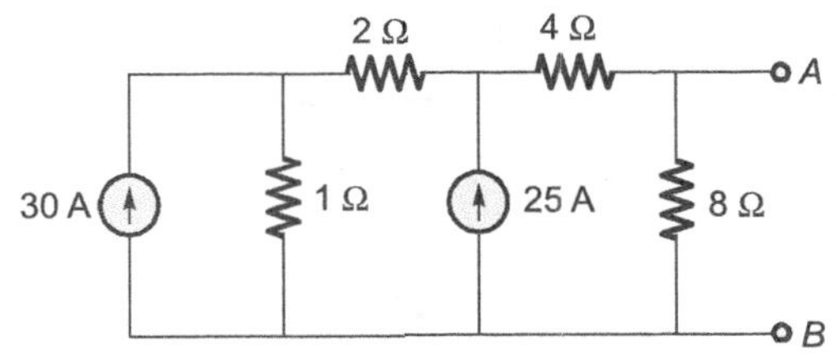

then we are going to change current source (30 A) to respective voltage source

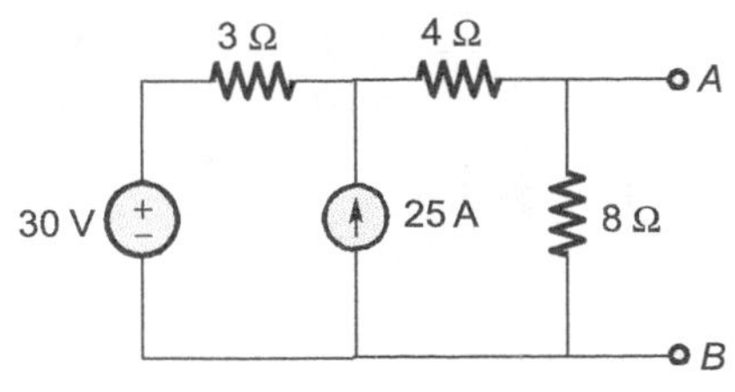

By changing (30 V) voltage source to respective current source

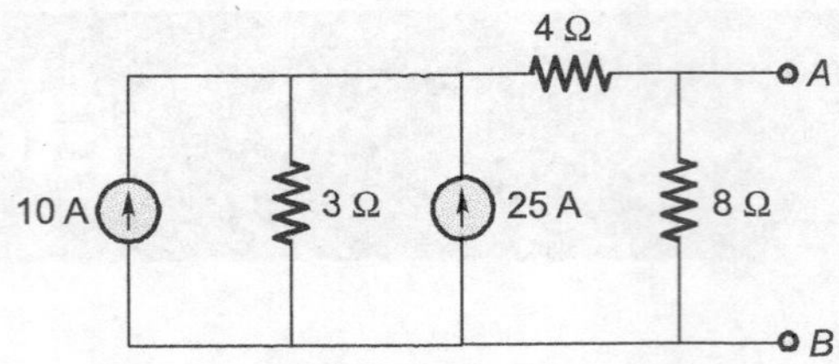

We combining current source

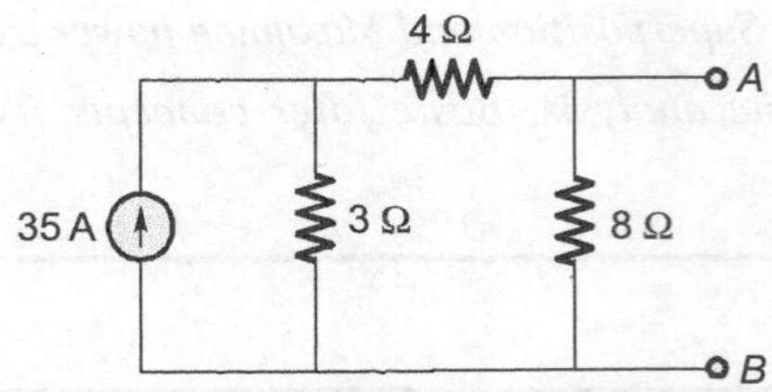

Now we make voltage source 35 A into current source

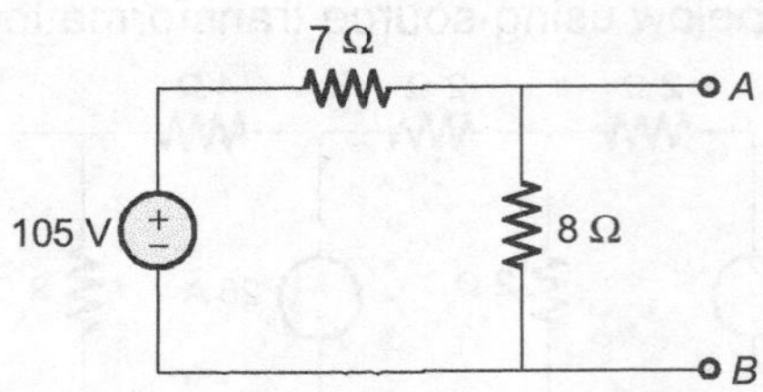

then by potential divider rule,

$$V_{AB} = \frac{105}{15} \times 8 = 56 \text{ V}$$

1.2 **Determine the value of I and R in the circuit shown in the figure.**

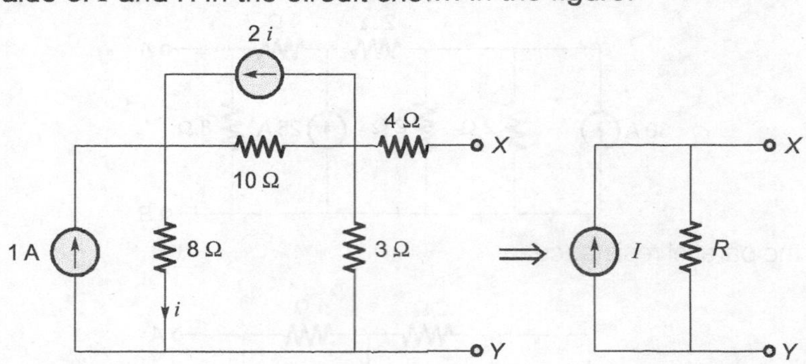

[14 marks : 2003]

Solution:

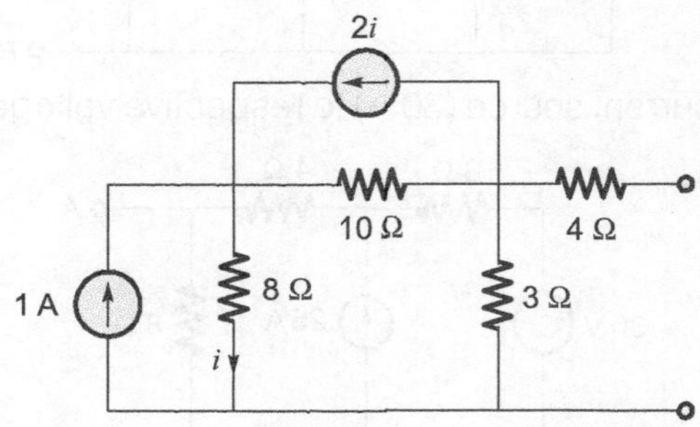

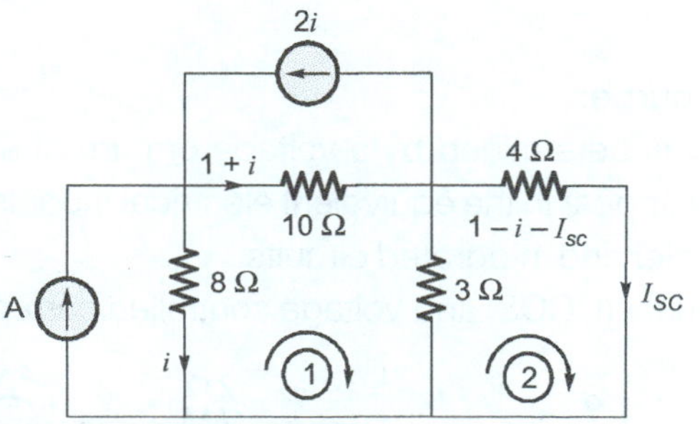

Using KVL in loop (1) we have,

$$-10(1+i) - 3(1 - i - I_{SC}) + 8i = 0$$
$$i + 3I_{SC} = 13$$

Using KVL in loop (2) we have,

$$-4I_{SC} + 3(1 - i - I_{SC}) = 0$$
$$7I_{SC} + 3i = 3$$
$$7I_{SC} + 3(13 - 3I_{SC}) = 3$$
$$2I_{SC} = 36$$
$$I_{SC} = 18 \text{ Amps.}$$

Now opening circuit the terminal xy, we have using KVL in the loop (1) we have

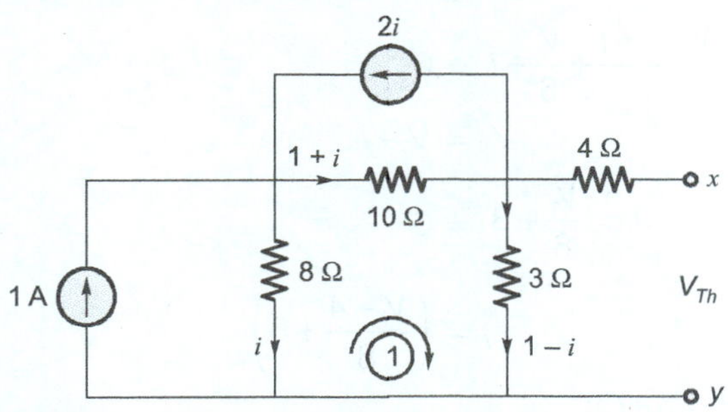

$$-10(1+i) - 3(1-i) + 8i = 0$$
$$13\text{ A} = i$$
$$V_{th} = (1-i)\cdot 3 = -36 \text{ volts}$$

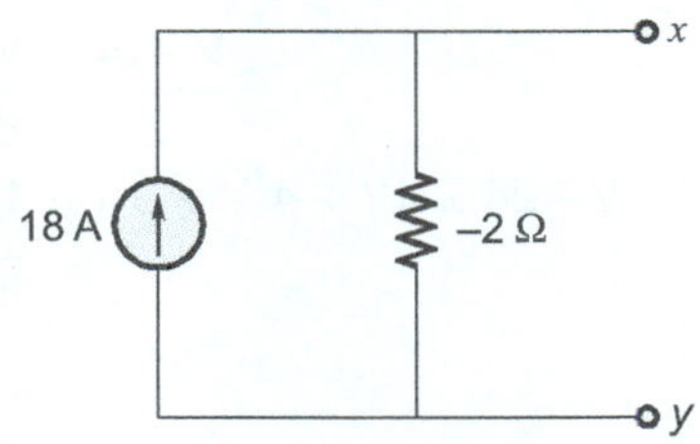

then, $$R_{th} = \frac{V_{th}}{I_{SC}} = \frac{-36}{18} = -2\ \Omega$$

and $$I = 18 \text{ Amps.}$$

Note: The negative value of the resistance tells us that the circuit in supplying power. Of course, the resistor in the circuit can not supply power (they absorb power). It is the dependent source that supplies the power. This is a circuit how dependent source and resistance could be used to simulate negative resistance.

1.3 Explain what you understand by voltage-dependent current source and current-dependent current source. For the circuit shown in figure determine the currents i_1, i_2 and i_3 using nodal analysis.

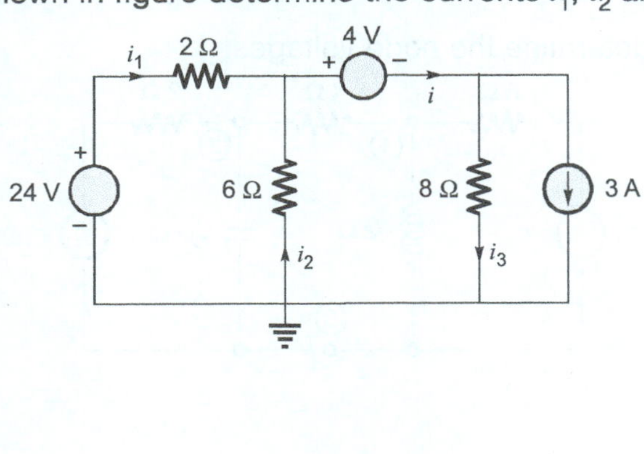

[12 marks : 2004]

Solution:

Dependent or controlled source:

In which the source quantity is determined by a voltage or current existing at some other location in the system source such as these appear in the equivalent electrical models for many electronic devices such as transistors, operational amplifier and integrated circuits.

Current controlled current source (CCCS) and Voltage controlled current source (VCCS)

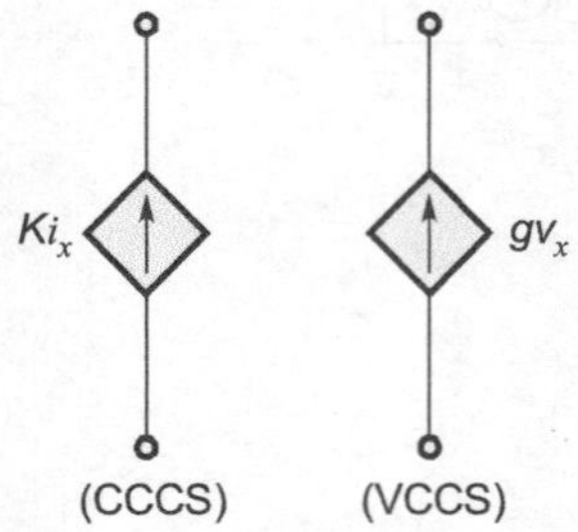

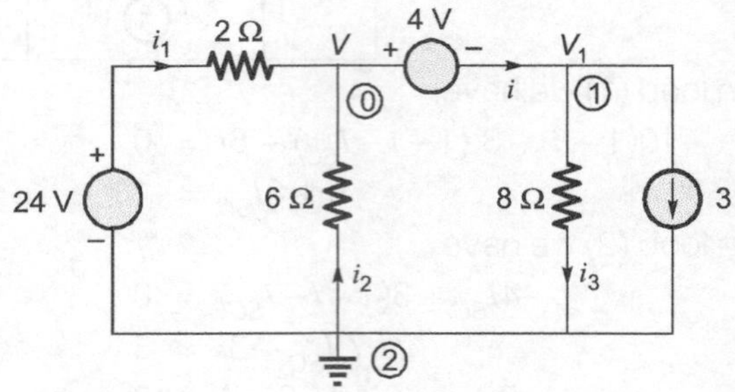

By nodal analysis, $(-i_1) + (-i_2) + i = 0$

$$-\left(\frac{24-V}{2}\right) + \left(-\frac{0-V}{6}\right) + i = 0$$

$$\frac{(V-24)}{2} + \frac{V}{6} + i = 0 \qquad ...(1)$$

and $V_1 = V - 4$

KCL at node 1, $-i + \frac{V_1}{8} + 3 = 0$

$$i = \left(\frac{V-4}{8} + 3\right) \qquad ...(2)$$

Combining both equation (1) and (2),

$$\therefore \quad \frac{V-24}{2} + \frac{V}{6} + \frac{V-4}{8} + 3 = 0$$

$12\,V - 24 \times 12 + 4\,V + 3\,V - 12 + 3 \times 24 = 0$

$$9\,V = 24\left(12 + \frac{1}{2} - 3\right)$$

$$\therefore \quad V = 12\text{ V}, V_1 = 8\text{ V},$$

$$i_1 = \frac{24-12}{2} = 6\text{ A}$$

$$i_2 = -\frac{12}{6} = -2\text{ A}$$

$$i_3 = 1\text{ A}$$

1.4 For the network shown, determine the node voltages.

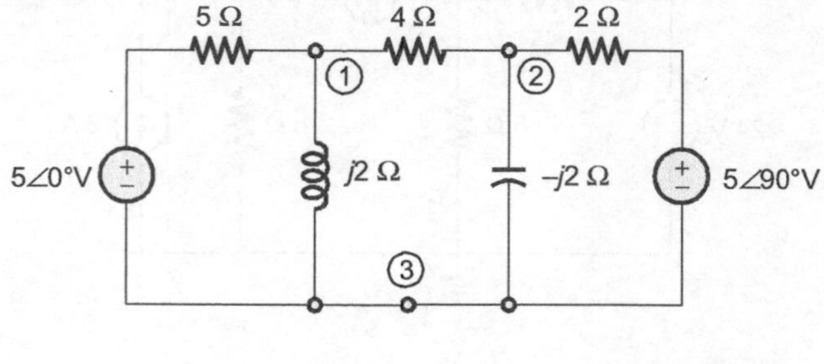

[12 marks : 2007]

Solution:

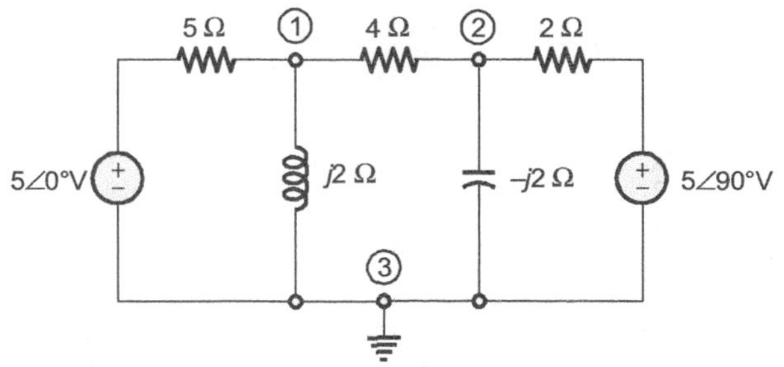

Let us take node (3) as reference node or (datum node)

By applying KCL at node (1),

$$\frac{V_1 - 5\angle 0^\circ}{5} + \frac{V_1}{2j} + \frac{V_1 - V_2}{4} = 0$$

$$0.67\angle{-48}\ V_1 - 0.25\ V_2 = 1\angle 0^\circ \qquad ...(1)$$

By applying KCL at node (2),

$$\frac{V_2 - V_1}{4} + \frac{V_2}{-2j} + \frac{V_2 - 5\angle 90^\circ}{2} = 0$$

$$0.9\angle 33.7^\circ\ V_2 - \frac{V_1}{4} = 2.5\angle 90^\circ \qquad ...(2)$$

$$\begin{bmatrix} 0.67\angle -48^\circ & -0.25 \\ -0.25 & 0.90\angle 33.7 \end{bmatrix} \begin{bmatrix} V_1 \\ V_2 \end{bmatrix} = \begin{bmatrix} 1\angle 0^\circ \\ 2.5\angle 90^\circ \end{bmatrix}$$

$$\Delta = \begin{vmatrix} 0.45 - j.5 & -0.25 \\ -0.25 & 0.75 + \frac{1}{2}j \end{vmatrix} = 0.525 - 0.15j = 0.54\angle -16^\circ$$

$$\therefore \qquad V_1 = \frac{1}{\Delta}\begin{bmatrix} 1 & -0.25 \\ j2.5 & 0.75 + j0.5 \end{bmatrix} = 2.476\angle 72^\circ \text{ V}$$

$$V_2 = \frac{1}{\Delta}\begin{bmatrix} 0.45 - j0.5 & 1 \\ -0.25 & j2.5 \end{bmatrix} = 3.43\angle 52^\circ \text{ V}$$

1.5 Find the power dissipated in the resistor R in the ladder network shown in the figure below.

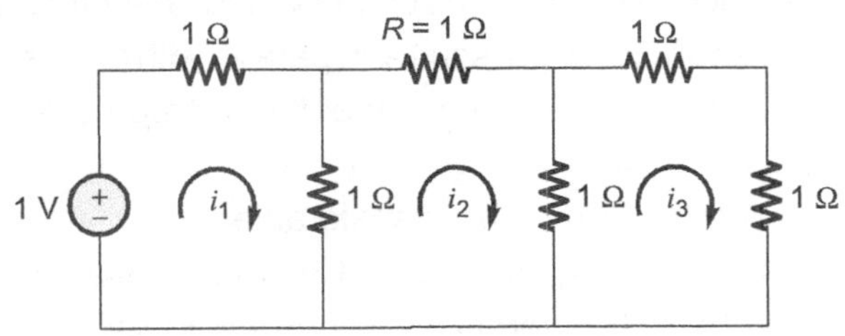

[12 marks : 2008]

Solution:

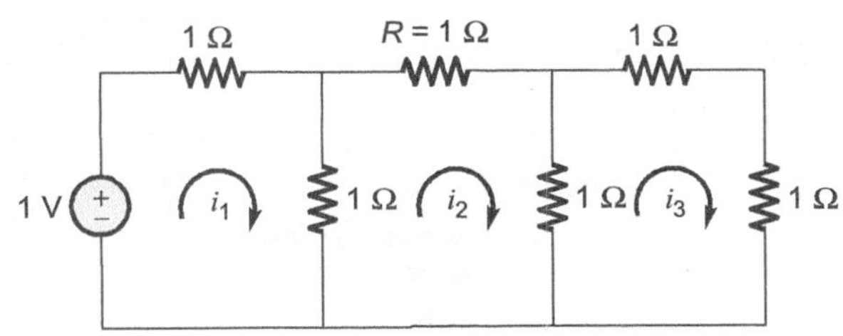

Using KVL in loop,

$$1 = 2i_1 - i_2 \quad \text{...(i)}$$

$$0 = 3i_2 - i_1 - i_3 \quad \text{...(ii)}$$

$$0 = 3i_3 - i_2 \quad \text{...(iii)}$$

$$\therefore \quad i_3 = \frac{i_2}{3}$$

By solving the equations, we get,

$$i_2 = \frac{3}{13}$$

$\therefore$ Power dissipated in the resistor $R = i^2 R = \dfrac{9}{169}$ W

1.6 State 'Voltage to current source transformation' theorem. It is required to replace network *N* in figure by a suitable equivalent network. Which of the networks of figure could be valid equivalent network (s)?

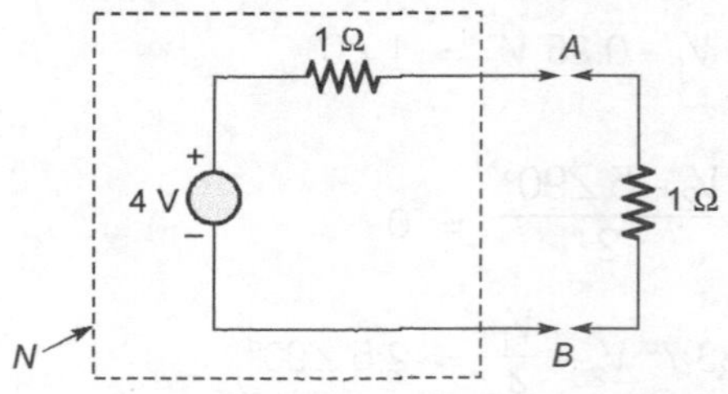

Fig. (a)

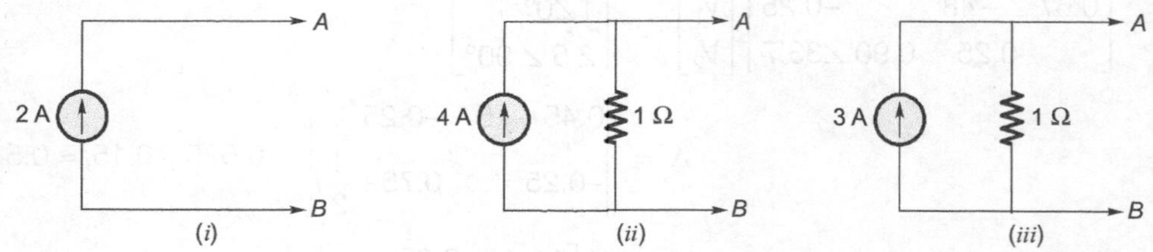

Fig. (b)

[12 marks : 2009]

Solution:

Source Transformation:

- Any voltage source consists of an ideal voltage source in series with an internal impedance (for ideal voltage source, this impedance being zero, the output voltage becomes independent of the load current) while any current source is an ideal current source in parallel with an internal impedance (for ideal current source, this parallel impedance is infinity such that the source current does not face any branching through this internal impedance path.
- The voltage and current sources are mutually transferable.
- For any voltage source, if the ideal voltage be V and internal resistance be R, the voltage source can be replaced by current source I with the internal resistance in parallel to the current source and the value of I is given by $I = V/R$.

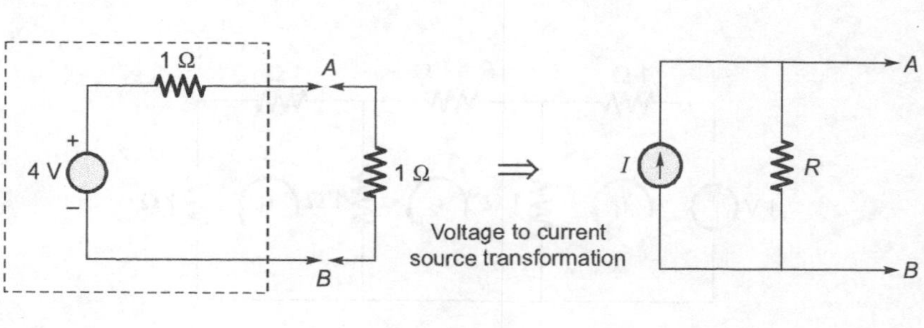

where, $R = 1\ \Omega$ and $I = \frac{V}{R} = \frac{4}{1} = 4A$

Hence,

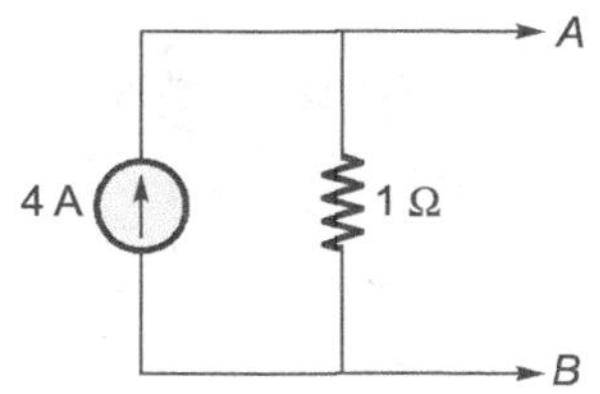

is the valid equivalent network.

1.7 **The following mesh equations pertain to a network**

$$8I_1 - 5I_2 - I_3 = 110$$
$$-5I_1 + 10I_2 + 0 = 0$$
$$-I_1 + 0 + 7I_3 = 115$$

(i) Draw the network showing each element.

(ii) Calculate the current in the 110 volts source. **[8 + 4 = 12 marks : 2010]**

Solution:

All the mesh equations are rearranged as follows:

$$8I_1 - 5I_2 - I_3 = 110$$

$$\Rightarrow \quad 5(I_1 - I_2) + (I_1 - I_3) + 2I_1 = 110 \quad \ldots(i)$$

$$-5I_1 + 10I_2 + 0 = 0$$

$$\Rightarrow \quad 5(I_2 - I_1) + 5I_2 = 0 \quad \ldots(ii)$$

$$-I_1 + 0 + 7I_3 = 115$$

$$(I_3 - I_1) + 6I_3 = 115 \quad \ldots(iii)$$

On the basis of equations (*i*), (*ii*) and (*iii*) the network is drawn below:

$$\begin{bmatrix} 8 & -5 & -1 \\ -5 & 10 & 0 \\ -1 & 0 & 7 \end{bmatrix} \begin{bmatrix} I_1 \\ I_2 \\ I_3 \end{bmatrix} = \begin{bmatrix} 110 \\ 0 \\ 115 \end{bmatrix}$$

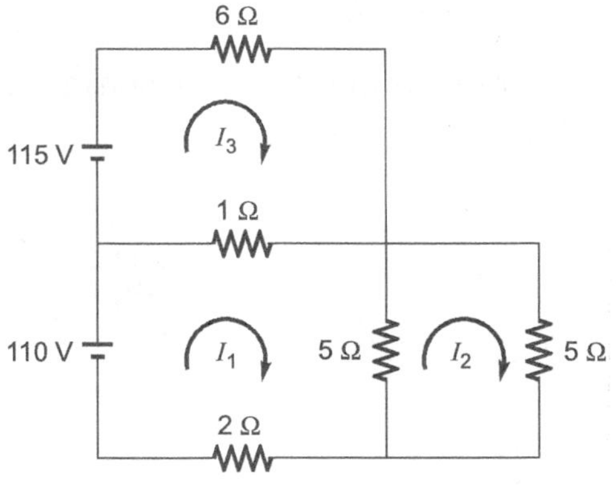

$$\Delta = \begin{vmatrix} 8 & -5 & -1 \\ -5 & 10 & 0 \\ -1 & 0 & 7 \end{vmatrix} = 375$$

$$\Delta_1 = \begin{vmatrix} 110 & -5 & -1 \\ 0 & 10 & 0 \\ 115 & 0 & 7 \end{vmatrix} = 8850$$

According to Cramer's rule, $I_1 = \frac{\Delta_1}{\Delta} = \frac{8850}{375} = 23.6\text{ A}$

(Current in the voltage source)

1.8 Find the current I_x which flows through the 3 Ω resistor in the circuit of the figure below:

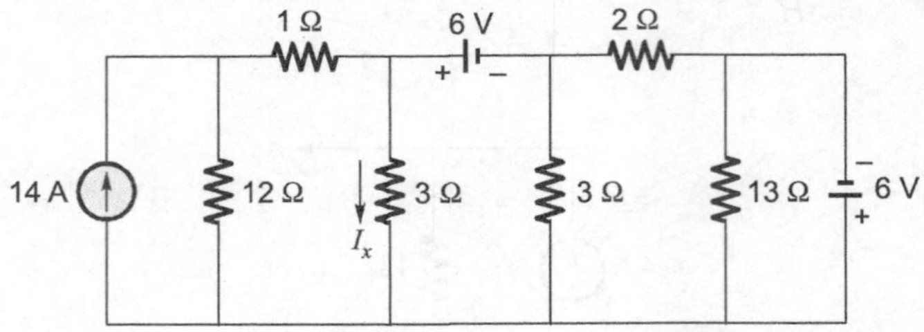

[8 + 4 = 12 marks : 2011]

Solution:

The 14 A current source is transformed to an equivalent voltage source and the circuit is redrawn.

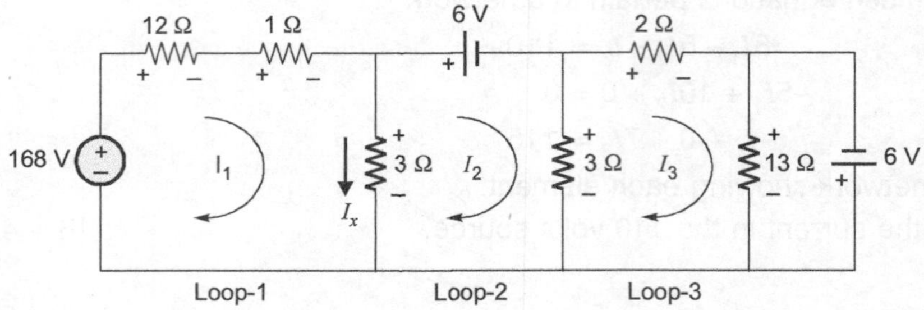

From the circuit,

$$I_x = I_1 - I_2 \quad ...(i)$$

Applying KVL in loop 1, $168 - 13I_1 - 3(I_1 - I_2) = 0$

$\Rightarrow$ $16I_1 - 3I_2 = 168$ ...(ii)

Applying KVL in loop 2, $3(I_1 - I_2) - 6 - 3(I_2 - I_3) = 0$

$\Rightarrow$ $3I_1 - 6I_2 + 3I_3 = 6$ ...(iii)

Applying KVL in loop 3, $3(I_2 - I_3) - 2I_3 + 6 = 0$

$-3I_2 + 5I_3 = 6$ ...(iv)

On solving equations (*ii*), (*iii*) and (*iv*) we get

$I_1 = 12$ A, $I_2 = 8$ A and $I_3 = 6$ A

From equation (*i*), $I_x = I_1 - I_2 = 12 - 8 = 4$ A

1.9 Calculate the power dissipation in each resistor of figure given below. Use KVL and KCL to solve the problem.

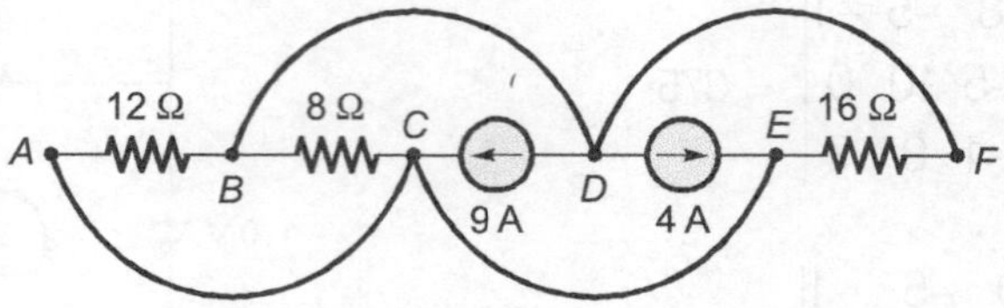

[10 marks : 2013]

Solution:

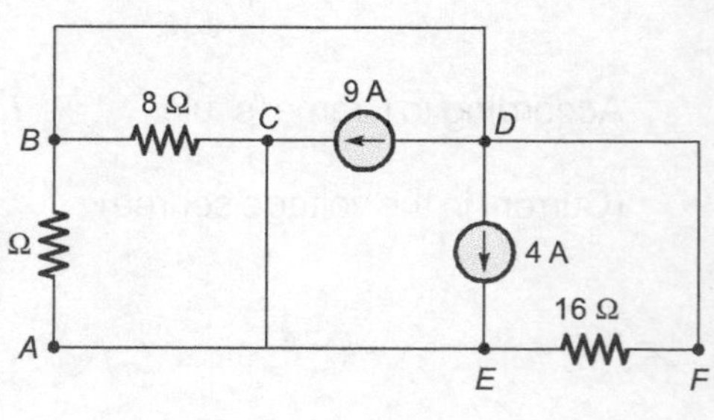

Let, $V_B = V_D = V_F = V_1$

and $V_A = V_C = V_E = 0$ V (reference)

Using KCL at node *B*,

$$\frac{V_B - V_A}{12} + \frac{V_B - V_C}{8} + \frac{V_F - V_E}{16} + 9 + 4 = 0$$

$$\frac{V_1}{3}+\frac{V_1}{2}+\frac{V_1}{4}+52 = 0$$

$$\frac{13\,V_1}{12} = -52$$

$$\Rightarrow \qquad V_1 = -48\text{ V}$$

$\because$ Resistors always dissipates the power irrespective of direction of current flowing through them.

$\therefore$ Power dissipated across 12 Ω resistor $= \frac{(48)^2}{12} = 192\text{ W}$

Power dissipated across 8 Ω resistor $= \frac{(48)^2}{8} = 288\text{ W}$

Power dissipated across 16 Ω resistor $= \frac{(48)^2}{16} = 144\text{ W}$

1.10 **Carry out the nodal analysis for the circuit given below and then compute the voltage across the 2-mho's conductance.**

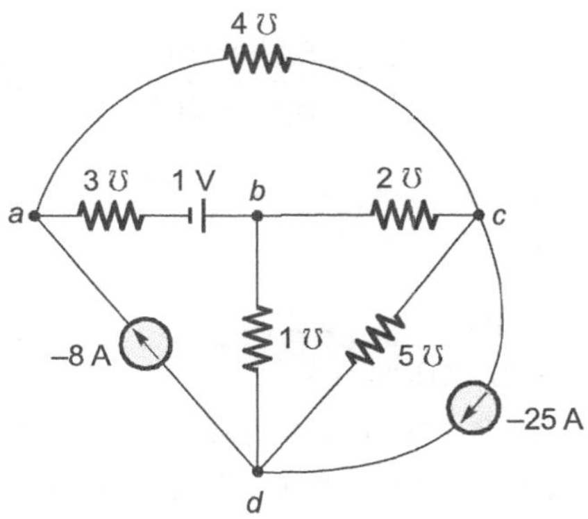

[10 marks : 2013]

Solution:

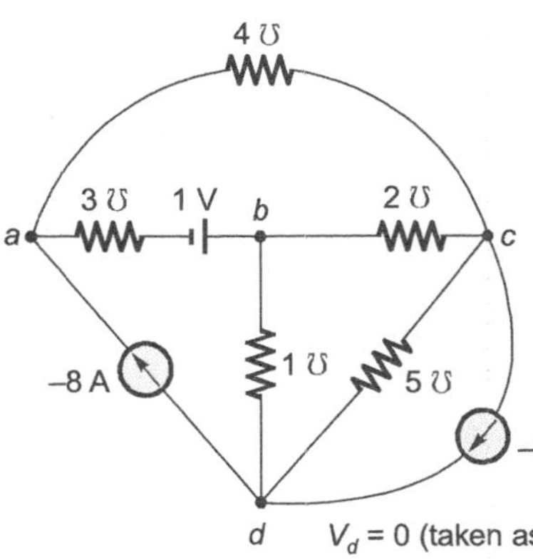

Nodal at *a*:

$$-(-8) + [V_a - (V_b - 1)]3 + (V_a - V_c)(4) = 0$$

$$7V_a - 3V_b - 4V_c = -11 \quad ...(i)$$

Nodal at *b*:

$$(V_b)1 + (V_b - V_c)(2) + (V_b - V_a)(3) = 0$$

$$-3V_a + 6V_b - 2V_c = 3 \quad ...(ii)$$

Nodal at *c*:

$$(V_c - V_b)(2) + (V_c - V_a)(4) + (V_c)(5) - 25 = 0$$

$$-4V_a - 2V_b + 11V_c = 25 \quad ...(iii)$$

From equations (*i*), (*ii*) and (*iii*),

$$V_a = 1\text{ V}, \quad V_b = 2\text{ V}, \quad V_c = 3\text{ V}$$

Hence, voltage across 2-mho's conductance $= |V_c - V_b| = |3-2| = 1\text{V}$

1.11 **Three capacitors C_1, C_2 and C_3 whose values are 10 μF, 5 μF and 2 μF respectively, have breakdown voltages of 10 volts, 5 volts and 2 volts respectively. For their interconnection shown below, find the maximum safe voltage that can be applied across the combination and the corresponding total charge stored in the effective capacitance across the terminals *AB*.**

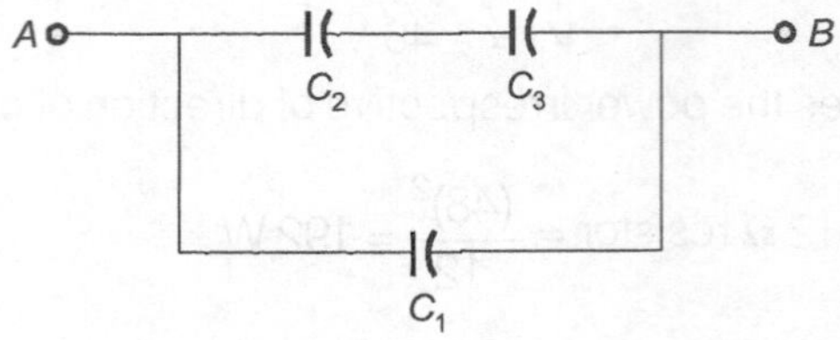

[12 marks : 2014]

Solution:

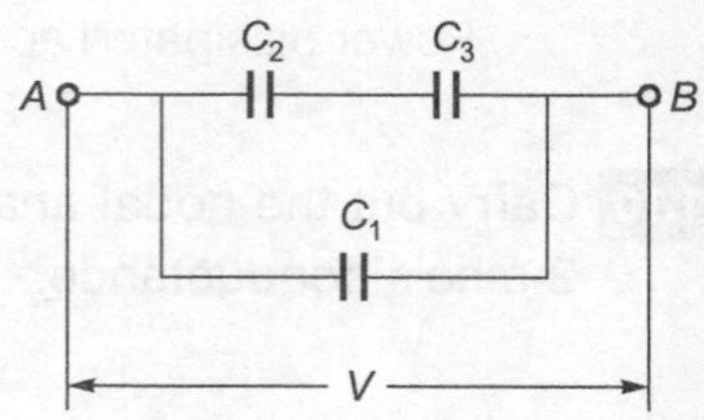

$$\text{Charge on } C_1 = 10\,\mu F \times 10\,V = 100\,\mu C$$
$$\text{Charge on } C_2 = 5\,\mu F \times 5 = 25\,\mu C$$
$$\text{Charge on } C_3 = 2\,\mu F \times 2 = 4\,\mu C$$

Since, the capacitors are connected in series = Charge is same (i.e.)

(Minimum charge value of C_2 or C_3) = 4 μC

$$\text{Effective capacitance of } C_2 \text{ and } C_3 = \frac{C_2 C_3}{C_2 + C_3} = \frac{5\mu F \times 2\mu F}{5\mu F + 2\mu F} = 1.4285\,\mu F$$

$$\Rightarrow \qquad \text{Voltage } (V) = \left(\frac{Q}{C}\right)$$

$$V = \frac{4\mu C}{1.4285 \times 10^{-6}} = 2.8\text{ V}$$

New charge on the capacitor (C_1) = 2.8 × 10 μF = 28 μC

The total charge stored in the effective capacitance = 28 μC + 4 μC = 32 μC

1.12 **Determine current-*I* through the 10 Ω resistances shown in figure below:**

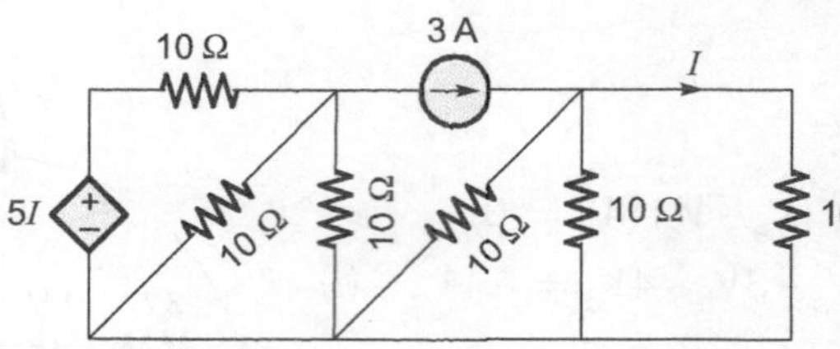

[10 marks, 2015]

Solution:

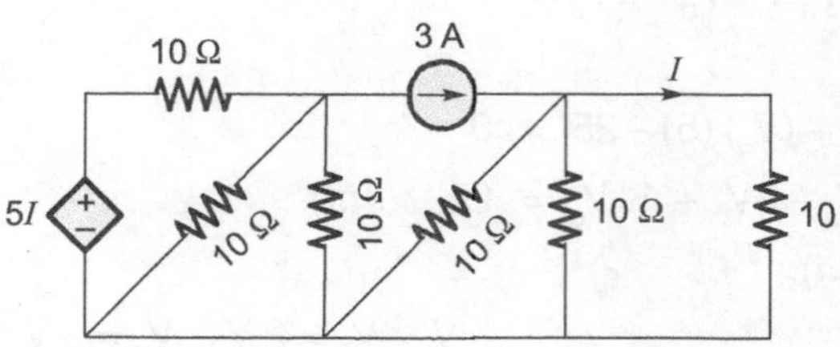

The given circuit above can be simplified as,

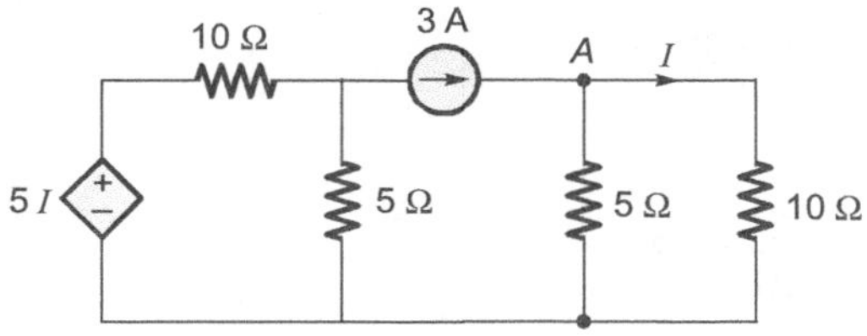

using current division on node A,

$$I = \frac{5}{5+10}\times 3 = 1\text{ A}$$

1.13 Calculate the equivalent resistance R_{ab} in the circuit given below:

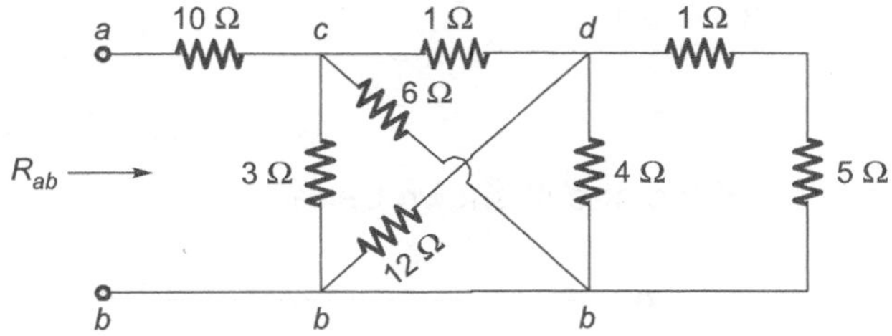

[10 marks : 2016]

Solution:

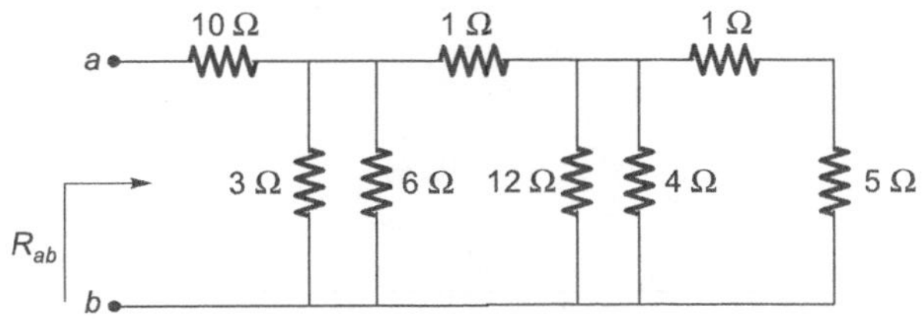

$$R_{ab} = \left[\{(5+1)\parallel(12\parallel 4)\}+1+\{(6\parallel 3)\}\right]+10$$

$$R_{ab} = \left[\{(6\parallel 3)+1\}\parallel 2\right]+10$$

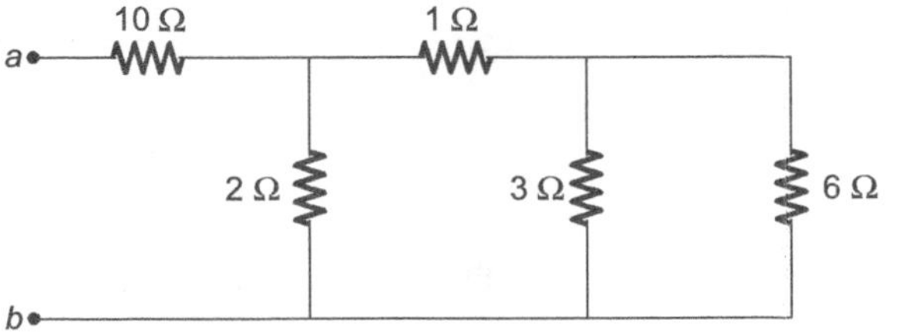

$$R_{ab} = \frac{\left(\frac{6\times 3}{6+3}+1\right)\times 2}{3+2}+10$$

$$R_{ab} = \frac{56}{5} = 11.2\ \Omega$$

1.14 Find the currents and voltages in the circuit shown in figure below.

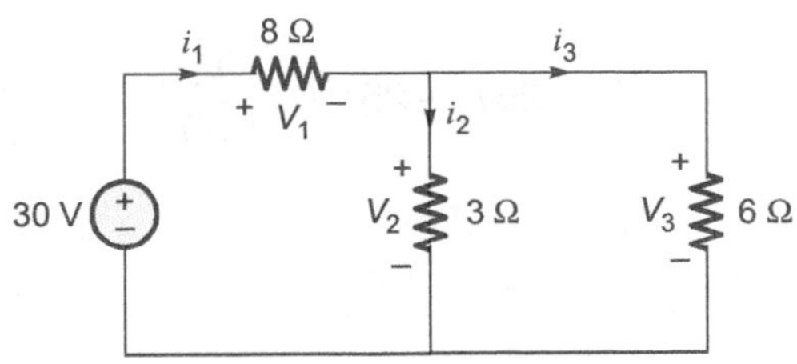

[10 marks : 2016]

Solution:

Total resistance,
$$R_T = (6 \parallel 3) + 8$$
$$R_T = 10\ \Omega$$
$$i_1 = \frac{30}{10} = 3\text{A};$$
$$V_1 = 8 \times 3 = 24\text{ V}$$
$$i_2 = 3 \times \frac{6}{9} = 2\text{A} \quad \text{(Current division)}$$
$$V_2 = 2 \times 3 = 6\text{ V}$$
$$i_3 = 3 \times \frac{3}{9} = 1\text{A} \quad \text{(Current division)}$$
$$V_3 = 6 \times 1 = 6\text{ V}$$

1.15 Find the input resistance R_{in}, of the circuit shown below.

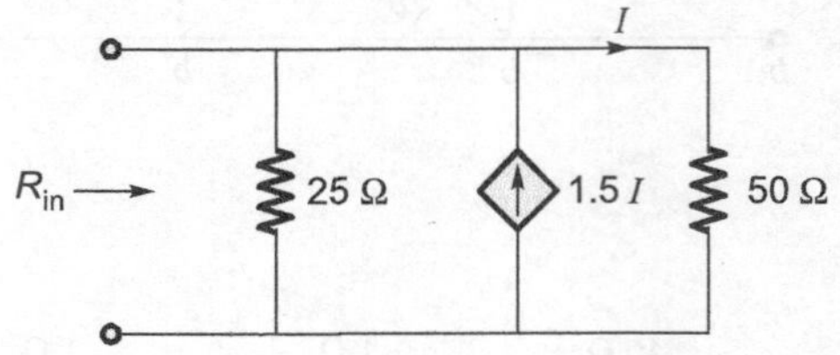

[10 marks : 2016]

Solution:

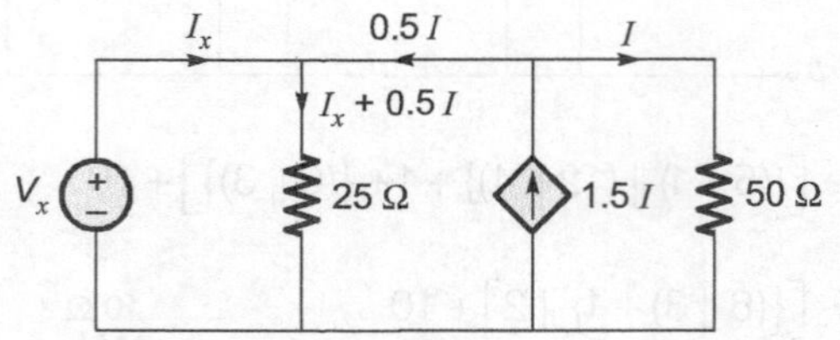

$$R_{in} = \frac{V_x}{I_x}$$

Write KVL equation in first loop,

$$V_x = 25\,(I_x + 0.5\,I) \quad \text{...(i)}$$
$$I = \frac{V_x}{50} \quad \text{...(ii)}$$

Substitute equations (ii) in (i),

$$V_x = 25 I_x + 25 \times 0.5\left(\frac{V_x}{50}\right)$$
$$\left(\frac{V_x}{I_x}\right) = R_m = 33.33\ \Omega$$

1.16 For the circuit shown in the figure,

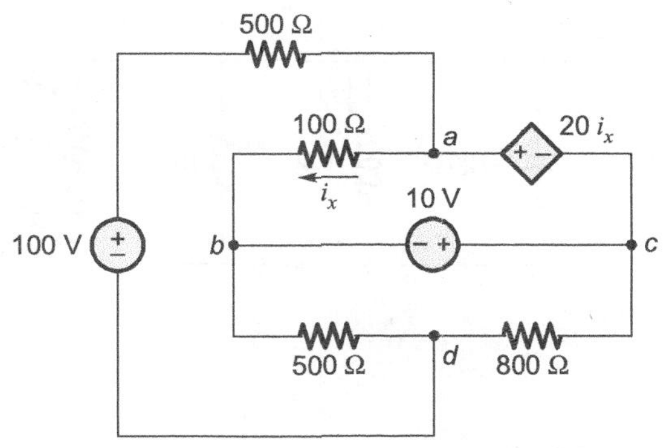

(i) Find the node voltages.

(ii) Power absorbed by the 800 Ω resistor. [20 marks : 2019]

Solution:

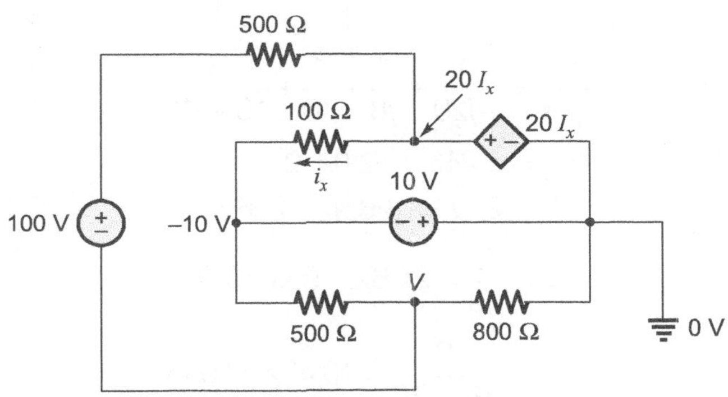

$$\frac{V+100-20I_x}{500}+\frac{V+10}{500}+\frac{V}{800} = 0$$

$$8V+800-160I_x+8V+80+5V = 0$$

$$21\,V-160I_x = -880 \qquad \text{... (i)}$$

Write KCL equation for I_x

$$\frac{20I_x+10}{100} = I_x$$

$$I_x = \frac{1}{8}\text{ A} \qquad \text{... (ii)}$$

$$\therefore \qquad 20\,I_x = 2.5$$

Substituting (ii) in (i),

$$21V-160\left(\frac{1}{8}\right) = -880$$

$$V = -40.95\text{ V}$$

Node voltages: –10 V, –40.95 V, 2.5 V

power consumed by 800 Ω = $\frac{V^2}{R}=\frac{(-40.95)^2}{800} = 2.096$ W

1.17 For the circuit shown in the figure given below, the total current I entering the circuit is $30.0 \angle -21°$ A. Determine the voltage V and the branch current I_2 and I_4.

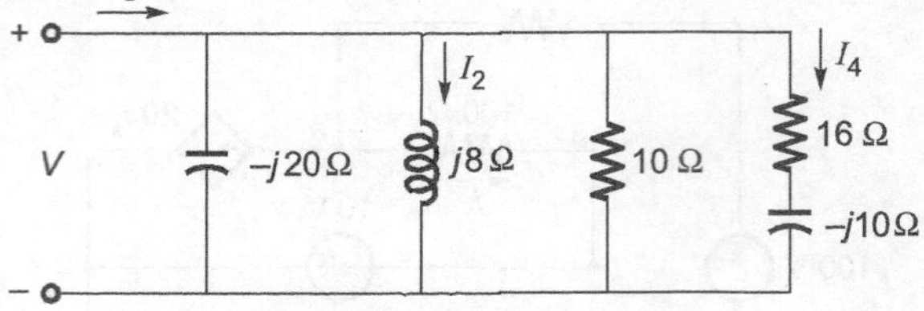

[20 marks : 2020]

Solution:

Given that,

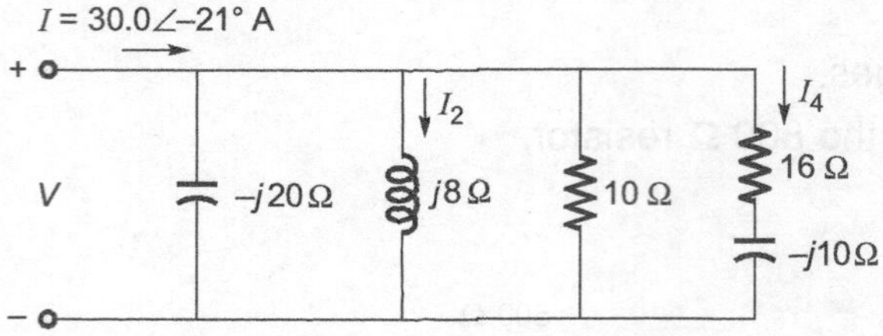

$$\frac{1}{Z_{eq}} = \frac{1}{-j20} + \frac{1}{j8} + \frac{1}{10} + \frac{1}{16 - j10}$$

$$Z_{eq} = 6.245 + j2.02\ \Omega$$

$$V = Z_{eq} I = 196.9 \angle -3.06\ \text{V}$$

$$I_2 = \frac{V}{j8} = 24.6 \angle -93.066\ \text{A}$$

$$I_4 = \frac{V}{16 - j10} = 10.43 \angle 28.945\ \text{A}$$

2. Network Theorems

2.1 Determine the value of R shown in the below figure such that the 6 Ω resistor consumes the maximum power.

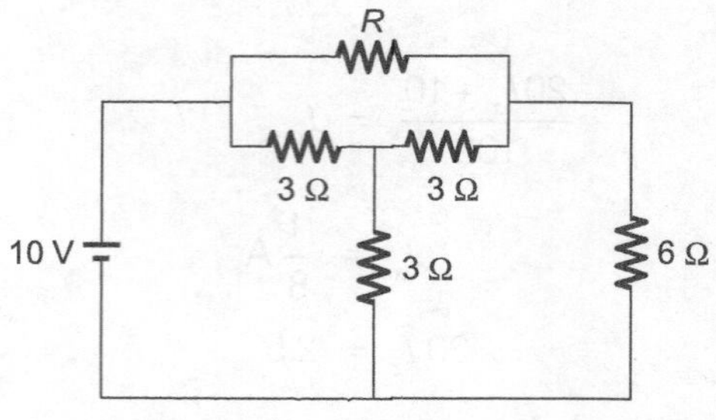

[8 marks : 2002]

Solution:

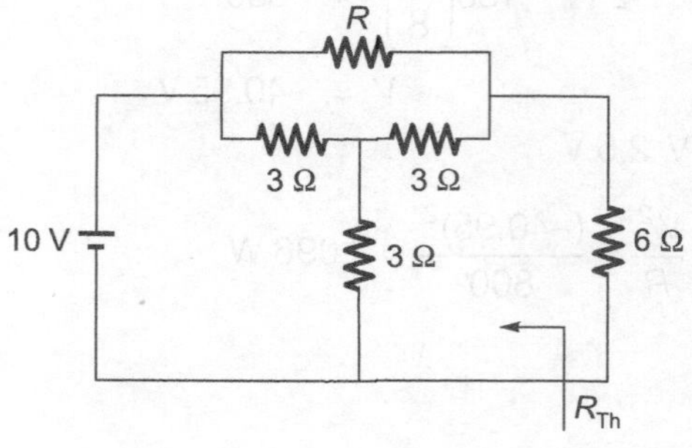

R_{Th} is calculated by shorting the voltage source and converting star into delta network,

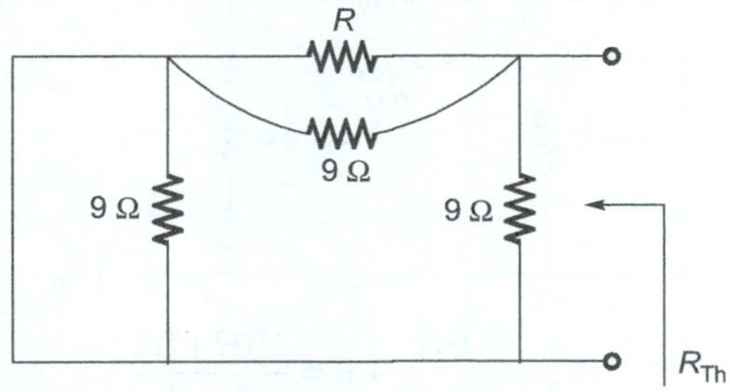

$$R_{th} = \frac{9R}{9+R} \parallel 9 = \frac{\left(\frac{9R}{9+R}\right)\times(9)}{\frac{9R}{9+R}+9} = \frac{9R}{2R+9}$$

According to maximum power transfer theorem,

$$\text{Load resistance} = R_{Th}$$

$$R_{th} = \frac{9R}{2R+9} = 6$$

$$9R = 12R + 54$$

$$R = -\frac{54}{3}\ \Omega$$

Resistance can not be negative.

If $R = 0$, voltage source will be directly connected to the load resistor (6 Ω), all the power will be absorbed by load resistor (6 Ω).

So, $R = 0$ for maximum power transfer.

2.2 Determine a non-negative value of R such that the power consumed by the 2 Ω resistor in the figure shown is maximum.

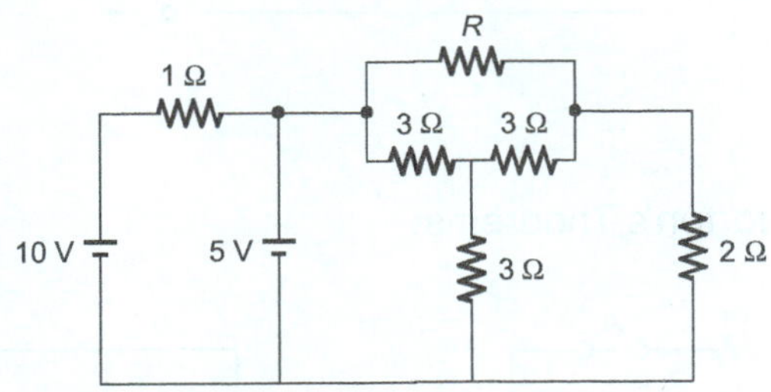

[10 marks : 2003]

Solution:

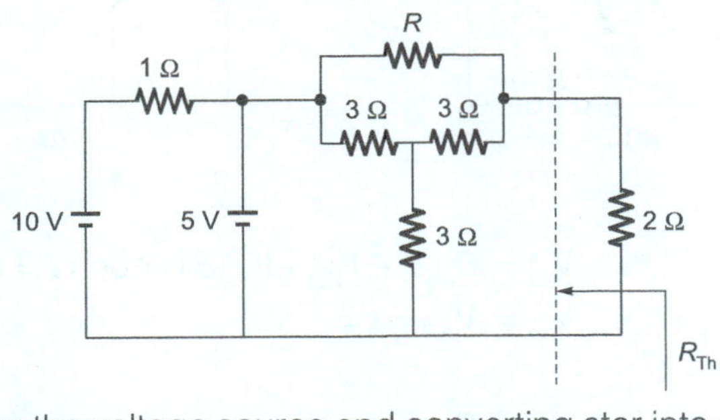

R_{Th} is calculated by shorting the voltage source and converting star into delta network,.

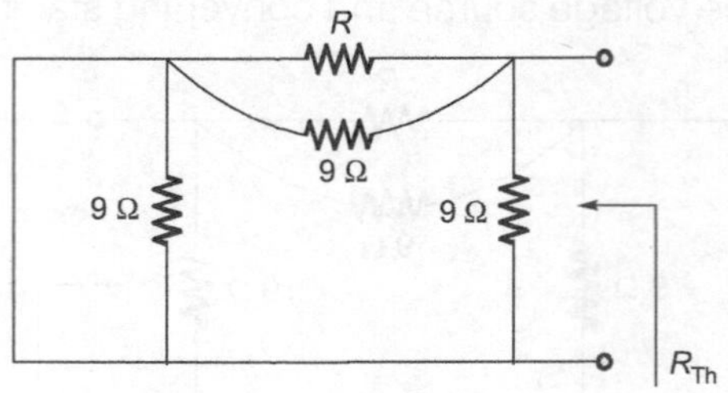

$$R_{Th} = \frac{9R}{9+R} + 9 = \frac{18R+81}{9+R}$$

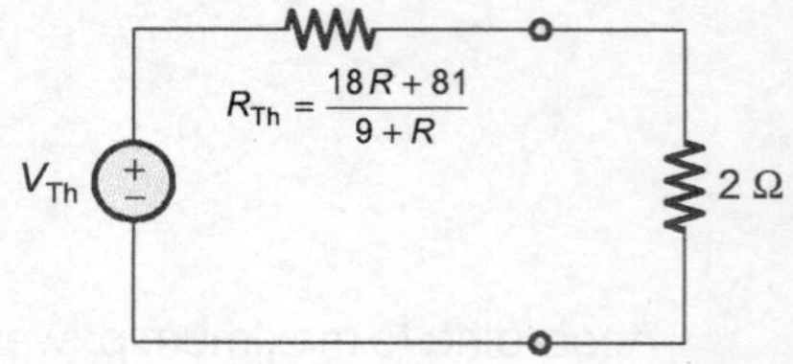

The above circuit is replaced by an equivalent Thevenin circuit. In the circuit load resistance is fixed and source resistance is variable, so we have to choose minimum internal resistance of the source that results maximum current flow through the fixed load which ultimately gives maximum power dissipation across the load.

$$R = 0 \quad \text{(Minimum value)}$$

2.3 Show that Thevenin's and Norton's theorems are dual to each other. Determine the voltage across 3-ohm resistor for the circuit shown in figure below using Thevenin's theorem.

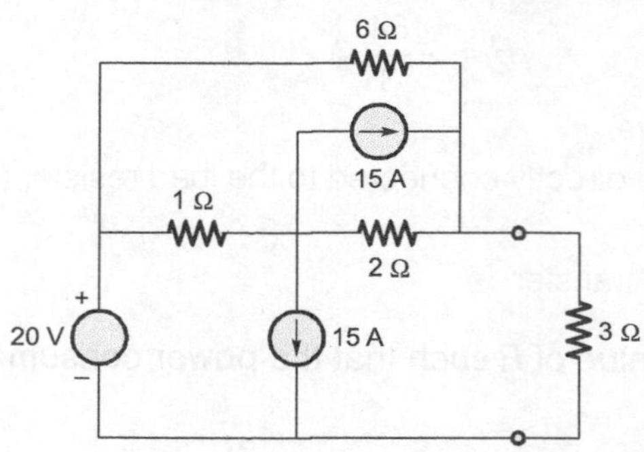

[10 marks : 2004]

Solution:

Duality of Thevenin's and Norton's Theorems:

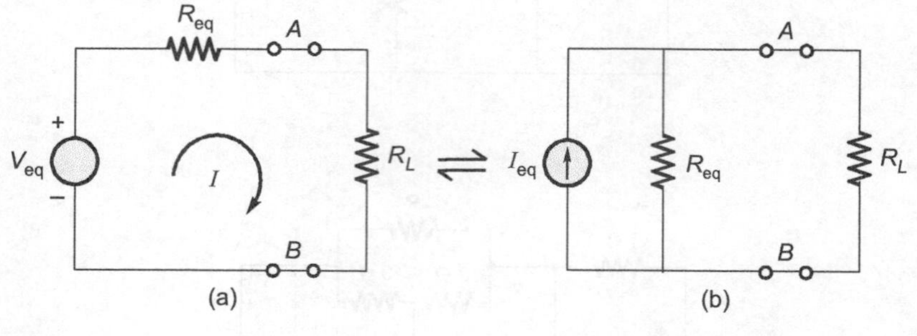

For circuit (a):

$$V_{eq} = R_{eq}I + V_{AB} \quad \text{If } AB \text{ is open } I = 0$$

∴

$$V_{eq} = V_{AB} = V_{OC}$$

V_{eq} is called Thevenin voltage V_{th}

If AB is shorted,

$$V_{eq} = R_{eq}I_{sc}$$

$$R_{eq} = \frac{V_{th}}{I_{SC}}$$

For circuit (b):

If AB is open,
$$I_{eq} = \frac{V_{AB}}{R_{eq}} = \frac{V_{OC}}{V_{OC}/I_{sc}}$$

$$I_{eq} = I_{sc}$$

I_{sc} is called Norton current I_N.

So, the current that flows by short circuiting the terminals of a Thevenin circuit is the Norton current. We also see that the two equivalent resistances are same, i.e. $R_{th} = R_N = R_{eq}$. So, a Thevenin circuit can be converted into a Norton circuit or vice-versa. So, a Norton equivalent circuit is the dual of Thevenin equivalent circuit. Voltage source with series equivalent resistance is transformed into a current source with the equivalent resistance in parallel.

For finding 'V_{th}'

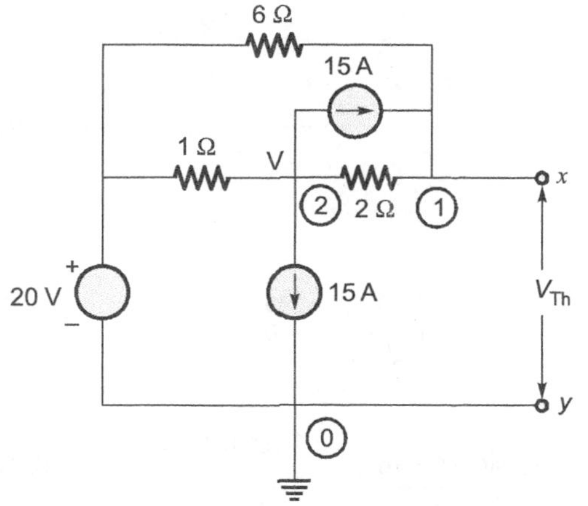

KCL at node (1),

$$\frac{V_{th}-V}{2}+\frac{V_{th}-20}{6}-15 = 0 \quad \text{...(1)}$$

KCL at node (2),

$$\frac{V-V_{th}}{2}+\frac{V-20}{1}+30 = 0 \quad \text{...(2)}$$

Solving we get,

$$3\,V_{th} - 3\,V + V_{th} - 20 - 90 = 0$$
$$4V_{th} - 3V = +110$$
$$V - V_{th} + 2\,V - 40 + 60 = 0$$
$$4\,V_{th} - 3V = 110$$
$$-\,V_{th} + 3V = -20$$

$\therefore$
$$V_{th} = 30\text{ V}$$

R_{th} is the calculated from the following figure,

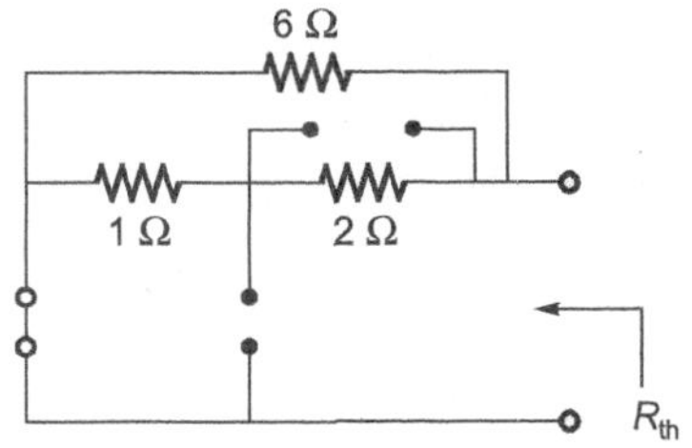

$R_{th} = 3||6 = 2\ \Omega$

Thevenin equivalent circuit across the AB terminals,

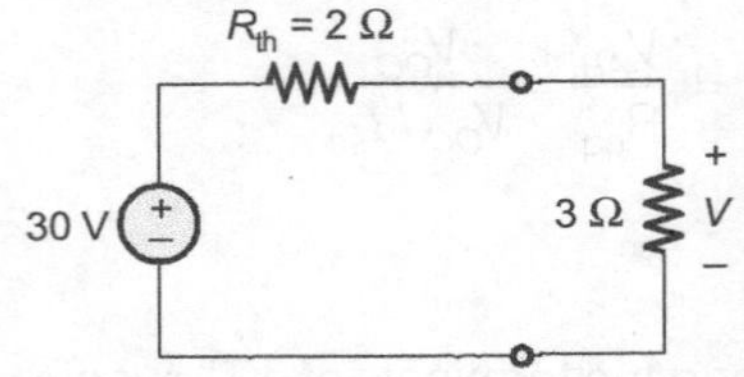

$$V = 30 \times \frac{3}{3+2} = 18 \text{ V}$$

2.4 **For the circuit shown in Figure, find the Thevenin and Norton equivalent circuits looking in at terminals '*a*' and '*b*'.**

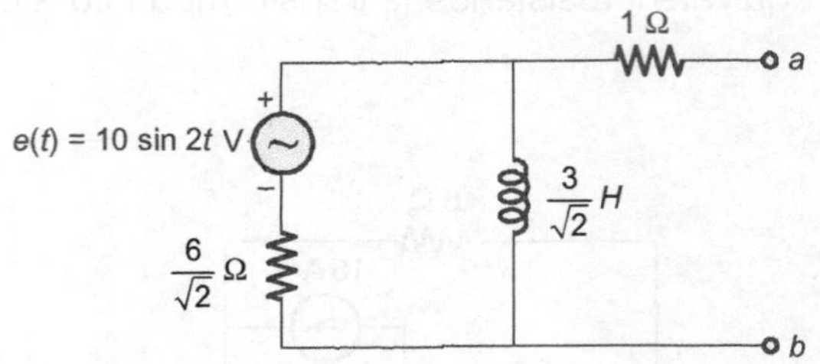

[10 marks : 2005]

Solution:

$$X_L = \omega L = 2 \times \left(\frac{3}{\sqrt{2}}\right) = \frac{6}{\sqrt{2}}$$

$$V_{th} = \frac{10\sin 2t \times j\omega L}{R + j\omega L} = \frac{10\sin 2t \times j\frac{6}{\sqrt{2}}}{\frac{6}{\sqrt{2}} + j\frac{6}{\sqrt{2}}} = \frac{10\sin 2t \times j1}{1 + j1} = 7.07\sin(2t + 45°)$$

For finding (R_{th}):

$$R_{th} = 1 + \frac{\frac{6}{\sqrt{2}} \times j\frac{6}{\sqrt{2}}}{\frac{6}{\sqrt{2}} + j\frac{6}{\sqrt{2}}} = 1 + \frac{j6 \times \frac{6}{\sqrt{2}}}{6 + j6} = 1 + \frac{j\left(\frac{6}{\sqrt{2}}\right)}{1 + j} = 3.12 + j2.12$$

$$I_N = \frac{V_{th}}{R_{th}} = \frac{7.07\sin(2t + 45°)}{3.12 + j2.12} = 1.875\sin(2t + 10°)$$

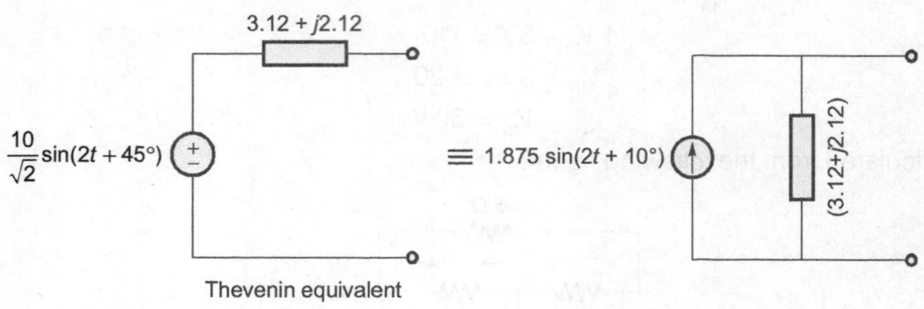

Thevenin equivalent

2.5 Calculate the current *I* following through the 10 Ω resistor in the circuit shown below. Apply Thevenin's theorem.

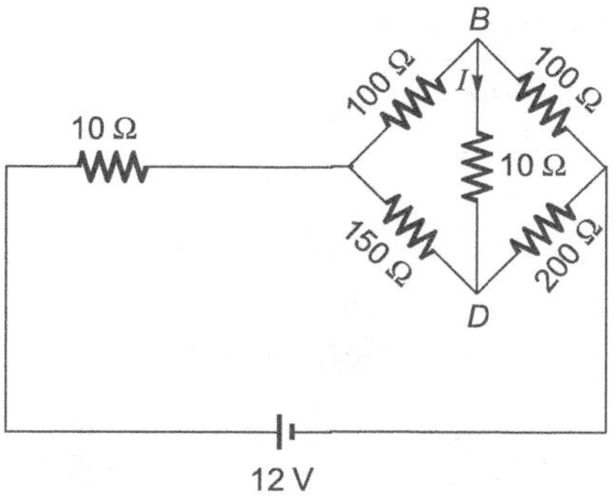

[12 marks : 2006]

Solution:

For finding '$R_{thevenin}$ (R_{th})',

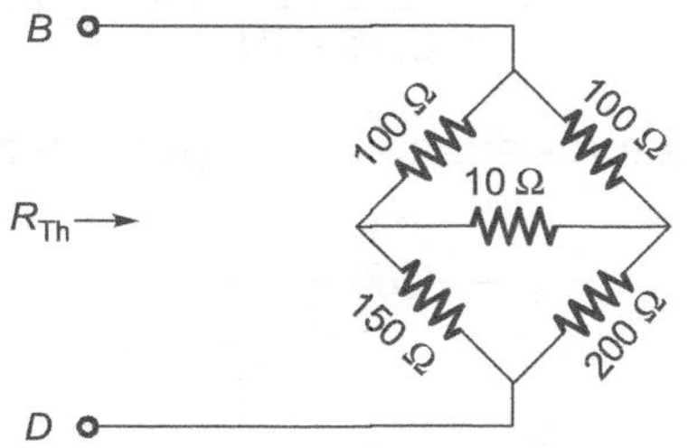

By delta-star conversion,

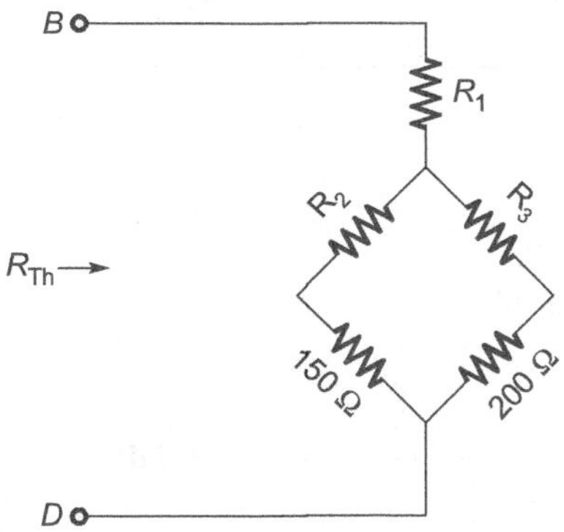

where,
$$R_1 = \frac{100 \times 100}{210} = 47.61$$

$$R_2 = \frac{1000}{210} = 4.761$$

$$R_3 = \frac{1000}{210} = 4.761$$

$$\therefore \quad R_{thevenin}(R_{th}) = 47.61 + (150 + 4.761) \,||\, (200 + 4.761)$$
$$= 135.76\ \Omega$$

For finding $V_{Thevenin}$ (V_{th}),

$$I' = \frac{12}{10 + \left(\frac{200 \times 350}{550}\right)} = 0.0874\ \text{A}$$

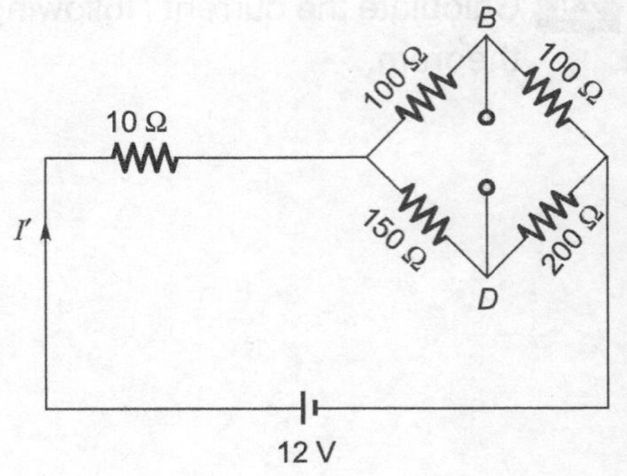

$$V_B = \frac{I' \times 350}{550} \times 100 = 5.56 \text{ V}$$

$$V_D = \frac{I' \times 200 \times 200}{550} = 6.35$$

$$V_{BD} = -0.79 \text{ V}$$

$$V_{th} = -0.79 \text{ V}$$

The Thevenin's equivalent circuit is

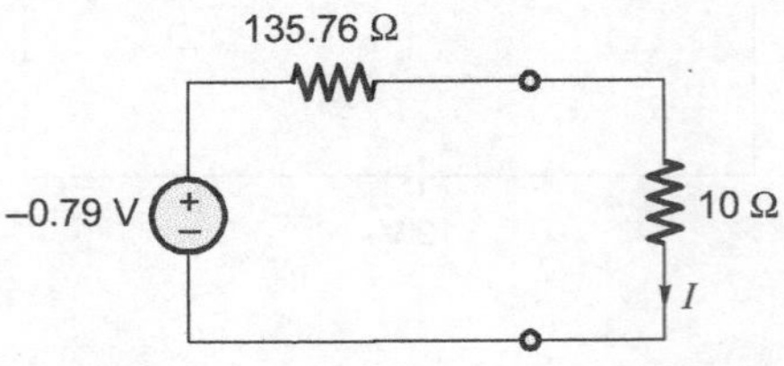

i.e., The current pass through 10 Ω resistor,

$$I = \frac{-0.79 \text{ V}}{135.76 + 10} = 5.419 \text{ mA} \quad \text{(direction opposite)}$$

2.6 **Calculate the voltage V across the resistance R in Figure by the principle of superposition.**

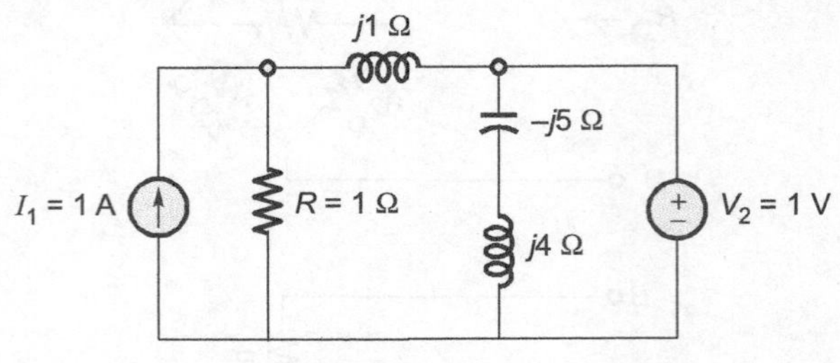

[12 marks : 2007]

Solution:

By applying super position theorem,
First short circuiting voltage source V_2,
we have,

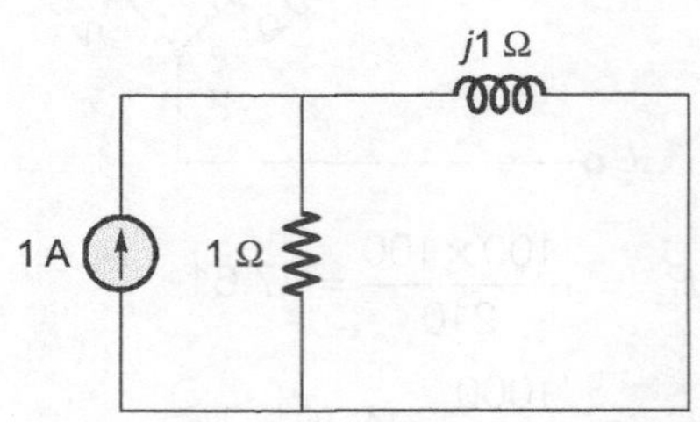

$$\therefore \quad V_{R1} = \frac{1 \times j}{1+j} = \frac{j}{1+j} = \frac{1}{2}(1+j)$$

Second, open circuit the current source.
We have,

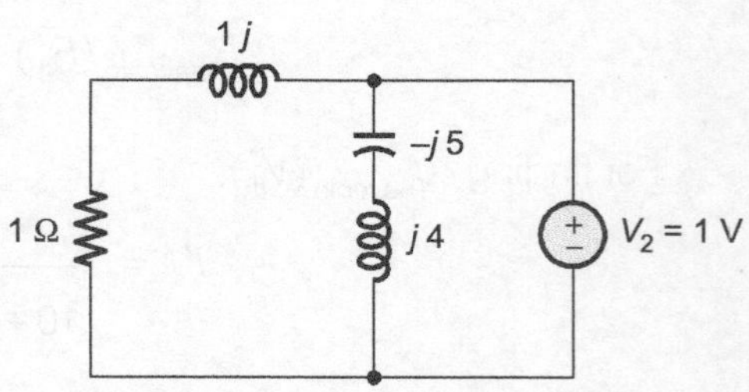

$$\therefore \quad V_{R2} = 1 \times \frac{1}{1+j} = \frac{1-j}{2}$$

Then,

Voltage across R, $\quad V_R = V_{R1} + V_{R2} = \frac{1+j}{2} + \frac{1-j}{2}$

Hence, $\quad V_R = 1 \text{ V}$

2.7 Draw the Thevenin's equivalent of the circuit at *aa*' and *bb*' as shown in the figure below and find the load current.

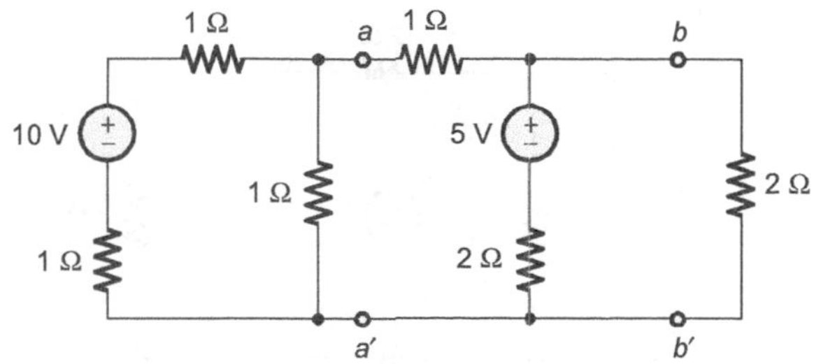

[12 marks : 2008]

Solution:

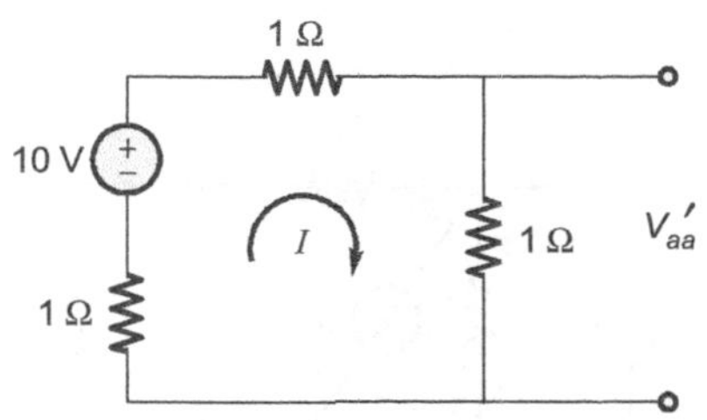

$$\therefore \quad V_{aa}' = \frac{10}{3}\times 1 = \frac{10}{3}\text{ V}$$

$$R_{aa}' = \frac{2\times 1}{2+1} = \frac{2}{3}\Omega$$

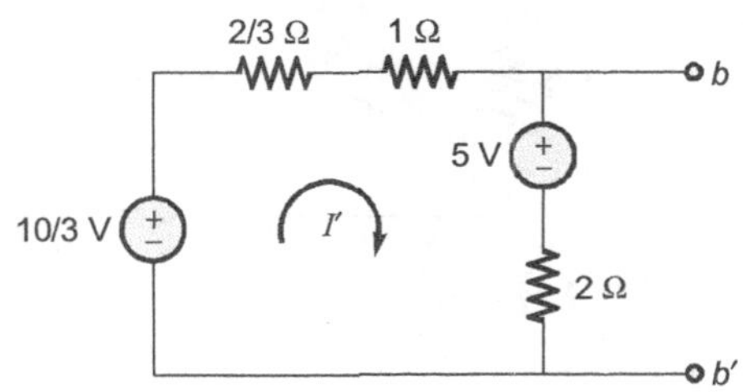

By using KVL,

$$\frac{10}{3} - \frac{2}{3}I' - I' - 5 - 2I' = 0$$

$$\therefore \quad I' = \frac{-5}{11}$$

$$\therefore \quad V_{bb}' = 5 + 2\,i' = 5 + 2\times\left(-\frac{5}{11}\right) = \frac{45}{11}\text{ V}$$

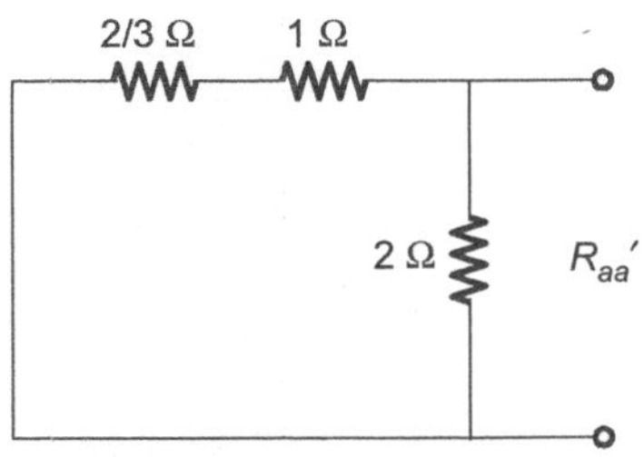

For finding R_{bb}', our circuit becomes

$$\therefore \quad R_{bb}' = \frac{5/3\times 2}{5/3+2} = \frac{10}{11}\Omega$$

So, that current from the equivalent circuit,

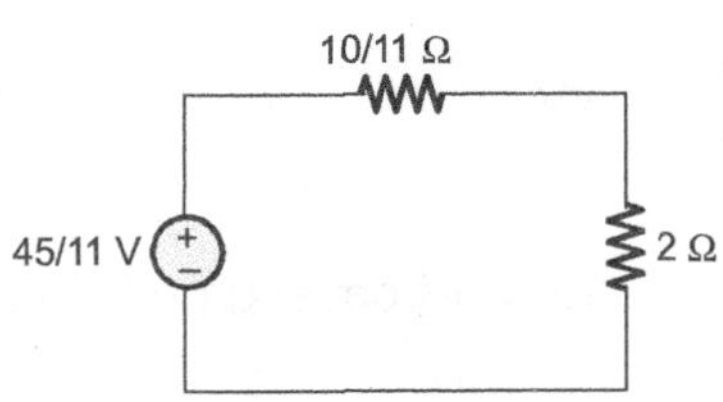

$$I_{\text{load}} = \frac{(45/11)}{(10/11+2)} = 1.406\text{ A}$$

2.8 **For the circuit shown, determine the voltage across 3 ohm resistor applying Thevenin's theorem. State the maximum power theorem for an ac circuit.**

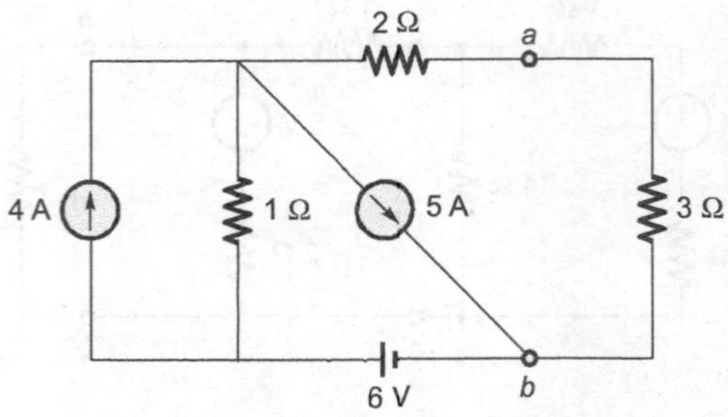

[12 marks : 2012]

Solution:

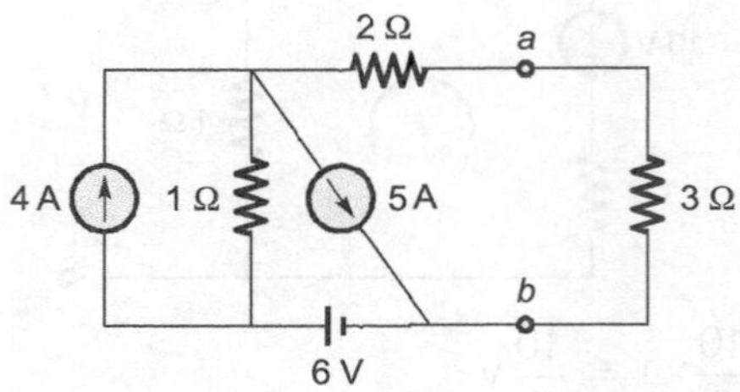

To find R_{Th}

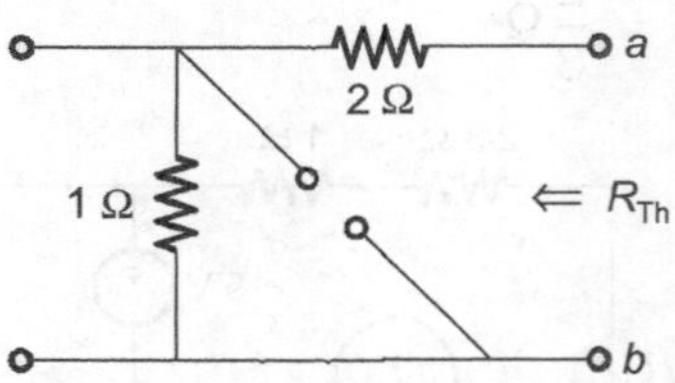

$$\Rightarrow \quad R_{Th} = 1 + 2 = 3\ \Omega$$

To find V_{Th},

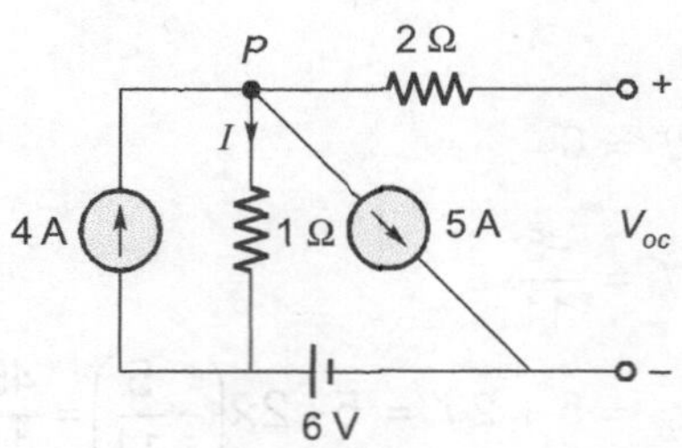

Applying KCL at node *P*

$$4 = I + 5$$

$$\Rightarrow \quad I = -1$$

$$\therefore \quad V_{oc} = 1 \cdot I + 6 = -1 + 6 = 5\text{ V}$$

$$\therefore \quad V_{ab} = 3 \cdot \frac{5}{6} = 2.5\text{ V}$$

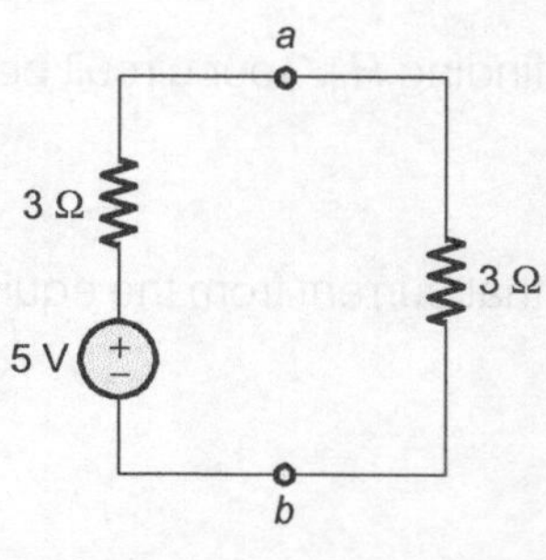

Maximum power theorem:

In order to transfer maximum amount of power from source to load, the load impedance must be complex conjugate of source impedance.

$$Z_L = Z_s^*$$

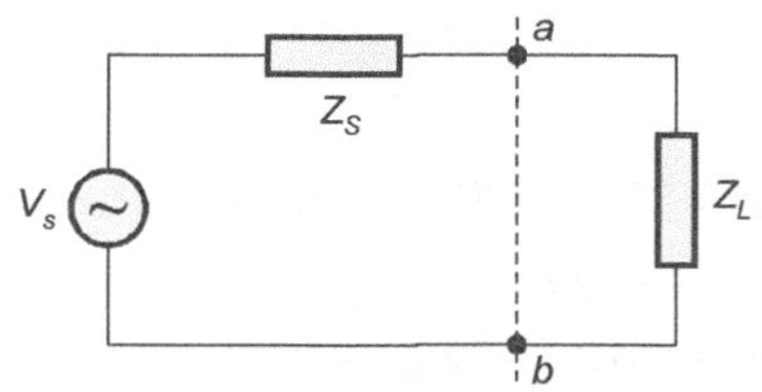

For maximum power transfer, if the load is inductive then the source impedance must be capacitive in nature and vice-versa.

2.9 **Determine the value of the variable resistor R in figure such that maximum power is absorbed by 2 ohm resistor.**

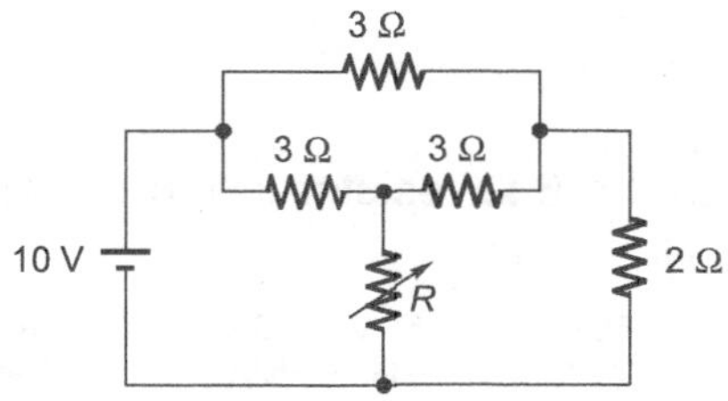

[10 marks, 2015]

Solution:

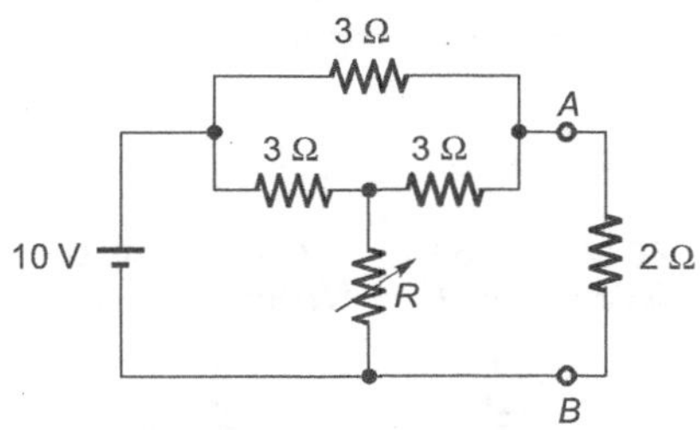

Maximum power transfer will be when R_{Th} across $AB = 2$.

and,

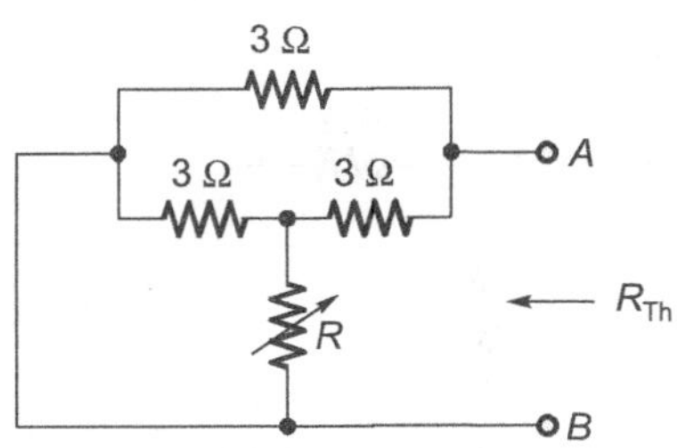

$$R_{Th} = 2 \qquad \text{when, } R = \infty$$

2.10 **Use mesh analysis to find the currents marked in the circuit shown below.**

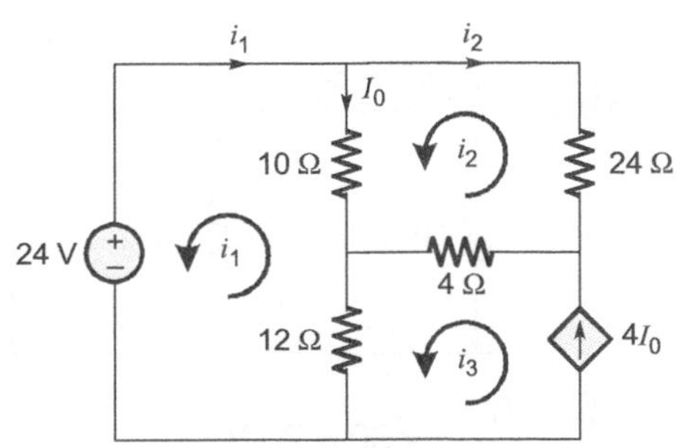

[5 marks : 2016]

Solution:

Using KVL in three loops, we get

$$24 = 22\, i_1 - 10\, i_2 - 12\, i_3 \qquad \text{...(i)}$$

$$0 = -10\, i_1 + 38\, i_2 - 4\, i_3 \qquad \text{...(ii)}$$

$$-4\, I_0 = -12\, i_1 - 4\, i_2 + 16\, i_3 \qquad \text{...(iii)}$$

$$I_0 = i_2 - i_1 \qquad \text{...(iv)}$$

$$4(i_2 - i_1) = -12\, i_1 - 4\, i_2 + 16\, i_3$$

$$8\, i_1 + 8\, i_2 - 16\, i_3 = 0 \qquad \text{...(v)}$$

Solve (i), (ii) and (iv), $i_1 = 2.25$ A; $i_2 = 0.75$ A; $i_3 = 1.5$ A

2.11 **Find the current flowing through the 6 Ω resistor in the current shown using Thevenin's theorem. Find also the power delivered to it.**

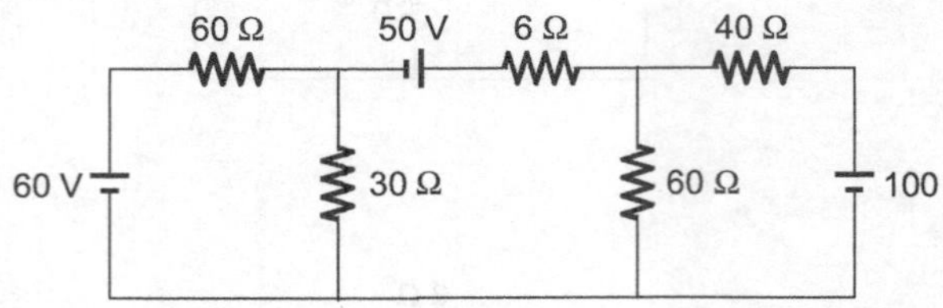

[20 Marks : 2017]

Solution:

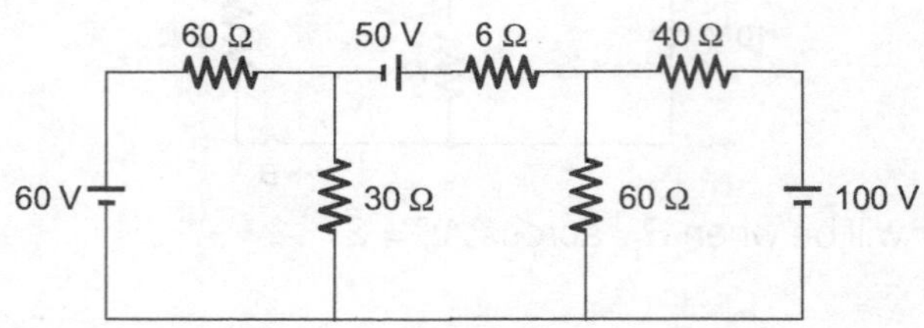

Shorting all voltage sources and open-circuit,

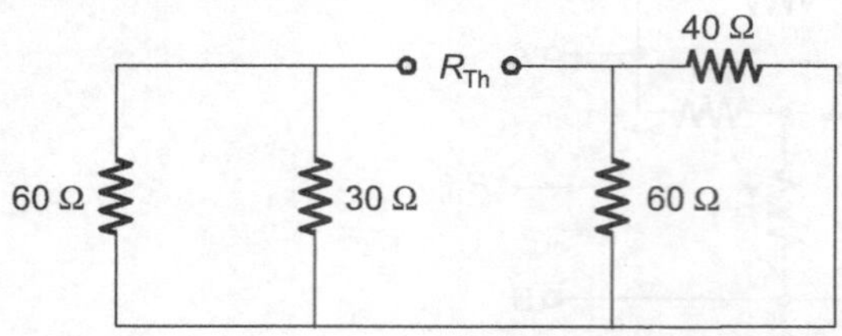

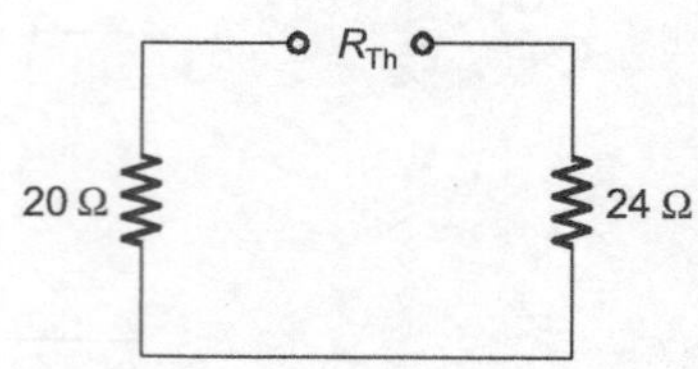

$$\therefore \qquad R_{Th} = 20 + 24 = 44\ \Omega$$

For V_{Th} calculation:

Open-circuit across 6 Ω

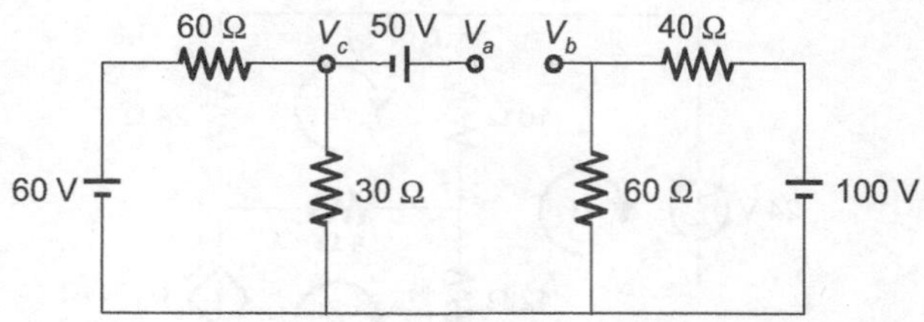

$$V_c = \left[\frac{60}{60+30}\right] \times 30 = 20\text{ V}$$

$$V_a - V_c = 50 \text{ V}$$

$$\therefore \quad V_a = 70 \text{ V}$$

$$V_b = \left(\frac{100}{60+40}\right) \times 60 = 60 \text{ V}$$

$$\therefore \quad V_{Th} = V_a - V_b = 70 - 60 = 10 \text{ V}$$

$$\therefore \quad i = \frac{V_{Th}}{R_{Th} + 6} = \frac{10}{44+6} = 0.2 \text{ A}$$

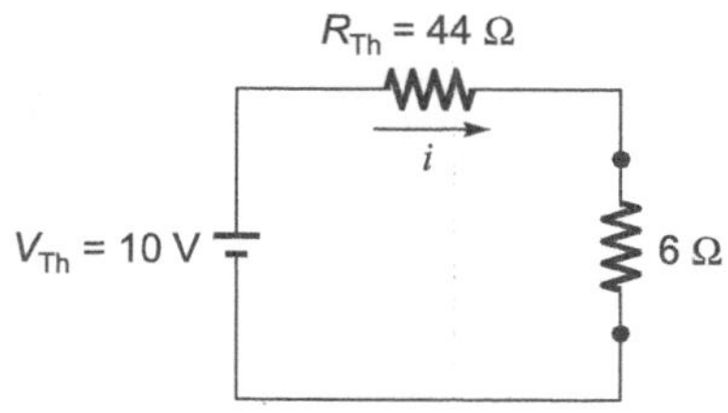

$$\text{Power delivered to 6 } \Omega \text{ resistor} = (i^2 \times 6) \text{ W} = 0.04 \times 6 = 0.24 \text{ W}$$

2.12 **Determine the value of load resistance R_L to be connected across terminals X-Y to receive maximum power. Also, obtain the value of this maximum power.**

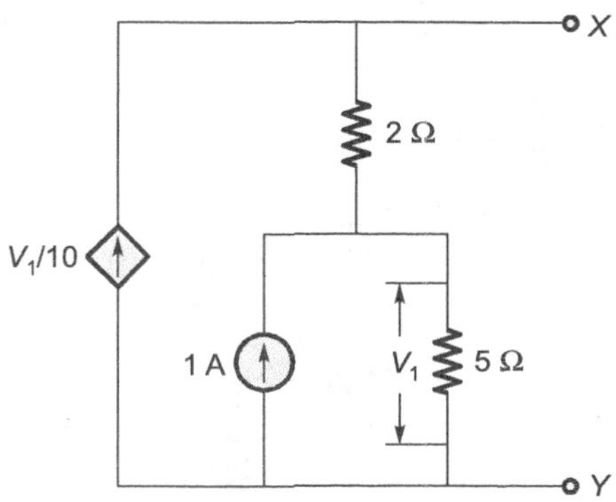

[20 marks : 2018]

Solution:

Lets find R_{Th} first by connecting 1 V voltage source across XY

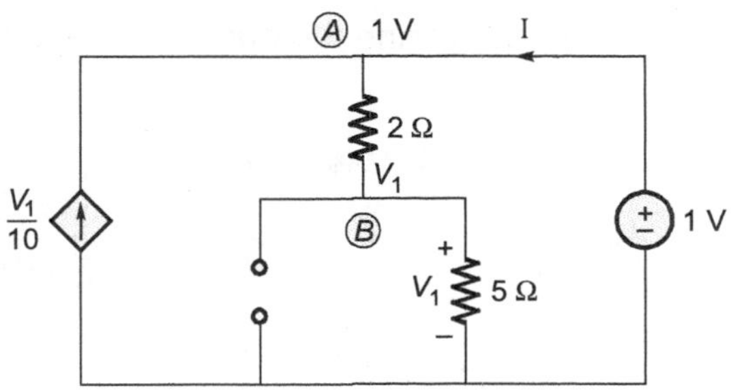

Applying voltage divider rule, $\quad V_1 = 1\text{ V} \times \frac{5}{5+2} = \frac{5}{7}\text{volts}$

Applying KCL at (A), $\quad \frac{V_1}{10} + I - \frac{1}{7} = 0$

$$I = -\frac{V_1}{10} + \frac{1}{7} = \frac{1}{14}\text{A}$$

$$\therefore \quad R_{Th} = \frac{1}{I} = 14\,\Omega$$

Hence using maximum power transfer theorem,

$$R_L = R_{Th} = 14\ \Omega$$

For V_{Th}:

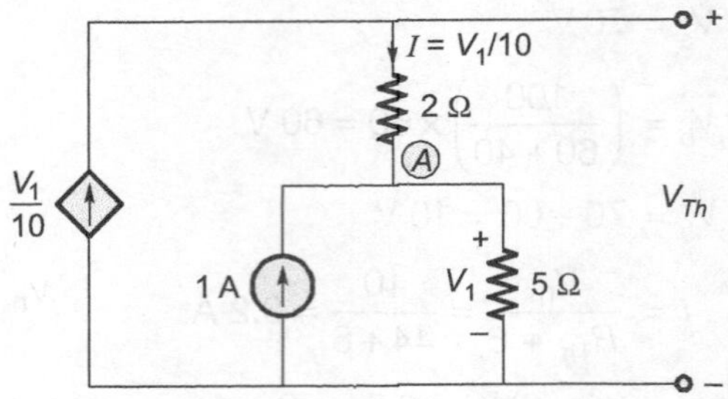

Applying KVL, $$V_{Th} = 2I + V_1$$

$$I = \frac{V_1}{10}$$

$$\therefore \quad V_{Th} = \frac{6V_1}{5} \quad \text{...(i)}$$

KCL at point (A), $$\frac{V_1}{10} + 1 - \frac{V_1}{5} = 0$$

$$V_1 = 10 \text{ volts}$$

$$\therefore \quad V_{Th} = 12 \text{ V from equation (i),}$$

Maximum power transferred =

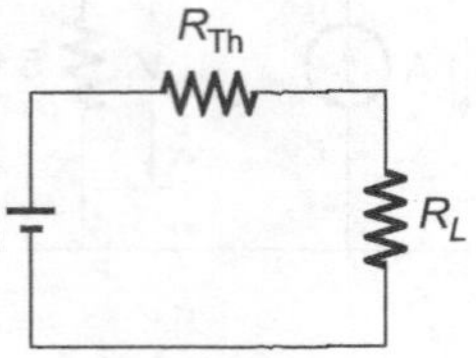

$$P_{max} = \frac{V_{Th}^2}{4R_L} = \frac{12^2}{4 \times 14} = 2.57 \text{ Watt}$$

2.13 **Using the principle of superposition, determine the current I in the 5 Ω resistor in the circuit shown in the figure.**

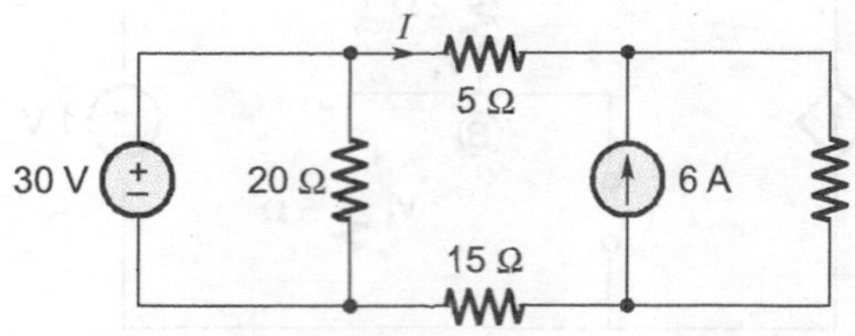

[12 marks : 2020]

Solution:

We let, $$I = I' + I''$$

Where I' and I'' are due to 6 A current source and 30 V voltage source respectively. To obtain I', we turn off the 30 V voltage source so that we have the circuit in figure,

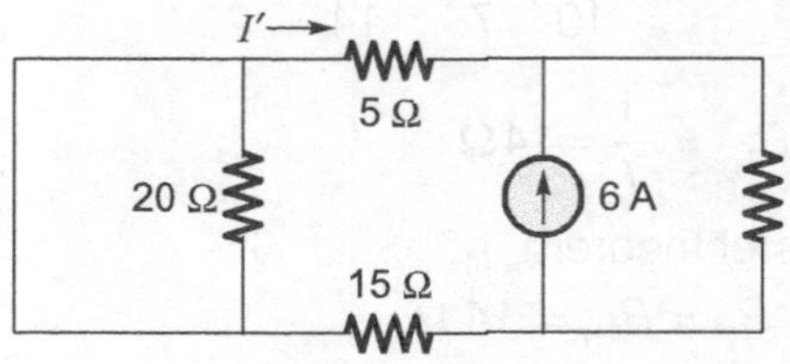

Circuit can be redrawn as

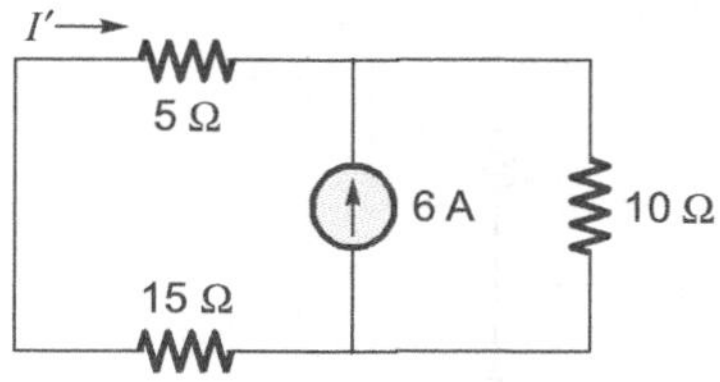

Applying current division rule,

$$I' = -6 \times \frac{10}{20+10} = -2\text{ A}$$

To obtain I'', we turn off the 6 A current source so that we have the circuit in figure

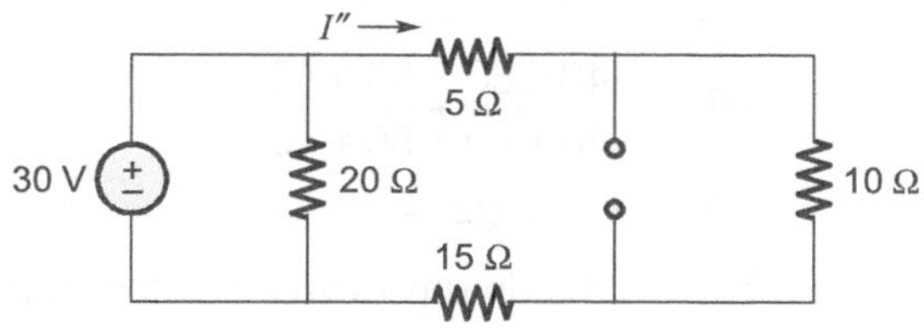

Write KVL equation,

$$30 - 5I'' - 10I'' - 15I'' = 0$$

$$I'' = \frac{30}{30} = 1\text{ A}$$

So, $\quad I = I' + I'' = 1 - 2 = -1\text{ A}$

2.14 **Determine the value of resistance '*R*' which will extract the maximum power from the circuit shown in the figure. Also calculate the value of maximum power.**

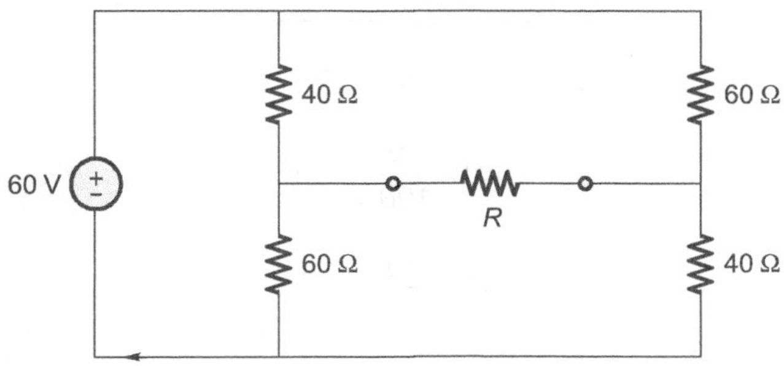

[20 marks : 2021]

Solution:

The given circuit can be replaced by Thevenin's equivalent circuit.

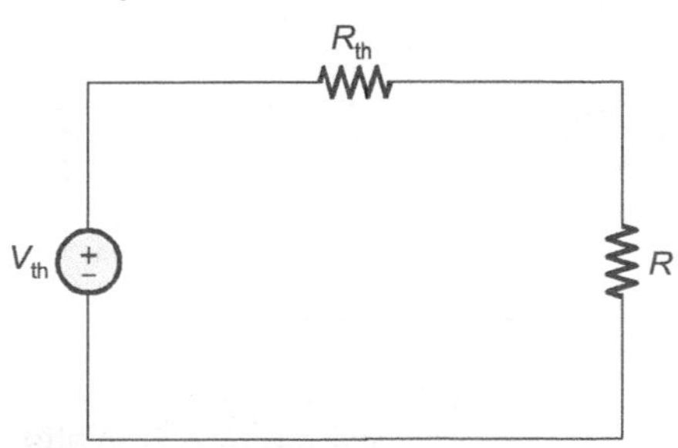

If R is equal to R_{th}, the maximum power will be transferred from source to load.

For R_{th}; $\quad V_s = 0$

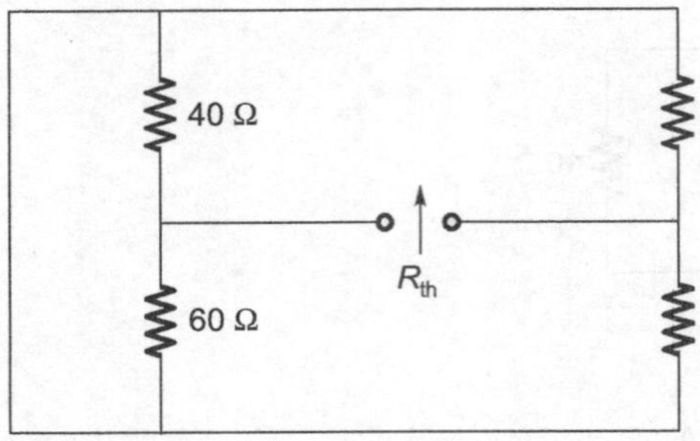

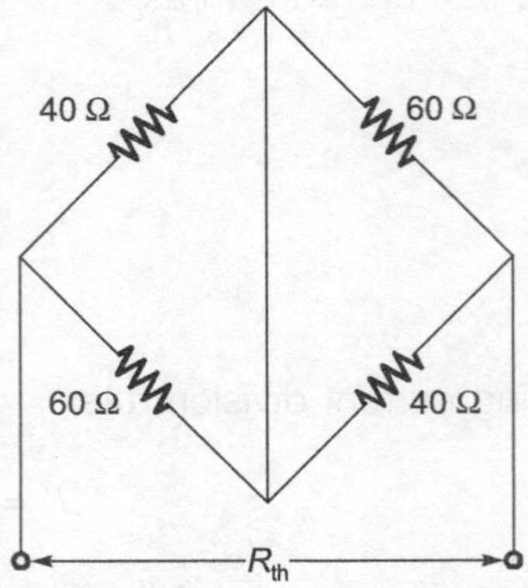

$$R_{th} = (40 \parallel 60) + (60 \parallel 40)$$

$$R_{th} = \frac{40\times 60}{40+60}+\frac{60\times 40}{60+40}$$

$$R_{th} = 24 + 24 = 48\ \Omega$$

If the resistance (R) value is 48 Ω, then maximum power transfer from source to load.

Find V_{th} :

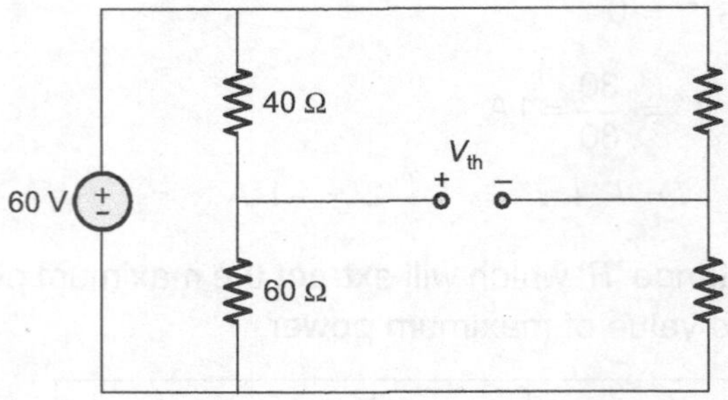

Apply KVL : $\quad V_{th} = V_{60} - V_{40}$

Using voltage division : $\quad V_{60} = 60\times\frac{60}{100} = 36\ \text{V}$

$$V_{40} = 60\times\frac{40}{100} = 24\ \text{V}$$

$$V_{th} = 36 - 24 = 12\ \text{V}$$

Equivalent Thevenin's circuit

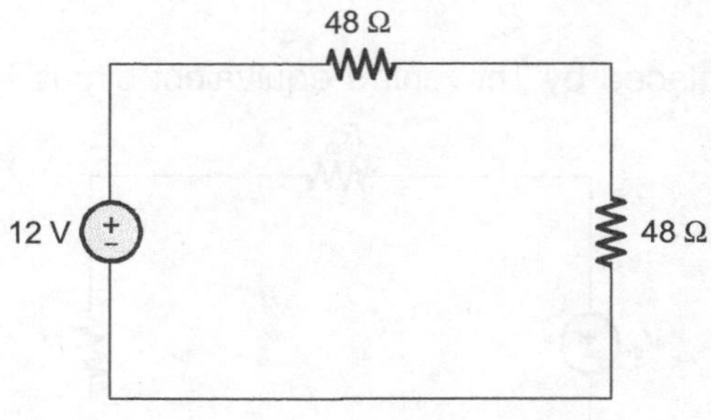

Maximum power to R is $\quad P_{max} = \frac{V_{th}^2}{4R_L} = \frac{12\times 12}{4\times 48} = \frac{3}{4}$ Watts

2.15 Obtain the Thevenin's equivalent at the terminals *a*, *b* for the figure shown. What will be the maximum power supplied by the circuit to a load resistance?

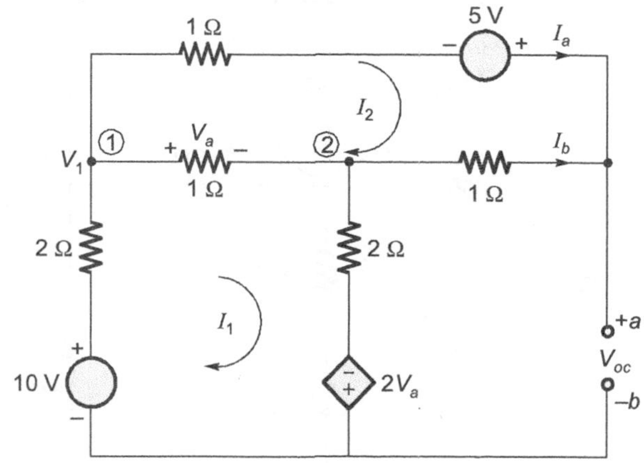

[20 marks : 2022]

Solution:

To Thevenin's, we find the V_{OC} across *ab* first

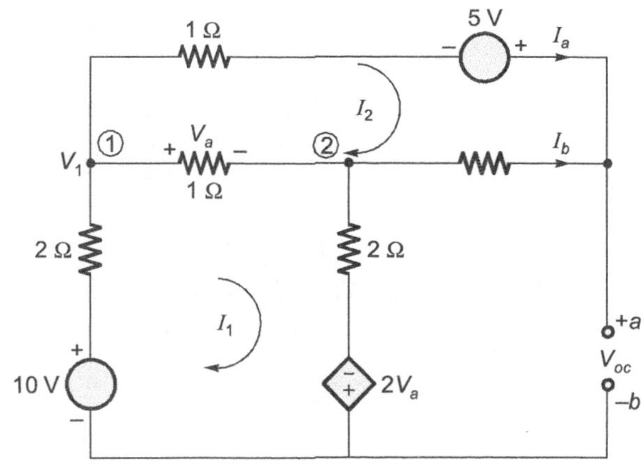

So writing KCL at V_2 first

$$\frac{V_2 - 5 - V_1}{2} + \frac{V_2 + 2V_a}{2} + \frac{V_2 - V_1}{1} = 0$$

$$V_2 - 5 - V_1 + V_2 + 2(V_1 - V_2) + 2V_2 - 2V_1 = 0$$

$$2V_2 - V_1 = 5 \quad \text{...(i)}$$

Writing KCL at V_1, we get

$$\frac{V_1 + 5 - V_2}{2} + \frac{V_1 - 10}{2} + \frac{V_1 - V_2}{1} = 0$$

$$V_1 + 5 - V_2 + V_1 - 10 + 2V_1 - 2V_2 = 0$$

$$4V_1 - 3V_2 = 5 \quad \text{...(ii)}$$

Using (i) and (ii), we get

$$V_1 = 5 \text{ V}, V_2 = 5 \text{ V}$$

So, I_1 in figure is

$$\frac{V_2 - 5 - V_1}{2} = -2.5 \text{ A}$$

So, $V_a = V_2 - I_1 \times 1\,\Omega = 5 - [-2.5] = 7.5$ V

So, $V_{OC} = 7.5$ V

Now, for R_{th}, all independent sources are replaced by their internal resistance.

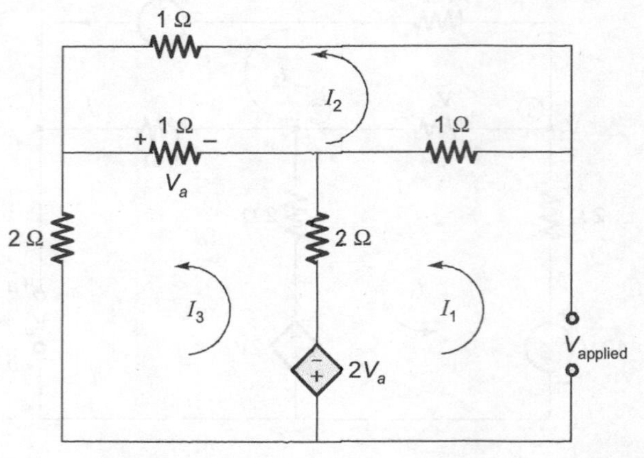

Doing Mesh Analysis :

$$V - (I_1 - I_2)1 - (I_1 - I_3)2 + 2(I_2 - I_3) = 0$$
$$V - I_1 + I_2 - 2I_1 + 2I_3 + 2I_2 - 2I_3 = 0$$
$$V - 3I_1 + 3I_2 = 0 \quad \text{...(i)}$$

For Loop I_2 :

$$I_2 - I_3 + I_2 - I_1 + I_2 = 0$$
$$3I_2 = I_3 + I_1 \quad \text{...(ii)}$$

For Loop 3 :

$$2(I_2 - I_3) + 2(I_3 - I_1) + I_3 - I_2 + 2I_3 = 0$$
$$2I_2 - 2I_3 + 2I_3 - 2I_1 + I_3 - I_2 + 2I_3 = 0$$
$$I_2 - 2I_1 + 3I_3 = 0 \quad \text{...(iii)}$$
$$I_2 = 2I_1 - 3I_3$$
$$3I_2 = I_3 + I_1$$

So,

$$6I_1 - 9I_3 = I_3 + I_1$$
$$5I_1 = 10I_3$$
$$I_3 = \frac{I_1}{2} \quad \text{...(iv)}$$

Using (iv) in (ii)

$$3I_2 = \frac{3I_1}{2}$$
$$I_2 = \frac{I_1}{2} \quad \text{...(v)}$$

Using (v) in (i)

$$V = 3I_1 - \frac{3I_1}{2}$$
$$V = \frac{3I_1}{2}$$
$$\frac{V}{I_1} = \frac{3}{2} = R_{int}$$

So, Thevenin's equivalent circuit is

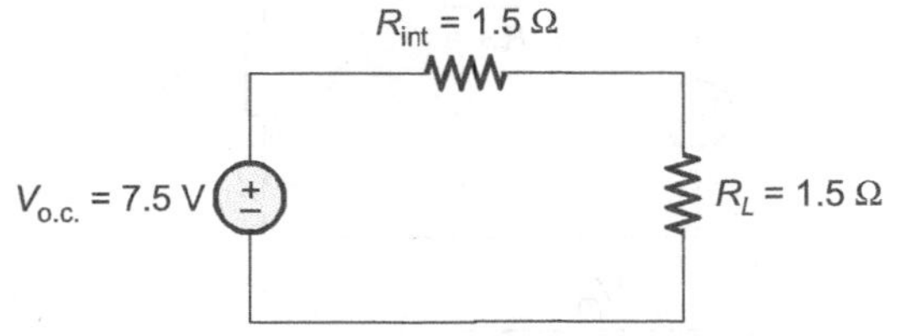

So, maximum power transfer when

$$R_{load} = 1.5\ \Omega$$

So,

$$P_{max} = \left(\frac{7.5}{3}\right)^2 \times 1.5\ \Omega$$

$$= 9.375\text{ W}$$

2.16 Two terminals shown in figure coming out of a network show a voltage of 150 V on open circuit and give a current of 3 A on short-circuit. If a resistor of 10 Ω is connected as a load between the two terminals the current falls to 2.65 A. Determine the equivalent voltage generator circuit and find whether the circuit is capacitive or inductive.

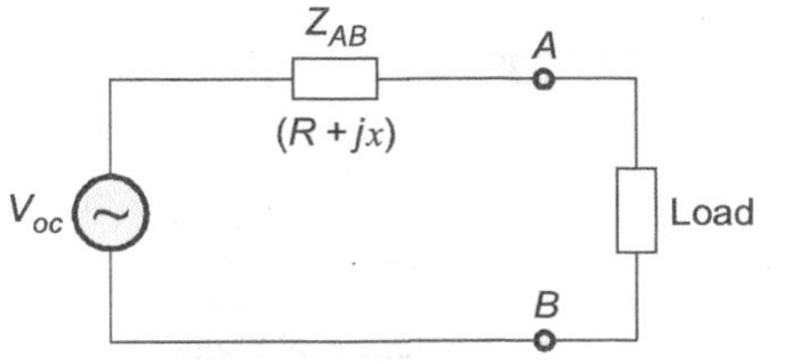

[12 marks : 2022]

Solution:

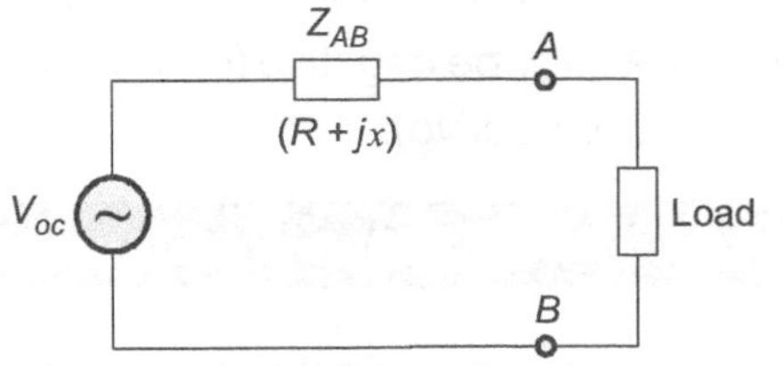

Given, Open circuit $= 150 = |V_{OC}|$

$$V_{OC} = 150\angle 0°\text{ V}$$

Now when the circuit is short circuited around points AB we get

$$I_{SC} = \frac{V_{OC}}{R + jx}$$

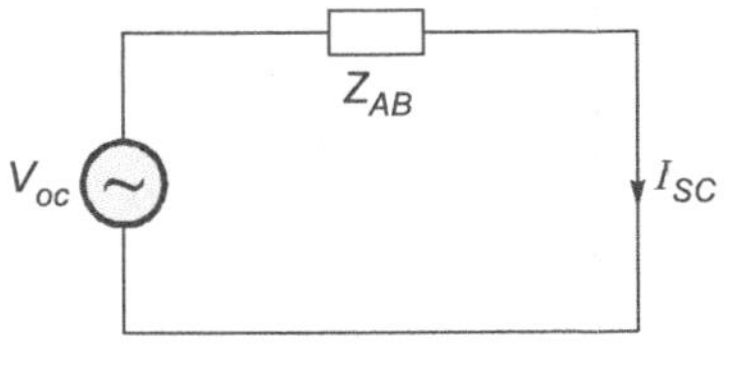

$$\Rightarrow \quad |I_{SC}| = \frac{|V_{OC}|}{\sqrt{R^2 + X^2}}$$

So,

$$\frac{150}{\sqrt{R^2 + X^2}} = 3$$

$$\Rightarrow \quad R^2 + X^2 = 2500 \quad \text{...(i)}$$

Now when load $R = 10\ \Omega$ applied across AB

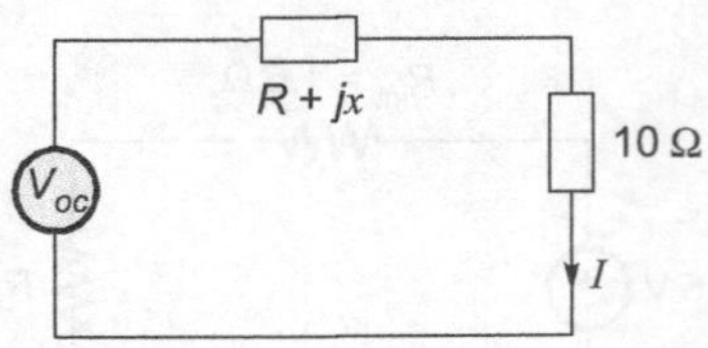

$$I = \frac{V_{OC}}{(R+10)+jX}$$

$$|I| = \frac{|V_{OC}|}{\sqrt{(R+10)^2+X^2}} = \frac{150}{2.65} = 56.6$$

$$(R+10)^2 + X^2 = 3203.987 \quad \text{...(ii)}$$

Using equation (i) in (ii),

$$(R+10)^2 + 2500 - R^2 = 3203.98$$

$$R^2 + 100 + 20R + 2500 - R^2 = 3203.98$$

$$20R = 3203.98 - 2500 - 100$$

$$20R = 603.98$$

$$R = 30.199\ \Omega$$

Now in (i), $R^2 + X^2 = 2500$

$$X^2 = 1587.998$$

$\Rightarrow$ $X = \pm 39.85\ \Omega$

So equivalent circuit is

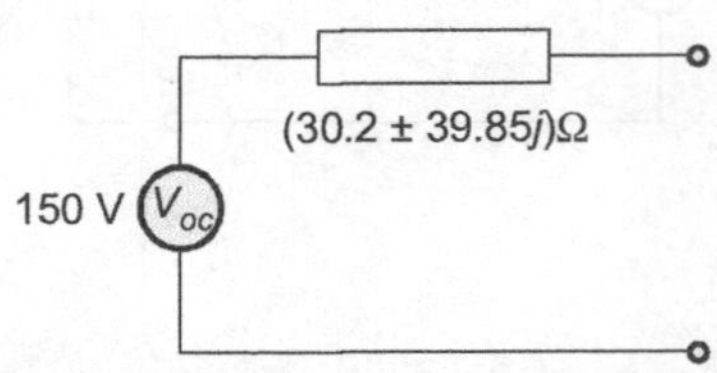

According to mathematical equation 'X' may be negative (or) positive then 'X' may be capacitive or inductive but for generator 'X' is always positive (inductive).

3. Transient Response of DC and AC Networks, Sinusoidal Steady State Analysis

3.1 **In the circuit below, determine the initial and final values of the current through the 1 F-capacitor. Make use of theorems.**

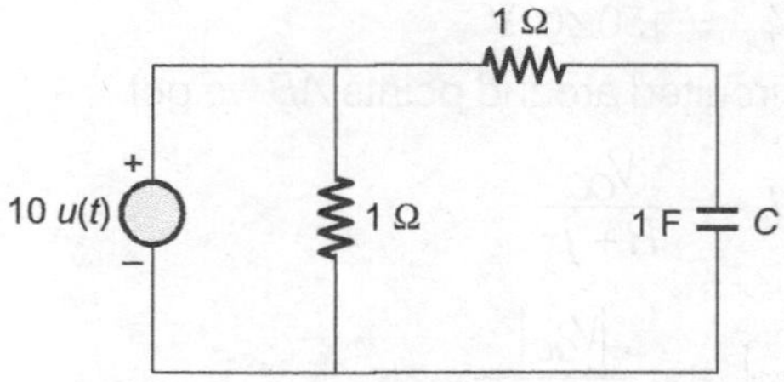

[12 marks : 2002]

Solution:

$$10 = I_C \times 1 + V_C$$

$$I_C = C\frac{dV_C}{dt}$$

$$10 = C\frac{dV_C}{dt} + V_C$$

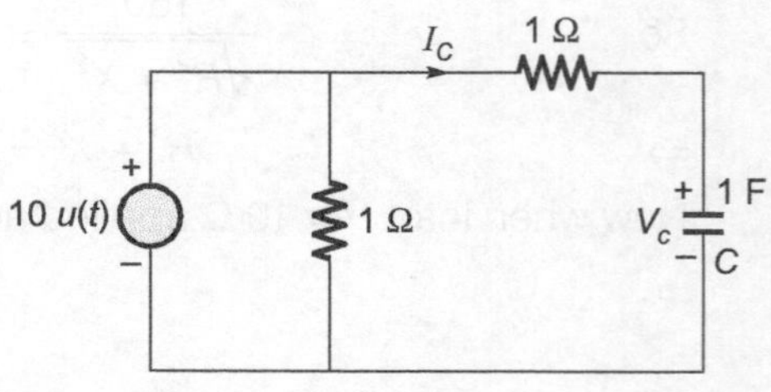

Taking Laplace transform,

$$\frac{10}{s} = (sC+1)V_C(s) = (s+1)V_C(s)$$

$$V_C(s) = \frac{10}{s(s+1)}$$

$$I_C(s) = \frac{V_C(s)}{1/Cs} = sV_C(s) = \frac{10s}{s(s+1)} = \frac{10}{s+1}$$

By, initial value theorem,

$$i_C(0) = \lim_{s\to\infty} sI_C(s) = \lim_{s\to\infty} \frac{10\times s}{s+1} = 10$$

Final value theorem,

$$i_C(\infty) = \lim_{s\to 0} sI_C(s) = \lim_{s\to 0} \frac{10\times s}{s+1} = 0$$

3.2 **For the circuit shown below, draw the phasor diagram, Derive the condition for the two branch currents, I_L and I_C to be in quadrature.**

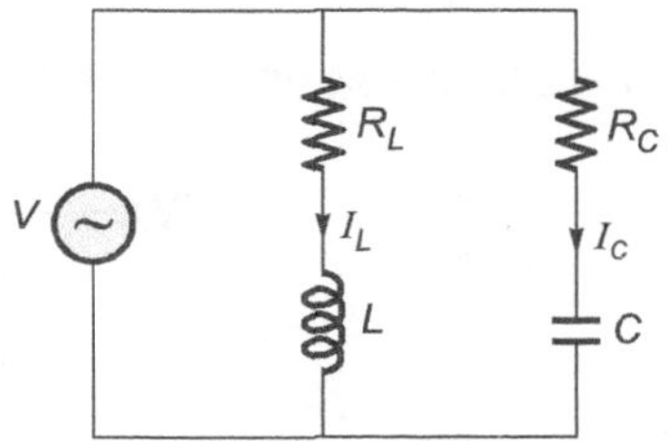

[12 marks : 2002]

Solution:

Taking V as the reference,

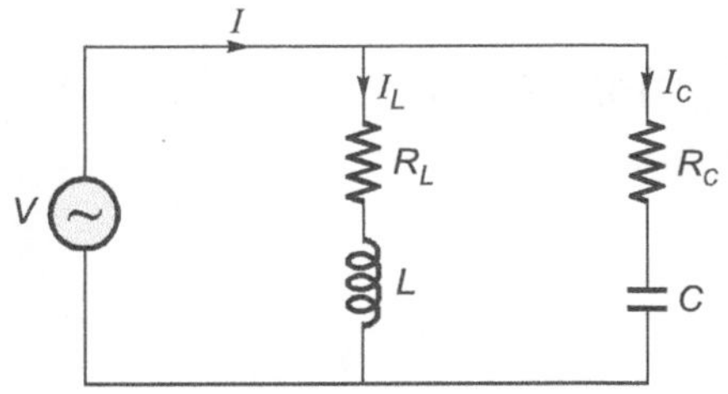

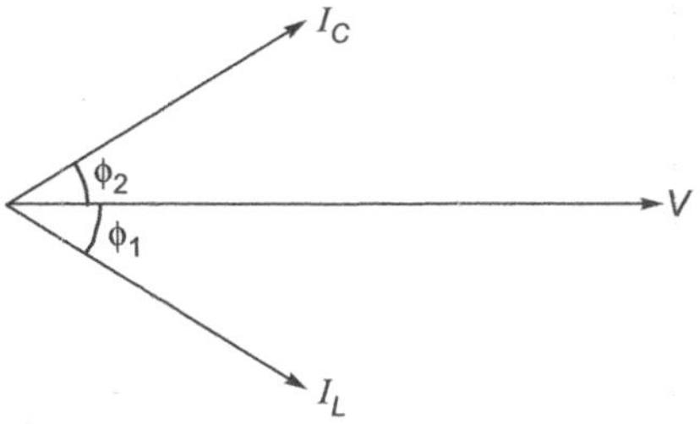

$$I_L = \frac{V}{R_L + j\omega L} = \frac{V}{\sqrt{R_L^2 + (\omega L)^2}} \angle -\tan^{-1}\left(\frac{\omega L}{R_L}\right)$$

$$\therefore \quad \phi_1 = \angle -\tan^{-1}\left(\frac{\omega L}{R_L}\right)$$

$$I_C = \frac{V}{R_C - j/c\omega} = \frac{V}{\sqrt{R_C^2 + (1/C\omega)^2}} \angle \tan^{-1}\frac{1}{\omega R_C C}$$

$$\therefore \quad \phi_2 = \angle \tan^{-1}\frac{1}{\omega R_C C}$$

For two branches to be in quadrature

$$\phi_1 + \phi_2 = 90^\circ$$

$$\left(\tan^{-1} + \frac{\omega L}{R_L} + \tan^{-1}\frac{1}{\omega R_c C}\right) = 90^\circ$$

$$\tan^{-1} A + \tan^{-1} B = \tan^{-1}\left[\frac{A+B}{1-AB}\right]$$

$$\tan^{-1}\left[\frac{\left(\frac{\omega L}{R_L} + \frac{1}{\omega R_c C}\right)}{\left[1 - \frac{L}{R_L R_c C}\right]}\right] = 90^\circ$$

$$1 - \frac{L}{R_L R_c C} = 0$$

$$\therefore \qquad L = R_L R_C C \Rightarrow \frac{L}{R_L} = R_c C$$

This is condition for two branch currents I_L and I_C to be in quadrature.

3.3 In the circuit shown in the figure, S is closed at time of $t = 0$. Determine $i_c(t)$ and the time constant.

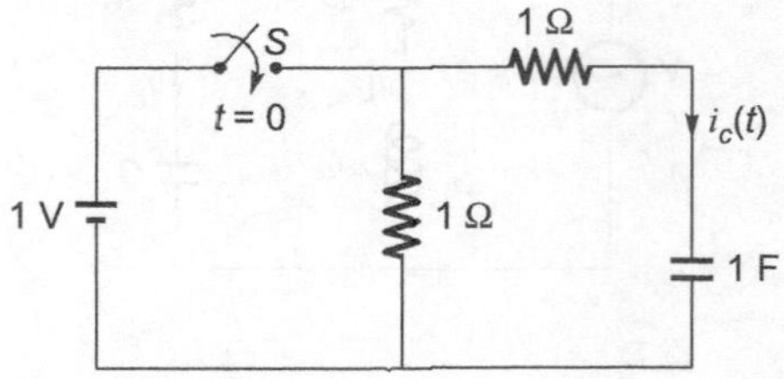

[8 marks : 2003]

Solution:

By KVL,
$$1 = i_C(t) \cdot R + 1/C \int_0^t i_c(t)dt + V_c(0^+)$$

Here, $V_c(0^+) = 0$

then, using Laplace transform, we get

$$\frac{1}{s} = I_C(s) \cdot [R + 1/Cs]$$

By putting the value of (R and C),

$$I_C(s) = \frac{1}{s+1}$$

By using inverse Laplace transform of both sides, we have

$$i_C(t) = e^{-t}$$

and, Time constant $= RC$

$= 1$ sec.

3.4 In the circuit shown below the switch is initially in position 1. Find the voltage across the coil at the instant at which the switch is changed to position 2.

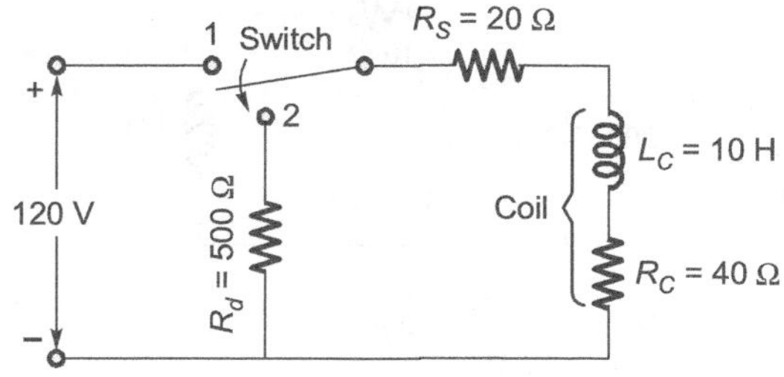

[12 marks : 2006]

Solution:

For $t < 0$, the circuit will be in steady

State at $t = 0^-$

$$i(0^-) = \frac{120}{20+40} = \frac{120}{60}$$

$$i(0^-) = 2\text{ A}$$

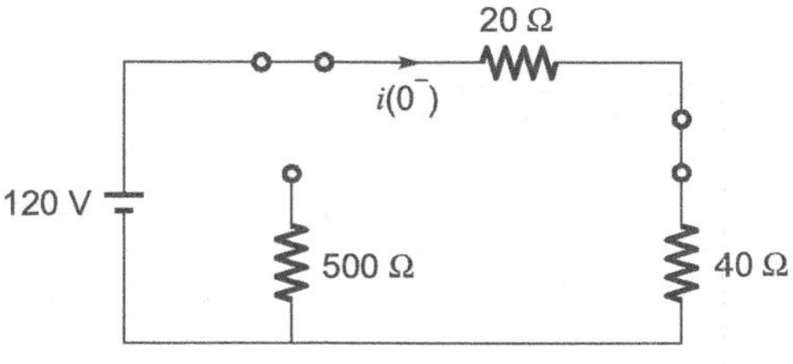

For $t \geq 0$, switch will be in position 2,

$$i(0^+) = i(0^-) = 2\text{A}$$

Applying KVL

$$\frac{L\,di(t)}{dt} + (500+40+20)\,i(t) = 0$$

$$L\frac{di(t)}{dt} + 560\,i(t) = 0 \quad \text{...(1)}$$

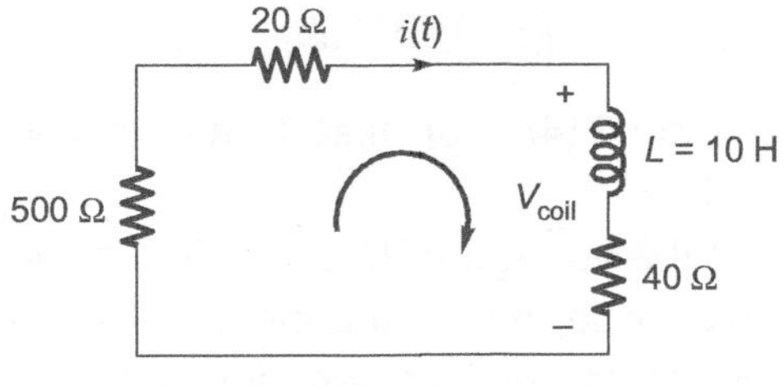

Voltage across the coil at the instant of switching $t = 0^+$

$$V_{coil}(0^+) = L\frac{di}{dt}(0^+) = -520\,i(0^+)$$

$$= -1040\text{ V}$$

3.5 The circuit in the figure below is initially under steady-state condition. The switch is moved from position 1 to position 2 at $t = 0$. Find the current after switching.

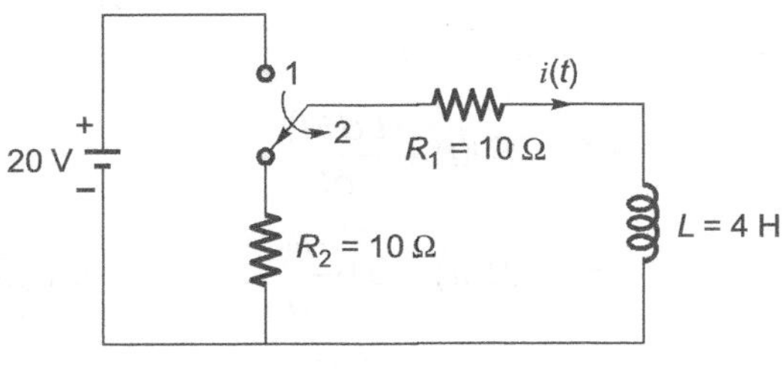

[12 marks : 2008]

Solution:

For $t < 0$, at steady state the inductor will be short circuited

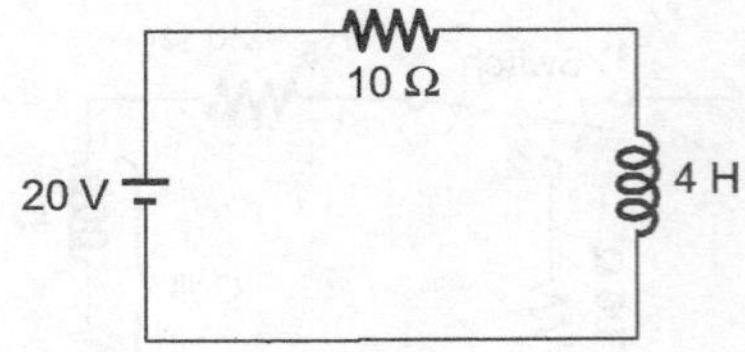

$$i(0^-) = \frac{20}{10} = 2\text{ A}$$

For $t \geq 0$

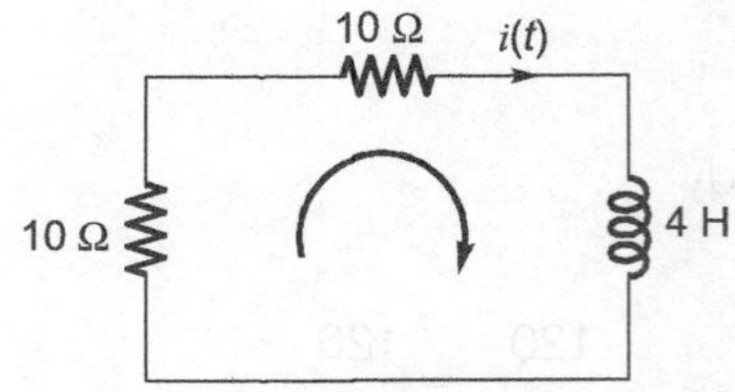

Applying KVL,

$$L\frac{di(t)}{dt} + (10+10)i(t) = 0$$

$$\frac{di(t)}{dt} + \frac{20}{4}i(t) = 0$$

Taking Laplace transform,

$$[sI(s) - i(0^-)] + 5\,I(s) = 0$$

$$[s+5]I(s) = i(0^-) = 2$$

$$I(s) = \frac{2}{s+5}$$

$$i(t) = 2e^{-5t}\,u(t)$$

3.6 **A sinusoidal voltage $e(t)$ = 141 sin 314t is applied to an initially de-energized series R-L circuit in which R = 100 Ω and L = 0.5 H.**

(i) If the switch which applied the voltage to R-L circuit is closed at the instant when $e(t)$ passing through zero with a positive slope, determine the initial value of the transient current.

(ii) Write the complete expression for the transient solution.

(iii) Write the expression for the complete solution of the current response.

(iv) At what instantaneous value of the applied voltage will the closing of the switch result in no transient component of the current.

[8 + 4 = 12 marks : 2010]

Solution:

(i) As the circuit is de-energized initially, $i(0^-) = 0$ and instantaneous value of the applied voltage is zero at the instant of switching.

Hence, initial value of the transient current is 0.

Applying KVL,

$$e(t) = R\,i(t) + \frac{L\,di(t)}{dt}$$

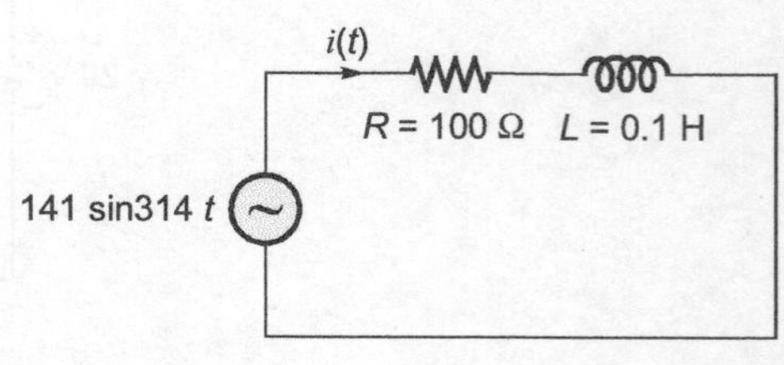

$$\Rightarrow \quad 141\sin 314t = 100\,i(t) + 0.5\frac{di(t)}{dt}$$

$$\Rightarrow \quad 282\sin 314t = 200\,i(t) + \frac{di(t)}{dt}$$

Applying Laplace transform,

$$\frac{282 \times 314}{s^2 + 314^2} = 200\, I(s) + s\, I(s)$$

$$\Rightarrow \qquad I(s) = \frac{282 \times 314}{(s+200)(s^2+314^2)}$$

(ii) Using partial fraction method,

$$\frac{A}{(s+200)} + \frac{Bs+C}{s^2+314^2} = \frac{282 \times 314}{(s+200)(s^2+314^2)}$$

$A(s^2 + 314^2) + (Bs + C)(s + 200) = 282 \times 314$

$$A + B = 0 \qquad \ldots(i)$$

$$200\,B + C = 0 \qquad \ldots(ii)$$

$$314^2 A + 200\,C = 282 \times 314 \qquad \ldots(iii)$$

On solving equation (*i*), (*ii*) and (*iii*), we get

$A = 0.6388,\ B = -0.6388,\ C = 127.77$

$$I(s) = \frac{0.6388}{s+200} - \frac{0.6388s}{s^2+314^2} + \frac{127.77}{s^2+314^2}$$

$$= \frac{0.6388}{s+200} - \frac{0.6388s}{s^2+314^2} + 0.4069 \times \frac{314}{s^2+314^2}$$

$$i(t) = L^{-1}[I(s)] = 0.6388\, e^{-200t} - 0.6388 \cos 314t + 0.4069 \sin 314t$$

(iii) Complete solution of current response,

$$i(t) = 0.6388\, e^{-200t} + 0.757 \sin(314t - 57.7)$$

where, $i(t)$ = transient response of the current = $0.6388\, e^{-200t}$

$i_{ss}(t)$ = steady state response of the current = $0.757 \sin(314t - 57.7)$

(iv) As the circuit is initially deenergized before switching, $i(0^-) = 0$ and inductor is present in the circuit.

so, $i(0^+) = i(0^-) = 0$

Hence, $i(t) = 0$ initial current is zero, at the instant of switching (a)

$$i(t) = i_r(t) + i_{ss}(t) \qquad \ldots(iv)$$

For transient free response $i_r(t) = 0$(c)

Using conditions (a) and (c), equation (*iv*) becomes,

$$0 = 0 + i_{ss}(t)$$

$$i_{ss}(t) = 0$$

$$0.757 \sin(314t - 57.7) = 0$$

$$\Rightarrow \qquad 314t = 57.7$$

so, switch should be closed at $314t = 57.7$ for transient free current.

Instantaneous value of the applied voltage at the instant of switching.

$$e(t) = 141 \sin 314t = 141 \sin 57.7 = 118.92 \text{ volt}$$

3.7 Compute the steady-state current in the 8 Ω inductor in the circuit given in figure.

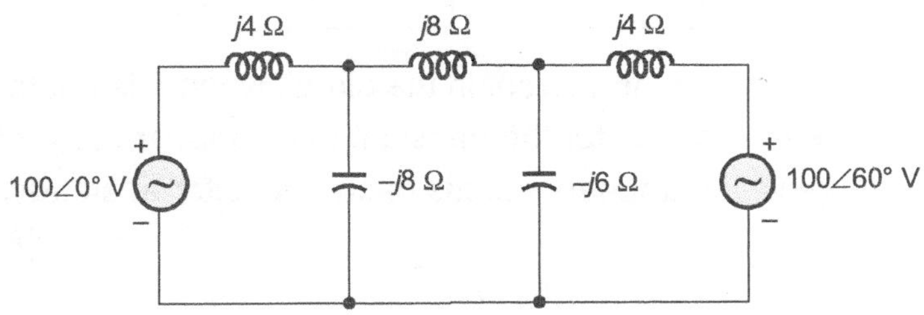

[8 + 4 = 12 marks : 2010]

Solution:

Using nodal analysis,

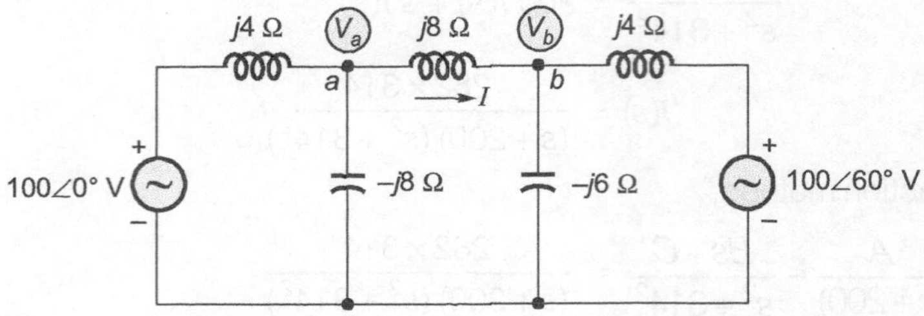

Applying KCL at note *a*,

$$\frac{V_a - 100\angle 0^\circ}{j4} + \frac{V_a - V_b}{j8} + \frac{V_a}{-j8} = 0$$

Multiplying by $j8$, $\quad 2[V_a - 100] + (V_a - V_b) - V_a = 0$

$\Rightarrow \quad V_b = 2V_a - 200 \quad \quad ...(i)$

Applying KCL at node *b*,

$$\frac{V_b - V_a}{j8} + \frac{V_b}{-j6} + \frac{V_b - 100\angle 60^\circ}{j4} = 0$$

Multiplying both sides by $j24$,

$$3[V_b - V_a] - 4V_b + 6[V_b - 100\angle 60^\circ] = 0$$

Putting value of V_b from equation (*i*),

$$-3V_a + 5[2V_a - 200] = 600\angle 60^\circ$$

$$7V_a = 1000 + 600\angle 60^\circ$$

$$V_a = 200\angle 21.78^\circ \text{ V}$$

Putting value of V_a into equation (*i*), $\quad V_b = 2V_a - 200$

$$= 2 \times 200\angle 21.78^\circ - 200$$

$\Rightarrow \quad V_b = 226.76\angle 40.88 \text{ V}$

Current through 8 Ω inductor, $\quad I = \dfrac{V_a - V_b}{j8} = \dfrac{200\angle 21.78^\circ - 226.76\angle 40.88}{j8}$

$$= 9.44\angle -169.11^\circ \text{ A}$$

3.8 In the circuit of the figure below, switch *S* has been in position 1 for a long time:

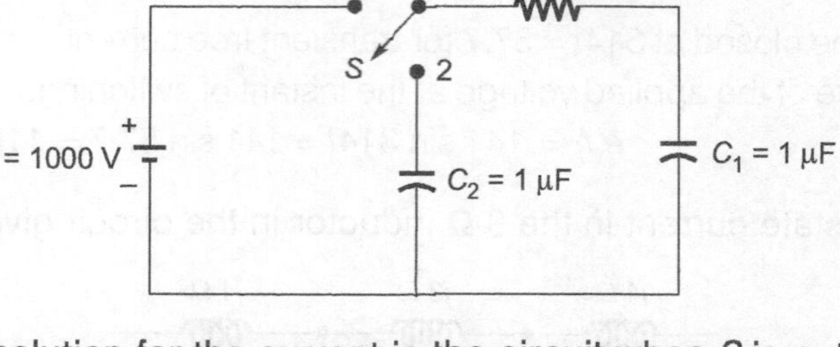

(i) Find the complete solution for the current in the circuit when *S* is put to position 2.

(ii) How long does it take in seconds for the transient to disappear (current to decay within 1%).

(iii) Determine the voltage which appears across each capacitor at steady state.

[8 + 4 = 12 marks : 2011]

Solution:

At $t = 0$, the switch was moved from position 1 to 2. The circuit before moving the switch.

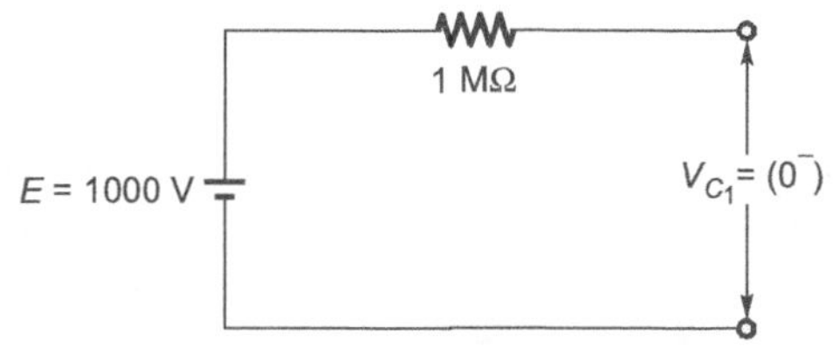

The capacitor C_1 acts as open circuit and get charged with 1000 V.

So, $V_{C_1}(0^-) = 1000\text{ V}$

(i) Equivalent circuit in s-domain, after moving the switch as $V_{C_1}(0^+) = V_{C_1}(0^-)$

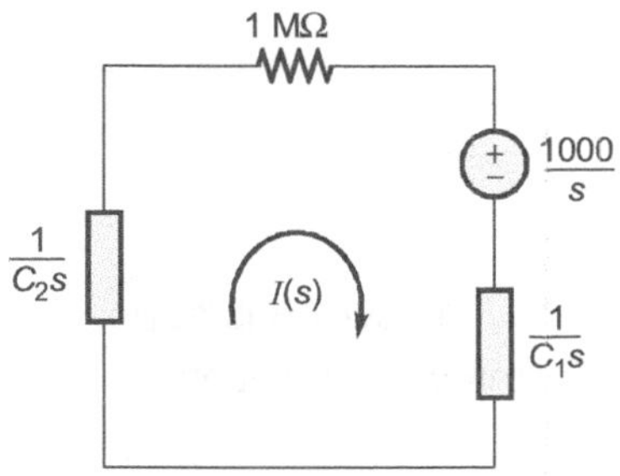

$$I(s) = \frac{\frac{1000}{s}}{\frac{1}{C_1 s} + \frac{1}{C_2 s} + R} = \frac{1000}{\left(\frac{1}{C_1} + \frac{1}{C_2}\right) + Rs}$$

$$= \frac{1000}{10^6 s + \left(\frac{1}{10^{-6}} + \frac{1}{10^{-6}}\right)} = \frac{10^{-3}}{s+2}$$

Taking Inverse Laplace,

$$i(t) = 10^{-3} e^{-2t}$$

(ii) Let, at $t = T$, the transient disappear (within 1%)

$$i(T) = 0.01 \times 10^{-3}$$

$$10^{-3} e^{-2T} = 0.01 \times 10^{-3}$$

$$e^{2T} = 100$$

$$T = \frac{1}{2} ln(100) = 2.3\text{ sec.}$$

(iii) As $C_1 = C_2 = 1\ \mu\text{F}$, $V_{C_1}(s) = V_{C_2}(s)$

Voltage across each capacitor is same and given by

$$V_{C_1}(s) = V_{C_2}(s) = \left(\frac{1}{C_1 s}\right) I(s) = \left(\frac{1}{C_1 s}\right) \times \left[\frac{10^{-3}}{s+2}\right]$$

$$\text{steady state value of voltages} = \underset{s\to 0}{\text{Lt}} [sV_1(s)]$$

$$= \underset{s\to 0}{Lt}\left[s \times \left(\frac{1}{C_1 s}\right)\left(\frac{10^{-3}}{s+2}\right)\right] = \frac{10^{-3}}{2C_1} = \frac{10^{-3}}{2 \times 10^{-6}} = 500\text{ V}$$

Hence, voltage across each capacitor is 500 V.

3.9 For the below circuit the switch has been in position 1 for a long time.

(i) Find the indicated currents.

(ii) Find the indicated currents and voltage immediately after the switch has been thrown to position 2.

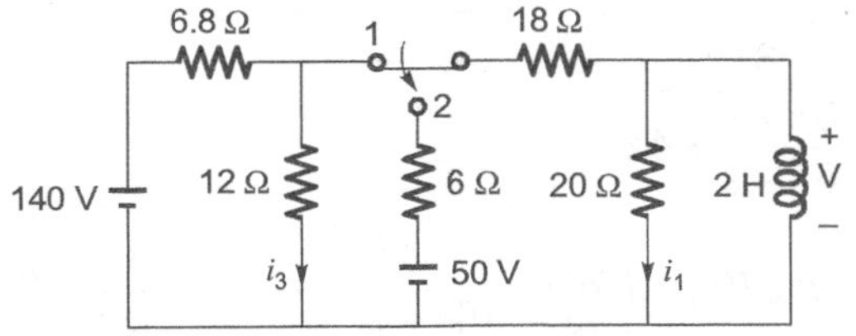

[12 marks : 2012]

Solution:

When switch is at position 1,

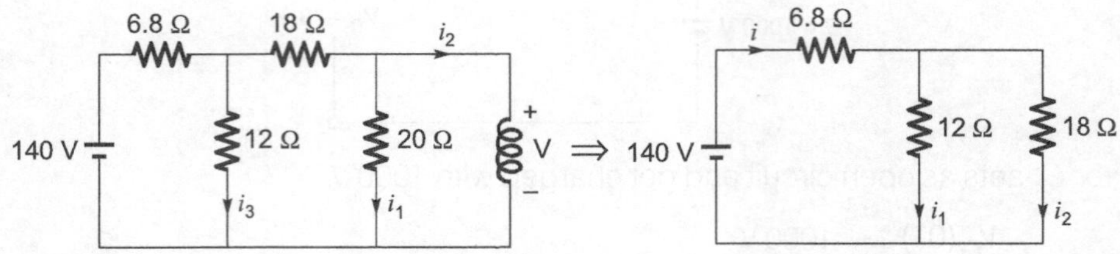

As switch at position 1 for long time, the inductor will behave like conductor i.e. short circuit

$\Rightarrow$ $$i_1 = 0$$

Current, $$i = \frac{140}{6.8+(12\|18)} = 10\text{ A}$$

$\Rightarrow$ $$i_2 = \frac{12}{30}\times 10 = 4\text{ A}$$

$$i_3 = 10-4 = 6\text{ A}$$

When switch is at position 2. The current through inductor will not change just after the switch has been thrown to position 2.

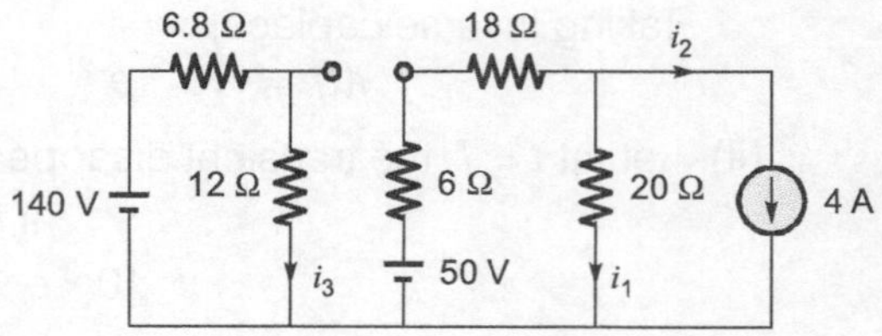

$$i_2 = 4\text{ A}$$

$$i_3 = \frac{140}{6.8+12} = 7.45\text{ A}$$

$$50 = (i_1+4)(6+18)+20\,i_1$$

$$50 = 44i_1 + 96$$

$$i_1 = -\frac{46}{44} = -1.045\text{ A}$$

$$v = 20\times i_1 = -20.91\text{ V}$$

3.10 **A parallel combination of R = 16 kΩ and L = 1 H is excited by a constant voltage source of 8 volts, having internal resistance of 4 kΩ. Obtain the current through inductance in steady state and the time constant of the circuit.**

[8 marks : 2014]

Solution:

Method-I:

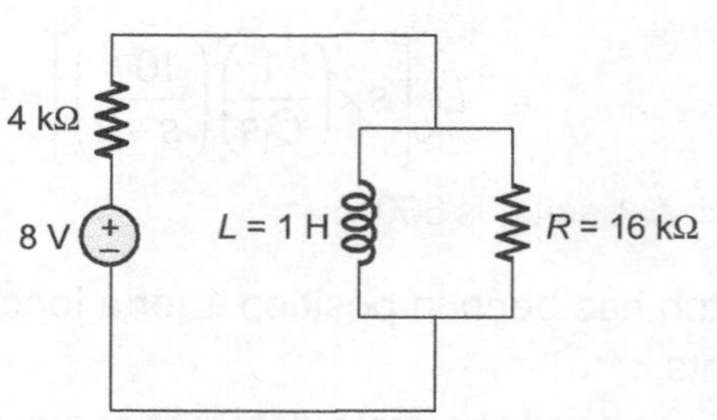

After converting into s-domain,

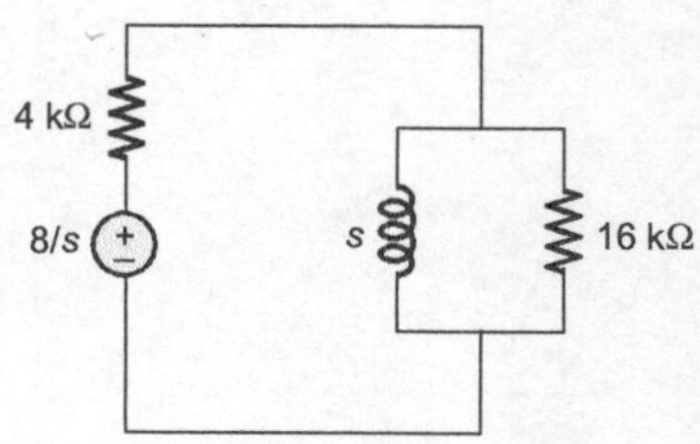

$$I(s) = \frac{8/s}{\frac{16k\times s}{16k+s}+4k}$$

At steady state, $$I(s)_{ss} = \lim_{s\to 0} sI(s) = \lim_{s\to 0} s\cdot\frac{8/s}{\frac{16\text{k}\times s}{16\text{k}+s}+4\text{k}} = \frac{8}{4\text{ k}\Omega} = 2\text{ mA}$$

Method-II:

At steady state, inductor behaves as short circuit. So,

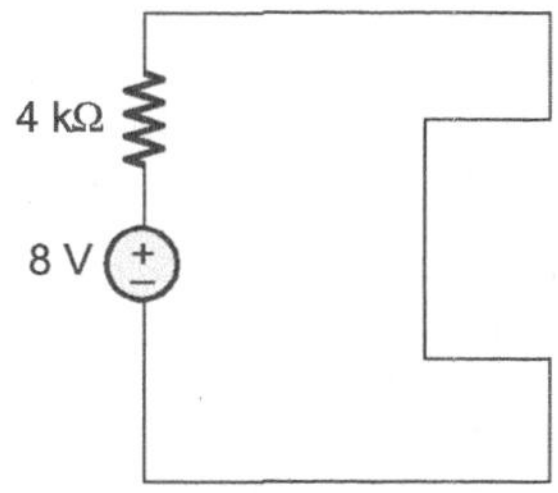

$$I_{ss} = \frac{8}{4\,k\Omega} = 2\,mA$$

$$R_{eq} = \frac{4\,k\Omega \times 16\,k\Omega}{20\,k\Omega} = 3.2\,k\Omega$$

$$L_{eq} = 1\,H$$

$$\tau = \frac{L_{eq}}{R_{eq}} = \left(\frac{1}{3.2\,k\Omega}\right) = 0.3125\,\text{msec.}$$

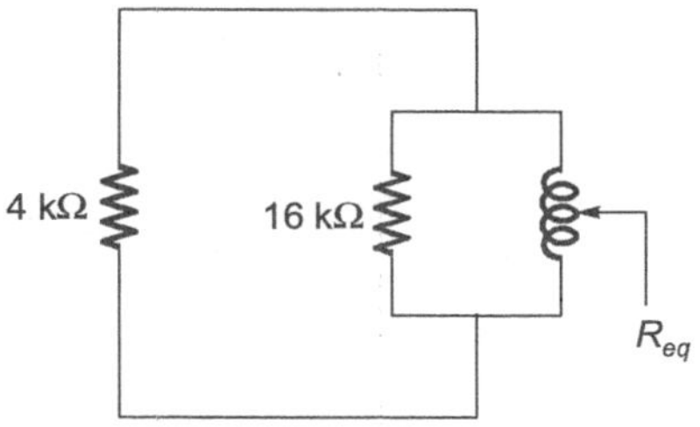

3.11 For the circuit shown below, the capacitor has initial voltage of V_0 volts as per polarity indicated. It is supplied by a constant voltage source of V volts. Draw s-domain equivalent circuit. Hence, obtain the expression for charging current $i(t)$ and the voltage across capacitance. Switch is closed at $t = 0$.

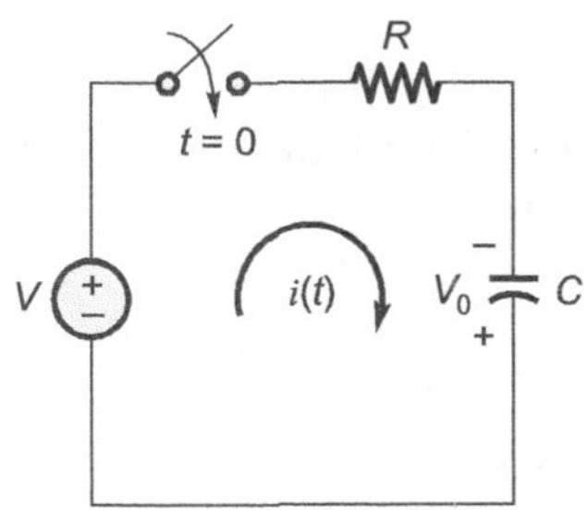

[8 marks : 2014]

Solution:

$$V = iR + \frac{1}{C}\int i\,dt$$

$$\frac{V}{s} = RI(s) + \frac{1}{Cs}[I(s)] - \frac{V_0}{s}$$

$$= \left(R + \frac{1}{Cs}\right)I(s) - \frac{V_0}{s}$$

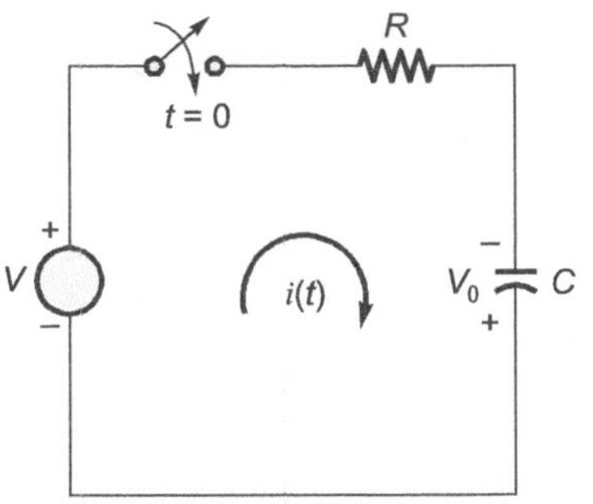

After chosing the switch at $t = 0$,

$$I(s) = \frac{\frac{V}{s} + \frac{V_0}{s}}{R + \frac{1}{sC}} = \frac{(V + V_0)C}{RCs + 1} = \frac{\frac{(V + V_0)C}{RC}}{s + \frac{1}{RC}}$$

$$i(t) = \frac{(V + V_0)}{R} e^{-t/RC}$$

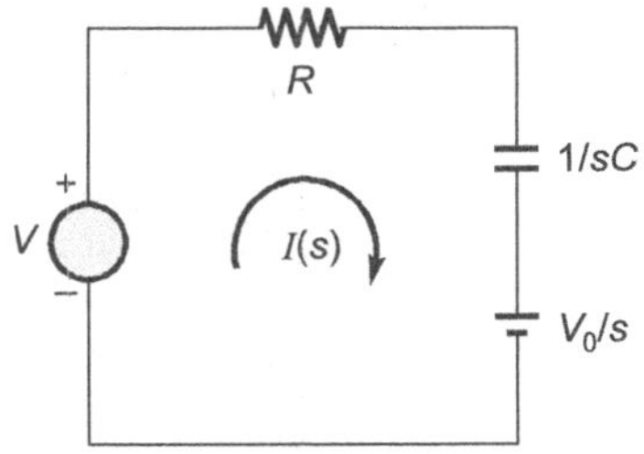

Voltage across capacitance = $I(s) \times \frac{1}{sC} = \frac{(V+V_0)C}{RCs+1} \times \frac{1}{sC}$

$$V_c(s) = \frac{(V+V_0)/RC}{s\left(s+\frac{1}{RC}\right)} \quad \text{...(1)}$$

or, $$\frac{(V+V_0)/RC}{s\left(s+\frac{1}{RC}\right)} = \frac{A}{s} + \frac{B}{s+\frac{1}{RC}}$$

equating coefficient of 's' and constant term, $A + B = 0$

$$\frac{A}{RC} = \frac{V+V_0}{RC}$$

$$A = V + V_0$$

so, $$B = -(V + V_0)$$

or, $$V_c(s) = \frac{V+V_0}{s} + \frac{-(V+V_0)}{s+\frac{1}{RC}}$$

$$V_c(t) = (V + V_0) - (V + V_0)\, e^{-t/RC} = (V + V_0)\, [1 - e^{-t/RC}]$$

3.12 The winding of an electromagnet has an inductance of 3 H and a resistance of 15 Ω. When it is connected to a 120 V, d.c. supply, calculate:

(i) the steady-state value of current flowing in the winding.

(ii) the time constant of the circuit,

(iii) the value of the induced e.m.f. after 0.1s,

(iv) the time for the current to rise to 85% of its final value, and

(v) the value of the current after 0.3s. [10 marks : 2016]

Solution:

(i) In steady state, inductor acts as s.c.

$$I = \frac{120}{15} = 8 \text{ A}$$

(ii) $$\text{Time constant } (\tau) = \frac{L}{R} = \frac{3}{15} = \frac{1}{5} \text{ sec}$$

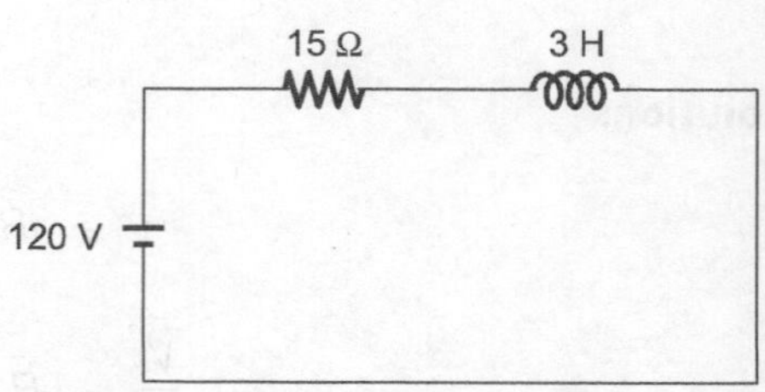

(iii) Voltage across inductor is '0' in steady state.

$$V(0^+) = 120 \text{ V},$$

$$V_L = V_L(\infty) + (V_L(0^+) - V_L(\infty))\, e^{-t/\tau}$$

$$V_L(t) = 0 + (120 - 0)e^{-5t}$$

$$V_L(t) = 120\, e^{-5t}$$

$$V_L(0.1) = 120\, e^{-5(0.1)} = 72.78 \text{ V}$$

(iv) $$i(0^+) = 0;$$

$$i(\infty) = 8 \text{ A}$$

$$i(t) = 8 + (0 - 8)\, e^{-5t}$$

$$i(t) = 8 - 8\, e^{-5t}$$

$$0.85 \times 8 = 8 - 8\, e^{-5t}$$

$$t = 0.3795 \text{ sec}$$

(v) $$i(0.3) = 8 - 8\, e^{-5(0.3)}$$

$$i(0.3) = 6.2149 \text{ Amp}$$

3.13 For the circuit shown in figure

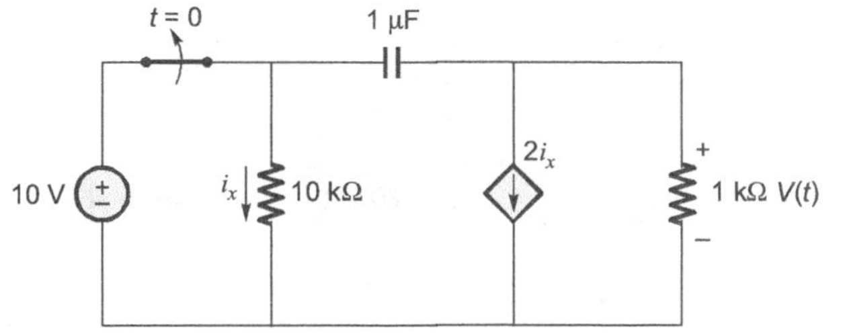

(i) Find the expression of $V(t)$, the voltage across 1 kΩ resistor when the switch is opened at time, $t = 0$.

(ii) Sketch $V(t)$ with respect to time (t) and marks the time constant t.

[20 marks : 2019]

Solution:

For $t < 0$,

Initially switch is closed,

$$V_C(0^-) = 10000\, i_x + 1000\,(2i_x) = 12000 i_x$$

$$= 12000 \times \frac{10}{10000} = 12 \text{ V}$$

Converting circuit in laplace domain,

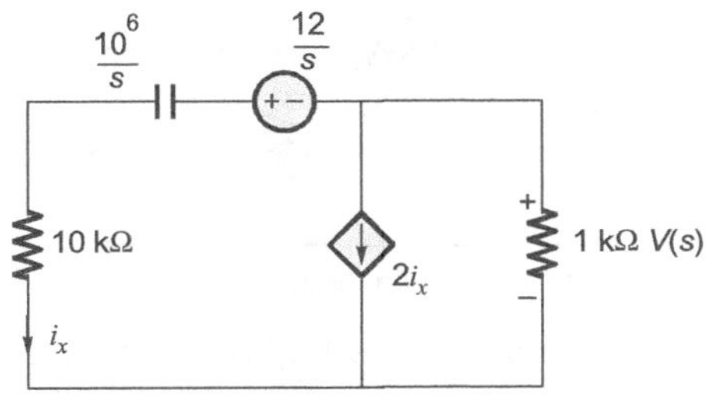

$$3i_x = -V(t)$$

$$10ki_x + 3ki_x + \frac{10^6}{s} i_x = \frac{12}{s}$$

$$s(13 \times 10^3\, i_x) + 10^6\, i_x = 12$$

$$i_x = \frac{12}{13 \times 10^6 s + 10^6} = \frac{\frac{12}{13} \times 10^{-3}}{s + \frac{10^3}{13}}$$

$$i_x(t) = 0.923 \times 10^{-3}\, e^{-76.92}$$

$$V(t) = -3i_x(t)$$

$$= -2.769\, e^{-76.92t} \text{ V}$$

At $t = 0$, $V(0) = -2.769$

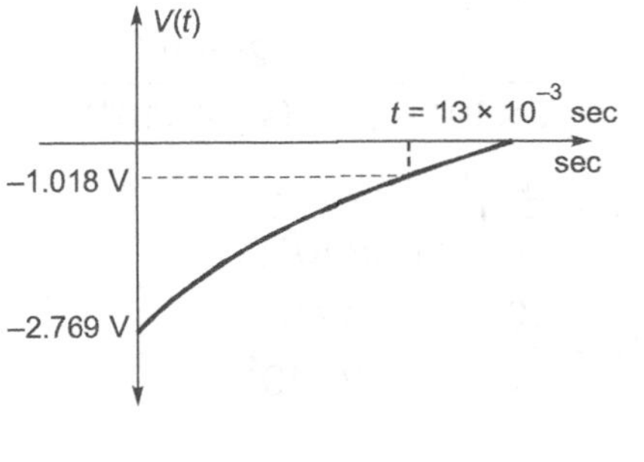

3.14 In the circuit shown in the figure given below, the switch is closed at $t = 0$. Obtain the current i and capacitor voltage v_c, for $t > 0$.

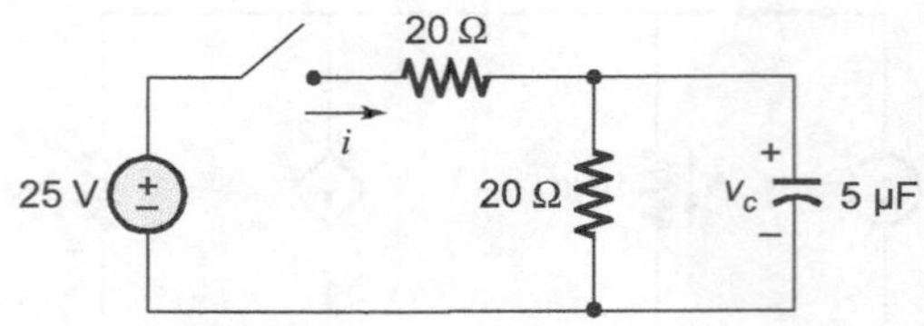

[20 marks : 2020]

Solution:

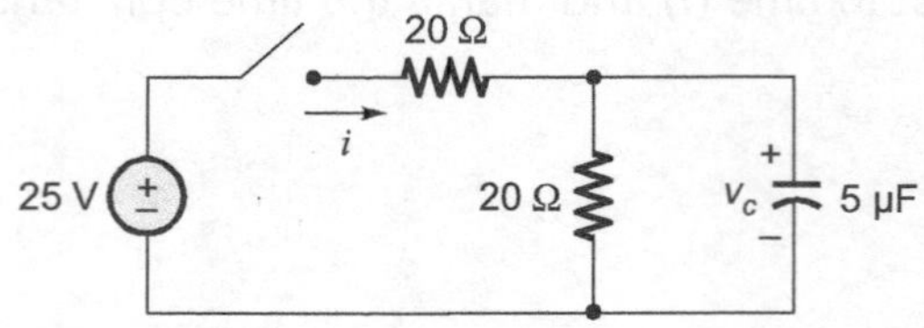

At $t = 0^-$

$$V_C = 0 \text{ V} \quad \text{(As no sources is connected with capacitor)}$$

$$\therefore \quad v_c(0^+) = v_c(0^-) = 0 \text{ V}$$

For $t > 0$

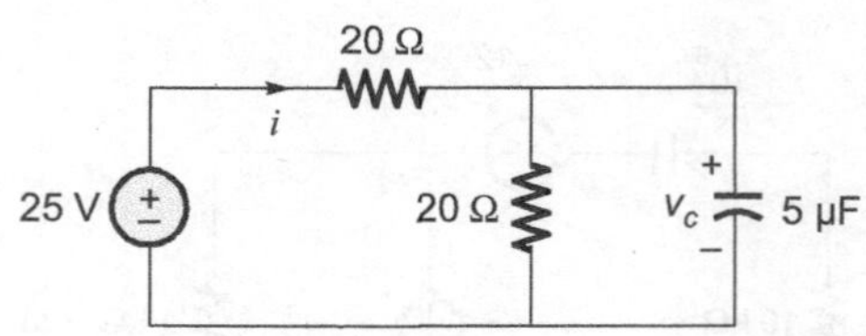

figure (i)

Transforming the circuit from time domain to the Laplace domain,

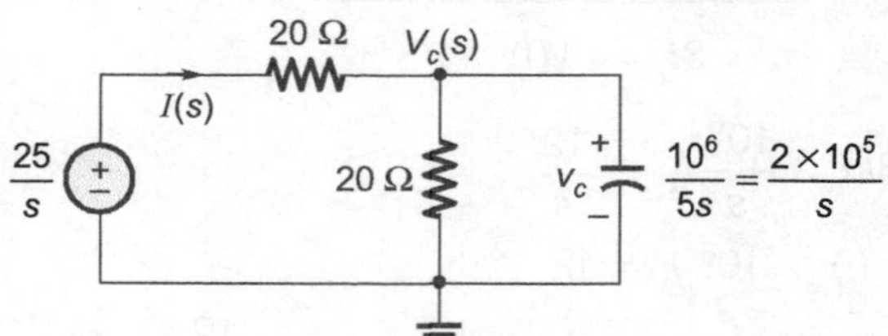

Apply KCL,

$$\frac{V_c(s) - \frac{25}{s}}{20} + \frac{V_c(s)}{20} + \frac{V_c(s)\cdot s}{20\times10^4} = 0$$

$$V_c(s)\left[10^4 + 10^4 + s\right] - \frac{25}{s}\times10^4 = 0$$

$$V_c(s) = \frac{\frac{25\times10^4}{s}}{2\times10^4 + s} = \frac{25\times10^4}{s(s+2\times10^4)}$$

$$V_c(s) = \frac{a_0}{s} + \frac{a_1}{s+2\times10^4}$$

$$V_c(s) = \frac{12.5}{s} + \frac{-12.5}{s+2\times10^4}$$

Taking inverse laplace transform,

$$v_c(t) = 12.5u(t) - 12.5e^{-2\times10^4 t}u(t)$$

$$v_c(t) = 12.5\left[1 - e^{-2\times10^4 t}\right]u(t)\text{ V}$$

From figure (i)

$$i = \frac{25 - v_c(t)}{20}$$

$$= \frac{25 - 12.5 + 12.5e^{-2\times10^4 t}}{20}$$

$$i(t) = \frac{12.5}{20}\left[1 + e^{-2\times10^4 t}\right]$$

$$i(t) = \frac{5}{8}\left[1 + e^{-2\times10^4 t}\right]\text{A} \qquad (t > 0)$$

3.15 **A circuit shown in the figure is initially being quiescent and excited by a current impulse. Find the expression for $V_c(t)$ for $t > 0$ and draw the voltage response. Calculate the value of voltage across the capacitor at $t = 50$ ms.**

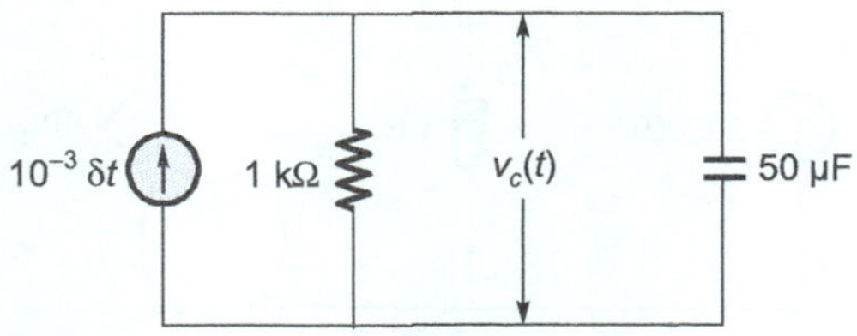

[20 marks : 2021]

Solution:

$t = 0^-$: Capacitor is not connected to any independent source.

$$V_C(0^-) = 0\text{ V}$$

$t > 0$: Circuit in s- domain :

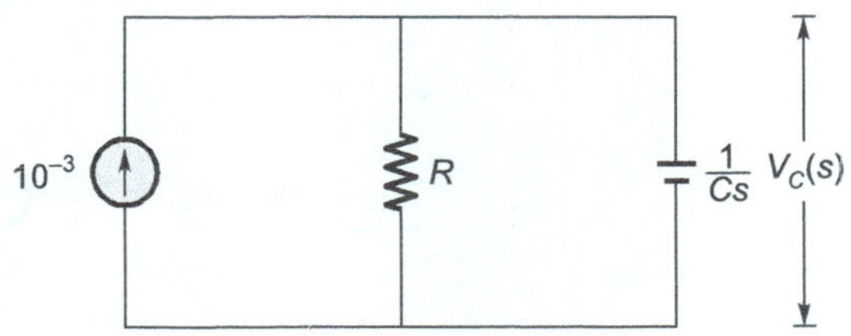

$$V_C(s) = \left(\frac{10^{-3}\times R}{R + \frac{1}{Cs}}\right)\times\frac{1}{Cs}$$

$$V_C(s) = 10^{-3}\left(\frac{RCs}{RCs+1}\right)\times\frac{1}{Cs}$$

$$V_C(s) = 10^{-3}\left(\frac{1/C}{s + \frac{1}{RC}}\right)$$

Substitute $R = 1$ kΩ; $C = 50 \times 10^{-6}$ F

$$V_C(s) = 10^{-3}\left[\frac{\frac{1}{50}\times10^{-6}}{s + \frac{1}{1\times10^3\times50\times10^{-6}}}\right]$$

$$V_C(s) = 10^{-3}\left[\frac{20000}{s+20}\right] = \frac{20}{s+20}$$

On taking inverse laplace transistor of above equation,

$$V_C(t) = (20e^{-20t}) \quad t \geq 0$$

$$V_C(50 \times 10^{-3}) = 20e^{-20(50\times10^{-3})} = 7.3575 \text{ V}$$

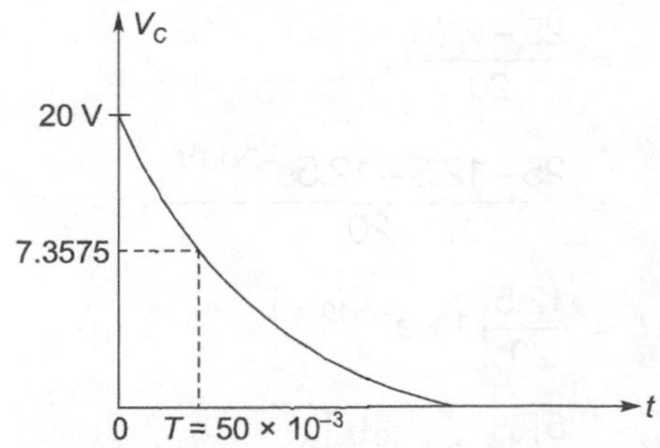

3.16 **(i) For the network shown in the following figure, compute $i_L(t)$ and $i_1(t)$, if the initial current through the inductor is 0 ampere :**

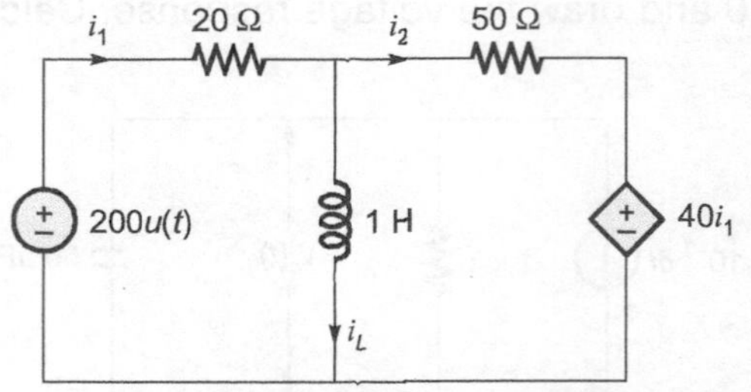

(ii) Determine the current at $t > 0$, if AC voltage V is applied, when switch S is moved from position 2 to position 1 at t = 0, for the network shown in the following figure. Assume steady-state current of 1 ampere in the network when the switch is at position 1 :

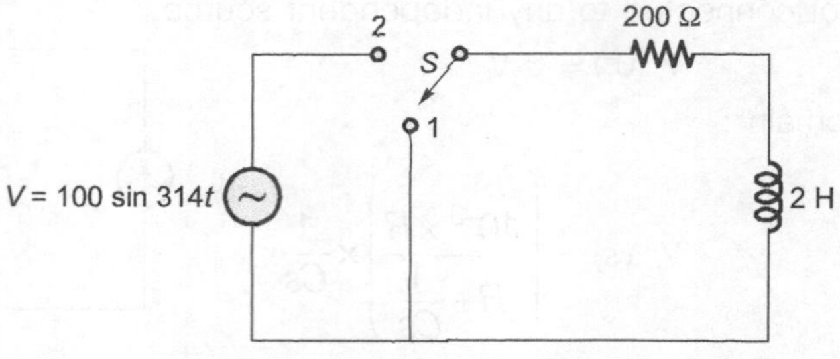

[10 + 10 marks : 2023]

Solution:

(i) For the network shown in the figure. Given : initial current in the inductor is zero,
To find : $i_L(t)$ and $i_1(t)$

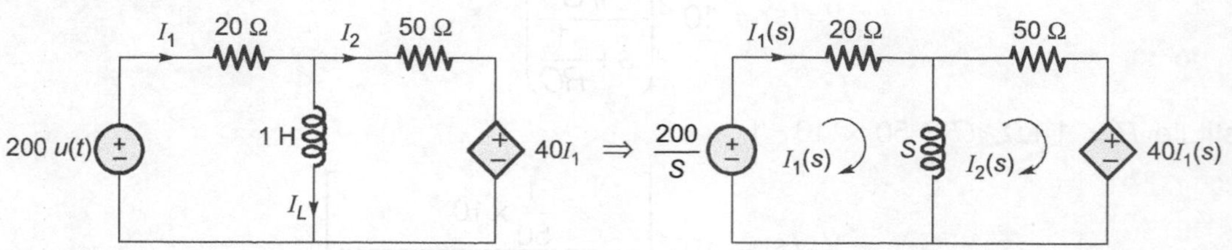

$$\frac{200}{s} = (20 + s)\,I_1(s) - sI_2(s) \qquad \text{...(i)}$$

$(50 + S)I_2(s) - sI_1(s) + 40I_1(s) = 0$ ...(ii)

From equation (ii),

$$I_2(s) = \frac{(s-40)I_1(s)}{(50+s)} \quad ...(iii)$$

Substitute equation (iii) in equation (i),

$$\frac{200}{s} = (20+s)I_1(s) - \frac{s(s-40)}{(s+50)}I_1(s)$$

$$\frac{(s+50)200}{s} = (110s + 1000)\,I_1(s)$$

$$I_1(s) = \frac{200s + 10^4}{110s^2 + 1000s} \quad ...(iv)$$

$$i_1(t) = L^{-1}I_1(s)$$

$$= \left[10 - \frac{90}{11}e^{\frac{-100}{11}t}\right]$$

$$I_L(s) = I_1(s) - I_2(s)$$

$$= I_1(s) - \frac{(s-40)}{(50+s)}I_1(s)$$

$$= \left(\frac{s+50-s+40}{s+50}\right)I_1(s)$$

$$I_L(s) = \frac{18000}{s[110s+1000]} = \frac{1800}{s[11s+100]}$$

$$i_L(t) = L^{-1}I_L(s)$$

$$= 18 - 18e^{\frac{-100}{11}t}$$

(ii) Determine the current at $t > 0$, when AC voltage is applied, when switch is moved from position-2 to position-1 at $t = 0$ sec. Assume steady state current of 1 A in the network, when the switch is at position-1.

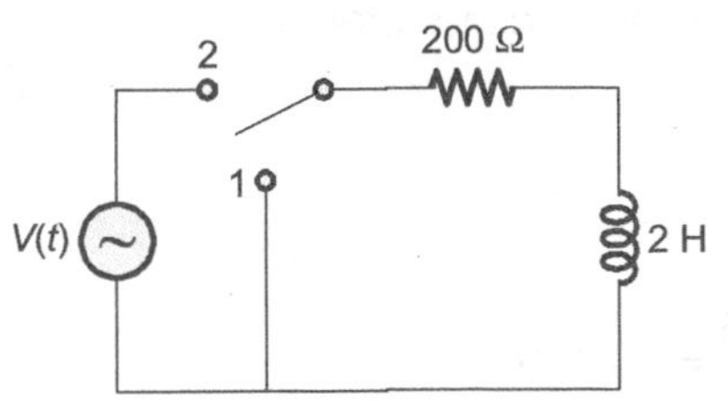

$$V(t) = 100\sin 314t$$

At $t > 0$

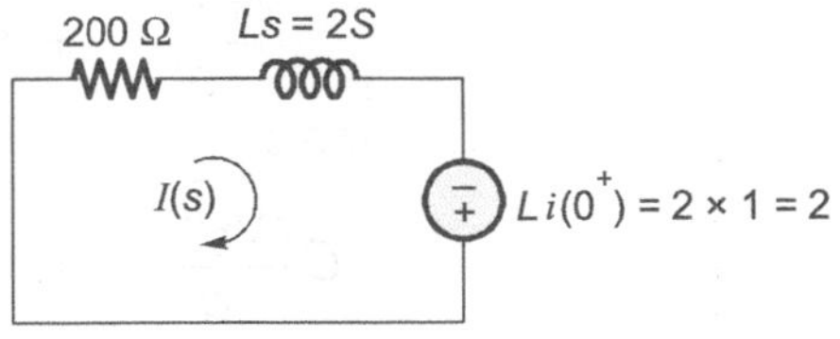

$$(200 + 2s)I(s) = 2$$

$$I(s) = \frac{2}{200 + 2s}$$

$$i(t) = L^{-1}I(s)$$

$$i(t) = [e^{-100t}]\,u(t)$$

4. Second Order RLC Circuit, Resonance

4.1 The network shown in the figure below has reached steady state when the switch *S* moves from *a* to *b*.

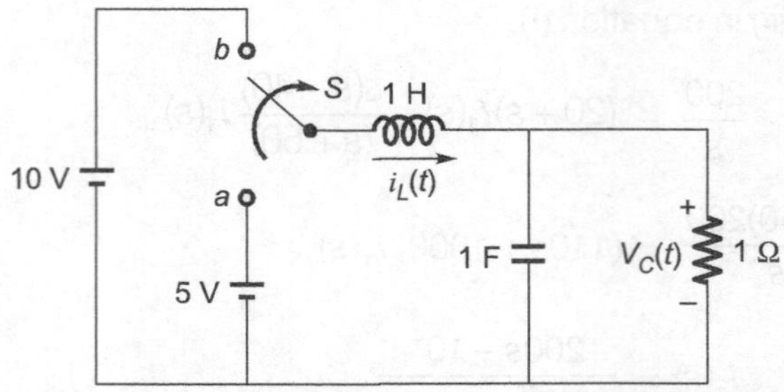

(a) Determine initial values for $i_L(t)$ and $V_C(t)$ and their first derivatives with switch in position *b*.
(b) Determine $V_C(t)$ for $t > 0$. Sketch $V_C(t)$ as a function of time
(c) Determine damping ratio, undamped and damped natural frequencies.

[6 + 20 + 6 = 32 marks : 2001]

Solution:

(a) For $t < 0$, the circuit will be in steady state

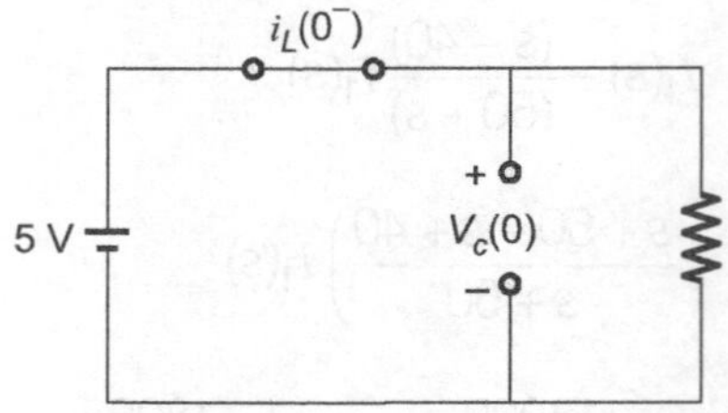

$$i_L(0^-) = 5\text{ A}$$
$$V_C(0^-) = 5\text{ V}$$

For $t \geq 0$, the switch will be in position *b*.
at $t = 0^+$, the circuit will be
Since the current in the inductor and voltage in the capacitor cannot change instantaneously.

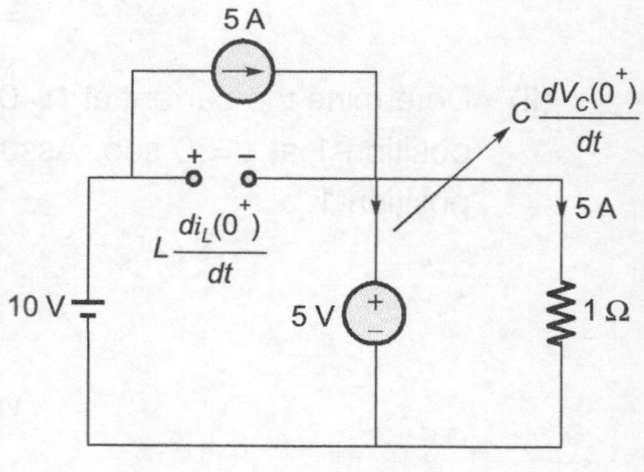

$$i_L(0^+) = i(0^-) = 5\text{ A}$$
$$V_C(0^+) = V_C(0^-) = 5\text{ V}$$

From the circuit, the voltage across the inductor at $t = 0^+$, will be

$$L\frac{di_L(0^+)}{dt} = 10 - 5 = 5$$

$$\frac{di_L(0^+)}{dt} = \frac{5}{L} = 5\text{ A/sec}$$

$$\frac{di_L(0^+)}{dt} = 5\text{ A/sec}$$

Current in the capacitor at $t = 0^+$

$$C\frac{dv_C(0^+)}{dt} = 0$$

$$\frac{dv_C(0^+)}{dt} = 0\text{ V/sec}$$

(b) Drawing the circuit in Laplace domain for $t \geq 0$

Applying nodal analysis,

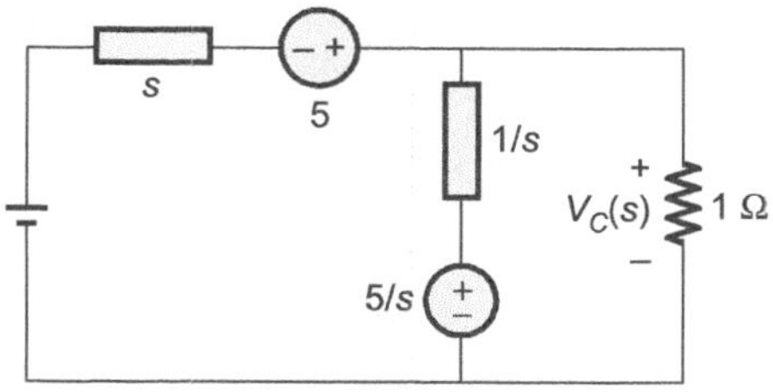

$$\frac{\frac{10}{s}+5-V_C(s)}{s} = \frac{V_C(s)-\frac{5}{s}}{1/s}+V_C(s)$$

$$\frac{10}{s}+5-V_C(s) = s^2V_C(s)-5s+sV_C(s)$$

$$(s^2+s+1)V_C(s) = \frac{10}{s}+5s+5$$

$$V_C(s) = \frac{\frac{10}{s}+5s+5}{s^2+s+1} = \frac{5(s^2+s+2)}{s(s^2+s+1)} = \frac{5s^2+5s+10}{s(s^2+s+1)}$$

$$V_C(s) = \frac{10}{s}-\frac{5s+5}{s^2+s+1}$$

$$V_C(s) = \frac{10}{s}-5\left[\frac{s+\frac{1}{2}}{\left(s+\frac{1}{2}\right)^2+\left(\frac{\sqrt{3}}{2}\right)^2}+\frac{1}{2}\frac{1}{\left(s+\frac{1}{2}\right)^2+\left(\frac{\sqrt{3}}{2}\right)^2}\right]$$

$$V_C(s) = \frac{10}{s}-5\left[\frac{s+\frac{1}{2}}{\left(s+\frac{1}{2}\right)^2+\left(\frac{\sqrt{3}}{2}\right)^2}+\frac{1}{\sqrt{3}}\frac{\frac{\sqrt{3}}{2}}{\left(s+\frac{1}{2}\right)^2+\left(\frac{\sqrt{3}}{2}\right)^2}\right]$$

Taking inverse Laplace transform,

$$V_C(t) = 10-5e^{-t/2}\left[\cos\frac{\sqrt{3}}{2}t+\frac{1}{\sqrt{3}}\sin\frac{\sqrt{3}}{2}t\right]$$

$$V_C(t) = 10-5e^{-0.5t}\left[\cos\frac{\sqrt{3}}{2}t+\frac{1}{\sqrt{3}}\sin\frac{\sqrt{3}}{2}t\right]$$

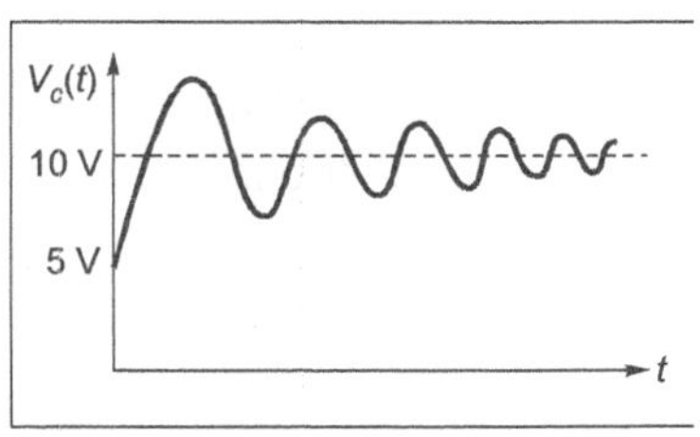

(c) From the above equation,

$$V_C(t) = 10-e^{-0.5t}\left[\cos\frac{\sqrt{3}}{2}t+\frac{1}{\sqrt{3}}\sin\frac{\sqrt{3}}{2}t\right]$$

Comparing with the response of a second order system subjected to step input

$$S(t) = 1-e^{-\xi\omega_n t}\sin(\omega_d t+\phi)$$

where, $\xi \rightarrow$ damping ratio

$$\omega_d = \omega_n\sqrt{1-\xi^2} \rightarrow \text{damped frequency}$$

$\omega_n \rightarrow$ undamped frequency

$$\therefore \qquad \xi\omega_n = 0.5 = \frac{1}{2}$$

$$\omega_d = \omega_n\sqrt{1-\xi^2} = \frac{\sqrt{3}}{2}$$

$$\omega_n^2-\omega_n^2\xi = \frac{3}{4}$$

$$\omega_n^2 - \left(\frac{1}{2}\right)^2 = \frac{3}{4}$$

$$\omega_n = \sqrt{\frac{3}{4} + \frac{1}{4}} = 1 = 1 \text{ rad/sec}$$

$$\xi = \frac{1}{2\omega_n} = \frac{1}{2}$$

Therefore, damping ratio $= \frac{1}{2}$

undamped frequency = 1 rad/sec

damped frequency $= \frac{\sqrt{3}}{2}$ rad/sec

4.2 In the circuit shown in the figure, S is closed at ($t = 0$). Find the current $i_C(t)$ through the capacitor at ($t = 0$).

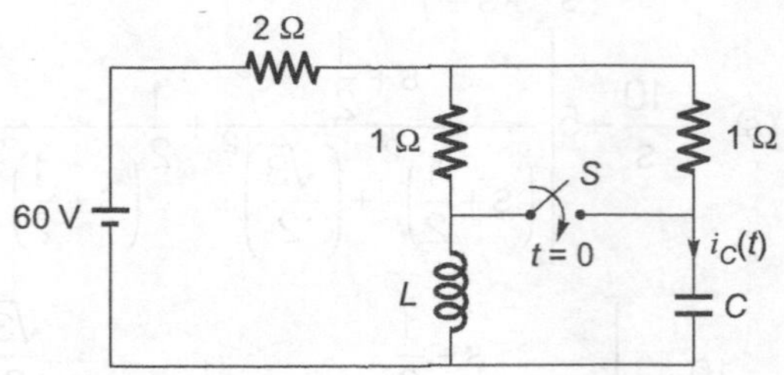

[12 marks : 2003]

Solution:

Before switch (S) is closed. Inductor 'L' is shorted and 'C' is open.

at ($t = 0^-$)

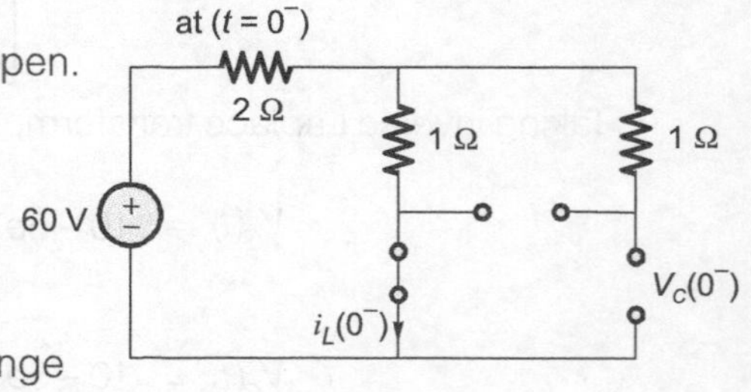

$$v_c(0^-) = v_c(0^+) = \frac{60}{3} \times 1 = 20 \text{ V}$$

$$i_L(0^-) = i_L(0^+) = \frac{60}{3} = 20 \text{ A}$$

At ($t = 0^+$), Inductor current and capacitor voltage never change instantaneously.

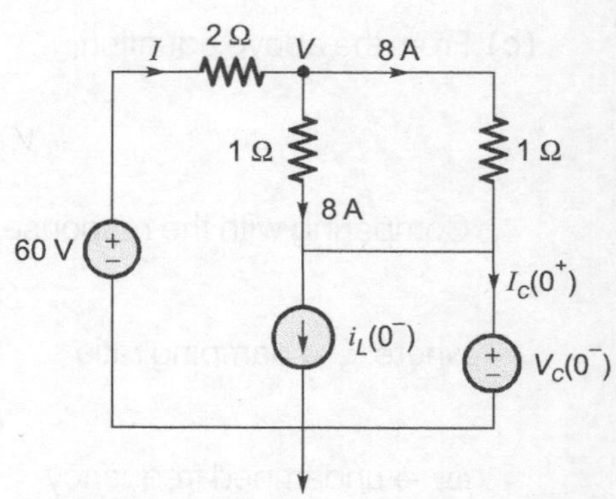

By KCL,
$$\frac{V-60}{2} + \frac{V-20}{0.5} = 0$$

$$V - 60 + 4V - 80 = 0$$

$$5V = 140$$

$\therefore$
$$V = 28 \text{ V}$$

$\therefore$
$$I(0^+) = \frac{60-28}{2} = 16 \text{ A}$$

Then capacitor current,
$$I_C(0^+) = I(0^+) - i_L(0^+) = 16 - 20$$
$$= -4 \text{ Amps.}$$

4.3 Show that a high-Q coil resonant circuit can be approximated as shown in figure.

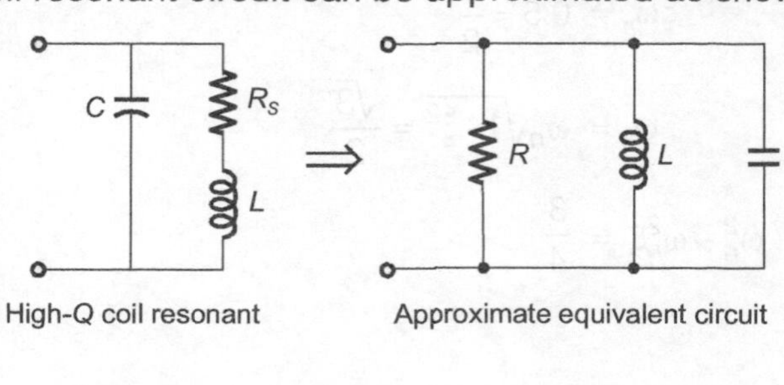

High-Q coil resonant | Approximate equivalent circuit

For a practical tank circuit shown in figure below, the resonance occurs at 1 MHz. Assuming a high Q-coil, find out the quality factor of high-Q coil at resonance frequency.

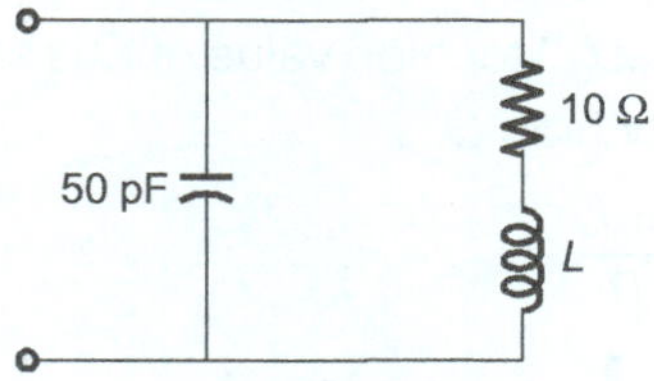

[12 marks : 2004]

Solution:

Circuit-1,

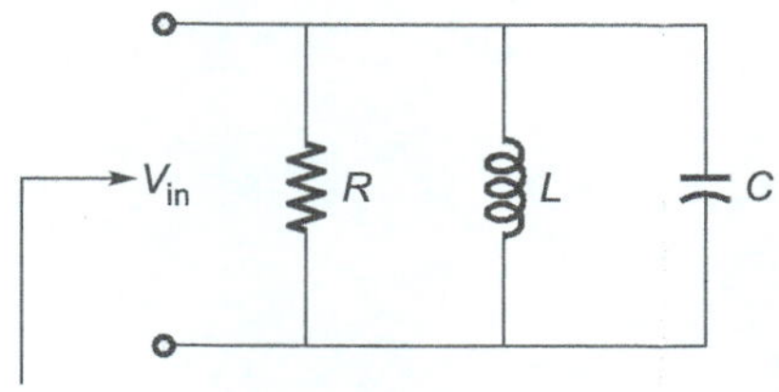

$$y_{in} = \frac{1}{R} + j\omega C + \frac{1}{j\omega L}$$

$$= \frac{1}{R} + j\left(\omega C - \frac{1}{\omega L}\right)$$

The frequency w_0, at which resonance occurs is

$$\omega_0 C - \frac{1}{\omega_0 L} = 0$$

$$\Rightarrow \qquad \omega_0 = \frac{1}{\sqrt{LC}}$$

Circuit-2,

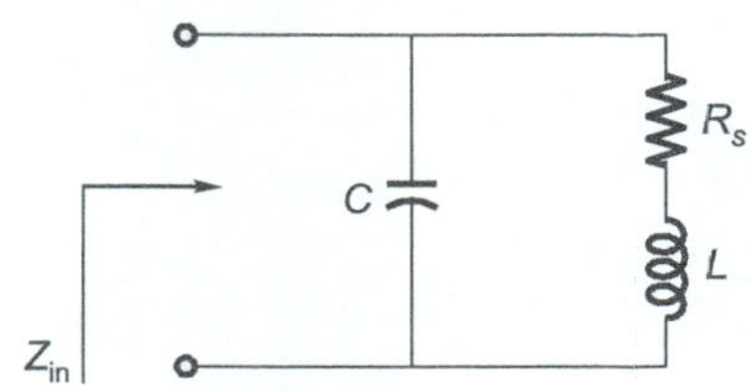

$$Z_{in} = \frac{(R_s + j\omega L) \times \left(\frac{1}{j\omega C}\right)}{R_s + j\omega L + \frac{1}{j\omega C}} = \frac{R + j\omega L}{1 + j\omega RC - \omega^2 LC}$$

$$= \omega L \times \left[\frac{\frac{R}{\omega L} + j}{1 + j\omega RC - \omega^2 LC}\right] \qquad \text{...(1)}$$

$$Q \text{ of the coil} = \frac{\omega L}{R} \qquad \text{...(2)}$$

Hence, equation (1) becomes,

$$Z_{in} = \omega L \times \left[\frac{\frac{1}{Q} + j}{1 + j\omega RC - \omega^2 LC}\right]$$

Since, Q is very high, equation (2) becomes,

$$Z_{in} = \frac{j\omega L}{1 + j\omega RC - \omega^2 LC} = \frac{1}{\frac{RC}{L} + j\left(\omega C - \frac{1}{\omega L}\right)} = \frac{\frac{RC}{L} - j\left(\omega C - \frac{1}{\omega L}\right)}{\left[\left(\frac{RC}{L}\right)^2 + \left(\omega C - \frac{1}{\omega L}\right)^2\right]}$$

For resonance at frequency ω_0, imaginary term is zero

$$\omega_0 C - \frac{1}{\omega_0 L} = 0$$

$$\Rightarrow \qquad \omega_0 = \frac{1}{\sqrt{LC}}$$

Resonance frequency ω_0 of the circuit (2) for high value of Q is same as that of circuit (1). Hence, circuit (2) can be approximated as circuit (1) for high Q.

$$Q = \frac{\omega_0 L}{R}$$

$$\omega_0 = \frac{1}{\sqrt{LC}} \text{ for high } Q$$

$$L = \frac{1}{\omega_0^2 C}$$

$$Q = \frac{\omega_0}{R} \times \left(\frac{1}{\omega_0^2 C}\right) = \frac{1}{\omega_0 CR}$$

$$= \frac{1}{2\pi \times 1 \times 10^6 \times 50 \times 10^{-12} \times 10} = 318$$

4.4 In the *R-L-C* circuit shown in figure, I_S = 10 A, R = 1 Ω, L = 1 H, C = 1 μF and $i_L(0^-) = 0$. Determine $V(0^+)$, $\frac{dV}{dt}(0^+)$ and $\frac{d^2V}{dt^2}(0^+)$ after the switch is closed at (t = 0).

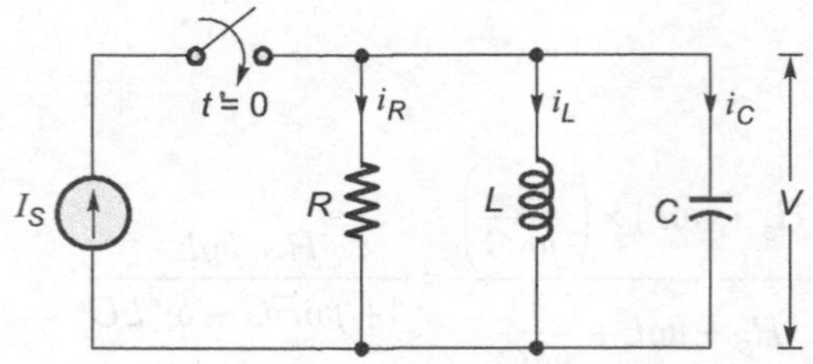

[10 marks : 2005]

Solution:

$$i_L(0^-) = 0 \text{ A}$$
$$V_c(0^+) = V_c(0^-) = 0 \text{ V}$$

Applying KCL at node A, we get

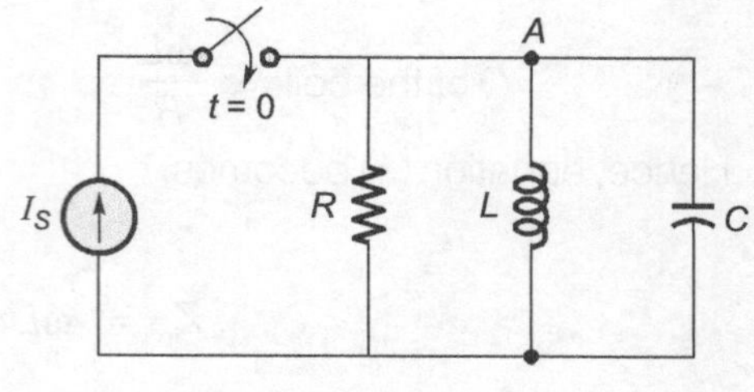

$$\frac{V}{R} + \frac{1}{L}\int V dt + C\frac{dV}{dt} = I$$

$$\frac{V}{1} + \int V dt + 1 \times 10^{-6}\frac{dV}{dt} = 10$$

By differentiations, we get

$$\frac{dV}{dt} + V + 1 \times 10^{-6}\frac{d^2V}{dt^2} = 0$$

$$10^{-6}\frac{d^2V}{dt^2} + \frac{dV}{dt} + V = 0$$

$$V(0^+) = 0 \text{ V}$$

$$10^{-6}\frac{d^2V}{dt^2} + \frac{dV}{dt} = 0 \qquad \text{...(1)}$$

Circuit at $t = 0^+$

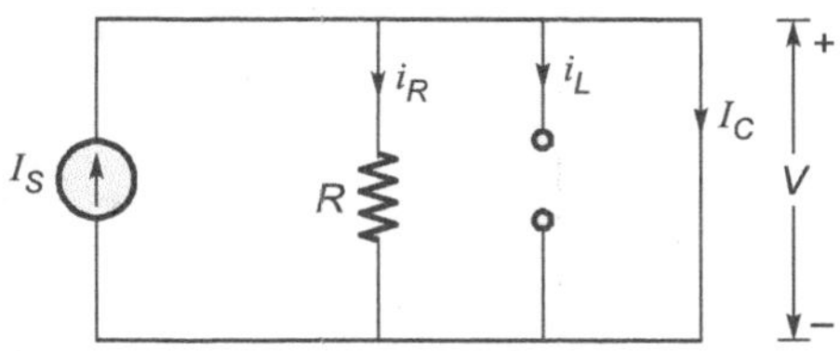

Thus, $$i_c(t) = C\frac{dV}{dt}$$

At $(t = 0)$ $$i_c(0^+) = 10\text{ A}$$

$$10 = 1\times10^{-6}\frac{dV}{dt}(0^+)$$

$$\frac{dV(0^+)}{dt} = 10\times 10^6 = 10^7\text{ V/S}$$

From equation (1), we get

$$10^{-6}\frac{d^2V}{dt^2}+10^7 = 0$$

$$\therefore \quad \frac{d^2V}{dt^2} = -10^{+7}\times 10^6 = -10^{13}\text{ V}^2/\text{sec}^2$$

4.5 Show that the resonant frequency ω_0 of an *R-L-C* series circuit is geometric mean of ω_1 and ω_2, the lower and upper half-power frequencies respectively.

[10 marks : 2005]

Solution:

In series *RLC* circuit, $$I = \frac{V}{R+j\left(\omega L-\frac{1}{C\omega}\right)}$$

$$|I| = \frac{V}{\sqrt{R^2+\left(\omega L-\frac{1}{\omega C}\right)^2}}$$

At half power points, $I = \frac{I_0}{\sqrt{2}}$ where, $I_0 = \frac{V}{R}$ (Current at resonance)

$$\frac{V}{\sqrt{2}\,R} = \frac{V}{\sqrt{R^2+\left(\omega L-\frac{1}{C\omega}\right)^2}}$$

$$\omega L-\frac{1}{\omega C} = \pm R$$

At lower half power point,

$$\omega_1 L-\frac{1}{\omega_1 C} = -R$$

$$\omega_1^2+\frac{R}{L}\omega_1-\frac{1}{LC} = 0$$

$$\omega_1 = \frac{-R}{2L}\pm\sqrt{\left(\frac{R}{2L}\right)^2+\frac{1}{LC}}$$

At upper half power point,

$$\omega_2 L-\frac{1}{\omega_2 C} = R$$

$$\omega_2^2-\frac{R}{L}\omega_2-\frac{1}{LC} = 0$$

$$\omega_2 = \frac{R}{2L} \pm \sqrt{\left(\frac{R}{2L}\right)^2 + \frac{1}{LC}}$$

$$\omega_1 \omega_2 = \left[-\frac{R}{2L} + \sqrt{\left(\frac{R}{2L}\right)^2 + \frac{1}{LC}}\right]\left[\frac{R}{2L} + \sqrt{\left(\frac{R}{2L}\right)^2 + \frac{1}{LC}}\right] = \frac{1}{LC}$$

$\therefore$ $$\omega_1 \omega_2 = \omega_0^2$$

$\therefore$ $$\omega_0 = \sqrt{\omega_1 \omega_2}$$

4.6 Find out resonant frequency of the circuit shown below with terms of circuit parameters.

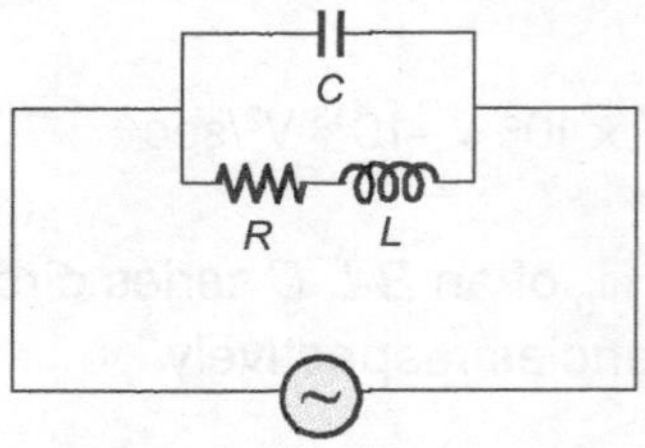

[12 marks : 2006]

Solution:

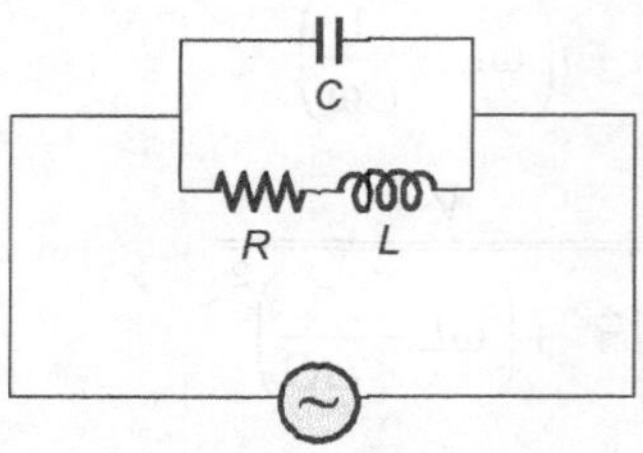

Admittance of the circuit Y_{in} is

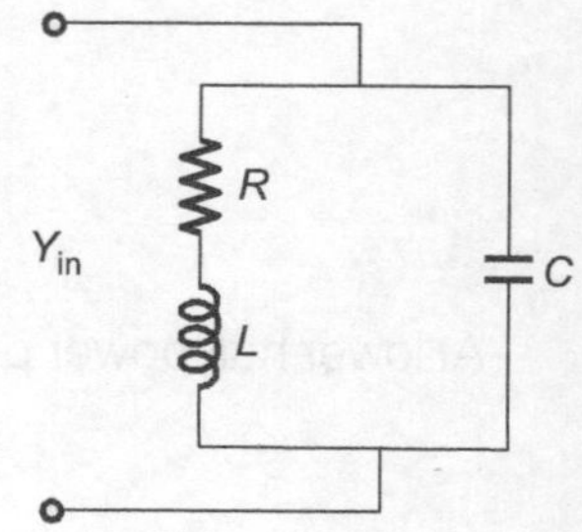

$$Y_{in} = \frac{1}{R + j\omega L} + j\omega C$$

$$= \frac{R - j\omega L}{R^2 + \omega^2 L^2} + j\omega C$$

At resonance, the imaginary term of Y_{in} will be zero

$$\text{imaginary } [Y_{in}]\Big|_{\omega=\omega_m} = \frac{-\omega_m L}{R^2 + \omega_m^2 L^2} + \omega_m C = 0$$

$$\frac{-L}{R^2 + \omega_m^2 L^2} + C = 0$$

$$L = CR^2 + \omega_m^2 CL^2$$

$$\omega_m^2 CL^2 = L - CR^2$$

$$\omega_m^2 = \frac{L - CR^2}{CL^2}$$

$$\omega_m = \sqrt{\frac{1}{LC} - \left(\frac{R}{L}\right)^2} \text{ rad/sec}$$

4.7 Compare the resonant frequency of the circuit shown in figure for $R = 0$ to that for $R = 50\ \Omega$.

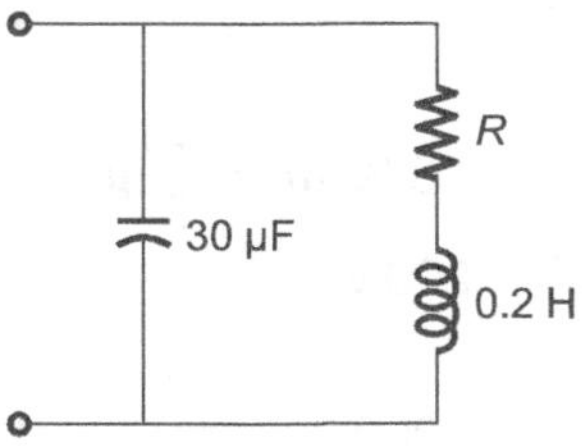

[12 marks : 2007]

Solution:

For $R = 0$, our circuit becomes simple (LC) network

$$\therefore \quad \omega_0 = \frac{1}{\sqrt{LC}} = \frac{1}{\sqrt{0.2 \times 30 \times 10^{-6}}} = 408.2 \text{ rad/sec} = 65 \text{ Hz}$$

For $R = 50\ \Omega$, then

$$y_{in} = j\omega C + \frac{1}{R + j\omega L} = \frac{R}{R^2 + (\omega L)^2} + j\left[\omega C - \frac{\omega L}{R^2 + (\omega L)^2}\right]$$

At resonance, Imaginary part of $y(j\omega) = 0$

$$\therefore \quad \omega_0 = \frac{1}{\sqrt{LC}}\sqrt{1 - R^2\frac{C}{L}} = 408.2 \times 0.79$$

$$= 322.7 \text{ rad/sec.} = 51.4 \text{ Hz}$$

4.8 In figure, $R = 200\ \Omega$, $L = 0.10$ H, $C = 13.33\ \mu$F, and $V_C(0^-) = 200$ V. Obtain the current transient, if the switch is closed at $t = 0$.

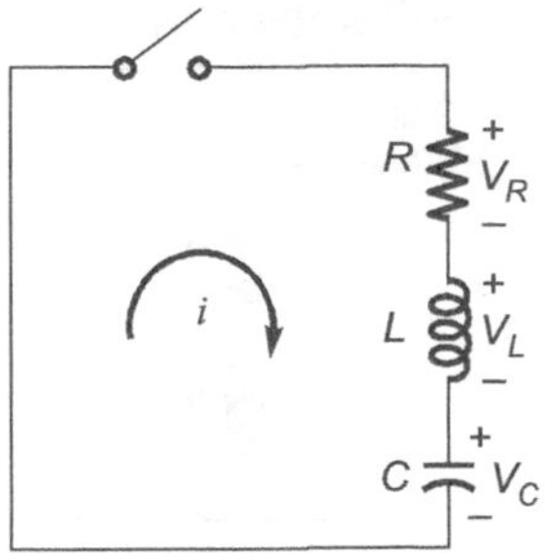

[12 marks : 2007]

Solution:

The circuit in Laplace domain will be

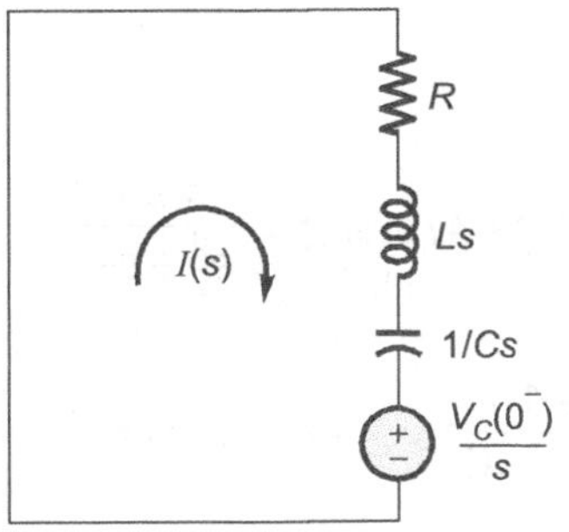

Applying KVL,

$$RI(s) + LsI(s) + \frac{I(s)}{Cs} + \frac{V_C(0^-)}{s} = 0$$

Putting the values:

$$R = 200\ \Omega$$
$$L = 0.1\text{ H}$$
$$C = 13.33\mu\text{F} = \frac{40}{3}\mu\text{F}$$
$$V_C(0^-) = 200\text{ V}$$

$$\left[200 + 0.1s + \frac{1}{\frac{40}{3}\times 10^{-6}s}\right]I(s) = -\frac{200}{s}$$

$$\left[200 + 0.1s + \frac{75\times 10^3}{s}\right]I(s) = -\frac{200}{s}$$

$$\frac{\left[0.1s^2 + 200s + 75\times 10^3\right]I(s)}{s} = -\frac{200}{s}$$

$$I(s) = \frac{-200}{0.1s^2 + 200s + 75\times 10^3} = \frac{-2000}{s^2 + 2000s + 75\times 10^4}$$

$$I(s) = \frac{-2000}{(s+500)(s+1500)} = -2\left[\frac{1}{(s+500)} - \frac{1}{(s+1500)}\right]$$

Taking inverse Laplace transform, $i(t) = 2\left[e^{-1500t} - e^{-500t}\right]u(t)$

$$= 2[e^{-1500t} - e^{-500t}] \quad t \geq 0$$

4.9 Obtain the impedance $Z_{in}(s)$ for the circuit shown in the figure at

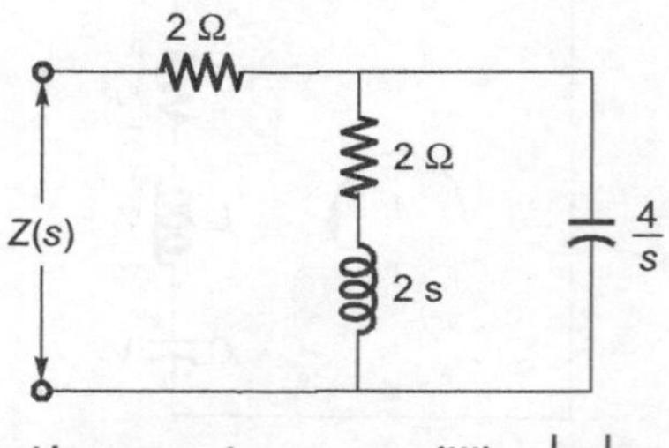

(i) $s = 0$, (ii) $s = j4$ rad/sec and (iii) $|s| = \infty$.

[12 marks : 2008]

Solution:

$$Z(s) = \left\{\frac{4}{s}\,\middle|\middle|\,(2+2s)\right\} + 2$$

$$= 2 + \frac{4}{s}\times(2+2s)\Big/2 + 2s + 4/s = 2 + \frac{4(s+1)}{s^2+s+2}$$

(i) $s = 0$, $Z(0) = 2 + 2 = 4\ \Omega$

(ii) $s = 4j$, $Z(4j) = 2 + \frac{4(1+4j)}{2+4j-16} = 2.33\angle{-29°}\ \Omega$

(iii) $s = \infty$, $Z(\infty) = 2 + \frac{4\left(\frac{1}{s}+\frac{1}{s^2}\right)}{1+\frac{1}{s}+\frac{1}{s^2}} = 2\ \Omega$

4.10 A series RLC circuit has $R = 25\ \Omega$, $L = 0.04$ H, $C = 0.01\ \mu$F. Calculate the resonant frequency. If a 1 volt source of the same frequency, as the frequency of resonance, is applied to this circuit calculate the frequencies at which the voltages across L and C are maximum. Calculate the voltages.

[12 marks : 2008]

Solution:

Resonant frequency, $$f_0 = \frac{1}{2\pi\sqrt{LC}} = \frac{1}{2\times 3.14\sqrt{0.04\times 0.01\times 10^{-6}}} = 7960 \text{ Hz}$$

At resonance current, $$I_0 = \frac{V}{R} = \frac{1}{25} = 0.04 \text{ A}$$

The frequency at which V_L is maximum,

$$f_L = \frac{1}{2\pi}\left[\sqrt{LC - \frac{C^2R^2}{2}}\right]^{-1} \approx 7960$$

The frequency at which (V_c) is maximum is

$$f_c = \left(\frac{1}{2\pi}\right)\sqrt{\frac{1}{LC} - \frac{R^2}{2L}} = 7960$$

∴ Voltage across the inductance

$$= I_0 \times \omega_0 L = 0.04 \times 50 \times 10^3 \times 0.04 = 80 \text{ V}$$

∴ Voltage across the capacitance

$$V_c = \frac{I_0}{\omega_0 C} = 80 \text{ V}$$

4.11 The circuit shown in the figure is in steady-state with switch S open. The switch is closed at time $t = 0$. Determine the value of R so that the current $i_s(0^+)$ through the switch is 1.2 A. Determine $i_s(\infty)$.

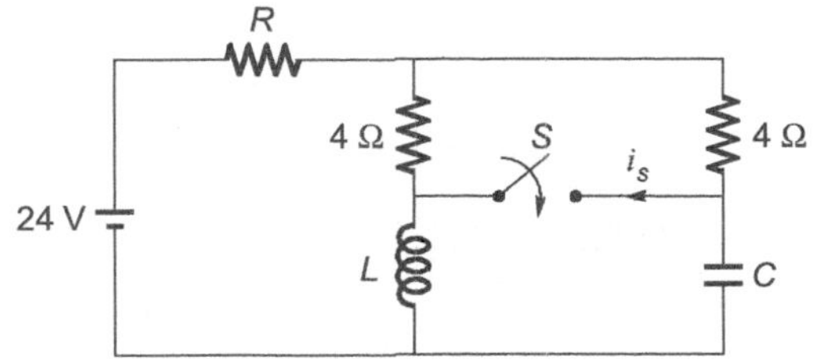

[12 marks : 2009]

Solution:

For $t < 0$, at steady state

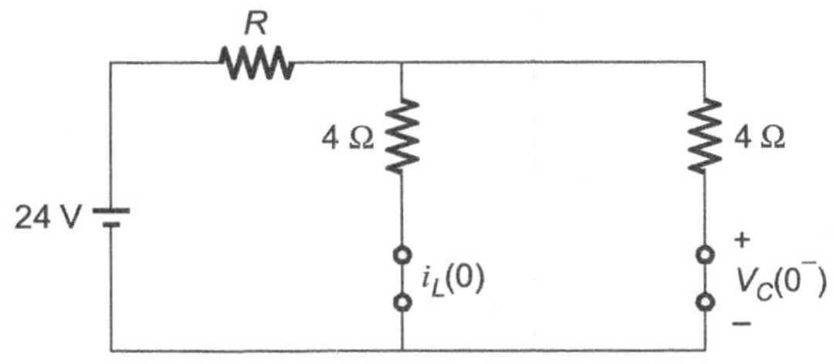

$$i_L(0^-) = \frac{24}{R+4}$$

$$V_C(0^-) = 4i_L(0^-) = \frac{24\times 4}{R+4}$$

For $t > 0$, circuit at $t = 0^+$

Since the current is the inductor and voltage in the capacitor cannot change instantaneously

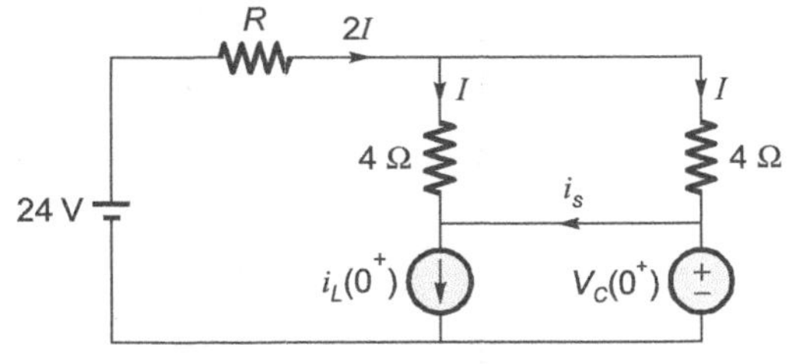

$$i_L(0^-) = i_L(0^+) = \frac{24}{R+4}$$

$$V_C(0^-) = V_C(0^+) = \frac{24\times 4}{R+4}$$

$$\text{Current,} \quad I = \frac{1}{2}\frac{24 - V_C(0^+)}{R + (4||4)} = \frac{24 - \frac{24\times 4}{R+4}}{2(R+2)}$$

$$i_L(0^+) = I + i_s$$

$$\frac{24}{R+4} = \frac{24 - \frac{24\times 4}{R+4}}{2(R+2)} + 1.2$$

$$\frac{24\times 2\times(R+2)}{R+4} = 24 - \frac{24\times 4}{R+4} + 2.4(R+2)$$

$$\frac{2(R+2)}{R+4} = 1 - \frac{4}{R+4} + 0.1(R+2)$$

$$\frac{2(R+2)+4}{R+4} = 1 + 0.1R + 0.2$$

$$0.1\,R = 0.8$$

$$R = 8\ \Omega$$

At $t = \infty$, the behaviour of the circuit is shown in figure.

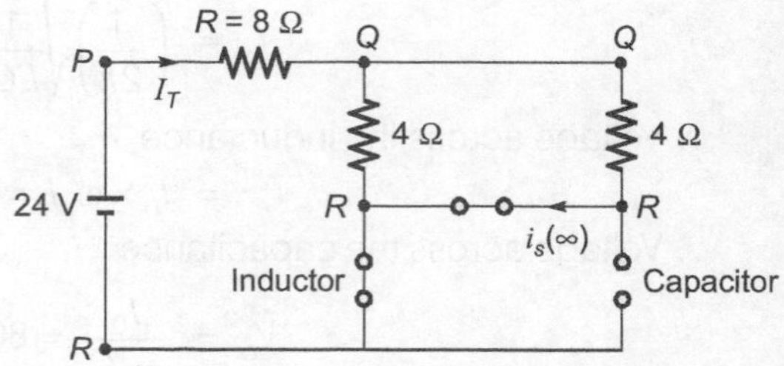

$$R_T = 8 + \frac{(4\times 4)}{4+4} = 10\ \Omega$$

$$I_T = \frac{24}{10} = 2.4\ \text{A}$$

$$i_s(\infty) = 2.4\times\frac{4}{4+4} = 1.2\ \text{A}$$

4.12 A series R-L-C circuit having $R = 25\ \Omega$, $L = 2$ H and $C = 30\ \mu$F is connected across an a.c. variable frequency source. At what frequencies will the phase angle of the circuit be (i) 45° lagging and (ii) 45° leading, the applied voltage?

[8 + 4 = 12 marks : 2010]

Solution:

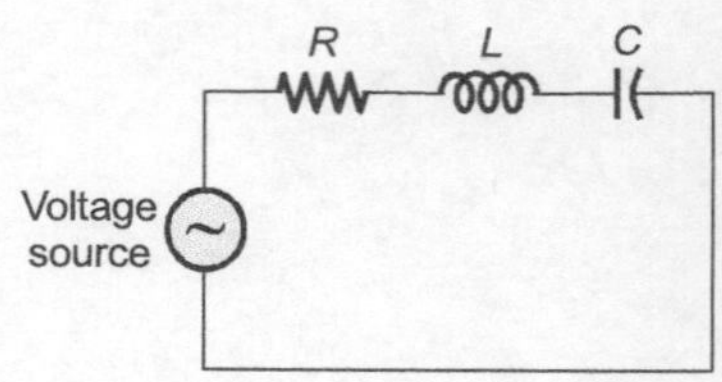

Where, $R = 25\ \Omega,\ L = 2$ H

and $C = 30\ \mu$F

$$\text{Impedance of the circuit,} \quad Z(j\omega) = R + j\omega L + \frac{1}{j\omega C} = R + j\left(\omega L - \frac{1}{\omega C}\right)\Omega$$

$$\theta = \tan^{-1}\left[\frac{\left(\omega L - \frac{1}{\omega C}\right)}{R}\right]$$

$$\Rightarrow \quad \tan\theta = \frac{\left(\omega L - \frac{1}{\omega C}\right)}{R}$$

(i) For phase angle leading or $\theta = -45°$ [current leads voltage by 45°]

$$\tan(-45°) = \frac{\left(2\omega - \frac{1}{\omega\times 30\times 10^{-6}}\right)}{25}$$

$$\Rightarrow \quad -25 = 2\omega - \frac{10^6}{30\,\omega}$$

$$\Rightarrow \quad -25 \times 30\,\omega = 2 \times 30\,\omega^2 - 10^6$$

$$\Rightarrow \quad 60\,\omega^2 + 750\,\omega - 10^6 = 0 \quad \ldots(i)$$

Roots of the equation (*i*),

$$\omega_1 = 123 \text{ rad/s and } \omega_2 = -135.5 \text{ rad/sec.}$$

$$\omega_2 = -135.5 \text{ rad/s is rejected as } \omega \text{ can be negative}$$

$$f = \frac{123}{2\pi} = 19.57 \text{ Hz}$$

For frequency of 19.57 Hz, phase angle of the circuit is 45° leading.

(ii) For phase angle 45° lagging, $\theta = 45°$ [current lages voltage by 45°]

$$\tan(45°) = \frac{\left(\omega L - \frac{1}{\omega C}\right)}{R} = \frac{\left(2\omega - \frac{10^6}{30\,\omega}\right)}{25}$$

$$\Rightarrow \quad 25 = 2\omega - \frac{10^6}{30\,\omega}$$

$$\Rightarrow \quad 60\,\omega^2 - 750\,\omega - 10^6 = 0 \quad \ldots(ii)$$

Roots of the equation (*ii*),

$\omega_1 = -123$ rad/sec and $\omega_2 = 135.5$ rad/sec.

$\omega_1 = -123$ rad/sec is rejected as ω can not be negative

$$f = \frac{135.5}{2\pi} = 21.56 \text{Hz}$$

For frequency of 21.56 Hz, phase angle of the circuit is 45° lagging.

4.13 **A fixed capacitance (X_C = 20 ohms) is placed in parallel with a series combination of resistance (R = 8 ohms) and variable inductance (X_L ohms), having negligible resistance. An alternating voltage of 120 volts is applied across the parallel combination. Show that the value of X_L which will produce unity power factor resonance is given by**

$$X_L = \frac{X_C}{2} \pm \sqrt{\frac{X_C^2}{4} - R^2}$$

Determine the minimum value of the current drawn from the supply.

[8 + 4 = 12 marks : 2011]

Solution:

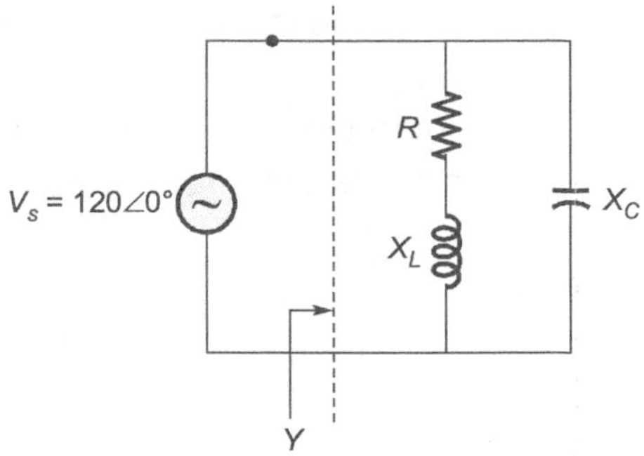

Y = equivalent admittance of the parallel circuit

$$= \frac{1}{R + jX_L} + \frac{1}{(-jX_C)} = \frac{(R - jX_L)}{(R + jX_L)(R - jX_L)} + \frac{j}{X_C}$$

$$= \frac{R - jX_L}{R^2 + X_L^2} + \frac{j}{X_C} = \frac{R}{R^2 + X_L^2} + j\left(\frac{1}{X_L} - \frac{X_L}{R^2 + X_L^2}\right)$$

For unity power factor resonance, imaginary term must be zero,

$$\frac{1}{X_C} - \frac{X_L}{R^2 + X_L^2} = 0$$

$$\Rightarrow \quad X_L^2 - X_C X_L + R^2 = 0 \quad \ldots(i)$$

Roots of the equation (*i*),

$$X_L = \frac{X_C}{2} \pm \sqrt{\frac{X_C^2}{4} - R^2} \quad ...(ii)$$

Putting values of R and X_C in equation (*ii*),

$$X_L = \frac{20}{2} \pm \sqrt{\frac{20^2}{4} - 8^2}$$

$$\Rightarrow \quad X_{L_1} = 16\ \Omega \text{ and } X_{L_2} = 4\ \Omega$$

$$Y = \frac{R}{R^2 + X_L^2} \quad \text{(Imaginary part is zero)}$$

Current drawn from the supply,

$$I_s = V_s Y = 120 \frac{R}{(R^2 + X_L^2)}$$

For $X_L = X_{L_1} = 16\Omega$

$$I_{S_1} = 120 \times \left(\frac{8}{8^2 + 16^2}\right) = 3\ \text{A}$$

For $X_L = X_{L_2} = 4\ \Omega$

$$I_{S_1} = 120 \times \left(\frac{8}{8^2 + 4^2}\right) = 12\ \text{A}$$

Minimum value of the current drawn from the supply $I_{S_1} = 3\ \text{A}$.

4.14 A circuit has the configuration depicted in the given figure.

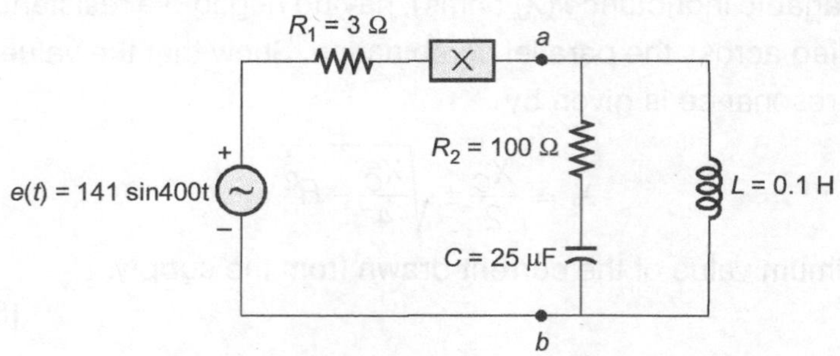

(i) Find the equivalent impedance appearing to the right of points *ab*.
(ii) Determine the value of the reactance *X* which makes the source current in phase with the source voltage.
(iii) Should the reactance *X* of part (ii) be inductive or capacitive? Find the required value of *L* or *C*.
(iv) Compute the effective value of the source current for the condition described in part (ii).

[8 + 4 = 12 marks : 2011]

Solution:

$$e(t) = 141 \sin 400t$$

$$\omega = 400 \text{ rad/sec.}$$

Let,

$$Z_a = R_2 + \frac{1}{j\omega C} = 100 - j\left(\frac{1}{400 \times 25 \times 10^{-6}}\right)$$

$$= 100 - j100\ \Omega$$

and

$$Z_b = j\omega L = j400 \times 0.1 = j40\ \Omega$$

(i) Equivalent impedance $= Z_{eq} = Z_a \parallel Z_b = \dfrac{(100-j100)(j40)}{[(100-j100)+(j40)]}$

$= 11.76 + j47.06\ \Omega$

(ii) Total impedance $= Z_T = Z_{eq} + R_1 + jX = 11.76 + j47.06 + 3 + jX$

$= 14.76 + j(47.06 + X)$

Imaginary part of Z_T should be zero to make the source current in phase with the voltage source

$$47.06 + X = 0$$

$$\Rightarrow \quad X = -47.06\ \Omega$$

(iii) Negative value of X indicates, reactance is capacitive

$$jX = \frac{1}{j\omega C}$$

$$\Rightarrow \quad -j47.06 = \frac{-j}{400 \times C} \Rightarrow C = 53.12\ \mu\text{F}$$

(iv) For the condition described in part (*ii*),

Total impedance becomes,

$$Z_T = 14.76\ \Omega$$

$$\text{Source current } i(t) = \frac{V(t)}{Z_T} = \frac{141 \sin 400t}{14.76} = 9.55 \sin 400t$$

$$\text{Maximum value of } i(t) = 9.55\ \text{A}$$

$$\text{Rms value of } i(t) = \frac{9.55}{\sqrt{2}} = 6.75\ \text{A}$$

4.15 **A voltage of $e(t) = 150 \sin 1000t$ is applied across a series RLC circuit where $R = 40\ \Omega$, $L = 0.13$ H and $C = 10\ \mu$F.**

(i) Compute the rms value of the S.S. current.

(ii) Find the rms voltage across the inductor.

(iii) Find the rms voltage across the capacitor.

(iv) Draw the complete phasor diagram showing all voltage components.

(v) Determine the reactive power supplied by the source.

Solution:

(i)

$$e(t) = 150 \sin(1000t)$$

$$jX_L = j\omega L = j1000 \times 0.13 = 130j\ \Omega$$

$$-jX_C = \frac{-j}{\omega C} = \frac{-j}{1000 \times 10 \times 10^{-6}} = -100j\ \Omega$$

$$Z = R + j(X_L - X_C) = 40 + j30\ \Omega$$

$$= 50\angle 36.87^\circ\ \Omega$$

$$I = \frac{e(t)}{Z} = 3 \sin(1000t - 36.87^\circ)$$

$\therefore$ rms value of steady state current $= \dfrac{3}{\sqrt{2}}$ A

(ii) rms voltage across inductor $= (130)\left(\dfrac{3}{\sqrt{2}}\right) = 275.77$ V

(iii) rms voltage across capacitor = $(100)\left(\frac{3}{\sqrt{2}}\right) = 212.132$ V

(iv)

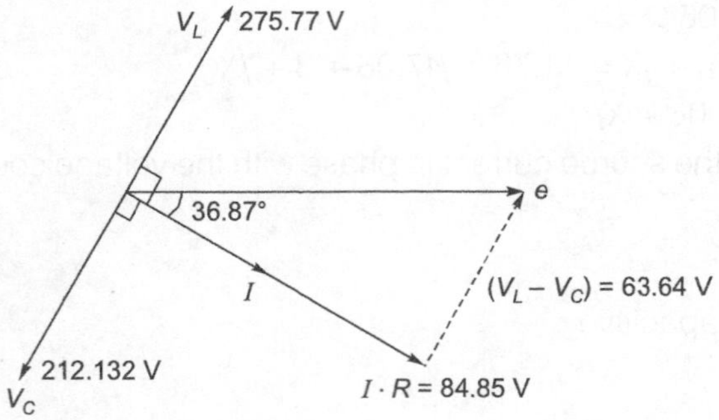

(v) Reactive power supplied = $e_{rms} \times I_{rms} \sin\theta = \left(\frac{150}{\sqrt{2}}\right)\left(\frac{3}{\sqrt{2}}\right)(0.6) = 135$ VAR

4.16 **Define bandwidth in a series RLC circuit. Show that bandwidth increases with the increase in the value of resistance. The voltage applied to a series RLC circuit is 8.5 V. The quality factor of the coil is 50 and the value of the capacitor is 320 pF. The resonance frequency of the circuit is 175 kHz. Find the value of inductance, current flowing in the circuit during resonance and voltage across the capacitor under resonance. Draw phasor diagram.**

[12 marks : 2012]

Solution:

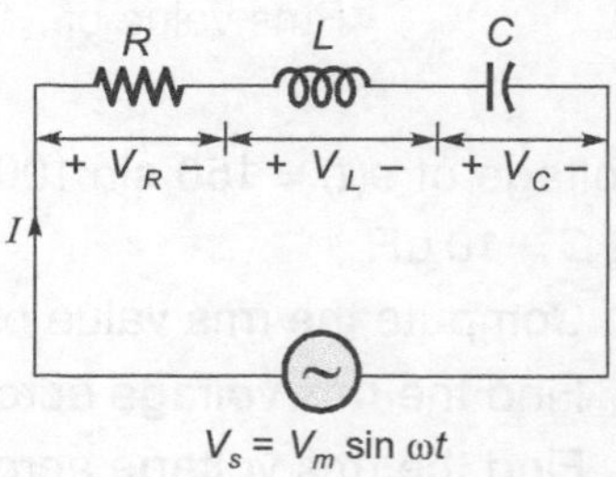

Current, $$I = \frac{V_m}{\sqrt{R^2 + \left(\omega L - \frac{1}{\omega C}\right)^2}}$$

At half power $$I = \frac{I_m}{\sqrt{2}} = 0.707 I_m$$

ω_1 and ω_2 are called cut-off frequencies or half power frequencies.
The distance between these two points is called bandwidth (BW)

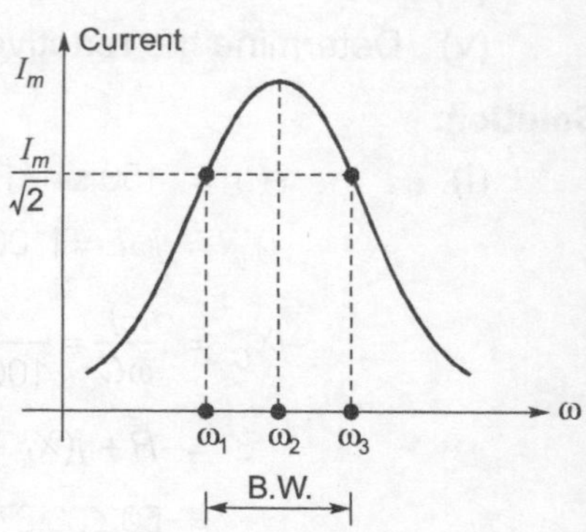

Quality factor, $$Q = \frac{|V_L|}{|V|} = \frac{|V_C|}{|V|} = \frac{\omega_r L}{R} = \frac{1}{\omega_r RC}$$

where, $\omega_r = \frac{1}{\sqrt{LC}}$

$$Q = \frac{\omega_r}{\text{B.W.}} \Rightarrow \text{B.W.} = \frac{\omega_r}{Q} = \frac{\omega_r}{\left(\frac{\omega_r L}{R}\right)}$$

$$\text{BW} = \frac{R}{L}$$

Hence, the value of bandwidth (BW) increases with increase in the value of resistance.

Given, $Q = 50, \quad C = 320 \text{ pF}, \quad f_r = 175 \text{ kHz}$

$$f_r = \frac{1}{2\pi\sqrt{LC}}$$

$$\Rightarrow \quad L = \frac{1}{f_r^2\, 4\pi^2 C} = \frac{1}{(175 \times 10^3)^2 \times 4\pi^2 \times 320 \times 10^{-12}} = 2.584 \text{ mH}$$

Current I at resonance,

$$I = \frac{V}{R} = \frac{V}{\left(\frac{1}{Q\omega_r C}\right)} \qquad \text{as} \quad Q = \frac{1}{\omega_r RC}$$

$$= \frac{8.5}{\frac{1}{50\times 2\pi\times 175\times 10^3\times 320\times 10^{-12}}}$$

$$I = 149.53 \text{ mA}$$

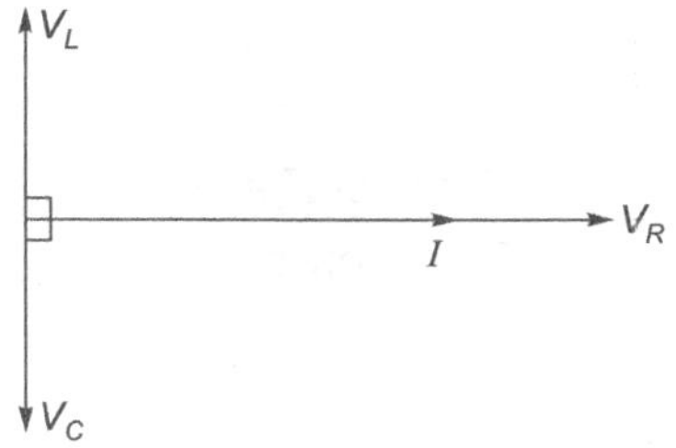

Voltage across capacitor,

$$|V_c| = Q\cdot|V| = 50\times 8.5 = 425 \text{ V}$$

4.17 **A series combination of resistance R and inductance L is connected across capacitance C. Determine values of R, L and C using following data:**

(i) poles of network function $z(s)$ are at $-0.5\pm j\frac{\sqrt{3}}{2}$.

(ii) zeros of $z(s)$ are at $s = -1$

(iii) $z(j0) = 1$

[12 marks : 2014]

Solution:

$$Z(s) = \frac{(R+sL)\left(\frac{1}{sC}\right)}{(R+sL)+\left(\frac{1}{sC}\right)} = \frac{R+sL}{RsC+s^2LC+1}$$

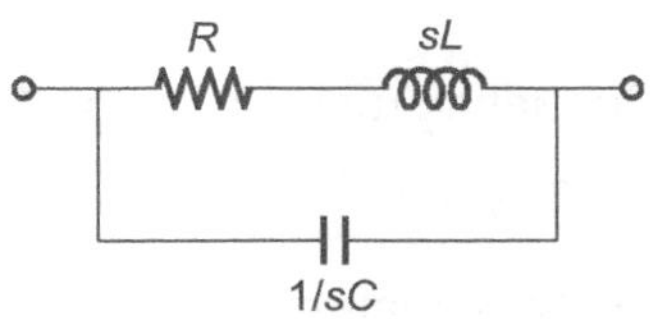

(i) Poles of $Z(s)$, $\quad s = -0.5\pm j\frac{\sqrt{3}}{2}$ ...(1)

C.E.

$$\Rightarrow \quad s = \frac{-RC\pm\sqrt{(RC)^2-4(LC)}}{2\cdot(LC)} = \frac{-RC\pm\sqrt{R^2C^2-4LC}}{2LC}$$

$$= -\frac{R}{2L}\pm\sqrt{\frac{R^2C^2}{4L^2C^2}-\frac{4LC}{4L^2C^2}}$$

$$s = -\frac{R}{2L}\pm\sqrt{\frac{R^2}{4L^2}-\frac{1}{LC}} \qquad ...(2)$$

$$\frac{R}{2L} = 0.5 \qquad ...(3)$$

$$\sqrt{\frac{1}{LC}-\frac{R^2}{4L^2}} = \frac{\sqrt{3}}{2} \qquad ...(4)$$

From equation (3) and (4),

$$\Rightarrow \quad \sqrt{\frac{1}{LC}-(0.5)^2} = \frac{\sqrt{3}}{2} \Rightarrow \frac{1}{LC}-(0.5)^2 = \frac{3}{4}$$

$$\Rightarrow \quad \frac{1}{LC} = \frac{3}{4} + (0.5)^2$$

$$\Rightarrow \quad \frac{1}{LC} = 1$$

$$\Rightarrow \quad LC = 1 \quad ...(5)$$

(ii) Zeros of $Z(s)$ are at $s = -1$ ...(6)

From network function,

$$\Rightarrow \quad R + sL = 0$$

$$\Rightarrow \quad s = -\frac{R}{L} = -1 \quad \text{[From equation (6)]}$$

$$\Rightarrow \quad R = L \quad ...(7)$$

From equation (5) and (7), $RC = 1$ ...(8)

(iii) $Z(j_0) = 1$

From network function, $Z(j_0) = \dfrac{R + j_0 L}{RC(j_0) + (j_0)^2 LC + 1} = 10$

putting, $LC = 1$ and $RC = 1$ from equation (5) and (8) respectively,

$$\Rightarrow \quad \frac{R + j_0 L}{j_0 - 1 + 1} = 1 \Rightarrow \frac{R + j_0 L}{j_0} = 1$$

$$\Rightarrow \quad R + j_0 L = j_0$$

equating real and imaginary part,

$R = 0$, $L = 0$, $C = 0$ the network is not synthesizable.

From equation (5) and (8), $LC = 1 \Rightarrow C = \dfrac{1}{L} = \dfrac{1}{1} = 1$

$$RC = 1 \Rightarrow C = \frac{1}{R} = \frac{1}{1} = 1$$

So, $R = 1\ \Omega$, $L = 1$ H, $C = 1$ F

4.18 **For the network shown below, obtain the source frequency w so that applied voltage V and current I shall be in phase.**

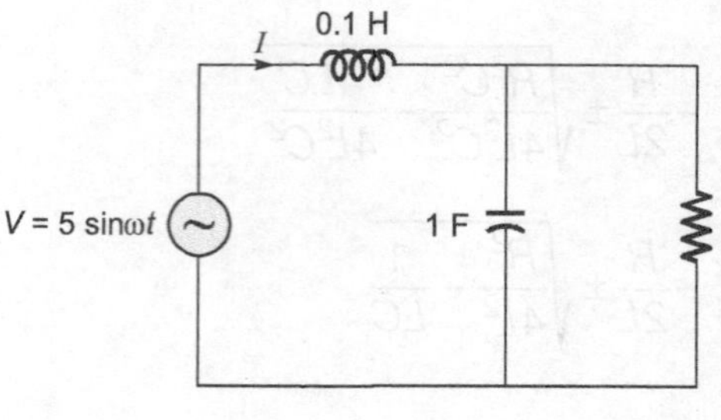

[8 marks : 2014]

Solution:

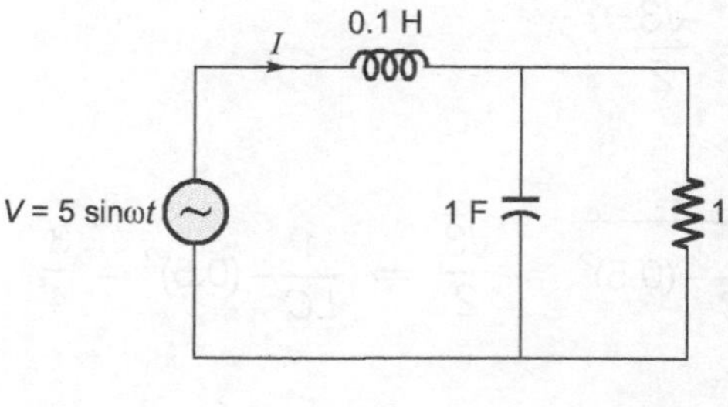

For, V and I in phase, imaginary part of $Z(s)$, where $s = j\omega$ should be zero.

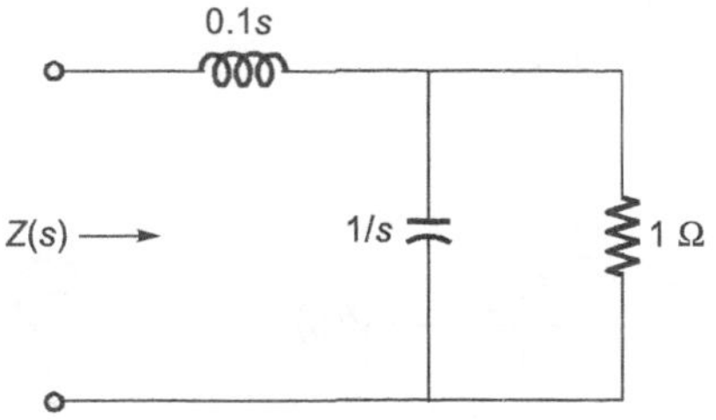

$$Z(s) = \frac{\frac{1}{s}\times 1}{\frac{1}{s}+1} + 0.1s = \frac{1}{1+s} + 0.1s = \frac{1+0.1s+0.1s^2}{s+1}$$

multiplying nominator and denominator by $(s-1)$,

$$Z(s) = \frac{(0.1s^2+0.1s+1)(s-1)}{(s+1)(s-1)}$$

$$= \frac{0.1s^3+0.1s^2+s-0.1s^2-0.1s-1}{s^2+s-s-1} = \frac{0.1s^3+0.9s-1}{s^2-1}$$

Put, $s = j\omega$

$$Z(j\omega) = \frac{0.1(j\omega)^3+0.9(j\omega)-1}{(j\omega)^2-1} = \frac{-0.1j\omega^3+0.9j\omega-1}{-\omega^2-1} = \frac{j(0.9\omega-0.1\omega^3)-1}{-\omega^2-1}$$

Equating imaginary part to zero

$$\Rightarrow \quad 0.9\omega - 0.1\omega^3 = 0$$

$$\Rightarrow \quad \omega(0.9-0.1\omega^2) = 0$$

$$\Rightarrow \quad \omega = 0$$

or $\quad \omega^2 = 9 \Rightarrow \omega = \pm 3$ rad/sec.

Taking positive non-zero value,

$$\omega = 3 \text{ rad/sec.}$$

4.19 **Show that the voltage V across R shown in figure is independent of R at radian frequency ω of voltage source when $\omega^2 L_1 C_1 = \omega^2 L_2 C_2 = 1$. Find also the expression of this voltage.**

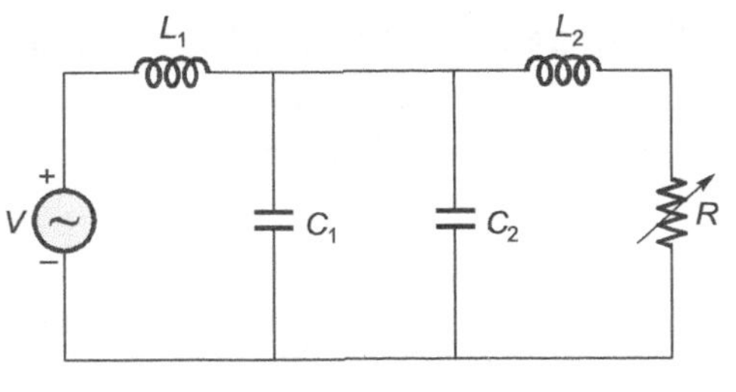

[10 marks : 2015]

Solution:

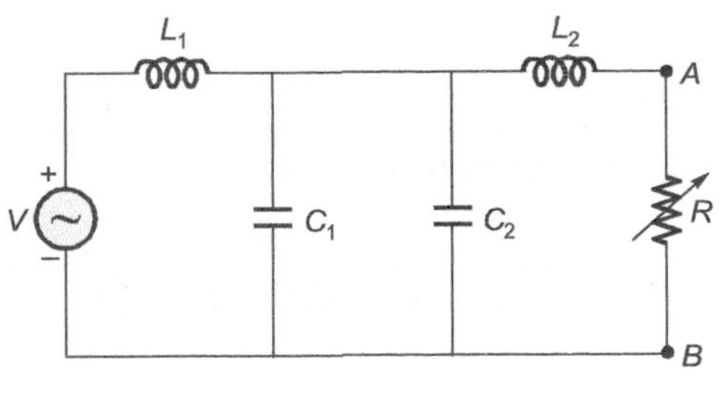

At radian frequency 'ω',

$$\omega^2 L_1 C_1 = \omega^2 L_2 C_2 = 1 \quad \text{...(1)}$$

From circuit diagram, using Thevenin's theorem,

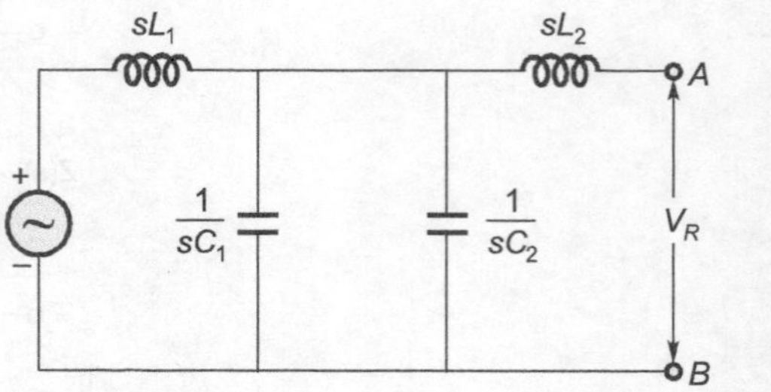

Here, $V_{AB} = V_{Th}$

$$V_{Th} = \frac{\frac{1}{s(C_1+C_2)}}{sL_1 + \frac{1}{s(C_1+C_2)}} \times V(s)$$

$$V_{Th} = \frac{V(s)}{1+s^2L_1(C_1+C_2)} = \frac{V(s)}{1+s^2L_1C_1+s^2L_1C_2} \quad \text{...(2)}$$

Similarly,

$$Z_{Th} = \frac{sL_1 \times \frac{1}{s(C_1+C_2)}}{sL_1 + \frac{1}{s(C_1+C_2)}} + sL_2$$

$$Z_{Th} = \frac{sL_1}{1+s^2L_1(C_1+C_2)} + sL_2 \quad \text{...(3)}$$

Using equation (1) in equation (2) and (3).

$$V_{Th} = \frac{V(s)}{s^2L_1C_2} \text{ and } Z_{Th} = \frac{1}{sC_2} + sL_2$$

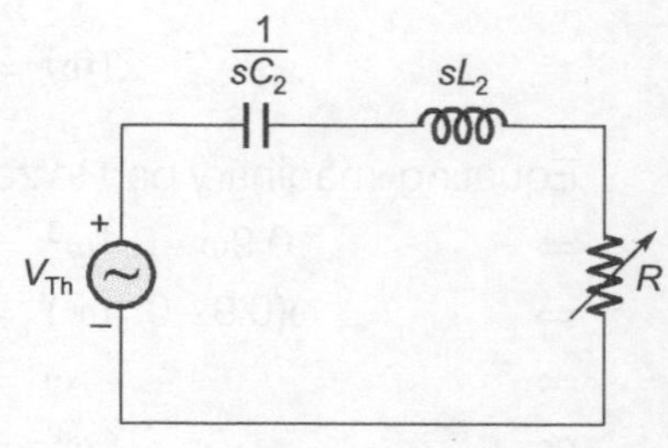

Since L_2 and C_2 are in resonance then all the supply voltage V_{Th} is applied across R and voltage V_{Th} is independent of 'R'.

$$V_R = V_{Th} = \frac{V(s)}{s^2L_1C_2} = \frac{V}{-\omega^2L_1C_2}$$

$$\boxed{V_R = \frac{V}{-\omega^2L_1C_2}}$$

4.20 A pure inductance of 150 mH is connected in parallel with a 40 μF capacitor across a 50 V variable frequency supply as shown in the circuit below. Determine:
(i) the resonant frequency of the circuit and
(ii) the current circulating in the capacitance and inductance at resonance.

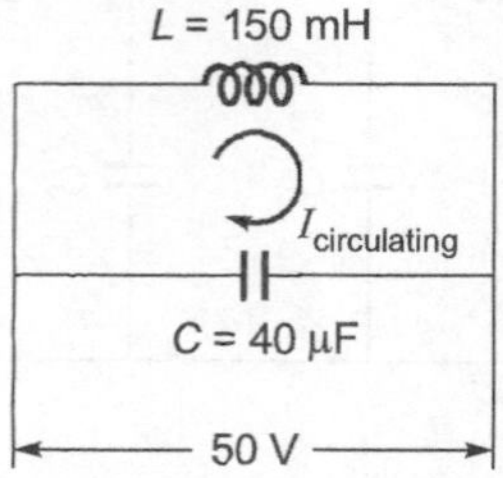

[5 marks : 2016]

Solution:

$$\omega_0 = \frac{1}{\sqrt{LC}} = \frac{1}{\sqrt{150\times10^{-3}\times40\times10^{-6}}} = 408.248 \text{ rad/sec}$$

$$\vec{I}_L = \frac{V}{jX_L} = \frac{V}{j\omega_0 L} = \frac{50}{j408.25\times150\times10^{-3}} = -j\,0.816$$

$$\vec{I}_C = \frac{V}{-jX_C} = jV.\omega.C$$

$$= j\,50 \times 408.25 \times 40 \times 10^{-6} = 0.816 \text{ A}$$

Circulating current is 0.816 A.

4.21 **A coil of inductance 100 mH and resistance 800 Ω is connected in parallel with a variable capacitor across a 12 V, 5 kHz supply. Determine for the condition when the supply current is minimum.**
(i) the capacitance of the capacitor,
(ii) the dynamic resistance,
(iii) the supply current, and
(iv) the *Q*-factor

[10 marks : 2016]

Solution:

At resonance, the supply current is minimum and impedance is maximum,

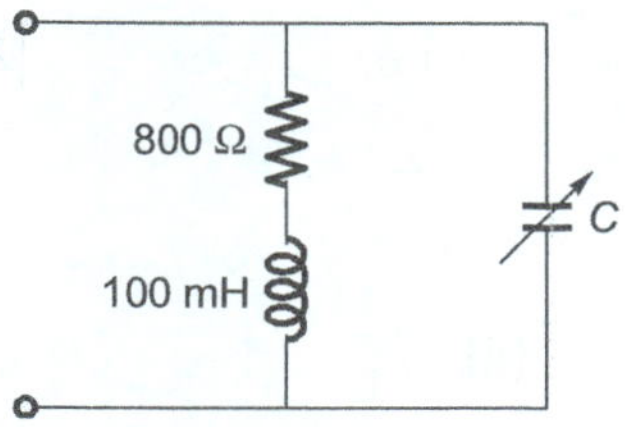

$$Y = \frac{1}{R + jX_L} \times \frac{R - jX_L}{R - jX_L} + \frac{1}{-jX_C}$$

$$Y = \frac{R - jX_L}{R^2 + X_L^2} + \frac{j}{X_C}$$

For resonance imaginary term = 0

$$\frac{X_L}{R^2 + X_L^2} = \frac{1}{X_C}$$

$$\frac{\omega L}{R^2 + \omega^2 L^2} = \omega_0 C,$$

$$C = \frac{100 \times 10^{-3}}{(800)^2 + (5 \times 10^3 \times 2\pi \times 100 \times 10^{-3})^2}$$

(i) $C = 9.515 \text{ nF}$

(ii) If resonance, $z = \frac{L}{RC}$ (dynamic resistance)

$$z = \frac{100 \times 10^{-3}}{800 \times 9.515 \times 10^{-9}} = 13.137 \text{ k}\Omega$$

(iii) Supply current $(I) = \frac{V}{z} = 0.913 \text{ mA}$

(iv) $\text{Q-factor} = \frac{\omega L}{R} = \frac{2\pi \times 5 \times 10^3 \times 100 \times 10^{-3}}{800} = 3926$

4.22 **A circuit is shown below.**

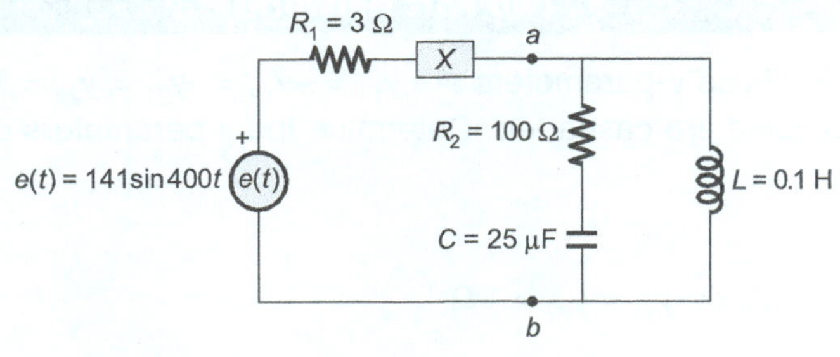

(i) Determine the value of the reactance X which makes the source current in phase with the source voltage.

(ii) Compute also the effective value of the current for the condition described in part (i).

[20 marks : 2017]

Solution:

(i) To make the source current in phase with source voltage, inductive reactance should be equal to capacitive reactance.

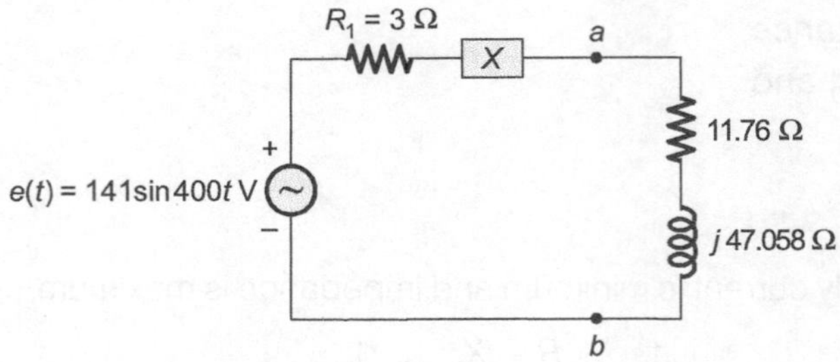

i.e. $$|X_C| = |X_L| = 47.058\ \Omega$$

∴ $$X_L - X_C = 0$$

$$X_C = 47.058\ \Omega$$

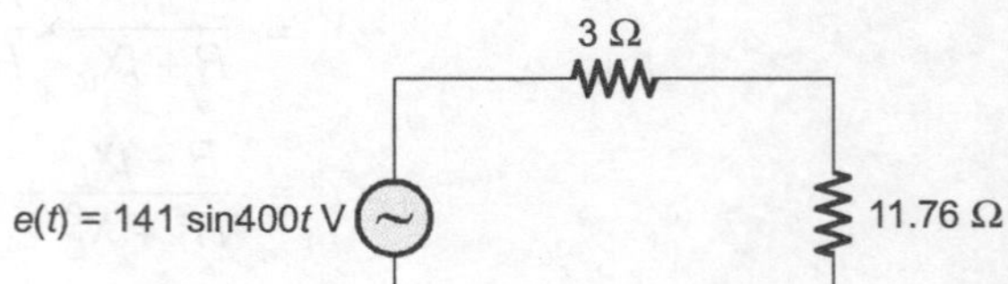

(ii) $$I_{rms} = \frac{141}{\sqrt{2}(3+11.76)} = 6.75\ \text{A}$$

4.23 For a series R-L-C circuit excited from an AC source, find the resonant frequency, bandwidth and quality factor, if $R = 100\ \Omega$, $L = 0.5$ H and $C = 0.4\ \mu$F.

[12 marks : 2023]

Solution:

For series RLC circuit, excited from AC source.

Given that :

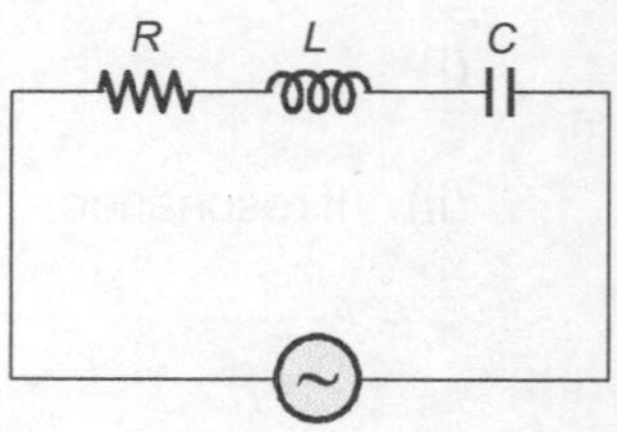

$$R = 100\ \Omega$$
$$L = 0.5\ \text{H}$$
$$C = 0.4\ \mu\text{F}$$

1. Resonant frequency, $$\omega_0 = \frac{1}{\sqrt{LC}} = 2.23 \times 10^3\ \text{rad/s}$$

2. Bandwidth, $$\text{B.W.} = \omega_2 - \omega_1 = \frac{R}{L} = 200\ \text{rad/s}$$

3. Quality factor, $$Q = \frac{\omega_0}{\text{B.W.}} = \frac{\omega_0}{\omega_2 - \omega_1} = 11.15$$

5. Two Port Networks

5.1 Draw a 2-port network whose y-parameters are $y_{11} = -y_{12} = -y_{21} = y_{22} = 1\ \Omega^{-1}$. If two such 2-port networks are cascaded. Determine the y-parameters of the overall network.

[10 marks : 2002]

Solution:

For 2-port network $y_{11} = -y_{12} = -y_{21} = y_{22} = 1\ \Omega^{-1}$

$$Y = \begin{bmatrix} y_{11} & y_{12} \\ y_{21} & y_{22} \end{bmatrix} = \begin{bmatrix} 1 & -1 \\ -1 & 1 \end{bmatrix}$$

Considering a π-network,

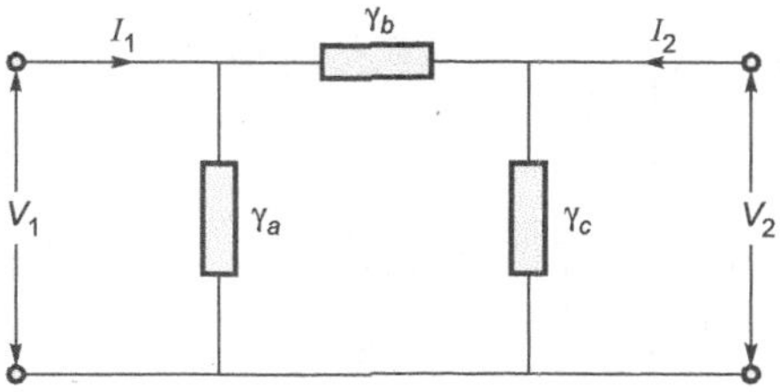

For a π-network, y-parameters,

$$y_{11} = Y_a + Y_b \qquad y_{12} = -Y_b$$
$$y_{21} = -Y_b \qquad y_{22} = Y_b + Y_c$$

As per problem,

$$y_{12} = -y_b = -1 \quad \Rightarrow \quad Y_b = 1$$
$$y_{11} = y_a + y_b = Y_A + 1 = 1 \quad \Rightarrow \quad Y_a = 0$$
$$y_{21} = y_b + Y_c = 1 + Y_c = 1 \quad \Rightarrow \quad Y_c = 0$$

Hence, the network,

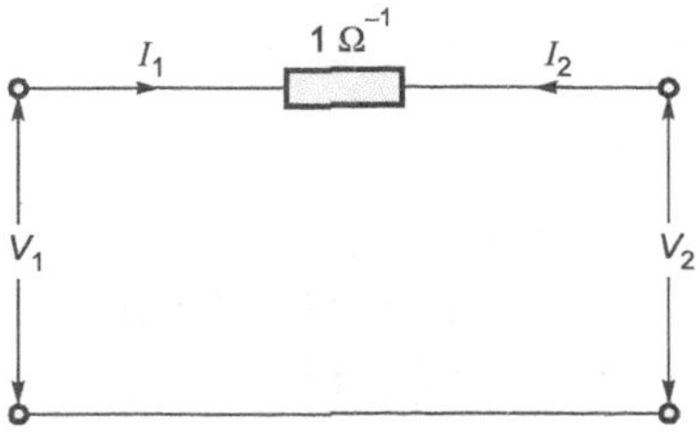

$$I_1 = y_{11}V_1 + y_{12}V_2 = V_1 - V_2$$
$$I_2 = y_{21}V_1 + y_{22}V_2 = -V_1 + V_2 \quad \Rightarrow \quad V_1 = V_2 - I_2$$
$$I_1 = -I_2$$
$$\begin{bmatrix} V_1 \\ I_1 \end{bmatrix} = \begin{bmatrix} 1 & 1 \\ 0 & 1 \end{bmatrix}\begin{bmatrix} V_2 \\ -I_2 \end{bmatrix} \Rightarrow [T] = \begin{bmatrix} 1 & 1 \\ 0 & 1 \end{bmatrix}$$

If two such system are cascaded,

$$T' = [T] \times [T] = \begin{bmatrix} 1 & 1 \\ 0 & 1 \end{bmatrix}\begin{bmatrix} 1 & 1 \\ 0 & 1 \end{bmatrix} = \begin{bmatrix} 1 & 2 \\ 0 & 1 \end{bmatrix}$$

$$\begin{bmatrix} V_1 \\ I_1 \end{bmatrix} = \begin{bmatrix} 1 & 2 \\ 0 & 1 \end{bmatrix}\begin{bmatrix} V_2 \\ -I_2 \end{bmatrix}$$
$$V_1 = V_2 - 2I_2$$
$$2I_2 = -V_1 + V_2$$
$$I_2 = -\frac{1}{2}V_1 + \frac{1}{2}V_2$$
$$I_1 = -I_2 = \frac{V_1}{2} - \frac{V_2}{2}$$
$$\begin{bmatrix} I_1 \\ I_2 \end{bmatrix} = \begin{bmatrix} 1/2 & -1/2 \\ -1/2 & 1/2 \end{bmatrix}$$

y-parameters,

$$y_{11} = -y_{12} = -y_{21} = y_{22} = \frac{1}{2}\Omega^{-1}$$

5.2 A 2-port network N_1 has $Z_{11} = Z_{12} = Z_{21} = Z_{22} = 1\ \Omega$. A similar 2-port network N_2 is connected in cascade to N_1. Determine the Z-parameters of the composite network.

[8 marks : 2003]

Solution:

$$Z\text{-parameters of Network } N_1 = \begin{bmatrix} 1 & 1 \\ 1 & 1 \end{bmatrix}$$

$$\begin{bmatrix} V_1 \\ V_2 \end{bmatrix} = \begin{bmatrix} 1 & 1 \\ 1 & 1 \end{bmatrix}\begin{bmatrix} I_1 \\ I_2 \end{bmatrix}$$

$$V_1 = I_1 + I_2$$

$$V_2 = I_1 + I_2$$

$$\therefore \quad V_1 = V_2$$

$$I_1 = V_2 - I_2$$

$$\begin{bmatrix} V_1 \\ I_1 \end{bmatrix} = \begin{bmatrix} 1 & 0 \\ 1 & 1 \end{bmatrix}\begin{bmatrix} V_2 \\ -I_2 \end{bmatrix}$$

$$T\text{-parameters of network } T_1 = \begin{bmatrix} 1 & 0 \\ 1 & 1 \end{bmatrix}$$

Similar two networks are in cascade so

Resultant, $\quad T = [T_1][T_2]$

Resultant, $\quad T = \begin{bmatrix} 1 & 0 \\ 1 & 1 \end{bmatrix}\begin{bmatrix} 1 & 0 \\ 1 & 1 \end{bmatrix} = \begin{bmatrix} 1 & 0 \\ 2 & 1 \end{bmatrix}$

$$\therefore \quad V_1 = V_2$$

$$\therefore \quad I_1 = 2V_2 - I_2$$

$$V_2 = \frac{I_1 + I_2}{2} \text{ and } V_1 = \frac{I_1 + I_2}{2}$$

$$\therefore \quad Z_{\text{Resultant}} = \begin{bmatrix} \frac{1}{2} & \frac{1}{2} \\ \frac{1}{2} & \frac{1}{2} \end{bmatrix}$$

5.3 Find the *ABCD* parameters for the circuit shown below.

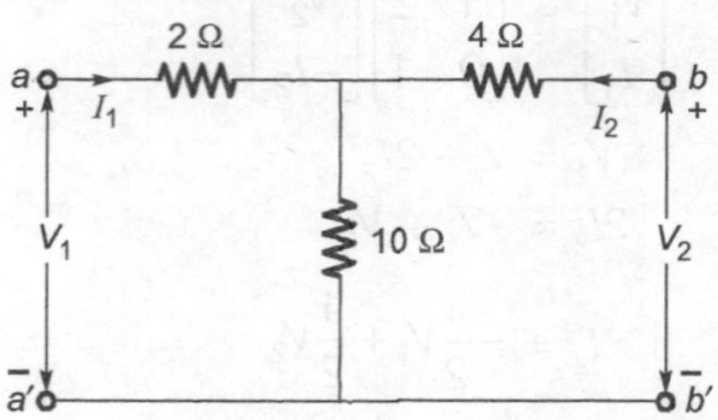

[10 marks : 2006]

Solution:

From this (T) network,

$$V_1 = 2I_1 + 10(I_1 + I_2) = 12I_1 + 10I_2$$

$$V_2 = 4I_2 + 10(I_1 + I_2) = 10I_1 + 14I_2$$

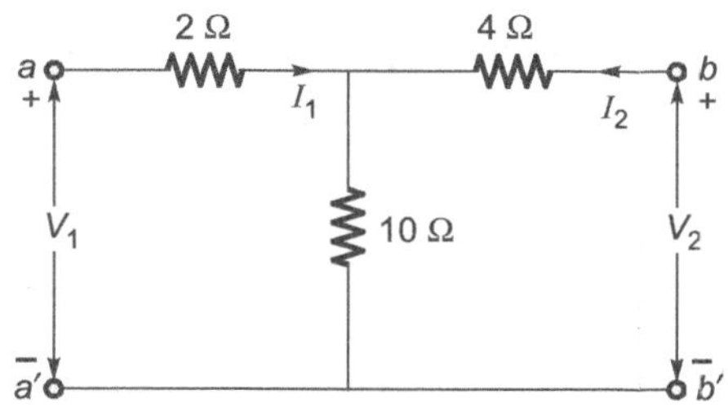

$$10 I_1 = V_2 - 14 I_2$$

$$\therefore \quad V_1 = 12\frac{(V_2 - 14 I_2)}{10} + 10 I_2$$

$$\therefore \quad V_1 = 1.2 V_2 - 1.4 \times 12 I_2 + 10 I_2$$

$$= 1.2 V_2 - 6.8 I_2 \qquad ...(i)$$

$$I_1 = 0.1 V_2 - 1.4 I_2 \qquad ...(ii)$$

$$\therefore \quad \begin{bmatrix} V_1 \\ I_1 \end{bmatrix} = \begin{bmatrix} 1.2 & 6.8 \\ 0.1 & 1.4 \end{bmatrix} \begin{bmatrix} V_2 \\ -I_2 \end{bmatrix}$$

The [*ABCD*] parameters of the network is

$$\therefore \quad T = \begin{bmatrix} 1.2 & 6.8 \\ 0.1 & 1.4 \end{bmatrix}$$

5.4 **Write the driving-point impedance of the ladder network in figure.**

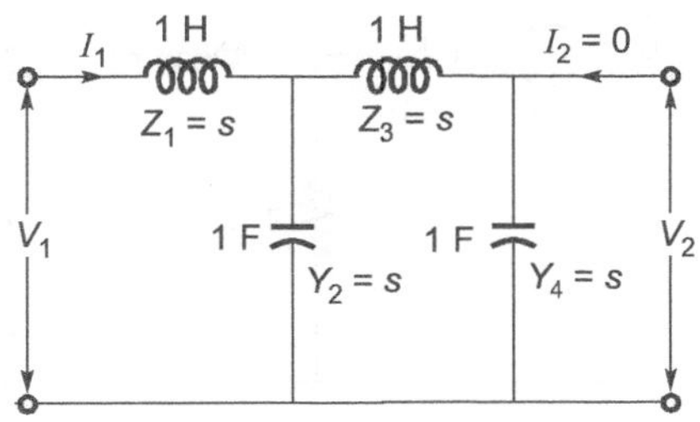

[12 marks : 2007]

Solution:

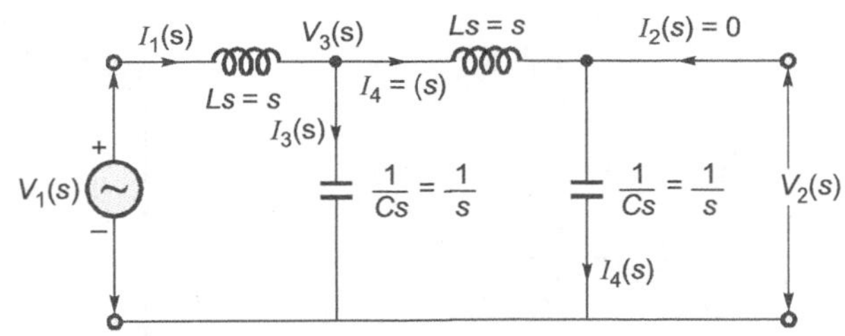

Let us transform the given circuit to *s*-domain.

Here, $$I_4(s) = \frac{V_2(s)}{1/Cs} = sV_2(s)$$

Also, $$V_3(s) = V_2(s) + I_4(s) \cdot Ls = V_2(s) + sV_2(s) \cdot Ls$$

$$= V_2(s)\,[s^2L + 1] = V_2(s)\,[1 + s^2]$$

$$I_1(s) = I_3(s) + I_4(s) = \frac{V_3(s)}{1/Cs} + sV_2(s)$$

$$= Cs\, V_2(s) \cdot (1 + s^2) + sV_2(s)$$

$$= V_2(s)\,[s^3 + 2s]$$

Now, $$V_1(s) = V_3(s) + I_1(s)\, Ls$$

$$= V_2(s)\,[1 + s^2] + V_2(s)\,[s^3 + 2s]s$$

$$= V_2(s)\,[1 + s^2 + s^4 + 2s^2] = V_2(s)\,[s^4 + 3s^2 + 1]$$

$$\therefore \quad \text{Driving point impedance} = \frac{V_1(s)}{I_1(s)} = \frac{V_2(s)\left[s^4 + 3s^2 + 1\right]}{V_2(s)\left[s^3 + 2s\right]}$$

The driving point impedance can be written as continued fraction

$$Z(s) = s+\cfrac{1}{s+\cfrac{1}{s+\cfrac{1}{s}}}$$

$$Z(s) = s+\cfrac{1}{s+\cfrac{1}{\frac{s^2+1}{s}}} = s+\cfrac{1}{s+\cfrac{s}{s^2+1}}$$

or, Driving point-impedance $= \dfrac{s^4+3s^2+1}{s^3+2s}$

5.5 A 2-port network has z-parameters Z_{11}, Z_{12}, Z_{21}, Z_{22}. It is to be represented by the equivalent circuit shown in the figure. Determine the values of Z_A, Z_B and β.

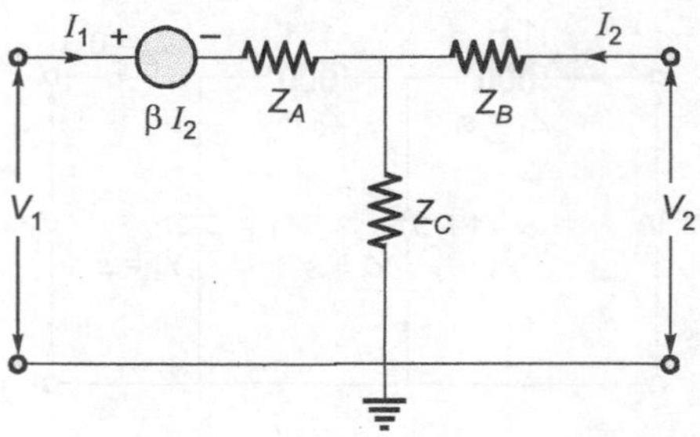

[12 marks : 2009]

Solution:

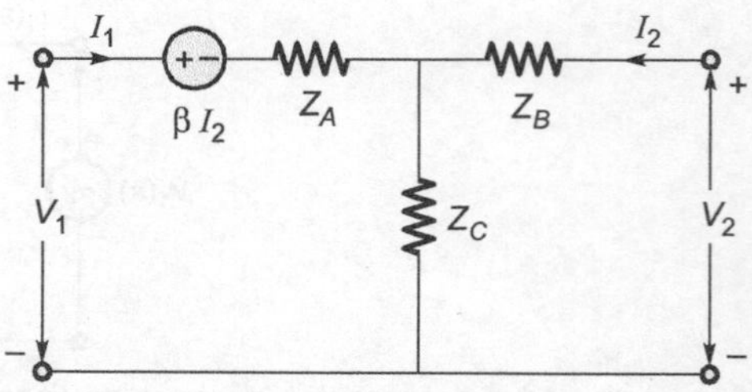

$$V_1 = Z_{11} I_1 + Z_{12} I_2$$

$$V_2 = Z_{21} I_1 + Z_{22} I_2$$

$$Z_{11} = \left.\frac{V_1}{I_1}\right|_{I_2=0},\quad Z_{22} = \left.\frac{V_2}{I_2}\right|_{I_1=0},$$

$$Z_{12} = \left.\frac{V_1}{I_2}\right|_{I_1=0} \text{ and } Z_{21} = \left.\frac{V_2}{I_1}\right|_{I_2=0}$$

∴ $V_1 = \beta I_2 + Z_A I_1 + Z_C (I_1 + I_2) = \beta I_2 + Z_A I_1 + Z_C (I_1 + I_2)$

$= I_2 (\beta + Z_C) + I_1 (Z_A + Z_C)$ ...(i)

∴ $V_2 = Z_B I_2 + (I_2 + I_1) Z_C = Z_B I_2 + Z_C I_2 + Z_C I_1$

$= (Z_B + Z_C) I_2 + Z_C I_1$ ...(ii)

$$Z_{11} = \left.\frac{V_1}{I_1}\right|_{I_2=0} = Z_A + Z_C$$

$$Z_{12} = \left.\frac{V_1}{I_2}\right|_{I_1=0} = \beta + Z_C$$

$$Z_{21} = \left.\frac{V_2}{I_1}\right|_{I_2=0} = Z_C$$

$$Z_{22} = \left.\frac{V_2}{I_2}\right|_{I_1=0} = Z_B + Z_C$$

$$Z_{11} = Z_A + Z_C \quad \text{...(A)}$$

$$Z_{12} = \beta + Z_C \quad \text{...(B)}$$

$$Z_{21} = Z_C \quad \text{...(C)}$$

$$Z_{22} = Z_B + Z_C \quad \text{...(D)}$$

From equation (A), $Z_{11} = Z_A + Z_C = Z_A + Z_{21}$

$\therefore$ $\boxed{Z_A = Z_{11} - Z_{21}}$

From equation (D), $Z_{22} = Z_B + Z_C = Z_B + Z_{21}$

$\therefore$ $\boxed{Z_B = Z_{22} - Z_{21}}$

From equation (B),

$\therefore$ $\boxed{\beta = Z_{12} - Z_{21}}$

5.6 When can a two-port circuit be declared as a reciprocal circuit? Is the two-port circuit shown in figure reciprocal? Justify the answer by a 15 V voltage source?

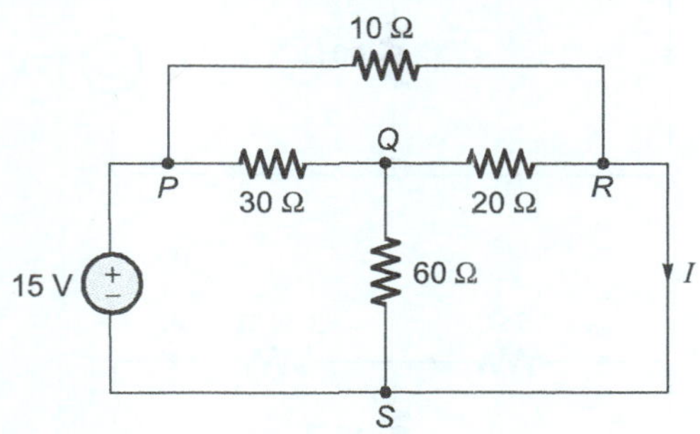

[8 + 4 = 12 marks : 2010]

Solution:

A two port network is termed to be reciprocal if the ratio of the response variable to the excitation variable remains identical even if the positions of the response and excitation in the network are interchanged.

Conditions for reciprocal network in various parameters.

Parameter	Condition for reciprocal
Z	$Z_{12} = Z_{21}$
Y	$y_{12} = y_{21}$
h	$h_{12} = -h_{21}$
$ABCD$	$AD - BC = 1$

15 V voltage source is connected at terminals 1 – 1′ and terminals 2 – 2′ are shorted.

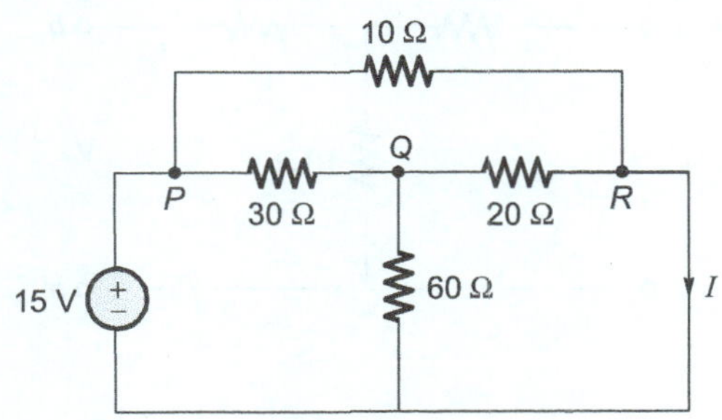

Converting Δ into Y-circuit,

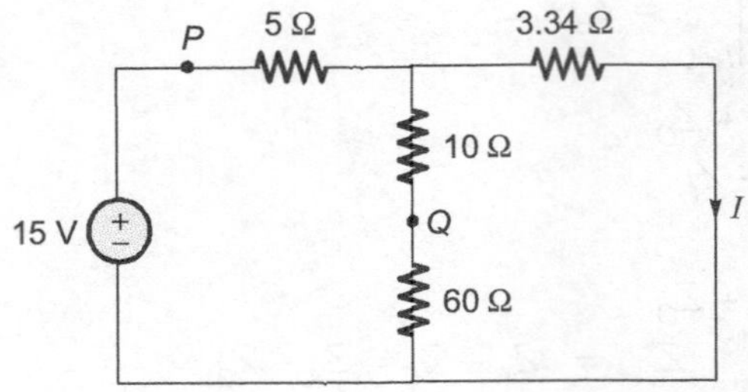

$$I = \frac{15}{5+(10+60)\|3.34} \times \frac{(10+60)}{(10+60)+3.34}$$

$$= 1.748 \text{ A}$$

Now 15 V voltage source is at 2 – 2′ terminals at terminals 1 – 1′ are shorted.

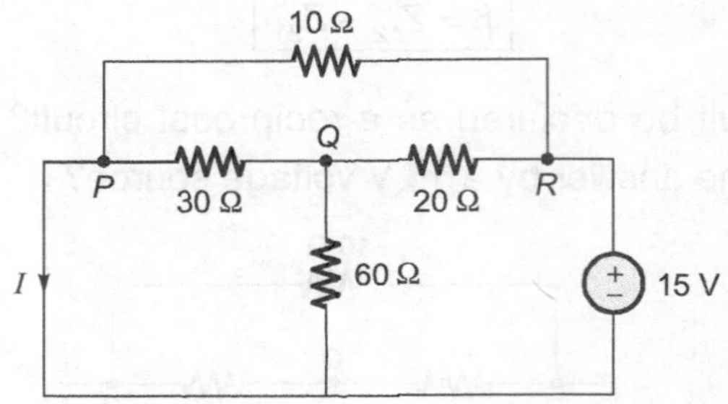

Converting Δ into Y-circuit,

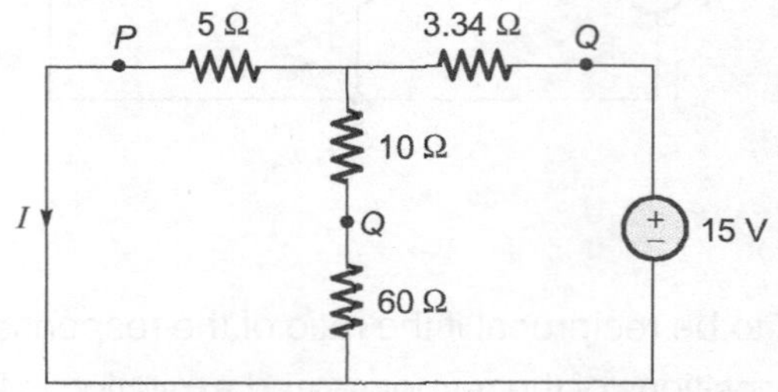

$$I = \frac{15}{(10+60)\|5+3.34} \times \frac{(10+60)}{(10+60)+5}$$

$$= 1.748 \text{ A}$$

For the same excitation (15 V voltage source) placed at both terminals the response (I = 1.748 A) remains identical. Hence the given network is reciprocal.

5.7 Find *ABCD* parameters for the two-port network shown in the figure below:

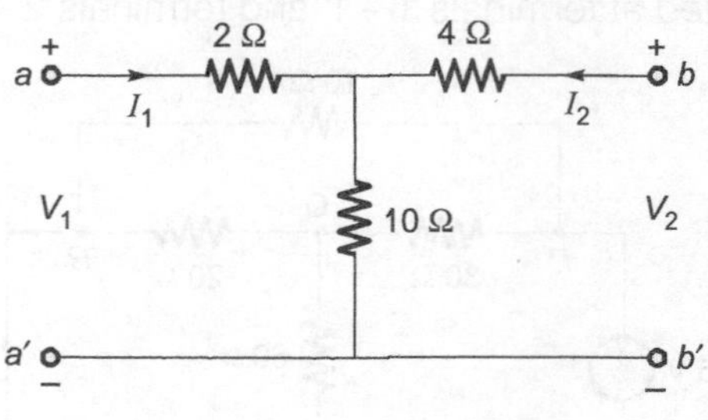

[8 + 4 = 12 marks : 2011]

Solution:

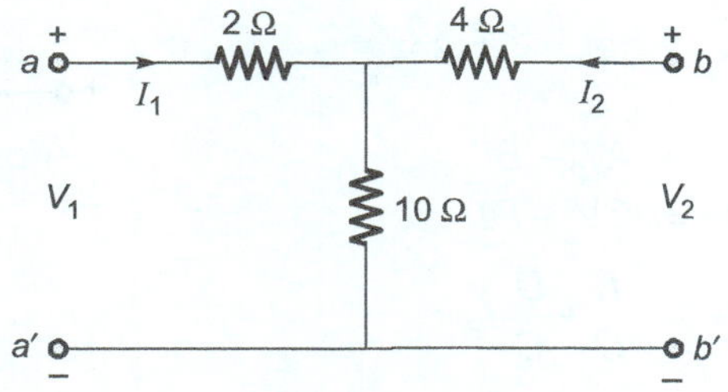

Applying KVL,

$$V_1 = 2I_1 + 10(I_1 + I_2)$$

$$= 12I_1 + 10I_2 \quad \text{...}(i)$$

$$V_2 = 4I_2 + 10(I_1 + I_2)$$

$$= 10I_1 + 14I_2 \quad \text{...}(ii)$$

$$\Rightarrow \quad 10I_1 = V_2 - 14I_2$$

$$I_1 = 0.1\,V_2 - 1.4I_2 \quad \text{...}(iii)$$

Putting value of I_1 in equation (i),

$$V_1 = 12\,(0.1\,V_2 - 1.4I_2) + 10I_2$$

$$= 1.2\,V_2 - 6.8I_2 \quad \text{...}(iv)$$

$$V_1 = 1.2\,V_2 - 6.8I_2$$

$$I_1 = 0.1\,V_2 - 1.4I_2$$

$$\begin{bmatrix} V_1 \\ I_1 \end{bmatrix} = \begin{bmatrix} 1.2 & 6.8 \\ 0.1 & 1.4 \end{bmatrix} \begin{bmatrix} V_2 \\ -I_2 \end{bmatrix}$$

Hence the *ABCD* parameters are

$$\begin{bmatrix} A & B \\ C & D \end{bmatrix} = \begin{bmatrix} 1.2 & 6.8 \\ 0.1 & 1.4 \end{bmatrix}$$

5.8 Obtain z-parameters of a two-port network in terms *ABCD* parameters. Obtain *z*-parameters for the circuit shown in the below figure.

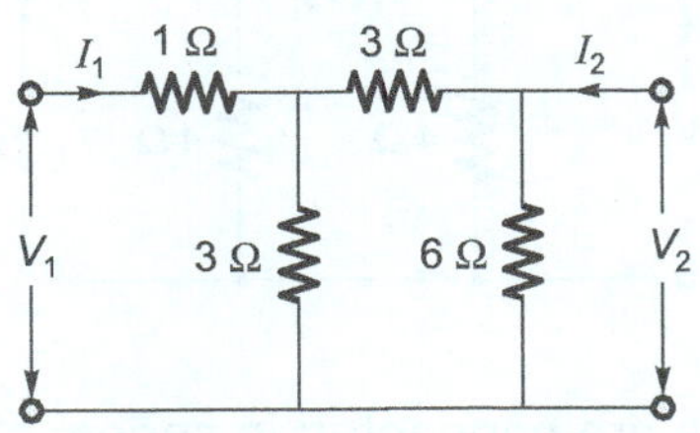

[12 marks : 2012]

Solution:

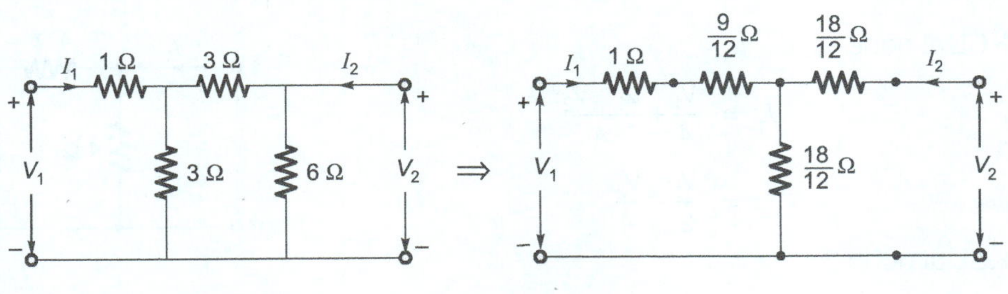

Z-parameters: $V_1 = z_{11} I_1 + z_{12} I_2$

$$V_2 = z_{21} I_1 + z_{22} I_2$$

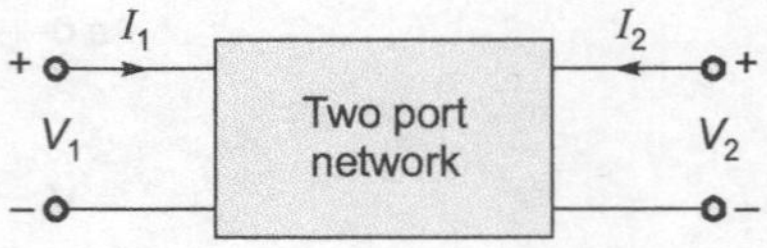

ABCD parameters:

$$V_1 = AV_2 - BI_2$$

$$I_1 = CV_2 - DI_2$$

$$V_2 = \frac{I_1}{C} + \frac{D}{C} I_2$$

$$\Rightarrow \quad V_1 = \frac{A}{C} I_1 + \left(\frac{AD}{C} - B\right) I_2$$

$$V_1 = \frac{A}{C} I_1 + \frac{(AD - BC)}{C} I_2$$

$$V_2 = \frac{1}{C} I_1 + \frac{D}{C} I_2$$

$$z_{11} = \frac{A}{C}, \; z_{12} = \frac{AD - BC}{C}$$

$$z_{21} = \frac{1}{C}, \; z_{22} = \frac{D}{C}$$

Applying KVL in above circuit,

$$V_1 = \left(1 + \frac{3}{4} + \frac{3}{2}\right) I_1 + \frac{3}{2} I_2 = \frac{13}{4} I_1 + \frac{3}{2} I_2$$

$$V_2 = \left(\frac{3}{2} + \frac{3}{2}\right) I_2 + \frac{3}{2} I_1 = \frac{3}{2} I_1 + 3I_2$$

$$\therefore \quad z_{11} = \frac{13}{4}, \; z_{12} = \frac{3}{2}$$

$$z_{21} = \frac{3}{2}, \; z_{22} = 3$$

5.9 For the π-network shown below:

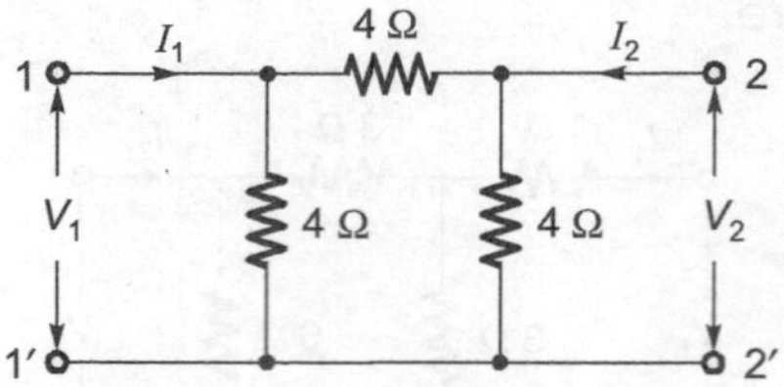

(i) Determine *y*-parameters.
(ii) If two such identical networks are connected in series, obtain the *z*-parameters of the resultant network.

[10 marks : 2013]

Solution:

(i) Using KCL at node 1,

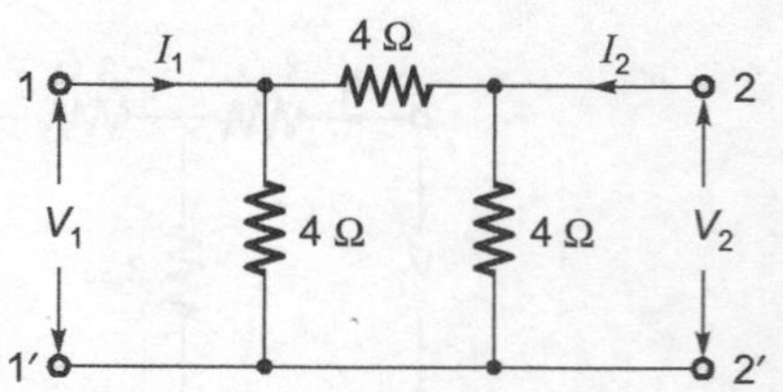

$$I_1 = \frac{V_1}{4} + \frac{V_1 - V_2}{4}$$

$$= \frac{V_1}{2} - \frac{V_2}{4} \quad \ldots(i)$$

Using KCL at node 2,

$$I_2 = \frac{V_2}{4} + \frac{V_2 - V_1}{4}$$

$$= -\frac{V_1}{4} + \frac{V_2}{2} \quad ...(ii)$$

Using equation (*i*) and (*ii*) we have,

$$y \text{ parameters} = \begin{bmatrix} 1/2 & -1/4 \\ -1/4 & 1/2 \end{bmatrix}$$

(ii)
$$[Z] = [Y]^{-1} = \frac{1}{1/4 - 1/16}\begin{bmatrix} 1/2 & 1/4 \\ 1/4 & 1/2 \end{bmatrix}$$

$$[Z] = \begin{bmatrix} 8/3 & 4/3 \\ 4/3 & 8/3 \end{bmatrix}$$

$\because$ For series connection,

$$[Z_{overall}] = [Z_1] + [Z_2]$$

$$\therefore \quad Z_{overall} = \begin{bmatrix} 8/3 & 4/3 \\ 4/3 & 8/3 \end{bmatrix} + \begin{bmatrix} 8/3 & 4/3 \\ 4/3 & 8/3 \end{bmatrix} = \begin{bmatrix} 16/3 & 8/3 \\ 8/3 & 16/3 \end{bmatrix}$$

5.10 **Determine the Z-parameters of the network shown in the below figure.**

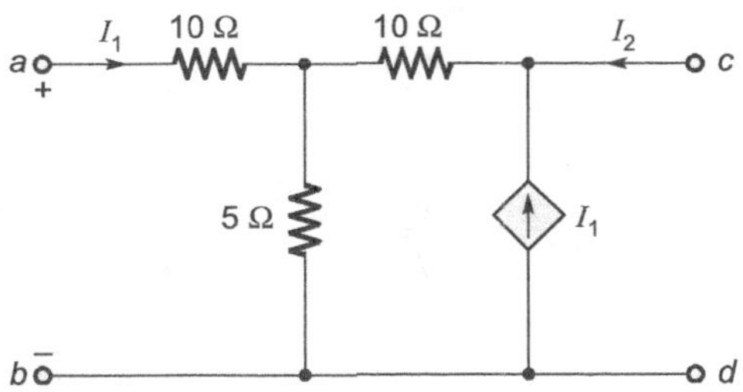

[8 marks : 2014]

Solution:

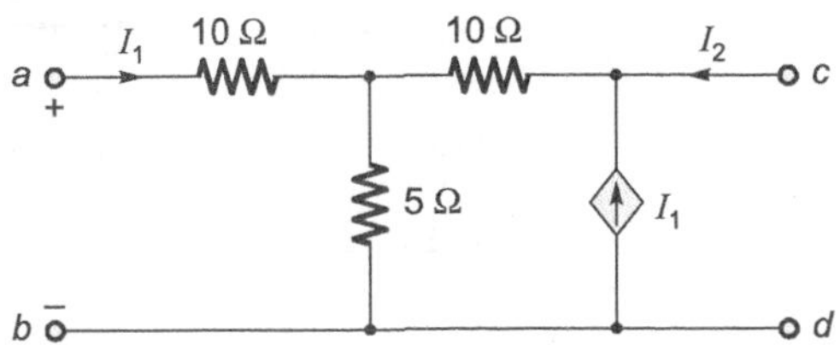

2-Parameter:

$$V_1 = Z_{11} I_1 + Z_{12} I_2$$
$$V_2 = Z_{21} I_1 + Z_{22} I_2$$

Case-I:

When,
$$I_2 = 0$$

$$Z_{11} = \frac{V_1}{I_1}$$

$$Z_{21} = \frac{V_2}{I_1}$$

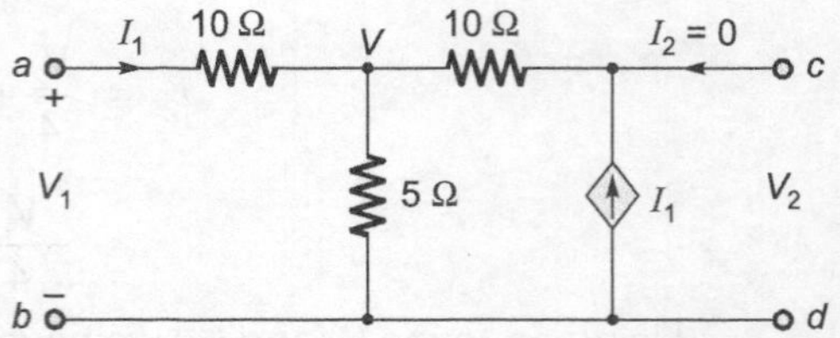

Applying KCL at V,

$$\Rightarrow \quad -I_1 + \frac{V}{5} - I_1 = 0$$

$$\Rightarrow \quad 2I_1 = \frac{V}{5}$$

$$V = 10I_1 \qquad \text{...(1)}$$

Also, $$V_1 - 10I_1 = V \qquad \text{...(2)}$$

Putting, $V = 10I_1$ from equation (1) into equation (2),

$$V_1 - 10I_1 = 10I_1 = 20I_1$$

$$\frac{V_1}{I_1} = 20\ \Omega = Z_{11}$$

Z_{21}:

$$V + 10I_1 = V_2 \qquad \text{...(3)}$$

Putting, $V = 10I_1$ from equation (1) into equation (3),

$$\Rightarrow \quad 10I_1 + 10I_1 = V_2$$

$$\Rightarrow \quad 20I_1 = V_2$$

$$\Rightarrow \quad \frac{V_1}{I_1} = Z_{21} = 20\ \Omega$$

Case-II:

When, $$I_1 = 0$$

$$Z_{12} = \frac{V_2}{I_2}$$

$$Z_{22} = \frac{V_2}{I_2}$$

For Z_{22},

Applying KCL in $ca'b'd$,

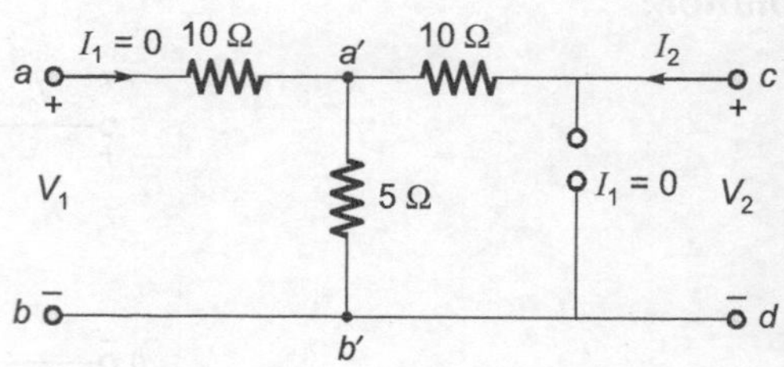

$$V_2 - 10I_2 - 5I_2 = 0$$

$$V_2 = 15I_2$$

$$\frac{V_2}{I_2} = Z_{22} = 15\ \Omega$$

For Z_{12}, $$V_1 = 5I_2$$

$$\frac{V_2}{I_2} = 5\ \Omega = Z_{12}$$

Second Method:

KVL to loop1,

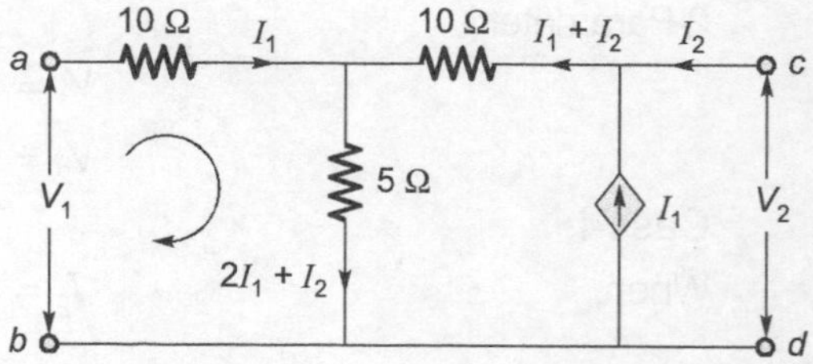

$$V_1 = 10I_1 + 5(2I_1 + I_2)$$

$$= 10I_1 + 10I_1 + 5I_1$$

$$\boxed{V_1 = 20I_1 + 5I_2}$$

KVL to loop2,

$$V_2 = 10(I_1 + I_2) + 5(2I_1 + I_2)$$

$$= 10I_1 + 10I_2 + 10I_1 + 5I_2 = 20I_1 + 15I_2$$

$$= \begin{bmatrix} 20 & 5 \\ 20 & 15 \end{bmatrix}$$

5.11 A two port network follows the following voltage-current relations:

$$I_1 = 2V_1 - V_2$$
$$2I_2 = -2V_1 + 4V_2$$

Determine z-parameters of the network and its T-equivalent circuit.

[10 marks : 2015]

Solution:

$$I_1 = 2V_1 - V_2 \quad ...(1)$$
$$2I_2 = -2V_1 + 4V_2 \quad ...(2)$$

Z-parameter,

$$V_1 = Z_{11} I_1 + Z_{12} I_2$$
$$V_2 = Z_{21} I_1 + Z_{22} I_2$$

From equation (2),

$$4V_2 = 2I_2 + 2V_1$$

$$\Rightarrow \quad V_2 = \frac{I_2}{2} + \frac{V_1}{2} \quad ...(3)$$

Put, V_2 from equation (3) into equation (1),

$$\Rightarrow \quad I_1 = 2V_1 - \frac{I_2}{2} - \frac{V_1}{2}$$

$$\Rightarrow \quad \left(2V_1 - \frac{V_1}{2}\right) = I_1 + \frac{I_2}{2}$$

$$\Rightarrow \quad \frac{3V_1}{2} = I_1 + \frac{I_2}{2}$$

$$\Rightarrow \quad V_1 = \frac{2I_1}{3} + \frac{I_2}{3} \quad ...(4)$$

From equation (4), $\quad Z_{11} = \frac{2}{3}, \quad Z_{12} = \frac{1}{3}$

From equation (1), $\quad 2V_1 = V_2 + I_1$

$$\Rightarrow \quad V_1 = \frac{V_2}{2} + \frac{I_1}{2} \quad ...(5)$$

Put V_1 from equation (5) into equation (2),

$$2I_2 = -2\left[\frac{V_2}{2} + \frac{I_1}{2}\right] + 4V_2$$

$$\Rightarrow \quad 2I_2 = -V_2 - I_1 + 4V_2$$
$$\Rightarrow \quad 3V_2 = I_1 + 2I_2$$

$$\Rightarrow \quad V_2 = \frac{I_1}{3} + \frac{2}{3} I_2 \quad ...(6)$$

From equation (6), $\quad Z_{21} = \frac{1}{3}, \quad Z_{22} = \frac{2}{3}$

For T-equivalent circuit,

$$V_1 = Z_{11} I_1 + Z_{12} I_2$$
$$V_2 = Z_{21} I_1 + Z_{21} I_2$$

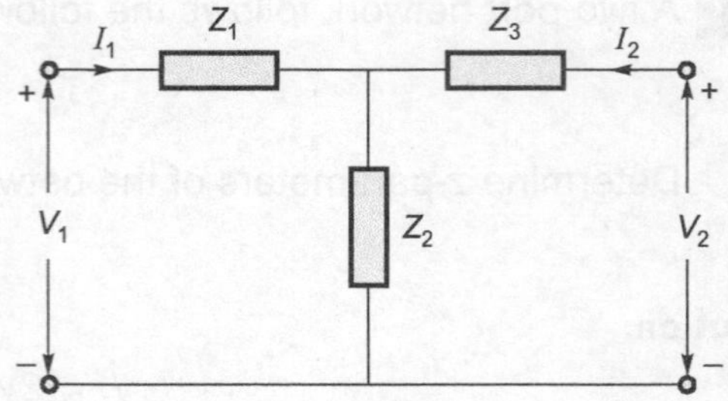

when, $I_2 = 0$, $$Z_{11} = \frac{V_1}{I_1} \text{ and } Z_{21} = \frac{V_2}{I_1}$$

and from figure,

$$V_1 = I_1(Z_1 + Z_2)$$

$$\Rightarrow \quad Z_1 + Z_2 = \frac{V_1}{I_1} = Z_{11} = \frac{2}{3}$$

or, $$Z_1 + Z_2 = \frac{2}{3} \quad \text{...(7)}$$

and, $$V_2 = I_1 Z_2$$

$$Z_2 = \frac{V_2}{I_1} = Z_{21} = \frac{1}{3}$$

or $$Z_2 = \frac{1}{3} \quad \text{...(8)}$$

From equation (7) and equation (8),

$$Z_1 = \frac{1}{3} \text{ and } Z_2 = \frac{1}{3}$$

when, $I_1 = 0$, $$Z_{12} = \frac{V_1}{I_2} \text{ and } Z_{22} = \frac{V_2}{I_2}$$

From figure, $$V_2 = I_2(Z_2 + Z_3) \Rightarrow Z_2 + Z_3 = \frac{V_2}{I_2} = Z_{22}$$

$$Z_2 + Z_3 = \frac{2}{3} \quad \text{...(9)}$$

and, $$V_1 = I_2 Z_2$$

$$\Rightarrow \quad \frac{V_1}{I_2} = Z_2 = Z_{12} = \frac{1}{3}$$

$$\Rightarrow \quad Z_2 = \frac{1}{3} \quad \text{...(10)}$$

From equaton (9) and (10), $$Z_3 = \frac{1}{3}, Z_2 = \frac{1}{3}$$

so, $$Z_1 = Z_2 = Z_3 = \frac{1}{3}$$

5.12 **Determine overall z-parameters when two identical 2-port networks with $z_{11} = z_{12} = z_{21} = z_{22} = 2\ \Omega$ are connected in cascade.**

[10 marks : 2015]

Solution:

$$V_1 = 2I_1 + 2I_2 \quad \text{...(1)}$$

$$V_2 = 2I_1 + 2I_2 \quad \text{...(2)}$$

ABCD parameter,

$$\begin{bmatrix} V_1 \\ I_1 \end{bmatrix} = \begin{bmatrix} A & B \\ C & D \end{bmatrix} \begin{bmatrix} V_2 \\ -I_2 \end{bmatrix}$$

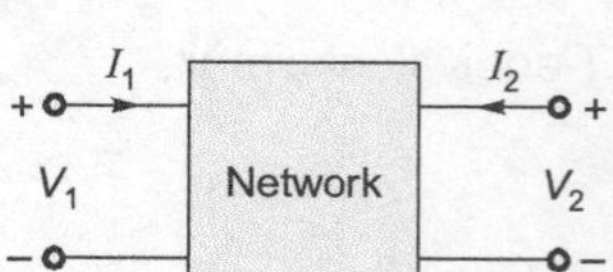

Put, value of I_1 from equation (2) into equation (1).

$$V_1 = 2\left[\frac{V_2 - 2I_2}{2}\right] + 2I_2$$

$$\Rightarrow \quad V_1 = V_2 - 2I_2 + 2I_2$$

$$\Rightarrow \quad V_1 = V_2 + 0 \cdot I_2 \quad \text{...(3)}$$

From equation, (3), $A = 1, \quad B = 0$

From equation, (2), $I_1 = \frac{V_2 - 2I_2}{2} = \frac{V_2}{2} - I_2$

$$I_1 = \frac{1}{2}V_2 - I_2 \quad \text{...(4)}$$

From equation (4), $C = \frac{1}{2}, \quad D = 1$

So, $A = 1, \quad B = 0, \quad C = \frac{1}{2}, \quad D = 1$

or, $\begin{bmatrix} A & B \\ C & D \end{bmatrix} = \begin{bmatrix} 1 & 0 \\ \frac{1}{2} & 1 \end{bmatrix}$

If two such network connected in cascade,

$$\begin{bmatrix} A_1 & B_1 \\ C_1 & D_1 \end{bmatrix}\begin{bmatrix} A_2 & B_2 \\ C_2 & D_2 \end{bmatrix} = \begin{bmatrix} 1 & 0 \\ \frac{1}{2} & 1 \end{bmatrix}\begin{bmatrix} 1 & 0 \\ \frac{1}{2} & 1 \end{bmatrix}$$

$$\begin{bmatrix} A & B \\ C & D \end{bmatrix} = \begin{bmatrix} 1 & 0 \\ 1 & 1 \end{bmatrix}$$

$$\begin{bmatrix} V_1 \\ I_1 \end{bmatrix} = \begin{bmatrix} 1 & 0 \\ 1 & 1 \end{bmatrix}\begin{bmatrix} V_2 \\ -I_2 \end{bmatrix}$$

$$V_1 = V_2$$

$$I_1 = V_2 - I_2$$

$$V_2 = I_1 + I_2$$

$$V_1 = I_1 + I_2$$

$$\begin{bmatrix} V_1 \\ V_2 \end{bmatrix} = \begin{bmatrix} 1 & 1 \\ 1 & 1 \end{bmatrix}\begin{bmatrix} I_1 \\ I_2 \end{bmatrix}$$

$$Z = \begin{bmatrix} 1 & 1 \\ 1 & 1 \end{bmatrix}\Omega$$

5.13 **For the two-port network shown, the y-parameter matrix is given.**

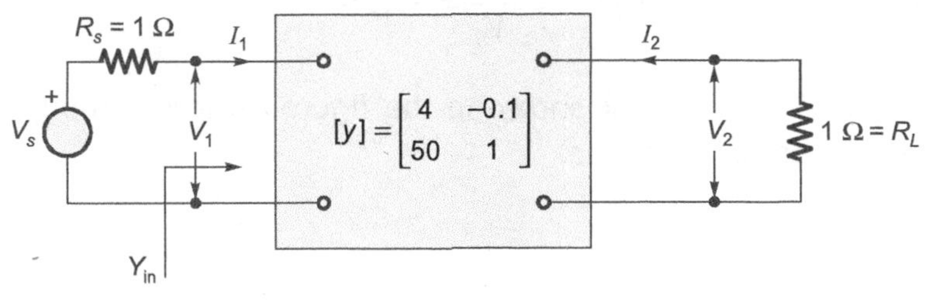

(i) Obtain the input admittance Y_{in} as shown in the figure.
(ii) Obtain voltage gain V_2/V_s.

[20 marks : 2018]

Solution:

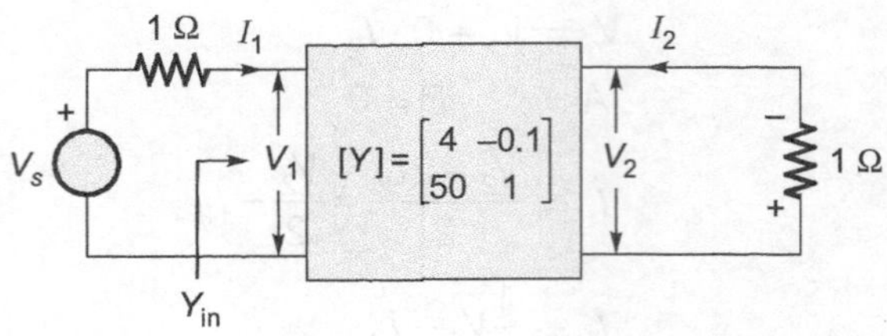

(i) KVL at input,

$$V_S = V_1 + I_1 R_S$$
$$V_S = V_1 + I_1$$

Using Y-parameters,

$$I_1 = 4\,V_1 - 0.1\,V_2 \quad \text{...(i)}$$
$$I_2 = 50\,V_1 + V_2 \quad \text{...(ii)}$$
$$Y_{in} = \frac{I_1}{V_1}$$

KVL at ouput,

$$V_2 = -I_2$$

$\therefore$ Putting in equation (ii),

$$-V_2 = 50\,V_1 + V_2$$
$$-2\,V_2 = 50\,V_1$$
$$V_2 = -25\,V_1 \quad \text{...(iii)}$$

Putting in equation (i),

$$I_1 = 4\,V_1 + 0.1 \times 25\,V_1$$
$$I_1 = 6.5\,V_1 \quad \text{...(iv)}$$
$$Y_{in} = \frac{I_1}{V_1} = 6.5\ \mho$$

(ii)

$$V_S = V_1 + I_1 \text{ from input loop}$$

and

$$I_1 = 6.5\,V_1 \text{ from equation (iv),}$$

$\therefore$
$$V_S = V_1 + 6.5\,V_1$$

$\therefore$
$$V_S = 7.5\,V_1 \quad \text{...(v)}$$

Now from equation (iii),

$$V_2 = -25\,V_1$$

$\therefore$
$$V_S = 7.5 \times \left(\frac{-V_2}{25}\right)$$

$\therefore$
$$\frac{V_2}{V_S} = \frac{-25}{7.5}$$

$\therefore$
$$\text{Voltage gain} = \frac{V_2}{V_S} = -3.33$$

5.14 Find the Z-parameter of the network shown in the figure. Is the network reciprocal? If so why? Assume the operational amplifier is ideal.

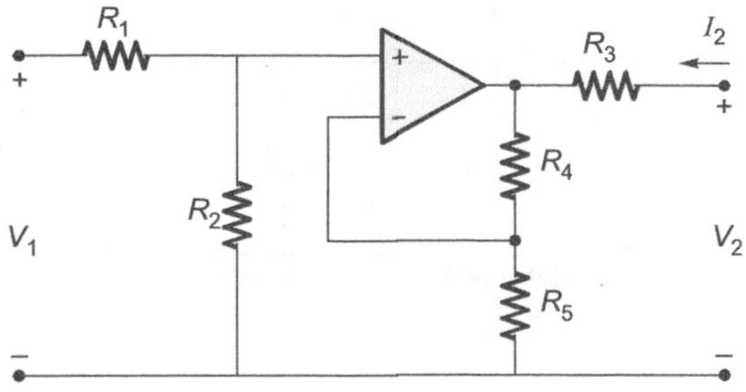

[12 marks : 2019]

Solution:

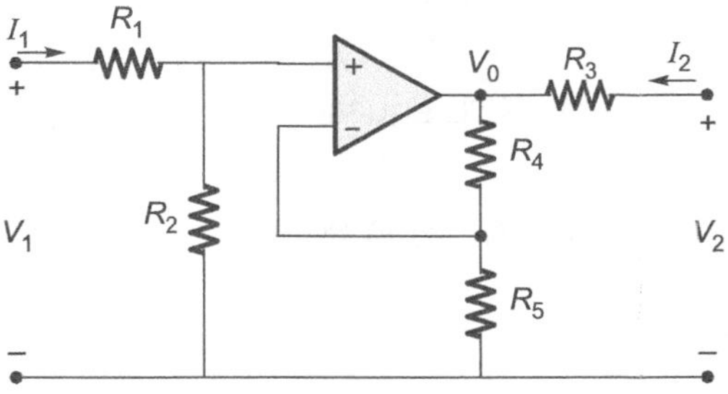

$$V_1 = I_1(R_1 + R_2) + 0I_2$$

$$\therefore \quad z_{11} = (R_1 + R_2)$$

$$z_{12} = 0$$

$$V_0 = \left(1+\frac{R_4}{R_5}\right)\times\frac{R_2}{R_1+R_2}\times V_1$$

$$= \left(1+\frac{R_4}{R_5}\right)\times\frac{R_2}{R_1+R_2}\times(R_1+R_2)I_1$$

$$= \left(1+\frac{R_4}{R_5}\right)\times R_2 I_1 \quad \text{... (i)}$$

$$I_2 = (V_2 - V_0)\times\frac{1}{R_3}$$

$$I_2 R_3 = V_2 - V_0$$

$$V_0 = V_2 - I_2 R_3 \quad \text{... (ii)}$$

Put equation (ii) in equation (i),

$$V_2 - I_2 R_3 = \left(1+\frac{R_4}{R_5}\right)\times R_2 I_1$$

$$V_2 = \left(1+\frac{R_4}{R_5}\right)\times R_2 I_1 + I_2 R_3$$

$$z_{21} = \left(R_2 + \frac{R_2 R_4}{R_5}\right)$$

and $z_{22} = R_3$

$$\begin{bmatrix} z_{11} & z_{12} \\ z_{21} & z_{22} \end{bmatrix} = \begin{bmatrix} (R_1+R_2) & 0 \\ \left(R_2+\frac{R_2R_4}{R_5}\right) & R_3 \end{bmatrix}$$

For a reciprocal network, $z_{12} = z_{21}$

But for above network, $z_{12} \neq z_{21}$

Hence it is not reciprocal network.

5.15 The equivalent ABCD constants of a network are obtained by cascade connection of two networks as shown in the figure. The value of constant B of equivalent network is $5 + j40$. Calculate the value of z_1.

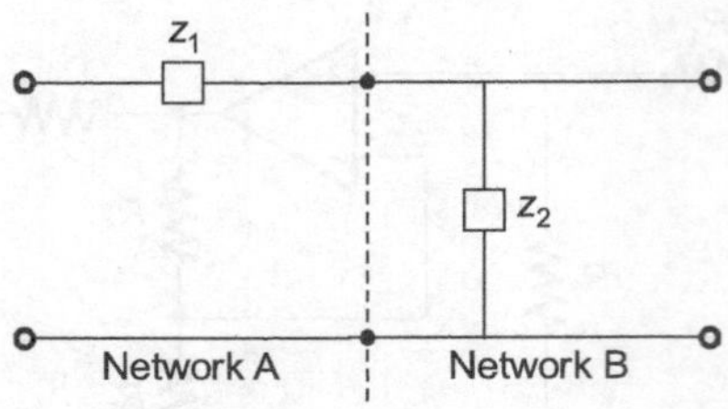

[12 marks : 2021]

Solution:

For network A:

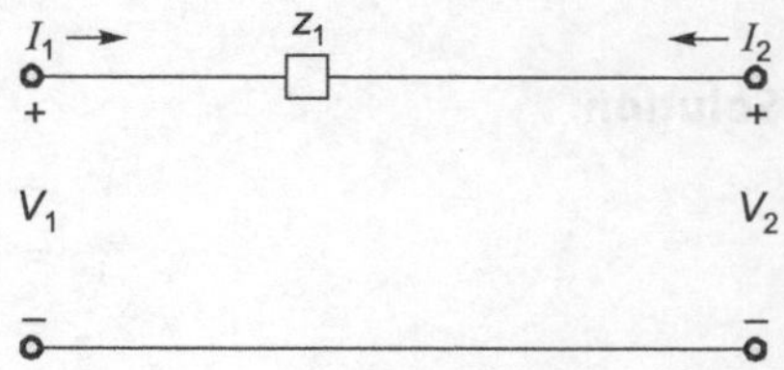

Write KVL equation,

$$V_1 = Z_1 I_1 + V_2 \quad ...(1)$$

Write KCL equation

$$I_1 = -I_2 \quad ...(2)$$

Sub. eq. (2) in eq. (1)

$$V_1 = Z_1(-I_2) + V_2$$

$$V_1 = V_2 - Z_1 I_2 \quad ...(3)$$

Compare eq. (2), and eq. (3) with

$$V_1 = AV_2 - BI_2$$

$$I_1 = CV_2 - DI_2$$

$$\therefore \quad \begin{bmatrix} A_1 & B_1 \\ C_1 & D_1 \end{bmatrix} = \begin{bmatrix} 1 & Z_1 \\ 0 & 1 \end{bmatrix}$$

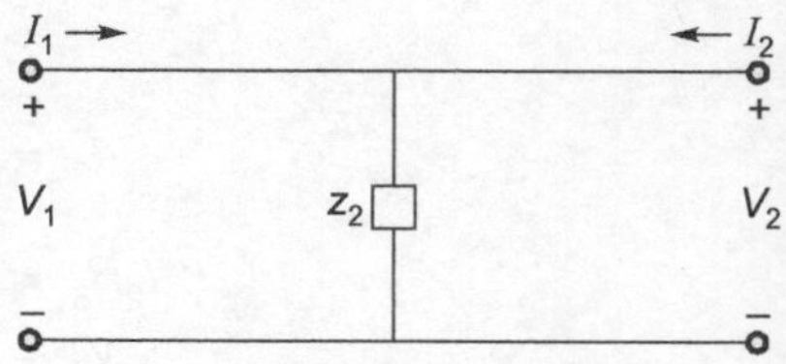

For network B, $ABCD$ parameters,

Write KVL equation

$$V_1 = V_2 \quad ...(4)$$

Write KCL equation

$$I_1 = \frac{V_2}{Z_2} - I_2 \quad ...(5)$$

From equation (4) and equation (5), $ABCD$ parameters are

$$\begin{bmatrix} A & B \\ C & D \end{bmatrix} = \begin{bmatrix} 1 & 0 \\ \frac{1}{Z_2} & 1 \end{bmatrix}$$

For cascade connection, multiply individual $ABCD$ parameters

$$\begin{bmatrix} A & B \\ C & D \end{bmatrix} = \begin{bmatrix} 1 & Z_1 \\ 0 & 1 \end{bmatrix} \begin{bmatrix} 1 & 0 \\ \frac{1}{Z_2} & 1 \end{bmatrix}$$

$$\begin{bmatrix} A & B \\ C & D \end{bmatrix} = \begin{bmatrix} 1+\frac{Z_1}{Z_2} & Z_1 \\ \frac{1}{Z_2} & 1 \end{bmatrix}$$

Constant $B = Z_1$, $\quad Z_1 = (5 + j40)\ \Omega$

5.16 Draw the equivalent z parameter circuit of a network as shown in the figure and determine whether the circuit is (i) reciprocal, and (ii) symmetrical.

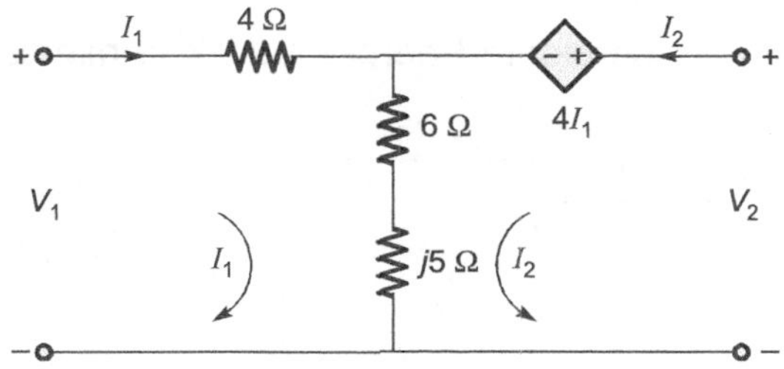

[20 marks : 2021]

Solution:

Write KVL equation in first loop.

$$V_1 = 4I_1 + 6(I_1 + I_2) + j5(I_1 + I_2)$$

$$V_1 = (10 + j5)I_1 + (6 + j5)I_2 \quad ...(1)$$

Write KVL equation in second loop

$$V_2 = 4I_1 + 6(I_1 + I_2) + j5(I_1 + I_2)$$

$$V_2 = 10I_1 + j5I_1 + (6 + j5)I_2$$

$$V_2 = (10 + j5)I_1 + (6 + j5)I_2 \quad ...(2)$$

Compare equation (1) and equation (2) with

$$V_1 = Z_{11}I_1 + Z_{12}I_2$$

$$V_2 = Z_{21}I_1 + Z_{22}I_2$$

$$\begin{bmatrix} Z_{11} & Z_{12} \\ Z_{21} & Z_{22} \end{bmatrix} = \begin{bmatrix} 10+j5 & 6+j5 \\ 10+j5 & 6+j5 \end{bmatrix}$$

Equivalent Circuit :

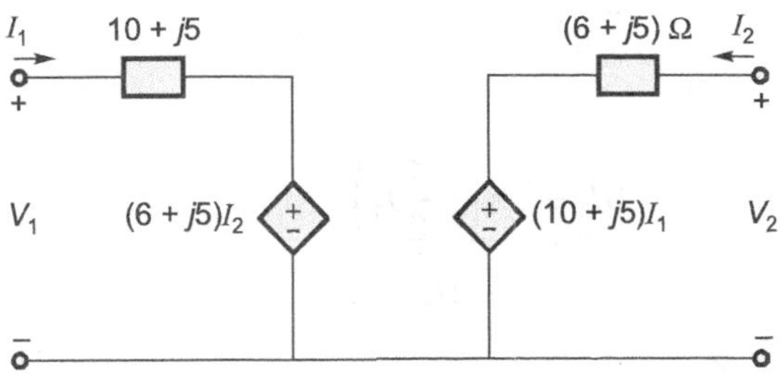

Condition for Reciprocity :

$$Z_{12} = Z_{21}$$

$$Z_{12} = (6 + j5)\ \Omega;\ Z_{21} = (10 + j5)\ \Omega$$

Since, $Z_{12} \neq Z_{21}$ So, given network is not reciprocal.

Condition for Symmetry :

$$Z_{11} = Z_{22}$$

$$Z_{11} = (10 + j5)\ \Omega;$$

$$Z_{22} = (6 + j5)\ \Omega$$

Since $Z_{11} \neq Z_{22}$, so it is not a symmetrical network.

5.17 The following equations refer to a two-port network

$$V_1 = 5I_1 + 2I_2$$
$$V_2 = 2I_1 + I_2$$

A load resistance of 3 ohms is connected across port-2 terminals as shown in figure. Calculate the input Impedance.

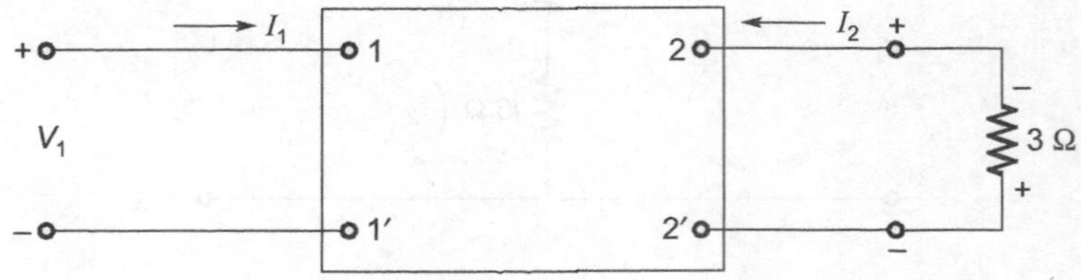

[12 marks : 2022]

Solution:

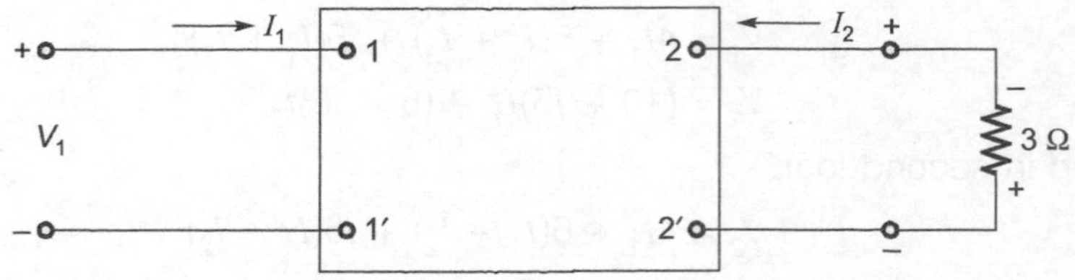

$$V_1 = 5I_1 + 2I_2 \quad \text{...(i)}$$
$$V_2 = 2I_1 + I_2 \quad \text{...(ii)}$$

As a load resistance of 3 Ω is connected across port-2

$$V_2 = -V_L = 3I_2$$
$$V_2 = -3I_2$$

Put the value of V_2 in eqn. (ii)

$$-3I_2 = 2I_1 + I_2$$
$$-4I_2 = 2I_1$$
$$I_2 = -\frac{1}{2}I_1$$

Now put the value of I_2 in eqn. (i)

$$V_1 = 5I_1 + 2\times\left(-\frac{1}{2}I_1\right)$$
$$= 5I_1 - I_1 = 4I_1$$

Input impedance, $\frac{V_1}{I_1} = 4\ \Omega$

Hence, Input impedance = 4 Ω

5.18 Find the hybrid parameters of the following circuit :

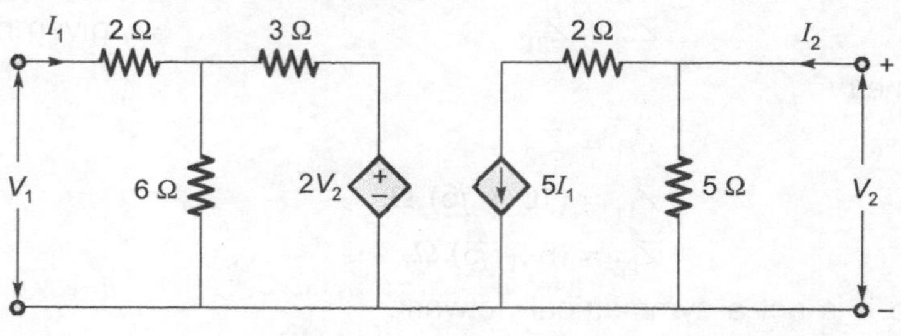

[12 marks : 2023]

Solution:

Hybrid parameter of the network

$$V_1 = h_{11}I_1 + h_{12}V_2$$
$$I_1 = h_{21}I_1 + h_{22}V_2$$

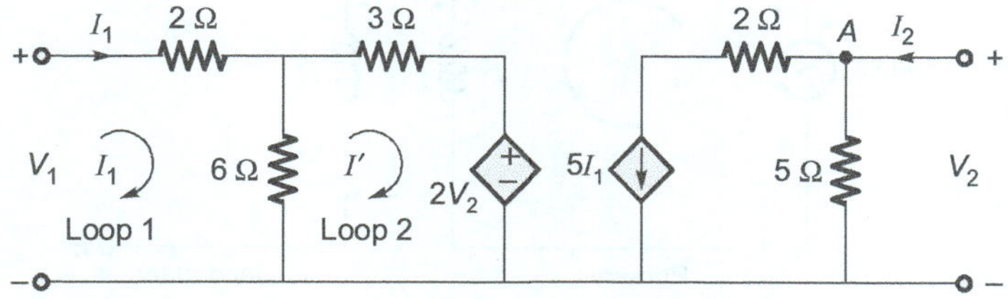

Applying KCL at node A,

$$5I_1 + \frac{V_2}{5} = I_2$$
$$h_{21}I_1 + h_{22}V_2 = I_2$$
$$\left.\begin{array}{l} h_{21} = 5 \\ h_{22} = 1/5\,℧ \end{array}\right.$$

Applying KVL in loop 1,

$$V_1 = (2+6)I_1 - 6I' \quad \text{...(i)}$$
$$(3+6)I' - 6I_1 + 2V_2 = 0$$
$$I' = \frac{6I_1 - 2V_2}{9} \quad \text{...(ii)}$$

Substitute equation (ii) in (i),

$$V_1 = 8I_1 - 6\left(\frac{6I_1 - 2V_2}{9}\right) = 8I_1 - \frac{12}{3}I_1 + \frac{4}{3}[V_2]$$

$\Rightarrow$
$$V_1 = 4I_1 + \frac{4}{3}V_2$$

We know,
$$V_1 = h_{11}I_1 + h_{12}V_2$$
$$\left.\begin{array}{l} h_{11} = 4\ \Omega \\ h_{12} = 4/3 \end{array}\right.$$

$$\begin{bmatrix} h_{11} & h_{12} \\ h_{21} & h_{22} \end{bmatrix} = \begin{bmatrix} 4 & \frac{4}{3} \\ 5 & \frac{1}{5} \end{bmatrix}$$

6. Magnetically Coupled Circuits

6.1 Obtain Thevenin's equivalent across terminals AB and hence obtain Z_L to be connected across AB to receive maximum power.

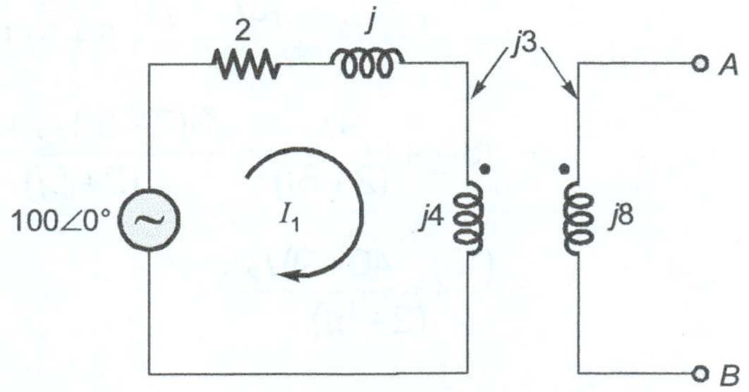

[10 marks : 2014]

Solution:

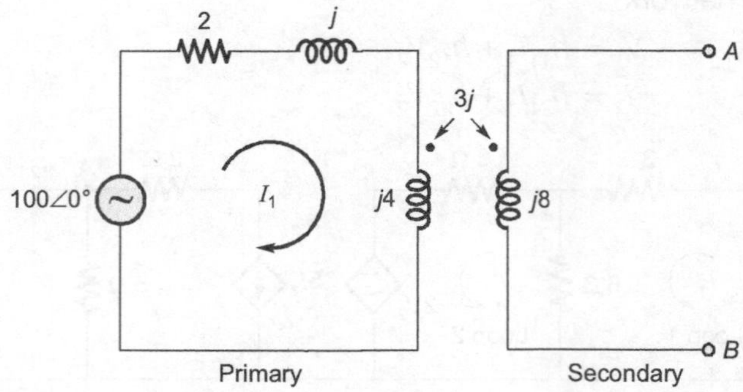

Thevenin's Voltage:

Since, the secondary winding is open. So, no current flows through the terminals *A* and *B*.

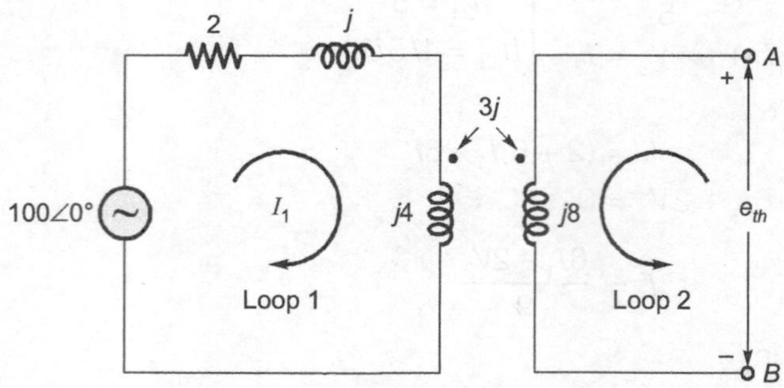

KVL around the loop1:

$$100\angle 0^\circ - (2+5j)\,I_1 = 0$$

$$100\angle 0^\circ = (2+5j)\,I_1$$

$$I_1 = \frac{100\angle 0^\circ}{(2+5j)}$$

KVL around the loop2:

$$e_{th} - (3j)I_1 = 0$$

$$e_{th} = 3j\,I_1 = \frac{3j\,100\angle 0^\circ}{(2+5j)}$$

$$= \frac{300\angle 90^\circ}{(2+5j)} = 55.708\angle 21.80^\circ$$

Thevenin's Resistance:

$$e_1 = j8I_2 + 3j\,I_1 = j8I_2 + j3\left(\frac{-3jI_2}{2+5j}\right)$$

$$= j8I_2 + \frac{9I_2}{(2+5j)} = \frac{j8(2+5j)I_2 + 9I_2}{(2+5j)}$$

$$= \frac{(16j - 40 + 9)I_2}{(2+5j)}$$

$$\left(\frac{e_1}{I_2}\right) = \left(\frac{-31+16j}{2+5j}\right) = 6.478\angle 84.5^\circ$$

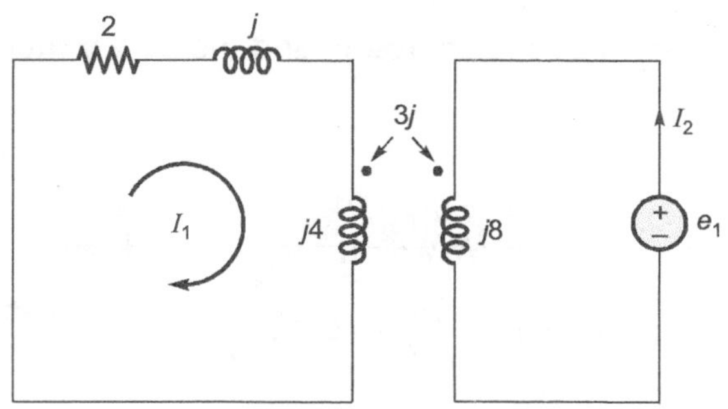

Thevenin's Equivalent:

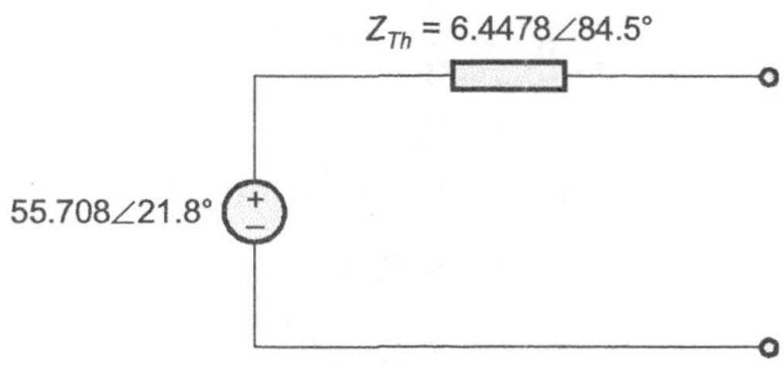

6.2 Two mutually coupled identical coils are connected in series having self inductance L = 4 mH and mutual inductance M = 2 mH, what is the maximum ratio of two possible values of effective inductances? Determine the coefficient of coupling between the two coils.

[10 marks : 2015]

Solution:

$$L_{eq_1} = L_1 + L_2 + 2M = (2 \times 4 + 2 \times 2)\text{ mH} = 12\text{ mH}$$

and, 

$$L_{eq_2} = L_1 + L_2 - 2M = (2 \times 4 - 2 \times 2)\text{ mH} = 4\text{ mH}$$

$$\text{Maximum ratio} = \frac{L_{eq_1}}{L_{eq_2}} = \frac{12\text{ mH}}{4\text{ mH}} = 3$$

Also,
$$M = K\sqrt{L_1 L_2}$$

$\Rightarrow$
$$K = \frac{M}{\sqrt{L_1 L_2}} = \frac{2\text{ mH}}{4\text{ mH}} = \frac{1}{2}$$

6.3 For the network shown, $V_S = 24\sqrt{2}\sin 100t$, coefficient of coupling $k = \frac{1}{3}$ between two coupled coils.

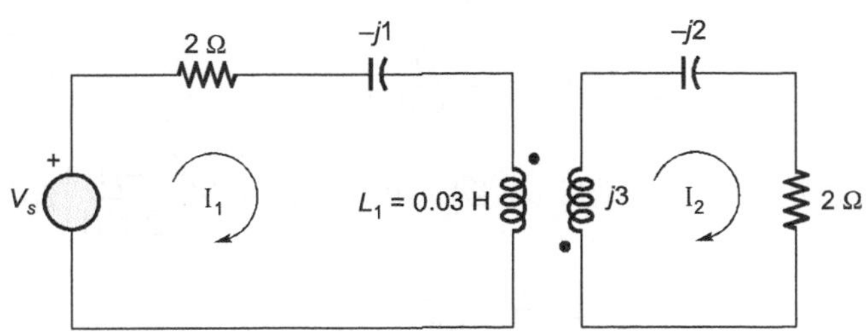

(i) Write loop equations in vector matrix for currents I_1 and I_2.

(ii) Obtain the impedance seen by the source v_s and the power factor of the source.

[20 marks : 2018]

Solution:

(i)

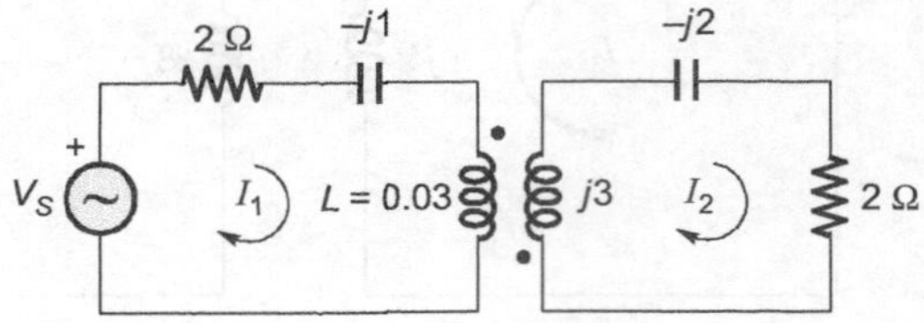

$$V_S = 24\sqrt{2}\sin 100t$$

$$K = \frac{1}{3} \rightarrow \text{coefficient of coupling}$$

$$\omega = 100$$

$$\therefore \quad \omega L = 100 \times 0.03 = 3\ \Omega$$

$$M = K\sqrt{L_1 L_2}$$

$$\Rightarrow \quad X_m = K\sqrt{X_{L1} X_{L2}}$$

$$\therefore \quad X_m = \frac{1}{3}\sqrt{3 \times 3} = 1\ \Omega$$

Using decoupled technique,

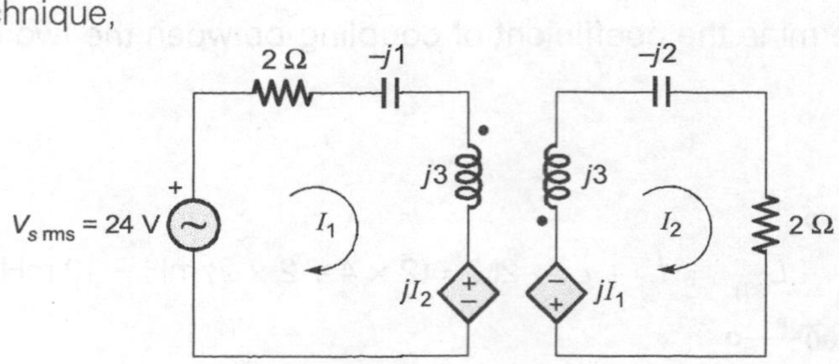

KVL in loop (1),

$$24 = (2 - j1 + j3)\, I_1 + j\, I_2$$

$$24 = (2 + j2)\, I_1 + j\, I_2 \quad \text{...(i)}$$

KVL in loop (2),

$$0 = jI_1 + (j3 - j2 + 2)\, I_2$$

$$0 = j\, I_1 + (2 + j1) I_2 \quad \text{...(ii)}$$

Writing in matrix vector form:

$$\begin{bmatrix} 2+j2 & j1 \\ j1 & 2+j1 \end{bmatrix} \begin{bmatrix} I_1 \\ I_2 \end{bmatrix} = \begin{bmatrix} 24 \\ 0 \end{bmatrix}$$

(ii) Impedance seen by the source V_S is,

$$Z_{\text{in}} = \frac{V_S}{I_1}$$

From equation (ii),

$$I_2 = \frac{-j1}{2+j1} I_1$$

Putting it in equation (i),

$$V_S = 24 = (2 + j2)I_1 + jX\left(\frac{-j1}{2+j1}\right) I_1$$

$$V_S = I_1\left[2 + j2 + \frac{1}{2+j1}\right]$$

$$Z_{in} = \frac{V_S}{I_1} = 2 + j2 + \frac{2 - j1}{5}$$

$$Z_{in} = \frac{12}{5} + j\frac{9}{5} = (2.4 + j1.8)\ \Omega$$

$$I_1 = \frac{24}{2.4 + j1.8} = 8\angle{-36.86}$$

$\therefore$ Power factor of source = cos (36.86)

p.f. = 0.800 lagging

6.4 (i) Two inductive coils having same self-inductance when connected in series carrying a current of I amperes store W joules of magnetic energy in their fields. When the connections of one of the coils are interchanged and the current is reduced to $\left(\frac{I}{3}\right)$ amperes, the stored energy remains the same. Calculate the ratio of mutual to self-inductance.

(ii) Determine the transfer function $\frac{V_o(s)}{V_g(s)}$ for the circuit shown below for R_1 = 500 Ω, R_2 = 50 Ω, L = 10 mH and C = 2 μF :

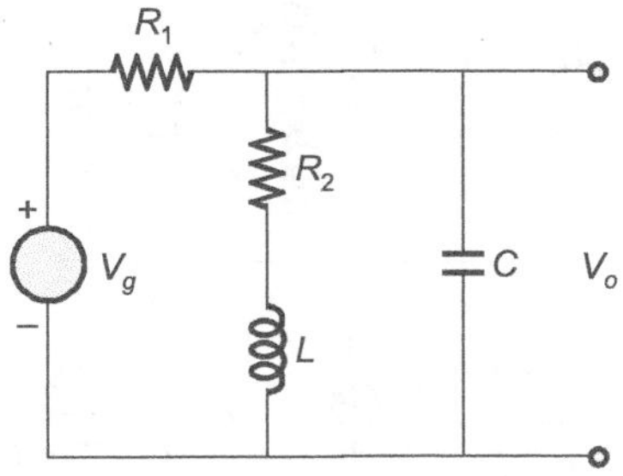

[10 + 10 = 20 marks : 2023]

Solution:

(i) Given :

$$L_1 = L_2 = L$$

Case (1) : Inductance are connecting in series opposing.

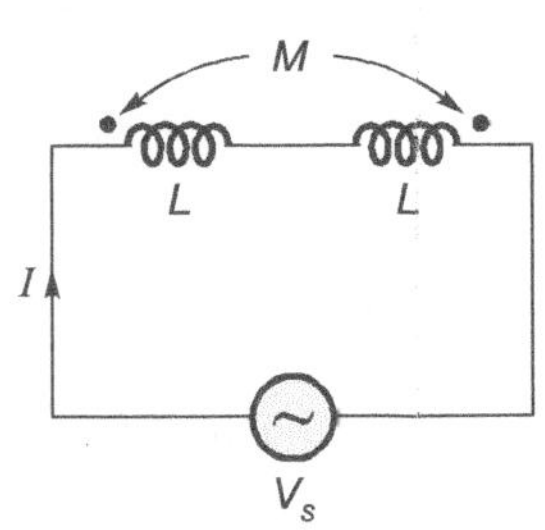

$$L_{eq} = L_1 + L_2 - 2M$$

$$W = \frac{1}{2}L_{eq} \cdot I^2$$

$$W = \frac{1}{2}[L + L - 2M]I^2$$

$$= \frac{1}{2}[2L - 2M]I^2 = (L - M)I^2 \quad \text{...(i)}$$

Case (2) : Inductance are connecting in series aiding.

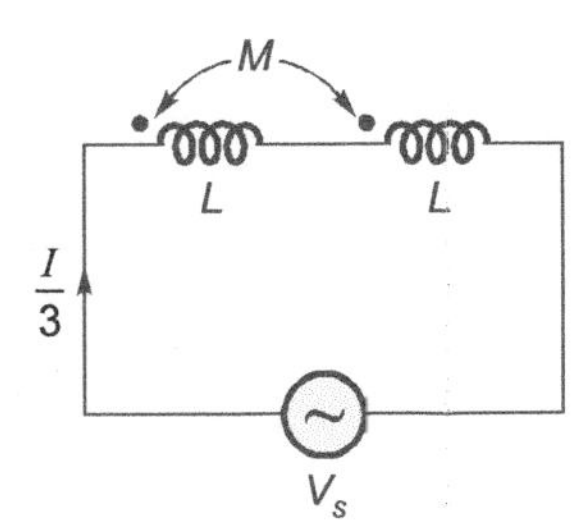

$$L_{eq} = L_1 + L_2 + 2M$$

$$W = \frac{1}{2}L_{eq}(I')^2$$

$$W = \frac{1}{2}[L_1 + L_2 + 2M]\left(\frac{I}{3}\right)^2$$

$$= \frac{1}{2}(2L + 2M)\left(\frac{I}{3}\right)^2 = (L + M)\frac{I^2}{9} \quad \text{...(ii)}$$

By equating (i) = (ii),

$$\frac{M}{L} = \frac{4}{5}$$

$$(L-M)I^2 = (L+M)\frac{I^2}{9}$$

$$8L - 9M = L + M$$

$$8L = 10M$$

$$\frac{M}{L} = \frac{8}{10} = \frac{4}{5}$$

(ii)

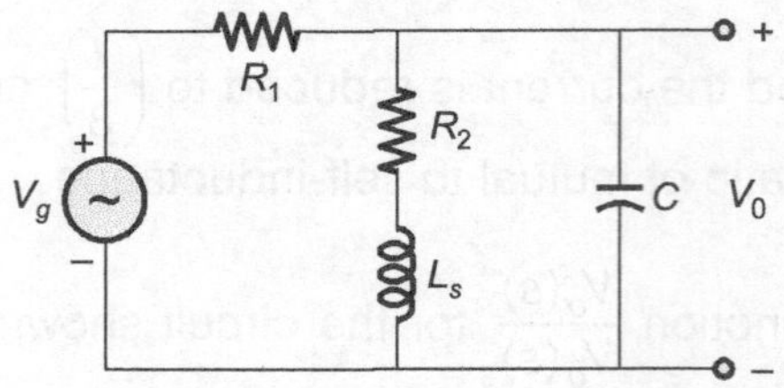

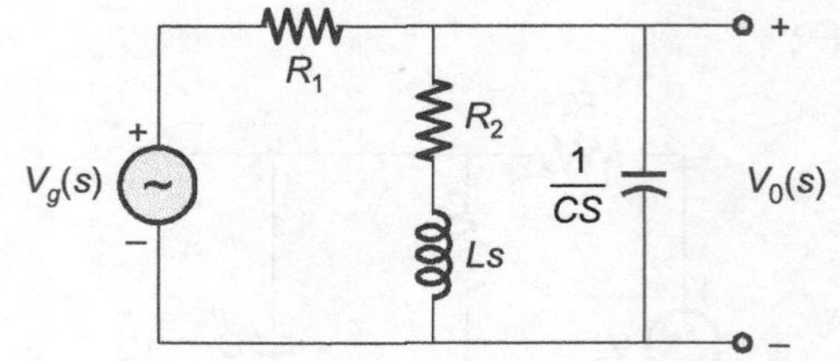

$$Z'(s) = \frac{(R_2 + Ls)\cdot\frac{1}{Cs}}{R_2 + Ls + \frac{1}{Cs}} \Rightarrow \frac{R_2 + Ls}{LCs^2 + R_2Cs + 1}$$

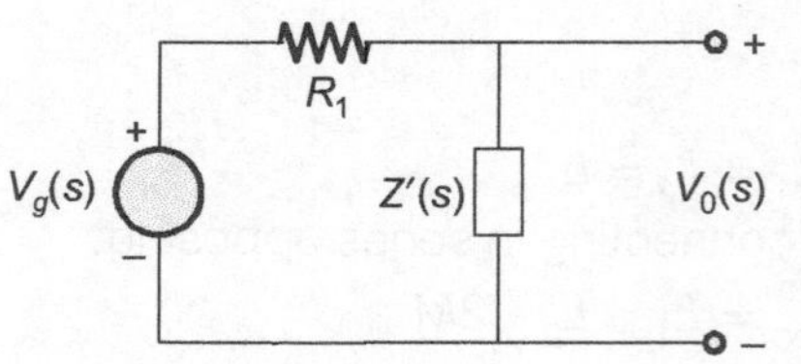

$$V_g(s) = [R, Z'(s)]I(s) \quad \text{...(1)}$$

$$V_o(s) = Z'(s)I(s) \quad \text{...(2)}$$

Dividing equation (2) by equation (1)

$$\frac{V_o(s)}{V_g(s)} = \frac{\frac{R_2 + Ls}{LCs^2 + R_2Cs + 1}}{R_1 + \frac{R_2 + Ls}{LCs^2 + R_2Cs + 1}}$$

$$\frac{V_o(s)}{V_g(s)} = \frac{R_2 + Ls}{R_1LCs^2 + R_1R_2Cs + R_1 + R_2 + Ls}$$

$$\frac{V_o(s)}{V_g(s)} = \frac{R_2 + Ls}{R_1LCs^2 + (R_1R_2C + L)s + R_1 + R_2}$$

$$\frac{V_o(s)}{V_g(s)} = \frac{50 + 10^{-2}s}{10^{-5}s^2 + 0.06s + 550}$$

$$= \frac{10^{-2}[s+5000]}{10^{-5}[s^2+6000s+55\times10^6]}$$

$$\frac{V_o(s)}{V_g(s)} = \frac{1000(s+5000)}{s^2+6000s+55\times10^6}$$

7. Network Graph, Functions and Synthesis

7.1 The driving point admittance of network is given by the following equation:

$$Y(s) = \frac{s^3+2s^2+3s+1}{s^3+s^2+2s+1}$$

(a) Perform a continued fraction of admittance function $Y(s)$ given in the above equation

(b) Realise a network from the continued fraction of $Y(s)$.

(c) With the help of Routh-Hurwitz criteria, show that neither the numerator, nor the denominator of $Y(s)$ given in the above equation has any zeros on the right half of s-plane.

[10+10+12 = 32 marks : 2001]

Solution:

(a) $$Y(s) = \frac{s^3+2s^2+3s+1}{s^3+s^2+2s+1}$$

Continued fraction of $Y(s)$ can be obtained as,

$$
\begin{array}{l}
s^3+s^2+2s+1)\,s^3+2s^2+3s+1\,(1 \rightarrow Y_1 \\
\qquad\qquad\quad \underline{s^3+s^2+2s+1} \\
\qquad\qquad\qquad s^2+s)\,s^3+s^2+2s+1\,(s \rightarrow Z_1 \\
\qquad\qquad\qquad\qquad \underline{s^3+s^2} \\
\qquad\qquad\qquad\qquad 2s+1)\,s^2+s\,(s/2 \rightarrow Y_2 \\
\qquad\qquad\qquad\qquad\qquad \underline{s^2+s/2} \\
\qquad\qquad\qquad\qquad\qquad s/2)\,2s+1\,(4 \rightarrow Z_2 \\
\qquad\qquad\qquad\qquad\qquad\quad \underline{2s} \\
\qquad\qquad\qquad\qquad\qquad\qquad 1)\,s/2\,(s/2 \rightarrow Y_3 \\
\qquad\qquad\qquad\qquad\qquad\qquad\quad \underline{s/2} \\
\qquad\qquad\qquad\qquad\qquad\qquad\quad \times
\end{array}
$$

(b) where,

$R_1 \rightarrow 1\ \Omega$

$L_1 \rightarrow 1\text{ H}$

$C_2 \rightarrow 1/2\text{ F}$

$R_2 = 4\ \Omega$

$C_3 = 1/2\text{ F}$

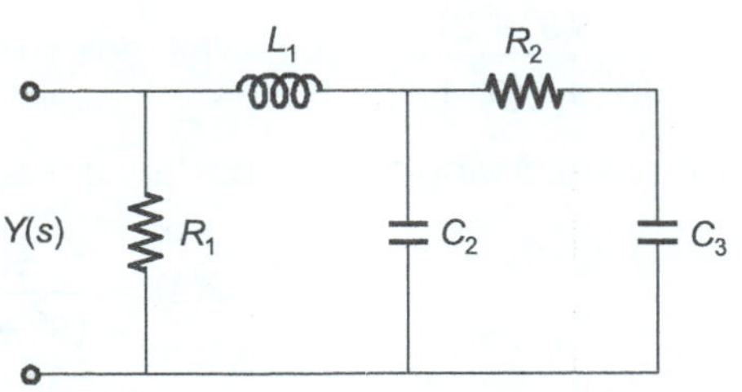

(c) $$Y(s) = \frac{N(s)}{D(s)}$$

where, $N(s) = s^3 + 2s^2 + 3s + 1$

and $D(s) = s^3 + s^2 + 2s + 1$

First, constructing Routh array for numerator

$$\begin{array}{c|cc} s^3 & 1 & 3 \\ s^2 & 2 & 1 \\ s^1 & 5/2 & \\ s^0 & 1 & \end{array}$$

Since all the elements in the first column are positive, therefore there is no zeros in right half s-plane.
Now, constructing Routh array for denominator.

$$D(s) = s^3 + s^2 + 2s + 1$$

$$\begin{array}{c|cc} s^3 & 1 & 2 \\ s^2 & 1 & 1 \\ s^1 & 1 & 0 \\ s^0 & 1 & \end{array}$$

As all the elements in the first column are positive, then there is no zero in RHS-plane.

7.2 Point out the necessary and sufficient condition for positive real function, Justify that the following function can be driving point impedance reactance network.

$$Z(s) = \frac{s(s^2+2)}{(s^2+1)(s^2+4)}$$

Give the realization of the network using first Foster form of this network.

[10 marks : 2004]

Solution:

The necessary and sufficient conditions of positive realness of a given rational function $[F(s)]$:

1. $F(s)$ must have simple poles on $j\omega$-axis with real and positive residues.
2. If $F(s) = \frac{P(s)}{Q(s)}$, then $P(s) + Q(s)$ must be Hurwitz.
3. If $F(s) = \frac{P(s)}{Q(s)} = \frac{M_1(s)+N_1(s)}{M_2(s)+N_2(s)}$, M being the even parts and N being the odd parts, then for positive realness $M_1 M_2 - N_1 N_2|_{s=j\omega} \geq 0$ for all ω

   Such that $\text{Re}[F(j\omega)] \geq 0$ for all ω.

$Z(s) = \frac{s(s^2+2)}{(s^2+1)(s^2+4)}$, since the poles and zeros alternatively on imaginary axis therefore $Z(s)$ can be realized as a driving point impedance reactive network.

$$Z(s) = \frac{s(s^2+2)}{(s^2+1)(s^2+4)}$$

$$Z(j\omega) = \frac{j\omega(2-\omega^2)}{(1-\omega^2)(4-\omega^2)} = jX(\omega)$$

$$X(\omega) = \frac{\omega(2-\omega^2)}{(1-\omega^2)(4-\omega^2)} \quad \text{...(i)}$$

(i) $\text{Re}\{Z(j\omega)\} = 0$

(ii) Zeros of $Z(s) = s = 0,\ s = \pm\sqrt{2},\ s \to \infty$

Poles of Z(s): $s = \pm j1,\ s = \pm j2$

Poles-zero diagram:

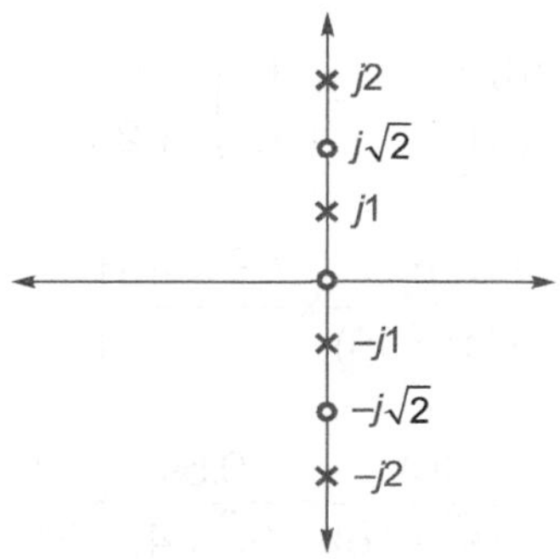

(iii) Poles and zeros are simple on $j\omega$-axis and alternating.

(iv) The first critical frequency should be either a pole (or) zero. Here it is zero.

(v) The last critical frequency should be either a pole (or) zero. Here it is zero.

(vi) $\dfrac{dX(\omega)}{d\omega} < 0$

This condition is satisfied for $X(\omega)$.

∴ The given driving point impedance is satisfying the conditions of LC reactive network.

$$\therefore \qquad \frac{Z(s)}{s} = \frac{(s^2+2)}{(s^2+1)(s^2+4)}$$

where, $(s^2 = x)$

$$\frac{Z(s)}{s} = \left(\frac{(x+2)}{(x+1)(x+4)}\right) = \frac{A}{x+1} + \frac{B}{x+4}$$

$$A = \left.\frac{x+2}{x+4}\right|_{x=-1} = \frac{1}{3}$$

$$B = \left.\frac{x+2}{x+1}\right|_{x=-4} = \frac{2}{3}$$

$$\frac{Z(s)}{s} = \frac{1}{3(s^2+1)} + \frac{2}{3(s^2+4)}$$

$$Z(s) = \frac{s/3}{s^2+1} + \frac{2s/3}{s^2+4} = \frac{1}{3s + \frac{3}{s}} + \frac{1}{\frac{3s}{2} + \frac{6}{s}}$$

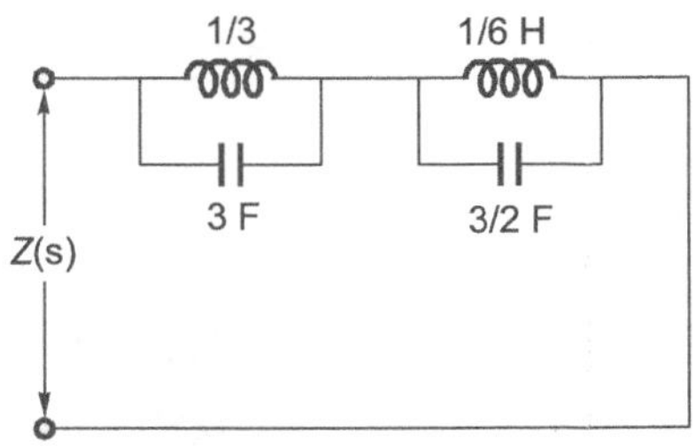

Now the network realized is shown figure.

7.3 Synthesize the impedance function $z(s) = \dfrac{s(s^2+10)}{(s^2+4)(s^2+16)}$ using any one of the Foster forms of network realization.

[8 marks : 2014]

Solution:

Foster-I form:

$$Z(s) = \frac{s(s^2+10)}{(s^2+4)(s^2+16)} = \frac{k_1 s}{(s^2+4)} + \frac{k_2 s}{(s^2+16)}$$

$$\Rightarrow \quad k_1 = \left.\frac{(s^2+10)}{(s^2+16)}\right|_{s^2=-4} = \left(\frac{6}{12}\right) = \left(\frac{1}{2}\right) = 0.5$$

$$\Rightarrow \quad k_2 = \left.\frac{s(s^2+10)}{(s^2+4)(s^2+16)} \times \frac{(s^2+16)}{s}\right|_{s^2=-16}$$

$$Z(s) = \frac{s(s^2+10)}{(s^2+4)(s^2+16)} = \frac{0.5s}{(s^2+4)} + \frac{0.5s}{(s^2+16)}$$

$$= \frac{s/2}{(s^2+4)} + \frac{s/2}{(s^2+16)} = \frac{1}{\left(2s+\frac{8}{s}\right)} + \frac{1}{\left(2s+\frac{32}{s}\right)}$$

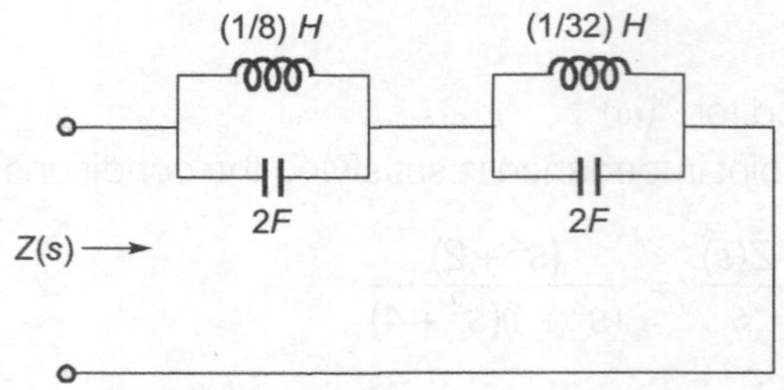

Foster-II form:

$$Y(s) = \frac{(s^2+4)(s^2+16)}{s(s^2+10)}$$

$Y(s)$ has got a pole at infinity.
So, remove the pole at infinity,

$$Y(s) = \frac{(s^4+20s^2+64)}{(s^3+10s)}$$

$$s^3+10s \overline{)\, s^4+20s^2+64 \,(} \; s$$
$$\underline{s^4+10s^2}$$
$$10s^2+64$$

$$Y(s) = s + \frac{10s^2+64}{s(s^2+10)} = s + \frac{A}{s} + \frac{Bs}{s^2+10}$$

i.e.,
$$\frac{10s^2+64}{s(s^2+10)} = s + \frac{A}{s} + \frac{Bs}{s^2+10}$$

$$A = \left.\frac{10s^2+64}{s(s^2+10)} \cdot s\right|_{s=0} = \left(\frac{64}{10}\right) = \left(\frac{32}{5}\right)$$

$$\Rightarrow \quad B = \left(\frac{10s^2+64}{s^2}\right)_{s^2=10} = \left(\frac{-100+64}{-10}\right) = \left(\frac{-36}{-10}\right) = \left(\frac{18}{5}\right)$$

$$Y(s) = s+\frac{32}{5s}+\frac{18s}{5(s^2+10)} = s+\frac{32}{5s}+\frac{1}{\left(\frac{5}{18}\right)s+\left(\frac{50}{18s}\right)}$$

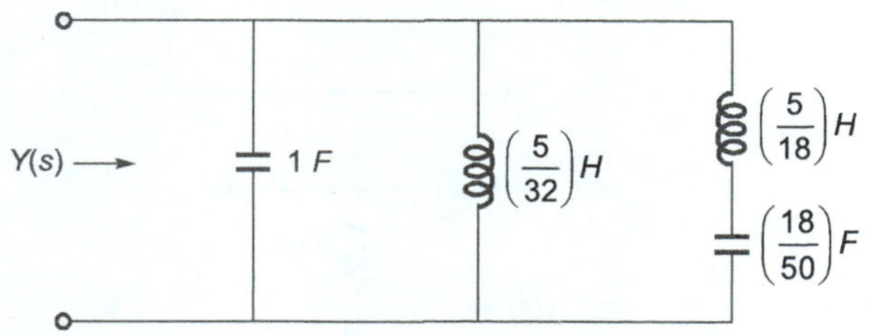

7.4 **State the conditions that are to be satisfied for a network function to be L-C impedance function. Test whether the following network function $z(s)$ represents L-C impedance function:**

$$z(s) = \frac{(s^2+4)(s^2+9)}{s(s^2+6)}$$

[8 marks : 2014]

Solution:

Conditions that are to be satisfied for a network function to be *L-C* impedance function, are:

1. The *LC* impedance function is a ratio of even to odd or odd to even polynomial.
2. The poles are zeros are simple, lie along the $j\omega$-axis and they alternate (or interlace).
3. There must be either a zero or a pole at the origin or at ∞.
4. The highest degree of numerator and denominator polynomial may different at the most by unity.
5. The lowest power of the numerator and denominator polynomial must also differ at the most by unity.

The given *L-C* impedance function is

$$Z(s) = \frac{(s^2+4)(s^2+9)}{s(s^2+6)}$$

Zeros: $s = \pm j2, \pm j3$

Poles: $s = 0, \pm j2.45$

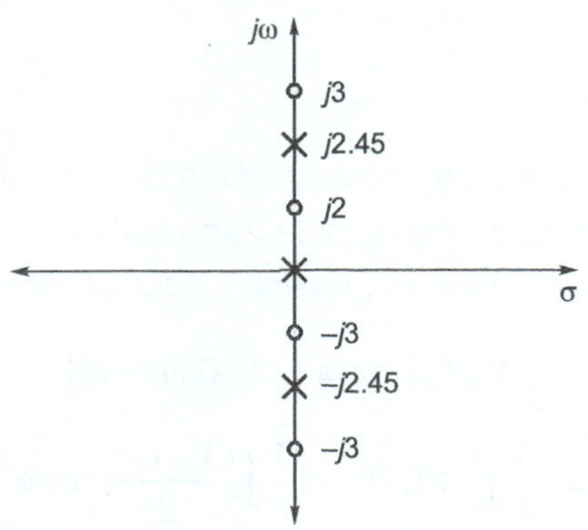

Yes, $Z(s)$ represents *L-C* impedance function.

7.5 **Find all the four canonical forms of the impedance $z(s) = s + 1$. State if you observe some speciality in these realisations.**

[10 marks : 2015]

Solution:

The given impedance function is $z(s) = s + 1$, we can see that the function cannot be realized using *L-C* as two components of the circuit. The circuit can be *R-L* or *RC*, since $z(s) = s + 1$ so it is *RL* function.

so (1) Foster form 1:

$$z(s) = s + 1 \quad \Rightarrow$$

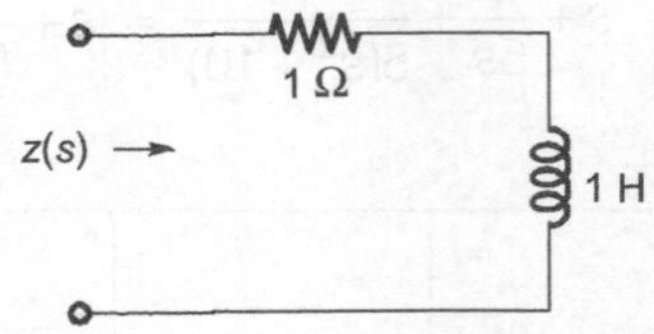

(2) Foster form 2:

$$y(s) = \frac{1}{s+1} \quad \Rightarrow$$

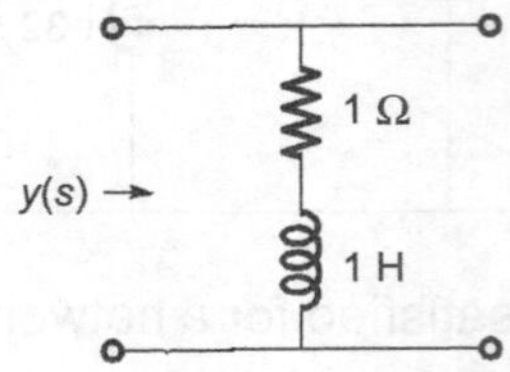

(3) Cauer form 1:

$$z(s) = s + 1 \Rightarrow$$

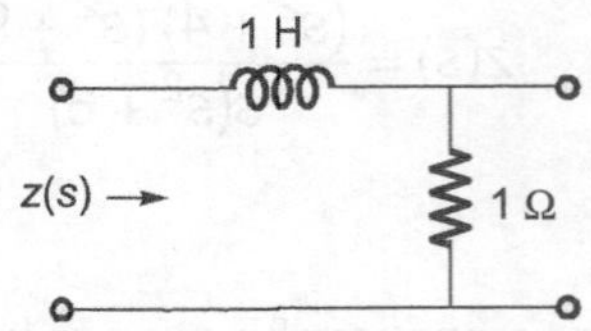

(4) Cauer form 2:

$$z(s) = \frac{1}{1+s} \quad \Rightarrow$$

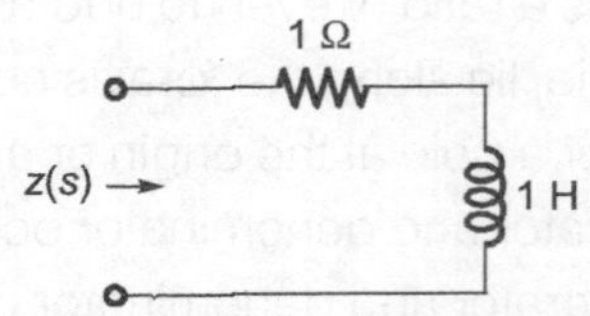

The above realisation show that all the four realization are same when there is no pole in the $z(s)$, also when the impedance function is simple then all the four forms have same structure.

8. Miscellaneous (Three-Phase Circuit + Filters)

8.1 Prove that the average power in an ac circuit is given by $W = VI\cos\phi$ where symbols have their usual meaning.

[12 marks : 2002]

Solution:

Let the voltage in the a.c. circuit be $v(t) = V_m \cos\omega t$ and the current $i(t) = I_m \cos(\omega t - \phi)$ where the phase angle may be positive or negative according leading or lagging current.

$$\text{Instantaneous power } p(t) = v(t)\, i(t) = V_m \cos\omega t \cdot I_m (\cos\omega t - \phi)$$

$$= \frac{1}{2} V_m I_m [\cos\phi + \cos(2\omega t - \phi)]$$

$$\text{Average power in a.c. circuit} = \frac{1}{T}\int_0^T p(t)\, dt = \frac{1}{T}\int_0^T \frac{V_m I_m}{2}[\cos\phi + \cos(2\omega t - \phi)]\, dt$$

$$P = \frac{1}{2T} V_m I_m \cos\phi [t]_0^T + \frac{1}{2T} V_m I_m \left[\frac{\sin(2\omega t - \phi)}{2\omega}\right]_0^T$$

The average of $\cos(2\omega t - \phi)$ is zero so that,

$$P = \frac{1}{2} V_m I_m \cos\phi$$

$$\text{Average power in a.c. circuit} = \frac{1}{2} V_m I_m \cos\phi = \left(\frac{V_m}{\sqrt{2}}\right)\left(\frac{I_m}{\sqrt{2}}\right)\cos\phi = VI\cos\phi$$

ϕ is phase difference between V and I.

8.2 A large consumer of electricity requires 10 kW, 230 V_{rms} of power at a power factor angle of 60° lagging through a transmission line of resistance 0.1 Ω. Determine the power produced by the utility to meet the load demand. If the consumer corrects his power factor from 0.5 to 0.9 lagging, what power the utility has to produce to meet the load demand?

[10 marks : 2004]

Solution:

$$VI\cos\phi = P$$

when, Load requirement $P = 10$ kW

and Power factor $= \cos\phi = \cos 60 = 0.5$

$$I = \frac{P}{V\cos\phi} = \frac{10\times10^3}{230\times0.5} = 86.96 \text{ A}$$

$$\text{Loss in line} = I^2R = (86.96)^2\times0.1$$

$$= 756.2 \text{ Watts} = 0.756 \text{ kW}$$

$$\text{Power produced by the utility} = \text{Load requirement} + \text{Loss}$$

$$= 10 + 0.756$$

$$= 10.756 \text{ kW}$$

For a power factor of 0.9,

$$I = \frac{P}{V\cos\phi} = \frac{10\times10^3}{230\times0.9} = 48.31 \text{ A}$$

$$\text{Loss in line} = I^2R = (48.31)^2\times0.1$$

$$= 233.38 \text{ Watts}$$

$$= 0.233 \text{ kW}$$

$$\text{Power produced by the utility} = \text{Load requirement} + \text{Loss}$$

$$= 10 + 0.233 = 10.233 \text{ kW}$$

8.3 An industrial consumer is operating a 3-phase 1.0 kW induction motor at a lagging power factor of 0.8 and at a source voltage of 400 V r.m.s. He wishes to raise the power factor to 0.95 lagging by connecting a circuit element in parallel to the load (motor). Indicate the type of circuit element (inductive or capacitive) and find its value if the operating frequency is 50 Hz. Give your comments on the new line and motor currents.

[10 marks : 2005]

Solution:

Assuming the induction motor is star connected.

As we know,

$$I_1 = \frac{1000}{(\sqrt{3})\times400\times0.8} = 1.8 \text{ A}$$

$$\cos\phi = 0.8$$

$$\therefore \quad \phi = 36.87°, \tan\phi = 0.75 = \frac{Q}{P} \qquad (\because P = 1.0 \text{ kW})$$

$$\therefore \quad Q = 750 \text{ VAR}$$

The consumer wishes to raise the power factor to 0.95 lagging by connecting capacitance element in parallel to the load.

$$\text{New p.f.} = 0.95$$

$$\cos\phi = 0.95$$

$$\phi = 18.2°$$

$$\tan\phi = 0.3286 = \frac{Q'}{P}$$

$$Q' = 328.6 \text{ VAR}$$

$$I_{new} = \frac{1000}{\sqrt{3}\times 400\times 0.95}$$

$$= 1.5193$$

$$Q_c = 750 - 328.6$$

$$= 421.4 \text{ V}$$

For per phase,

$$\frac{Q_c}{3} = \frac{V_{ph}^2}{X_C}$$

$$\frac{421.4}{3} = \left(\frac{400}{\sqrt{3}}\right)^2 \Big/ X_C$$

$$X_C = 379.68$$

$\therefore$

$$X_C = \frac{1}{\omega C}$$

$$379.68 = \frac{1}{C\times 2\times\pi\times 50}$$

$\therefore$

$$C = 8.38\ \mu\text{F}$$

So, after the power factor improvement, the net current drawn by combination of (*IM*) and capacitor has been reduced.

8.4 **A 415 V, three-phase generator supplies power to both a delta and a star-connected load in the manner shown in Figure. All the phase impedances are identical and specifically equal to (5 + *j*8.66) Ω. Compute the total generator current in each line.**

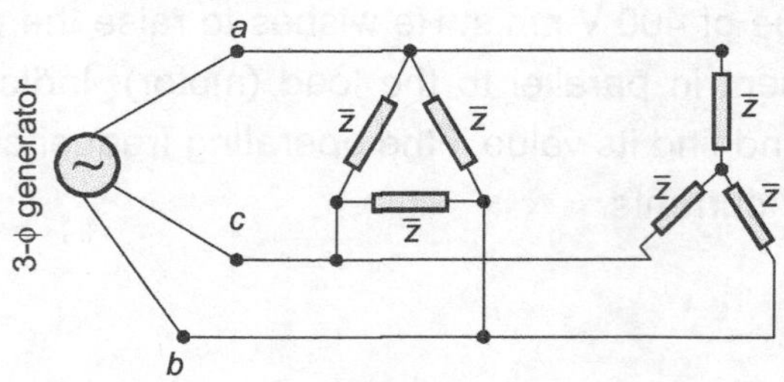

[10 marks : 2005]

Solution:

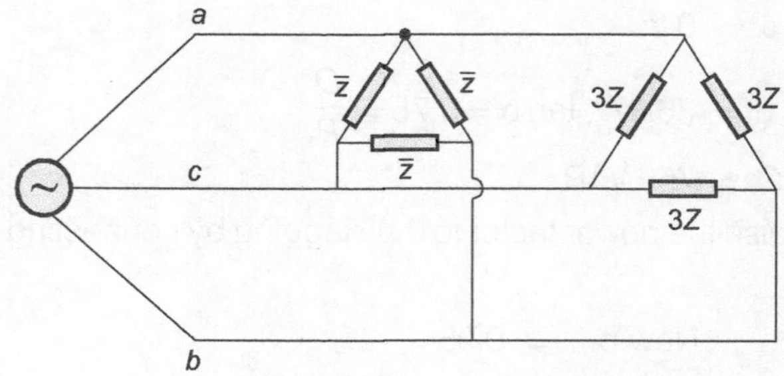

By solving,

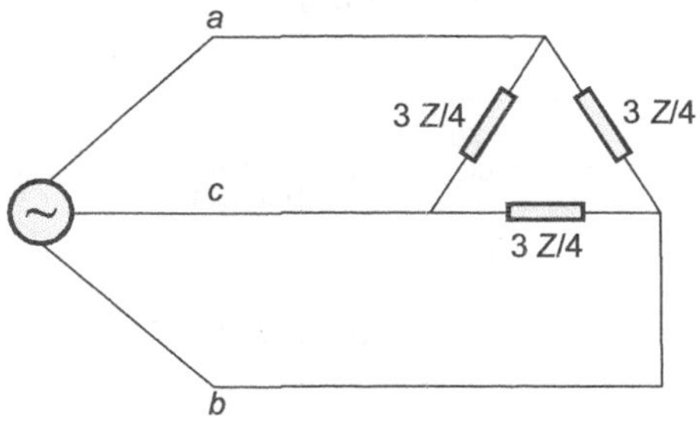

Equivalent star connection,

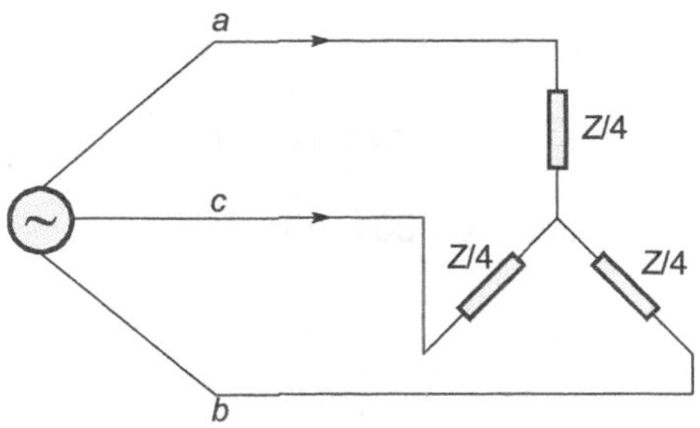

$$\text{(line current)} = \frac{415 \times 4}{\sqrt{3} \times Z} = \frac{415 \times 4}{(\sqrt{3}) \times (5 + 8.66j)}$$

$$= 47.92 - 83\,j = 95.84\angle -60^\circ$$

$$I_{al} = 95.84\angle -60^\circ$$

$$I_{bl} = 95.84\angle -180^\circ$$

$$I_{cl} = 95.84\angle 60^\circ$$

8.5 **An inductive load of resistance 20 W and inductance 0.1 H is switched on to an ac voltage $v = 100 \sin(\omega\tau + \alpha)$.**
Find the switching angle α such that there is no transient at $\omega = 314$ rad/s.

[12 marks : 2006]

Solution:

Applying KVL,

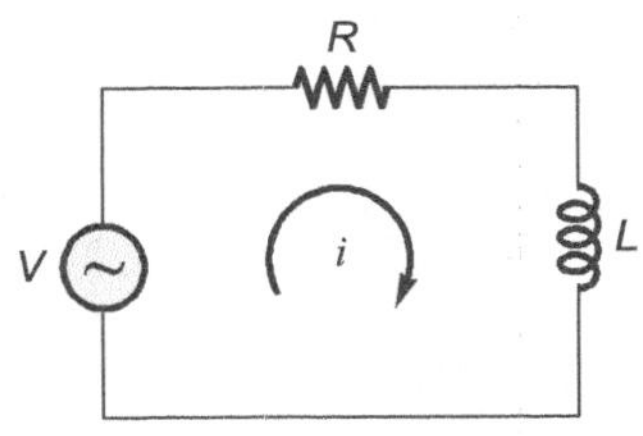

$$L\frac{di}{dt} + Ri = V$$

$$L\frac{di}{dt} + Ri = 10 \sin(\omega t + \alpha)$$

$$= 10\,[\cos\alpha \cdot \sin\omega t + \sin\alpha \cdot \cos\omega t]$$

Taking Laplace transform

$$(Ls + R)\,I(s) = \frac{100}{s^2 + \omega^2}[\omega \cos\alpha + s \sin\alpha]$$

$$I(s) = \frac{100[\omega \cos\alpha + s \sin\alpha]}{(s^2 + \omega^2)(R + Ls)}$$

The solution of current will have two parts, one particular (i_p) and the other complementary (i_c) solution. The particular solution is responsible for steady state part while the complementary part is responsible for the transient part.

$$I(s) = \frac{100[\omega\cos\alpha + s\sin\alpha]}{(s^2+\omega^2)(R+Ls)}$$

$$= \underbrace{\frac{As+B}{s^2+\omega^2}}_{\substack{\text{Particular solution part}\\ \text{Steady state}}} + \underbrace{\frac{C}{R+LS}}_{\substack{\text{Complementary solution part}\\ \text{Transient}}}$$

Finding the coefficients A, B and C:

$(AL + C)s^2 + (AR + BL)s + (BR + C\omega^2) = 100\ (\omega \cos \alpha + s \sin \alpha)$

$$AL + C = 0 \quad \text{...(i)}$$

$$AR + BL = 100 \sin \alpha \quad \text{...(ii)}$$

$$BR + C\omega^2 = 100\ \omega \cos \alpha \quad \text{...(iii)}$$

Putting the value of A from equation (*i*) to (*ii*),

$$-\left(\frac{R}{L}\right)C + BL = 100 \sin \alpha] \times R \quad \text{...(iv)}$$

$$BR + C\omega^2 = 100\ \omega \cos\alpha] \times L \quad \text{...(v)}$$

$L(v) - R(iv) \Rightarrow$

$$C\left[L\omega^2 + \frac{R^2}{L}\right] = 100[\omega L\cos \alpha - R \sin \alpha]$$

$$C = \frac{100[\omega L\cos\alpha - R\sin\alpha]}{R^2+\omega^2L^2}L$$

The complementary response is, $I(s) = \dfrac{C}{R+sL} = \dfrac{C}{R(s+R/L)}$

Taking inverse Laplace transform,

$$i(t) = \frac{C}{L}e^{-Rt/L}$$

This is a decreasing exponential and will be zero only when the coefficient C is zero.

$$C = 0 = \frac{100[\omega L\cos\alpha - R\sin\alpha]}{R^2+\omega^2L^2}L$$

$$\omega L \cos \alpha - R \sin \alpha = 0$$

$$\tan \alpha = \frac{\omega L}{R}$$

$$\alpha = \tan^{-1}\frac{\omega L}{R}$$

Putting the values, $\alpha = \tan^{-1}\left(\dfrac{314\times 0.1}{20}\right) = 57.50° = 1 \text{ rad.}$

8.6 **An unbalanced star connected load is fed from a three-phase three wire balanced 400 V supply. The load impedance in the three phases are:**

$$Z_R = (4 + j\,8)\ \Omega;\ Z_Y = (3 + j\,4)\ \Omega;\ Z_B = (15 + j\,20)\ \Omega$$

Assuming the phase sequence to be *RYB*, calculate the total power absorbed by load.

[14 marks : 2006]

Solution:

The given load is star connected unbalanced 3-wire load. For this we can use any following method to calculate each line/phase current.

(i) Nodal analysis (ii) Mesh/loop analysis
(iii) Millman's theorem (iv) Star to delta transformation

By using mesh analysis,

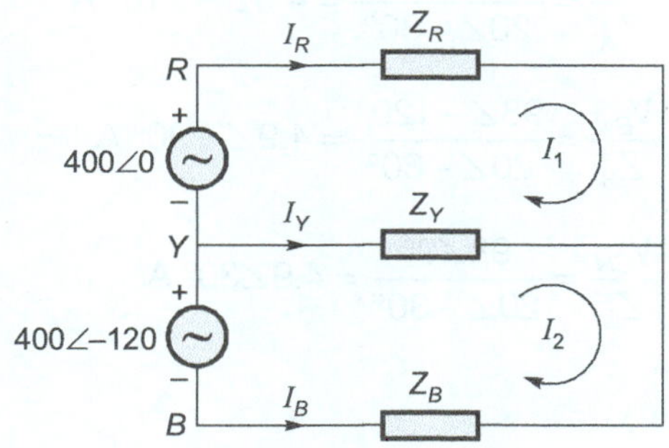

$$400\angle 0^\circ = (Z_R + Z_Y)I_1 - Z_Y I_2 \quad \text{...(i)}$$
$$400\angle -120^\circ = -Z_Y I_1 + (Z_Y + Z_B) I_2 \quad \text{...(ii)}$$

Substitute Z_R, Z_Y and Z_B values in above equation and solve the equation to get

$$I_1 = 28.44\angle -69$$
$$I_2 = 13.02\angle -152.4$$
$$I_R = I_1 = 28.44\angle -69$$
$$I_Y = I_2 - I_1 \Rightarrow 29.88\angle 136.6$$
$$I_B = I_2 = 13.02\angle -152.4$$

$$P = |I_R|^2 \times 4 + |I_Y|^2 \times 3 + |I_B|^2 \times 15$$
$$= (28.44)^2 \times 4 + (29.88)^2 \times 3 + (13.02)^2 \times 15$$
$$= 8456 \text{ W}$$

8.7 **A three-phase, 4-wire CBA system, with an effective line voltage of 169.7 V, has three impedance of 20∠–30° Ω in a Y-connection (Figure). Determine the line currents and draw the voltage-current phasor diagram.**

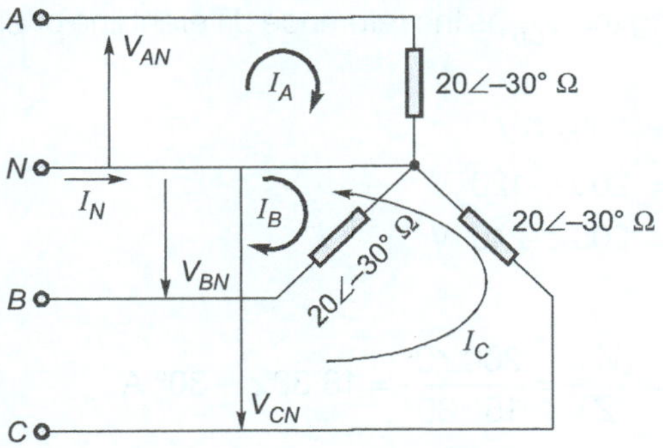

[12 marks : 2007]

Solution:

Line voltage is 169.7 V.

Hence, $$|V_{CN}| = |V_{BN}| = |V_{AN}| = \frac{169.7}{\sqrt{3}} \simeq 98\ V$$

Phase sequence is *CBA* and V_{CN} is taken as the reference.

$V_{CN} = 98\angle 0^\circ$ V, $V_{BN} = 98\angle -120^\circ$ V and $V_{AN} = 98\angle -240^\circ$

$Z_A = Z_B = Z_C = 20\angle{-30°}$

Line currents are

$$I_A = \frac{V_{AN}}{Z_A} = \frac{98\angle-240°}{20\angle-30°} = 4.9\angle-210° \text{ A}$$

$$I_B = \frac{V_{BN}}{Z_B} = \frac{98\angle-120°}{20\angle-30°} = 4.9\angle-90° \text{ A}$$

$$I_C = \frac{V_{CN}}{Z_C} = \frac{98\angle0°}{20\angle-30°} = 4.9\angle30° \text{ A}$$

Voltage current phasor diagram,

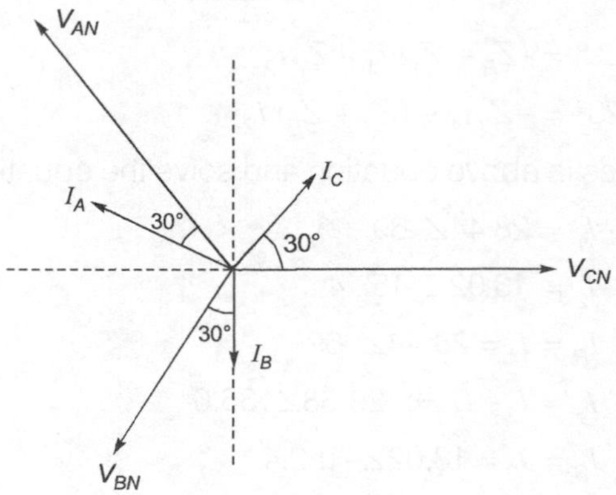

8.8 A three-phase, three-wire *CBA* system, with an effective line voltage 200 V, has a balanced Δ-connected load with impedance $Z = 15\angle30°\ \Omega$. Obtain the line and phase currents by the single-line equivalent method.

[12 marks : 2008]

Solution:

$$Z_P = \text{Impedance of the phase} = 15\angle30°$$

Phase sequence is *CBA*. Taking the V_{CB} as the reference different line or phase voltages of the delta connected load.

$$V_{CB} = 200\angle0 \text{ V}$$
$$V_{BA} = 200\angle-120° \text{ V}$$
$$V_{AC} = 200\angle-240° \text{ V}$$

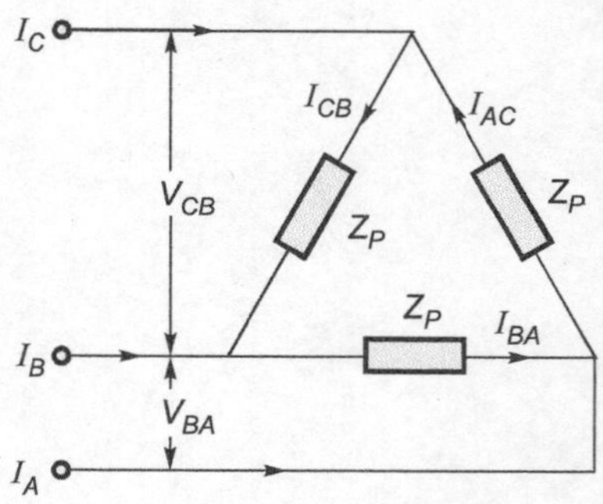

Different phase currents,

$$I_{CB} = \frac{V_{CB}}{Z_P} = \frac{200\angle0°}{15\angle30°} = 13.33\angle-30° \text{ A}$$

Similarly, $I_{BA} = 13.33\angle-150°$ A

$I_{AC} = 13.33\angle-270°$ A

Different line currents,

$$I_A = I_{AC} - I_{BA} = 13.33\angle-270° - 13.33\angle-150° = 23.08\angle60° \text{ A}$$

$$I_B = I_{BA} - I_{CB} = 13.33\angle-150° - 13.33\angle-30° = 23.08\angle180° \text{ A}$$

$$I_C = I_{CB} - I_{AC} = 13.33\angle-30° - 13.33\angle-270° = 23.08\angle-60° \text{ A}$$

8.9 A star-connected load is connected across a 3-phase balanced a-c supply as shown in the figure. The lamps L_B and L_Y have equal resistance. If the phase-sequence is $\overrightarrow{RYB}$, draw a neat phasor diagram with V_{YB} as the reference phasor, and hence find out which lamp will glow higher.

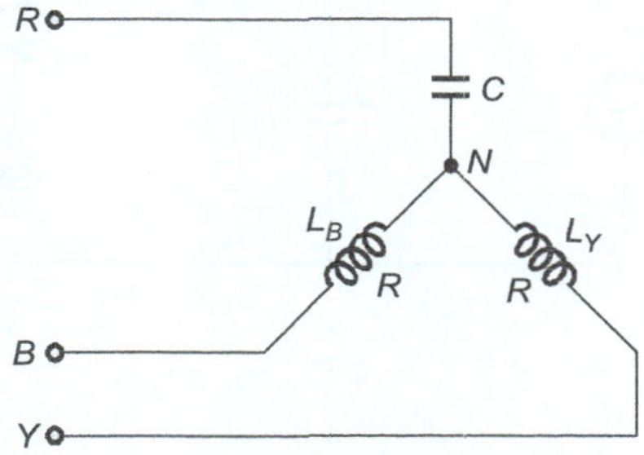

[12 marks : 2009]

Solution:

The given circuit is star connected unbalanced load so we use mesh analysis method to find the currents. Now rearranging the circuit where V_{YB} is reference.

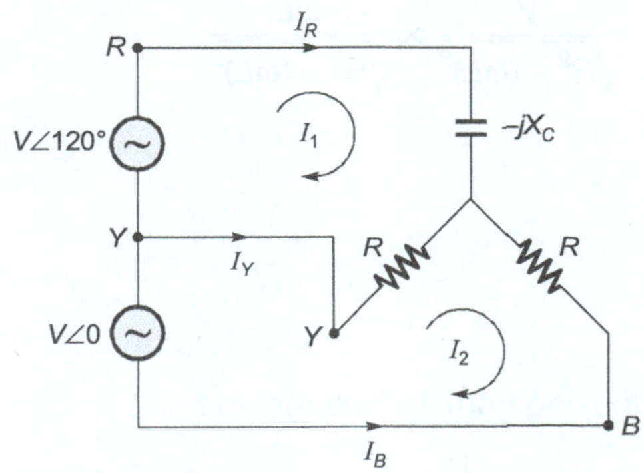

$$V_{YB} = V\angle 0°$$
$$V_{BR} = V\angle -120°$$
$$V_{RY} = V\angle 120°$$

Applying KVL in each loop:

$$V\angle 120° = I_1(R - jX_C) - I_2R \quad ...(1)$$

$$V\angle 0° = I_1R + 2I_2R \quad ...(2)$$

From equation (2),

$$I_2 = \frac{V}{2R} - \frac{I_1}{2} \quad ...(3)$$

Substitute (3) in (1),

$$V\angle 120° = I_1(R - jX_C) - R\left(\frac{V}{2R} + \frac{I_1}{2}\right)$$

$$I_1 = \frac{V(j1.732)}{R - j2X_C} \quad ...(4)$$

Substitute (4) in (3), $I_2 = \dfrac{V}{2R} \cdot \left(\dfrac{R - j(1.732R + 2X_C)}{R - j2X_C}\right)$

$$I_B = -I_2 = -\frac{V}{2R}\left(\frac{R + j(1.732R + 2X_C}{R - j2X_C}\right)$$

$$I_Y = I_2 - I_1 = \frac{V}{2R}\left(\frac{R - j(5.196R + 2X_C}{R - j2X_C}\right)$$

From the numerators of I_B and I_Y, it can be seen that $|I_Y| > |I_B|$.

$\therefore$ Lamp L_Y glow brighter.

8.10 **For the parallel circuit shown in the figure draw a neat phasor diagram and determine an expression for the value of the capacitor C for the circuit to exhibit the unity power factor.**

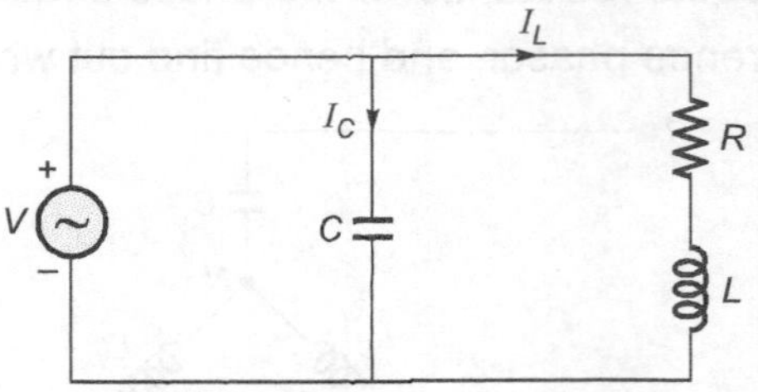

[8+4 = 12 marks : 2009]

Solution:

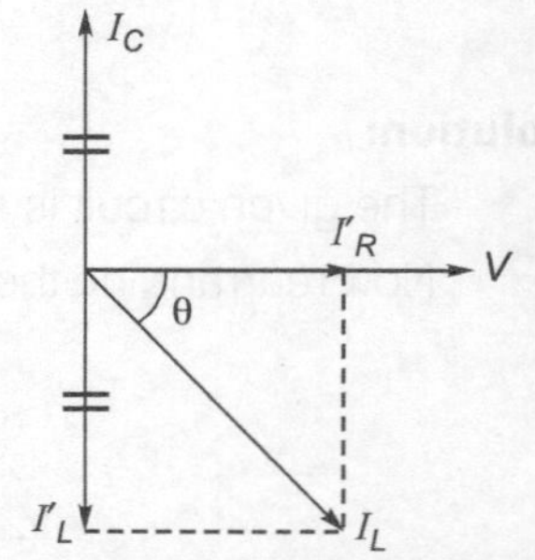

Phase diagram of the network

where, $\tan\theta = \left(\dfrac{\omega L}{R}\right)$

From the phase diagram,
For unity pf ($I'_L = I_C$)

where, $I'_L = I_L \times \sin\theta = \dfrac{V}{\sqrt{R^2+(\omega L)^2}} \times \dfrac{\omega L}{\sqrt{R^2+(\omega L)^2}}$

and $I_C = V \times \omega C$

$$\omega C = \frac{\omega L}{R^2+(\omega L)^2}$$

$$C = \frac{L}{R^2+(\omega L)^2}$$

Let us consider the admittance of network looking from the voltage source.

$$y = \frac{1}{Z_C} + \frac{1}{Z_L} = jC\omega + \frac{1}{R + j\omega L}$$

$$= jC\omega + \frac{(R - j\omega L)}{R^2+(\omega L)^2}$$

For unity pf, imaginary term of admittance (y) should be equal to zero.

$$jC\omega - \frac{j\omega L}{R^2+(\omega L)^2} = 0$$

$\therefore$ $$C = \frac{L}{R^2+(\omega L)^2}$$

8.11 **Three identical coils are connected in star to a 200 V, 3-phase supply and each takes 400 W. The p.f. is 0.8 lagging. What will be the magnitude of line currents and total power if the same coils are connected in delta to the same supply? If the total power in delta-connected coils is measured by two-wattmeter method, what will be the readings of the wattmeters?**

[8+4 = 12 marks : 2010]

Solution:

Let resistance of the phase is R and reactance is X.

$$V_{l-l} = 200\text{ V}$$

$$V_{l-N} = \frac{200}{\sqrt{3}} = 115.47\text{ V}$$

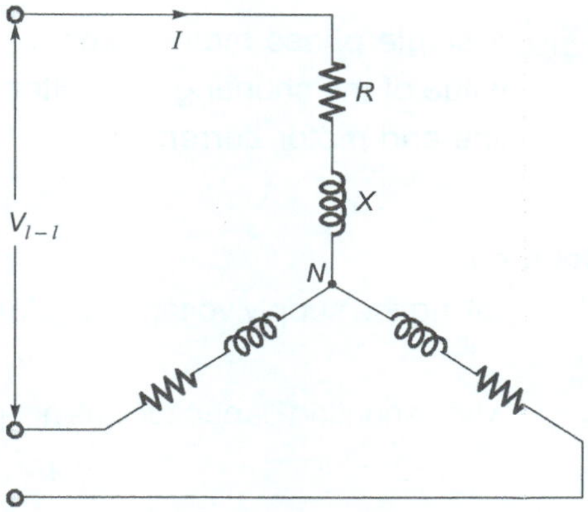

Power consumed by each phase = 400 W

$$V_{l-n} \times I \times \cos\phi = 400$$
$$115.47 \times I \times 0.8 = 400$$
$$I = 4.33 \text{ A}$$
$$I^2R = 400$$
$$\Rightarrow \quad 4.33^2 R = 400$$
$$R = 21.33\ \Omega$$
$$Z = \frac{V_{l-n}}{I} = \frac{115.47}{4.33} = 26.66\ \Omega$$
$$X = \sqrt{Z^2 - R^2}$$
$$= \sqrt{26.66^2 - 21.33^2} = 16\ \Omega$$

If same coils are connected in delta.

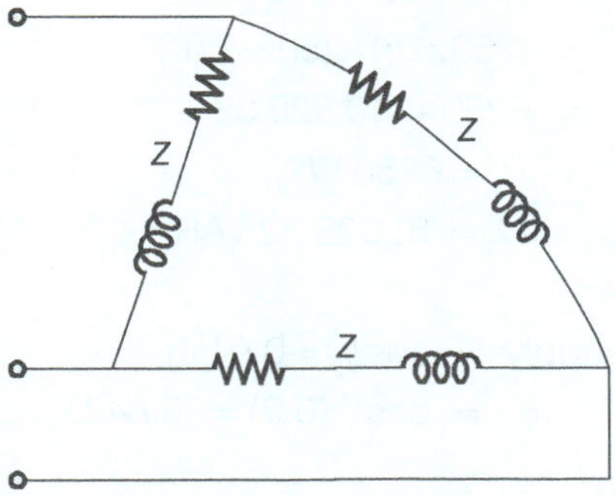

where, $Z = R + jX = 21.33 + j16$

Converting the delta connection into equivalent star connection.

where, $$Z' = \frac{Z}{3} = \frac{21.33 + j16}{3}$$
$$Z' = R' + jX' = 7.11 + j5.33$$
$$= 8.89\angle 36.87°$$
$$R' = 7.11\ \Omega$$
$$I' = \frac{V_{l-n}}{Z'} = \frac{115.47}{8.89\angle 36.87} \simeq 13\angle -36.87 \text{ A}$$

Magnitude of line current = $|I'|$ = 13 A and $\phi = -36.87°$

p.f. = cos 36.87 = 0.8 lagging

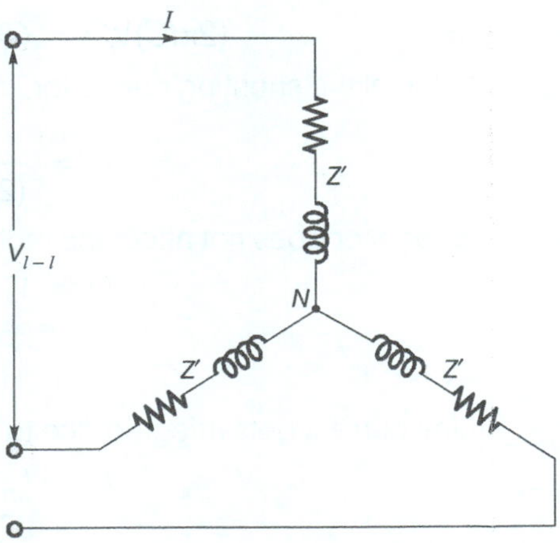

$$\text{Total power} = 3 \times V_{l-n} \times I' \times \cos 36.87$$
$$= 3 \times 115.47 \times 13 \times 0.8$$
$$= 3602.66 \text{ Watts}$$

P_1 and P_2 are power measured by two different wattmeters in two-wattmeter method

$$P_1 = V_{l-l} \times I' \times \cos(30° + \phi)$$
$$= 200 \times 13 \times \cos(30° + (-36.87°))$$
$$= 2581.33 \text{ Watts}$$

and $$P_2 = V_{l-l} \times I' \times \cos(30° - \phi)$$
$$= 200 \times 13 \times \cos(30° - (-36.87°))$$
$$= 1021.32 \text{ Watts}$$

8.12 A single phase motor takes 50 A at a p.f. of 0.5 lag from a 250 V, 50 Hz supply. What must be the value of the shunting capacitor to raise the overall p.f. to 0.9 lag? How does the capacitor effect the line and motor currents?

[8+4 = 12 marks : 2011]

Solution:

Taking the supply voltage as reference,

$$V_s = 250\angle 0° \text{ V}$$

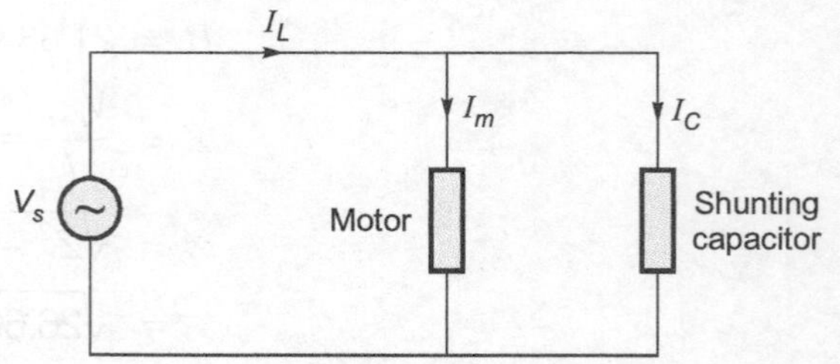

when shunting capacitor was not present,

$$\text{p.f.} = \cos\phi_1 = 0.5 \text{ lag}$$

$$\Rightarrow \quad \phi_1 = \cos^{-1}(0.5) = 60°$$

$$\text{Motor current} = I_m = 50\angle -60° \text{ A}$$

$$\text{Line current} = I_L = I_m = 50\angle -60° \text{ A}$$

$$\text{Power supplied} = S_1 = P_1 + jQ_1 = V_s I^*_L$$

$$= (250\angle 0°) \times (50\angle -60°)^*$$

$$= 6250 + j10{,}825.32$$

where,

$$\text{Active power} = P_1 = 6250 \text{ W}$$

$$\text{Reactive power} = Q_1 = 10{,}825.32 \text{ VAR}$$

when shunting capacitor is present,

$$\text{p.f. of the supply} = \cos\phi_2 = 0.9 \text{ lag}$$

$$\phi_2 = \cos^{-1}(0.9) = 25.842°$$

As we know,

$$\tan\phi_2 = \frac{Q_2}{P_2}$$

Only reactive power is affected by the shunting capacitor, so active power remains same $P_2 = P_1 = 6250$ W

$$Q_2 = P_2 \tan\phi_2 = 6250 \times \tan(25.842°) = 3027.02 \text{ VAR}$$

Shunting capacitor supplies reactive power (Q_C) to the system

$$Q_C = Q_1 - Q_2 = 10{,}825.32 - 3027.02 = 7798.3 \text{ VAR}$$

If X_C is reactance of the shunting capacitor,

$$Q_C = \frac{V_s^2}{X_C}$$

where,

$$X_C = \frac{1}{\omega C} = \frac{1}{2\pi f C}$$

$$\Rightarrow \quad (2\pi f C)V_s^2 = Q_C$$

Value of the shunting capacitor,

$$C = \frac{7798.3}{(250)^2 \times 2 \times \pi \times 50} = 397.16 \ \mu\text{F}$$

Capacitor does not affect the motor current.

$$I_C = V_s \times (j2\pi f C)$$

$$= 250 \times 2 \times \pi \times 50 \times 397.16 \times 10^{-6}\angle 90°$$

$$\Rightarrow \quad I_C = 31.2\angle 90° \text{ A}$$

Line current gets affected due to current taken by the shunting capacitor and whose new value is

$$I_L = I_m + I_C = 50\angle -60° + 31.20\angle 90°$$

$$= 27.77\angle -25.84° \text{ A}$$

8.13 Using star-delta transformation, find the currents I_1, I_2 and I_3 for the given circuit. All the resistances are in ohm.

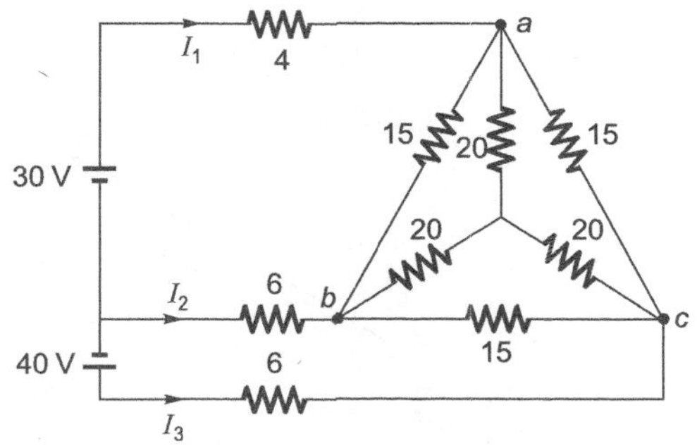

[12 marks : 2012]

Solution:

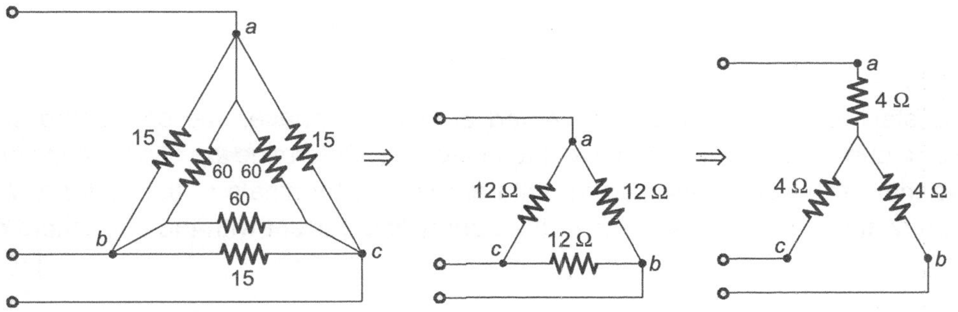

The original circuit becomes,

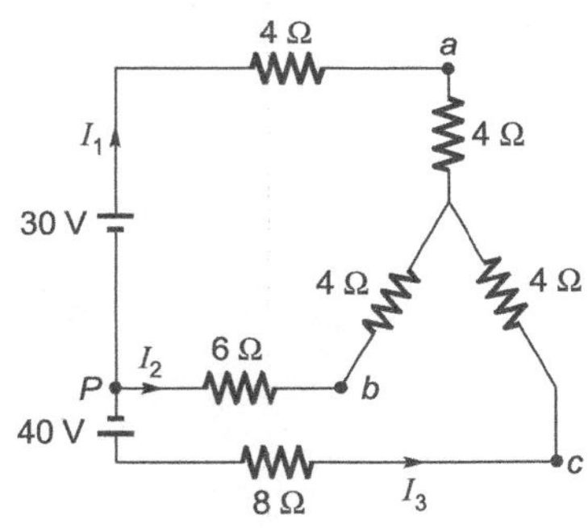

By applying KCL at node P

$$I_1 + I_2 + I_3 = 0 \quad \ldots(i)$$

By applying KLV in loop $P_{ab'}$

$$30 = (4+4)I_1 - (4+6)I_2$$
$$30 = 8I_1 - 10I_2$$
$$4I_1 - 5I_2 = 15 \quad \ldots(ii)$$

By applying KVL in loop $P_{bc'}$

$$40 = (8+4)I_3 - (4+6)I_2$$
$$20 = 6I_3 - 5I_2 \quad \ldots(iii)$$
$$20 = 6(-I_1 - I_2) - 5I_2$$
$$20 = -6I_1 - 11I_2$$
$$\Rightarrow \quad 6I_1 + 11I_2 = -20 \quad \ldots(iv)$$

After solving equation (*ii*) and (*iv*),

$$I_1 = +\frac{65}{74}\text{ A},\ I_2 = -\frac{85}{37}\text{ A},\ I_3 = \frac{85}{37} - \frac{65}{74} = \frac{105}{74}\text{ A}$$

8.14 A balanced star-connected load is supplied from a 3-phase, 50 Hz, 415 V supply. Current in each phase is 20∠–30° A, the 30° angle being w.r.t. phase voltage.
Determine:
(i) the total power supplied,
(ii) the overall p.f. and
(iii) draw the phasor diagram.

[5 marks : 2013]

Solution:

(i) Total power supplied $= 3\,V_{ph} I_{ph} \cos\phi = (3)\left(\frac{415}{\sqrt{3}}\right)(20)\cos(30°) = 12450$ W

(ii) From phasor diagram, angle between V_{RY} and I_R is 60°.

$\Rightarrow$ p.f. = cos60° = 0.5 lag

(iii) Taking V_R as reference

Phase diagram:

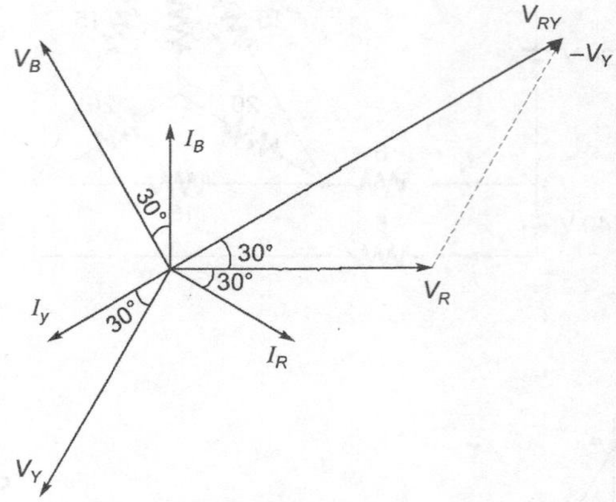

8.15 A 3-phase star connected motor, connected across a 3-phase, star-connected supply of 400 V, 50 Hz takes a current of 20 A at 0.8 power factor lagging. Determine the capacitance of the capacitors per phase that are to be connected in delta across the terminals of the motor to raise the power factor to unity and also the new value of the supply line current with the capacitors connected.

[10 marks : 2013]

Solution:

$$Q_{reactive} = \sqrt{3}\, V_L I_L \sin\phi = (\sqrt{3})(400)(20)(0.6) = 8313.84 \text{ VAR}$$

So, VAR supplied by capacitor bank = 8313.84 VAR

So, $$3\frac{V_L^2}{X_c} = 8313.84$$

$$\Rightarrow \quad C = \frac{8313.84}{(3)(400)^2 \times 2 \times \pi \times 50} = 55.13 \ \mu\text{F/phase}$$

$$P_{active} = \sqrt{3}\, V_L I_L \cos\phi = (\sqrt{3})(400)(20)(0.80) = 11085.125 \text{ W}$$

Now for unity power factor, $\phi = 0°$ but P_{active} will remain same.

$$P_{active} = \sqrt{3}\, V_L I_{new} \cos(0°) = \sqrt{3}\, V_L I_{new}$$

$$I_{new} = \frac{11085.125}{\sqrt{3} \times 400} = 16 \text{ A}$$

8.16 A balanced three-phase, 208 Volts generator supplies a unit of 1800 Watts at a line current of 10 Amp. When three identical impedances are arranged in Y-connection across the line terminals of the generator calculate the resistive and reactive components of each phase impedance. Sketch also the phase and magnitude relations between phase and line voltages of a Y-connected three-phase system.

[20 marks : 2017]

Solution:

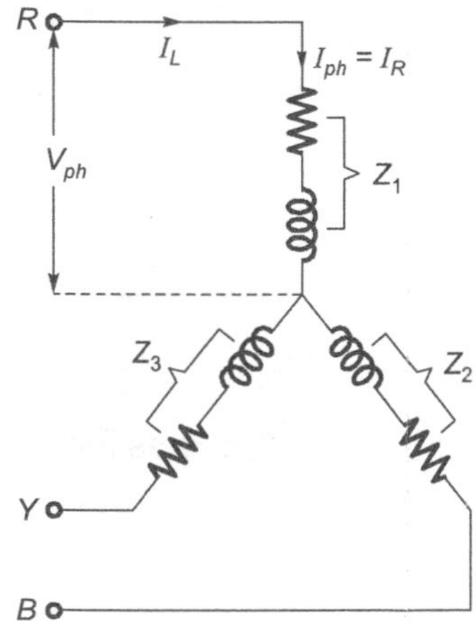

Given that,

Generator terminal voltage, $V = 208\angle 0°\text{ V}$

Line current, $I_L = 10\text{ A}$

Power supplies, $P = 1800\text{ W}$

$$P = \sqrt{3} V_L I_L \cos\phi$$

$$\cos\phi = \frac{1800}{\sqrt{3} \times 208 \times 10} = 0.50$$

Phase angle, $\phi = 60°$

In star connected load, $I_L = I_{ph} = 10\text{ A}$

and $V_{ph} = \frac{V_L}{\sqrt{3}} = \frac{208}{\sqrt{3}}\text{ V}$

Given, $Z_1 = Z_2 = Z_3 = Z$

Impedance, $Z = \frac{V_{ph}}{I_{ph}} = \frac{(208/\sqrt{3})\angle 0°}{10\angle -60°} = 12\angle 60°\ \Omega$

$$= (6.0 + j10.4)\ \Omega$$

Resistive components, $R = 6.0\ \Omega$

Reactive component, $X = 10.4\ \Omega$

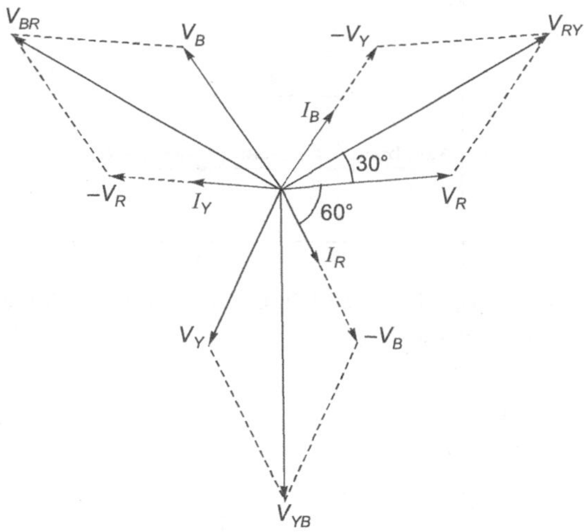

8.17 What is a filter? State the ideal characteristics of a filter.

[10 marks : 2022]

Solution:

A filter is a electronic circuit that removes any unwanted component or features from a signal. In other words, it is a circuit that rejects certain band of frequencies and allows others to pass through it. Filters are mainly classified into two categories :

(1) Active filter; (2) Passive filter

Active Filters : This filter circuit consists of active components like transistors and op-amps in addition to RLC components.

Passive Filters : These filters only consists of passive element (RLC).

Filters are further categorized based on the operating frequencies of a particular circuit.

(1) **Low Pass Filter :** This filter attenuates all the frequencies above the cut-off frequencies. Ideal low pass filter provides a constant gain from zero to cut-off frequency.

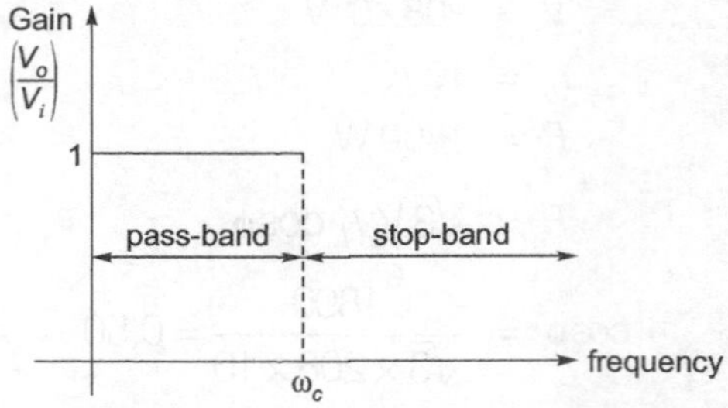

(2) **High Pass Filter :** Filter which attenuates all the frequency components below the cut-off frequency.

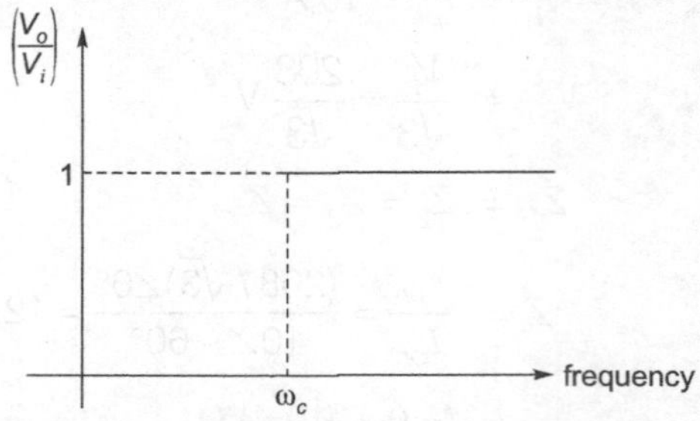

(3) **Band Pass Filters :** This filter pass all the frequency components in a particular band of frequencies and reject all the frequency components outside the pass band.

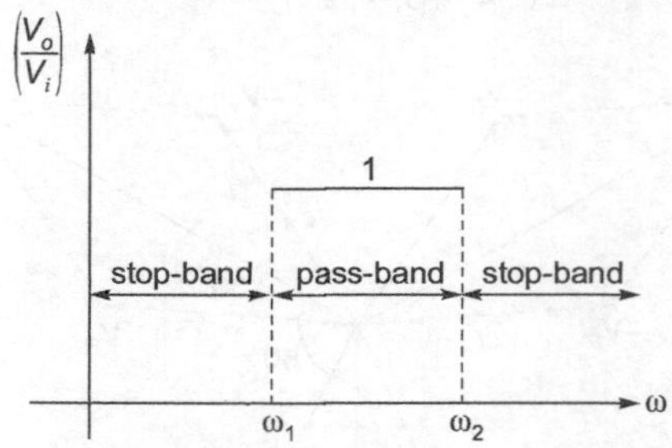

(4) **Band Reject Filters :** This filter reject all the frequency components lies in a particular band of frequencies and pass all other frequency components.

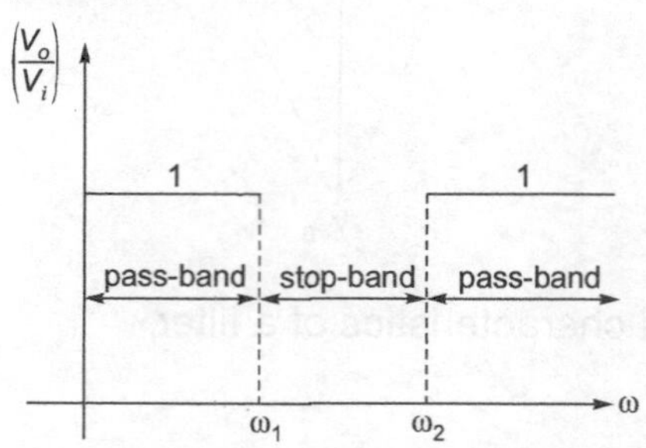

(5) **All Pass Filters :** This filter pass all frequency components of a signal. It is also known as phase-shifter, time-delay filter.

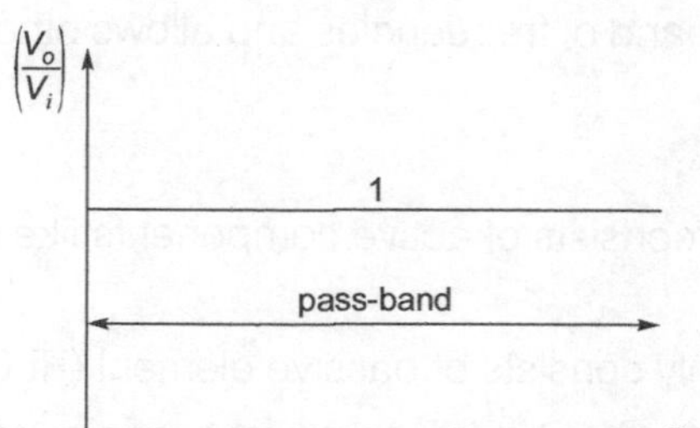

■■■■

3 Electromagnetic Theory

Revised Syllabus of ESE: *Gauss theorem, electric field and potential due to point, line, plane and spherical charge distributions, Ampere's and Biot-Savart's laws; inductance, dielectrics, capacitance; Maxwell's equations.*

---

1. Coordinate Systems & Vector Calculus

1.1 "A hand 'curl meter' in the form of a pin wheel is used to indicate curl of a vector field." Justify the statement.

[12 marks : 2001]

Solution:

A pin or paddle wheel as a 'curl meter': The force is exerted on the each blade of the paddle wheel, the force being proportional to the component of the field normal to the surface of that blade. To test and field for curl we dip one paddle wheel into the field with the axis of the paddle wheel lined up with the direction of the component of curl desired. No rotation means no curl, larger angular velocities mean greater values of the curl a reverse in the direction of spin means a reversal in the sign of the curl. In order find the direction of vector curl, we should place one paddle wheel in the field and hunt around for the orientation which produces the greatest torque. The direction of the curl is then along the axis of the paddle wheel.

1.2 Explain the concept of gradient. Determine the gradient of the given field:

$$V = \rho z \sin\phi + z^2 \cos^2\phi + \rho^2$$

[5 marks : 2013]

Solution:

Gradient of a Scalar:

Gradient V = Grad $V = \nabla V$ ... always a vector quantity.

∇V

Magnitude $|\nabla V|$ — Direction

Magnitude represents the rate of change of scalar quantity V with respect to various co-ordinates.

Direction represents the direction in which this rate of change is maximum.

$$\vec{E} = -\nabla V = -\text{Grad}\,V$$

$$V = -\int_c \vec{E}\cdot d\vec{l}$$

Numerical:

$$V = \rho z \sin\phi + z^2 \cos^2\phi + \rho^2$$

Given system is in cylindrical coordinate system.

Hence, $$\nabla V = \frac{\partial V}{\partial \rho}\hat{a}_\rho + \frac{1}{\rho}\frac{\partial V}{\partial \phi}\hat{a}_\phi + \frac{\partial V}{\partial z}\hat{a}_z$$

Now, $$\frac{\partial V}{\partial \rho} = \frac{\partial}{\partial \rho}(\rho z \sin\phi + z^2\cos^2\phi + \rho^2) = z\sin\phi + 0 + 2\rho$$

$$\frac{\partial V}{\partial \rho} = z\sin\phi + 2\rho$$

Similarly, $$\frac{\partial V}{\partial \phi} = \left[z(\cos\phi) + z^2(2\cos\phi \times (-\sin\phi))\right] + 0 = z\cos\phi - 2z^2\sin\phi\cos\phi$$

Similarly, $$\frac{\partial V}{\partial z} = \frac{\pi}{\rho}\sin\phi + 2z\cos^2\phi$$

So, $$\nabla V = (z\sin\phi + 2\rho)\hat{a}_\rho + \frac{(z\cos\phi - 2z^2\sin\phi\cos\phi)}{\rho}\hat{a}_\phi + \left(\frac{\pi}{\rho}\sin\phi + 2z\cos^2\phi\right)\hat{a}_z$$

1.3 State and explain Divergence theorem. Verify Divergence theorem for the vector A in spherical co-ordinates for the volume enclosed between $r = 1$ and $r = 2$, where $A = 5r^2\,\hat{a}_r$.

[10 marks : 2013]

Solution:

Divergence Theorem:

$$\boxed{\oint\!\!\!\oint \vec{A}\cdot d\vec{s} = \iiint_V \nabla\cdot\vec{A}\,dV}$$

Statement: Closed surface integral of any vector $\vec{A}$ integrated over any closed surface area S is always equal to the volume integral of the divergence of the vector $\vec{A}$ integrated over the volume V which is enclosed by the closed surface area 'S'. The theorem holds good, irrespective of

- the shape of closed surface area 'S'
- the type of the coordinate system
- the type of the vector $\vec{A}$

Numerical:

Given, $$\vec{A} = 5r^2\,\hat{a}_r$$

From $$r = 1 \text{ to } 2$$

$\Rightarrow$ $$\nabla\cdot\vec{A} = \frac{1}{r^2}\cdot\frac{\partial}{\partial r}(r^2\vec{A}_r\hat{a}_r)$$

$$= \frac{1}{r^2}\times\frac{\partial}{\partial r}(r^2\times 5r^2) = \frac{1}{r^2}[5\times 4r^3] = 20\,r$$

$$dV = r^2\sin\theta\,dr\,d\theta\,d\phi$$

$\therefore$ $$\int_V(\nabla\cdot\vec{A})\,dV = \int_0^{2\pi}\int_0^{\pi}\int_1^{2} r^2\sin\theta\times 20r\,dr\,d\theta\,d\phi$$

$$= \int_0^{2\pi}\int_0^{\pi}\int_1^{2} 20r^3 \sin\theta \, dr\, d\theta\, d\phi = \int_0^{2\pi}\int_0^{\pi} 20 \times \left[\frac{r^4}{4}\right]_1^2 \sin\theta\, d\theta\, d\phi$$

$$= \int_0^{2\pi} 20 \times \left(\frac{16-1}{4}\right)[-\cos\theta]_0^{\pi} d\phi$$

$$= 20 \times \frac{15}{4} \times 2 \times 2\pi = 300\pi \quad \ldots(i)$$

$$\oint_s \vec{A}\cdot ds = \int_0^{2\pi}\int_0^{\pi} Ar^2 \cdot r^2 \sin\theta\, d\theta d\phi$$

$$= Ar^4 \times \sin\theta\, d\theta\, d\phi = 5 \times (15) \times 2 \times 2\pi$$

$$= 300\,\pi \quad \ldots(ii)$$

From equations (*i*) and (*ii*), $\boxed{\oiint \vec{A}\cdot d\vec{s} = \iiint_V \nabla\cdot\vec{A}\, dV}$

Hence, divergence theorem is verified.

1.4 **State Gauss divergence theorem. Let R be the region bounded by the closed cylinder $x^2 + y^2 = 4$, $z = 0$ and $z = 2$. Verify this theorem, if $\vec{F} = 3x^2\hat{i} + y^2\hat{j} + z\hat{k}$.**

[20 marks : 2023]

Solution:

Divergence Theorem : If $\vec{F}$ is differentiable vector point function in a closed surface S having volume V, then

$$\iint_S \vec{F}\cdot\hat{n}dS = \iiint_V \bar{\nabla}\cdot\vec{F}dV$$

Verify : We need to prove that

$$\iint_S \vec{F}\cdot\hat{n}ds = \iiint \bar{\nabla}\cdot\vec{F}dV$$

Consider LHS :

$$\iint_S \vec{F}\cdot\hat{n}ds = \iint_{S_1} \vec{F}\cdot\hat{n}ds + \iint_{S_2} \vec{F}\cdot\hat{n}ds + \iint_{S_3} \vec{F}\cdot\hat{n}ds$$

For S_1 : $Z = 0,\ \hat{n} = -\hat{k}$

$$\iint_{S_1} \vec{F}\cdot\hat{n}ds = \iint_{S_1} (-z)ds = 0 \quad (\because z = 0)$$

For S_2 : $Z = 2,\ \hat{n} = \hat{k}$

$$\iint_{S_2} \vec{F}\cdot\hat{n}ds = \iint_{S_2} z\cdot ds = 2 \text{ (Area of circle)}$$

$$= 2(\pi \times 2^2)$$

$$= 8\pi$$

For S_3 : It is curved surface of cylinder. Project the surface onto XZ-plane.

$$\iint_{S_3} \vec{F}\cdot\hat{n}ds = \iint_R \vec{F}\cdot\hat{n}\cdot\frac{dxdz}{\hat{n}\cdot\hat{j}}$$

$$\hat{n} = \frac{xi + yj}{2}$$

$$\vec{F}\cdot\hat{n} = (3x^2 i + y^2 j + zk)\cdot\frac{(xi + yj)}{2} = \frac{3x^3 + y^3}{2}$$

and $$\hat{n}\cdot\hat{j} = \frac{y}{2}$$

$$\iint_{S_3} \vec{F}\cdot\hat{n}ds = \iint_R \frac{3x^3 + y^3}{2}\cdot\frac{dxdz}{\frac{y}{2}}$$

$$= \iint_R \left(3\frac{x^3}{y} + y^2\right)dxdz$$

Put $y = 2\sin\theta$, $x = \cos\theta$, $dx = -2\sin\theta d\theta$

$$= \int_{z=0}^{2}\int_{\theta=0}^{2\pi}\left(3\left(\frac{8\cos^3\theta}{2\sin\theta}\right) + 4\sin^2\theta\right)(-2\sin\theta)d\theta dz$$

$$= \int_{z=0}^{2}\int_{0}^{2\pi}\left(-24\cos^3\theta - 8\sin^3\theta\right)d\theta dz$$

$$= \int_{z=0}^{2}\left(-24\int_0^{2\pi}\cos^3\theta d\theta\right) - \left(8\int_0^{2\pi}\sin^3\theta d\theta\right)dz$$

$$= \int_{z=0}^{2}\left[(-24)\times 2\int_0^{\pi}\cos^3\theta d\theta - 0\right]dz$$

$$= -48 \times 0 = 0$$

$$\therefore \quad \iint_S \vec{F}\cdot\hat{n}ds = \iint_{S_1}\vec{F}\cdot\hat{n}ds + \iint_{S_2}\vec{F}\cdot\hat{n}ds + \iint_{S_3}\vec{F}\cdot\hat{n}ds$$

$$= 0 + 8\pi + 0$$

$$= 8\pi$$

Consider RHS :

$$\iiint \overline{\nabla}\cdot\vec{F}dV = \iint_R\int_{z=0}^{2}(6x + 2y + 1)dzdydx$$

$$= \int_{x=-2}^{2}\int_{y=-\sqrt{4-x^2}}^{\sqrt{4-x^2}}(6x + 2y + 1)(z)_0^2\, dy\, dx$$

$$= \int_{x=-2}^{2}\left[(6x+1)y + y^2\right]_{-\sqrt{4-x^2}}^{\sqrt{4-x^2}} dx$$

$$= 2\int_{x=-2}^{2}\left[(6x+1)\sqrt{4-x^2} + (4-x^2) - (4-x)^2\right]dx$$

$$= 2\int_{-2}^{2}](6x+1)y + 0]_{-\sqrt{4-x^2}}^{\sqrt{4-x^2}} dx = 2\times 2\int_{-2}^{2}(6x+1)\sqrt{4-x^2}dx$$

$$= 4\left[\int_{-2}^{2} 6x\sqrt{4-x^2}\,dx + \int_{-2}^{2}\sqrt{4-x^2}\,dx\right]$$

$$= 4\left[0 + 2\int_{-2}^{2}\sqrt{4-x^2}\,dx\right] = 4\left[2\times\frac{1}{4}\pi(2)^2\right] = 8\pi$$

L.H.S. = R.H.S.

Hence verified.

2. Electrostatic Fields

2.1 **State and explain Gauss's law. A spherical volume charge distribution ρ is given by**

$$\rho = \rho_0\left(1-\frac{r^2}{100}\right) \qquad \text{for } r \le 10 \text{ mm}$$

$$= 0 \qquad \text{for } r > 10 \text{ mm}$$

Show that the maximum value of electric field intensity E occurs at r = 7.45 mm. Obtain the value of E at r = 7.45 mm.

[12 marks : 2001]

Solution:

Gauss's Law:

Surface integral of the normal component of the electric vector D over any closed surface is equal to the total charge enclosed by the surface.

The total flux through the closed surface,

$$\Psi = \int d\Psi = \oint \bar{D}_s\cdot\bar{ds} = \int_{Vol}\rho_v\,dv$$

$$\oint \bar{D}\cdot\bar{ds} = \int_{Vol}\rho_v dv$$

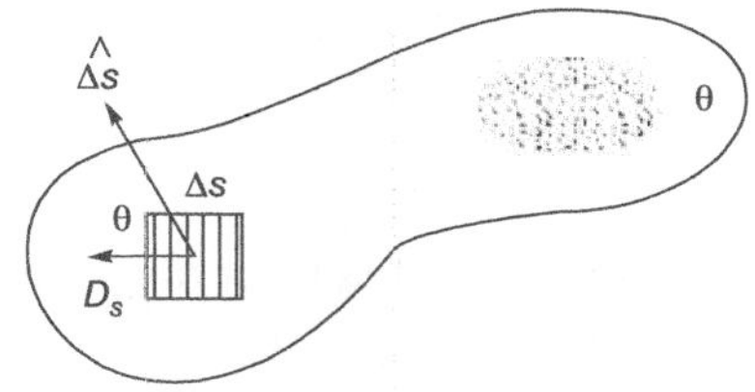

Applying Gauss's law to the spherical surface at ($r \le 10$)

$$E'(4\pi r^2) = \frac{Q'}{\epsilon_0} = \frac{1}{\epsilon_0}\left[4\pi\rho_0\int\left(1-\frac{r^2}{100}\right)r^2dr\right]$$

$$= \frac{4\pi\rho_0 r^2}{\epsilon_0}\left[\frac{r}{3}-\frac{1}{5}\frac{r^3}{100}\right] = \frac{\rho_0}{\epsilon_0}\left[\frac{r}{3}-\frac{1}{5}\frac{r^3}{100}\right]$$

As clearly seen from the above expression, (E') varies with 'r' and for maximum value of (E')

$$\frac{dE'}{dr} = 0$$

Differentiating with respect to r and equating to zero

$$\frac{\rho_0}{\epsilon_0}\left(\frac{1}{3}-\frac{3}{5}\frac{r^2}{10^2}\right) = 0$$

$$\frac{r^2}{10^2} = \frac{5}{3}\times\frac{1}{3}$$

$$\therefore \qquad r = 7.45 \text{ mm}$$

$$\frac{d^2E'}{dr^2} = \frac{\rho_0}{\epsilon_0}\left(\frac{-3}{5}\cdot\frac{2r}{10^2}\right)$$

At $r = 7.45$ mm, $\quad \dfrac{d^2E'}{dr^2} < 0$

Therefore, maximum E occurs at $r = 7.45$ mm

$$E'_{max} = \frac{\rho_0}{\epsilon_0}\left[\frac{7.45}{3} - \frac{1}{5}\times\frac{7.45^3}{10^2}\right] = \frac{\rho_0}{\epsilon_0}\times 1.6563 \text{ V/mm}$$

2.2 **Two infinite and conducting cones both on z-axis, one is $\theta = \theta_1 = 45°$ (constant) cone and the other is $\theta = \theta_2 = 150°$ constant cone. The region between them is characterized by ϵ, $\rho_v = 0$, $V = 0$ at θ_1 and $V = 10$ V at θ_2. Find expression for V between $45° < \theta < 150°$.**

[10 marks : 2001]

Solution:

Consider the coaxial cone as shown in figure. Here V depends only on θ, so Laplace's equation on spherical coordinate becomes.

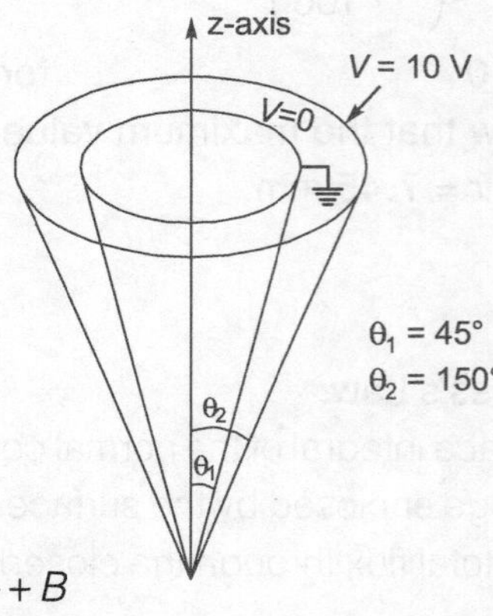

$$\nabla^2 V = \frac{1}{r^2\sin\theta}\frac{d}{d\theta}\left[\sin\theta\frac{dV}{d\theta}\right] = 0$$

or
$$\frac{d}{d\theta}\left[\sin\theta\frac{dV}{d\theta}\right] = 0$$

Integrating once we get

$$\frac{dV}{d\theta} = \frac{A}{\sin\theta}$$

Integrating this result we get

$$V = A\int\frac{d\theta}{\sin\theta} + B = A\int\frac{d\theta}{2\cos\theta/2\sin\theta/2} + B$$

$$V = A\int\frac{1}{2}\frac{\sec^2\theta/2}{\tan\theta/2}d\theta + B = A\int\frac{d(\tan\theta/2)}{\tan\theta/2} + B$$

$$V = A\,ln\tan(\theta/2) + B$$

Now apply the boundary condition to determine the value of A and B

$V(\theta = \theta_1) = 0$

$$O = A\,ln\tan\left(\frac{45°}{2}\right) + B$$

$$B = -A\,ln\tan(\theta_1/2)$$

$$\therefore \quad V = A\,ln\left[\frac{\tan(\theta/2)}{\tan(\theta_1/2)}\right]$$

Now at $\quad \theta = \theta_2,\ V = V_0 = 10$ V

$$\therefore \quad 10 = A\,ln\left[\frac{\tan(\theta_2/2)}{\tan(\theta_1/2)}\right]$$

$$\therefore \quad V = \frac{10\,ln\left[\dfrac{\tan(\theta/2)}{\tan(\theta_1/2)}\right]}{ln\left[\dfrac{\tan(\theta_2/2)}{\tan(\theta_1/2)}\right]} = \frac{10}{2.198}\,ln\left[\frac{\tan(\theta/2)}{\tan(\theta_1/2)}\right]$$

$$V = 4.548\,ln\left[\frac{\tan(\theta/2)}{\tan(\theta_1/2)}\right] \text{ for } 45° < \theta < 150°$$

2.3 A sheet of charge, ρ_s = 2 nc/m^2 is present at the plane x = 3 in free space, and a line charge, ρ_L = 20 nC/m is located at x = 1, z = 4. Find

(i) the magnitude of the electric field intensity $\bar{E}$ at the origin

(ii) the direction of $\bar{E}$ at (4, 5, 6)

(iii) the force per meter length on the line charge. [12 marks : 2002]

Solution:

(i) A sheet charge is a surface charge distribution at a rate ρ_s = 2 nc/m^2 located at the plane (x = 3). Field due to this at origin

$$E_s = \frac{\rho_s}{2\varepsilon_0}(-a_x) = \frac{2\times 10^{-9}}{2\times\varepsilon_0}(-a_x) = -113\,\bar{a}_x$$

Field produced by the line charge is (at origin),

$$|E_L| = \frac{\rho_L}{2\pi\varepsilon_0 \times r}\hat{n}$$

where, $\rho_L = 20 \times 10^{-9}$ C/m, $\vec{r} = -\hat{a}_x - 4\hat{a}_z$

$$|\vec{r}| = \sqrt{1^2+4^2} = \sqrt{17}$$

$$E_L = \frac{20\times 10^{-9}}{2\pi\varepsilon_0\times\sqrt{17}}\times\frac{-\hat{a}_x-4\hat{a}_z}{\sqrt{17}} = \frac{20\times 10^{-9}\times 18\times 10^{9}}{17}(-\hat{a}_x-4\hat{a}_z)$$

$$E_{Resultant} = E_S + E_L = -21.17\,\hat{a}_x - 84.70\hat{a}_z - 113\hat{a}_x = 134\hat{a}_x - 84.70\hat{a}_z$$

$$\text{Magnitude, } |E_{Resultant}| = |-134\,\hat{a}_x - 84.70\,\hat{a}_z| = 158.425 \text{ V/m}$$

(ii) At point (4, 5 ,6)

Electric field due to surface charge,

$$E_{(s)} = +113\hat{a}_x \quad (\because \text{ point is on the right side of sheet of charge})$$

Electric field due to line charge at point (4, 5, 6) due to line charge located at (x = 1 and z = 4).

Vector distance between this the two point

$$\vec{r}_1 = (4-1)\hat{a}_x + 5\hat{a}_y + 2\hat{a}_z \quad \left(\because \vec{r}_1 = (4-1)\hat{a}_x + 5\hat{a}_y + (6-4)\hat{a}_z\right)$$

$$\vec{E}_l = \frac{\rho_L}{2\pi\varepsilon_0\times|\vec{r}_1|}\hat{n} = \frac{20\times 10^{-9}\times 18\times 10^{9}}{\sqrt{38}}\times\frac{3\hat{a}_x+5\hat{a}_y+2\hat{a}_z}{\sqrt{38}}$$

$$\vec{E}_l = 9.474(3\hat{a}_x + 5\hat{a}_y + 2\hat{a}_z)$$

$$E_{Resultant} = 28.421\,\hat{a}_x + 47.368\hat{a}_y + 18.949\hat{a}_z \quad (\because \vec{E}_{resultant} = \vec{E}_s + \vec{E}_l)$$

(*iii*)

$$\text{Force/Length} = \bar{E}_s \times \rho_L$$

$$= (113\,\hat{a}_x)\times\rho_L = (+2.26\times 10^{-6})a_x \text{ N/m}$$

2.4 If V in the free space is given by

$V = 60 \sin\theta/r^2$, and a point P is located at r = 3 m, θ = 60°, ϕ = 25°, find at P.

(*i*) V_P (*ii*) $\bar{E}_P$ (*iii*) dV/dN

(*iv*) $\bar{a}_N$ (*v*) ρ_v

[12 marks : 2002]

Solution:

(*i*)

$$V_p = \frac{60\sin 60°}{3^2} = 5.7735 \text{ V} \quad (\because \theta = 60°,\ r = 3\text{ m})$$

(ii)

$$\vec{E} = -\nabla V = -\left(\frac{\partial V}{\partial r}\vec{a}_r + \frac{1}{r}\times\frac{\partial V}{\partial \theta}\vec{a}_\theta\right) = -\left[-\frac{120\sin\theta}{r^3}\vec{a}_r + \frac{60\cos\theta}{r^3}\vec{a}_\theta\right]$$

$$\bar{E}_p = \left[\frac{120\sin 60}{3^3}\hat{a}_r - \frac{60\cos 60}{3^3}\vec{a}_\theta\right] = 3.849\,\bar{a}_r - 1.111\,\vec{a}_\theta \text{ V/m}$$

(iii)

$$\left.\frac{dV}{dN}\right|_p = |E_p| = \sqrt{(3.849)^2 + (1.111)^2} = 4.06 \text{ V/m}$$

(iv)

$$\left.\vec{a}_n\right|_p = \frac{-\vec{E}_p}{|E_p|} = \frac{-(3.849\,\hat{a}_r - 1.111\,\vec{a}_\theta)}{4.006} = -0.96\,\vec{a}_r + 0.277\,\vec{a}_\theta$$

(v)

$$\vec{D} = \epsilon_0\,\vec{E} = \epsilon_0\left[\frac{120\sin\theta}{r^3}\hat{a}_r - \frac{60\cos\theta}{r^3}\hat{a}_\theta\right]$$

Now,

$$\rho_v = \nabla\cdot\vec{D}$$

$$= \frac{1}{r^2\sin\theta}\left[\frac{\partial}{\partial r}\left(r^2\sin\theta\times\frac{120\sin\theta}{r^3}\right) + \frac{\partial}{\partial\theta}r\sin\theta\left(\frac{-60\cos\theta}{r^3}\right)\right]$$

$$\rho_v = \frac{1}{r^2\sin\theta}\times\epsilon_0\left[-\frac{120\sin^2\theta}{r^2} - \frac{60\cos 2\theta}{r^2}\right]$$

ρ_v at P

$$= \frac{1}{3^2\times\sin 60}\left[-\frac{120\sin^2 60}{3^2} - \frac{60\cos 120}{3^2}\right]$$

ρ_v at P

$$= \epsilon_0\frac{1\times 2}{3^2\times\sqrt{3}}\left[-\frac{120}{9}\times\frac{3}{4} + \frac{60}{9}\times\frac{1}{2}\right]$$

$$= 8.85\times 10^{-12}\times(-0.8553)$$

$$= -7.569\times 10^{-12}\text{ C/m}^2$$

At P,

$$\rho_v = -7.569 \text{ pC/m}^2$$

2.5 Conducting cylinders at ρ_1 = 1.6 cm and ρ_2 = 5 cm in free space are held at potentials of 80 V and –40 V, respectively. Find

(i) V and $\bar{E}$ at ρ = 2 cm

(ii) the surface at which $V = 0$

(iii) capacitance per meter length between the conducting cylinders.

[12 marks : 2002]

Solution:

Cylindrical surfaces are equipotential. The voltages at ρ = 1.6 cm and 5 cm are 80 and –40 V respectively. The potential is constant with ϕ and z.

The Laplace equation gets reduced to ($\nabla^2 V = 0$)

($\nabla^2 V = 0$)

$$\frac{1}{r}\frac{d}{dr}\left[r\frac{dv}{dr}\right] = 0$$

Integrating,

$$r\frac{dv}{dr} = C_1[\text{constant}]$$

$$\frac{dv}{dr} = \frac{C_1}{r}$$

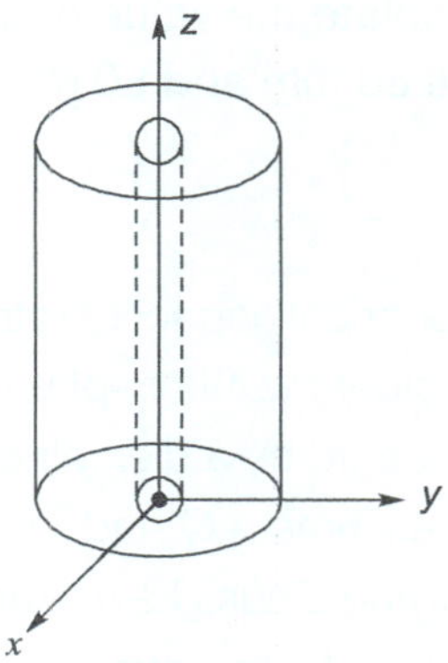

Integrating again, $V = C_1 \, ln\, r + C_2$ (another constant)

Applying boundary condition,

$$V \text{ at } (1.6 \text{ cm}) = 80 = C_1 \, ln\,(1.6 \times 10^{-2}) + C_2$$

$$V(5 \text{ cm}) = -40 = C_1 \, ln\,(5 \times 10^{-2}) + C_2$$

Solving, $C_1 = -105.315$ and $C_2 = -355.5$

$\therefore$ $V = -105.315 \ln \rho - 355.5$

(i) $V_{(\rho = 2 \text{ cm})} = 56.49 \text{ V}$

Now, $$E = -\nabla V = \frac{105.315}{\rho} \hat{a}_\rho \text{ V/m}$$

$$\bar{E} = -\nabla V = \frac{105.315}{2} \hat{a}_\rho = 52.6575 \, \hat{a}_\rho \text{ V/cm} \qquad \because \rho = 2 \text{ cm}$$

$$= 5265.75 \, \hat{a}_\rho \text{ V/m}$$

(ii) The surface at which ($V = 0$)

$$V = 0 = -105.315 \, ln\, \rho_0 - 355.5$$

$\therefore$ $\rho_0 = 0.03419 \text{ m} = 3.419 \text{ cm}$

(iii) Total change on this cylinder $= \rho_s \times 2\pi \, \rho h$

$$\rho_s = D = \epsilon_0 E = \frac{\epsilon_0 \times 105.315}{\rho}$$

$\therefore$ $$Q = \frac{105.315 \epsilon_0}{\rho} \times 2\pi \, \rho h = 105.315 \times 2\pi \times \varepsilon_0 \times h$$

Potential difference between the cylinder = 80 –(–40) = 120 V

Capacitance, $$C = \frac{105.315 \times 2\pi \times \epsilon_0 \, h}{120} = 48.824 \, h \text{ pF}$$

$\therefore$ Capacitance/meter = 48.823 pF = C/h = 48.824 pF/m

2.6 Explain dipole moment. A dielectric slab of flat surface with relative permittivity of 5 is disposed with its normal to a uniform field with flux density 2.0 coulombs per m^2. The volume of the slab is 0.1 m^3. It is uniformly polarized. Calculate its dipole moment.

[12 marks : 2003]

Solution:

When an electric field is applied, the positive atomic nuclei and the electrons are pulled in opposite direction in the material. Consequently, their centres are displaced in the direction of the field. This displacement is represented as an electric dipole moment here.

$$\text{Polarization } \bar{p} = \bar{D}\left(1 - \frac{1}{\varepsilon_r}\right) = 2\left(1 - \frac{1}{5}\right) = 1.6 \text{ C/m}^2$$

[$\bar{P}$ = (Number of dipoles/m^3) × dipole moment]

Then, dipole moment $\vec{d} = \bar{p} \cdot v = (1.6) \times 0.1 = 0.16$ Cm (taking magnitude energy)

or polarization → dipole moment per unit volume

$\therefore$ dipole moment = polarization × volume

2.7 **Calculate the value of capacitance of a spherical capacitor consisting of two concentric spheres of radii 60 mm and 80 mm with air as dielectric medium between the two. Derive the formula used.**

[12 marks : 2003]

Solution:

A spherical capacitor is the case of two concentric spherical conductors. Consider the inner sphere of radius a and outer sphere of radius $b\,(b > a)$ separated by a dielectric medium with permittivity $\in$ as shown in figure. We assume $+Q$ and $-Q$ charges on the inner and outer spheres. By applying Gauss Law to an arbitrary gaussion spherical surface of radius $r(a < r < b)$ we have

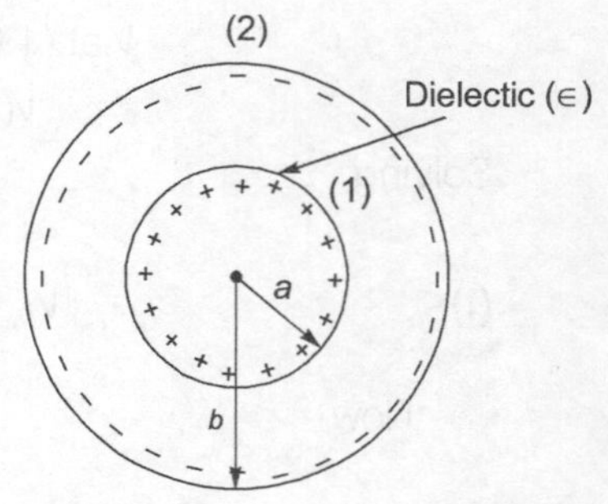

$$Q = \in \oint \vec{E} \cdot \overrightarrow{dS} = \in E_r 4\pi r^2$$

$$E_r = \frac{Q}{4\pi \in r^2}$$

The potential difference between the conductors is

$$V = \int_2^1 \vec{E} \cdot \vec{dl} = -\int_b^a \frac{Q}{4\pi \in r^2} \hat{a}_r \cdot dr \cdot \hat{a}_r$$

$$V = -\int_b^a \frac{Q}{4\pi \in r^2} dr = \frac{Q}{4\pi \in}\left[\frac{1}{r}\right]_b^a = \frac{Q}{4\pi \in}\left[\frac{1}{a} - \frac{1}{b}\right] = \frac{Q}{4\pi \in}\left(\frac{b-a}{ab}\right)$$

$\therefore$ Charge, $C = Q/V$

$$C = \frac{4\pi \in ba}{b-a}$$

From the data,

$$b = 80 \text{ mm} = 80 \times 10^{-3} \text{ m}$$

$$a = 60 \text{ mm} = 60 \times 10^{-3} \text{ m}$$

$\in_r = 1$ (Vacuum or air) assume

$$\therefore \quad C = \frac{4\pi \times 8.85 \times 10^{-12} \times 60 \times 10^{-6} \times 80}{20 \times 10^{-3}}$$

$$C = 26690.97 \times 10^{-15} \text{ F}$$

$$C = 26.69097 \times 10^{-12} \text{ F} = 26.69 \text{ pF}$$

2.8 **Prove that the capacitance between two lines is given by**

$$C = \frac{\pi \varepsilon_0}{ln(d/r)}$$

where d is the distance of separation between these lines and r is radius of each line.

[12 marks : 2004]

Solution:

We will assume that $+\lambda$ and $-\lambda$ are the charges in coulomb per meter length of lines A and B. Spaced D meteres apart and radius of each wire is r meter. Let us consider $(D >> r)$. If 'E_x' is field intensity at any point 'p' on this line with distance 'x' from A and $(D - x)$ from B, the

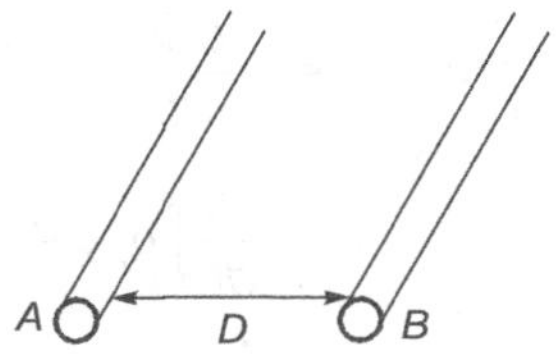

Parallel Lines

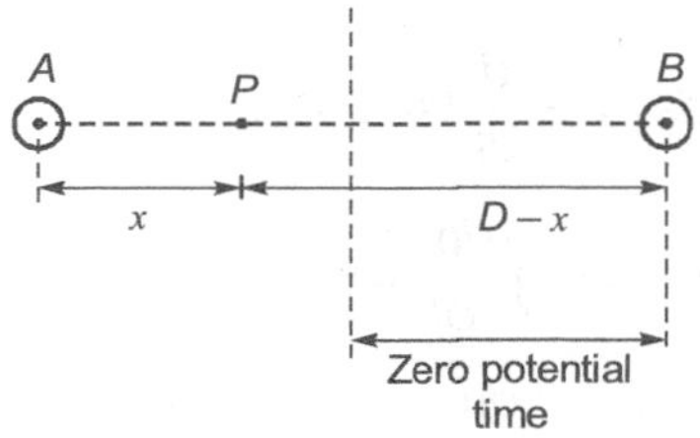

$$V_{BA} = \text{Potential rise from } B \text{ to } A$$

$$V_{BA} = -\int_B^A E_x\, dx$$

The field at point '*p*' is the vector sum of the field due to *A* and *B*. The field due to a charge of λ per meter length at a distance *x* is $\left(\dfrac{\lambda}{2\pi\varepsilon_0 x}\right)$, directed from *A* to *B*.

$$E_{resultant}(P) = \frac{\lambda}{2\pi\varepsilon_0 x} + \frac{\lambda}{2\pi\varepsilon_0 (D-x)}$$

$$\therefore \qquad V = -\frac{\lambda}{2\pi\varepsilon_0}\int_{D-r}^{r}\left[\frac{1}{x}+\frac{1}{D-x}\right]dx = -\frac{\lambda}{2\pi\varepsilon_0}\times\left[ln\frac{r}{D-r}+ln\frac{r}{D-r}\right]$$

$$= \frac{\lambda}{2\pi\varepsilon_0}\times 2\times ln\frac{D-r}{r}$$

$$= \frac{\lambda}{\pi\varepsilon_0}\times ln\frac{D-r}{r} \cong \frac{\lambda}{\pi\varepsilon_0}\times ln\frac{D}{r} \qquad \because D >> r$$

Hence
$$V = \frac{\lambda}{\pi\varepsilon_0} ln\frac{D}{r}$$

The capacitance between the two lines *A* and *B* therefore given by

$$C_{AB} = \frac{\lambda}{V} = \frac{\pi\varepsilon_0}{ln\dfrac{D}{r}}$$

2.9 Explain relaxation time constant for a metal. Derive an expression for the continuity of current equation for dynamic current. Show that the volume charge density is a solution of $\dfrac{\partial\rho_v}{\partial t} + (\sigma/\varepsilon)\rho_v = 0$.

[12 marks : 2005]

Solution:

In metal consider the relation between collision time of an electron with its probability of collision. The probability of an electron to collide with the lattice during a short interval (*dt*) be given by (dt/τ_e) where (τ_e) – average relaxation time constant or relaxation time constant. For isotropic materials, the mean time of collision is the same as the relaxation time, that is $\tau_c = \tau_r$ (where τ_r - relaxation time).

$\therefore$ $$\text{current} = -\frac{dQ}{dt}$$

$$I = \frac{-d}{dt}\int \rho_V \, dV \qquad \rho_V = \text{volume charge density}$$

We know that,
$$I = \oint J \cdot ds$$

$$\oint_s J \cdot ds = -\frac{d}{dt}\int \rho_V \, dV$$

$$\int_V \nabla \cdot J dV = -\int \frac{d\rho_V}{dt} dV \qquad \text{(Divergence theorem)}$$

$$\nabla \cdot J = -\frac{d\rho_V}{dt} \qquad \text{[Continuity equation]}$$

where,
$$\oint_s J \cdot ds = \int_{Vol} (\nabla \cdot J) dv$$

$$\nabla \cdot J + \frac{\partial \rho_V}{\partial t} = 0$$

put,
$$J = \sigma E$$

$$\nabla \cdot (\sigma E) + \frac{\partial \rho_V}{\partial t} = 0$$

$$\frac{\sigma}{\in}(\nabla \cdot D) + \frac{\partial \rho_V}{\partial t} = 0$$

$$(\sigma/\in)\rho_V + \frac{\partial \rho_V}{\partial t} = 0 \qquad (\because \nabla \cdot D = \rho_V \text{ from Maxwell's } 1^{st} \text{ equation})$$

$\therefore$ Volume charge density is the solution of this equation.

2.10 A line charge of length of 2 m has a linear charge density of ρ_L (cm^{-1}). Show that the Electric field at a distance r perpendicular to the line charge from its middle point as shown in the below figure is

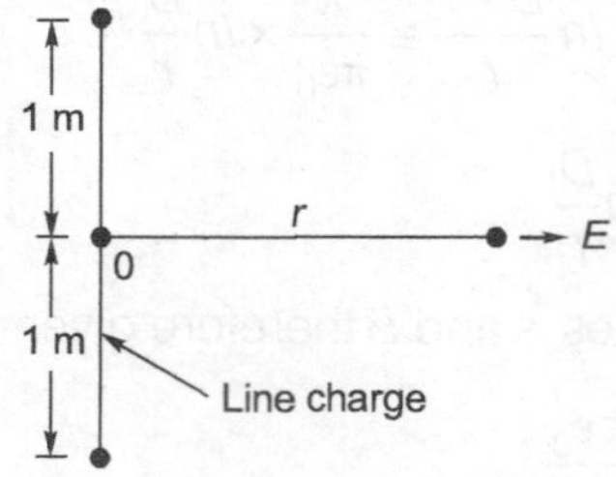

$$E_r = \frac{\rho_L}{2\pi \varepsilon_0 r \sqrt{\left(\frac{r}{m}\right)^2 + 1}}$$

[12 marks : 2006]

Solution:

Consider a uniformly charged wire (L) with the charge assuming to be uniformly distributed on it at the rate of 'ρ_L' Coulomb per meter length. Let 'P' be any point at which field intensity has to be determined.
Let dE be the field due to charge element ($\rho_L\, dx$).
It has component dE_x and dE_y

$$dE_x = dE \cos\theta$$
$$dE_y = dE \sin\theta$$

$$dE = \frac{\rho_L \, dx}{4\pi \in_0 xr^2}$$

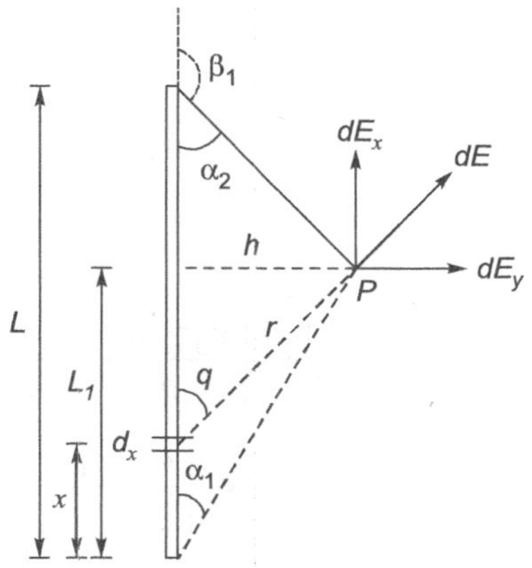

Thus, $$dE_x = \frac{\rho_L \cos\theta dx}{4\pi \in_0 r^2} \quad \text{differentiation}$$

$$-dx = -h \operatorname{cosec}^2 \theta \, d\theta \quad [L_1 - x = h\cot\theta]$$

$$r = h \operatorname{cosec} \theta$$

$$E_x = \frac{\rho_L}{4\pi \in_0 h} \cos\theta \, d\theta$$

$\therefore$ $$E_x = \int_{\theta=\alpha_1}^{\beta_1} dE_x = \frac{\rho_L}{4\pi \in_0 h} \int_{\alpha_1}^{\pi-\alpha_2} \cos\theta \, d\theta$$

$\therefore$ $$E_x = \frac{\rho_L}{4\pi \in_0 h}(\sin\alpha_2 - \sin\alpha_1)$$

Similarly, $$E_y = \frac{\rho_L}{4\pi\varepsilon_0 h}(\cos\alpha_1 + \cos\alpha_2)$$

Here, point 'P' is along the perpendicular bisector of wire, it follows that $\alpha_1 = \alpha_2 = \alpha$ therefore, field 'E_x' vanishes. 'x' components neutralize each other. However 'r'-components always add to each other.

$$E = \frac{\rho_L}{2\pi \in_0 r} \times \cos\alpha = \frac{\rho_L}{2\pi \in_0 r} \times \frac{1}{\sqrt{\left(\frac{r}{m}\right)^2 + 1}} \qquad \because \left[\cos\alpha = \frac{m}{\sqrt{r^2 + m^2}}\right]$$

2.11 **A parallel capacitor of width 'w', Length 'l', and separation 'd' has a solid dielectric slab of permittivity ε in the space between the plates. The capacitor is charged to a voltage V_0 by a battery, as shown in figure below. Assuming that dielectric slab is withdrawn to the position shown. Determine the force acting on the slab.**

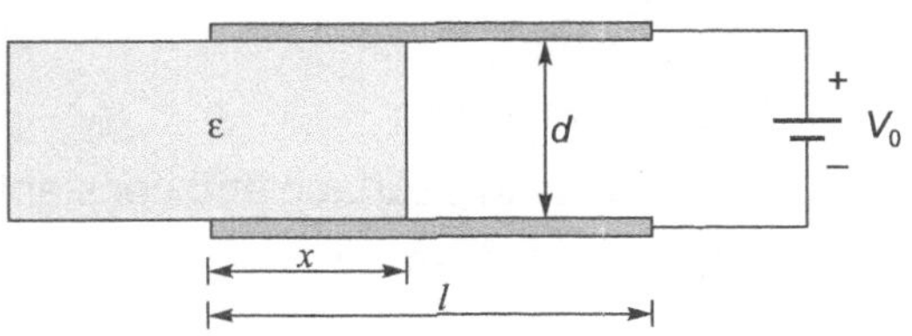

[12 marks : 2007]

Solution:

Initially, $$C = \frac{\in_0 \in_r wl}{d} \quad \ldots(i)$$

Now, $$C_1 = \frac{\in_0 \in_r wx}{d}, \quad \in_2 = \frac{\in_0 w(l-x)}{d}$$

$$C_{eq} = C_1 + C_2 = \frac{\in_0 \in_r wx}{d} + \frac{\in_0 w(l-x)}{d}$$

$$= \frac{\in_0 w}{d}[\in_r x + l - x] \quad \ldots(ii)$$

Initially energy stored $$= \frac{1}{2}CV^2 \quad \ldots(iii)$$

Now,

$$\text{energy stored} = \frac{1}{2}C_{eq}V^2$$

$$\text{Change in energy stored} = \text{work done}$$

$$\frac{1}{2}CV^2 - \frac{1}{2}C_{eq}V^2 = Fx^*(l-x) \qquad \because (l-x \text{ is displacement})$$

$$\frac{1}{2}V^2\left[\frac{\epsilon_0\epsilon_r wl}{d} - \frac{\epsilon_0 w}{d}\{\epsilon_r x + l - x\}\right] = Fx(l-x)$$

$$\frac{1}{2}V^2\left[\frac{\epsilon_0 w}{d}\{\epsilon_r l - \epsilon_r x - (l-x)\}\right] = Fx(l-x)$$

$$\frac{1}{2}V^2\frac{\epsilon_0 w}{d}\left[\epsilon_r(l-x) - (l-x)1\right] = Fx(l-x)$$

$$\therefore \qquad Fx = \frac{1}{2}V^2\frac{\epsilon_0 w}{d}[\epsilon_r - 1]$$

2.12 A dielectric sphere of radius 'a' and permittivity 'ϵ' is situated in a vacuum and is charged throughout its volume by a volume charge density of free charges $\rho = \frac{\rho_0 a}{r}$, r being the distance from the centre of the sphere, show that the electrostatic energy of the system is

$$We = \pi\rho_0^2 a^5\left[\frac{1}{6\epsilon} + \frac{1}{2\epsilon_0}\right]$$

[10 marks : 2008]

Solution:

Volume charge density is given as

$$\rho_v = \begin{cases} \dfrac{\rho_o a}{r}, & 0 \le r \le a. \\ 0\ , & r > a \end{cases}$$

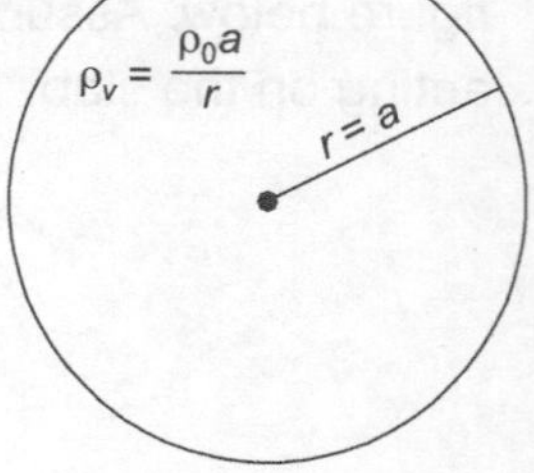

Now let us assume a spherical surface of radius $r \le a$

$\therefore$ flux lines will be directed radially outside to this surface and constant at every point on this surface so according to Gauss law:

$$\oint \vec{D}.\vec{dS} = \int_V \rho_v dV \qquad \left(\because Q_{\text{inclosed}} = \int_V \rho_v\, dV\right)$$

$$\int_s D\hat{a}r.r^2\sin\theta d\phi\hat{a}_r = \int_0^r\int_0^{\pi}\int_0^{2\pi}\frac{\rho_o a}{r}r^2\sin\theta dr d\theta d\phi$$

$$D\int_0^{\pi}\int_0^{2\pi} r^2\sin\theta d\theta d\phi = \int_0^r\int_0^{\pi}\int_0^{2\pi}\frac{\rho_o a}{r}r^2\sin\theta dr d\theta d\phi$$

$$Dr^2.2.2\pi = \rho_0 a 2.2\pi\int_0^r rdr$$

$$D = \frac{\rho_0 a}{2}, \quad r \le a$$

$$\epsilon E_1 = \frac{\rho_0 a}{2}, \quad r \le a$$

$$\therefore \qquad E_1 = \frac{\rho_0 a}{2\in}, \quad r \le a$$

Let as assume a spherical surface of radius $r > a$ flux lines will be radially outside this surface and value of D will be constant throughout this surface.

$$\oint_s \vec{D}.\vec{dS} = \int_V \rho_v dV$$

$$\int_0^{\pi}\int_0^{2\pi} Dr^2 \sin\theta\, d\theta\, d\phi = \int_0^{a}\int_0^{\pi}\int_0^{2\pi} \frac{\rho_0 a}{r} r^2 \sin\theta\, d\theta\, d\phi\, dr$$

$$D.r^2 4\pi = \rho_0 a 4\pi \int_0^a r dr$$

$$D 4\pi r^2 = \rho_0 a 4\pi \left[\frac{r^2}{2}\right]_0^a$$

$$D = \frac{\rho_0 a^3}{2r^2}, \quad r > a$$

$$\in_0 E_2 = \frac{\rho_0 a^3}{2r^2}, \quad r > a$$

$$\therefore \qquad E_2 = \frac{\rho_0 a^3}{2\in_0 r^3}, \quad r > a$$

Now electrostatic energy stored in the system

$$U = \int_V \text{energy density } dV$$

where, $\quad$ energy density $= \frac{1}{2}\in_0 E_2^2 + \frac{1}{2}\in E_1^2$

$$\therefore \qquad U = \int_V \left(\frac{1}{2}\in_0 E_2^2 + \frac{1}{2}\in E_1^2\right) dV$$

$$U = \int_a^{\infty}\int_0^{\pi}\int_0^{2\pi} \frac{1}{2}\in_0 E_2^2 dV + \int_0^{a}\int_0^{\pi}\int_0^{2\pi} \frac{1}{2}\in E_1^2 r^2 \sin\theta\, d\theta \phi\, dr$$

$$= \int_a^{\infty}\int_0^{\pi}\int_0^{2\pi} \frac{1}{2}\in_0 \left[\frac{\rho_0 a^3}{2\in_0 r^2}\right]^2 r^2 \sin\theta\, d\theta\, d\phi\, dr + \int_0^{a}\int_0^{\pi}\int_0^{2\pi} \frac{1}{2}\in \left[\frac{\rho_0 a}{2\in}\right]^2 r^2 \sin\theta\, d\theta\, d\phi\, dr$$

$$= \frac{4\pi\rho_0^2 a^6}{8\in_0}\int_a^{\infty} \frac{1}{r^2} dr + \frac{\rho_0^2 a^2}{8\in} 4\pi \int_0^a r^2 dr$$

$$= \frac{\pi\rho_0^2 a^6}{2\in_0}\left[-\frac{1}{r}\right]_a^{\infty} + \frac{\rho_0^2 a^2 \pi}{2\in}\left[\frac{r^3}{3}\right]_0^a = \frac{\pi\rho_0^2 a^5}{2\in_0} + \frac{\rho_0^2 a^5 \pi}{6\in} = \pi\rho_0^2 a^5\left[\frac{1}{6\in} + \frac{1}{2\in_0}\right]$$

2.13 A certain linear, homogeneous isotropic, dielectric material has a relative permittivity of 1.8. If the applied voltage in material is –4000 V in the y-axis direction, determine electric field intensity $\vec{E}$, electric flux density $\vec{D}$ and polarization $\vec{P}$. Also find the volume charge density r_v.

[10 marks : 2009]

Solution:

Given $\in_r = 1.8$

Let us consider d be thickness of homogeneous, linear isotropic dielectric material.

$\therefore$ medium is homogeneous so electric field will be same at all points and directed towards y-direction.

$\therefore$ $\vec{E}$ electric field is always towards decreasing potential.

$$\therefore \quad \vec{E} = \frac{4000}{d}\hat{a}_y \text{ V/m}$$

$$\therefore \quad \vec{D} = \in \vec{E} = \frac{8.85\times10^{-12}\times1.8\times4000}{d}\hat{a}_y \text{ C/m}^2$$

$$\vec{D} = \frac{15.93\times4\times10^{-9}}{d}\hat{a}_y \text{ C/m}^2$$

$$\vec{D} = \frac{63.72\times10^{-9}}{d}\hat{a}_y \text{ C/m}^2$$

Polarization,
$$\vec{P} = \vec{D}\left(1-\frac{1}{\in_r}\right)$$

$$\vec{P} = \frac{63.72}{d}\times10^{-9}\left(1-\frac{1}{1.8}\right)\hat{a}_y$$

$$\vec{P} = \frac{28.32}{d}\hat{a}_y \text{ nC/m}^2$$

Now
$$\rho_v = \nabla\cdot\vec{D} \text{ (according Maxwell's 1}^{\text{st}}\text{ equation)}$$

$$\therefore \quad \nabla\cdot\left(\frac{63.72}{d}\hat{a}_y n \text{ C/m}^2\right) = \rho_v$$

$$\frac{\partial}{\partial y}\left(\frac{63.72}{d}\times10^{-9}\right) = \rho_v$$

Where d is variable and it represents y-coordinate.

$$\therefore \quad \rho_v = -\frac{63.72}{d^2}\times10^{-9} \text{ C/m}^3$$

$$\rho_v = -\frac{63.72}{d^2} \text{ nC/m}^3$$

2.14 **A long co-axial cable with radii a and b with $b > a$ has uniform charge density ρ_l C/m. Find:**

(i) energy stored per unit length for $a < r < b$

(ii) capacitance per unit length.

[10 marks : 2013]

Solution:

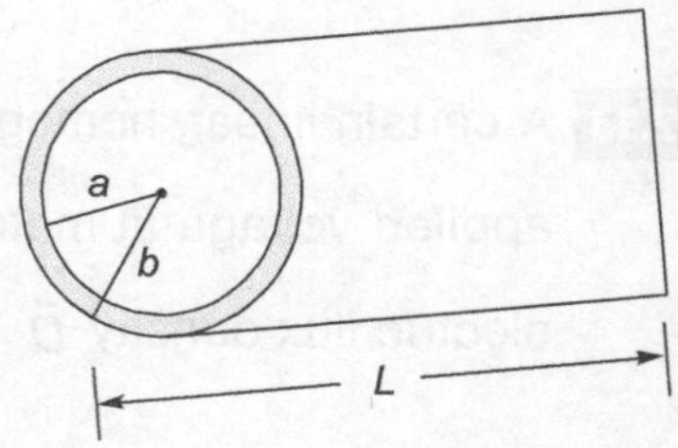

A coaxial capacitor is essentially a coaxial cable or coaxial cylindrical capacitor. Consider length L of two coaxial conductors of inner radius a and outer radius b ($b > a$) as shown in figure. Let the space between the conductors be filled with a homogeneous dielectric with permittivity $\in$. We assume that conductors 1 and 2, respectively carry charges $+Q$ and $-Q$ uniformly distributed on them. By applying Gauss's law to an arbitrary Gaussian cylindrical surface of radius r($a < r < b$), we obtain

$$Q = \in \oint E \cdot dS = \in E_\rho 2\pi rL$$

Hence, $$E = \frac{Q}{2\pi \in rL}\hat{a}_r$$

Neglecting flux fringing at the cylinder ends,

$$V = -\int_2^1 E \cdot dr = -\int_b^a \left[\frac{Q\hat{a}_r}{2\pi \in rL}\right] \cdot dr\,\hat{a}_r = \frac{Q}{2\pi \in L} ln\frac{b}{a}$$

(i) The capacitance per unit length of a coaxial cylinder is given by

$$\boxed{C = \frac{Q}{V} = \frac{2\pi \in}{ln\frac{b}{a}}}$$

(ii) Energy stored $= \frac{1}{2}CV^2 = \frac{1}{2}\frac{Q^2}{C} = \frac{1}{2}\frac{(\rho_l L)^2}{\frac{2\pi \in L}{ln\frac{b}{a}}} = \frac{\rho_l^2 L}{4\pi \in} ln\frac{b}{a}$

$\therefore$ Energy stored per unit length $= \frac{\rho_l^2}{4\pi \in} ln\frac{b}{a}$

2.15 Derive Coulomb's law, starting from Gauss theorem. State any resonable assumptions which you think are necessary for the derivation.

[8 marks : 2014]

Solution:

$$Q = \int_s D_s \cdot ds$$

Assumptions:

1. D_s is everywhere either normal (or) tangential to the closed surface, so that $D_s \cdot ds$ becomes either $D_s \cdot ds$ (or zero, respectively.
2. On that portion of the closed surface for which $D_s \cdot ds$ is not zero, D_s = constant.
   Consider a charge (Q) enclosed in sphere of radius 'r'.
   Consider small element 'ds' with sides $r\sin\theta\, d\phi$ and $r\, d\theta$.

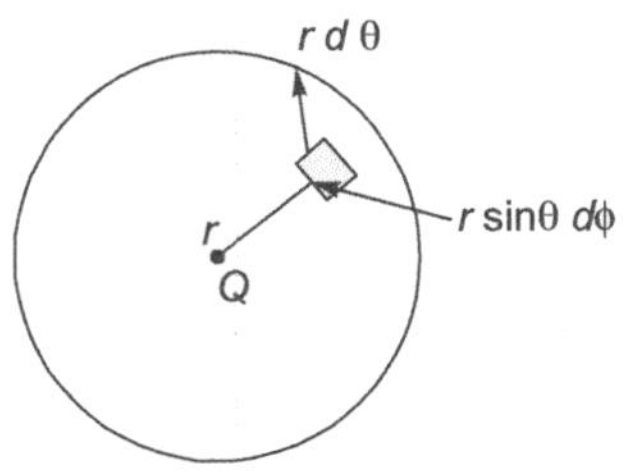

To cover the entire sphere $\int_0^{\pi}\int_0^{2\pi} D_s(r^2\sin\theta)\, d\theta\, d\phi = Q$

$$\int_0^{2\pi} D_s r^2 d\phi \int_0^{\pi} \sin\theta\, d\theta = 0$$

$$\int_0^{2\pi} D_s r^2 d\phi (2) = Q$$

$$D_s = \frac{Q}{4\pi r^2}$$

$\because \quad D = \in E$ where $\in = \varepsilon_0 \varepsilon_r$

$\Rightarrow$ $$\boxed{E = \frac{Q}{4\pi\varepsilon r^2}}$$

2.16 Determine the E field using Gauss's law caused by a spherical cloud of electron with a volume charge density

$$P_v(r, \theta, \phi) = P_0 \frac{r}{a};\ 0 < r < a$$
$$= 0\ ;\ a \le r < \infty$$

[10 marks : 2015]

Solution:

$$P_v(r, \theta, \phi) = P_0 \frac{r}{a}; 0 < r < a$$
$$= 0;\ a \le r \le \infty$$

(i) For $0 < r < a$: using Gauss's law,

$$\oint\!\!\oint \vec{D}.\vec{ds} = Q_{\text{enclosed}} = \iiint P_v \cdot dV$$

$$D \oint\!\!\oint r^2 \sin\theta\, d\theta\, d\phi = \iiint P_0 \cdot \frac{r}{a} \cdot r^2 \sin\theta\, dr\, d\theta\, d\phi$$

$$D \cdot \int_{\theta=0}^{\pi} \int_{\phi=0}^{2\pi} r^2 \sin\theta \cdot d\theta d\phi = P_0 \int_{r=0}^{r} \int_{\theta=0}^{\pi} \int_{\phi=0}^{2\pi} \frac{r^3}{a} \cdot \sin\theta \cdot dr \cdot d\theta \cdot d\phi$$

$$D \cdot r^2 [-\cos\theta]_0^{\pi} \times [\phi]_0^{2\pi} = \frac{P_0}{a}\left[\frac{r^4}{4}\right]_0^r \times [-\cos\theta]_0^{\pi} \times [\phi]_0^{2\pi}$$

$$D \cdot 4\pi r^2 = \frac{P_0}{a} \times \frac{r^4}{4} \times 4\pi$$

$$D = \frac{P_0}{a} \cdot \frac{r^2}{4}, \quad \boxed{\vec{D} = \frac{P_0}{4} \cdot \frac{r^2}{a} \cdot \hat{a}_r}$$

$$\boxed{\vec{E} = \frac{P_0 \cdot r^2}{4\varepsilon_0 a} \cdot \hat{a}_r}$$

(ii) For $a \le r < \infty$

$$E_r = \frac{1}{4\pi\varepsilon_0 r^2} \int_{r=0}^{a} \int_{\theta=0}^{\pi} \int_{\phi=0}^{2\pi} \frac{P_0 \cdot r^3}{a} \cdot \sin\theta \cdot dr d\theta \cdot d\phi = \frac{1}{4\pi} \cdot \frac{P_0 \cdot a^4}{\varepsilon_0 a \cdot r^2} \times \frac{4\pi}{4}$$

$$E_r = \frac{P_0 \cdot a^4}{4\varepsilon_0 r^2}; \quad \boxed{\bar{E} = \frac{P_0 a^3}{4\varepsilon_0 r^2} \cdot \hat{a}_r}$$

2.17 If we consider yz-plane of a Cartesian co-ordinate system as an interface between two dielectric mediums having $\epsilon_1 = 5\epsilon_0\ (x > 0)$ and $\epsilon_2 = 3\epsilon_0\ (x < 0)$, then find the expressions for D_1 and E_1 in medium 1 and D_2 in medium 2 with known electric field in medium 2 is $\bar{E}_2 = 10\bar{i}_x + 20\bar{i}_z$.

[10 marks : 2015]

Solution:

Let assume,
$$\vec{E}_1 = E_{x_1}\hat{i}_x + E_{y_1}\hat{i}_y + E_{z_1}\hat{i}_z$$

$x > 0$	$x = 0$	$x < 0$
medium (1)		medium (2)
$\varepsilon_1 = 5\varepsilon_0$		$\varepsilon_2 = 3\varepsilon_0$
		$\bar{E}_2 = 10\hat{i}_x + 20\hat{i}_z$

Using Boundary conditions:

1. For tangential component,

$$E_{y_1} = E_{y_2} \quad \text{and} \quad E_{z_1} = E_{z_2}$$

Therefore, $$E_{y_1} = 0 \quad \text{and} \quad E_{z_1} = 20$$

2. For perpendicular components,

$$D_{x_1} - D_{x_2} = \rho_s$$

$\because$ $$D_{x_1} - D_{x_2} = 0, \quad \text{as} \quad \rho_s = 0$$

Hence $$\varepsilon_1 E_{x_1} = \varepsilon_2 E_{x_2}$$

and $$E_x = \frac{3\varepsilon_0}{5\varepsilon_0} \times 10 = 6$$

Now, $$\boxed{\vec{E}_1 = 6\hat{i}_x + 20\hat{i}_z}$$

and $$\vec{D}_1 = 5\varepsilon_0(6\hat{i}_x + 20\hat{i}_z)$$

$$\boxed{\vec{D}_1 = \varepsilon_0(30\hat{i}_x + 100\hat{i}_z)}$$

and $$D_{n_1} = D_{n_2} = 30\epsilon_0\, i_x$$

$$D_t = \epsilon_2 E_{t_1} = 3\epsilon_0\,(20)\hat{i}_z = 60\epsilon_0\,\hat{i}_z$$

so, $$D = D_{n_1} + D_{t_1}$$

$$= \epsilon_0(30\hat{i}_x + 60\hat{i}_z)$$

2.18 What is the difference between the coaxial cable used for power line and the one used in an electronic circuit ? Derive an expression of capacitance for such a cable per unit length.

[10 marks : 2015]

Solution:

In electronic circuit co-axial cable mainly used for high frequency, (kHz to MHz) low power (m Watt) signal or data transfer they have small conductor size due to small current density. They are mechanically less stronger than powerline co-axial cable.

Example: Transfer of video signal.

Where as power line co-axial cable used for low frequency (50 Hz, 60 Hz) high power (M Watt) transfer and they are having large cross section area due to large power transfer. They are mechanically much stronger than electronic cable.

Example: Under ground cable.

Expression of capacitance of coaxial cable:

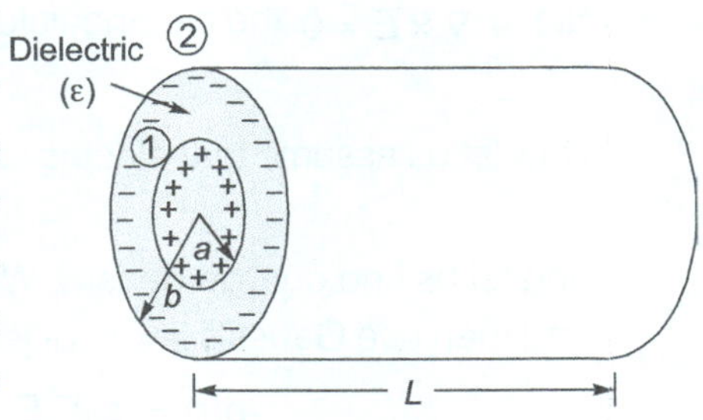

Consider a coaxial cable of length L of inner radius 'a' and outer radius 'b' ($b > a$) as shown in figure. Let the space between the conductors be filled with a homogenous dielectric with permittivity 'ε'. We assume that conductors 1 and 2 respectively carry $+Q$ and $-Q$ uniformly distributed on them. By applying Gauss's law to an arbitrary Gaussian cylinderical surface of radius $\rho(a < \rho < b)$, we obtain,

$$Q = \varepsilon\oint E.ds = \varepsilon E\cdot 2\pi\cdot\rho\cdot L$$

Hence, $$E = \frac{Q}{2\pi\varepsilon\rho L}\hat{a}_\rho$$

Neglecting flux fringing at the cylinder ends,

$$V = -\int_b^a E\cdot dl = -\int_b^a\left[\frac{Q}{2\pi\varepsilon\rho L}\hat{a}_\rho\right]\cdot d\rho\cdot\hat{a}_\rho = \frac{Q}{2\pi\varepsilon L}\ln\frac{b}{a}$$

Thus the capacitance of a coaxial cylinder is given by

$$C = \frac{Q}{V} = \frac{2\pi\varepsilon L}{\ln\frac{b}{a}}$$

Capacitance per unit length $\boxed{C = \frac{2\pi\varepsilon}{\ln(b/a)}}$

2.19 **The electric field is given by $E(r) = E_0\cos\left(\frac{z}{z_0}\right)\exp\left(\frac{-r}{r}\right)\hat{r}$, where z_0 and r_0 are consonant. Check whether or not the electric field is conservative. If it is conservative find the potential, if it is not suggest how it may be possible to find the electrostatic part of the electric field (if present) and the corresponding electrostatic potential $V(r)$.**

[10 marks : 2016]

Solution:

First we need to test whether or not the field is purely electrostatic, i.e., whether or not it is conservative. If $\nabla\times E = 0$ then E is conservative. First write the field in spherical coordinates

$$E(r,\theta,\phi) = E_0\cos\left(\frac{r\cos\theta}{z_0}\right)\exp\left(\frac{-r}{r_0}\right)\hat{r}$$

and use $$\nabla\times A = \frac{1}{r\sin\theta}\left[\frac{\partial}{\partial\theta}(\sin\theta A_\phi) - \frac{\partial A_\theta}{\partial_\phi}\right]\hat{r} + \frac{1}{r}\left[\frac{1}{\sin\theta}\frac{\partial A_r}{\partial\phi} - \frac{\partial}{\partial r}(rA_\phi)\right]\hat{\theta} + \frac{1}{r}\left[\frac{\partial}{\partial r}(rA_\theta) - \frac{\partial A_r}{\partial\theta}\right]\hat{\phi}$$

$\therefore$ $$\nabla\times E = -\frac{1}{r}\frac{\partial E_r}{\partial\theta}\hat{\phi}$$

$$= -\frac{E_0}{r}\exp\left(\frac{-r}{r_0}\right)\frac{\partial}{\partial\theta}\cos\left(\frac{r\cos\theta}{z_0}\right)\hat{\phi} = -\frac{E_0}{r}\exp\left(\frac{-r}{r_0}\right)\left[-\sin\left(\frac{r\cos\theta}{z_0}\right)\right]\left[\frac{-r\sin\theta}{z_0}\right]\hat{\phi}$$

$$= -\frac{E_0}{z_0}\exp\left(\frac{-r}{r_0}\right)\left[-\sin\left(\frac{r\cos\theta}{z_0}\right)\right]\sin\theta\hat{\phi}$$

Since $\nabla\times E \neq 0$ the electric field is not purely electrostatic.

Now let us assume that electric field is given by $E(r) = E_0\cos\left(\frac{z}{z_0}\right)\exp\left(\frac{-r}{r}\right)\hat{r}$, where z_0 and r_0 are constants

and let us find charge density. We will need to write E in terms of either Cartesian or spherical coordinates, and then use Gauss's law in differential form. Choosing spherical coordinates E is in the radial direction,

$$\rho(r) = \varepsilon_0\nabla.E$$

$$= \varepsilon_0 E_0 \frac{1}{r^2}\frac{\partial}{\partial r}\left[r^2 \cos\left(\frac{r\cos\theta}{z_0}\right)\exp\left(-\frac{r}{r_0}\right)\right]$$

$$= \frac{\varepsilon_0 E_0}{r^2}\left[2r\cos\left(\frac{r\cos\theta}{z_0}\right)\exp\left(-\frac{r}{r_0}\right) - r^2\sin\left(\frac{r\cos\theta}{z_0}\right)\frac{\cos\theta}{z_0}\exp\left(-\frac{r}{r_0}\right)\right.$$

$$\left. + r^2\cos\left(\frac{r\cos\theta}{z_0}\right)\exp\left(-\frac{r}{r_0}\right)\left(\frac{-1}{r_0}\right)\right]$$

$$= \frac{\varepsilon_0 E_0}{r}\cos\left(\frac{r\cos\theta}{z_0}\right)\exp\left(-\frac{r}{r_0}\right)\left[2 - \tan\left(\frac{r\cos\theta}{z_0}\right)\frac{r\cos\theta}{z_0} - \frac{r}{r_0}\right]$$

$$= \frac{\varepsilon_0 E_0}{r}\cos\left(\frac{z}{z_0}\right)\exp\left(-\frac{r}{r_0}\right)\left[2 - \tan\left(\frac{z}{z_0}\right)\frac{z}{z_0} - \frac{r}{r_0}\right]$$

So the space has non-zero charge density and so there must be an electrostatic component of electronic field. The electric field and potential can be computed from ρ using coulomb law.

2.20 An air-spaced coaxial line has inner and outer conductors with radii *a* and *b* respectively as shown in figure below. Show that the breakdown voltage of the line is highest when $ln(a/b) = 1$.

[5 marks : 2016]

Solution:

We have an air spaced coaxial line with inner and outer conduction with radii *a* and *b*.

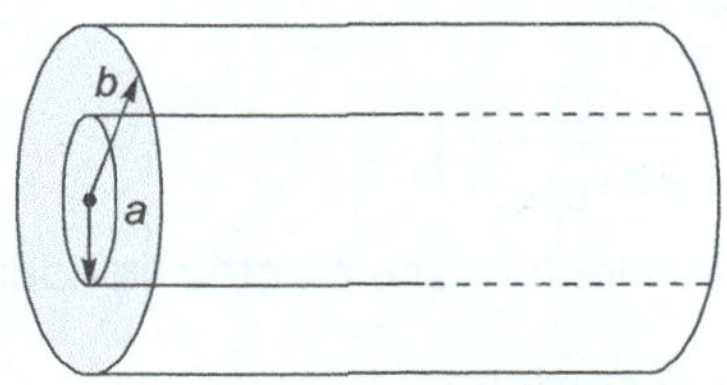

Let us assume that potential difference between the two conductors (inner and outer cylinder is ΔV) and charge per unit length is

Using gauss law we can get that,

$$E = \left(\frac{\lambda}{2\pi\varepsilon_0 \pi}\right)\hat{a}r$$

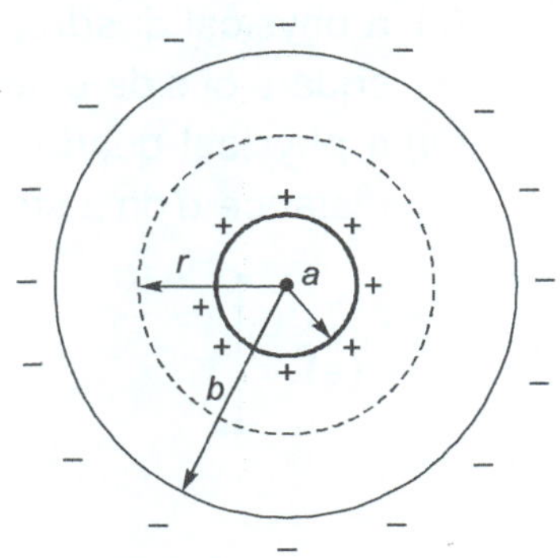

Now, $$\Delta V = \int_a^b E.dr = \frac{\lambda}{2\pi\varepsilon_0}\int_a^b \frac{1}{r}dr = \frac{\lambda}{2\pi\varepsilon_0} ln(b/a)$$

or we can say that, $$\frac{\lambda}{2\pi\varepsilon_0} = \frac{\Delta V}{ln(b/a)}$$

So, we get that, $$E = \left(\frac{\Delta V}{ln(b/a)r}\right)$$

and '*E*' is maximum at $r = a$

Now from this we can analyze that if E_0 is breakdown electric field of the line then breakdown voltage will be

$$\Delta V = \left(E_0 . ln(b/a)a\right)$$

So breakdown voltage will be maximum when $ln(b/a)$ is highest

2.21 **Figure below shows coaxial cable with two layers of dielectric material. Find an expression for the electric field and point within the dielectric.**

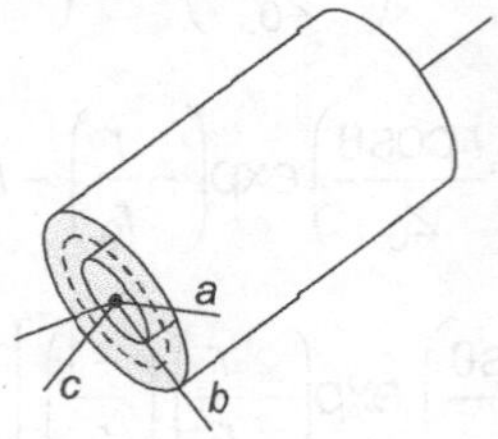

[10 marks : 2016]

Solution:

The given coaxial cable contain two dielectrics and we need to find expression of electric field at any point.
The charge will remain same on all three plates, Q.
So to find electric field.
For $a < r < b$, if we draw a gaussion surface of radius r we get that,

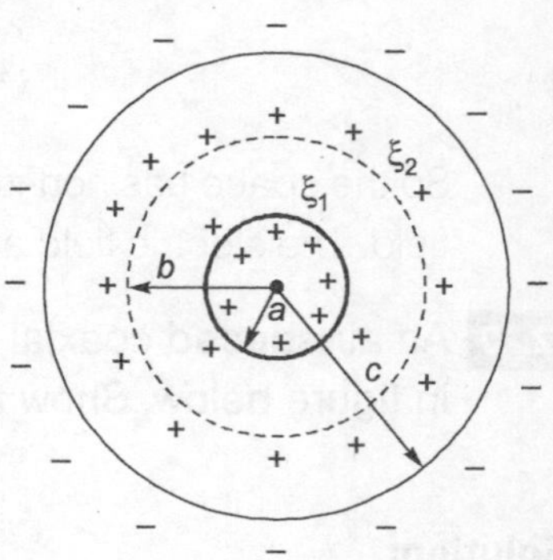

$$D = \frac{Q}{2\pi rl}$$

Where Q is charge on inner plate, r is radius of gaussian surface and l is length of cylinder.

So, $$E = \left(\frac{Q}{2\pi\varepsilon_0\varepsilon_{r_1} rl}\right)$$

Here, ε_{r1} is dielectric constant of 1st dielectric.
The electric field in 2nd dielectric will be

$$E = \left(\frac{Q}{2\pi\varepsilon_0\varepsilon_{r2} rl}\right)$$

Here ε_{r2} is dielectric constant of 2nd dielectric. The electric flux density will be same in both the dielectric.

2.22 **How much work must be done to assemble:**

(i) a physical dipole made of charge $+q$ and charge $-q$ separated by distance d.

(ii) a physical quadrupole made up of four charges $+q$, $-q$, $+q$ and $-q$ on successive corners of a square of side d, and

(iii) a physical quadrupole made up of four charges $-q$, $+q$, $+q$ and $-q$ equally spaced apart by distance d on a straight line (see diagram below).

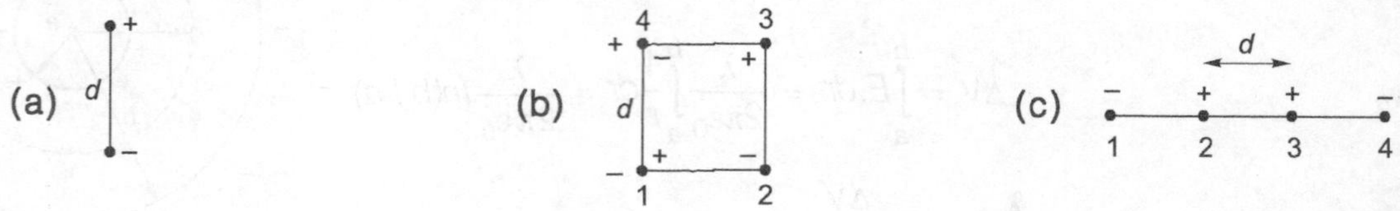

[10 marks : 2016]

Solution:

The work done to bring together a group of N charges is

$$W = \frac{1}{2}\sum_{i=1}^{N} q_i V(r_i)$$

(a) $$W = \frac{1}{2}\left((-q)\frac{(+q)}{4\pi\varepsilon_0 d} + (+q)\frac{(-q)}{4\pi\varepsilon_0 d}\right) = -\frac{q^2}{4\pi\varepsilon_0 d}$$

(b)
$$V_1 = V_3 = \frac{q}{4\pi\varepsilon_0}\left(\frac{-1}{d}+\frac{\sqrt{2}}{d}+\frac{-1}{d}\right) = \frac{q}{4\pi\varepsilon_0 d}\left(\sqrt{2}-2\right)$$

$$\therefore \quad q_1V_1 = q_3V_3 = \frac{q^2}{4\pi\varepsilon_0 d}\left(\sqrt{2}-2\right)$$

$$V_2 = V_4 = \frac{q}{4\pi\varepsilon_0}\left(\frac{1}{d}+\frac{-\sqrt{2}}{d}+\frac{1}{d}\right) = \frac{q}{4\pi\varepsilon_0 d}\left(2-\sqrt{2}\right),$$

$$\therefore \quad q_2V_2 = q_4V_2 = \frac{q^2}{4\pi\varepsilon_0 d}\left(\sqrt{2}-2\right)$$

Hence,
$$W = \frac{1}{2}\times 4\times\frac{q^2}{4\pi\varepsilon_0}(\sqrt{2}-2) = -\frac{q^2}{2\pi\varepsilon_0 d}\left(2-\sqrt{2}\right)$$

(c)
$$V_1 = V_4 = \frac{q}{4\pi\varepsilon_0}\left(\frac{1}{d}+\frac{1}{2d}+\frac{-1}{3d}\right) = \frac{7}{6}\frac{q}{4\pi\varepsilon_0 d}$$

$$\therefore \quad q_1V_1 = q_4V_4 = -\frac{7}{6}\frac{q^2}{4\pi\varepsilon_0 d}$$

$$V_2 = V_3 = \frac{q}{4\pi\varepsilon_0}\left(\frac{-1}{d}+\frac{1}{d}+\frac{-1}{2d}\right) = -\frac{1}{2}\frac{q}{4\pi\varepsilon_0 d}$$

$$\therefore \quad q_2V_2 = q_3V_3 = -\frac{1}{2}\frac{q^2}{4\pi\varepsilon_0 d}$$

Hence,
$$W = -\frac{1}{2}\frac{q^2}{4\pi\varepsilon_0 d}\left(-\frac{7}{6}-\frac{1}{2}-\frac{1}{2}-\frac{7}{6}\right) = -\frac{10}{3}\frac{q^2}{4\pi\varepsilon_0 d}$$

2.23 Four charges $q_1 = -2.0 \times 10^{-9}$ Coulomb, $q_2 = 3.0 \times 10^{-9}$ Coulomb, $q_3 = 2.0 \times 10^{-9}$ Coulomb and $q_4 = 1.0 \times 10^{-9}$ Coulomb are placed at four corners of a square of 2.0 metre side. Find out the potential at the centre of square.

[12 marks : 2017]

Solution:

Net potential at, the center,

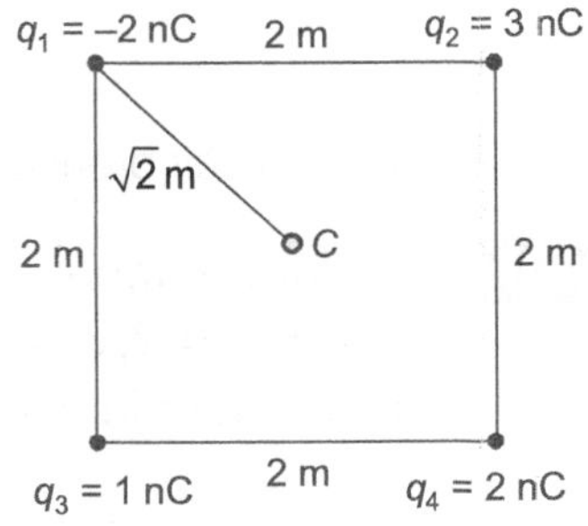

$$C,\ V_C = \frac{1}{4\pi\xi_0}\frac{q_1}{r}+\frac{1}{4\pi\xi_0}\frac{q_2}{r}+\frac{1}{4\pi\xi_0}\frac{q_3}{r}+\frac{1}{4\pi\xi_0}\frac{q_4}{r}$$

$$= [-2\times 10^{-9}+3\times 10^{-9}+10^{-9}+2\times 10^{-9}]\times\frac{1}{4\pi\xi_0}\times\frac{1}{\sqrt{2}}\ \text{V}$$

$$V_C = \frac{4\times 10^{-9}\times 9\times 10^{9}}{\sqrt{2}}\ \text{V} = 18\sqrt{2}\ \text{V}$$

2.24 Two parallel plates are separated 0.25 cm apart. The gap is partially filled with 0.23 cm thick sheet of fibre with $\epsilon_r = 5$. A potential difference of 2.5 kV is applied across the plate. Determine the electric field strength in the fibre and air film between the fibre and the plate. Will the air break down? Dielectric strength of air can be taken as 30 kV/cm.

[12 marks : 2017]

Solution:

Given that:

Distance between two parallel plate = 0.25 cm = 0.25×10^{-2} m

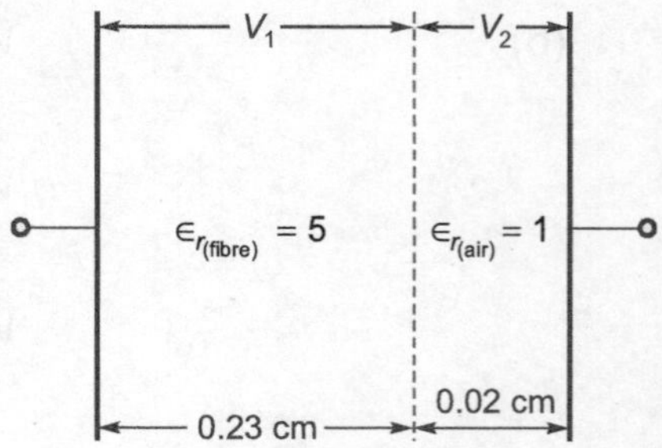

Relative permittivity of fibre, $\epsilon_{r(\text{fibre})} = 5$

Relative permittivity of air, $\epsilon_{r(\text{air})} = 1$

It is equivalent to two capacitors in series of capacitances

$$C_1 = \frac{\epsilon_0 \epsilon_1 A}{d_1} \text{ and } C_2 = \frac{\epsilon_0 \epsilon_2 A}{d_2}$$

$$\epsilon_1 = \epsilon_{\text{fibre}} = 5 \text{ and } \epsilon_2 = \epsilon_{\text{air}} = 1$$

and
$$C_{eq} = \frac{C_1 C_2}{C_1 + C_2}$$

For capacitors connected in series, charge being same

i.e.,
$$C_1 V_1 = C_2 V_2$$

So,
$$\frac{V_2}{V_1} = \frac{C_1}{C_2} = \frac{\epsilon_1 d_2}{\epsilon_2 d_1} \Rightarrow V_1 = \frac{\epsilon_1 d_2}{\epsilon_2 d_1} \cdot V_1 \quad \text{...(i)}$$

and
$$V = V_1 + V_2 = 2.5 \text{ kV} = 2500 \text{ V} \quad \text{...(ii)}$$

Substitute the value of ϵ_{r_1}, ϵ_{r_2}, d_1 and d_2 in equation (i), we get

$$V_2 = \frac{5 \times 0.02 \times 10^{-2}}{1 \times 0.23 \times 10^{-2}} \cdot V_1 = 0.4348\, V_1$$

From equation (ii), we get

$$V_1 + 0.4348\, V_1 = 2500$$

$$1.4348\, V_1 = 2500$$

$$V_1 = \frac{2500}{1.4348} = 1742.403 \text{ V}$$

and
$$V_2 = 0.4348\, V_1 = 0.4348 \times 1742.403 = 757.597 \text{ V}$$

Now, electric field strength,

$$E_{(\text{fibre})} = E_1 = \frac{V_1}{d_1} = \frac{1742.403}{0.23} = 7575.665 \text{ V/cm or } 7.5756 \text{ kV/cm}$$

and
$$E_{(\text{air})} = E_2 = \frac{V_2}{d_2} = \frac{757.597}{0.02} = 37879.85 \text{ V/cm or } 37.8797 \text{ kV/cm}$$

so,
$$E_{\text{air}} = 37.8798 \text{ kV/cm} > 30 \text{ kV/cm}$$

Hence, the air breakdown will be occur.

2.25 An infinite number of charges each equal to 'q Coulombs' are placed in a free space along the line at $x = 1$, $x = 2$, $x = 4$, $x = 8$, $x = 16$ and so on. Find the expression for potential and electric field intensity at point $x = 0$, due to these systems of charges. Assume that values of x are in metres.

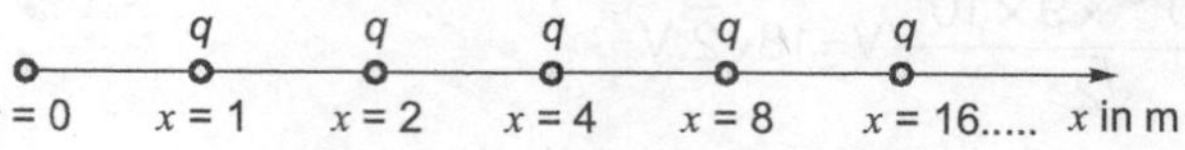

[12 marks : 2018]

Solution:

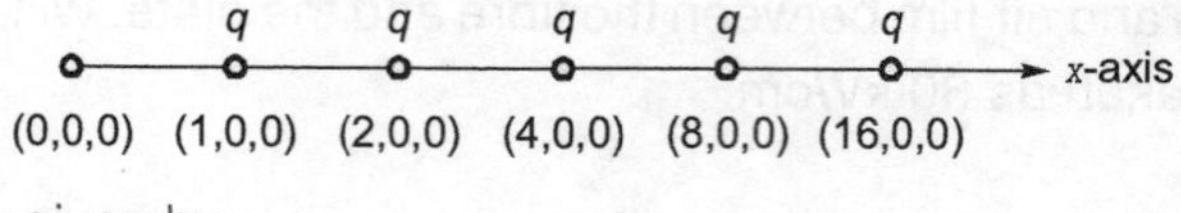

Total potential at origin is given by:

$$V = \frac{q}{4\pi\varepsilon_0(2^0)} + \frac{q}{4\pi\varepsilon_0(2^1)} + \frac{q}{4\pi\varepsilon_0(2^2)} + \ldots..$$

$$V = \frac{q}{4\pi\varepsilon_0}\left[\frac{1}{2^0}+\frac{1}{2^1}+\frac{1}{2^2}+\ldots.\right]$$

$$V = \frac{q}{4\pi\varepsilon_0}\left[1+\frac{1}{2}+\frac{1}{2^2}+\ldots.\right]$$

Now we know that summation of an infinite G.P. is given as:

$[a + ar + ar^2 + ar^3 + \ldots. = \dfrac{a}{1-r}$ if $|r| < 1$]

$$V = \frac{q}{4\pi\varepsilon_0}\left[\frac{1}{1-1/2}\right] = \frac{q}{2\pi\varepsilon_0}\text{ V}$$

Electric field at origin is given by,

$$\vec{E} = \frac{q}{4\pi\varepsilon_0}\frac{\vec{R}_{12}}{|\vec{R}_{12}|^3}$$

$$\text{Total } \vec{E} = \frac{q}{4\pi\varepsilon_0}\frac{(-1\hat{a}_x)}{1^3}+\frac{q}{4\pi\varepsilon_0}\frac{(-2\hat{a}_x)}{2^3}+\frac{q}{4\pi\varepsilon_0}\left(\frac{-2^2\hat{a}_x}{(2^2)^3}\right)+\ldots.$$

$$= \frac{q}{4\pi\varepsilon_0}\left[1+\frac{1}{2^2}+\frac{1}{(2^2)^2}+\frac{1}{(2^3)^2}+\ldots\right](-\hat{a}_x)$$

$$\vec{E} = \frac{q}{4\pi\varepsilon_0}\left[\frac{1}{\left(1-\frac{1}{2^2}\right)}\right](-\hat{a}_x) = \frac{q}{4\pi\varepsilon_0}\left[\frac{4}{3}\right](-\hat{a}_x)$$

$$= -\frac{q}{3\pi\varepsilon_0}\hat{a}_x \text{ V/m}$$

2.26 A parallel plate capacitor consisting of two dielectric materials is shown in figure. The middle dielectric slab is placed symmetrically with respect to the plates. If the potential difference between one of the plates and nearest surface of dielectric interface is 4 volts, determine ϵ_1/ϵ_2. Assume parallel plate capacitor has an electrode area of A m^2.

[12 marks : 2018]

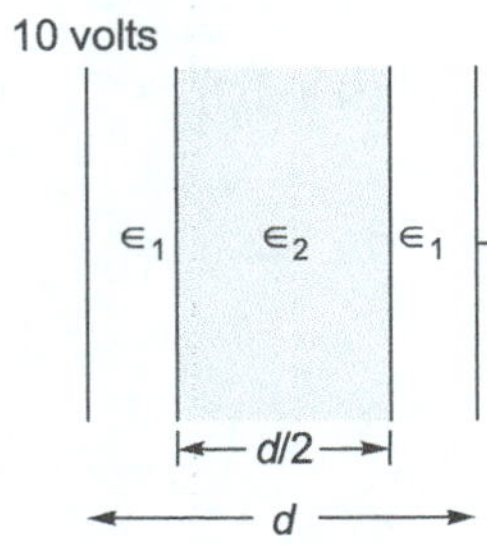

Solution:

According to given condition:

All the three dielectrics are connected in series, so charge will be same throughout the dielectrics.

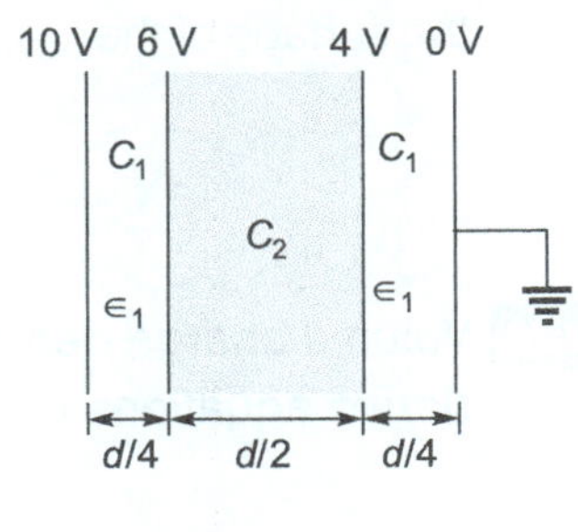

$$C_1 = \frac{\epsilon_0\epsilon_1 \times A}{d/4} = \frac{4\,\epsilon_0\epsilon_1 \times A}{d}$$

$$C_2 = \frac{\epsilon_0\epsilon_1 A}{d/2} = \frac{2\,\epsilon_0\epsilon_2 A}{d}$$

$$Q_1 = Q_2$$

$$\therefore \quad C_1V_1 = C_2V_2$$

$$\therefore \quad \frac{4\epsilon_0\epsilon_1 A}{d}\times 4 = \frac{2\epsilon_0\epsilon_2 A}{d}\times 2$$

$$\Rightarrow \quad \frac{\epsilon_1}{\epsilon_2} = \frac{1}{4}$$

2.27 **Find the force with which the plates of a parallel-plate capacitor attract each other. Also determine the pressure on the surface of the plate due to the field.**

[12 marks : 2019]

Solution:

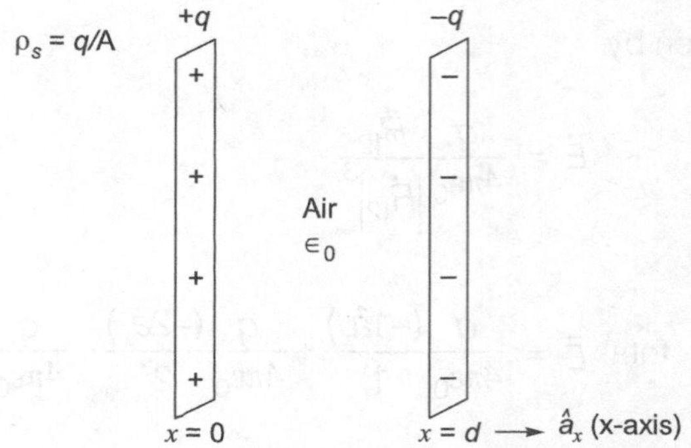

$$r_s = \frac{q}{A} = \text{Surface charge density (C/m}^2)$$

A - Area of the plate.

Electric field intensity due to single plate having r_s (C/m²) surface charge density is

$$\vec{E} = \frac{\rho_s}{2\epsilon_0}\hat{a}_x \text{ for } x > 0 \qquad \text{... (i)}$$

Force on f charge due to $\vec{E}$ is, $\vec{F} = Q\vec{E}$ ... (ii)

Force on $-q$ charge due to $\vec{E}$ is, $\vec{F} = -q\vec{E}$ ... (iii)

Put (i) in (iii), $\vec{F} = -q\frac{\rho_s}{2\epsilon_0}\hat{a}_x$ where, $\rho_s = \frac{q}{A}$

$$= -q\left(\frac{q}{2A\epsilon_0}\right)\hat{a}_x$$

$$\vec{F} = \frac{-q^2}{2A\epsilon_0}\hat{a}_x \qquad \text{... (iv)}$$

Force on $-q$ charge plate is towards $+q$ charge plate indicating both plates attract each other. Pressure on the surface of the plate

$$= P = \frac{F}{A} = \frac{q^2}{2\epsilon_0 A^2}\ (\text{N/m}^2)$$

2.28 **Volume charge density is the same as the divergence of the electric flux density. Using Gauss's law, derive equations to prove it.**

[12 marks : 2019]

Solution:

Gauss law states that the total electric flux crossing any closed surface is equal to the total charge enclosed by that closed surface.

$$\psi \text{ crossing closed surface} = Q_{\text{enclosed}} \quad \text{... (i)}$$

Given $\vec{D}$ electric flux density then electric flux crossing closed surface is

$$\psi \text{ crossing closed surface} = \oint\!\!\oint \vec{D}\cdot d\vec{S} \quad \text{... (ii)}$$

Put (ii) in (i),

$$\Rightarrow \quad \oint\!\!\oint \vec{D}\cdot d\vec{S} = Q_{\text{enclosed}}. \quad \text{... (iii)}$$

Inside a closed surface volume is present. Total charge in a volume inside closed surface is

$$Q_{\text{enc}} = \iiint \rho_v dv \quad \text{... (iv)}$$

Put (iv) in (iii),

$$\oint\!\!\oint \vec{D}\cdot d\vec{S} = \iiint \rho_v dv \quad \text{... (v)}$$

From divergence theorem $\oint\!\!\oint \vec{D}\cdot d\vec{S} = \iiint (\vec{\nabla}\cdot\vec{D}) dv$ ... (vi)

Put (vi) in (v),

$$\iiint (\vec{\nabla}\cdot\vec{D})\, dv = \iiint \rho_v dv$$

Compare both sides, $\vec{\nabla}\cdot\vec{D} = \rho_v$ ... (viii)

equation (viii) says divergence of electric flux density is equal to volume charge density.

2.29 Determine the total charge for the following:
(a) on a line $0 \le x \le 5$ m if $\rho_L = 12x^2$ milii C/m
(b) on a cylinder $\rho = 3$, $0 \le z \le 4$ m if $\rho_s = \rho z^2$ nC/m^2

[10 marks : 2019]

Solution:

(a)

$$Q = \int \rho_l dl = \int_{x=0}^{5} 12x^2(10^{-3}) = 12\times10^{-3}\int_{x=0}^{5} x^2 dx$$

$$= 12\times10^{-3}\left(\frac{x^3}{3}\right)_{x=0}^{5} = 12\times10^{-3}\left(\frac{5^3}{3}\right) = 4\times10^{-3}(125)$$

$$= 0.5 \text{ Coulomb}$$

(b) On a cylinder $\rho = 3$; $0 \le z \le 4$ m if $\rho_s = \rho z^2$ nC/m^2

$$Q = \iint \rho_s ds = \iint \rho z^2(\times10^{-9})\rho\, d\phi\, dz\Big|_{p=3 \text{ surface}}$$

$$= (3)^2\times10^{-9}\int_{\phi=0}^{2\pi} d\phi \int_{z=0}^{4} z^2 dz = 9\times10^{-9}(\phi)_{\phi=0}^{2\pi}\left(\frac{z^2}{3}\right)_{z=0}^{4}$$

$$= 18\pi\times10^{-9}\left(\frac{4^3}{3}\right) = 1.206\ \mu\text{C}$$

2.30 Explain and derive continuity of current equation using the principle of conservation of charge.

[12 marks : 2020]

Solution:

From the principle of charge conservation, the time rate of decrease of charge within a given volume must be equal to the net outward current flow through the closed surface of the volume.

Mathematically, $$I_{out} = \frac{-dQ_{in}}{dt}$$

$$I = \frac{-dQ}{dt} \quad \text{...(i)}$$

Where Q is the total charge enclosed by the closed surface

I is current going out of the volume, which is enclosed by the closed surface,

$$Q = \iiint \rho_v dv \quad \text{...(ii)}$$

$$I = \oiint \vec{J} \cdot d\vec{s} \quad \text{...(iii)}$$

Put (ii), (iii) in (i),

$$\oiint \vec{J} \cdot d\vec{s} = \frac{-d}{dt}\left(\iiint \rho_v dv\right) \quad \text{...(iv)}$$

From the divergence theorem,

$$\oiint \vec{J} \cdot d\vec{s} = \iiint (\vec{\nabla} \cdot \vec{J}) dv \quad \text{...(v)}$$

Put (v) in equation (iv),

$$\iiint (\vec{\nabla} \cdot \vec{J}) dv = \iiint \left(-\frac{\partial \rho_v}{\partial t}\right) dv \quad \text{...(vi)}$$

Compare both sides

$$\vec{\nabla} \cdot \vec{J} = -\frac{\partial \rho_v}{\partial t} \quad \text{...(vii)}$$

equation (vii) is called the continuity of current equation.

2.31 Prove that insulation resistance of a cable is inversely proportional to its length. Define insulation resistance.

[12 marks : 2020]

Solution:

The cable conductor is provided with an insulation of suitable thickness so as to avoid leakage of current. The path for leakage current is radial, as shown in figure (a), through the insulating material. The opposition offered by the insulation to the leakage current is called the insulation resistance of the cable.

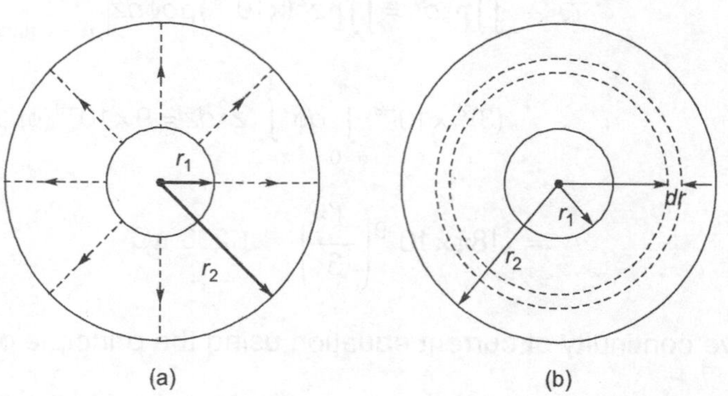

(a) (b)

Consider a single core cable of conductor radius r_1, internal sheath radial r_2, length l and insulation material resistivity ρ.

Let an elementary cylindrical section of the insulation of radius r and thickness dr be considered. Now the length through which the leakage current will flow is dr and area of cross-section is $2\pi rl$.

The insulation resistance offered to the leakage current by elementary cylindrical section of the insulation under consideration is $\frac{\rho dr}{2\pi rl}$.

Insulation resistance of the cable,

$$R_{INS} = \int_{r_1}^{r_2} \frac{\rho dr}{2\pi rl} = \frac{\rho}{2\pi l}[\log_e r]_{r_1}^{r_2} = \frac{\rho}{2\pi l}\log_e \frac{r_2}{r_1}$$

i.e., the insulation resistance of the cable varies inversely as the length of the cable $\left(R_{INS} \alpha \frac{1}{l}\right)$.

2.32 An electric dipole is placed in a uniform electric field of strength 7 × 10⁶ V/m as shown in the figure. The charge on each dipole is opposite in nature and each equal to the charge of an electron. Both charges are separated by a distance of 2 angstrom. Both, field and dipoles are in same plane. Calculate the magnitude and direction of the dipole moment and torque.

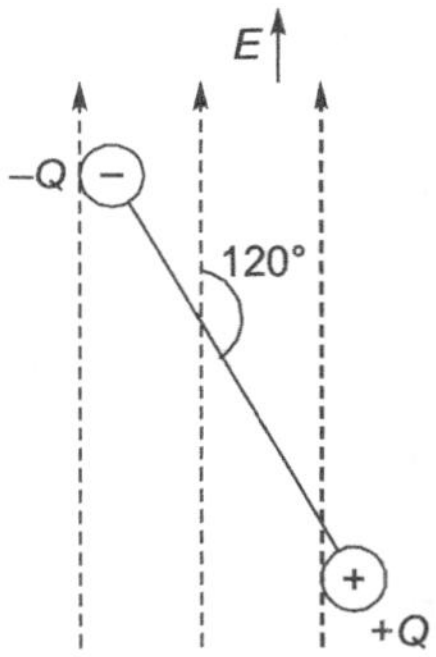

[12 marks : 2021]

Solution:

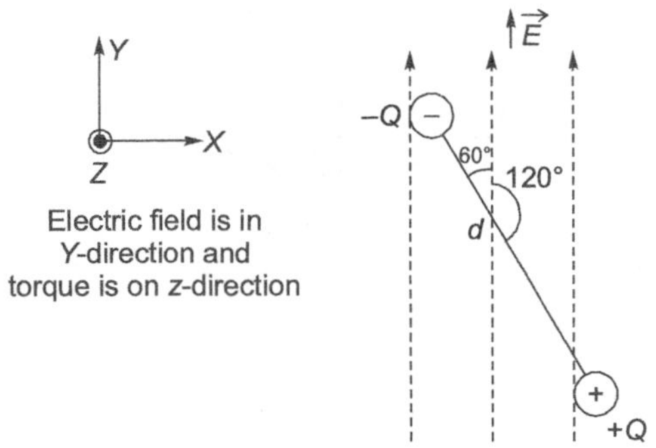

Electric field strength,

$$\vec{E} = 7\times10^6 \hat{a}_y \text{ V/m}$$

Charge on each dipole, $Q = 1.60 \times 10^{-19}$ C (charge of an electron)

Both charges are separated by a distance,

$$d = 2 \text{ Å} = 2 \times 10^{-10} \text{ m}$$

Dipole moment is given by $\vec{P} = Q\times\vec{d}$

$$\vec{P} = Q\left[d\sin 60°\hat{a}_x - d\cos 60°\hat{a}_y\right]$$

$$= 1.60\times10^{-19}\times2\times10^{-10}\times\left(\frac{\sqrt{3}}{2}\hat{a}_x - \frac{1}{2}\hat{a}_y\right)$$

$$\vec{P} = (2.77\hat{a}_x - 1.60\hat{a}_y)\times10^{-29} \text{ C-m}$$

Torque, $\vec{\tau} = \vec{P}\times\vec{E}$

$$= (2.77\hat{a}_x - 1.60\hat{a}_y)\times10^{-29}\times7\times10^{6}\hat{a}_y \text{ V/m}$$

$$= 19.39\times10^{-23}\hat{a}_z \text{ N-m}$$

Magnitude of torque, $|\vec{\tau}| = 19.39\times10^{-23}$ N-m

2.33 (i) Evaluate value of frequency at which conduction current density and displacement current density are equal

(a) distilled water, where, $\sigma = 2.0\times10^{-4}$ s/m and $\epsilon_r = 81$.

(b) sea water where $\sigma = 4.0$ s/m and $\epsilon_r = 1$.

(ii) In spherical co-ordinates $V = -35$ V on a conductor at $r = 3$ cm and $V = 150$ V at $r = 40$ cm. The space between conductors is a dielectric for which $\epsilon_r = 3.12$. Evaluate surface charge densities on the conductors.

[10 + 10 marks : 2022]

Solution:

(i) Conduction current density,

$$J_C = \sigma E$$

Displacement current density

$$J_D = \varepsilon_o\frac{\partial D}{\partial t}$$

$$= \varepsilon_o\varepsilon_r\omega E$$

(a) For distilled water, $\sigma = 2\times10^{-4}$ s/m, $\varepsilon_r = 81$

$$J_C = J_D$$

$$\sigma E = \omega\epsilon_r\epsilon_o E$$

$$\sigma = \omega\epsilon_r\epsilon_o$$

$$\omega = \frac{\sigma}{\epsilon_r\epsilon_o} = \frac{2\times10^{-4}}{81\times8.88\times10^{-12}}$$

$$= 2.78\times10^{5} \text{ rad/sec}$$

$$f = \frac{\omega}{2\pi} = 44.27 \text{ kHz}$$

(b) Sea water, $\sigma = 4$ s/m, $\epsilon_r = 1$

$$\omega = \frac{\sigma}{\epsilon_r\epsilon_o} = \frac{4}{1\times8.88\times10^{-12}}$$

$$= 4.504\times10^{11} \text{ rad/sec}$$

$$f = 7.17\times10^{10} \text{ Hz}$$

(ii) $V = -35$ V, $r = 3$ cm

$V = 150$ V, $r = 40$ cm

Spherical surface are equipotential. The potential is constant with θ and ϕ. The Laplace equation gets reduced to

$$\nabla^2 V = 0$$

$$\frac{\partial}{\partial r} r^2 \frac{\partial V}{\partial r} = 0$$

$$r^2 \frac{\partial V}{\partial r} = A$$

$$\frac{\partial V}{\partial r} = \frac{A}{r^2}$$

$$\int \partial V = \int \frac{A}{r^2} \partial r$$

$$V = -\frac{A}{r} + B$$

Put the given value to find out the value of A and B.

$$-35 = -\frac{A}{3 \times 10^{-2}} + B \quad \text{...(1)}$$

$$150 = -\frac{A}{40 \times 10^{-2}} + B \quad \text{...(2)}$$

By solving equation (1) and (2), we get

$$A = 6, B = 165$$

$$V(r) = -\frac{6}{r} + 165$$

Now,

$$\vec{E} = -\nabla V$$

$$= -\frac{\partial V}{\partial r} \hat{a}_r = -\frac{6}{r^2} \hat{a}_r$$

$$\vec{D} = \varepsilon_r \varepsilon_o E = -\frac{6\varepsilon_r \varepsilon_o}{r^2} \hat{a}_r \text{ c/m}^2$$

Surface charge densities,

$$\rho_s = |\vec{D}| = \frac{6\varepsilon_r \varepsilon_o}{r^2} = \frac{18.72\varepsilon_o}{r^2} \text{ C/m}^2$$

Surface charge density on inner sphere (r = 3 cm) is

$$\rho_s(r = 3 \text{ cm}) = \frac{6 \times 8.854 \times 10^{-12} \times 3.12}{(3 \times 10^{-2})^2} = 184.16 \text{ nC/m}^2$$

Surface charge density on the outer sphere (r = 40 cm) ($\varepsilon_r = 1$)

$$\rho_s(r = 40 \text{ cm}) = \frac{6\varepsilon_0(1)}{(0.40)^2} = 0.332 \text{ nC/m}^2$$

2.34 **Show that the capacitor of figure (a) has capacitance $C_{eq} = \frac{\epsilon_0 \epsilon r_1 A_1}{d} + \frac{\epsilon_0 \epsilon r_2 A_2}{d}$ and capacitor of figure (b) has reciprocal capacitance $\frac{1}{C_{eq}} = \frac{1}{\frac{\epsilon_0 \epsilon r_1 A}{d_1}} + \frac{1}{\frac{\epsilon_0 \epsilon r_2 A}{d_2}} = \frac{1}{C_1} + \frac{1}{C_2}$.**

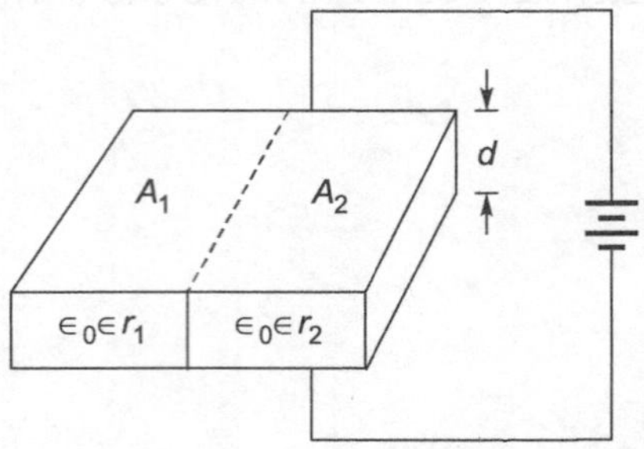

Figure (a)

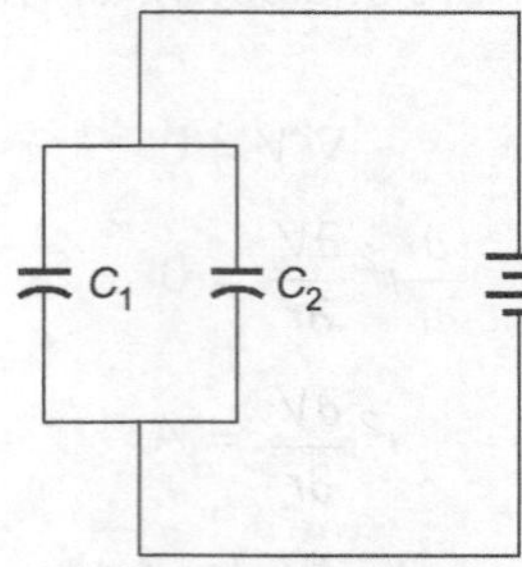

Figure (b)

[10 marks : 2022]

Solution:

Since flux density is continuous and $\vec{E}$ is discontinuous therefore given configuration is two parallel connected capacitors

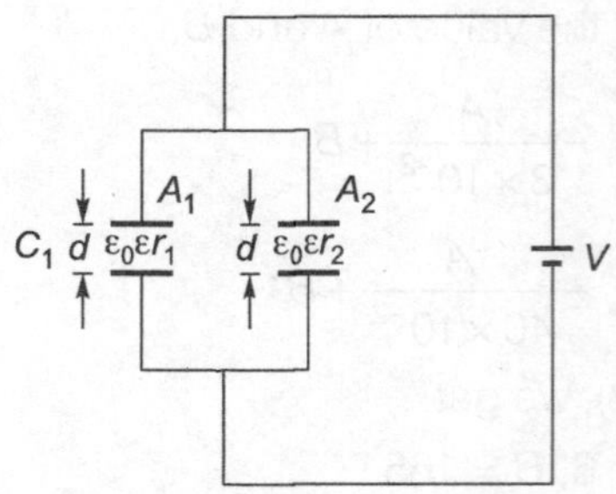

$$C_1 = \frac{\varepsilon_0 \cdot \varepsilon_{r1} A_1}{d}$$

and $$C_2 = \frac{\varepsilon_0 \cdot \varepsilon_{r2} A_2}{d}$$

When two capacitor are connected in parallel then equivalent capacitance is given as

$$C_{eq} = C_1 + C_2 = \frac{\varepsilon_0 \cdot \varepsilon_{r1} A_1}{d} + \frac{\varepsilon_0 \cdot \varepsilon_{r2} A_2}{d}$$

From figure (b) as the result needed, given capacitors must be connected in series, so we are providing solution accordingly,

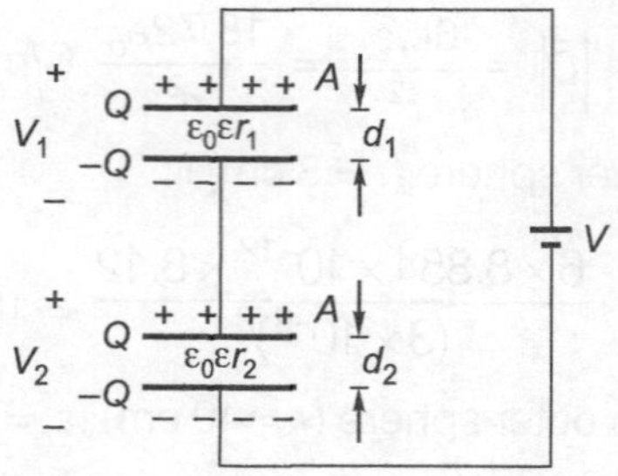

where, $$C_1 = \frac{\varepsilon_0 \cdot \varepsilon_{r1} A}{d_1}$$

and $$C_2 = \frac{\varepsilon_0 \cdot \varepsilon_{r2} A}{d_2}$$

$$V = V_1 + V_2$$

$$= \frac{Q}{C_1} + \frac{Q}{C_2}$$

$$\frac{V}{Q} = \frac{1}{C_1} + \frac{1}{C_2}$$

$$\frac{1}{C_{eq}} = \frac{1}{C_1} + \frac{1}{C_2} = \frac{1}{\frac{\varepsilon_0 \varepsilon_{r_1} A}{d_1}} + \frac{1}{\frac{\varepsilon_0 \varepsilon_{r_2} A}{d_2}}$$

2.35 Derive an expression for capacitance (C) of concentric spheres having radii a and b ($a < b$) respectively with single dielectric.

[12 marks : 2023]

Solution:

We know that, $$V = \int \vec{E} \cdot \vec{dl}$$

$$\vec{E}\Big|_{\text{sphere}} = \frac{Q}{4\pi \in r^2} \hat{a}_r$$

$$\vec{dl} = dl\hat{a}_r$$

$$\therefore \quad V = -\int_a^b \frac{Q}{4\pi \in r^2} dr = \frac{Q}{4\pi \in} \frac{1}{r}\Big|_a^b = \frac{Q}{4\pi \in}\left[\frac{1}{b} - \frac{1}{a}\right]$$

$$= -\frac{Q}{4\pi \in}\left[\frac{1}{a} - \frac{1}{b}\right]$$

Negative sign signifies that in the direction of $\overline{dl}$ from $a \to b$, V decreases. Hence, taking absolute value of V, we get

$$C = \frac{Q}{|V|} = \frac{4\pi \in}{\frac{1}{a} - \frac{1}{b}} \text{ F}$$

2.36 What is the electric field intensity $\vec{E}_1$ V/m due to an infinite sheet of uniform charge density σ C/m^2?

(i) Derive the electric field intensity $\vec{E}_2$ V/m at P contributed by the circular portion of this infinite sheet charge of radius b metre on the perpendicular axis at a metre from the sheet as shown in the figure below :

(ii) Find b, if $a = 0.5$ m and $\vec{E}_2 = \frac{\vec{E}_1}{2}$.

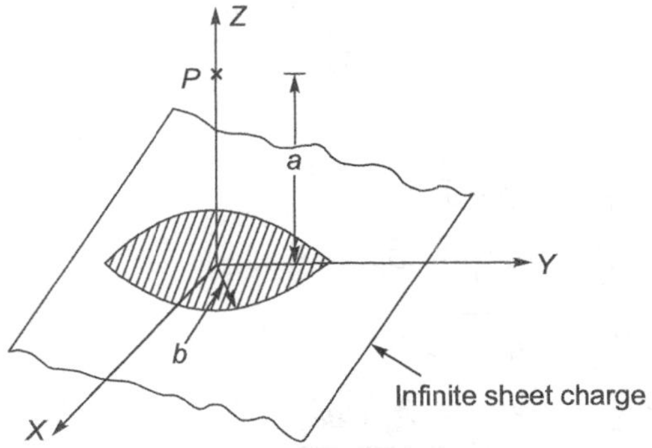

[20 marks : 2023]

Solution:

From Gauss law,

$$\oint \bar{D}\cdot d\bar{s} = Q_{enc}$$

$$Q_{enc} = \int \sigma ds = \sigma \int ds = \sigma A$$

$$\oint \bar{D}\cdot d\bar{s} = D\left[\underbrace{\int ds}_{top} + \underbrace{\int ds}_{bottom}\right]$$

$$= D[A + A]$$

$$= D[2A]$$

$\therefore$ $$D2A = \sigma A$$

$\Rightarrow$ $$D = \frac{\sigma}{2}$$

In terms of vector, $$\bar{D} = \frac{\sigma}{2}\hat{a}_n$$

$\hat{a}_n$ = Unit normal vector from surface charge and taken towards point of interest.

$\Rightarrow$ $$\bar{E}_1 = \frac{\sigma}{2\epsilon_o}\hat{a}_n \text{ V/m}$$

(i) The electric field is given as

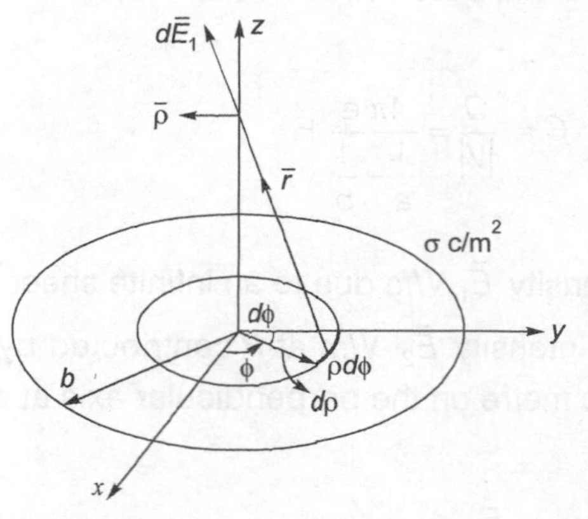

$$d\bar{E}_2 = \frac{Kdq}{r^2}\hat{a}_r\,; \quad K = \frac{1}{4\pi\epsilon_o}$$

Here, $$dq = \sigma ds = \sigma\rho d\rho d\phi$$

$\hat{a}_r$:

$$\bar{r} = -\bar{\rho} + \bar{a}$$

$\Rightarrow$ $$\bar{r} = -\rho\hat{a}_\rho + a\hat{a}_z$$

$\Rightarrow$ $$|\bar{r}| = \sqrt{\rho^2 + a^2}$$

$\therefore$ $$\hat{a}_r = \frac{\bar{r}}{|\bar{r}|} = \frac{-\rho\hat{a}_\rho + a\hat{a}_z}{\sqrt{a^2+\rho^2}}$$

Hence, $$d\bar{E}_2 = \frac{K\sigma\rho d\rho d\phi}{(\rho^2+a^2)^{3/2}}(-\rho\hat{a}_\rho + a\hat{a}_z)$$

The sum of the contribution along ρ gives zero due to symmetry.

Hence, $$d\bar{E}_2 = \frac{K\sigma\rho d\rho d\phi a}{(\rho^2+a^2)^{3/2}}\hat{a}_z$$

$$\Rightarrow \quad \bar{E}_2 = \int d\bar{E} = K\sigma a\int\frac{\rho d\rho}{(\rho^2+a^2)^{3/2}}\int d\phi\hat{a}_z$$

Let $\rho^2 + a^2 = t \Rightarrow 2\rho d\rho = dt \Rightarrow \rho d\rho = \frac{dt}{2}$

At $\rho = 0$, $t = a^2$

At $\rho = b$, $t = b^2 + a^2$

$$\therefore \quad \bar{E}_2 = \frac{K\sigma a}{2}\int_{t=a^2}^{a^2+b^2}\frac{dt}{t^{3/2}}\int_{\phi=0}^{2\pi} d\phi\hat{a}_z = \frac{K\sigma a}{2}\times\frac{-2}{\sqrt{t}}\Bigg|_{a^2}^{a^2+b^2}\cdot 2\pi\hat{a}_z$$

$$\Rightarrow \quad \bar{E}_2 = -K\sigma a\left[\frac{1}{\sqrt{a^2+b^2}} - \frac{1}{a}\right]2\pi\hat{a}_z$$

$$\Rightarrow \quad \bar{E}_2 = \frac{1}{4\pi\epsilon_o}\sigma a 2\pi\left[\frac{1}{a} - \frac{1}{\sqrt{a^2+b^2}}\right]\hat{a}_z$$

$$\Rightarrow \quad \bar{E}_2 = \frac{\sigma a}{2\epsilon_o}\cdot\frac{1}{a}\left[1 - \frac{a}{\sqrt{a^2+b^2}}\right]\hat{a}_z$$

Hence, $$\bar{E}_2 = \frac{\sigma}{2\epsilon_o}\left[1 - \frac{a}{\sqrt{a^2+b^2}}\right]\hat{a}_z \text{ V/m}$$

(ii) Given : $$\bar{E}_2 = \frac{\bar{E}_1}{2}$$

$$\Rightarrow \quad \frac{\sigma}{2\epsilon_o}\left[1 - \frac{a}{\sqrt{a^2+b^2}}\right] = \frac{\sigma}{2\epsilon_o}\times\frac{1}{2}$$

$$\Rightarrow \quad 1 - \frac{a}{\sqrt{a^2+b^2}} = \frac{1}{2}$$

$$\Rightarrow \quad \frac{a}{\sqrt{a^2+b^2}} = \frac{1}{2}$$

$$\Rightarrow \quad 2a = \sqrt{a^2+b^2}$$

$$\Rightarrow \quad 4a^2 = a^2 + b^2$$

$$\Rightarrow \quad 3a^2 = b^2$$

$$\Rightarrow \quad b = \sqrt{3}a = 0.5\sqrt{3}$$

$$\therefore \quad b = 0.5\sqrt{3}\text{ m}$$

2.37 **In the region between the two coaxial cones with insulated vertices as shown in the figure below, the voltage at θ_1 = 30° is 0 volt and a θ_2 = 45° is 125.5 volts :**

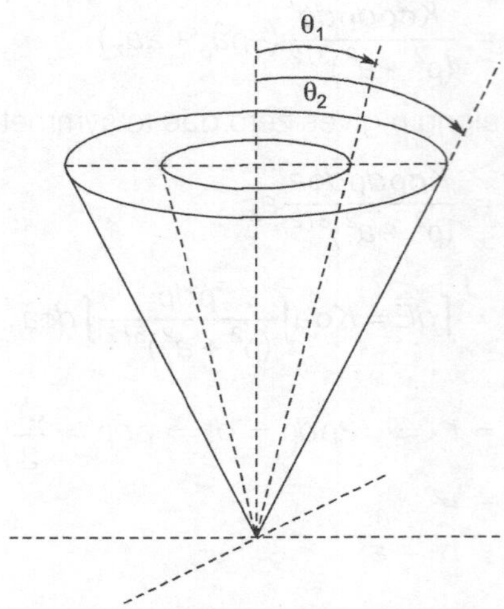

(i) Calculate the angle θ at which the voltage is 75 volts. Assume air as the dielectric in the region between the two coaxial cones.

(ii) Find the charge distribution on the conducting plane at $\theta_2 = 90°$.

[20 marks : 2023]

Solution:

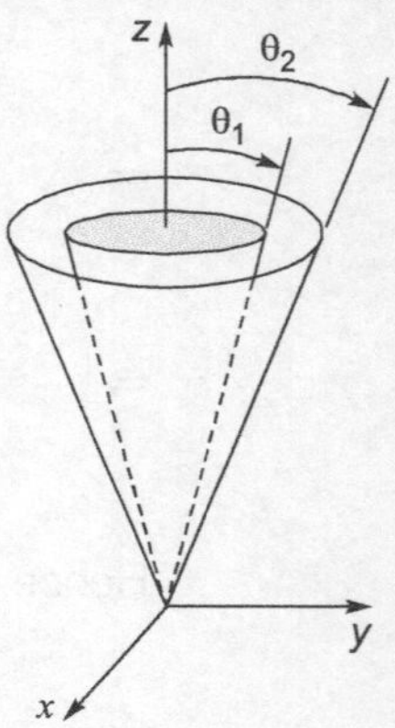

Given data : $V(\theta_1 = 30°) = 0$ V

$\Rightarrow \quad V(\theta = \theta_1) = 0$

$V(\theta_2 = 45°) = 125.5$ V

$\Rightarrow \quad V(\theta = \theta_2) = V_0$

$\Rightarrow \quad V = f(\theta)$

(i) As we know that, $\rho_V = 0$

∴ Laplace equation, $\nabla^2 V = 0$

Sp. co. sy :

$$\nabla^2 V = \frac{1}{r^2 \sin\theta}\left[\frac{\partial}{\partial\theta}\left(\frac{r\sin\theta}{r}\cdot\frac{\partial v}{\partial\theta}\right)\right] = 0$$

$$\Rightarrow \quad \frac{\partial}{\partial\theta}\left[\sin\theta\cdot\frac{\partial V}{\partial\theta}\right] = 0$$

$$\Rightarrow \quad \sin\theta\frac{dV}{d\theta} = 0$$

$$\Rightarrow \quad \frac{dV}{d\theta} = \frac{A}{\sin\theta}$$

$$\Rightarrow \quad V = A\int\frac{d\theta}{\sin\theta} + B = A\int\frac{d\theta}{2\sin\frac{\theta}{2}\cos\frac{\theta}{2}} + B$$

$$\Rightarrow \quad V = \frac{A}{2}\int\frac{d\theta}{\dfrac{\sin\frac{\theta}{2}}{\cos\frac{\theta}{2}}\cdot\cos^2\frac{\theta}{2}} + B$$

$$= \frac{A}{2}\int\frac{\sec^2\frac{\theta}{2}}{\tan\frac{\theta}{2}}d\theta + B$$

$$\Rightarrow \qquad V = A\int \frac{\frac{1}{2}\sec^2\frac{\theta}{2}d\theta}{\tan\frac{\theta}{2}} + B$$

$$= A\int \frac{d\left(\tan\frac{\theta}{2}\right)}{\tan\frac{\theta}{2}} + B = A\left(\tan\frac{\theta}{2}\right) + B$$

At $\theta = \theta_1$: $\qquad V = 0$

$$\Rightarrow \qquad 0 = A\ln\left[\tan\frac{\theta_1}{2}\right] + B$$

$$\Rightarrow \qquad B = -A\ln\left[\tan\frac{\theta_1}{2}\right]$$

At $\theta = \theta_2$: $\qquad V = V_o$

$$\Rightarrow \qquad V_o = A\ln\left[\tan\frac{\theta_2}{2}\right] - A\ln\left[\tan\frac{\theta_1}{2}\right]$$

$$\Rightarrow \qquad V_o = A\ln\left[\frac{\tan\left(\frac{\theta_2}{2}\right)}{\tan\left(\frac{\theta_1}{2}\right)}\right]$$

$$\Rightarrow \qquad A = \frac{V_o}{\ln\left[\frac{\tan\left(\frac{\theta_2}{2}\right)}{\tan\left(\frac{\theta_1}{2}\right)}\right]}$$

and

$$B = \frac{-V_o}{\ln\left[\frac{\tan\left(\frac{\theta_2}{2}\right)}{\tan\left(\frac{\theta_1}{2}\right)}\right]}\ln\left[\tan\frac{\theta_1}{2}\right]$$

Hence,

$$V = \frac{V_o}{\ln\left[\frac{\tan\frac{\theta_2}{2}}{\tan\frac{\theta_1}{2}}\right]}\cdot\ln\left[\tan\frac{\theta_2}{2}\right] - \frac{V_o}{\ln\left[\frac{\tan\frac{\theta_2}{2}}{\tan\frac{\theta_1}{2}}\right]}\ln\left[\tan\frac{\theta_1}{2}\right]$$

$$\Rightarrow \qquad V = \frac{V_o}{\ln\left[\frac{\tan\frac{\theta_2}{2}}{\tan\frac{\theta_1}{2}}\right]}\cdot\ln\left[\frac{\tan\frac{\theta}{2}}{\tan\frac{\theta_1}{2}}\right]$$

Putting the value, we get

$$V = \frac{125.5}{\ln\left[\frac{\tan 22.5°}{\tan 15°}\right]} \ln\left[\frac{\tan\frac{\theta}{2}}{\tan 15°}\right]$$

$$\Rightarrow \quad V = 288.118\left[\ln\left(\tan\frac{\theta}{2}\right) + 1.316\right]$$

Now at $\theta = \theta$, $\quad V = 75$

$$\Rightarrow \quad 75 = 288.118\left[\ln\left(\tan\frac{\theta}{2}\right) + 1.316\right]$$

$$\Rightarrow \quad \tan\frac{\theta}{2} = 0.348 \Rightarrow \theta = 38.36°$$

(ii) At $\theta_2 = 90°$ and $\theta_1 = 30°$

$$V = \frac{125.5}{\ln\left[\frac{\tan 45°}{\tan 15°}\right]} \ln\left[\frac{\tan\frac{\theta}{2}}{\tan 15°}\right]$$

$$= 95.295\left[\ln\left(\tan\frac{\theta}{2}\right) + 1.316\right]$$

Now,

$$\bar{E} = -\nabla V = -\frac{1}{r}\cdot\frac{dV}{d\theta}\hat{a}_\theta = \frac{-95.295}{(r\sin\theta)}\hat{a}_\theta$$

$$\Rightarrow \quad \bar{D} = \epsilon_o \bar{E} = \frac{-95.295\,\epsilon_o}{r\sin\theta}\hat{a}_\theta$$

Now, on the plane $\theta = 90°$,

$$\therefore \quad \rho_s = |\bar{D}| = \frac{95.295\,\epsilon_o}{r}\ \text{C/m}^2$$

$$\Rightarrow \quad \rho_s = \frac{8.43 \times 10^{-10}}{r}\ \text{C/m}^2$$

3. Magnetostatic Fields

3.1 A coaxial cable carries uniformly distributed current in the inner conductor and $-I$ in the outer conductor. Determine magnetic field intensity distributions within and outside the coaxial cable by using Ampere's circuital law.

[12 marks : 2002]

Solution:

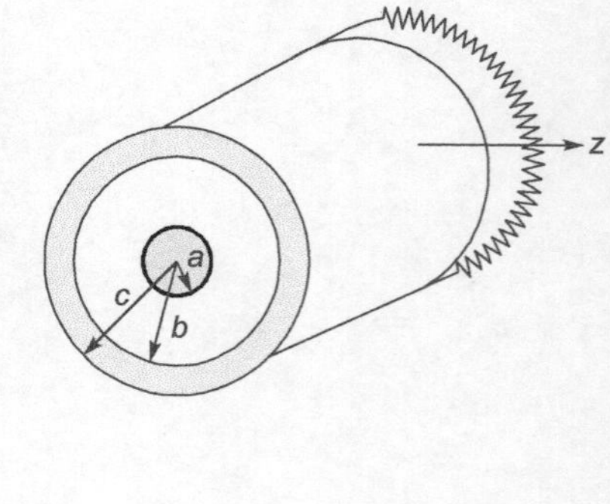

Consider an infinite long coaxial transmission line carrying a uniformly distributed total current (I) in the center conductor and ($-I$) in the outer conductor. Symmetry shows that H is not a function of ϕ and z.

Let us consider a circular path of radius 'ρ' where 'ρ' is larger than the radius of the inner conductor but less than the inner radius of the outer conductor, from ampere's circuital law,

$$|H_\phi| = \frac{I}{2\pi\rho}(a < \rho < b)$$

If we choose ρ smaller than the radius of the inner conductor, the current enclosed is

$$I_{enclosed} = I\left(\frac{\rho^2}{a^2}\right)$$

{$\because J$ is constant so $J = I/\pi a^2$. Now current through circular cross-section of area $\pi\rho^2$ is $= J\pi\rho^2 = I(\rho^2/a^2)$}

From ampere's circuital law,

$$2\pi\rho H_\phi = I\frac{\rho^2}{a^2}$$

$$\therefore \qquad H_\phi = \frac{I\rho}{2\pi a^2} \quad (\rho < a)$$

If the radius 'ρ' is larger than the outer radius of the outer conductor no current is enclosed.

$$\therefore \qquad \text{Net current} = I + (-I) = 0$$

$$H_\phi = 0 \ (\rho > c)$$

Finally if the path lies within the outer conductor

From ampere's circuital law,

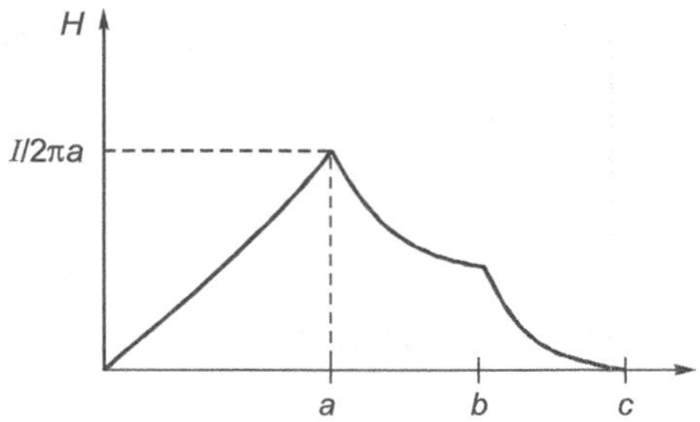

$$2\pi\rho\, H_\phi = I - I\left(\frac{\rho^2 - b^2}{\rho^2 - a^2}\right)$$

$$H_\phi = \frac{I}{2\pi\rho}\left[\frac{b^2 - a^2}{\rho^2 - a^2}\right] b < \rho < c$$

3.2 **Two long thin parallel conductors perpendicular to the plane of the paper and parallel to the surface of an infinite iron plate of high permeability, each carrying a steady current of 100 A in opposite directions are shown in figure.**

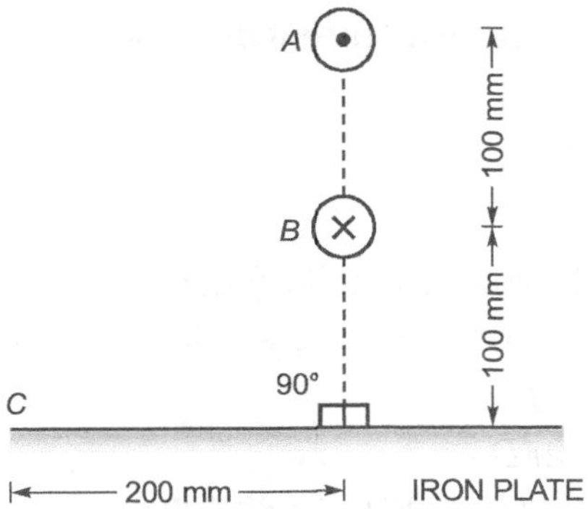

Calculate the magnetic field strength at a point *C*, 200 mm away from the bottom point on iron plate below the conductors. Also calculate the mechanical force per unit length on the conductor *A*.

[12 marks : 2003]

Solution:

Magnetic strength, $\qquad \bar{H}_A = \frac{I}{2\pi d_{CA}}; \qquad \bar{H}_B = \frac{I}{2\pi d_{CB}}$

Magnetic field strength due to A and A' at C will be

$$\bar{H}_{AA'} = 2\left|\bar{H}_A\right| \sin\theta_2\,(\hat{a}_x)$$

$$H_{BB'} = 2\left|\bar{H}_B\right| \sin\theta_1\,(-\hat{a}_x)$$

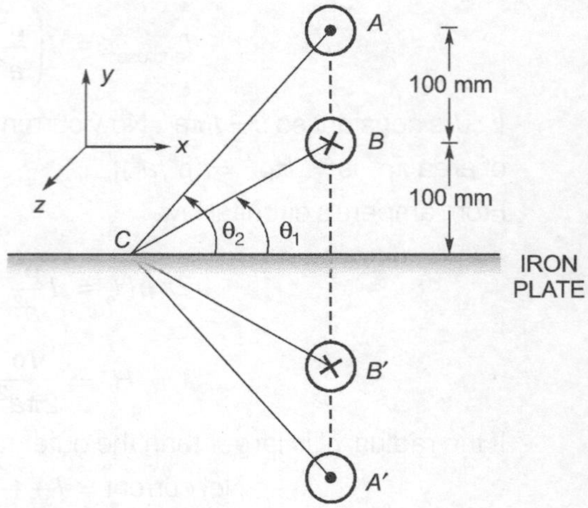

Then resultant magnetic field strength $\bar{H}_r = \bar{H}_{AA'} + \bar{H}_{BB'}$

$$= \left[2|H_A|\sin\theta_2 - 2|\bar{H}_B|\sin\theta_1\right](\hat{a}_x)$$

$$= \left[\frac{2I}{2\pi d_{CA}}\sin\theta_2 - \frac{2I}{2\pi d_{CB}}\sin\theta_1\right]\hat{a}_x$$

$$= \frac{I}{\pi}\left[\frac{\sin\theta_2}{d_{CA}} - \frac{\sin\theta_1}{d_{CB}}\right]\hat{a}_x$$

where, $d_{CA} = \sqrt{200^2 + 200^2}$ mm

$d_{CB} = \sqrt{200^2 + 100^2}$ mm

$$= \frac{100}{\pi}\left[\frac{1}{\sqrt{2}}\times\frac{1}{200\times\sqrt{2}} - \frac{1}{\sqrt{5}}\times\frac{1}{\sqrt{5}\times 100}\right] = \frac{1}{\pi}\left[\frac{1}{4} - \frac{1}{5}\right]\hat{a}_n = \frac{1}{20\pi}\hat{a}_x \text{ (A/mm)}$$

Mechanical force on the conductor A,

$$\bar{F} = I\,\overline{dl}\times\bar{B}$$

$$\bar{F} = I\,\overline{dl}\,(\bar{a}_z)\times B(\hat{a}_x)$$

Force/length on the conductor A,

$$F/dl = I\times\frac{\mu_0 I}{2\pi d_{AB}} = \frac{100\times 4\pi\times 10^{-7}\times 100}{2\pi\times 100\times 10^{-3}}$$

$$\therefore \quad F/dl = (2\times 10^{-2}\text{N/m})\,\hat{a}_y$$

3.3 **Two parallel wires each of 3 metres length having separation of 4 mm. Calculate the forces exerted on each of these wires when they carry a current of 5 A in the same direction and opposite direction. Assume $\mu_0 = 4\pi\times 10^{-7}$ H/m.**

[12 marks : 2004]

Solution:

Consider two straight, long parallel line situated in space D meter. Consider the conductor (1) produces a field around and its value (B_1) at the location of conductor (2) is given by

$$\therefore \quad B_1 = \mu_0\frac{I_1}{2\pi D}$$

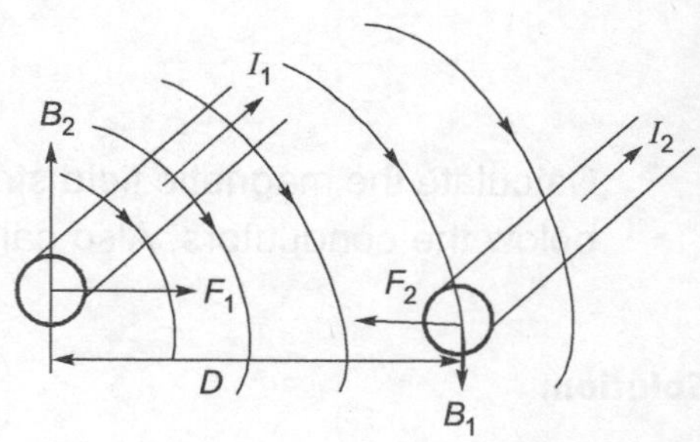

While conductor (2) carries a current (I_2) and is situated in a field with flux density B_1.

$$dF = I\,\overline{dl}\times\bar{B}$$

We can write, for a l length

$$\therefore \quad F_2 = l\,I_2\times B_1$$

As the conductor (2) and B_1 are mutually perpendicular.

$$F_2 = l[(-kI_2)\times(-jB_1)] = -i(l\,I_2\,B_1)$$

Force is along x-axis in the –ve direction. So, these is force of attraction between the conductor if both are carrying current in the same direction.

The force of attraction is given by

$$F = l\times I_2\times B_1 = l\,I_2\left(\mu_0\frac{I_1}{2\pi D}\right) = \frac{\mu_0 I_1 I_2}{2\pi D}l$$

If the direction of current (I_2) is reversed the direction of (F) is also reversed consequently there is force of repulsion instead of attraction between the conductors and magnitude is same for both case. Thus there is no change is magnitude of current

$$\text{Force of attraction} = \frac{\mu_0 I^2 l}{2\pi D} \qquad [I_1 = I_2 = I]$$

$$= \frac{4\pi\times10^{-7}\times25\times3}{2\pi\times4\times10^{-3}} = 3.75\times10^{-3}\text{ N}$$

3.4 Write the Lorentz force equation and explain its significance. Calculate the maximum torque of a rotor shown in the below figure when I = 60 A, r = 8 cm, B = 0.1 Wb/m^2 and length L = 12 cm.

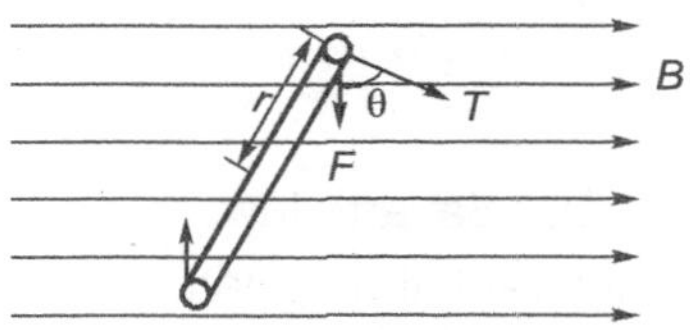

[10 marks : 2006]

Solution:

In a uniform magnetic field with flux-density 'B' and an electric charge element is moving in that region then there should have some force exerted on the moving charge. This force is called Lorentz force.

Let 'V' be the velocity of the charge-element (dQ).

The force experienced is given by

$$|dF| = dQ\, vB \sin\theta$$

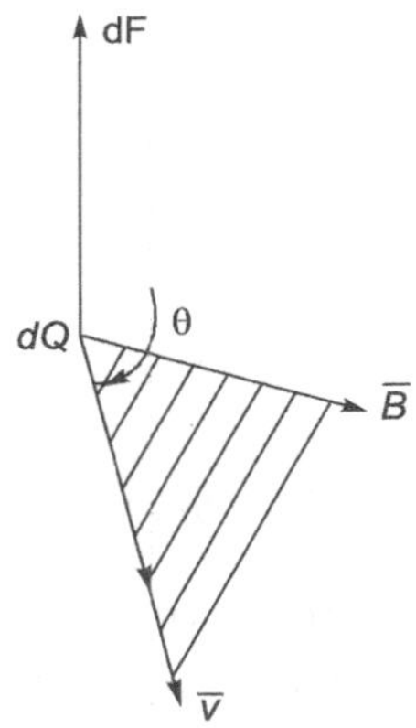

Relative direction of velocity of charge field B and force on charge-element.

$$\vec{dF} = dQ(\vec{v}\times\vec{B})$$

The direction of the force is perpendicular to the plane containing the vectors v and B is given by applying the right hand rule.

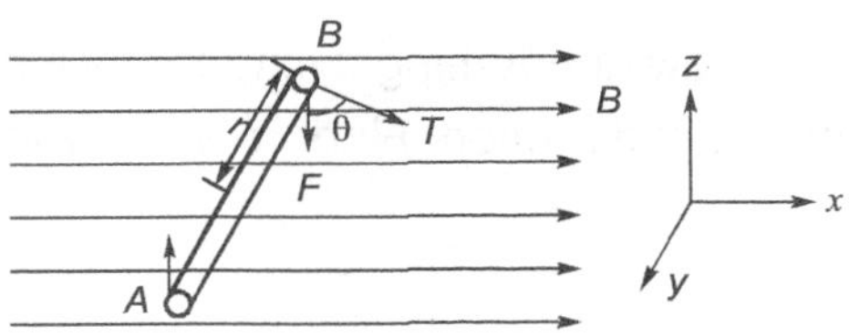

At A, $\bar{F}_A = \bar{I}\, i\times\bar{B} = (-60)\,(0.1)\,(0.12)\,(-\bar{a}_z)\times\bar{a}_x = 0.72\,\bar{a}_y$

$\bar{F}_B = \bar{I}\, i\times\bar{B} = (-60)\times(0.1)\times(1.2)\,(\bar{a}_z)\times(\bar{a}_x) = -0.72\,\bar{a}_y$

Torque at A, $\bar{J}_A = \overline{OA}\times\bar{F} = r(\sin\theta\,\bar{a}_x + \cos\theta\,\bar{a}_y)\times0.72\,a_y$

Torque at B, $\bar{J}_B = \overline{OA}\times\bar{F} = (\sin\theta\,\bar{a}_x + \cos\theta\,a_y)\,0.72\,\bar{a}_y = 0.72\,r\sin\theta\,\hat{a}_z$

$$\text{Total torque} = \bar{J}_A + \bar{J}_B = 1.44\times0.08\times\sin\theta\,\hat{a}_z$$

(J) will be maximum at $\theta = 90°$

$$\bar{J} = 1.44\times0.08\,\hat{a}_z = 0.1152\,\bar{a}_z\text{ N-m}$$

3.5 Give five properties of static magnetic field intensity. What are the different methods by which it can be calculated? Write a Maxwell's equation relating this in integral and differential forms.

[12 marks : 2011]

Solution:

If the charges are moving with constant velocity a static magnetic field is produced.

- A static magnetic field is produced by a constant current flow (or direct current).
- It is vector quantity and the direction of magnetic field intensity can be determined by the right hand rule with the right thumb pointing in the direction of current and right hand figures encircling the wire in the direction of *dH*.
- Flux line is closed and has no beginning or end.
- Magnetic flux lines are closed and do not cross each other regardless of current distribution.
- Magnetic flux lines always close upon themselves, this is because it is not possible to have isolated magnetic poles (or magnetic charges).

  It can be calculated using:
  1. Biot-Savart's law
  2. Ampere's circuit law

Here ampere's law is a special case of Biot-Savart's law.

- Biot-savart's law states that the differential magnetic field intensity dH produced by the differential current element Idl is proportional product of IdL and sine of angle α between the element and the line joining the point to the element and is inversely proportional to the square of the distance R between the point and the element.
- Ampere's circuit law states line integral of H around a closed path is the same as the net current $I_{inc.}$ enclosed by the path.

$$\oint H \cdot dl = I_{inc.}$$

Maxwell's equation for static magnetic field.

$$\nabla \times \vec{H} = \vec{J_C} \text{ is differential form}$$

$$\oint \vec{H} \cdot \vec{dl} = \int_S \vec{J_c} . \vec{dS}$$

Maxwell's equation in integral form to relating magnetic field intensity to conduction current density.

3.6 **State Ampere's circuital law. A hollow conducting cylinder has inner radius *a* and outer radius *b* and carries current *I* along the positive *z*-direction. Find $\vec{H}$ everywhere.**

[12 marks : 2020]

Solution:

Ampere's circuital law states that the closed line integral of magnetic field intensity is equal to the total current enclosed by that closed surface.

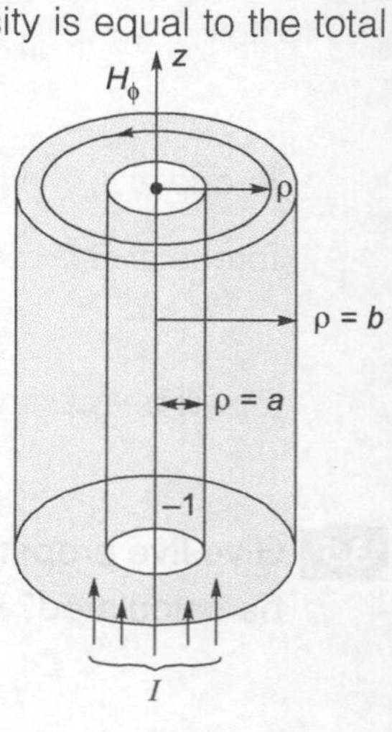

Mathematically $$\oint \vec{H} \cdot \vec{dl} = I_{enclosed} \quad \text{...(i)}$$

Derivation of $\vec{H}$ for hollow conducting cylinder :

Case-1: $\vec{H}$ in the region $0 < \rho < a$

$$\oint \vec{H} \cdot \vec{dl} = I_{enclosed}$$

$$\downarrow \quad \downarrow \quad \downarrow$$

$$H_\phi (2\pi\rho) = 0$$

$$H_\phi = 0 \quad \text{...(ii)}$$

Case-2: $\vec{H}$ in the region $a < \rho < b$

$$\oint \vec{H}\cdot d\vec{l} = I_{\text{enclosed}}$$

$$\downarrow \quad \downarrow \qquad \downarrow$$

$$H_\phi(2\pi\rho) = I\frac{(\rho^2 - a^2)}{(b^2 - a^2)}$$

$$H_\phi = \frac{I}{2\pi\rho}\left[\frac{\rho^2 - a^2}{b^2 - a^2}\right] \qquad \text{...(iii)}$$

Case-3: $\vec{H}$ in the region $b < \rho < \infty$

$$\oint \vec{H}\cdot d\vec{l} = I_{\text{enclosed}}$$

$$\downarrow \quad \downarrow \qquad \downarrow$$

$$H_\phi(2\pi\rho) = I$$

$$H_\phi = \frac{I}{2\pi\rho} \qquad \text{...(iv)}$$

$$\vec{H} = \begin{cases} 0 \text{ for } 0 < \rho < a \\ \dfrac{I}{2\pi\rho}\left[\dfrac{\rho^2 - a^2}{b^2 - a^2}\right]\hat{a}_\phi \textit{for } a < \rho < \beta \\ \dfrac{I}{2\pi\rho}\hat{a}_\phi \text{ for } b < \rho < \infty \end{cases}$$

3.7 **A solenoid of length l and radius a consists of N turns wire carrying current I. Derive an expression for H at point P along its axis. Also find expression for H if $l >> a$.**

[10 marks : 2022]

Solution:

Consider the cross-section of the solenoid as shown in figure,

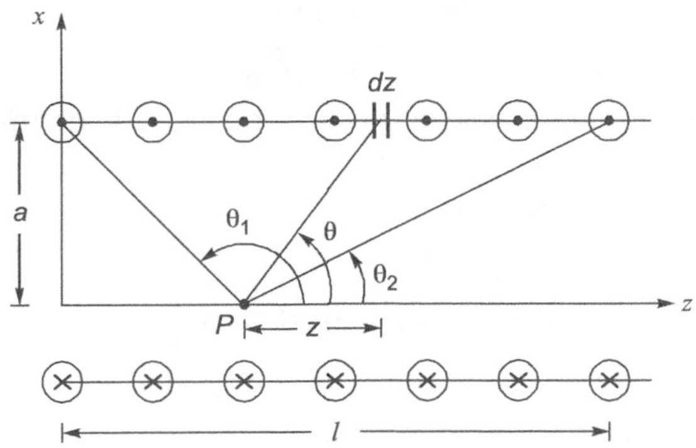

Since the solenoid consists of circular loops, therefore the contribution to the magnetic field H at point P by an element of the solenoid of length dz is

$$dH_z = \frac{I\,dl a^2}{2[a^2 + z^2]^{3/2}} = \frac{Ia^2 ndz}{2[a^2 + z^2]^{3/2}}$$

where, $$dl = ndz = \left(\frac{N}{l}\right)dz$$

From the figure, $$\tan\theta = \frac{a}{z}$$

$\therefore \quad dZ = -a\,\text{cosec}^2\theta\, d\theta = \dfrac{-[z^2+a^2]^{3/2}}{d^2}\sin\theta\cdot d\theta$

Therefore, $\quad dH_z = \dfrac{-nI}{2}\sin\theta\, d\theta$

$$H_z = \frac{-nI}{2}\int_{\theta_1}^{\theta_2}\sin\theta\, d\theta$$

$$\vec{H} = \frac{nI}{2}(\cos\theta_2 - \cos\theta_1)\hat{a}_z$$

Substituting, $\quad n = \dfrac{N}{l}$, gives

$$\vec{H} = \frac{NI}{2l}(\cos\theta_2 - \cos\theta_1)\hat{a}_z$$

If $l >> a$ or $\quad \theta_2 \approx 0°$

and $\quad \theta_1 \approx 180°$

$$\vec{H} = \frac{NI}{l}\hat{a}_z$$

4. Time Varying Electromagnetic Fields

4.1 Made statements about tangential components of *E* and *H* and normal component of *B* and *D* at any surface of discontinuity. Derive Maxwell's equation for harmonically varying fields in integral and differential form.

For coaxial capacitor (having outer radius *b*, inner radius a and length *l*) find displacement current flowing across surface at a radius *r* between *a* and *b* assume $V = V_{max}\sin\omega t$.

[10 marks : 2004]

Solution:

In case of dielectric to dielectric boundary condition,

$$E_{t_1} = E_{t_2} \text{ and } B_{n_1} = B_{n_2}$$

i.e. tangential component of electric field is same on the two side normal component of flux density is same on both the side.

In case of dielectric to conductor boundary,

$$E_{t_2} = B_{n_2} = 0$$

$\therefore$ Electric field and magnetic field do not exist inside a conductor.

$\therefore \quad E_{t_1} = 0$

$$B_{n_1} = 0$$

So at the boundary dielectric and conductor only normal component of electric field and tangential component of magnetic field exists.

Similarly, $\quad D_{n_1} - D_{n_2} = \rho_s$

$$H_{t_1} - H_{t_2} = \hat{a}_{n_{12}} \times \vec{J}_s$$

So in case there is no charge density on the surface then charge density will be induced inorder to support D_{n_1}. Similarly if there is no surface current density at the surface it will be induced inorder to support the tangential component of magnetic field.

For harmonically varying fields,

$$\frac{\partial}{\partial t} = j\omega$$

So Maxwell's equation,

(*i*) $$\nabla \cdot \vec{D} = \rho_v \Leftrightarrow \oint_s \vec{D} \cdot \overrightarrow{dS} = \int_v \rho_v \, dV$$

(*ii*) $$\nabla \cdot \vec{B} = 0 \Leftrightarrow \oint_s \vec{B} \cdot \overrightarrow{dS} = 0$$

(*iii*) $$\nabla \times \vec{E} = -\frac{\partial \vec{B}}{\partial t} \Leftrightarrow \oint_c \vec{E} \cdot \vec{dl} = -\int_s \frac{\partial \vec{B}}{\partial t} \cdot \overrightarrow{dS}$$

$$\Rightarrow \quad \nabla \times \vec{E} = -j\omega \vec{B}$$

$$\oint_c \vec{E} \cdot \vec{dl} = -\int_s j\omega \vec{B} \cdot \overrightarrow{dS}$$

(*iv*) $$\nabla \times \vec{H} = \vec{J}_c + \frac{\partial}{\partial t}\vec{D} \Leftrightarrow \oint_c \vec{H} \cdot \vec{dl} = \int_s \left(\vec{J}_c + \frac{\partial}{\partial t}\vec{D}\right) \cdot \overrightarrow{dS}$$

$$\nabla \times \vec{H} = \sigma \vec{E} + j\omega \in \vec{E} \Leftrightarrow \oint_c \vec{H} \cdot \vec{dl} = \int_s (\vec{J}_c + j\omega \in \vec{E}) \cdot \overrightarrow{dS}$$

$$\oint \vec{H} \cdot \vec{dl} = \int_s (\sigma \vec{E} + j\omega \in \vec{E}) \cdot \overrightarrow{dS}$$

Given that, $V = V_{max} \sin \omega t$

Let space between the conductors in filled with homogeneous dielectrics with permittivity $\in$. We assume that conductors 1 and 2 respectively carry $+Q$ and $-Q$ charge uniformly distributed on them.

Considering an arbitrary gaussion surface (cylindrical) of radius r $(a < r < b)$ and applying Gauss's law we have

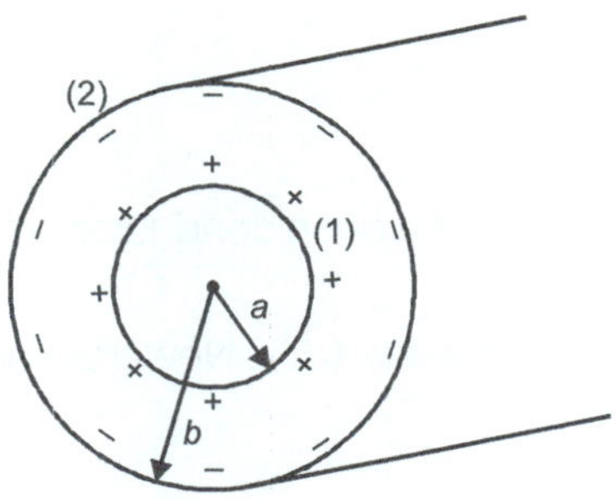

$$Q = \in \oint \vec{E} \cdot \overrightarrow{dS} = \in E_\rho 2\pi\rho L$$

$$\vec{E} = \frac{Q}{2\pi \in \rho L} \hat{a}_\rho \quad \ldots(i)$$

Neglecting flux fringing at cylinder ends

$$V = -\int_2^1 \vec{E} \cdot \vec{dl} = -\int_b^a \left[\frac{Q\hat{a}_\rho}{2\pi \in \rho L}\right] \cdot d\rho \, \hat{a}_\rho$$

$$V = \frac{Q}{2\pi \in L} \ln(b/a) \quad \ldots(ii)$$

From equation (*i*) and (*ii*), $\vec{E} = \dfrac{2\pi \in L}{\ln(b/a)} V \dfrac{1}{2\pi \in \rho L} \hat{a}_\rho$

$$\vec{E} = \frac{1}{\rho \ln(b/a)} V \hat{a}_\rho = \frac{1}{\rho \ln(b/a)} V_{max} \sin\omega t \, \hat{a}_\rho$$

∴ Displacement current density

$$\vec{J}_D = \frac{\partial}{\partial t}\vec{D} = \frac{\partial}{\partial t} \in \vec{E}$$

$$= \in \frac{\partial}{\partial t} \frac{1}{\rho \ln(b/a)} V_{max} \sin\omega t \vec{E} = \frac{\in}{\rho \ln(b/a)} V_{max} \, \omega \cos\omega t \, \hat{a}_\rho$$

$$\vec{J}_D = \frac{\in \omega V_{max} \cos\omega t\, \hat{a}_\rho}{\rho\, ln(b/a)}$$

This is displacement current density at a distance ρ from the centre of cylinder.

∴ Displacement current at the surface

$$I = \left|\vec{J}_D\right| \cdot A = \frac{\in \omega V_{max}}{\rho\, ln(b/a)} \cos\omega t \cdot 2\pi r l$$

For l length of cylinder.

4.2 **Write the Maxwell's equation in a good conductor for time harmonic fields. Show that the conduction current density J in a good conductor must satisfy. $\nabla(\nabla \times J) - \nabla^2 J = j\omega\mu\sigma J$.**

[12 marks : 2005]

Solution:

We know that Maxwell's equation are

$$\nabla \cdot \vec{D} = \rho_V \quad \text{...(i)}$$

$$\nabla \cdot \vec{B} = 0 \quad \text{...(ii)}$$

$$\nabla \cdot \vec{H} = \vec{J}_c + \frac{\partial}{\partial t}\vec{D} \quad \text{...(iii)}$$

$$\nabla \times \vec{H} = -\frac{\partial \vec{B}}{\partial t} = -j\omega\vec{B} = -j\omega\mu\vec{H} \qquad \because \frac{\partial}{\partial t} \sim j\omega \quad \text{...(iv)}$$

taking curt of equation (iv) we get

$$\nabla \times (\nabla \times \vec{E}) = \nabla \times (-j\omega\mu\vec{H})$$

$$\nabla(\nabla \cdot \vec{E}) - \nabla^2\vec{E} = j\omega\mu(\nabla \times \vec{H}) \quad \text{...(v)}$$

$$\vec{E} = \frac{\vec{J}_c}{\sigma}$$

$$\nabla \times \vec{H} = \vec{J}_c$$

∵ In a good conductor displacement current density $\left(\frac{\partial}{\partial t}\vec{D}\right)$ will be negligible compare to conduction current density $(\vec{J}_c)$. Now from equation (v),

$$\nabla\left(\nabla \cdot \frac{\vec{J}_c}{\sigma}\right) - \nabla^2\left(\frac{\vec{J}_c}{\sigma}\right) = -j\omega\mu\vec{J}_c$$

$$\frac{1}{\sigma}\nabla(\nabla \cdot \vec{J}_c) - \frac{1}{\sigma}\nabla^2(\vec{J}_c) = -j\omega\mu\vec{J}_c$$

$$\nabla(\nabla \cdot \vec{J}_c) - \nabla^2(\vec{J}_c) = -j\omega\mu\sigma\vec{J}_c$$

Proved.

4.3 **Show that in a source-free region ($\vec{J} = 0$, $\rho_v = 0$), Maxwell's equations can be reduced to two Identify the two all-embracing equations.**

[10 marks : 2009]

Solution:

Maxwell's equations are,

$$\nabla \times D = r$$

$$\nabla \cdot B = 0$$

⇒

$$\nabla \cdot H = 0$$

$$\nabla \times E = \frac{-\partial B}{\partial t}$$

$$\nabla \times H = J + \frac{\partial D}{\partial t}$$

Applying boundary conditions of the source force region.

$$\nabla \cdot D = 0 \quad \text{...(i)}$$

$$\Rightarrow \quad \nabla . E = 0 \quad \text{...(ii)}$$

$$\nabla \cdot B = 0 \quad \text{...(iii)}$$

$$\nabla \times E = \frac{-\partial B}{\partial t} \quad \text{...(iv)}$$

$$\nabla \times H = \frac{\partial D}{\partial t} \quad \text{...(v)}$$

These are first order differential equations. To find a characteristic solution, one of the field vectors E or H, has to eliminated. Eliminating H and applying the curl to (iv) $B = \mu_0 H$ and $D = \epsilon_0 E$, $J = \sigma E$.

$$\nabla \times (\nabla \times E) = \nabla \times \left(\frac{-\partial B}{\partial t}\right) = -\mu \frac{\partial}{\partial t}(\nabla \times H)$$

Applying the vector identity,

$$\nabla \times (\nabla \times E) = \nabla(\nabla \cdot E) - \nabla^2 E = -\mu \frac{\partial}{\partial t}\left(\sigma E + \epsilon \frac{\partial E}{\partial t}\right)$$

$$= 0 - \nabla^2 E = -\mu\sigma \frac{\partial E}{\partial t} - \mu \epsilon \frac{\partial^2 E}{\partial t^2} \quad (\because \nabla . E = 0)$$

$$\therefore \quad \nabla^2 E - \mu\sigma \frac{\partial E}{\partial t} - \mu \epsilon \frac{\partial^2 E}{\partial t^2} = 0 \quad \text{...}(i)$$

Similarly, taking (v), and eliminating E,

$$\nabla \times (\nabla \times H) = \nabla(\nabla \cdot H) - \nabla^2 H = \nabla \times \left(J + \frac{\partial D}{\partial t}\right)$$

$$-\nabla^2 H = \sigma(\nabla \times E) + \epsilon \frac{\partial}{\partial t}(\nabla \times E) \quad (\because \nabla \cdot H = 0)$$

$$-\nabla^2 H = -\mu\sigma \frac{\partial H}{\partial t} - \mu \epsilon \frac{\partial^2 H}{\partial t^2}$$

$$\therefore \quad \nabla^2 H - \mu\sigma \frac{\partial H}{\partial t} - \mu \epsilon \frac{\partial^2 H}{\partial t^2} = 0 \quad \text{...}(ii)$$

These equations for E and H are second order differential equation which they (E and H) must satisfy. They are known as wave equations, also called Helmholtz equations. So all the Maxwell's equation are reduced to two equation (i) and (ii).

4.4 State Ampere's circuit law. A hollow conducting cylinder has inner radius *a* and outer radius *b*, and carries a current *I* along the positive z-direction. Find $\vec{H}$ everywhere.

[12 marks : 2011]

Solution:

Ampere's Circuit Law: Ampere's circuit law states that the line integral of H around a closed path is the same as the net current $I_{inc.}$ enclosed by the path. Where H is magnetic field intensity.

In other word, the circulation of H equals $I_{inc.}$, i.e.

$$\oint H \cdot dl = I_{inc.}$$

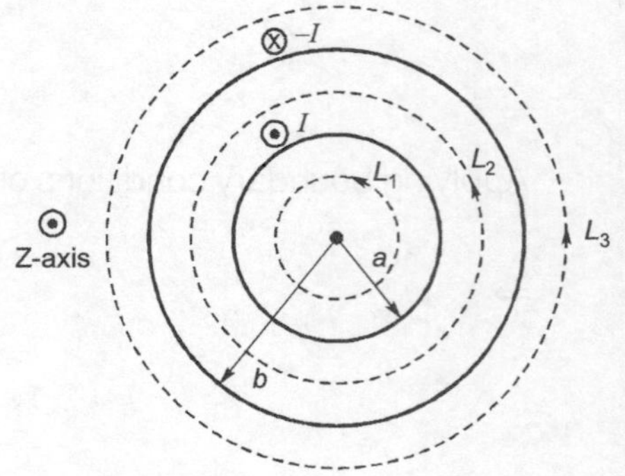

Let us consider the hollow cylinder with inner radius a and outer radius b and cross-section of this cylinder is shown above.

The inner conductor has radius a and carries current I, while outer conductor has radius b and carries return current I. Since current distribution is symmetrical, we apply Ampere's law along the Amperian path for each of the three possible region

$$0 < \rho \leq a, \quad a \leq \rho \leq b, \rho \geq b.$$

For region $0 \leq \rho \leq a$ we apply Ampere's law to path L_1, we have

$$\oint \vec{H} . \vec{dl} = I_{inc.} = \int_s \vec{J} . \vec{dS} \qquad ...(i)$$

Since the current is uniformly distributed over the cross-section

$$\vec{J} = \frac{I}{\pi a^2} a_z, \quad \vec{dS} = \rho d\phi d\rho a_z$$

$$I_{inc.} = \int_s \vec{J} . \vec{dS} = \frac{I}{\pi a^2} \int_0^{2\pi} \int_0^{a} \rho d\phi \, d\rho = \frac{I}{\pi a^2} \cdot \pi \rho^2 = \frac{I \rho^2}{a^2}$$

∴ From equation (i),

$$H_\phi \int_{L_1} dl = H_d 2\pi\rho = \frac{I\rho^2}{a^2}$$

$$H_\phi = \frac{I\rho}{2\pi a^2} \qquad \text{for} \quad 0 \leq \rho \leq a$$

$$\vec{H}_\phi = \frac{I\rho}{2\pi a^2} \hat{a}_d \qquad \text{for} \quad 0 \leq \rho \leq a$$

For region $a \leq \rho \leq b$ we use path L_2 as the Amperian path

$$\oint_{L_2} \vec{H} \cdot \vec{dl} = I_{inc} = I$$

$$H_d 2\pi\rho = I$$

$$H_\phi = \frac{I}{2\pi\rho} \qquad \text{for} \quad a \leq \rho \leq b$$

$$\vec{H}_\phi = \frac{I}{2\pi\rho} \hat{a}_d \qquad \text{for} \quad a \leq \rho \leq b$$

For region $\rho \geq b$ we use path L_3

$$\oint_L \vec{H} \cdot \vec{dl} = I - I = 0$$

$$H_d 2\pi\rho = 0$$

$$H_\phi = 0 \qquad \text{for} \quad \rho \geq b$$

4.5 **A parallel-plate capacitor with plate area of 10 cm^2 and plate separation of 6 mm has a voltage 50 sin($10^3 t$) volt applied to its plates. Determine the displacement current, assuming $\varepsilon = 2\varepsilon_0$.**

[10 marks : 2011]

Solution:

Given that, Plate area

$$A = 10 \text{ cm}^2 = 10 \times 10^{-4} \text{ m}^2$$

$A = 10^{-3}\ m^2$

Plate separation, $d = 6$ mm

$d = 6 \times 10^{-3}$ m

Applied voltage, $V = 50 \sin(10^3 t)$ volt

∴ electric field between the plate at every point

$$|\vec{E}| = \frac{V}{d} = \frac{50\sin(10^3 t)}{6\times10^{-3}} = \frac{5\times10^4}{6}\sin(10^3 t) = 8.33 \times 10^3 \sin(10^3\ t)\ \text{V/m}$$

From +ve plate to –ve plate.

Given, $\in = 2\in_0$

Now, $|\vec{D}| = \in|\vec{E}|$

$$|\vec{D}| = 2\in_0 \times 8.33 \times 10^3 \sin(10^3 t)$$

D — Electric flux density.

$$|\vec{D}| = 2 \times 8.85 \times 10^{-12} \times 8.33 \times 10^3 \sin(10^3 t)$$

$$= 14.75 \times 10^{-8} \sin(10^3 t)\ \text{c/m}^2$$

Now displacement current density,

$$|\vec{J}_d| = \left|\frac{d}{dt}\vec{D}\right| = \frac{d}{dt}|\vec{D}|$$

$$= \frac{d}{dt}14.75\times10^{-8}\sin(10^3 t) = 14.75 \times 10^{-5} \cos(10^3\ t)\ \text{A/m}^2$$

∴ Displacement current, $I_d = |\vec{J}_d| \times A = 14.75\times10^{-5}\cos(10^3 t)\times10^{-3}$

$$I_d = 14.75 \times 10^{-8} \cos(10^3\ t)\text{A}$$

$$(I_d)_{max} = 14.75 \times 10^{-8}\ \text{A}$$

4.6 Discuss the force exerted on a current element due to a magnetic field $\vec{B}$ in another current element.

[12 marks : 2012]

Solution:

A magnetic field can exert force only on a moving charge. The magnetic force $\vec{F}_m$ experienced by a charge Q moving with velocity $\vec{u}$ in magnetic field $\vec{B}$

$$\vec{F}_m = Q(\vec{u}\times\vec{B}) \quad ...(i)$$

To determine the force on a current element $I\,d\vec{l}$ of a current-carrying conductor due to magnetic field $\vec{B}$.

$$\vec{J} = \rho_v \vec{u}$$

$$Id\vec{l} = \vec{J}\,dv = \rho_v \vec{u}\,dv = dQ\vec{u}$$

An elemental charge da moving with velocity $\vec{u}$ is equivalent to a conduction current $Id\vec{l}$. Thus the force on a current element $Id\vec{l}$ in a magnetic field $\vec{B}$ is found by equation (i) by merely replacing $Q\vec{u}$ by $Id\vec{l}$

$$d\vec{F} = Id\vec{l}\times\vec{B}$$

∴ $$\vec{F} = \oint_L Id\vec{l}\times\vec{B}$$

If the magnetic field is produced by on another current element then

$$d\vec{B} = \frac{\mu_o}{4\pi} \cdot \frac{I_2 d\vec{l}_2 \times \hat{a}_{r21}}{r_{21}^2}$$

So force $d(d\vec{F}_1)$ on element $I_1 d\vec{l}_1$ due to the field $d\vec{B}_2$ produced by element $I_2 d\vec{l}_2$

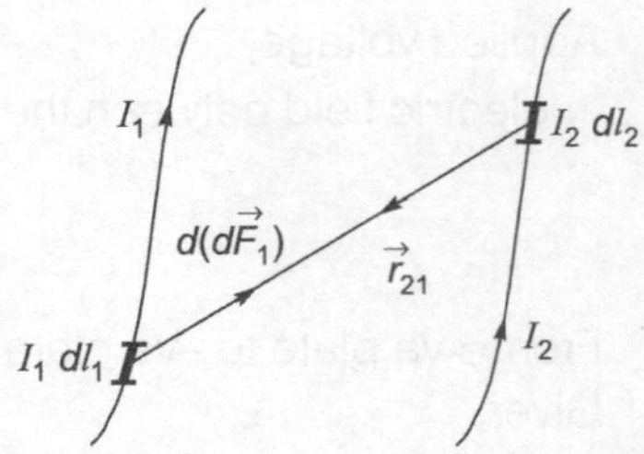

$$d(d\vec{F}_1) = I_1 d\vec{l}_1 \times d\vec{B}_2 = \frac{\mu_o I_1 d\vec{l}_1 \times (I_2 d\vec{l}_2 \times \hat{a}_{r_{21}})}{4\pi\, r_{21}^2}$$

Total force,
$$\vec{F}_1 = \frac{\mu_o I_1 I_2}{4\pi} \oint_{L_1} \oint_{L_2} \frac{d\vec{l}_1 \times (d\vec{l}_2 \times \hat{a}_{r_{21}})}{r_{21}^2}$$

4.7 **In free space, the magnetic flux density is**

$$B = y^2\,\hat{a}_x + z^2\,\hat{a}_y + x^2\hat{a}_z \text{ Wb/m}^2$$

(i) Show that B is a magnetic field.

(ii) Find the magnetic flux through $x = 1,\ 0 < y < 1$ and $1 < z < 4$.

(iii) Calculate current density (J). **[10 marks : 2013]**

Solution:

$$B = y^2\hat{a}_x + z^2\hat{a}_y + x^2\hat{a}_z$$

(i)
$$\nabla \cdot B = \frac{\partial}{\partial x}(B_x) + \frac{\partial}{\partial y}(B_y) + \frac{\partial}{\partial z}(B_z)$$

$$= \frac{\partial}{\partial x}(y^2) + \frac{\partial}{\partial y}(z^2) + \frac{\partial}{\partial z}(x^2) = 0$$

It is a solenoidal field in nature. As we know that in case of magnetic fields, isolated charge can't exist, so net magnetic flux emerging through any closed surface must be equal to zero viz. Varified from above calculations. Hence B is a magnetic field.

(ii)
$$\text{Magnetic flux} = \int_1^4\int_0^1 (y^2\,\hat{a}_x + z^2\,\hat{a}_y + x^2\,\hat{a}_z)\cdot(\hat{a}_x\, dy\, dz)$$

$$= \int_1^4\int_0^1 y^2 dy\, dz = \int_1^4 \left(\frac{y^3}{3}\right)_0^1 dz = \frac{1}{3}\int_1^4 1\, dz = 1\, wb$$

(iii)
$$J = \nabla \times B = \nabla \times \left(\frac{H}{\mu_o}\right)$$

$$= \frac{1}{\mu_o}\begin{vmatrix} \hat{a}_x & \hat{a}_y & \hat{a}_z \\ \frac{\partial}{\partial x} & \frac{\partial}{\partial y} & \frac{\partial}{\partial z} \\ y^2 & z^2 & x^2 \end{vmatrix} = \left(\frac{1}{\mu_o}\right)(-2z\,\hat{a}_x - 2x\,\hat{a}_y - 2y\hat{a}_z)$$

$$\boxed{J = \frac{-2}{\mu_o}(z\hat{a}_x + x\hat{a}_y + y\hat{a}_z)}$$

4.8 How the four quantities *E*, *D*, *H* and *B* are related to each other in static and time varying fields? Name the relation as laws. Give the value of divergence and curl of each quantity.

[10 marks : 2015]

Solution:

Four quantities *E*, *D*, *H* and *B* are related by the Maxwell's equations to each other in static and time varying fields.

Maxwell's equations:

1. **Gauss's Law of electric field:**

(a) Integral form: $$\oiint_s \vec{D}.\vec{ds} = \iiint \rho_V.dV \quad ...(1)$$

(b) Point form: $$\boxed{\nabla\cdot\vec{D} = \rho_V} \quad ...(2)$$

or we can write $$\nabla\cdot\vec{E} = \frac{\rho_V}{\varepsilon} \quad ...(3)$$

2. **Gauss's law of magnetic field:**

(a) Integral form: $$\oiint B\cdot ds = 0 \quad ...(4)$$

(b) Point form: $$\nabla\cdot\vec{B} = 0 \quad ...(5)$$

or we can write, $$\nabla\cdot\vec{H} = 0 \quad ...(6)$$

3. **Faraday's law of electromagnetic induction:**

(a) Integral form: $$\oint \vec{E}\cdot\vec{dl} = -\iint\left(\frac{\partial B}{\partial t}\right)ds \quad ...(7)$$

(b) Point form: $$\nabla\times\vec{E} = -\frac{\partial B}{\partial t} \quad ...(8)$$

$$\nabla\times\vec{D} = -\varepsilon\left(\frac{\partial B}{\partial t}\right) \quad ...(9)$$

4. **Ampere's circuital law:**

(a) Integral form: $$\oint \vec{H}.\vec{dl} = \iint (J_c + J_d)ds = \iint\left(\sigma\vec{E} + \frac{\partial\vec{D}}{\partial t}\right)ds$$

(b) Point form: $$\nabla\times\vec{H} = J_c + J_d = \sigma\vec{E} + \frac{\partial\vec{D}}{\partial t}$$

$$\nabla\times\vec{B} = \mu\left(\sigma\vec{E} + \frac{\partial\vec{D}}{\partial t}\right)$$

- **For static fields:** $\frac{\partial E}{\partial t} = \frac{\partial D}{\partial t} = \frac{\partial B}{\partial t} = \frac{\partial H}{\partial t} = 0$

  then, Maxwell's equations are given below:

  (a) $\nabla\cdot\vec{D} = \rho_V$ or $\nabla\cdot\vec{E} = \frac{\rho_V}{\varepsilon}$ (b) $\nabla\times\vec{E} = 0$ or $\nabla\times\vec{D} = 0$

  (c) $\nabla\cdot\vec{B} = 0$ or $\nabla\cdot\vec{H} = 0$ (d) $\nabla\times\vec{H} = \sigma\vec{E}$ or $\nabla\times\vec{B} = \sigma\mu\vec{E}$

- **For time varying fields:**

  Let, $$\vec{D} = D_o e^{j\omega t} \text{ and } \vec{B} = B_o e^{j\omega t}$$

  $$\frac{\partial\vec{D}}{\partial t} = j\omega D_o.e^{j\omega t} \text{ and } \frac{\partial\vec{B}}{\partial t} = j\omega B_o.e^{j\omega t}$$

$$\frac{\partial \vec{D}}{\partial t} = j\omega\vec{D} \text{ and } \frac{\partial \vec{B}}{\partial t} = j\omega\vec{B}$$

Then, new Maxwell' equations are as follows,

(a) $\nabla \cdot \vec{D} = \rho_V$ or $\nabla \cdot \vec{E} = \frac{\rho_V}{\varepsilon}$

(b) $\nabla \cdot \vec{B} = 0$ or $\nabla \cdot \vec{H} = 0$

(c) $\nabla \times \vec{E} = -\frac{\partial B}{\partial t} = -j\omega\vec{B}$ or $\nabla \times \vec{D} = -j\omega\varepsilon.\vec{B}$

(d) $\nabla \times \vec{H} = \sigma E + \frac{\partial D}{\partial t} = \sigma\vec{E} + j\omega\varepsilon\vec{E}$ or $\nabla \times \vec{B} = \mu\left(\sigma\vec{E} + \frac{\partial \vec{D}}{\partial t}\right) = \mu \times \left(\sigma\vec{E} + j\omega\varepsilon\vec{E}\right)$

4.9 **Show that in a source-free region ($J = 0, \rho_v = 0$), Maxwell's equations can be reduced to two. Identify the two all-embracing equations.**

[10 marks : 2019]

Solution:

Let the medium is homogeneous, Isotropic, source free $(\vec{J} = 0; \rho_V = 0)$ then Maxwell equation are:

$$\vec{\nabla} \times \vec{E} = -\frac{\partial \vec{B}}{\partial t} = -\frac{\partial \mu \vec{H}}{\partial t} \Rightarrow \vec{\nabla} \times \vec{E} = -\mu\frac{\partial \vec{H}}{\partial t} \quad \text{... (i)}$$

$$\vec{\nabla} \times \vec{H} = \vec{J} + \frac{\partial \vec{D}}{\partial t} = 0 + \frac{\partial (\in \vec{E})}{\partial t} \Rightarrow \vec{\nabla} \times \vec{H} = \in \frac{\partial \vec{E}}{\partial t} \quad \text{... (ii)}$$

$$\vec{\nabla} \cdot \vec{D} = \rho_V \Rightarrow \vec{\nabla} \cdot \in \vec{E} = 0 \Rightarrow \in (\vec{\nabla} \cdot \vec{E}) = 0 \Rightarrow \vec{\nabla} \cdot \vec{E} = 0 \quad \text{... (iii)}$$

$$\vec{\nabla} \cdot \vec{B} = 0 \Rightarrow \vec{\nabla} \cdot \mu\vec{H} = 0 \Rightarrow \mu(\vec{\nabla} \cdot \vec{H}) = 0 \Rightarrow \vec{\nabla} \cdot \vec{H} = 0 \quad \text{... (iv)}$$

Take curl of equation (i) both sides,

$$\vec{\nabla} \times (\vec{\nabla} \times \vec{E}) = \vec{\nabla} \times \left(-\mu\frac{\partial \vec{H}}{\partial t}\right)$$

$$(\vec{\nabla} \cdot \vec{E})\vec{\nabla} - (\vec{\nabla} \cdot \vec{\nabla})\vec{E} = -\mu\frac{\partial}{\partial t}(\vec{\nabla} \times \vec{H}) \quad \text{... (v)}$$

Put (ii), (iii) in (v),

$$0 - \nabla^2\vec{E} = -\mu\frac{\partial}{\partial t}\left(\in \frac{\partial \vec{E}}{\partial t}\right) \quad \text{... (vi)}$$

Equation (vi) is called electric wave equation. Take curl of equation (ii) both sides,

$$\vec{\nabla} \times (\vec{\nabla} \times \vec{H}) = \vec{\nabla} \times \left(\in \frac{\partial \vec{E}}{\partial t}\right)$$

$$(\vec{\nabla} \cdot \vec{H})\vec{\nabla} - (\vec{\nabla} \cdot \vec{\nabla})\vec{H} = \in \frac{\partial}{\partial t}(\vec{\nabla} \times \vec{E}) \quad \text{... (vii)}$$

Put (i), (iv) in (vii), $$0 - \nabla^2\vec{H} = \in \frac{\partial}{\partial t}\left(-\mu\frac{\partial \vec{H}}{\partial t}\right)$$

$\Rightarrow$ $$\nabla^2\vec{H} = \mu \in \frac{\partial^2 \vec{H}}{\partial t^2} \quad \text{... (viii)}$$

Equation (viii) is called magnetic wave equation.

Equation (vi) and (viii) are similar equations. Solving equations (vi) and (viii) we get solutions of electric and magnetic field of EM waves. Equation (vi) and (viii) are called wave equations or Helmholtz equations.

4.10 Show that the ratio of the amplitudes of conduction current density and displacement current density is $\frac{\sigma}{\omega \in}$ for the applied field $E = E_m \sin \omega t$. Assume $\mu = \mu_0$.

What is the amplitude ratio if the applied field is $E = E_m e^{-t/\tau}$, where t is real?

[20 marks : 2020]

Solution:

$$J_C = \sigma E \quad \text{...(i)}$$

where $J_C \rightarrow$ Conduction current density (A/m²)

σ is conductivity (1/Ω–m)

E is applied electric field intensity (volt/m)

$$J_d = \frac{\partial D}{\partial t} = \frac{\partial}{\partial t}(\in E) = \in \frac{\partial E}{\partial t} \quad \text{...(ii)}$$

where $J_d \rightarrow$ Displacement flux density (A/m²)

Displacement flux density (C/m²)

$$D = \in E$$

$\in = \in_0 \in_r$ = permittivity (F/m)

Given, $$E = E_m \sin \omega t \quad \text{...(iii)}$$

Put (iii) in (i),

$$J_C = \sigma E_m \sin \omega t$$

Amplitude of J_C is σE_m

Put (iii) in (ii) then $J_D = \in \frac{d}{dt}(E_m \sin \omega t) = \in E_m \cos(\omega t)\, \omega$

Amplitude of J_D is $\omega \in E_M$.

$$\frac{\text{amplitude of } J_C}{\text{amplitude of } J_d} = \frac{\sigma E_m}{\omega \in E_m} = \frac{\sigma}{\omega \in} \quad \text{...(iv)}$$

Given, $$E = E_m e^{-t/\tau} \quad \text{...(v)}$$

$$J_C = \sigma E = \sigma E_m e^{-t/\tau} \quad \text{...(vi)}$$

Amplitude of J_C is σE_m

$$J_d = \in \frac{\partial E}{\partial t} = \in \frac{\partial}{\partial t}\left(E_m e^{-t/\tau}\right) = \in E_m e^{-t/\tau}\left(-\frac{1}{\tau}\right) \quad \text{...(vii)}$$

$$J_d = \frac{-\in E_m}{\tau} e^{-t/\tau}$$

Amplitude of J_d is $\frac{\in E_m}{\tau}$

$$\frac{\text{Amplitude } J_c}{\text{Amplitude of } J_d} = \frac{\sigma E_m}{\left(\frac{\in E_m}{\tau}\right)} = \frac{\sigma \tau}{\in} \quad \text{...(viii)}$$

where τ is time constant.

5. Miscellaneous

5.1 **Define Poynting's vector and Poynting's theorem, Show that ratio of Poynting's vector to energy density is ≤ 3 × 10⁸ m/s.**

[12 marks : 2001]

Solution:

Poynting's theorem and poynting vector.

Consider the flow of electromagnetic energy out of a closed surface. ...(i)

$$\text{div}(E\times H) = H\cdot \text{curl } E - E\cdot \text{curl } H$$

where,
$$\text{Curl } E = -\frac{\partial B}{\partial t} = -\mu\frac{\partial H}{\partial t}$$

$$\text{Curl } H = J + \in\frac{\partial E}{\partial t}$$

Substituting in equation (i),

$$\text{div}(E\times H) = -H\left(\mu\frac{\partial H}{\partial t}\right) - E.J - E\left(\in\frac{\partial E}{\partial t}\right) \quad ...(ii)$$

By divergence theorem,

$$\text{div}(E\times H) = \text{div }\rho$$

$$\iiint_{\text{volume}} \text{div } s = \iint_{\text{surface}} \rho\cdot\hat{n}\, da$$

Applying volume integral on both sides in equation (ii),

Thus, $$\iint \rho\cdot\hat{n}\, da + \frac{\partial}{\partial t}\iiint\left(\frac{\mu H^2}{2} + \frac{\varepsilon E^2}{2}\right)dv = \iiint E.J dv$$

The Poynting vector '$(s = \bar{E}\times\bar{H})$' may be regards as the vector representing the amount of energy per unit time crossing unit area at any point. The direction of power flow is in the direction of $(\bar{E}\times\bar{H})$.

$$\frac{\text{Poynting vector}}{\text{Energy density}} \le 3\times 10^8 \text{ m/s}$$

($\therefore$ rate of flow of energy can't be greater than the velocity of light)

5.2 **Explain travelling waves on a transmission line and define Standing Wave Ratio (SWR). A high frequency lossless transmission line has a characteristic impedance of 600 Ω. Calculate the value of current SWR when the load is (500 + j300) Ω.**

[12 marks : 2001]

Solution:

For a long transmission line terminals at load impedance (Z_L).

Here,
$$dI = y\,dx\,V \quad \text{and} \quad dV = ZI\,dx$$

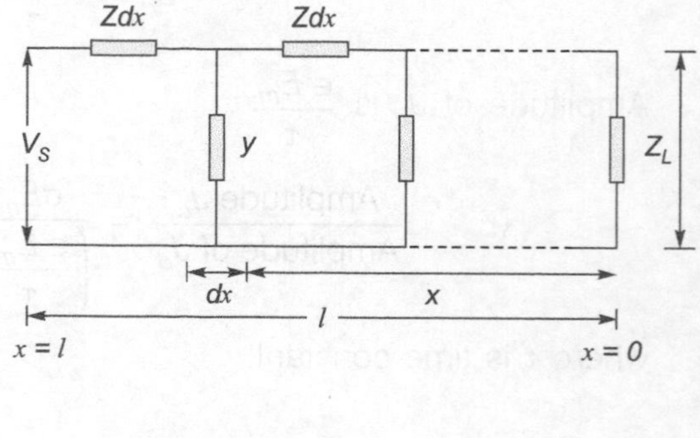

$$\therefore \quad \frac{dV}{dx} = ZI \quad \text{and} \quad \frac{dI}{dx} = yV$$

$$\therefore \quad \frac{d^2V}{dx^2} = z\frac{dI}{dx} = ZyV$$

$$\therefore \quad \frac{d^2V}{dx^2} - yZ\,V = 0$$

Let $yZ = r^2$ then, $\frac{d^2V}{dx^2} - r^2 V = 0$

Solution of differential equation,

$$V = V_+ e^{-rx} + V_- e^{rx}$$

$$I = I_+ e^{-rx} + I_- e^{rx}$$

At $(x = 0)$

$$Z_L = \frac{V}{I} = \frac{V_+ + V_-}{I_+ + I_-} = \frac{Z_0(I_+ - I_-)}{I_+ + I_-}$$

$$\therefore \quad \frac{I_-}{I_+} = \frac{Z_0 - Z_L}{Z_0 + Z_L} = \Gamma_R = \frac{V_-}{V_+}$$

Where, $|\Gamma_R|$ = Reflection coefficient

Z_0 = Characteristic impedance

Standing Wave Ratio: (SWR) → Ratio of maximum value of minimum value of amplitudes of the voltage or current waves.

$$\text{SWR} = \frac{1 + |\Gamma_R|}{1 - |\Gamma_R|}$$

where $Z_L = 500 + j300\ \Omega$

$Z_0 = 600\ \Omega$

then, $\Gamma_R = \frac{Z_L - Z_0}{Z_L + Z_0} = \frac{-100 + 300j}{1100 + 300j} = 0.277 \angle 93°$ $\quad (\therefore |\Gamma_R| = 0.277)$

$$\text{SWR} = \frac{1 + |\Gamma_R|}{1 - |\Gamma_R|} = \frac{1.27}{0.72} = 1.77$$

5.3 In a region where $\varepsilon_R = \mu_R = 1$ and $\sigma = 0$, $\bar{A} = 10^{-3} y \cos 3 \times 10^8 t \cos Z \bar{a}_z$ Wb/m, $V = 3 \times 10^5 y \sin 3 \times 10^8 t \sin Z$ V, Find $\bar{E}$ and $\bar{H}$. [12 marks : 2002]

Solution:

$$\bar{A} = 10^{-3} y \cos 3 \times 10^8 t \cos z\ \hat{a}_z$$

and $V = 3 \times 10^5 y \sin 3 \times 10^8 t \sin z\ v$

$$E = -\nabla V - \frac{\partial \bar{A}}{\partial t} = -\left[\frac{\partial V}{\partial x}\bar{a}_x + \frac{\partial V}{\partial y}\bar{a}_y + \frac{\partial V}{\partial z}\bar{a}_z\right] - \frac{\partial \bar{A}}{\partial t}$$

$$= -3 \times 10^5 \sin(3 \times 10^8 t)\left[\sin z\ \hat{a}_y + y \cos z\ \hat{a}_z\right] - 10^{-3} \times y \times (-3 \times 10^8) \sin(3 \times 10^8 t) \cos z\ \hat{a}_z$$

$$\vec{E} = -3 \times 10^5 \sin z \sin(3 \times 10^8)\ \hat{a}_y \text{ v/m}$$

$$\bar{B} = \bar{\nabla} \times \bar{A}$$

$$= \begin{vmatrix} \hat{a}_x & \hat{a}_y & \hat{a}_z \\ \frac{\partial}{\partial x} & \frac{\partial}{\partial y} & \frac{\partial}{\partial z} \\ 0 & 0 & 10^{-3} \times y \cos 3 \times 10^8 t \cos z \end{vmatrix} = 10^{-3} \times \cos(3 \times 10^8)\ t \cos z\ \hat{a}_x$$

$$\bar{H} = \frac{\vec{B}}{\mu_0 \mu_r} = \frac{\vec{B}}{\mu_0 \times 1} = 795.77472 \cos(3 \times 10^8)\ t \cdot \cos z\ \hat{a}_x \text{ A/m}$$

5.4 A transmission line operating at 500 M rad/sec has L = 0.5 μH/m, C = 32 pF/m, G = 100 μ℧/m and R = 25 Ω/m. Calculate γ, α, β, v, λ and z_0.

[12 marks : 2002]

Solution:

$$\omega = 500 \times 10^6 \text{ rad/sec}$$
$$L = 0.5 \times 10^{-6} \text{ H/m}$$
$$C = 32 \times 10^{-12} \text{ F/m}$$
$$G = 100 \times 10^{-6} \text{ S/m}$$
$$R = 25 \ \Omega\text{/m}$$
$$r(\text{Propagation constant}) = \sqrt{(R+j\omega L)(G+j\omega C)}$$
$$= \left[(25+j500\times10^6\times0.5\times10^{-6})(100\times10^{-6}+j500\times10^6\times32\times10^{-12})\right]^{1/2}$$
$$= \sqrt{-3.9975+0.425\,j} = (4.02\angle173°)^{1/2}$$
$$\therefore \quad r = \sqrt{4.02}\angle(173/2)°$$
$$= 0.1224 + 2j$$
$$\alpha(\text{Attenuation constant}) = 0.1224 \text{ Np/m}$$
$$\beta(\text{Phase constant}) = 2 \text{ rad/m}$$
$$(\text{Velocity of Propagation}) \ V = \frac{1}{\sqrt{LC}} = \frac{1}{\sqrt{0.5\times10^{-6}\times32\times10^{-12}}} = 2.5\times10^8 \text{ m/s}$$
$$\lambda(\text{Wavelength}) = \frac{2\pi}{\beta} = \frac{2\pi}{\beta} = 3.14 \text{ m}$$
$$\text{Characteristic impedance} = \sqrt{\frac{R+j\omega L}{G+j\omega C}}$$
$$= 125.3\angle-2.6762°$$

5.5 Three parallel transmission wires are charged with linear densities (+) 2 D, (–) D and (–) D in coulomb per meter length respectively. Obtain the condition for points on equipotential surfaces inside the parallel wires. Hence or otherwise show that, if three wire sections occupy the corners of an equilateral triangle then along the centroid axis over the whole length potential V = 0. Explain briefly the formula used.

[12 marks : 2003]

Solution:

Potential at any point in the region of space is given

$$V = \frac{-1}{2\pi\varepsilon_0}\times 2D\, ln\, r_1 - \lambda\, ln\, r_2 - \lambda\, ln\, r_3 \qquad \left(\text{where, } \lambda = \frac{D}{2\pi\varepsilon_0}\right)$$

$$= \frac{-D}{2\pi\varepsilon_0}[2\, ln\, r_1 - ln\, r_2 - ln\, r_3] = \frac{-D}{2\pi\varepsilon_0}\times\left[ln\, r_1^2 - ln\, r_2 - ln\, r_3\right]$$

$$= \frac{-D}{2\pi\varepsilon_0}\times ln\frac{r_1^2}{r_2\, r_3}$$

Now D and ε_0 are constant and so for any constant value of V – for that is the condition for any point to lie-on an equipotential surface it is necessary that the logarithmic part on the R.H.S. of equation should be constant. If r_1 and r_2 and r_3 are equal $\frac{r_1^2}{r_2 r_3} = 1$ making R.H.S. of equation equal to zero.

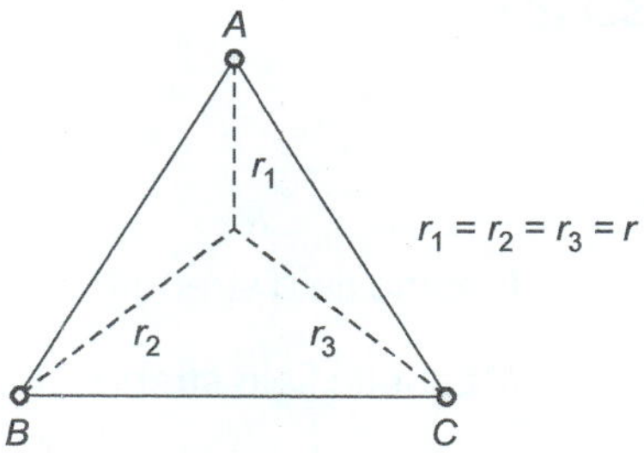

$$\because \quad ln\,1 = 0$$

Thus $$V = 0$$

5.6 The electric field intensity associated with a plane wave travelling in a perfect dielectric medium ($\mu_r = 1$) is given by $E_x(z, t) = 14.14 \cos\left(2\pi \times 10^7 t - \frac{\pi z}{10}\right) \hat{a}_x$ in volts per meter. Calculate the values of velocity of propagation and intrinsic impedance.

[12 marks : 2003]

Solution:

From Maxwell equation $\nabla \times \bar{E} = -\frac{\partial \bar{B}}{\partial t}$

$$\begin{vmatrix} \hat{a}_x & \hat{a}_y & \hat{a}_z \\ \frac{\partial}{\partial x} & \frac{\partial}{\partial y} & \frac{\partial}{\partial z} \\ E_m \cos(\omega t - \beta z) & 0 & 0 \end{vmatrix} = -\frac{\partial \bar{B}}{\partial t}$$

$$\frac{\partial}{\partial z}[E_m \cos(\omega t - \beta z)]\bar{a}_y - \frac{\partial}{\partial y}[E_m \cos \omega t - \beta z]\bar{a}_z = \frac{\partial \bar{B}}{\partial t}$$

$$\therefore \quad E_m \sin(\omega t - \beta z) \times \beta = -\frac{\partial \bar{B}}{\partial t}$$

$$\bar{B} = \frac{E_m \beta}{\omega} \cos(\omega t - \beta z)$$

$$\bar{H} = \frac{E_m \beta}{\omega \mu_0} \times \cos(\omega t - \beta z)$$

Here, $$\omega = 2\pi \times 10^7 \text{ and } \beta = \frac{\pi}{10}$$

$$\therefore \quad \text{Velocity of propagation} = C = \frac{\omega}{\beta} = 2 \times 10^8 \text{ m/s}$$

$$\therefore \quad \text{Intrinsic impedance} = \frac{\bar{E}}{\bar{H}} = \frac{E_m \omega \mu_0}{E_m \times \beta} = \frac{\omega \mu_0}{\beta}$$

$$= 2 \times 10^8 \times 4\pi \times 10^{-7} = 251.2\ \Omega$$

5.7 Derive wave equations for a conducting medium. A concentric cable has a d.c. voltage V between conductor and steady current I flowing in inner and outer conductors. Power is being transferred to a load resistance R along the concentric cable. Inner conductor has radius a and outer conductor has radius b. Find value of magnetic field strength H directed in circles about the axis. Also find the electric field strength E directed radially. Show that total power flow along cable is given by the integration of Poynting vector over any cross-sectional area.

[12 marks : 2004]

Solution:

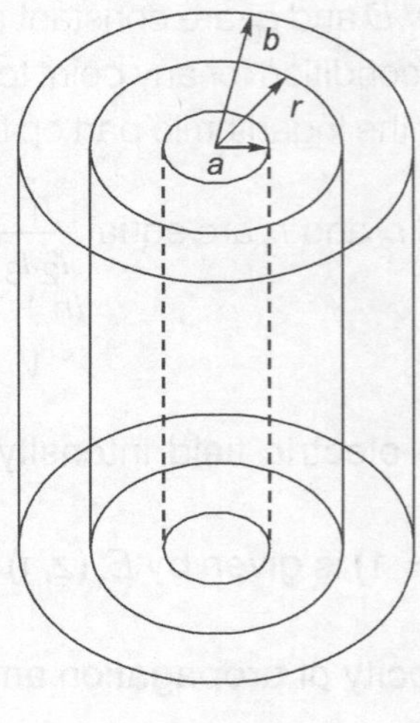

$$\bar{J} = \frac{I}{\pi(b^2 - r^2)}\hat{a}_z$$

Electric field strength, $$\bar{E}_z = \frac{\bar{J}}{\sigma} = \frac{\hat{a}_z}{\sigma} \times \frac{I}{\pi(b^2 - r^2)}$$

Magnetic field strength (H) from ampere's circuital law

$$\oint H \cdot 2\pi r = I_{\text{enclosed}}$$

$$\oint H \cdot 2\pi r = \frac{I(b^2 - r^2)}{(b^2 - a^2)}$$

$$H_\phi = \frac{I \times (b^2 - r^2)}{2\pi r(b^2 - a^2)}\hat{a}_\phi$$

Poynting vector, $$\bar{P} = \bar{E} \times \bar{H}$$

$$= (\hat{a}_z \times \hat{a}_\phi) \cdot \frac{I^2(b^2 - r^2)}{\sigma 2\pi^2 r(b^2 - a^2)(b^2 - r^2)} = \hat{a}_r \frac{I^2}{2\pi^2\ \sigma r\,(b^2 - a^2)}$$

then, $$\oint \bar{P} \cdot \overline{ds} = \oint \frac{I^2 \times 2\pi\, rl}{2\pi^2 \sigma r(b^2 - a^2)} ds$$

Where, $$\overline{ds} = 2\pi r L(\hat{a}_r)$$ $\left[\text{Where } R = \dfrac{l}{\sigma\pi(b^2 - a^2)}\right]$

so, $$\text{Power} = \oint \bar{P} \cdot \overline{ds} \quad \text{at radius } r = z^2 L$$

Total power flow along cable is given by the integration of poynting vector over any cross-sectional area.

5.8 For transmission lines equations $d^2V/dZ^2 = r^2V$ and $d^2I/dZ^2 = r^2I$.
Find the solution in exponential and hyperbolic function form. Prove that $Z_{SC} \cdot Z_{OC} = Z_0^2$.
How impedance matching is done by means of stub lines?

[12 marks : 2004]

Solution:

$$\frac{d^2V}{dx^2} = r^2 V$$

$$\frac{d^2I}{dx^2} = r^2 I$$

These are two linear differential equation whose general solution can be written as

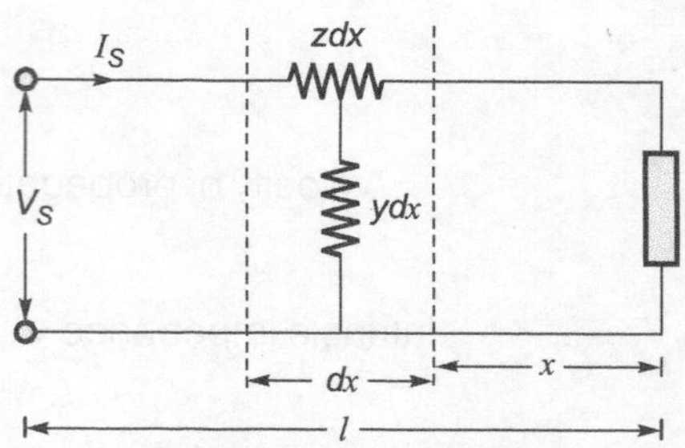

Schematic diagram of a long line

$$V = C_1e^{+rx} + C_2e^{-rx}$$

$$I = C_3e^{+rx} + C_4re^{-rx}$$

$\therefore$ $$I = \frac{1}{Z_C}(C_1e^{rx} - C_2e^{-rx})$$

The constant C_1 and C_2 are evaluating by using boundary condition or end condition.

at ($x = 0$), $$V = V_R \text{ and } I = I_R$$

$$V_R = 1(C_1 + C_2), I_R = \frac{I}{Z_C}(C_1 - C_2)$$

Solving,
$$C_1 = \frac{1}{2}(V_R + Z_C I_R)$$

$$C_2 = \frac{1}{2}(V_R - Z_C I_R)$$

$\therefore$
$$V = \left(\frac{V_R + Z_C I_R}{2}\right)e^{rx} + \left(\frac{V_R - Z_C I_R}{2}\right)e^{-rx}$$

$$I = \left(\frac{V_R / Z_C + I_R}{2}\right)e^{rx} - \frac{(V_R / Z_C - I_R)}{2}e^{-rx}$$

Where, Z_C = Characteristic impedance

r = propagation constant

$$V = V_R\left(\frac{e^{rx} + e^{-rx}}{2}\right) + I_R\, z_c\left(\frac{e^{rx} - e^{rx}}{2}\right)$$

$$I = \frac{V_R}{Z_C}\left(\frac{e^{rx} - e^{-rx}}{2}\right) + I_R\left(\frac{e^{rx} + e^{-rx}}{2}\right)$$

There can be written by hyperbolic functions

$$V = V_R \cos hrx + I_R Z_C \sin hrx$$

$$I = I_R \cos hrx + \frac{V_R}{Z_C} \sin h\, rx$$

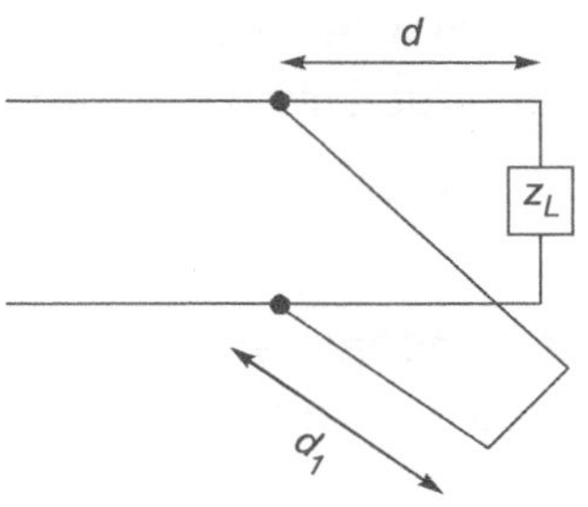

where, Z_0 = Characteristic impedance

at ($x = l$), input impedance,

$$z_i = z_0 \times \frac{z_L + z_0 \tan hrl}{z_0 + z_L \tan h\, rl}$$

where, Z_0 = characteristic impedance

Open circuit line $z_L \to \infty$, $z_{oc} = z_0 \cot hrl$

Short circuit line $z_L = 0$, $z_{sc} = z_0 \tan hrl$

Then,
$$z_{oc} \cdot z_{sc} = z_0^2$$

Impedance matching using stub line:

To match load to z_L line by placing a short circuit-stub of length d, a distance 'd' from the load. Stub line has same characteristic impedance as the main line. The input impedance of the line 'd' containing the load. The resultant input impedance must be $(1 + 0j)$. The input impedance of the length d containing load $(1 + jb_{in})$ for the addition of the input admittance of the stub (jb_{stub}) to produce total admittance $(1 + aj)$.

5.9 Derive an expression for characteristic impedance of a coaxial transmission line with inner and outer radius are a and b respectively.

[12 marks : 2005]

Solution:

The potential difference between inner and outer conductor.

$$V = -\int_a^b E \cdot dI = -\int_a^b \frac{\rho_L}{2\pi \in x}\, dx$$

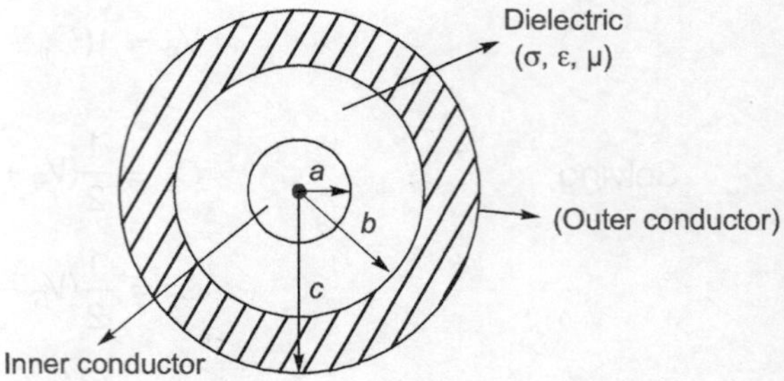

$$V = -\frac{\rho_L}{2\pi \in} \times ln(b/a)$$

$$|V| = \frac{\rho_L}{2\pi \in} ln(b/a) \text{ V}$$

$\therefore$ Capacitance/unit length $= \dfrac{\rho_L}{V} = \dfrac{2\pi \in}{ln(b/a)}$ F/m

$\therefore$ Total external flux, $\phi = \dfrac{\mu_0 I}{2\pi} ln(b/a)$ weber

$\therefore$ $L = \dfrac{\mu_0}{2\pi} ln(b/a)$ H/metre

$\therefore$ Characteristic impedance $= \sqrt{\dfrac{L}{C}} = \dfrac{1}{2\pi} \times \left[\sqrt{\dfrac{\mu}{\varepsilon}}\right] \times ln(b/a)\ \Omega$

5.10 **Explain the concept of wave impedance for plane waves travelling in a lossless medium and show how this may be extended to cover the case of a conductive medium. Show that the reflection coefficient for plane waves travelling in a vacuum and falling with normal incidence upon a medium having wave impedance η is given by $(\eta - \eta_0)/(\eta + \eta_0)$.**
Show that when a plane wave is incident normally upon a medium with permeability μ_0, permittivity ε_0 and conductivity σ the fraction of power absorbed is approximately $2(2\,\omega\varepsilon_0/s)^{1/2}$.

[12 marks : 2005]

Solution:

Wave impedance: Wave impedance of a medium is defined as the impedance offered by the medium for propagation of wave and mathematically, can be represented as

$$\eta = \frac{E_x}{H_y} = \sqrt{\frac{j\omega\mu}{\sigma + j\omega \in}}$$

Also called as intrinsic impedance for lossless medium $\sigma = 0$

$$\eta = \sqrt{\frac{\mu}{\in}} = \eta_0$$

For conductive medium, $\dfrac{\sigma}{\omega\in} >> 1$

$\Rightarrow$ $\sigma >> \omega\in$

$\therefore$ $\eta = \sqrt{\dfrac{j\omega\mu}{\sigma}} = \sqrt{\dfrac{\omega\mu}{\sigma}}\angle 45° = \sqrt{\dfrac{\omega\in}{\sigma}}\sqrt{\dfrac{\mu}{\in}}\angle 45°$

$$\eta = \eta_0 \cdot \sqrt{\frac{\omega \in}{\sigma}} \angle 45°$$

So η_0 may be extended to cover the wave impedance of conductive medium η

$$E_i = \sqrt{\frac{\mu_1}{\in_1}} H_i$$

$$E_r = -\sqrt{\frac{\mu_1}{\in_1}} H_r$$

The transmitted wave has the associated electric field given by

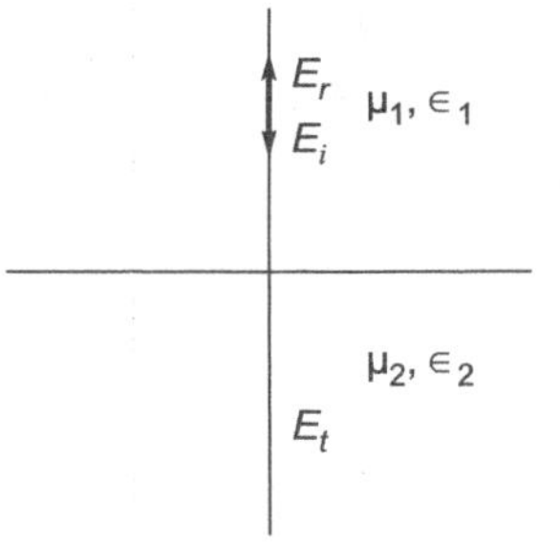

Normal incidence

$$E_t = \sqrt{\frac{\mu_2}{\epsilon_2}} H_t$$

$$H_t = \sqrt{\frac{\epsilon_1}{\mu_1}}(E_i - E_r) = \sqrt{\frac{\epsilon_2}{\mu_2}} E_t$$

and $E_t = E_i + E_r$

Simplifying, we get,

$$\frac{E_r}{E_i} = \frac{\eta_2 - \eta_1}{\eta_2 + \eta_1}$$

where, $\eta_2 = \eta, \eta_1 = \eta_0$

and

$$\frac{H_r}{H_i} = \frac{\eta_1 - \eta_2}{\eta_1 + \eta_2}$$

5.11 **Show that for metals,**

$$\eta = \sqrt{\frac{\omega\mu}{2\sigma}}(1+j)$$

$$k = \sqrt{\frac{\omega\mu\sigma}{2}}(1-j)$$

where η is intrinsic wave impedance of the medium and k is the wave number.

[12 marks : 2006]

Solution:

We know that,

For wave, wave number $k = \gamma^*$

γ is propagation constant, $\gamma = \sqrt{j\omega\mu(\sigma + j\omega\in)} = \sqrt{j\omega\mu \cdot j\omega\in(\sigma/j\omega\in + 1)}$

But for metals, $\frac{\sigma}{\omega\in} >> 1$

$$\gamma = \sqrt{j\omega\mu \cdot j\omega\in \frac{\sigma}{j\omega\in}} = \sqrt{\omega\mu\sigma j} = \sqrt{\omega\mu\sigma}\angle 45°$$

$$\gamma = \sqrt{\omega\mu\sigma}\frac{(1+j)}{\sqrt{2}} = \sqrt{\frac{\omega\mu\sigma}{2}}(1+j)$$

$$\therefore \quad K = \gamma^* = \sqrt{\frac{\omega\mu\sigma}{2}}(1-j)$$

$$\therefore \quad \text{Intrinsic impedance} = \frac{E}{H} = \eta = \sqrt{\frac{j\omega\mu}{\sigma + j\omega\varepsilon}}$$

In the case of a good conductor (metal)

$$\eta_{metal} = \sqrt{\frac{j\omega\mu}{\sigma}} = \sqrt{\frac{\omega\mu}{\sigma}}\angle 45° = \sqrt{\frac{\omega\mu}{2\sigma}}(1+j)$$

5.12 Draw a sketch of Electric and Magnetic field lines for a two conductor transmission line and mark the field cell.

Show that the cell impedance (intrinsic impedance of medium) for air or vacuum medium is $Z_{cell} = \sqrt{\mu_0/\varepsilon_0} = 377\ \Omega$.

[10 marks : 2006]

Solution:

Travelling waves in transmission circuits,

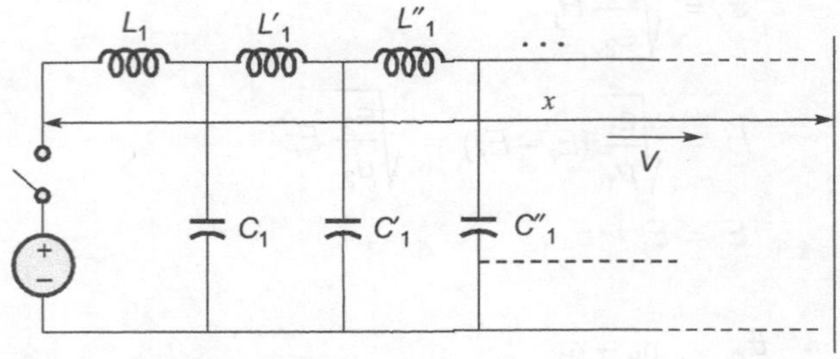

Transmission circuit is composed of elements inductors and capacitors.

Rate of change flux linkages is equal to the potential difference at that point.

Thus, $$V_x = I_x L_1 \frac{dx}{dt} = I_x L_1 V$$

Similarly the current, $$I_x = V_x C_1 \frac{dx}{dt} = V_x C_1 V$$

Intrinsic impedance, $$Z_n = \frac{V_x}{I_x} = \left(\frac{L_1}{C_1}\right) \times \frac{I_x}{V_x}$$

$$\therefore \quad Z_n = \sqrt{\frac{L_1}{C_1}}$$

From Maxwell equation-(free space),

$$\nabla \times E_s = -j\omega \mu_0 H_s$$

Which is greatly simplified for a single E_{xs} components varying only with Z.

$$\frac{dE_{xs}}{dz} = -j\omega \mu_0 H_{ys}$$

$$H_{ys} = -\frac{1}{j\omega\mu_0}(-jk_0)E_{x0}e^{-jk_0 z} = E_x \times \sqrt{\frac{\epsilon_0}{\mu_0}} e^{-jk_0 z}$$

In real instantaneous form,

$$H_y(z,t) = E_{x0}\sqrt{\frac{\epsilon_0}{\mu_0}} \cos(\omega t - k_0 z)$$

Therefore x directed E field that propagates in the positive z-direction is accompanied by a y-directed H-field.

$$\frac{E_x}{H_y} = \sqrt{\frac{\mu_0}{\epsilon_0}} = \sqrt{\frac{4\pi \times 10^{-7}}{\frac{1}{36\pi \times 10^9}}} = 377\,\Omega$$

5.13 A uniform plane wave with the field components E_x and H_y has an electric field amplitude of 100 $\angle 30°$ V/m and propagates at $f = 2$ MHz in a conductive region having the parameters $\mu = \mu_0$, $\varepsilon = 9\,\varepsilon_0$ and $(\sigma/\omega\varepsilon) = 0.5$.

(a) Find the values of attenuation constant, phase constant, phase velocity, wavelength, skin depth and intrinsic impedance for this wave.

(b) Express the electric and magnetic fields in both their complex and real time forms, with the values of (a) inserted.

[12 marks : 2007]

Solution:

Given:
$$\omega = 2\pi f$$
$$= 2\pi \times 2 \times 10^6 = 4\pi \times 10^6 \text{ rad/sec}$$

(a) Attenuation constant
$$= \alpha = \omega\sqrt{\frac{\mu \in}{2}\left(\sqrt{1+\left(\frac{\sigma}{\omega \in}\right)^2} - 1\right)}$$
$$= 2\pi\times2\times10^6\sqrt{\frac{\mu_0\times9\times\in_0}{2}}\left(\sqrt{1.25}-1\right)$$
$$= 2\pi\times2\times10^6\sqrt{0.53\times4\pi\times10^{-7}\times8.85\times10^{-12}}$$
$$= 3.05 \times 10^{-2} \text{ Np/m}$$

Phase constant,
$$\beta = \omega\sqrt{\frac{\mu\in}{2}\sqrt{1+\left(\frac{\sigma}{\omega\varepsilon}\right)^2}+1}$$
$$= 4\pi\times10^6\sqrt{\frac{\mu_0\times9\times\in_0}{2}\sqrt{1.25}+1} = 12.94 \times 10^{-2} \text{ rad/m}$$

Phase velocity
$$= \frac{\omega}{\beta} = \frac{4\pi\times10^6}{12.94\times10^{-2}} = 9.7 \times 10^7 \text{ m/s}$$

Wave length,
$$\lambda = \frac{2\pi}{\beta} = \frac{2\pi}{0.1294} = 48.56 \text{ m}$$

Skin depth,
$$\delta = \frac{1}{\sqrt{\pi f\mu\sigma}}$$
$$= \frac{1}{\sqrt{\pi\times2\times10^6\times4\pi\times10^{-7}\times0.5\times9\varepsilon_0\times4\pi\times10^6}} = 15.9 \text{ meter}$$

(n) Intrinsic impedance
$$= \sqrt{\frac{j\omega\mu}{\sigma+j\omega\varepsilon}} = \sqrt{\frac{j\mu/\varepsilon}{(\sigma/\omega\varepsilon)+j}}$$
$$= \sqrt{\frac{j\mu_0/9\varepsilon_0}{0.5+j}} = \sqrt{14104.69\angle 26.56} = 118.76\angle 13.28^\circ$$

(b) In complex from:
$$\bar{E}_x = E_{x_0}e^{-\gamma z}\hat{a}_x$$
$$\bar{E}_x = 100\angle 30\cdot e^{-(\alpha+j\beta)z}\hat{a}_x$$
$$\bar{E}_x = 100e^{-\alpha z}e^{j(\pi/6-\beta z)}\hat{a}_x$$
$$\bar{E}_x = 100e^{-(3.05\times10^{-2})z}\cdot e^{j(\pi/6-12.94\times10^{-2}z)}\hat{a}_x$$
$$\frac{E_{x_0}}{H_{y_0}} = \eta$$
$$\Rightarrow \quad H_{y_0} = \frac{E_{x_0}}{\eta} = \frac{100\angle 30}{118.76\angle 13.28^\circ} = 0.842\angle 16.72^\circ$$
$$\therefore \quad \vec{H}_y = H_{y_0}e^{-\gamma z}\hat{a}_y$$
$$\vec{H}_y = 0.842\angle 16.72^\circ e^{-(\alpha+j\beta)z}\hat{a}_y$$

$$\vec{H}_y = 0.842e^{-(3.05\times10^{-2}z)}\cdot e^{j(16.72°-12.94\times10^{-2}z)}\hat{a}_y$$

∴ Real instantaneous form is

$$\bar{E}_x = 100e^{-3.05\times10^{-2}z}\cos(\omega t+\pi/6-12.94\times10^{-2}z)\hat{a}_x \text{ V/m}$$

$$\vec{H}_y = 0.842e^{-3.05\times10^{-2}z}\cos(\omega t+16.72°-12.94\times10^{-2}z)\hat{a}_y \text{ A/m}$$

5.14 Region 1($z < 0$) is free space, whereas region 2($z > 0$) is a material medium characterized by $\sigma = 10^{-4}$(s/m), $\varepsilon = 5\,\varepsilon_0$, and $\mu = \mu_0$. For a uniform plane wave having the electric field:

$$\bar{E}_i = E_0 \cos(3\pi\times10^5\, t - 10^{-3}\pi z)\,\bar{a}_x \text{ (V/m)}$$

Incident normally on the interface z = 0 from region 1, obtain the expression for the transmitted wave electric field.

[10 marks : 2007]

Solution:

Given that medium 1 is free space

∴ $\eta_1 = 120\pi = 377\ \Omega$

$\sigma = 10^{-4}$ s/m, $\in = 5\in_0$, $\mu = \mu_0$ for medium 2

∴ $$\frac{\sigma}{\omega\in} = \frac{10^{-4}}{3\times\pi\times10^5\times8.85\times10^{-12}\times5}$$

Here, $\omega = 3\pi\times10^5$

∵ $\vec{E}_i = E_0\cos(3\pi\times10^5 t - 10^{-3}\pi z)$

Comparing with standard EM wave we get

$\omega = 3\pi\times10^5$ rad/sec

$\beta_1 = 10^{-3}\pi$ for free space

∴ $$\frac{\sigma}{\omega\in} = 2.3978 > 1$$

∴ this medium is behaving as good conductor.

For this medium,

∴ $$\alpha_2 = \beta_2 = \sqrt{\frac{\omega\mu\sigma}{2}} = \sqrt{\frac{3\pi\times10^5\times4\pi\times10^{-7}\times10^{-4}}{2}} = \pi\sqrt{6\times10^{-6}}$$

$\alpha_2 = \beta_2 = 2.449\times\pi\times10^{-3}$ rad/m or np/m

Now, transmission coefficient $$= \frac{E_{t_0}}{E_0} = \frac{2\eta_2}{\eta_1+\eta_2}$$

$$\eta_2 = \sqrt{\frac{j\omega\mu}{\sigma+j\omega\in}} = \sqrt{\frac{j\omega\mu_0}{\sigma+j\omega\times5\in_0}}$$

$$= \sqrt{\frac{j\times3\pi\times10^5\times4\pi\times10^{-7}}{10^{-4}+j\times3\pi\times10^{-7}\times4\pi\times10^5\times5\times8.85\times10^{-12}}}$$

$$\eta_2 = \sqrt{10930.24\angle67.35} = 104.54\angle33.7°\ \Omega$$

∴ Transmission coefficient $$= \frac{E_{t_0}}{E_0} = \frac{2\times104.54\angle33.7°}{377\angle0°+104.5\angle33.7°} = 0.447\angle26.57°$$

$$\therefore \qquad \frac{E_{t_0}}{E_0} = 0.447\angle 26.57°$$

$$E_{t_0} = 0.447\angle 26.57°\, E_0$$

$\therefore$ Transmitted wave equation $= E_{t_0} e^{-\gamma_2 z}\hat{a}_x$

Real instantaneous form will be

$$E(t) = E_{t_0} e^{-\alpha_2 z}\cos(\omega t - \beta_2 z)\hat{a}_x$$

$$E(t) = 0.447\, E_0\angle 26.57°\; e^{-\alpha_2 z}\cos(\omega t - \beta_2 z)\hat{a}_x$$

$$E(t) = 0.447 E_0 e^{-2.449\times\pi\times 10^{-3} z}\cos(3\pi\times 10^5 t - 2.44\pi\times 10^{-3} z + 26.57°)\hat{a}_x \text{ V/m}$$

5.15 **For the transmission line given in the figure below, determine (i) the input impedance, (ii) the reflection coefficient at input and (iii) the average power absorbed by each resistor.**

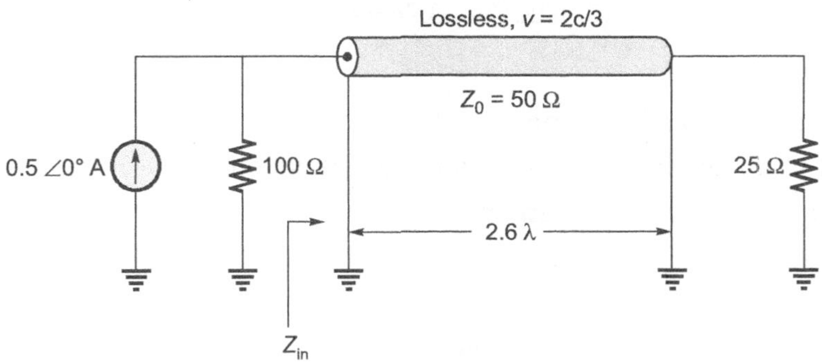

[10 marks : 2008]

Solution:

(i)
$$Z_{in} = Z_0 \times \frac{Z_L + jZ_0\tan\beta d}{Z_0 + jZ_L\tan\beta d}$$

$$= 50\times\frac{25 + j50\tan\left(\frac{2\pi}{\lambda}\times 2.6\lambda\right)}{50 + j25\tan\left(\frac{2\pi}{\lambda}\times 2.6\lambda\right)}$$

$$= 50\times\frac{1 + 2j\tan(5.2\pi)}{2 + j\tan(5.2\pi)} = 33.74 + j24.07\ \Omega$$

(ii) Reflection coefficient at input end

$$\Gamma_i = \frac{Z_L - Z_0}{Z_L + Z_0}$$

But here, $Z_L = 100\ \Omega$

$Z_0 = 50\ \Omega$

$$\therefore \qquad \Gamma_i = \frac{100-50}{100+50} = \frac{50}{150} = \frac{1}{3}$$

(iii) Net input impedance seen from the transmission end of line is

$$Z_{in} = 33.74 + j\,24.07\ \Omega$$

$\therefore$ Net equivalent diagram of line is

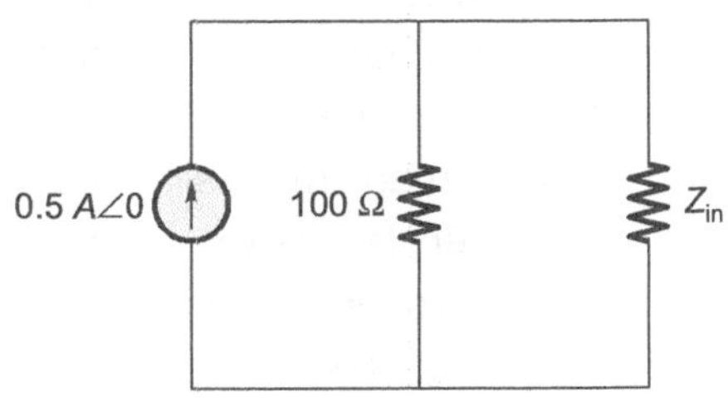

$\therefore$ Current through 100 Ω resistor is

$$= \frac{Z_{in}}{100 + Z_{in}} 0.5\angle 0°$$

$$i = \frac{33.74 \times 0.5}{133.74 + j24.07}$$

$$i = 0.122 - 0.02198\,j = 0.124\angle -10.21°$$

Current through Z_{in} $\quad i_1 = 0.5 - i$

$$= 0.378 + 0.02198\,j = 0.3786\angle 3.327°$$

So power dissipated in 100 Ω resistance is

$$P_1 = |i|^2 R$$

$$= (0.124)^2 \times 100 = 1.537 \text{ Watt}$$

Power dissipated is 25 Ω resistance is equal to power dissipated in Z_{in} (∵ line is lossless)

$$\therefore \quad P_2 = |i|^2 \times \text{Real part of } Z_{in}$$

$$P_2 = (0.3786)^2 \times 33.74 = 4.836 \text{ W}$$

5.16 Explain the following?

(i) Poynting vector and its significance

(ii) Loss tangent of dielectrics as used in wave propagation

(iii) Intrinsic impedance of a wave medium

[10 marks : 2011]

Solution:

(i) Energy can be transported from one point to another point by means of EM waves. The rate of such energy transportation can be calculated from Maxwell's equation.

- The quantity $\vec{E} \times \vec{H}$ is known as the poynting vector P, measured in watts per square meter (W/m^2) that is

$$\vec{P} = \vec{E} \times \vec{H}$$

- It represents the instantaneous power density vector associated with the EM field at a given point. The integration of poynting vector over any closed surface gives the net power flowing out of the surface.
- It also represents the direction of flow of EM waves.

(ii) **Loss tangent:** It is ratio of magnitude of the conduction current density J_c to that the displacement current density $\vec{J}_d$ in a lossy medium

$$\frac{|J_c|}{|J_d|} = \frac{|\sigma E|}{|J\omega \in E|} = \frac{\sigma}{\omega \in} = \tan\theta$$

$$\tan\theta = \frac{\sigma}{\omega \in}$$

- Where tan θ is known as loss tangent and θ is loss angle of the medium.
- Although a line of demarcation between good conductors and lossy dielectric is not easy to make; tan θ or θ may be used to determine, how lossy a medium is
- A medium is said to be good dielectric is tan θ is very small ($\sigma << \omega\in$) or a good conductor if tanθ is very large ($\sigma >> \omega\in$).
- So characteristic behaviours of a medium not only depends on σ, ∈ and μ but also on frequency of operation.

(iii) η for EM wave travelling is a medium,

$$\eta = \frac{E_x}{H_y} = \frac{E_y}{H_x}$$

η is known a intrinsic impedance of the medium and is a complex quantity.
It is also given by

$$\eta = \sqrt{\frac{j\omega\mu}{\sigma + j\omega \in}} = |\eta|\angle\theta_n = |\eta|e^{j\theta_n}$$

with
$$|\eta| = \frac{\sqrt{\mu/\in}}{\left[1+\left(\frac{\sigma}{\omega \in}\right)^2\right]^{1/4}}, \quad \tan 2\theta_n = \frac{\sigma}{\omega \in}$$

5.17 **What is a distortionless line? How do achieve distortionless condition on the line? Derive the necessary equations.**

[10 marks : 2011]

Solution:

A distortion line is one in which the attenuation constant α is frequency independent while phase constant β is linearly dependent on frequency from the general expression for α and β

$$\gamma = \alpha + j\beta = \sqrt{(R+j\omega L)(G+j\omega C)}$$

a distortionless line results if the line parameters are such that

$$\frac{R}{L} = \frac{G}{C}$$

Thus for a distortionless line,
$$\gamma = \sqrt{RG\left(1+\frac{j\omega L}{R}\right)\left(1+\frac{j\omega C}{G}\right)} = \sqrt{RG}\left(1+\frac{j\omega C}{G}\right) = \alpha + j\beta$$

$\therefore$
$$\alpha = \sqrt{RG},\ \beta = \sqrt{RG}\frac{\omega C}{G} = \omega\sqrt{\frac{R}{G}}C$$

or,
$$\alpha = \sqrt{RG} \text{ — independent of frequency}$$

$$\beta = \omega\sqrt{\frac{L}{C}}\cdot C$$

$$\beta = \omega\sqrt{LC} \text{ — Linearly dependent on frequency.}$$

5.18 **For a lossless two-wire transmission line, show that:**

(i) the phase velocity $\mu = c = \frac{1}{\sqrt{LC}}$;

(ii) the characteristic impedance $Z_0 = \frac{120}{\sqrt{\varepsilon_r}}\cosh^{-1}\frac{d}{2a}$

[10 marks : 2011]

Solution:

For a transmission line propagation constant,

$$\gamma = \sqrt{(R+j\omega L)(G+j\omega C)}$$

R— Resistance/m of Transmission line
L— Inductance/m of Transmission line
G— Conductance/m of Transmission line
C— Capacitance/m of Transmission line
For lossless Transmission line $R = G = 0$

$$\therefore \quad \gamma = \sqrt{j\omega L . j\omega C}$$

$$\gamma = j\omega\sqrt{LC}$$

We know, $$\alpha + j\beta = \gamma = j\omega\sqrt{LC}$$

$$\therefore \quad \alpha = 0, \beta = \omega\sqrt{LC} \quad ...(i)$$

and phase velocity of EM wave

$$\mu = \frac{\omega}{\beta} = \frac{\omega}{\omega\sqrt{LC}} = \frac{1}{\sqrt{LC}} \quad \text{Proved.}$$

5.19 Define intrinsic wave impedance for a medium and derive the equation for intrinsic impedance for a lossy dielectric medium (consider $E_x = 0, E_y = 0, E_z = E_0 e^{-\nu y}$)

[12 marks : 2012]

Solution:

Intrinsic impedance is the impedance offered by the loseless medium for propagation of wave. For a uniform plane wave propagation in a given medium, intrinsic impedance mathematically can be expressed as the ratio of electric field to magnetic field. The units of this ratio has units of ohms.

As given in question wave is travelling in $+a_y$ direction with electric field of

$$E_s = E_{zs}\, a_z = E_o\, e^{-\nu y}$$

the corresponding magnetic field is found from the source free Maxwell's equations

$$\nabla \times E_s = -j\omega\mu H_s$$

$$H_s = -\frac{1}{j\omega\mu}\nabla \times E_s = \frac{1}{j\omega\mu}\left[\frac{\partial E_{zs}}{\partial y}a_x - \frac{\partial E_{zs}}{\partial x}a_y\right] = -\frac{1}{j\omega\mu}\left[\frac{\partial}{\partial y}(E_o e^{-\nu y})a_x\right]$$

$$= -\frac{1}{j\omega\mu}(E_o e^{-\nu y} \times -\nu)a_x = \frac{\upsilon}{j\omega u}E_o e^{-\upsilon y}\, a_x = H_{xs}\, a_x$$

As the intrinsic impedance of the wave is defined as the ratio of the electric field and magnetic field phasors

$$\eta = \frac{E_{zs}}{H_{xs}} = \frac{E_o e^{-\nu y}}{\frac{\nu}{j\omega\mu}E_o e^{-\nu y}} = \frac{j\omega\mu}{\nu}$$

Here ν is a propagation constant of the medium.

$$\nu = \sqrt{j\omega\mu(\sigma + j\omega E)}$$

$$\therefore \quad \eta = \sqrt{\frac{j\omega\mu}{(\sigma + j\omega E)}} = |\eta| e^{j\theta_n}$$

5.20 In a non-magnetic medium $\vec{E} = 4\sin(2\pi \times 10^7 t - 0.8x)\vec{a}_z$ V/m. Find:
(i) ε_r, η and (ii) The time-average power carried by the wave.

[10 marks : 2012]

Solution:

(i) Since $\alpha = 0$ and $\beta \neq \omega/C$, the medium is not free space but a lossless medium

$$\vec{E} = 4\sin(2\pi\times 10^7 t - 0.8x)\,\hat{a}_x \text{ V/m}$$

$$\beta = 0.8, \quad \omega = 2\pi\times 10^7, \mu = \mu_o = \text{(non magnetic)}$$

$$\in = \in_o \in_r$$

Hence, $$\beta = \omega\sqrt{\mu \in} = \frac{\omega}{c}\sqrt{\in_r}$$

$\Rightarrow$ $$\sqrt{\in_r} = \frac{c\beta}{\omega} = \frac{3\times 10^8 \times 0.8}{2\pi\times 10^7} = \frac{12}{\pi}$$

$\Rightarrow$ $$\in_r = 14.59$$

$$\eta = \sqrt{\frac{\mu}{\in}} = \sqrt{\frac{\mu_o}{\in_o \in_r}} = \frac{120\pi}{\sqrt{\in_r}} = 120\pi\times\frac{\pi}{12} = 10\pi^2 = 98.7\ \Omega$$

(ii) Poynting vector, $$\vec{P} = \vec{E}\times\vec{H} = \frac{E_o^2}{\eta}\sin^2(\omega t - \beta x)\,\hat{a}_x$$

The time-average power carried by the wave.

$$P_{ave} = \frac{1}{T}\int_0^T P dt = \frac{E_0^2}{2\eta}\hat{a}_x = 81\hat{a}_x \text{ mW/m}^2$$

5.21 Define intrinsic impedance.
A uniform plane wave in free space is given by $E_s = 200\angle 30°\ e^{-j250z}\cdot a_x$ V/m.
Find:

(i) phase constant (β)
(ii) angular frequency (ω)
(iii) wavelength (λ)
(iv) intrinsic impedance (η)
(v) magnetic field (H_s)

[10 marks : 2013]

Solution:

Intrinsic impedance: It is resistance offered by the medium to the propagation of EM wave. Intrinsic impedance is a complex quantity.

$$E_s = E_m \angle\theta\ e^{j\omega t - z(\alpha + j\beta/a_x)}\cdot a_x \text{ V/m}$$

$\therefore$ on comparing,

(i) $\beta = 250$ rad/meter

(ii) $\omega = 0$ rad/sec

(iii) $\lambda = \dfrac{2\pi}{\beta} = \dfrac{2\pi}{250} = 0.0251\text{m}$

(iv) $\eta = \sqrt{\dfrac{\mu_o}{\in_o}} = 120\pi = 377\ \Omega$

(v) Direction of $\hat{H}_s$ = Propagation directional electrical field

$$\hat{a}_x\times\hat{a}_z = \hat{a}_y$$

$\because$ $$H_s = \frac{E_s}{\eta}$$

$\therefore$ $$H_s = \frac{200\angle 30°}{120\pi}e^{-j250z}\,\hat{a}_y \text{ A/m}$$

$$\boxed{H_s = 0.53\angle 30° e^{-j250z}\,\hat{a}_y \text{ A/m}}$$

5.22 **State and explain Poynting vector. A plane 200 MHz wave is travelling in a medium for which conductivity $\sigma = 0$, $\mu_r = 2$, $\varepsilon_r = 4$. If the average Poynting vector is 5 W/m², find E_{rms}, H_{rms}, phase velocity and impedance of the medium.**

[10 marks : 2013]

Solution:

Poynting's theorem states that the net power flowing out of a given volume v is equal to the time rate of decrease in the energy stored within *v* minus the ohmic losses.

Mathematically,

$$\underbrace{\oint_s (\vec{E} \times \vec{H}) d\vec{S}}_{\text{Total power leaving the volume}} = \underbrace{-\frac{\partial}{\partial t}\int_v \left[\frac{1}{2}\epsilon E^2 + \frac{1}{2}\mu H^2\right] dV}_{\text{Rate of decrease in energy stored in electric and magnetic fields}} - \underbrace{\int_v \sigma E^2 dV}_{\text{Ohmic power dissipated}}$$

The various terms in equation are identified using energy conservation arguments of EM fields.

- The quantity $\vec{E} \times \vec{H}$ on the left-hand side of the equation is known as Poynting vector $\vec{P}$, measured in watts per square meter (W/m²); that is,

$$\boxed{\vec{P} = \vec{E} \times \vec{H}}$$

- It should be noted that $\vec{P}$ is normal to both $\vec{E}$ and $\vec{H}$ and is therefore along the direction of wave propagation $\hat{a}_k$ for uniform plane waves.

  Thus, $$\boxed{\hat{a}_k = \hat{a}_E \times \hat{a}_H}$$

The fact that $\vec{P}$ points along $\hat{a}_k$ causes $\vec{P}$ to be regarded as a "pointing" vector.

Numerical:

Given,

$$P = 5 \text{ W/m}^2$$
$$\epsilon_r = 4$$
$$\mu_r = 2$$
$$f = 200 \text{ MHz}$$

Intrinsic impedance, $$\eta = 377\sqrt{\frac{\mu_r}{\epsilon_r}} = 377\sqrt{\frac{2}{4}} = 266.58\ \Omega$$

Phase velocity, $$v_p = \frac{c}{\sqrt{\epsilon_r \mu_r}} = \frac{3 \times 10^8}{\sqrt{4 \times 2}} = 1.06 \times 10^8 \text{ m/s}$$

$$E_{rms} = \sqrt{\eta P} = \sqrt{266.58 \times 5} = 36.51 \text{ V/m}$$

$$H_{rms} = \frac{E_{rms}}{\eta} = \frac{36.51}{266.58} = 0.137 \text{ A/m}$$

5.23 **Describe the boundary condition in magnetic fields. The *xy*-plane serves as interface between two different media. Medium 1 ($z < 0$) has $\mu_r = 6$ and medium 2($z > 0$) has $\mu_r = 4$. If the interface carries a surface current density of: $J_s = \frac{2}{\mu_o}\hat{a}_y$ and $B_2 = 5\hat{a}_x + 8\hat{a}_z$, find B_1.**

[10 marks : 2013]

Solution:

Boundary Relations for Magnetic Fields:

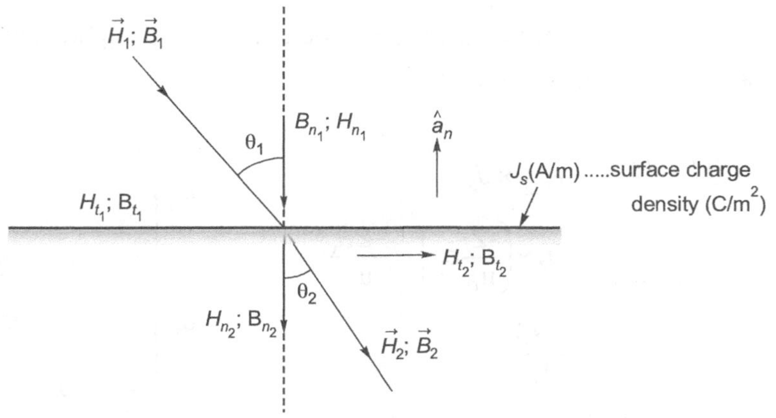

Tangential Component Relations:

$$H_{t_1} - H_{t_2} = J_s$$
$$\frac{B_{t_1}}{\mu_1} - \frac{B_{t_2}}{\mu_2} = J_s$$
$$\hat{a}_n \times (\vec{H}_1 - \vec{H}_2) = \vec{J}_s$$

where,
$$\left|\hat{a}_n \times \vec{H}_1\right| = H_{t_1}$$
$$\left|\hat{a}_n \times \vec{H}_2\right| = H_{t_2}$$

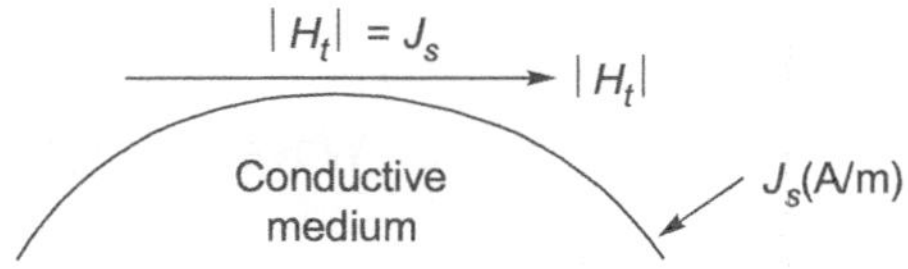

- The tangential component of magnetic field intensity $\vec{H}$ is discontinuous by a factor of J_s where J_s represents the surface current density on the common boundary separating two different magnetic media.
- When any of the two media is a conducting medium then the magnetic field inside a conducting medium is zero. Therefore, on any conducting surface only the tangential component of magnetic field will exist such that the magnitude of the tangential component of $\vec{H}$ field is always equal to the surface current density present on the conducting surface.

Normal Component Relation:

$$B_{n_1} = B_{n_2}$$
$$\mu_1 H_{n_1} = \mu_2 H_{n_2}$$
$$\hat{a}_n \cdot (\vec{B}_1 - \vec{B}_2) = 0$$

where,
$$\hat{a}_n \cdot \vec{B}_1 = B_{n_1}$$
$$\hat{a}_n \cdot \vec{B}_2 = B_{n_2}$$

- The normal component of magnetic flux density $\vec{B}$ is always continuous across a common boundary which separates two different magnetic media.

- The result is independent of the surface current density $\vec{J}_s$ present on the common boundary.
- Inside a conducting medium since the magnetic field does not exist, therefore the normal component of magnetic flux density $\vec{B}$ is zero across a common boundary separating two different media when one of them or both media are conducting media.

Numerical:

$$(H_{t_1} - H_{t_2}) = \hat{a}_{n_{12}} \times J_s$$

$$H_{t_1} - \frac{5}{\mu_o \times 4} = \hat{a}_z \times \left(\frac{2}{\mu_o}\hat{a}_y\right) = \frac{-2\hat{a}_x}{\mu_o}$$

$$H_{t_1} = \frac{-3\hat{a}_x}{4\mu_o}$$

$$B_n = \mu_o \mu_r H_n = -4.5\,\hat{a}_y$$

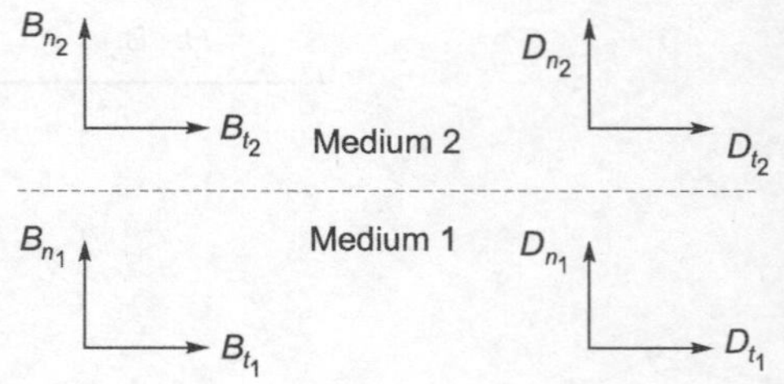

5.24 For a uniform plane wave in air, the magnetic field is given by: $H = i_x\, 2\exp\left\{j\left(\omega t - \frac{\pi}{20}z\right)\right\}$

Calculate:
(i) the wavelength
(ii) the frequency
(iii) the value of E at $t = \frac{1}{15}\,\mu$ sec, $z = 5$ m **[8 marks : 2014]**

Solution:

$$H = i_x\, 2\exp\left\{j\left(\omega t - \frac{\pi}{20}Z\right)\right\} = i_x \exp j(\omega t - \beta z)$$

(i) $$\text{Wavelength } (\lambda) = \frac{2\pi}{\beta}$$

where, $$\beta = \text{phase constant} = \left(\frac{\pi}{20}\right) \text{ rad/m}$$

$$\lambda = \frac{2\pi}{(\pi/20)} = 40 \text{ meters}$$

(ii) The frequency, $$(f) = \left(\frac{C}{\lambda}\right)$$

$$C = \text{Velocity in light} = 3 \times 10^8 \text{ m/sec.}$$

$$f = \left(\frac{C}{\lambda}\right) = \frac{3\times 10^8 \text{ m/sec}}{40} = \left(\frac{3}{4}\right)\times 10^7 \text{ Hz}$$

$$= 0.75 \times 10^7 \text{ Hz} = 7.5 \text{ MHz}$$

(iii) $$\frac{E_y}{H_x} = \sqrt{\frac{\mu_0}{\varepsilon_0}} = \eta$$

$\therefore$ E and H are transverse in nature,

$$\frac{E_y}{H_x} = \sqrt{\frac{4\pi\times 10^{-7}}{8.854\times 10^{-12}}} = 376.73$$

$$E_y = (376.73)\left(2\exp\left\{j\left(\omega t - \frac{\pi}{20}z\right)\right\}\right)$$

$$E = 753.46\exp\left(j\left(\omega t - \frac{\pi}{20}z\right)\right)\hat{i}_y$$

$$= 753.46\exp\left(j\left(2\times\pi\times 7.5\times 10^6 \times \frac{1}{15}\times 10^{-6} - \frac{\pi}{20}\times 5\right)\right)\hat{i}_y$$

$$= 753.46\exp\left(j\left(\pi - \frac{\pi}{4}\right)\right)\hat{i}_y$$

$$E = 753.46\exp\left(j\left(\frac{3\pi}{4}\right)\right)\hat{i}_y$$

$|E|$ at $t = \left(\frac{1}{15}\right)\mu$sec, $Z = 5$ m, 753.46 V/m

5.25 **What is the term σ/ωϵ called in the study of EM waves? Find general expression of attenuation constant a and relate with skin depth δ. Prove that the skin depth δ is independent of frequency when σ/ωϵ = 1 (poor conductor) and decreases with frequency when σ/ωϵ ?1 (good conductor).**

[10 marks : 2015]

Solution:

According Maxwell's equation,

$$\nabla\times\vec{H} = \vec{J_c} + \vec{J_d} = \sigma\vec{E} + \frac{\partial\vec{D}}{\partial t}$$

or,
$$\nabla\times\vec{H} = \sigma\vec{E} + j\omega\varepsilon\vec{E}$$

we can write;

$$\nabla\times\vec{H} = j\omega\varepsilon\vec{E}\left[1 - j\frac{\sigma}{\omega\varepsilon}\right] \quad ...(1)$$

or,
$$\nabla\times\vec{H} = j\omega\varepsilon'\vec{E}$$

where,
$$\varepsilon' = \varepsilon\left[1 - j\frac{\sigma}{\omega\varepsilon}\right]$$

This complex permittivity ε′ is having imaginary term $\left(\frac{\sigma}{\omega\varepsilon}\right)$ which is nothing but loss component. So we define, loss tangent $\boxed{\tan\delta = \frac{\sigma}{\omega\varepsilon}}$;

In general for a lossy medium the propagation constant 'γ' is defined by

$$\gamma = \alpha + j\beta, \qquad \text{where } \gamma = \sqrt{j\omega\mu(\sigma + j\omega\varepsilon)}$$

or,
$$\gamma^2 = j\omega\mu(\sigma + j\omega\varepsilon)$$

or,
$$-\text{Real part of }(\gamma^2) = \beta^2 - \alpha^2 = \omega^2\mu\varepsilon \quad ...(a)$$

and
$$|\gamma^2| = \beta^2 + \alpha^2 = \omega\mu\sqrt{\sigma^2 + \omega^2\varepsilon^2} \quad ...(b)$$

From equation (a) and (b), we obtain,

$$\alpha = \omega\sqrt{\frac{\mu\varepsilon}{2}\cdot\left[\sqrt{1+\left(\frac{\sigma}{\omega\varepsilon}\right)^2}-1\right]};$$

$$\beta = \omega\sqrt{\frac{\mu\varepsilon}{2}\left[\sqrt{1+\left(\frac{\sigma}{\omega\varepsilon}\right)^2}+1\right]}$$

"Skin depth 'δ' is a measure of the depth to which an EM wave can penetrate the medium."

Skin depth is given by: $\boxed{\delta = \frac{1}{\alpha}}$; i.e. skin depth is inverse of attenuation constant 'α'.

For the case of poor conductor or good dielectric;

$$\frac{\sigma}{\omega\,\varepsilon} << 1;$$

So the value of $\alpha = \omega\sqrt{\frac{\mu\varepsilon}{2}\left[\sqrt{1+0}-1\right]} = 0$

Hence, skin depth $\delta = \frac{1}{0} \to \infty.$

i.e., for poor conductors skin depth approaches to infinity. So there is no effect of frequency over it.

For the case of Good conductor or poor dielectric

$$\frac{\sigma}{\omega\,\varepsilon} >> 1;$$

$$\alpha = \omega\sqrt{\frac{\mu\varepsilon}{2}\left[\sqrt{\left(\frac{\sigma}{\omega\,\varepsilon}\right)^2}-1\right]} = \sqrt{\frac{\mu\sigma\omega}{2}} = \sqrt{\pi f\mu\sigma}$$

skin depth $\delta = \frac{1}{\sqrt{\pi f\mu\sigma}}$; therefore skin depth decreases with increase in frequency.

5.26 The reflection and transmission of plane electromagnetic waves at plane dielectric boundaries normal to the direction of propagation can be treated by using transmission line theory in which the electric and magnetic field strengths are represented by the voltage and current and the wave impedance is represented by the characteristic impedance. Use this method to estimate the attenuation of an electromagnetic wave at 2 GHz as it passes through a brick wall 120 mm thick assuming that the relative permittivity is 4.0 and that losses can be neglected.

[10 marks : 2016]

Solution:

Given that: Frequency = 2 GHz

Thickness of wall is 120 mm, relative permittivity of wall is 4.0.

The whole system can be treated as transmission line.

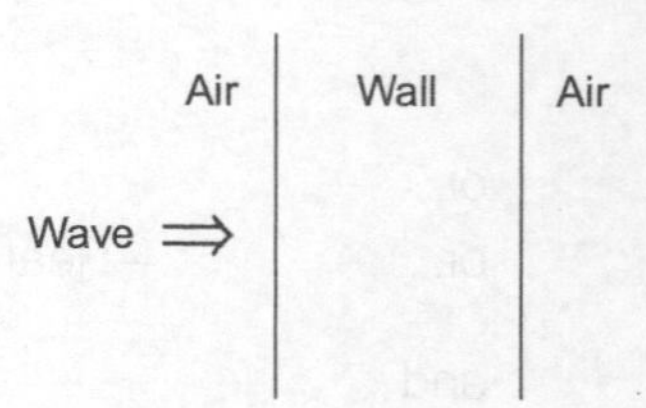

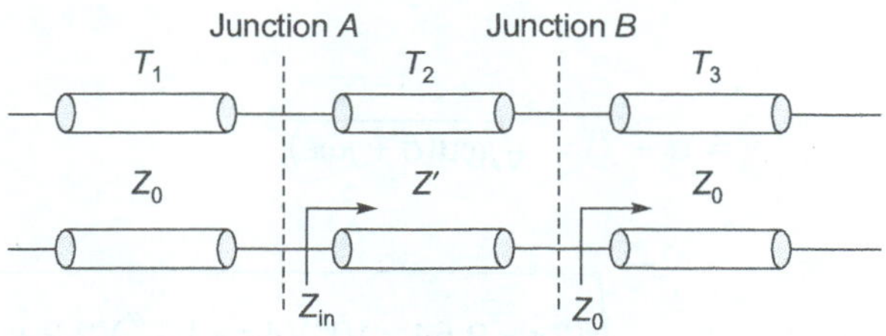

T_1 and T_3 are transmission line representing air, with characteristic impedance,

$$Z_0 = \eta_0$$

T_2 is transmission line representing brick wall with characteristic impedance,

$$Z' = \frac{\eta_0}{2}$$

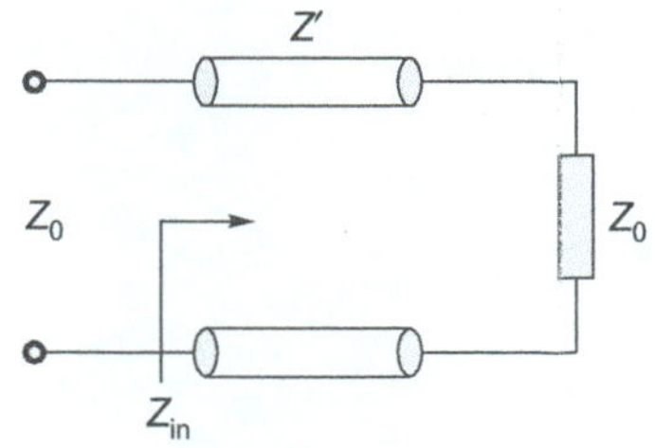

The thickness of brick wall is 120 mm and λ of wave is 75 mm.

The scenario can be represented as,

The wave will be reflected at junction *A* and *B*

$$Z_{in} = Z'\left(\frac{Z_L + jZ'\tan\beta L}{Z' + jZ_L\tan\beta L}\right)$$

Since, $$Z' = \frac{Z_0}{2}, \quad Z_L = Z_0$$

$$= \frac{Z_0}{2}\left[\frac{Z_0 + j\frac{Z_0}{2}\tan\left(\frac{2\pi}{75}\times 120\right)}{\frac{Z_0}{2} + jZ_0\tan\left(\frac{2\pi}{75}\times 120\right)}\right] = \frac{Z_0}{2}\left[\frac{1 + j0.363}{0.5 + j0.762}\right]$$

$$Z_{in} = 220\angle -36.77^\circ\ \Omega$$

So, reflection coefficient at junction *A*,

$$\frac{Z_{in} - Z_0}{Z_{in} + Z_0} = 0.422\angle -133.3^\circ$$

and reflection coefficient at junction *B*,

$$\left(\frac{Z_0 - Z_0/2}{Z_0 + Z_0/2}\right) = \frac{1}{3}$$

so, the wave that reach at load passes through reflection at junction *A* and *B*.

So, if power of input wave is *P* then power transmitted through junction *A* will be $P - P(\tau_{junctionA})^2$ and power reaching load will be $P(1-\tau_{junctionA}{}^2)\,(1-\tau_{junctionB}{}^2)$

$$P(1-0.422^2)\,(1-0.33^2) = 0.7185\,P$$

so, wave attenuate by 28.14%.

5.27 **The material parameters of a certain food item are given by σ = 2.2 s/m, $\epsilon = 48\epsilon_0$ and $\mu = \mu_0$ at the operating frequency, f = 2.54 GHz of a microwave oven. Determine α, β, γ, v_p and $\bar{\eta}$. Take $\epsilon_0 = \frac{10^{-9}}{36\pi}$ and $\mu_0 = 4\pi \times 10^{-7}$.**

[12 marks : 2022]

Solution:

Now for any material

$$\gamma = \alpha + j\beta = \sqrt{j\omega\mu(\sigma + j\omega\varepsilon)}$$

Given the values :

$$\gamma = \sqrt{j(2\pi \times 2.54 \times 10^9)(4\pi \times 10^{-7})(2.2 + j \times (2\pi \times 2.54 \times 10^9)(48)\left(\frac{10^{-9}}{36\pi}\right)}$$

$$\gamma = \sqrt{142825.15\angle 162.00}$$

$$\gamma = 377.922\angle 81° = 59.10 + j373.27 = \alpha + j\beta$$

So, $\alpha = 59.10$ Nep/m, $\beta = 373.27$ rad/m

Now, $$\lambda = \frac{2\pi}{\beta} = \frac{2\pi}{373.27} = 0.0168 \text{ m}$$

$$V_P = \frac{1}{\sqrt{\mu \in}} = \frac{1}{\sqrt{\mu_o \times 48\varepsilon_o}}$$

$$V_P = \frac{1}{\sqrt{4\pi \times 10^{-7} \times 48 \times \frac{10^{-9}}{36\pi}}} = 4.33 \times 10^7 \text{ m/s}$$

Now, for $\bar{\eta}$ $$= \left|\sqrt{\frac{j\omega\mu}{(\sigma + j\omega \in)}}\right|$$

$$\bar{\eta} = \left|\sqrt{\frac{200.55 \times 10^2 j}{(2.2 + 6.773)}}\right| = \left|\sqrt{2817.31\angle 18.00}\right|$$

$$\bar{\eta} = \sqrt{2817.31} = 53.08$$

■■■■

4 Electrical & Electronic Measurements

Revised Syllabus of ESE: *Principles of measurement, accuracy, precision and standards; Bridges and potentiometers; moving coil, moving iron, dynamometer and induction type instruments, measurement of voltage, current, power, energy and power factor, instrument transformers, digital voltmeters and multi-meters, phase, time and frequency measurement, Q-meters, oscilloscopes, potentiometric recorders, error analysis, Basics of sensors, Transducers, basics of data acquisition systems.*

---

1. Characteristics of Instruments and Measurement Systems

1.1 What is the dimension of a quantity? Derive dimensions of capacitance, resistance and inductance in electrostatic and electromagnetic systems. Find a relation between two systems of units.

[8 marks : 2001]

Solution:

Dimensions: Every quantity has a quality which distinguishes it from all other quantities. This unique quality is called Dimension. The dimension is written in a characteristic notation, as for example $[L]$ for length, $[T]$ for time etc.

Dimension in electrostatic system of units:

In electrostatic system of units

$$\text{Dimension of charge } [Q] = [\epsilon^{1/2} M^{1/2} L^{3/2} T^{-1}]$$

$$\text{emf } [E] = [\epsilon^{-1/2} M^{1/2} L^{1/2} T^{-1}]$$

and

$$[I] = [\epsilon^{1/2} M^{1/2} L^{3/2} T^{-2}]$$

Capacitance:

Capacitance, $$C = \frac{Q}{E}$$

Therefore, dimensions of capacitance are,

$$[C] = \frac{[Q]}{[E]} = \frac{[\epsilon^{1/2} M^{1/2} L^{3/2} T^{-1}]}{[\epsilon^{-1/2} M^{1/2} L^{1/2} T^{-1}]} = [\epsilon L]$$

Resistance:

Resistance, $$R = \frac{E}{I}$$

Therefore, dimensions of resistance are,

$$[R] = \frac{[E]}{[I]} = \frac{[\epsilon^{-1/2} M^{1/2} L^{1/2} T^{-1}]}{[\epsilon^{1/2} M^{1/2} L^{3/2} T^{-2}]} = [\epsilon^{-1} L^{-1} T]$$

Inductance:

Inductance, $$L = \frac{\text{emf}}{\text{rate of change of current}} = \frac{E}{dI/dt}$$

Therefore, dimensions of inductance are,

$$[L] = \frac{[E]}{[I]/[T]} = \frac{[E][T]}{[I]} = \frac{[\epsilon^{-1/2} M^{1/2} L^{1/2} T^{-1}][T]}{[\epsilon^{1/2} M^{1/2} L^{3/2} T^{-2}]}$$

$$= [\epsilon^{-1} L^{-1} T^2]$$

Dimensions in electromagnetic system:

In electromagnetic system,

Dimension of pole strength

$$[Q] = [\mu^{-1/2} M^{1/2} L^{1/2}]$$

$$[E] = [\mu^{1/2} M^{1/2} L^{3/2} T^{-2}]$$

$$[I] = [\mu^{-1/2} M^{1/2} L^{1/2} T^{-1}]$$

Capacitance:

Capacitance, $C = \dfrac{Q}{E}$

Therefore, dimensions of capacitance are

$$[C] = \frac{[Q]}{[E]} = \frac{[\mu^{-1/2} M^{1/2} L^{1/2}]}{[\mu^{1/2} M^{1/2} L^{3/2} T^{-2}]} = [\mu^{-1} L^{-1} T^2]$$

Resistance:

Resistance, $R = \dfrac{E}{I}$

Therefore, dimensions of resistance are,

$$[R] = \frac{[E]}{[I]} = \frac{[\mu^{1/2} M^{3/2} L^{1/2} T^{-2}]}{[\mu^{-1/2} M^{1/2} L^{1/2} T^{-1}]} = [\mu L T^{-1}]$$

Inductance:

Inductance, $L = \dfrac{\text{emf}}{\text{rate of change of current}} = \dfrac{E}{dI/dt}$

Therefore, dimensions of inductance are,

$$[L] = \frac{[E]}{[I]/[T]} = \frac{[E][T]}{[I]} = \frac{[\mu^{1/2} M^{1/2} L^{3/2} T^{-2}][T]}{[\mu^{-1/2} M^{1/2} L^{1/2} T^{-1}]} = [\mu L]$$

Relationship between Electrostatic and Electromagnetic System of units: We observe that two different dimensional equations are obtained for the same quantity in e.m. and e.s. systems of units. Any one quantity should have the same dimensions whatever may be the system.

The dimensions of charge in e.m.u. system are:

$$[Q] = [\mu^{-1/2} M^{1/2} L^{1/2}]$$

and in e.s.u. system are:

$$[Q] = [\epsilon^{1/2} M^{1/2} L^{3/2} T^{-1}]$$

Equating the above two dimensions,

$$[\mu^{-1/2} M^{1/2} L^{1/2}] = [\epsilon^{1/2} M^{1/2} L^{3/2} T^{-1}]$$

We have:

$$[\mu^{-1/2} \epsilon^{-1/2}] = [L T^{-1}]$$

Now dimensions $[L T^{-1}]$ are those of velocity,

∴ Equating the dimensional equations of other quantities, we always get the same result.

1.2 **Describe briefly the primary and secondary standards of mass and length.**

[4 marks : 2001]

Solution:

Primary Standard of Mass: The primary unit of mass is a prototype kilogramme kept at National Physical Laboratories of every country. This has an accuracy of 1 part in 10^8 and is occasionally verified against the standard kept at the International Bureau.

Secondary Standard of Mass: The secondary standards of mass are kept by industrial laboratories. These standards have an accuracy of 1 ppm and are checked against the primary standards.

International Units of Length: The international unit of length, the metre, was defined in 1960 in terms of an optical standard. This was the orange-red radiation of krypton atom. The internationally recognised krypton-86 discharge lamp when excited and observed under well defined conditions, emits orange light whose wavelength constitutes the basic unit of length. This can be measured to an accuracy of 1 ppm. The metre is equal to 1650763.73 wavelengths in vacuum of the orange-red light radiation of the krypton-86 atom in its transition between levels $2_{p_{10}}$ and 5_{d_5}. In terms of time the length travelled by light in vacuum in a time interval of 1/299792458 second.

Working Standards of Length: These are usually precision Gauge Blocks made up of steel. These blocks have two parallel surfaces and the distance between the two surfaces is specified. The accuracy tolerance of this distance is usually 0.5 – 0.25 micron. The most outstanding advantages gained by using of gauge blocks are their precision, low cost and accuracy which is 1 part per million (1 ppm).

1.3 **If $\mu = \dfrac{\pi r^4 (P_1 - P_2)}{8Ql}$, determine the dimensions of μ. (r and l are radius and length P_1, P_2 are pressures, Q is flow).**
If, $r = (0.5 \pm 0.01)$ mm, $P_1 = (200 \pm 3)$ kPa, $P_2 = (150 \pm 2)$ kPa, $Q = 4 \times 10^{-7}$ m³/sec. Calculate the absolute error in μ.

[12 marks : 2003]

Solution:

Dimensions of $r = [L]$; $Q = [L^3\,T^{-1}]$,

$$P_1 \text{ and } P_2 = \frac{[MLT^{-2}]}{[L^2]} = [ML^{-1}T^{-2}] = [M\,L^{-1}\,T^{-2}]$$

π and 8 are dimensionless. Hence,

$$\text{Dimension of } \mu = \frac{\pi\, r^4\,(P_1 - P_2)}{8Ql} = \frac{[L]^4[ML^{-1}T^{-2}]}{[L^3T^{-1}][L]} = [M\,L^{-1}\,T^{-1}]$$

Assuming per unit length, as the data of l is provided in the question,

$$\mu = \frac{\pi\, r^4\,(P_1 - P_2)}{8Q} = \frac{\pi \times (0.5 \times 10^{-3})^4\,(200 - 150)10^3}{8 \times 4 \times 10^{-7}} = 0.3068 \times 10^{-2}$$

On taking extreme positive values of r, P_1 and P_2,

$$\mu = \frac{\pi\, r^4\,(P_1 - P_2)}{8Q} = \frac{\pi \times (0.51 \times 10^{-3})^4\,(203 - 152)10^3}{4 \times 8 \times 10^{-7}}$$

$$= 0.3387 \times 10^{-2}$$

$$\text{Absolute error} = (0.3387 \times 10^{-2}) - (0.3068 \times 10^{-2})$$

$$= 0.0319 \times 10^{-2}$$

1.4 The following 10 observations were recorded when measuring a voltage:

1	2	3	4	5	6	7	8	9	10
41.7	42	41.8	42	42.1	41.9	42.5	42	41.9	41.8

Find:
(a) Mean,
(b) Standard deviation,
(c) Probable error of one reading.

[10 marks : 2003]

Solution:

(a) Mean of the Voltage:

$$\overline{X} = 41 + \frac{0.7+1+0.8+1+1.1+0.9+1.5+1+0.9+0.8}{10}$$

$$\overline{X} = 41 + \frac{9.7}{10}$$

$$\overline{X} = 41.97$$

(b) Standard Deviation:

X	$\overline{X}$	$d(X-\overline{X})$	d^2
41.7	41.97	−0.27	0.0729
42	41.97	+0.03	0.0009
41.8	41.97	−0.17	0.0289
42	41.97	+0.03	0.0009
42.1	41.97	+0.13	0.0169
41.9	41.97	−0.07	0.0049
42.5	41.97	+0.53	0.2809
42	41.97	+0.03	0.0009
41.9	41.97	−0.07	0.0049
41.8	41.97	−0.17	0.0289
			0.4410

$$\sigma = \sqrt{\frac{\sum_{i=1}^{10} d_i^2}{n-1}} = \sqrt{\frac{0.4410}{9}} = 0.2214 \text{ V}$$

(c) Probable Error:

$$\text{Probable error} = 0.6745\,\sigma = 0.2214 \times 0.6745$$
$$= 0.1493$$

1.5 A 100.00 V reference d.c. source is used for calibration of a digital multimeter in the 200 V range. The uncertainity quoted by a lab of higher echelon in national calibration network in the calibration certificate of reference standard is ±0.01 volt. The observations made during calibration are 100.2, 100.3, 100.2 and 100.1 volts respectively. Find the assigned value and uncertainity associated with measurements. (Assume a confidence level 95% and corresponding student factor, $t = 2.78$).

[10 marks : 2004]

Solution:

The four observations including uncertainty of ±0.01 volts is 100.2 ± 0.01, 100.3 ± 0.01, 100.2 ± 0.01 and 100.1 ± 0.01.

That is

100.21	100.19
100.31	100.29
100.21	100.19
100.11	100.09

$$\text{Mean} = 100 + \frac{0.84}{4} = 100.21 \text{ and } = 100 + \frac{0.76}{4} = 100.19$$

Assuming, $\text{mean } \overline{X} = 100.21$

X	$\overline{X}$	d	d^2
100.21	100.21	0	0
100.31	100.21	+0.10	0.01
100.21	100.21	0	0
100.11	100.21	–0.10	0.01

$$\sum_{n=1}^{4} d^2 = +0.01 + 0.01 = 0.02$$

Variance, $$V = \frac{\Sigma d^2}{n-1} = \frac{0.02}{3}$$

Standard deviation, $$s = \sqrt{V} = \sqrt{\frac{0.02}{3}} = 0.082$$

For a confidence level of 95%, student factor $t = 2.78$

True mean of the voltage being measured lies between the limits

$$= \overline{X} \pm 2.78\,s = 100.21 \pm 2.78 \times 0.082 = 100.21 \pm 0.228 \text{ V}$$

The assigned value is 100.21 V and the uncertainity limits lie between 99.982 V and 100.438 V.

1.6 What is meant by absolute measurement of electrical quantities? Describe the method of absolute measurement of ohm.

[10 marks : 2005]

Solution:

In order to establish a satisfactory system of measurements we must determine the absolute values of electrical quantities. Determination of absolute values involves the measurement of a unit in terms of physical quantities which are measured practically. Thus if we can determine the units of, say, current and emf by practical measurements, standards for all other electrical quantities like resistance etc can be set up in terms of current and emf.

Absolute Measurement of ohm: The resistance has the dimensions [μ LT^{-1}] in the e.m. system of units. Taking the permeability as dimensionless, the dimensions of resistance are [LT^{-1}] i.e. those of velocity. Thus the absolute measurement of resistance involves the measurement of length and time.

We describe here the Lorenz method of absolute measurement of resistance.

Figure shows the experimental set up used for this method. There is circular metallic disc which is mounted concentrically inside a solenoid. The disc is rotated at a uniform speed. A steady current is passed through the solenoid. R is a low resistance whose absolute value is to be determined. The resistance R is connected across the disc with the help of leads and two brushes. One brush presses the edge of the disc and the other brush makes a contact with the disc near its centre.

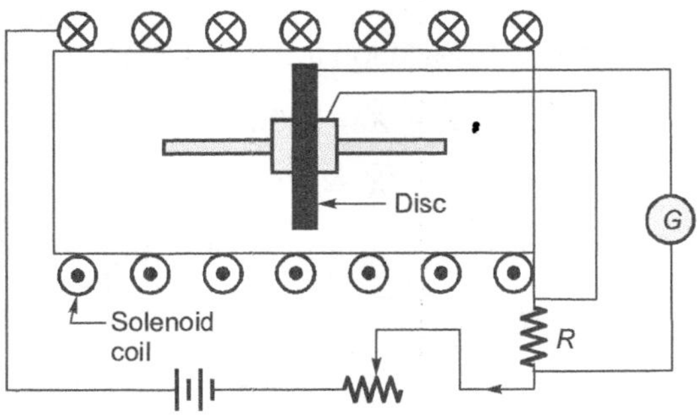

A sensitive galvanometer G is included in one of the leads. The magnetising current I for the solenoid is supplied by a battery. This current passes through resistance R also.

When we pass current through the solenoid, it produces a magnetic field. As the disc rotates an emf is induced in it. The current flowing through R producess a voltage drop across it. The emf generated in the disc is balanced against the voltage drop in resistance R, the balance being indicated by the galvanometer G. The speed of the disc adjusted till a balance is obtained.

Let, M = mutual inductance between disc and solenoid in henry

But flux linkages of disc = number of turns × flux linking the disc

= flux linking of the disc ($N = 1$)

Therefore, we conclude that the value of flux passing perpendicularly through the disc surface is MI weber if a current I ampere passes through the solenoid.

Let, n = speed of the disc in r.p.s. when the deflection of galvanometer is zero

∴ Emf induced in the disc = rate of change of flux linkage

$$= MI \times n$$

$$= MIn \text{ volt} \quad \ldots(i)$$

Voltage drop in resistance = IR volt,

where I and R are expressed in ampere and ohm respectively.

Thus when the deflection of the galvanometer is zero,

$$MIn = IR$$

or

$$R = Mn \text{ ohm} \quad \ldots(ii)$$

The value of mutual inductance M is obtained from the physical dimensions of the solenoid and the disc. Hence the value of R is determined solely in terms of linear dimensions of solenoid and disc, number of turns of solenoid and speed of rotation and is thus independent of other electrical units.

1.7 **During the measurement of a capacitor, following ten reading were obtained:**
1.002, 0.998, 1.005, 1.009, 0.995, 0.997, 1.004, 1.008, 1.003, 0.994 μF
Calculate:
(a) the arithmetic mean,
(b) the deviation from the mean,
(c) average deviation and
(d) standard deviation.

[12 marks : 2005]

Solution:

(a) The arithmetic mean of readings is,

$$\overline{X} = \frac{\Sigma x}{n}$$

$$\overline{X} = \frac{1.002 + 0.998 + 1.005 + 1.009 + 0.995 + 0.997 + 1.004 + 1.008 + 1.003 + 0.994}{10}$$

$$\overline{X} = 1.0015 \text{ μF}$$

(b) The deviation from the mean,

$$d_1 = x_1 - \overline{X} = 1.002 - 1.0015 = 0.0005 \text{ μF}$$

$$d_2 = x_2 - \overline{X} = = 0.998 - 1.0015 = -0.0035 \text{ μF}$$

$$d_3 = x_3 - \overline{X} = = 1.005 - 1.0015 = 0.0035 \text{ μF}$$

$$d_4 = x_4 - \overline{X} = = 1.009 - 1.0015 = 0.0075 \text{ μF}$$

$$d_5 = x_5 - \overline{X} = = 0.995 - 1.0015 = -0.0065 \text{ μF}$$

$$d_6 = x_6 - \overline{X} = = 0.997 - 1.0015 = -0.0045\ \mu F$$

$$d_7 = x_7 - \overline{X} = = 1.004 - 1.0015 = 0.0025\ \mu F$$

$$d_8 = x_8 - \overline{X} = = 1.003 - 1.0015 = 0.0015\ \mu F$$

$$d_9 = x_9 - \overline{X} = = 1.008 - 1.0015 = 0.0065\ \mu F$$

$$d_{10} = x_{10} - \overline{X} = = 0.994 - 1.0015 = -0.0075\ \mu F$$

(c) Average deviation is

$$\overline{D} = \frac{\Sigma|d|}{n}$$

$$\overline{D} = \frac{0.0005 + 0.0035 + 0.0035 + 0.0075 + 0.0065 + 0.0045 + 0.0025 + 0.0015 + 0.0065 + 0.0075}{10}$$

$$\overline{D} = 0.0044\ \mu F$$

(d) Since the number of readings is 10 which is less than 20, then standard deviation is,

$$s = \sqrt{\frac{\Sigma d^2}{n-1}}$$

$$s = \sqrt{\frac{(0.0005)^2 + (-0.0035)^2 + (0.0035)^2 + (0.0075)^2 + (-0.0065)^2 + (-0.0045)^2 + (0.0025)^2 + (0.0015)^2 + (0.0065)^2 + (-0.0075)^2}{(10-1)}} = 0.00527\ \mu F$$

1.8 The limiting errors for a four dial resistance box are:
Units : ±0.2%
Tens : ±0.1%
Hundreds : ±0.05%
Thousands : ±0.02%
If the resistance value is set at 4325 Ω calculate the limiting error for this value.

[10 marks : 2007]

Solution:

Thousand is set at 4000 Ω and error

$$= \pm 4000 \times \frac{0.02}{100} = \pm 0.8\ \Omega$$

$$\text{For hundred error} = \pm 300 \times \frac{0.05}{100} = \pm 0.15\ \Omega$$

Similarly,

$$\text{For ten error} = \pm 20 \times \frac{0.1}{100} = \pm 0.02\ \Omega$$

and

$$\text{For unit error} = \pm 5 \times \frac{0.2}{100} = \pm 0.01\ \Omega$$

Hence,

$$\text{Total error} = \pm(0.8 + 0.15 + 0.02 + 0.01)\ \Omega = \pm 0.98\ \Omega$$

$$\%\ \text{Relative error} = \frac{0.98}{4325} \times 100 = 0.0226\%$$

1.9 **Explain the advantages of electronic voltmeters over conventional type voltmeters. Describe the circuit diagram of a differential amplifier type electronic voltmeter using two FETs.**

[12 marks : 2007]

Solution:

Almost all electronic voltmeters make use of the rectifying properties of diodes whether vaccum tubes or metal rectifiers or semi-conductor diodes.

Advantage:

(a) **Detection of Low Level Signals:** Any measurement using a PMMC movement must draw a current 50 μA from the measured quantity for its operation for full scale deflection if conventional voltmeters are used. This would produce great loading effects especially in electronic and communication circuits. Electronic voltmeters avoid the loading errors by supplying power required for measurement by using external circuits like amplifiers.

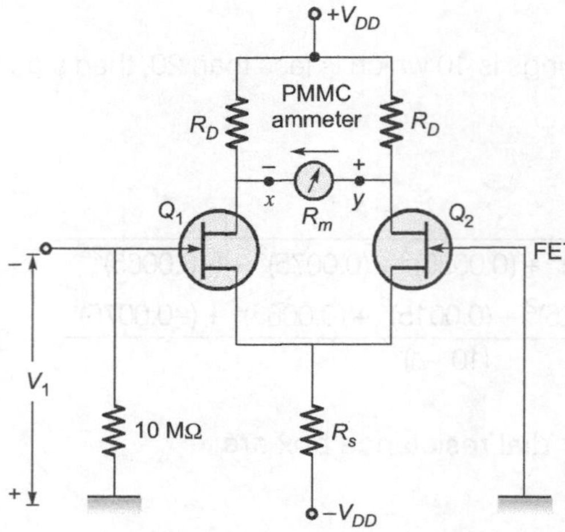

The amplifiers not only supply power for the operation but make it possible for low level signals, which produce a current less than 50 μA for full scale deflection, to be detected which otherwise cannot be detected in the absence of amplifiers.

Therefore, it is possible to measure currents below 50 μA, voltage below 10 mV and keep down the power drainage below 6.5 μW by using electronic voltmeters through use of amplifiers which is otherwise not possible with conventional types of meters using PMMC movement.

For the case of a.c. measurements, the use of an amplifier for detection of low level signals is ever more necessary for sensitive movements.

(b) **Low Power Consumption:** A conventional PMMC voltmeter is a rugged and accurate instrument, but it suffers from certain disadvantage. The principle problem is that it lacks both high sensitivity and high input resistance. It has a sensitivity of 20 kΩ/V with a 0 - 0.5 V range and has an input resistance of only 10 kΩ at its 0.5 V range with the result it has a full scale current of 50 μA which loads the measureand considerably.

In electronic and communication circuits as even this low value of current may not be tolerable on account of the fact that these circuits have very low operating currents.

The **Electronic Voltmeter (EVM)**, on the other hand, can have input resistances ranging from 10 MΩ to 100 MΩ with the input resistance remaining constant over all ranges instead of being different at different ranges, the EVM gives for less loading effects. Electronic voltmeters utilize the amplifying properties of vaccum tubes and transistors and therefore the power required for operating the deflecting element of PMMC instruments can be supplied from an auxiliary source.

Thus, while the circuit whose voltage is being measured controls the sensing element of the voltmeter, the power drawn from the circuit under measurement is very small or even negligible.

(c) **High Frequency Range:** The most important feature of electronic voltmeters is that their response can be made practically independent of frequency within extremely wide limits. Some electronic voltmeters permit the measurement of voltage from direct current to frequencies of the order of hundreds of MHz. The high frequency range may also be attributed to low input capacitance of most electronic devices. The capacitance may be of the order of a few pF.

Differential Amplifier type Electronic Voltmeter using Two FETs:

Field Effect Transistors (FETs) are used for increasing the input resistance of voltmeters. The use of FETs isolates the low resistance voltmeters from the circuit under test, thereby avoiding errors on account of loading effects.

The circuit shown consists of two FETs and these should be reasonably matched for current gain to ensure the thermal stability of the circuit. Therefore, an increase in the source current in one of FETs is offset by corresponding decrease in the source current of the other FETs.

The circuit is balanced when identical FETs are used so that for zero input no current flows through the ammeter. If a negative d.c. voltage is applied to the gate of left FET, a current would flow through the PMMC ammeter in the direction shown in figure.

The magnitude of the current depends upon the magnitude of input voltage. Through proper design of the circuit, the magnitude of current can be made directly proportional to the d.c. voltage across the input. Therefore, the meter can be made to read the input voltage directly in terms of volt.

1.10 What do you mean by dimensions of a quantity? The energy stored in a parallel-plate capacitor per unit volume is given by

$$W = K \in^a V^b d^c, \quad \text{where}$$

K = constant

$\in$ = permittivity of medium

V = Voltage between plates

d = distance between plates

Find the values of *a*, *b* and *c*.

[10 marks : 2008]

Solution:

Dimensions: Every quantity has a quality which distinguishes it from all other quantities. This unique quality is called Dimensions. The dimensions written in a characteristics notation.

For example: $[L]$ for length, $[T]$ for time etc.

Writing the dimensions of various quantities in LMTI system.

Energy density, $$W = \frac{\text{Energy}}{\text{Volume}}$$

$$[W] = \frac{[ML^2 T^{-2}]}{[L^3]} = [ML^{-1} T^{-2}]$$

Now, Force, $$F = \frac{Q_1 Q_2}{\in d^2}$$

$\therefore$ Permittivity, $$\in = \frac{Q_1 Q_2}{Fd^2}$$

or $$\in = \frac{[I^2 T^2]}{[MLT^{-2}][L^2]} = [M^{-1} L^{-3} I^2 T^4]$$

and $$\text{Voltage} = [V] = [I^{-1} ML^2 T^{-3}]; \; d = [L]$$

we get, $[ML^{-1}T^{-2}] = [I^2 M^{-1} L^{-3} T^4]^a [I^{-1} ML^2 T^{-3}]^b [L]^c$

$$= [I]^{2a-b} [M]^{-a+b} [L]^{-3a+2b+c} [T]^{4a-3b}$$

balancing the dimensions on bothe sides, we have,

$$0 = 2a - b$$

$$\Rightarrow \quad b = 2a \quad ...(i)$$

$$1 = -a + b \quad ...(ii)$$

$$-1 = -3a + 2b + c \quad ...(iii)$$

$$-2 = 4a - 3b \quad ...(iv)$$

from equation (*i*) and (*ii*),

$$1 = -a + 2a$$

$$\Rightarrow \quad a = 1 \,;\; b = 2$$

and

$$-1 = -3 + 4 + c$$

$$\Rightarrow \quad c = -2$$

From above; $a = 1$; $b = 2$ and $c = -2$

Hence, energy stored per unit volume is

$$\omega = K \in V^2 d^{-2} = \frac{K \in V^2}{d^2}$$

1.11 Define the terms 'Units', 'Absolute Units', 'Fundamental Units' and 'Derived Units'. Give suitable examples to support your answer.

[10 marks : 2008]

Solution:

Units: The result of a measurement of a physical quantity must be defined both in kind and magnitude. The standard measure of each kind of physical quantity is called a unit.

In general, $\text{Unit} = \dfrac{\text{Magnitude of physical quantity}}{\text{Numerical ratio}}$

Absolute Units: An absolute system of units is defined as a system in which the various units are all expressed in terms of a small number of fundamental units. Absolute measurements do not compare the measured quantity with arbitrary units of the same type but are made in terms of fundamental units.

Fundamental Units and Derived Units: The units for the fundamental dimensions are called the fundamental or base units. The fundamental unit are metre, kilogram, second, ampere, kelvin, candela and mole. The supplementary units are radian and steradian.

Derived Units: All other units which can be expressed in terms of fundamental units with the help of physical equations are called derived units. Every derived unit originate from some physical law an equation which defines that unit.

Example: Degree Celsius (°C), Hertz, Newton, Joule, Watt, Pascal, Coulomb, Volt, Farad, Ohm, Weber, etc.

1.12 Two sets of large number of 20 kΩ and 30 kΩ resistors are used to make a large number of 12 kΩ and 50 kΩ resistors choosing one from each group. If the standard deviations of the two sets of resistors of 20 kΩ and 30 kΩ are respectively 5% and 10%. Find the standard deviations of the combined resistor-sets of 12 kΩ and 50 kΩ?

[10 marks : 2010]

Solution:

Let, $R_1 = 20$ kΩ and $R_2 = 30$ kΩ

$$\sigma_{R_1} = \text{standard deviation of 20 k}\Omega\text{ resistor} = 20 \times \frac{5}{100}\text{k}\Omega = 1\text{k}\Omega$$

$$\sigma_{R_2} = \text{standard deviation of 30 k}\Omega\text{ resistor} = 30 \times \frac{10}{100}\text{ k}\Omega = 3\text{ k}\Omega$$

R_3 = 50 kΩ resistor is made by series combination of R_1 and R_2, so

$$R_3 = R_1 + R_2 = 20 + 30\text{ k}\Omega$$
$$= 50\text{ k}\Omega$$

Similarly, R_4 = 12 kΩ resistor is made by parallel combination of R_1 and R_2

so,
$$R_4 = \frac{R_1 R_2}{R_1 + R_2} = \frac{20 \times 30}{20 + 30} = 12\text{ k}\Omega$$

σ_{R_3} = standard deviation of R_3 and

σ_{R_4} = standard deviation of R_4

Standard deviation of R_3,

$$\sigma_{R_3} = \sqrt{\left(\frac{\partial R_3}{\partial R_1}\right)^2 \sigma_{R_1}^2 + \left(\frac{\partial R_3}{\partial R_2}\right)^2 \sigma_{R_2}^2}$$

where,
$$\frac{\partial R_3}{\partial R_1} = \frac{\partial (R_1 + R_2)}{\partial R_1} = 1$$

and
$$\frac{\partial R_3}{\partial R_2} = \frac{\partial (R_1 + R_2)}{\partial R_2} = 1$$

so,
$$\sigma_{R_3} = \sqrt{1^2 \times 1^2 + 1^2 \times 3^2}\text{ k}\Omega = \sqrt{10}\text{ k}\Omega = 3.16\text{ k}\Omega = \frac{3.16}{50} \times 100 = 6.32\%$$

Standard deviation of R_4,

$$\sigma_{R_4} = \sqrt{\left(\frac{\partial R_4}{\partial R_1}\right)^2 \sigma_{R_1}^2 + \left(\frac{\partial R_4}{\partial R_2}\right)^2 \sigma_{R_2}^2}$$

where,
$$\frac{\partial R_4}{\partial R_1} = \frac{\partial}{\partial R_1}\left(\frac{R_1 R_2}{R_1 + R_2}\right) = \frac{R_2}{R_1 + R_2} - \frac{R_1 R_2}{(R_1 + R_2)}$$

$$= \frac{R_2^2}{(R_1 + R_2)^2} = \left(\frac{R_2}{R_1 + R_2}\right)^2 = \left(\frac{30}{20 + 30}\right)^2 = 0.36$$

Similarly,
$$\frac{\partial R_4}{\partial R_2} = \frac{\partial}{\partial R_1}\left(\frac{R_1 R_2}{R_1 + R_2}\right) = \left(\frac{R_1}{R_1 + R_2}\right)^2 = \left(\frac{20}{20 + 30}\right)^2 = 0.16$$

$$\sigma_{R_4} = \sqrt{\left(\frac{\partial R_4}{\partial R_1}\right)^2 \sigma_{R_1}^2 + \left(\frac{\partial R_4}{\partial R_2}\right)^2 \sigma_{R_2}^2} = \sqrt{(0.36)^2 \times 1^2 + (0.16)^2}\ k\Omega$$

$$= \sqrt{0.36} = 0.6\text{ k}\Omega = \frac{0.6}{10} \times 100 = 6\%$$

1.13 The power factor of a circuit is determined by $\cos\phi = P/VI$ where P is the power in watt, V is the voltage in volt and I is the current in ampere. The relative error in power, current and voltage are respectively ±0.5%, ±1% and ±1%. Calculate the relative error in power factor. Also calculate the uncertainty in power factor if the errors were specified as uncertainties.

[10 marks : 2011]

Solution:

$$\cos\phi = \frac{P}{VI} = PV^{-1}I^{-1}$$

Relative error in measurement of P, V and I

$$\frac{\delta P}{P} = 0.5\%;\ \frac{\delta V}{V} = \pm 1.0\% \text{ and } \frac{\delta I}{I} = 1\%$$

Relative error in measurement of power factor ($\cos\phi$)

$$\frac{\delta \cos\phi}{\cos\phi} = \pm\left(\frac{\delta P}{P} + \frac{\delta V}{V} + \frac{\delta I}{I}\right) = (0.5 + 1 + 1)\% = \pm 2.5\%$$

Relative uncertainty in measurement of P, V and I

$$\frac{\omega_P}{P} = \pm 0.5\%,\ \frac{\omega_V}{V} = \pm 1\% \text{ and } \frac{\omega_I}{I} = \pm 1\%$$

Relative uncertainty in measurement of power factor ($\cos\phi$)

$$\frac{\omega_{\cos\phi}}{\cos\phi} = \sqrt{\left(\frac{\omega_P}{P}\right)^2 + \left(\frac{\omega_V}{V}\right)^2 + \left(\frac{\omega_I}{I}\right)^2} = \sqrt{(0.5)^2 + (1)^2 + (1)^2} = \pm 1.5\%$$

1.14 Explain, with a histogram, the normal distribution of errors. [4 marks : 2012]

Solution:

When a number of multisample observations are taken experimentally there is a scatter of data about central value. One method presenting test results in the form of a histogram. The normal or gaussian law of errors is the basis for the major part of study of random effects. The law of probability states the normal occurrence of deviations from average value of an infinite number of measurement or observations can be expressed by

$$y = \frac{h}{\sqrt{\pi}} \exp(-h^2 x^2)$$

Where x = Magnitude of deviation from mean

y = Number of at any deviation x, (the probability of occurrence of deviation x)

h = precision index

Interms of standard deviation σ

$$y = \frac{1}{\sigma\sqrt{2\pi}} \exp\left(-\frac{x^2}{2\sigma^2}\right)$$

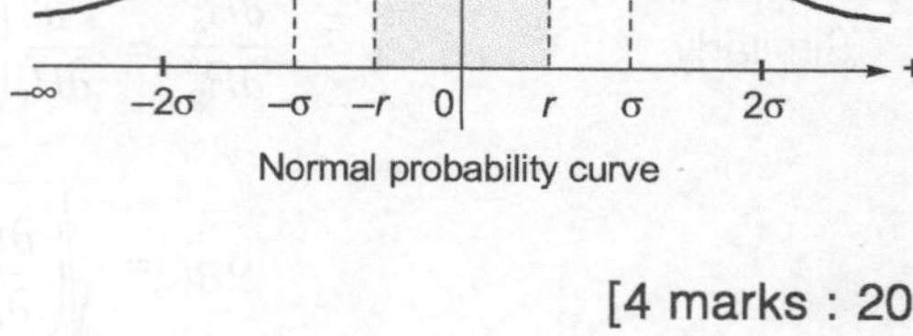

Normal probability curve

1.15 What is probable error and what is its significance? [4 marks : 2012]

Solution:

The probable error of a quantity is a value describing the probability distribution of that quantity. It defines the half-range of an interval about a central point for the distribution, such that half of the values from the distribution lie within the interval and half outside. Thus it is equivalent to half the interquartile range. The probable error (r) is a fixed multiple of the standard deviation σ.

$$r = 0.6745\,\sigma$$

Where the multiplying factor derives from the normal distribution.

1.16 **Explain units, system of units, standard and standard for absolute ampere.**

[4 marks : 2012]

Solution:

Unit: The result of a measurement of a physical quantity must be defined both in kind and magnitude. The standard measure of each kind of physical quantity is called a unit. Measurement implies comparison with a standard value.

A system of measurement is a set of units which can be used to specify anything which can be measured and were historically important, regulated and defined because of trade and internal commerce.

A ***standard*** is a physical representation of a unit of measurement. The term standard is applied to a piece of equipment having a known measure of physical quantity.

The international system of units (S.I.) defines the ampere as the constant current which, if maintained in two straight parallel conductors of infinite length and negligible circular cross-section placed 1 m apart in a vaccum, will produce between these conductor a force equal to 2×10^{-7} N/m.

1.17 **What are normal mode and common mode signals? How are they reduced?**

[10 marks : 2012]

Solution:

A normal-mode signal is any type (other than common mode) that appears between a pair of wires or on a single wire referenced to (or returned through) the earth, chasis or shield. Normal mode signals are read between two wire in a balanced or unbalanced transmission path.

For a balanced 2-wire path, one wire is driven positive while the other is driven negative by an equal amount, both with respect to a static or no-signal condition in which both lines assume the same voltage level relative to circuit common.

A common mode signal appears equally (with respect to local circuit common) on both lines of a 2-wire cable not connected to earth, shield, or local common. Usually, not always, this is an unwanted signal that should be rejected by the receiving circuit. Common mode voltage V_{cm} is expressed mathematically as the average of the two signal voltages with respect to local ground or common.

$$V_{cm} = \frac{\vec{V}_A + \vec{V}_B}{2}$$

Common mode signal (V_{cm}) must be rejected in the receiving circuit that rejection is easily accomplished when the receiving circuit is passive, transformer coupled etc.

1.18 **A circuit was tuned for resonance by eight different students, and the values of resonant frequency in kHz were recorded as 532, 548, 543, 535, 546, 531, 543 and 536.**

Calculate:

(i) the arithmetic mean, **(ii) deviations from mean,**

(iii) the average deviation, **(iv) the standard deviation, and**

(v) variance

[10 marks : 2013]

Solution:

(i)

$$\text{Mean} = \frac{d_1 + d_2 + d_3 + d_4 + d_5 + d_6 + d_7 + d_8}{8}$$

$$= \frac{532 + 548 + 543 + 535 + 546 + 531 + 543 + 536}{8} = 539.25 \text{ kHz}$$

(ii)

$$d_1 = -7.25$$

$$d_2 = 8.75$$

$$d_3 = 3.75$$
$$d_4 = -4.25$$
$$d_5 = 6.75$$
$$d_6 = -8.25$$
$$d_7 = 3.75$$
$$d_8 = -3.25$$

(iii) Mean deviation $= \dfrac{7.25 + 8.75 + 3.75 + 4.25 + 6.75 + 8.25 + 3.75 + 3.25}{8} = 5.75$

(iv) $$SD = \sqrt{\frac{(d_1)^2 + (d_2)^2 + (d_3)^2 + (d_4)^2 + (d_5)^2 + (d_6)^2 + (d_7)^2 + (d_8)^2}{n-1}}$$

$$= \sqrt{\frac{(7.25)^2 + (8.75)^2 + (3.75)^2 + (4.25) + (6.75)^2 + (8.25)^2 + (3.75)^2 + (3.25)^2}{7}} = 6.54$$

(v) Variance $= SD^2 = 42.785$

1.19 The following readings were observed when measuring a voltage.

S.No.	1	2	3	4	5	6	7	8
Volts	532	548	543	535	546	531	543	536

Calculate:
(i) Average deviation
(ii) Standard deviation
(iii) Probable error of one reading. **[10 marks : 2015]**

Solution:

S.No.	1	2	3	4	5	6	7	8
Volts	532	548	543	535	546	531	543	536

$$\text{Average} = \frac{532 + 548 + 543 + 535 + 546 + 531 + 543 + 536}{8} = 539.25$$

Deviations,

$$d_1 = 532 - 539.25 = -7.25$$
$$d_2 = 548 - 539.25 = 8.75$$
$$d_3 = 543 - 539.25 = 3.75$$
$$d_4 = 535 - 539.25 = -4.25$$
$$d_5 = 546 - 539.25 = 6.75$$
$$d_6 = 531 - 539.25 = -8.25$$
$$d_7 = 543 - 539.25 = 3.75$$
$$d_8 = 536 - 539.25 = -3.25$$

(i) Average deviation,

$$\bar{d}_{avg.} = \frac{|d_1| + |d_2| + |d_3| + |d_4| + |d_5| + |d_6| + |d_7| + |d_8|}{8} = 5.75$$

(ii) Standard deviation,

$$\sigma = \sqrt{\frac{(d_1)^2 + (d_2)^2 + (d_3)^2 + (d_4)^2 + (d_5)^2 + (d_6)^2 + (d_7)^2 + (d_8)^2}{n-1}} = 6.541$$

(iii) Probable error $= 0.6745\sigma = 4.4119$

1.20 **When 100 volts is applied across a pure resistance, it draws a current of 5 amperes. The uncertainties in measurements of voltage and current are ±0.2 volts and ±0.06 A respectively. Calculate the power dissipated in resistance and uncertainty in Power in% when Power *P* is calculated as *P* = *VI*.**

[12 marks : 2017]

Solution:

Given that: Applied voltage, $V = 100\text{ V}$

Current, $I = 5\text{ A}$

Power, $P = VI$

$= 100 \times 5 = 500\text{ W}$

Now, $P = VI$

$$\frac{\partial P}{\partial V} = I = 5$$

and

$$\frac{\partial P}{\partial I} = V = 100$$

$$W_V = 0.2 \text{ and } W_I = 0.06$$

$$\text{Uncertainty in power} = \sqrt{\left(\frac{\partial P}{\partial V}\right)^2 W_V^2 + \left(\frac{\partial P}{\partial I}\right)^2 \cdot W_I^2}$$

$$= \sqrt{(5)^2 \times (0.2)^2 + (100)^2 \times (0.06)^2}$$

$$= \sqrt{1+36} = \sqrt{37} = \pm 6.083\text{ W}$$

Power dissipated in resistance, $P = (500 \pm 6.083)\text{ W}$

$$\text{Uncertainty in power in \%} = \pm\frac{6.083}{500} \times 100 = \pm 1.2166\%$$

1.21 **The limiting errors for a four-dial resistance box are :**

Units : ± 0.2% Hundreds : ± 0.05%

Tens : ± 0.1% Thousands : ± 0.02%

If the resistance value is set at 3525 Ω, calculate the limiting error in the resistance value.

[5 marks : 2023]

Solution:

Given that Limiting errors for a four-dial resistance box :

Units : ± 0.2% Hundreds : ± 0.05%

Tens : ± 0.1% Thousands : ± 0.02%

Given resistance value : 3525 Ω

$$3525\ \Omega \Rightarrow \quad 3000 \pm 0.02\% \Rightarrow 3000 \pm 0.6$$

$$500 \pm 0.05\% \Rightarrow 500 \pm 0.25$$

$$20 \pm 0.1\% \Rightarrow 20 \pm 0.02$$

$$5 \pm 0.2\% \Rightarrow 5 \pm 0.01$$

$$3525 \pm 0.088 \Rightarrow \text{Error Value}$$

$$\Rightarrow \quad 3525 \pm 0.88 \Rightarrow \text{Error Value}$$

$$\text{\% Limiting errors} = \frac{0.88}{3525} \times 100 = 0.024\%$$

$$\Rightarrow \quad 3525 \pm 0.024\%$$

$$\text{Limiting errors} = \pm 0.024\%$$

2. Measurement of Resistance and Potentiometers

2.1 Resistors R_1 and R_2 have the nominal value 3 Ω and 6 Ω and tolerance of ±10% and ±5% respectively. What would be the tolerance in the equivalent resistance value when R_1 and R_2 are connected in parallel?

[12 marks : 2002]

Solution:

When, $R_1 = 3\ \Omega$ and $R_2 = 6\ \Omega$

Equivalent resistance in parallel,

$$R_{eq} = \frac{R_1 R_2}{R_1 + R_2} = \frac{3 \times 6}{3+6} = 2\ \Omega$$

Value of R_1 will be between 2.7 Ω to 3.3 Ω

So, taking maximum value of R_1 = 3.3 Ω

Similarly, R_2 will be between 5.7 Ω to 6.3 Ω

So, taking maximum value of R_2 = 6.3 Ω

$$R_{eq} = \frac{R_1 R_2}{R_1 + R_2} = \frac{6.3 \times 3.3}{6.3+3.3} = 2.165\ \Omega$$

$$\text{tolerance in equivalent circuit} = \pm\frac{2.165-2}{2} \times 100 = \pm 8.25\%$$

2.2 For measurement by ammeter and voltmeter method, what should be the accuracy of the instruments so that the overall measurement error is minimum?

[10 marks : 2006]

Solution:

Ammeter Voltmeter Method: The two types of connections employed for ammeter-voltmeter method are shown in figure (a) and (b). In both the cases, if readings of ammeter and voltmeter are taken, then the measured value of resistance is given by:

$$R_m = \frac{\text{Voltmeter reading}}{\text{Ammeter reading}} = \frac{V}{I}$$

The measured value of resistance R_m, would be equal to the true value, R, if the ammeter resistance is zero and the voltmeter resistance is infinite, so that the conditions in the circuit are not distrubed. However in practice this is not possible and hence both the methods give inaccurate results.

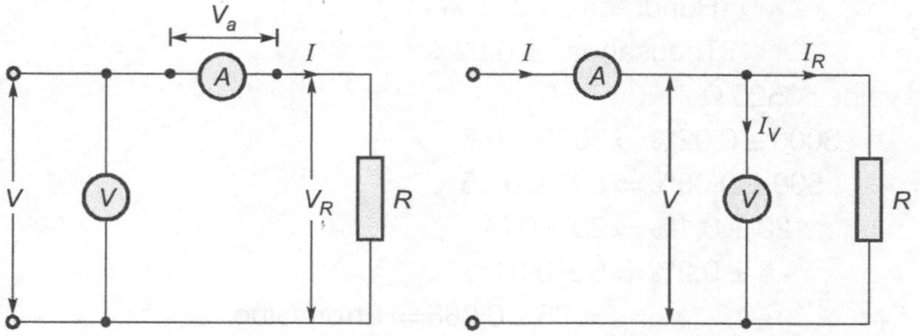

(a) (b)

Consider Circuit of Figure (a): In this circuit the ammeter measures the true value of the current through the resistance but the voltmeter does not measure the true voltage across the resistance. The voltmeter indicates the sum of the voltages across the ammeter and the measured resistance. Let R_a be the resistance of the ammeter.

∴ Voltage across the ammeter, $V_a = I R_a$

Now, measured value of resistance,

$$R_{m_1} = \frac{V}{I} = \frac{V_R + V_a}{I} = \frac{IR + IR_a}{I} = R + R_a \quad ...(i)$$

∴ True value of resistance,

$$R = R_{m_1} - R_a \quad ...(ii)$$

$$= R_{m_1}\left(1 - \frac{R_a}{R_{m_1}}\right) \quad ...(iii)$$

Thus the measured value of resistance is higher than the true value. It is also clear from above that the true value is equal to the measured only if the ammeter resistance, R_a is zero.

Relative error,
$$\in_r = \frac{R_{m_1} - R}{R} = \frac{R_a}{R} \quad ...(iv)$$

It is clear from equation (v) that the error in measurements would be small if the value of resistance under measurement is large as compared to the internal resistance of the ammeter. Therefore, the circuit of figure (a) should be used when measuring high resistance values.

Consider the Circuit of Figure (b): In this circuit the voltmeter measures the true value of voltage but the ammeter measures the sum of currents through the resistance and the voltmeter.

Let R_V be the resistance of the voltmeter.

∴ Current through the voltmeter,

$$I_V = \frac{V}{R_V}$$

Now, measured value of resistance,

$$R_{m_2} = \frac{V}{I} = \frac{V}{I_R + I_V} = \frac{V}{\frac{V}{R} + \frac{V}{R_V}} = \frac{R}{1 + \frac{R}{R_V}} \quad ...(v)$$

∴ True value of resistance,

$$R = \frac{R_{m_2} R_V}{R_V - R_{m_2}} = R_{m_2}\left(\frac{1}{1 - \frac{R_{m_2}}{R_V}}\right) \quad ...(vi)$$

From equation (vi) it is clear that the true value of resistance is equal to the measured value only if the resistance of voltmeter, R_V, is infinite. However, if the resistance of voltmeter is very large as compared to the resistance under measurement:

or
$$R_V >> R_{m2},$$

and therefore $\frac{R_{m_2}}{R_V}$ is very small.

or
$$R = R_{m_2} + \left(1 + \frac{R_{m_2}}{R_V}\right) \quad ...(vii)$$

Thus the measured value of resistance is smaller than the true value.

Relative error,
$$\in_r = \frac{R_{m_2} - R}{R} = \frac{R_{m_2}^2}{R_V R} \quad ...(viii)$$

The value of R_{m_2} is approximately equal to R,

$$\therefore \quad \epsilon_r = -\frac{R}{R_V} \qquad \ldots(ix)$$

It is clear from equation (*ix*) that the error in measurements would be small if the value of resistance under measurement is very small as compared to the resistance of the voltmeter. Hence, the circuit of figure (b) should be used when measuring low resistance values.

2.3 Describe the construction of a resistance standard of 1 Ω value. Discuss the techniques used to minimize the errors in this.

[10 marks : 2007]

Solution:

Figure shows a primary resistor of 1 Ω value. The standard resistor is coil of wire made of manganin mounted on a silk insulated brass spool. After wound resistance coils are impregnated with a shellac varnish and given an accelerated ageing by baking at a temperature of 150°C for about 48 hours. Ageing makes the coil resistance stable with time. The coil is supported from the hard rubber top by means of a thermometer tube in which a thermometer can be inserted from outside. The copper lead wires, which are silver soldered to the ends of resistance coil, are soft soldered to copper binding posts. The hard rubber top is screwed into the metal container which is filled with a good quality light mineral oil. The binding posts, thermometer well, and the threads by which the hard rubber top is connected to the container are all sealed with shellac which is not soluble in oil.

Techniques used to minimize the errors:

1. Due to skin effect when they are used at high frequencies. These error are minimized by using wires of small diameter.
2. Due to stray inductance and capacitances. The stray inductance can be reduced by arranging the adjacent wires to carry current in the opposite direction so as to reduce the resultant magnetic field. The stray capacitance can be reduced by arranging the adjacent wires to be widely spaced and to have minimal potential difference between them.

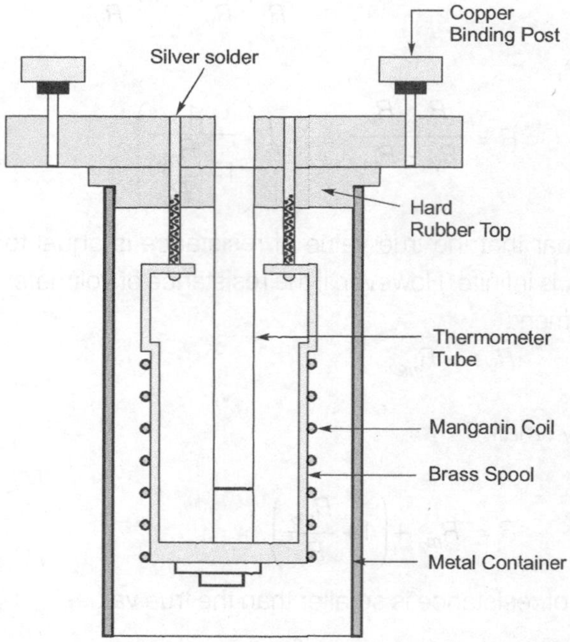

3. Due to contact resistances of the terminals in case of low value standards. The contact resistance is reduced by using a four terminal construction, in which current is passed through the two outer terminals.

2.4 **What are the different problems associated with measurement of low resistances? Explain the principle of working of Kelvin's Double Bridge and explain how the effect of contact resistance of leads is eliminated.**

[10 marks : 2007]

Solution:

Problem associated with the measurement of low resistances: Resistances having a value under 1 Ω is known as low resistances. The resistance of leads and contacts, though small, are appreciable in comparison in the case of low resistances.

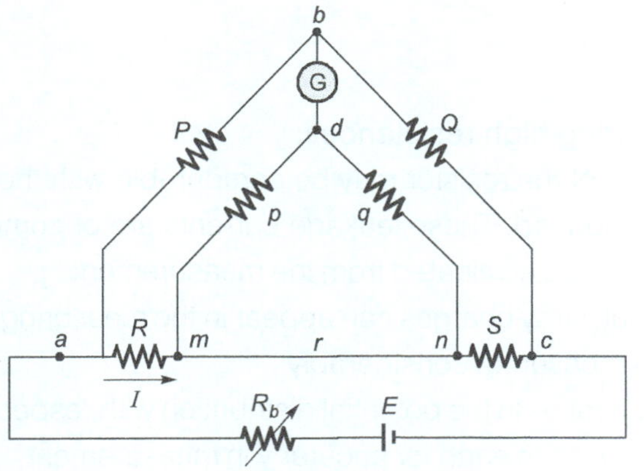

Kelvin's Double Bridge :

The Kelvin's Double bridge incorporates the idea of a second set of ratio arms – hence the name double bridge and the use of four terminal resistors for the low resistance arms. The first of ratio arms is P and Q. The second set of ratio arms p and q is used to connect the galvanometer to a point d at the appropriate potential between points m and n to eliminate the effect of connecting lead of resistance r between the known resistance R, and the standard resistance S.

The ratio p/q is made equal to P/Q. Under balance conditions there is no current through the galvanometer, which means that the voltage drop between a and b, E_{ab} is equal to the voltage drop E_{amd} between a and c.

Now,
$$E_{ab} = \frac{P}{P+Q}\cdot E_{ac}$$

and
$$E_{ac} = I\left[R+S+\frac{(p+q)r}{p+q+r}\right] \quad ...(i)$$

and
$$E_{amd} = I\left[R+\frac{p}{p+q}\left\{\frac{(p+q)r}{p+q+r}\right\}\right]$$

$$= I\left[R+\frac{pr}{p+q+r}\right] \quad ...(ii)$$

For zero galvanometer deflection,

or
$$\frac{P}{P+Q}\cdot I\left[R+S+\frac{(p+q)r}{p+q+r}\right] = I\left[R+\frac{pr}{p+q+r}\right]$$

or
$$R = \frac{P}{Q}\cdot s+\frac{qr}{p+q+r}\left[\frac{P}{Q}-\frac{p}{q}\right] \quad ...(iii)$$

Now if $\frac{P}{Q}=\frac{p}{q}$ equation (*iii*) becomes

$$R = \frac{P}{Q}\cdot S$$

as the expression of R does not involve r hence the effect of r (contact resistance of leads) is eliminated.

2.5 **Draw the circuit of a kelvin double ratio arm bridge and show how the effect of lead resistance is eliminated.**

[12 marks : 2009]

Solution:

(Refer Question No. 2.4).

2.6 **Discuss the difficulties faced in measuring high resistances. Explain the procedures adopted in the (laboratory for measuring high resistance).**

[12 marks : 2010]

Solution:

Difficulties faced in measuring high resistances:

- The insulation resistance of the resistor may be comparable with the actual value of the resistor. Thus leakage currents are produced. These leakage currents are of comparable magnitude to the current being measured and must be eliminated from the measurement.
- Due to electrostatic effect, stray charges can appear in the measuring circuit causing errors. Alternating field can also affect the measuring considerably.
- In order to obtain definite ratios in the potential distribution with respect to surroundings, one point of the circuit may be connected to the earth for accuracy in measurement.
- In measurement of insulation resistance, the specimen often has considered capacitance. On application of a direct voltage a large charging current flows initially which gradually decays down after a short interval.
- Fairly high voltages are used in toss in order to raise the currents to reasonable values in order to be measured. So normally a sensitive galvanometer or micro-ammeter is required and adequate steps have to be taken to prevent damage to these delicate instruments.

Measurement of High Resistance:

The direct deflection method is used for measuring insulation resistance of a cable.

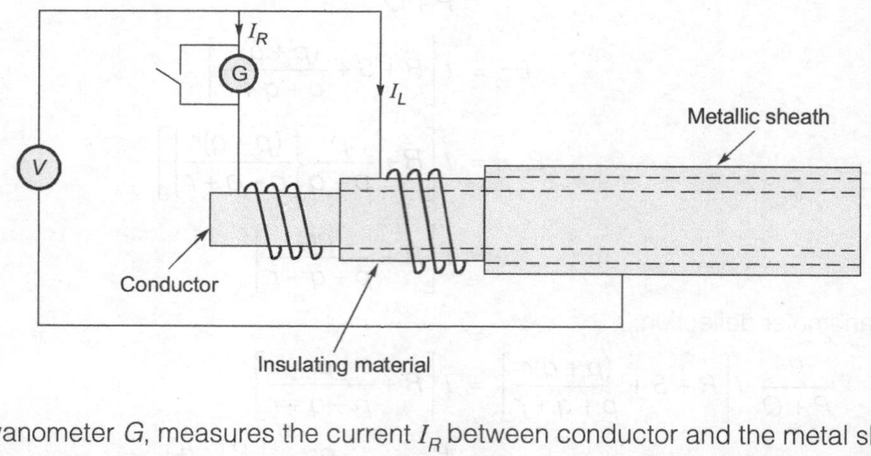

- The galvanometer G, measures the current I_R between conductor and the metal sheath. The leakage current I_L, over the insulating material is carried by the guard wire wound on the insulation and therefore does not flow through the galvanometer.
- Cables without metal sheaths can be tested in a similar way if the cable, except the end or ends on which connections are made, as immersed in water in a tank. The water and the tank then form the return path for the current. The cable is immersed is slightly water for about 24 hours and the temperature is kept constant and then the measurement is taken.

  Insulation resistance of the cable $R = V/I_R$

2.7 **Explain with the diagram the four-point probe method to measure the conductivity of semiconductor.** **[10 marks : 2013]**

Solution:

Four Probe Method: The 4-point probe setup consists of four equally spaced tungsten metal tips with finite radius. Each tip is supported by springs on the ends to minimize sample damage during probing. The four metal tips are part of an auto mechanical stage which travels up and down during measurements. A high impedance current source is used to supply current through the outer two probes, a voltmeter measures the voltage across the inner two probes in order to determine the sample resistivity and conductivity.

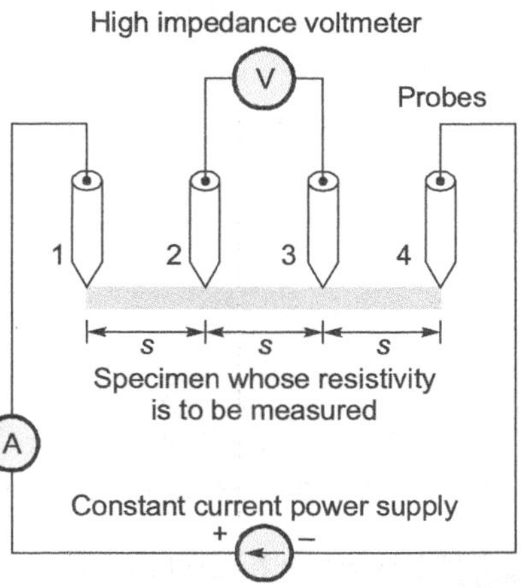

These inner probes draw no current because of the high input impedance voltmeter in the circuit. Thus unwanted voltage drop (IR drop) at Point-B and Point-C caused by contact resistance between probes and the sample is eliminated from the potential measurements.

The electric current carried through the two outer probes, sets up an electric field in the sample. In the figure shown below, the electric field lines are drawn solid and the equipotential lines are drawn broken. The two inner probes measure the potential difference between Point-B and Point-C.

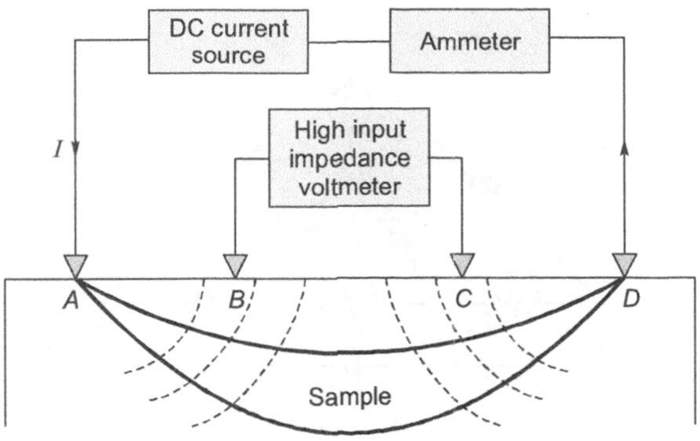

In order to use four probe method in semiconductor material it is necessary to assume that:

1. The resistivity of the material is uniform in the area of measurement.
2. If there is minority carrier injection into the semiconductor by the current carrying electrodes most of the carriers recombine near the electrodes, so that their effect on the conductivity is negligible.
3. The surface on which the probes rest is flat with no surface leakage.
4. The diameter of the contact between the metallic probes and the semiconductor should be small compared to the distance between probes.

Conductivity for Intrinsic Semiconductor: The total electrical conductivity is the sum of the conductivities of the valence and conduction band carriers, which are holes and electrons, respectively.

It can be expressed as

$$\sigma = e(n_e \mu_e + n_h \mu_h) \quad \text{...}(i)$$

where, n_e, μ_e are the electron's concentration and mobility. Similarly, n_h, μ_h are the hole's concentration and mobility, respectively.

The mobility is a quantity that directly relates the drift velocity v_d of charge carriers to the applied electric field E across the material, i.e.

$$\mu = \frac{v_d}{E} \quad \text{...}(ii)$$

In the intrinsic region the number of electrons is equal to the number of holes, $n_e = n_h = n_i$ so equation (*i*) implies that

$$\sigma = n_i e(\mu_e + \mu_h) \quad ...(iii)$$

Temperature Dependance of Conductivity: The electron density (electron/volume) in the conduction band is obtained by integrating from the bottom to top of the conduction band. The detailed calculation reveal that

$$n_i = NT^{3/2} \exp\left(-\frac{E_g}{2kT}\right) \quad ...(iv)$$

where N is some constant, substituting (*iv*) and (*iii*),

$$\sigma = Ne(\mu_e + \mu_h)T^{3/2} \exp\left(-\frac{E_g}{2kT}\right) \quad ...(v)$$

equation (*i*) shows that the electrical conductivity of intrinsic semiconductor depends on temperature and decreases exponentially with decrease in temperature.

2.8 Draw the circuit of a Kelvin's Double-Bridge used to measure low resistance. Derive the condition for balance. Calculate insulation resistance of a cable in which the voltage falls from 100 to 80 V in 20 s. The capacitance is 300 pF.

[10 marks : 2013]

Solution:

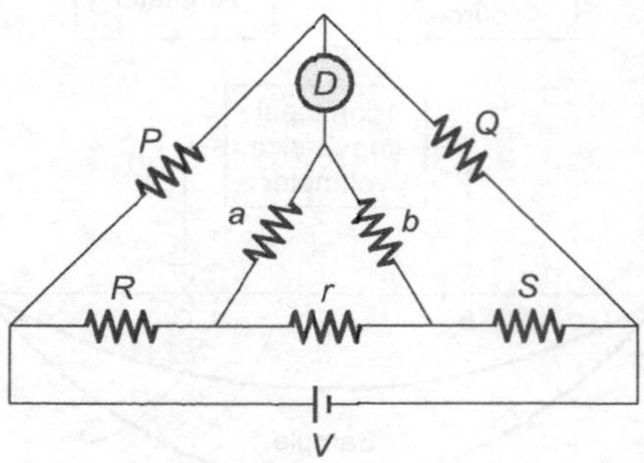

Using star-delta conversion:

$$X_1 = \frac{ar}{a+b+r}$$

$$X_2 = \frac{br}{a+b+r}$$

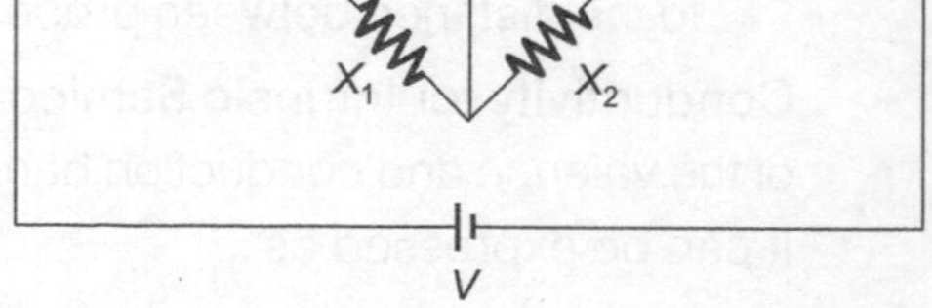

Using,

$$Z_1 Z_4 = Z_2 Z_3$$

$$P\left(S + \frac{br}{a+b+r}\right) = Q\left(R + \frac{ar}{a+b+r}\right)$$

$$\frac{PS}{Q} + \frac{P}{Q}\frac{br}{a+b+r} = \frac{ar}{a+b+r} + R$$

$$\boxed{R = \frac{PS}{Q} + \frac{br}{a(a+b+r)}\left(\frac{P}{Q} - \frac{a}{b}\right)}$$ Condition of balance for Kelvin's double bridge.

Numerical:

Given,

$$V = 100\text{ V}$$
$$C = 300\text{ pF}$$

$$t = 20 \text{ sec.}$$
$$V_c = 80 \text{ V}$$
$$R = \frac{0.4343t}{C\log_{10}(V/V_c)} = \frac{0.4343 \times 20}{300 \times 10^{-12}\log_{10}\left(\frac{100}{80}\right)} = 30.48 \times 10^4 \text{ M}\Omega$$

2.9 **Why and how are volt-ratio boxes utilized along with d.c. potentiometers? How should the value of the volt-ratio box resistance be chosen?**

[12 marks : 2020]

Solution:

Volt-Ratio Box: General purpose potentiometers cover a low voltage range. (approximately 0–1.6 V in case of American Instruments and 0–1.8 V in case of British Instruments). If higher voltage ranges have to be measured a precision potential divider, called a volt-ratio box, is used. The volt-ratio box provides multiple ranges as shown in figure.

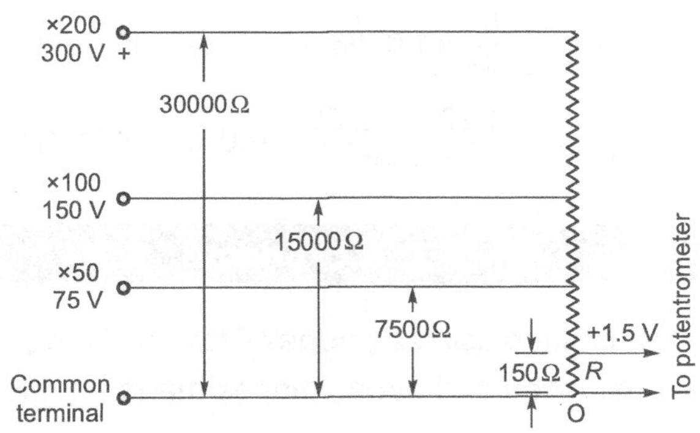

The voltage to be measured is connected to the appropriate binding post. Sometimes the volt-ratio box is provided with selector switch for voltage ranges. Supposing the voltage to be measured is connected between common terminal and 300 V switch. If the reading of the potentiometer is 0.826, then the value of unknown voltage is 0.826 × (300/1.5) = 165.2 V.

It was stated earlier that the potentiometer consumes no power from the source under balance conditions. However, when a volt-ratio box is used, this statement is no longer true. An examination of figure above indicates that a current must flow through resistance R, in order to produce a voltage drop so that a voltage measurement be made by the potentiometer in order that the unknown voltage be determined. Therefore, the resistance of the volt-ratio box represents a load on the unknown voltage source and thus consumes power. The value of volt-ratio box resistance should be as high as possible in order that the power consumption is low. However, low values of resistance are preferred since they have a greater degree of stability. Also low values of resistance give greater galvanometer sensitivity and also minimize the effects of high resistance leakage paths around the terminals.

Thus the choice of volt-ratio box resistance involves a compromise between low and high values. Volt-ratio boxes are usually constructed with resistance of 100 Ω/V or 200 Ω/V of the nominal voltage range, allowing a maximum current of 10 mA or 5 mA respectively from the measured source. Figure shows a volt-ratio box with a resistance of 100 Ω/V and a maximum current of 10 mA.

2.10 **A Standard Cell of 1.0186 volts used with a conventional slide wire potentiometer balances at 45 cm. Calculate :**

(*a*) the emf of the cell which balances at 65 cm and

(*b*) the percentage error in voltmeter which balances at 57 cm when reading 1.27 volts.

[10 marks : 2022]

Solution:

Given, $E_1 = 1.0186$ V and $l_1 = 45$ cm

(*a*) Let the emf of cell is E_2 which balances at $l_2 = 65$ cm,
As we know, from the potentiometer relation,

$$E \propto l$$

Therefore,
$$\frac{E_2}{E_1} = \frac{l_2}{l_1}$$

$$E_2 = \frac{65}{45} \times 1.0186 = 1.4713 \text{ Volt}$$

Therefore emf of cell is 1.4713 volt.

(*b*) Again let the cell emf is E_3 which balances at $l_3 = 57$ cm

Therefore,
$$E_3 = \frac{l_3}{l_1} \times E_1$$

$$E_3 = \frac{57}{45} \times 1.0186 = 1.2902 \text{ Volt}$$

Percentage error in voltmeter $= \dfrac{1.27 - 1.2902}{1.2902} \times 100 = -1.5677\%$

3. A.C. Bridges

3.1 **How can frequency be determined using a bridge? Draw this bridge and derive condition for balance. Why and how are two resistances and capacitances made equal?**

[12 marks : 2001]

Solution:

Measurement of Frequency Using Bridge: Some bridges have balance equations which involve frequency directly even if the performance of individual bridge elements is independent of frequency. These bridges may be used for determination of frequency in terms of values of various bridge elements.

Wien's Bridge: The Wien's bridge is primarily used as a frequency determining bridge.

Figure shows the Wien's Bridge under balance conditions.

At balance,

$$\left(\frac{R_1}{1 + j\omega C_1 R_1}\right) R_4 = \left(R_2 - \frac{j}{\omega C_2}\right) R_3$$

or
$$\frac{R_4}{R_3} = \frac{R_2}{R_1} + \frac{C_1}{C_2} + j\left(\omega C_1 R_2 - \frac{1}{\omega C_2 R_1}\right)$$

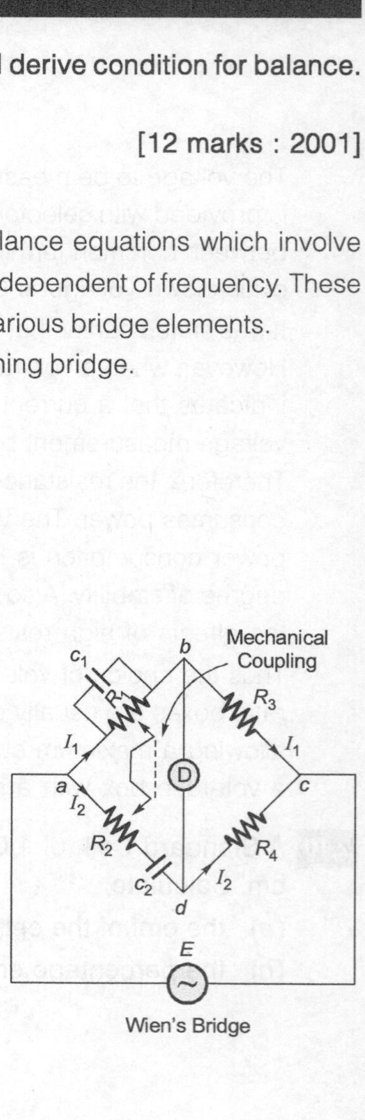

Wien's Bridge

Equating the real and imaginary parts,

or
$$\frac{R_4}{R_3} = \frac{R_2}{R_1} + \frac{C_1}{C_2} \quad \ldots (i)$$

and $\omega C_1 R_2 - \dfrac{1}{\omega C_2 R_1} = 0$ from which

$$\omega = \frac{1}{\sqrt{R_1 R_2 C_1 C_2}}$$

and frequency,
$$f = \frac{1}{2\pi\sqrt{R_1 R_2 C_1 C_2}} \text{ Hz} \quad \ldots (ii)$$

In most Wien bridges, the components are so chosen that,

$$R_1 = R_2 = R$$

and $$C_1 = C_2 = C$$

Then equation (*i*) reduces to $R_4/R_3 = 2$ and equation (*ii*) reduces to

$$f = \frac{1}{2\pi RC} \quad \ldots(iii)$$

Switches for resistors R_1 and R_2 are mechanically linked so as to fulfil the condition $R_1 = R_2$. As long as C_1 and C_2 are fixed capacitors equal in value and $R_4 = 2R_3$, the wien's bridge may be used as a frequency determining device, balanced by a single control.

3.2 **In the circuit shown below when $I = 0$, show that,**

$$\frac{\omega L_1}{R_1} = \frac{1}{\omega C_4 R_4}$$

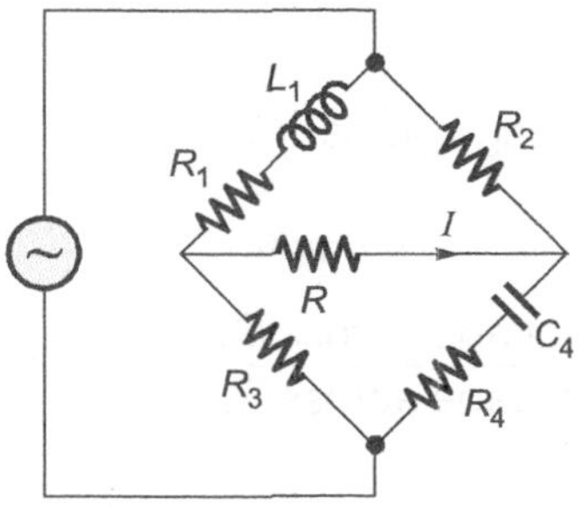

[10 marks : 2002]

Solution:

In a bridge circuit the magnitudes $Z_1 Z_4 = Z_2 Z_3$ and phase angles.

$$\angle\theta_1 + \angle\theta_4 = \angle\theta_3 + \angle\theta_2 \text{ should be equal.}$$

Let, $$Z_1 = R_i + j\omega L, \quad Z_2 = R_2$$

$$Z_3 = R_3; \quad Z_4 = R_4 + \frac{1}{j\omega C_4}$$

Taking phase angles,

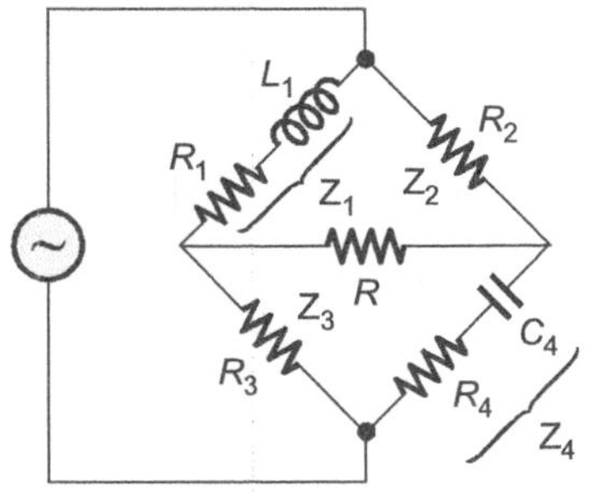

$$\tan\angle\theta_1 = \frac{j\omega L_1}{R_1}; \ \theta_2 = 0$$

$$\tan\angle\theta_4 = \frac{1}{j\omega C_4 R_4}; \ \theta_3 = 0$$

$$\frac{j\omega L_1}{R_1} + \left(-\frac{j}{\omega C_4 R_4}\right) = 0$$

or $$\frac{\omega L_1}{R_1} = \frac{1}{\omega C_4 R_4}$$

that is for bridge balance, the inductive phase angle must equal the capacitive phase angle.

3.3 **In a Hay's bridge, the four arms are $R_1 - L_1$, R_2, $R_3 - C_3$, R_4 connected in clockwise order, show that, under the 'phase null' condition, Q of the coil is given by**

$$Q = \frac{\text{Voltage across } C_3}{\text{Voltage across } R_3}$$

[12 marks : 2003]

Solution:

At balance the phase angles should be equal

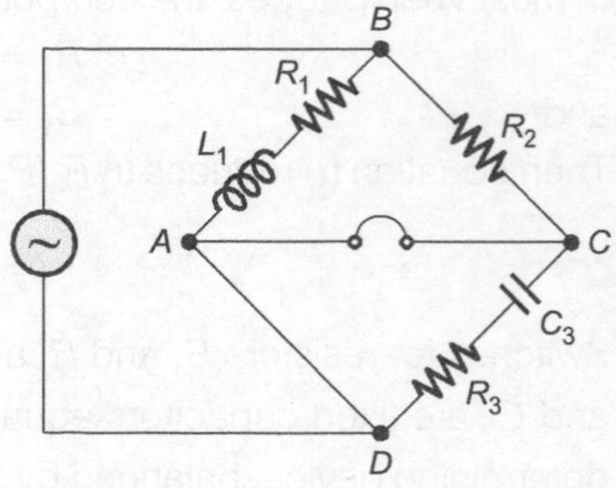

$$\therefore \qquad \tan\theta_L = \frac{X_{L_1}}{R_1} = \frac{\omega L_1}{R_1} = Q$$

$$\tan\theta_C = \frac{X_{C_3}}{R_3} = \frac{1}{\omega C_3 R_3}$$

$$\tan\theta_L = \tan\theta_C$$

Under phase null condition,

$$\text{or} \qquad Q_X = \frac{1}{\omega C_3 R_3}$$

The current I passing through arm CD is the same for both C_3 and R_3.

$$\therefore \qquad Q_X = I \times \frac{1}{\omega C_3} \times \frac{1}{I R_3}$$

$$I \times \frac{1}{\omega C_3} = \text{Voltage across } C_3$$

$$I R_3 = \text{Voltage across } R_3$$

$$\therefore \qquad Q_X = \frac{\text{Voltage across } C_3}{\text{Voltage across } R_3}$$

3.4 **In a test on a Bakelite sample at 20 kV, 50 Hz by schering bridge, having a standard capacitor of 106 pF, balance was obtained with a capacitance of 0.35 μF in parallel with a non-inductive resistance of 318 ohms, the non-inductive resistance in the remaining arm of the bridge is 130 ohms. Calculate the power factor and equivalent series resistance of the capacitor. Derive also the balance condition of the bridge.**

[16 marks : 2004]

Solution:

For derivation of the balance condition of the bridge. At balance,

$$\left(R_x + \frac{1}{j\omega C_x}\right)\left(\frac{R_1}{1 + j\omega C_1 R_1}\right) = \frac{1}{j\omega C_s} \cdot R_2$$

$$\text{or} \qquad \left(R_x + \frac{1}{j\omega C_x}\right) R_1 = \frac{R_2}{j\omega C_s}(1 + j\omega C_1 R_1)$$

$$\text{or} \qquad R_1 R_x - \frac{jR_1}{\omega C_x} = -j\frac{R_2}{\omega C_s} + \frac{C_1 R_1 R_2}{C_s}$$

Equating the real and imaginary terms, $R_x = \dfrac{C_1}{C_s} \cdot R_2$ $\qquad \left(C_x = \dfrac{C_s}{R_2} \cdot R_1\right)$

Series resistance of the capacitor,

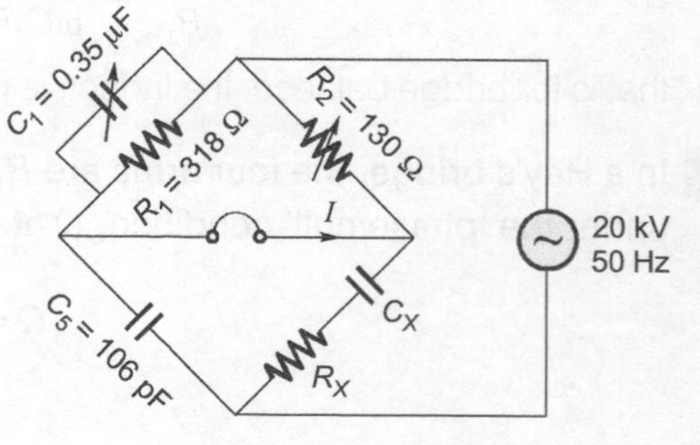

$$R_x = \frac{C_1}{C_s} R_2 = \frac{0.35 \times 10^{-6}}{106 \times 10^{-12}} \times 130$$

$$= 0.0033 \times 130$$

$$= 0.429 \times 10^6\ \Omega$$

$$= 429\ \text{k}\Omega$$

Power factor of capacitance,

$$\cos\theta_x = \frac{R_x}{|Z_x|} \text{ where}$$

$$|Z_x| = \sqrt{R_x^2 + \left(\frac{1}{\omega C_x}\right)^2}$$

as $$\left(\frac{1}{\omega C_x}\right)^2 >> R_x$$

$$|Z_x| = \frac{1}{\omega C_x}$$

$$\text{Power factor} = \frac{R_x}{Z_x} = R_x\,\omega\, C_x$$

$$C_x = \frac{R_1}{R_2} C_s$$

$$\text{pf} = \omega \times \frac{C_1}{C_s} \times R_2 \times \frac{R_1}{R_2} \times C_s = \omega\, C_1\, R_1$$

$$= 2\pi \times 50 \times 0.35 \times 10^{-6} \times 318$$

$$= 34965.9 \times 10^{-6} = 34966 \times 10^{-6} = 0.035$$

3.5 **A balanced bridge has the following arm parameters:**

Arm $AB - R_1$ = 1500 Ω in parallel with
C_1 = 0.03 μF
Arm $BC - R_2$ = 1876 Ω in series with
C_2 = 0.03 μF
Arm CD – unknown
Arm $DA - C_3$ = 0.5 μF

If an a.c. voltage of frequency 1 kHz is applied across the junction BD, find the parameters of arm CD.

[12 marks : 2005]

Solution:

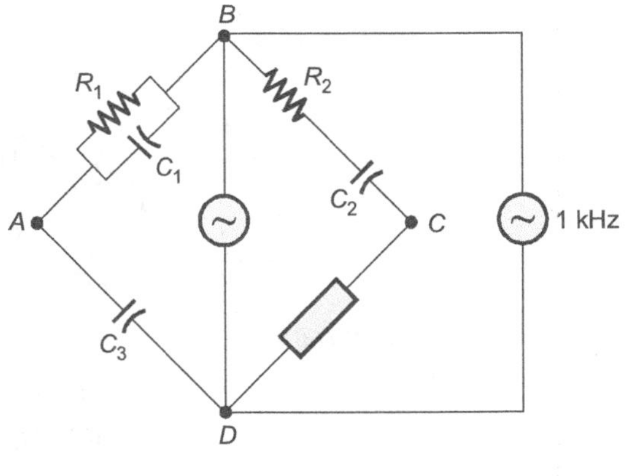

$$Z_x = \frac{Z_2 Z_3}{Z_1} = \left(R_2 + \frac{1}{j\omega C_2}\right) \times \left(\frac{1}{j\omega C_3}\right) \times \left(\frac{1}{R_1} + j\omega C_1\right)$$

$$= \left(\frac{R_2}{j\omega C_3} + \frac{1}{j^2\omega^2 C_2 C_3}\right)\left(\frac{1}{R_1} + j\omega C_1\right)$$

$$= \frac{-jR_2}{\omega C_3 R_1} + \frac{R_2 C_1}{C_3} - \frac{1}{\omega^2 C_2 C_3 R_4} - \frac{jC_1}{\omega C_2 C_3}$$

$$= \left\{\frac{R_2 C_1}{C_3} - \frac{1}{\omega^2 C_2 C_3 R_1}\right\} - j\left\{\frac{R_2}{\omega C_3 R_1} + \frac{C_1}{\omega C_2 C_3}\right\}$$

$$= \left\{\frac{\omega^2 C_1 C_2 R_1 R_2 - 1}{\omega^2 R_1 C_2 C_3}\right\} - j\left\{\frac{R_2 C_2 + R_1 C_1}{\omega R_1 C_2 C_3}\right\}$$

On putting values of the parameters,

$$Z_x = 112.56 - j716$$

Negative value of imaginary part indicates, there is capacitor.

So, $$Z_x = R_x + \frac{1}{j\omega C_x} = 112.56 - j716$$

$$R_x = 112.56\ \Omega$$

and $$\frac{1}{\omega C_x} = 716$$

$$\Rightarrow \quad C_x = \frac{1}{(2\times\pi\times1000\times716)} = 0.222\ \mu\text{F}$$

3.6 A capacitor bushing forms the arm *AB* of a high voltage schering bridge, a standard capacitor of 500 mmF and negligible loss forms the arms *AD*. Arm *BC* consists of a non-inductive resistance of 300 Ω. The bridge is balanced with the arm *CD* having a resistance of 72.6 Ω and capacitance of 0.148 μF in parallel. 50 Hz supply is connected between *A* and *C*. Calculate from first principles the capacitance and dielectric loss angle of bushing.

[12 marks : 2006]

Solution:

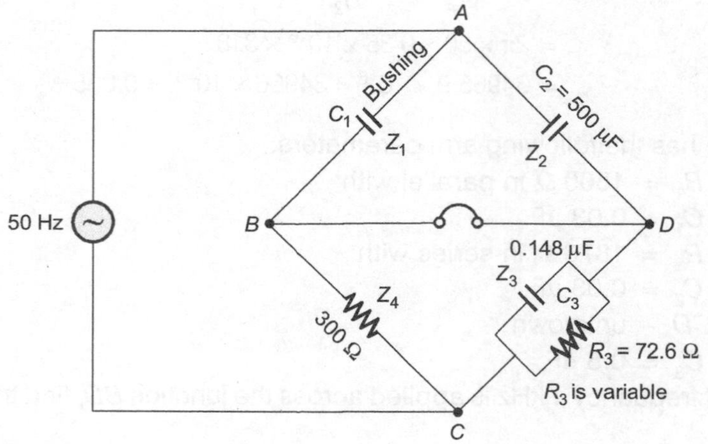

$$Z_1 Z_3 = Z_2 Z_4$$

$$Z_1 = Z_2 Z_4 Y_3$$

$$\frac{1}{j\omega C_1} = \frac{1}{j\omega C_2}\times R_4 \times\left(\frac{1}{R_3} + j\omega C_3\right)$$

$$-\frac{j}{\omega C_1} = -\frac{j}{\omega C_1}\times\frac{R_4}{R_3} + \frac{\omega C_3 R_4}{\omega C_2} \qquad \ldots(i)$$

Equating imaginary parts,

$$-\frac{j}{\omega C_1} = -\frac{1}{\omega C_2}\frac{R_4}{R_3}$$

$$C_1 = C_2\cdot\frac{R_3}{R_4} = 500\times10^{-12}\times\frac{72.6}{300} = 121\times10^{-12}\ \text{F} = 121\ \text{pF}$$

Let the loss of bushing be represented as R_1 in series with C_1.

The dielectric loss angle,

$$\tan\delta = \omega C_1 R_1 = \omega\times\left(C_2\frac{R_3}{R_4}\right)\times R_1 \qquad \ldots(ii)$$

Equation (i) is modified, due to loss in capacitance,

$$R_1 - \frac{j}{\omega C_1} = -\frac{j}{\omega C_2} \times \frac{R_4}{R_3} + \frac{C_3 R_4}{C_2}$$

Equating real parts, $R_1 = R_4 \dfrac{C_3}{C_2}$

Putting in equation(ii),

loss angle, $\tan\delta = \omega C_2 \cdot \dfrac{C_3}{C_2} \times R_4 \times \dfrac{C_3}{C_2} = \omega R_3 C_3$

$$= 2\pi \times 50 \times 72.6 \times 0.148 \times 10^{-6}$$
$$= 3375.58 \times 10^{-6} = 0.00337558$$
$$\delta = 0.19^\circ$$

3.7 **A bridge circuit for measurement of effective resistance and inductance of an iron-cored coil consists of**
Arm *AB*, the unknown impedance
Arm *BC*, a pure resistance R_3
Arm *CD*, a loss less capacitor C_4
Arm *DA*, a capacitor C_2 in series with a resistance R_2.
Under balanced condition,
$R_3 = 10\ \Omega$, $R_2 = 842\ \Omega$, $C_2 = 0.135\ \mu F$ and $C_4 = 1\ \mu F$.
Calculate the value of effective resistance and self-inductance at a supply frequency of 100 Hz. Derive the equations of balance and draw the phasor diagram under balanced condition.
[10 marks : 2008]

Solution:

Let R_1 and L_1 be the effective resistance and inductance of the specimen respectively,

At balance, $Z_1 Z_4 = Z_2 Z_3$

$$(R_1 + j\omega L_1)\frac{1}{j\omega C_4} = R_3\left(R_2 + \frac{1}{j\omega C_2}\right)$$

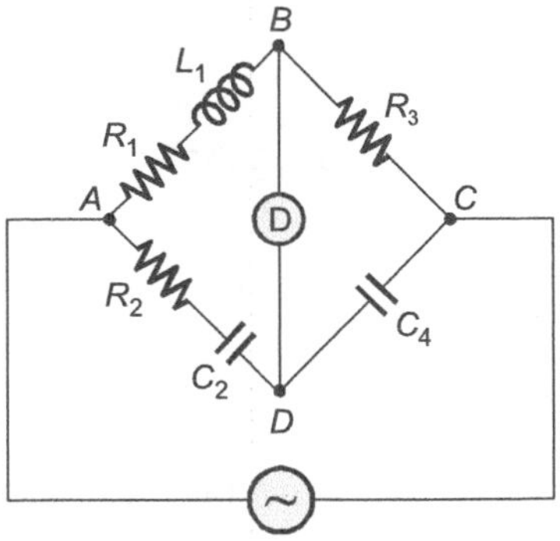

Separating and comparing real and imaginary parts,

$$L_1 = R_2 R_3 C_4$$
$$= 842 \times 10 \times 1 \times 10^{-6}\ \text{H} = 8.42\ \text{mH}$$

and $R_1 = R_3 \dfrac{C_4}{C_2} = 10 \times \dfrac{1 \times 10^{-6}}{0.135 \times 10^{-6}}\ \Omega$

$$R_1 = 74.07\ \Omega$$

Phasor diagram,

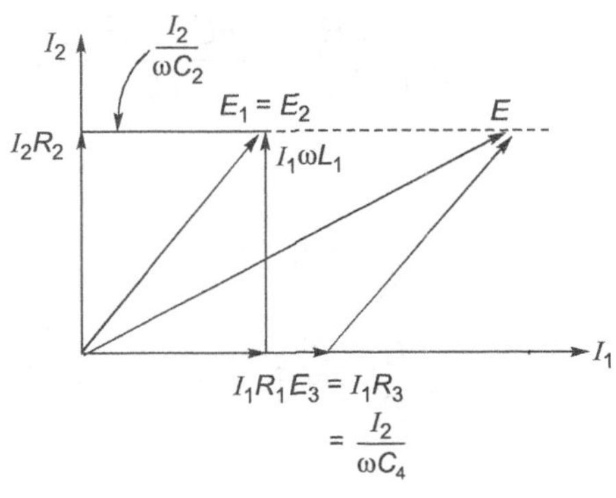

3.8 **Draw the circuit of Anderson bridge. Derive the null conditions. Show that Maxwell bridge is the special case of this bridge.**

[12 marks : 2009]

Solution:

Anderson's Bridge: This bridge, in fact, is a modification of the Maxwell's inductance-capacitance bridge. In this method, the self-inductance is measured in terms of a standard capacitor. This method is applicable for precise measurement of self-inductance over a very wide range of values.

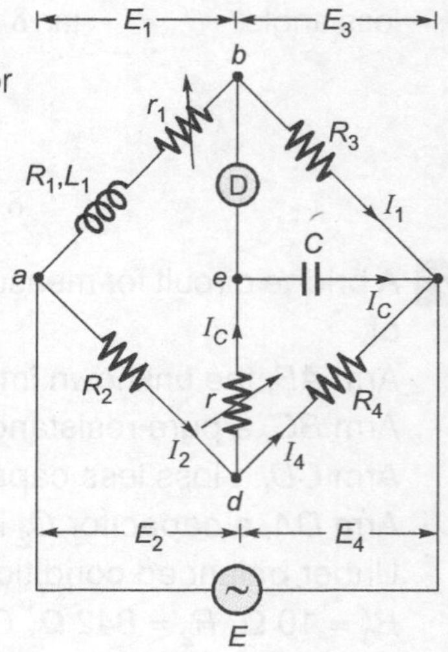

Let,
L_1 = self-inductance to be measured
R_1 = resistance of self-inductor
r_1 = resistance connected in series with self-inductor
and r, R_2, R_3, R_4 = known non-inductive resistances
C = fixed standard capacitor

At balance, $I_1 = I_3$ and $I_2 = I_c + I_4$

Now,
$$I_1 R_3 = I_C \times \frac{1}{j\omega C}$$

∴
$$I_c = I_1 j\omega C R_3$$

Writing the other balance equations,
$$I_1 (r_1 + R_1 + j\omega L_1) = I_2 R_2 + I_c r$$

and
$$I_c\left(r + \frac{1}{j\omega C}\right) = (I_2 - I_c) R_4$$

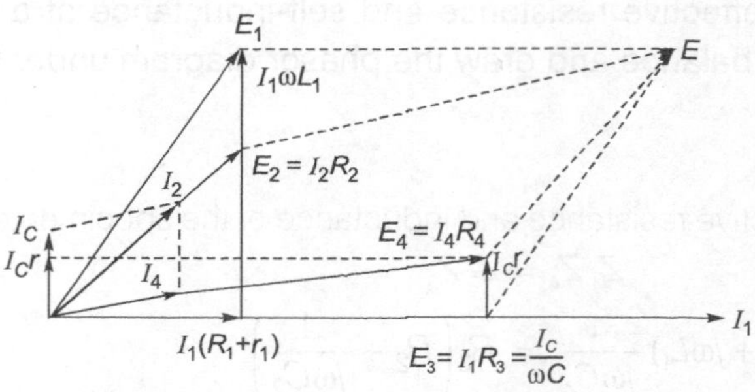

Substituting the value of I_c in the above equations,
$$I_1(r_1 + R_1 + j\omega L_1) = I_2 R_2 + I_1 j\omega C R_3 r$$

or
$$I_1(r_1 + R_1 + j\omega L_1 - j\omega C R_3 r) = I_2 R_2 \quad \ldots(i)$$

and
$$j\omega C R_3 I_1\left(r + \frac{1}{j\omega C}\right) = (I_2 - I_1 j\omega C R_3) R_4$$

or
$$I_1 (j\omega C R_3 r + j\omega C R_3 R_4 + R_3) = I_2 R_4 \quad \ldots(ii)$$

From equations (*i*) and (*ii*),
$$I_1(r_1 + R_1 + j\omega L_1 - j\omega C R_3 r) = I_1\left(\frac{R_2 R_3}{R_4} + \frac{j\omega C R_2 R_3 r}{R_4} + j\omega C R_3 R_2\right)$$

Equating the real and the imaginary parts:
$$R_1 = \frac{R_2 R_3}{R_4} - r_1$$

and
$$L_1 = C\frac{R_3}{R_4}\left[r(R_4 + R_2) + R_2 R_4\right]$$

3.9 **With the help of circuit diagram, explain the principle and operation of Owen's bridge for the measurement of incremental inductance.**

[10 marks : 2010]

Solution:

The Owen's bridge measures the value of a self-inductance in terms of a capacitance,

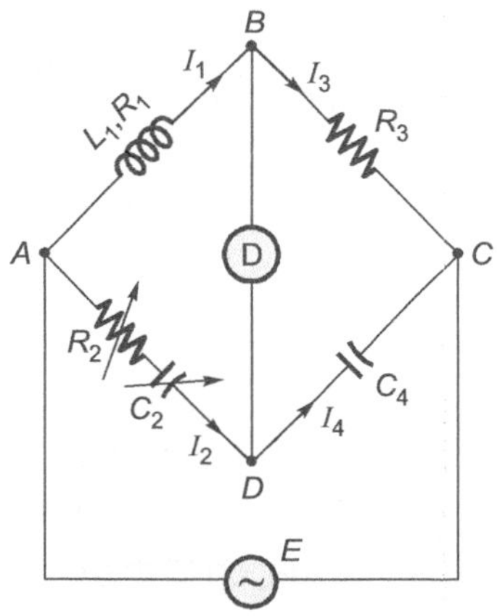

Owen's Bridge

Let, $Z_1 = R_1 + j\omega L_1;\quad Z_2 = R_2 + \dfrac{1}{j\omega C_2}$

$Z_3 = R_3;\quad Z_4 = \dfrac{1}{j\omega C_4}$

where, L_1 = Unknown self-inductance of resistance R_1

R_2 = Variable non-inductive resistance

R_3 = Standard non-inductive resistance

C_2 = Standard variable capacitance

C_4 = Standard fixed capacitance

At balance, we have,

$$Z_1Z_4 = Z_2Z_3$$

$$(R_1 + j\omega L_1)\left(\frac{1}{j\omega C_4}\right) = \left(R_2 + \frac{1}{j\omega C_2}\right)(R_3)$$

$$\Rightarrow \quad \frac{R_1}{j\omega C_4} + \frac{L_1}{C_4} = R_2R_3 + \frac{R_3}{j\omega C_2}$$

Separating and equating real and imaginary terms

$$\frac{L_1}{C_4} = R_2R_3 \Rightarrow L_1 = R_2R_3C_4$$

$$\frac{R_1}{j\omega C_4} = \frac{R_3}{j\omega C_2} \Rightarrow R_1 = \frac{C_4}{C_2}R_3$$

3.10 **Explain, with a diagram, how wien's bridge can be used for experimental determination of frequency. Derive the expression for frequency in terms of bridge parameters.**

[10 marks : 2011]

Solution:

- The wien's bridge shown in figure has a series RC combination in one arm and a parallel combination in the adjoining arm.

$$z_1 = R_1 - j/\omega C_1 \quad z_2 = R_2$$

$$y_3 = \frac{1}{R_3} + j\omega C_3 \quad \text{and} \quad z_4 = R_4$$

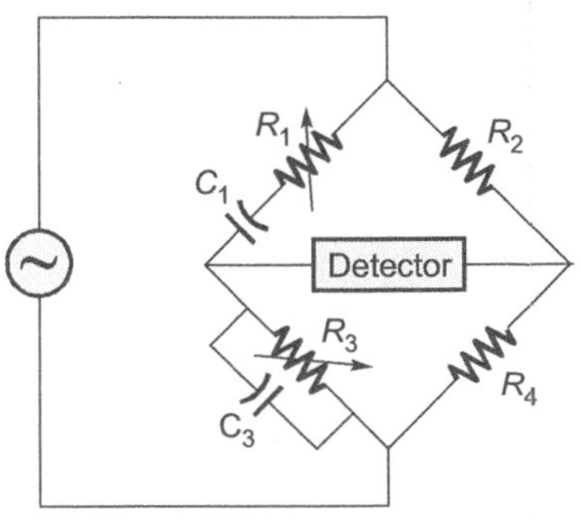

Using the bridge balance equation, we have

$$z_1 z_4 = z_2 z_3 = \frac{z_2}{y_3}$$

$$\Rightarrow \quad z_2 = z_1 z_4 y_3$$

$$\Rightarrow \quad R_2 = \left(R_1 - \frac{j}{\omega C_1}\right) \times R_4 \times \left(\frac{1}{R_3} + j\omega C_3\right)$$

$$= \frac{R_1R_4}{R_3} - \frac{jR_4}{\omega C_1 R_3} + j\omega C_3 R_1 R_4 + \frac{C_3R_4}{C_1}$$

$$\Rightarrow \qquad R_2 = \left(\frac{R_1 R_4}{R_3} + \frac{C_3 R_4}{C_1}\right) - j\left(\frac{R_4}{\omega C_1 R_3} - \omega C_3 R_1 R_4\right)$$

Equating the real and imaginary terms we have

$$R_2 = \frac{R_1 R_4}{R_3} + \frac{C_3 R_4}{C_1} \text{ and } \frac{R_4}{\omega C_1 R_3} - \omega C_3 R_1 R_4 = 0$$

Therefore,
$$\frac{R_2}{R_4} = \frac{R_1}{R_3} + \frac{C_3}{C_1}$$

and
$$\frac{1}{\omega C_1 R_3} = \omega C_3 R_1 \Rightarrow \omega^2 = \frac{1}{C_1 R_1 R_3 C_3} = \frac{1}{\sqrt{C_1 R_1 C_3 R_3}}$$

$$f = \frac{1}{2\pi\sqrt{C_1 R_1 C_3 R_3}} \qquad \ldots(i)$$

Using equation (*i*), the frequency of the applied voltage is determined.

- In most wien's bridge circuit, the components are chosen such that $R_1 = R_3 = R$ and $C_1 = C_3 = C$. Equation (*i*) therefore reduces to

$$f = \frac{1}{2\pi RC}$$

3.11 **Draw a circuit diagram of De Sauty Bridge for the measurement of capacitance and obtain an expression for the unknown capacitance. What are the defects of this bridge?**

[10 marks : 2012]

Solution:

At balance,

$$\left(R_1 + r_1 + \frac{1}{j\omega C_1}\right) R_4 = \left(R_2 + r_2 + \frac{1}{j\omega C_2}\right) R_3$$

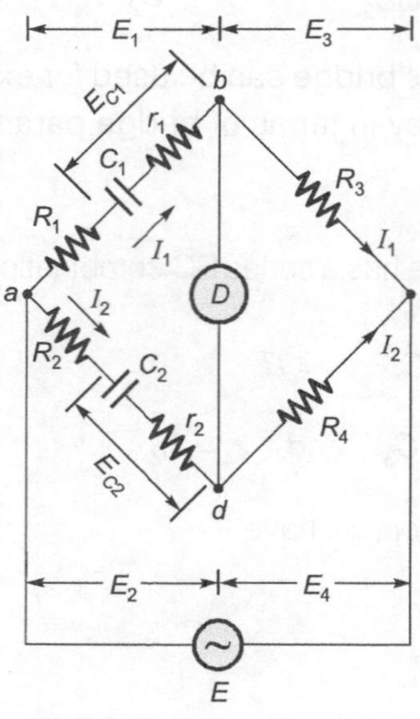

Comparing imaginary and real part on both side

$$(R_1 + r_1) R_4 = (R_2 + r_2) R_3$$

and
$$\frac{R_4}{\omega C_1} = \frac{R_3}{\omega C_2}$$

$$\Rightarrow \quad \frac{C_1}{C_2} = \frac{R_4}{R_3} = \frac{(R_2 + r_2)}{(R_1 + r_1)}$$

The phasor diagram of the bridge under balance condition

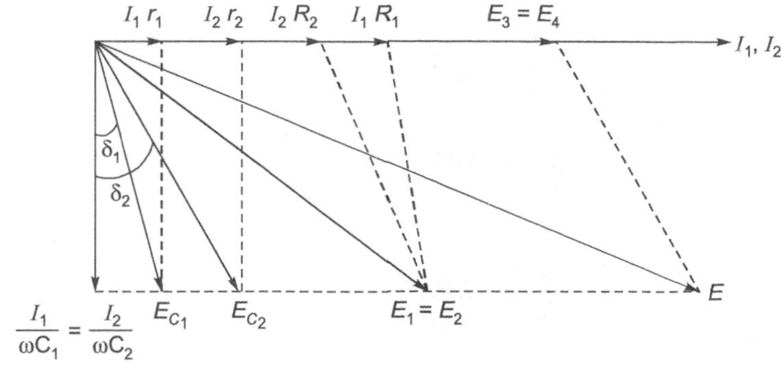

The angle δ_1 and δ_2 are the phase angles of capacitor C_1 and C_2 respectively.

Dissipation factor for the capacitors are

$$D_1 = \tan\delta_1 = \omega C_1 r_1$$

and
$$D_2 = \tan\delta_2 = \omega C_2 r_2$$

$$\therefore \quad D_2 - D_1 = \omega C_2\left(\frac{R_1 R_4}{R_3} - R_2\right)$$

Thus, this method does not give accurate results for dissipation factor since its value depends on difference of quantities $R_1\,R_4/R_3$ and R_2. These quantities are moderately large and their difference is very small and since this difference cannot be known with a high degree accuracy the dissipation factor can not be determined accurately.

3.12 In Hay's bridge, the four arms, namely, $R_1 - L_1$, R_2, $R_3 - C_3$ and R_4 are connected in clockwise order. Show that under balanced condition the quality factor Q of the coil is given by

$$Q = \frac{\text{Voltage across } C_3}{\text{Voltage across } R_3}$$

[8 marks : 2014]

Solution:

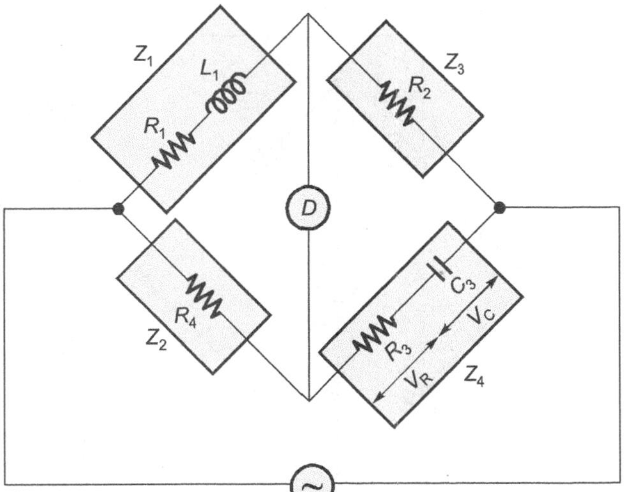

$$Q = \frac{\text{Voltage across } C_3}{\text{Voltage across } R_3}$$

as we known, $$Q = \frac{\omega L_1}{R_1} \quad ...(i)$$

From the bridge balanced condition,

$$Z_1 Z_4 = Z_2 Z_3$$

$$\Rightarrow \quad (R_1 + j\omega L_1)\left(R_3 + \frac{1}{j\omega C_3}\right) = R_4 R_2$$

$$\Rightarrow \quad R_1 R_3 + j\omega L_1 R_3 + \frac{R_1}{j\omega C_3} + \frac{L_1}{C_3} = R_4 R_2$$

equating real and imaginary path,

$$R_1 R_3 + \frac{L_1}{C_3} = R_4 R_2 \quad ...(ii)$$

$$\omega L_1 R_3 - \frac{R_1}{\omega C_3} = 0$$

$$\Rightarrow \quad \omega L_1 R_3 = \frac{R_1}{\omega C_3}$$

$$\Rightarrow \quad \frac{\omega L_1}{R_1} = \frac{1}{\omega C_3 \times R_3} \quad ...(iii)$$

From equation (*i*) and (*iii*), $$Q = \frac{\omega L_1}{R_1} = \frac{1/\omega C_3}{R_3} = \frac{1/C}{R_3} = \frac{1/C}{V_R}$$

3.13 **A bridge circuit for measurement of parameters of an iron-core coil consists of,**
Arm *AB*, iron-core with resistance R_1 and inductance L_1
Arm *BC*, a pure resistance R_3
Arm *CD*, lossless capacitor C_4
Arm *DA*, capacitor C_2 in series with resistance R_2
At balance, $R_3 = 10\ \Omega$, $R_2 = 842\ \Omega$, $C_2 = 0.135\ \mu F$ and $C_4 = 1\ \mu F$
Calculate power factor of coil at supply frequency of 1000 Hz. Draw the phasor diagram under balanced condition

[8 marks : 2014]

Solution:

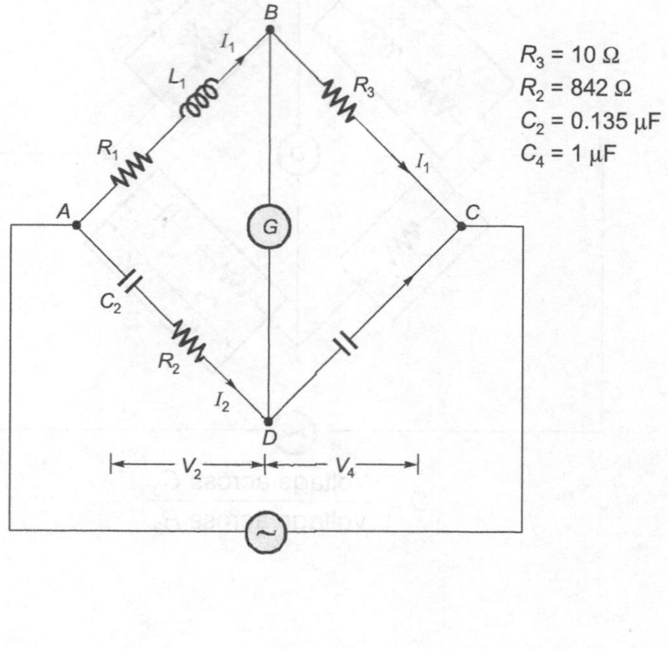

$$\tan\phi = \frac{\omega L_1}{R_1} \Rightarrow \phi = \tan^{-1}\left(\frac{\omega L_1}{R_1}\right)$$

At balance condition, $$(R_1 + j\omega L_1)\left(\frac{1}{j\omega C_4}\right) = R_3 \cdot \left(R_2 + \frac{1}{j\omega C_2}\right)$$

equating real and imaginary part,

$$\frac{L_1}{C_4} = R_2 R_3$$

$$L_1 = R_2 R_3 C_4 \Rightarrow L_1 = 842 \times 10 \times 1 \times 10^{-6} = 8.42 \text{ mH}$$

and $$-\frac{R_1}{\omega C_4} = -\frac{R_3}{\omega C_2}$$

$\Rightarrow$ $$R_1 = \frac{C_4 R_3}{C_2} = \frac{1\times 10^{-6} \times 10}{0.135 \times 10^{-6}} = 74.07\ \Omega$$

$$\tan\phi = \frac{\omega L_1}{R_1} = \frac{1000 \times 8.42 \times 10^{-3}}{74.07}$$

$$\phi = \tan^{-1}\left(\frac{8.42}{74.07}\right) = 6.4853°$$

$$\cos\phi = \cos(6.4853°) = 0.9936$$

$$\text{Power factor} = \cos\phi = 0.9936$$

Phasor diagram,

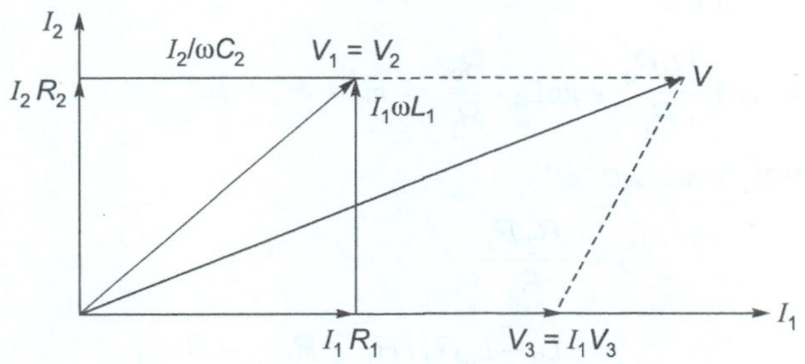

$$V_4 = \frac{I_2}{\omega C_4}$$

$$V_1 = I_1 R_1 + jI_1 \omega L_1 \quad \text{...(i)}$$

$$V_2 = I_2 R_2 + \frac{I_2}{\omega C_4}$$

$$V = V_1 + V_3 \text{ or } V_2 + V_4$$

3.14 **For the Bridge circuit shown, obtain the expression for mutual inductance *M*, when the bridge is at balance, in terms of circuit parameters.**

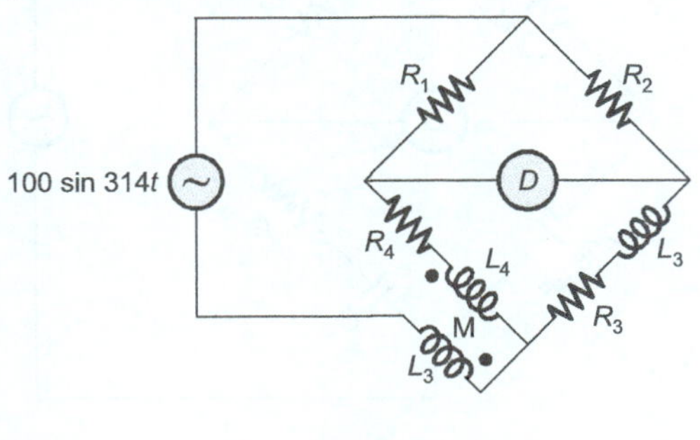

[20 marks : 2017]

Solution:

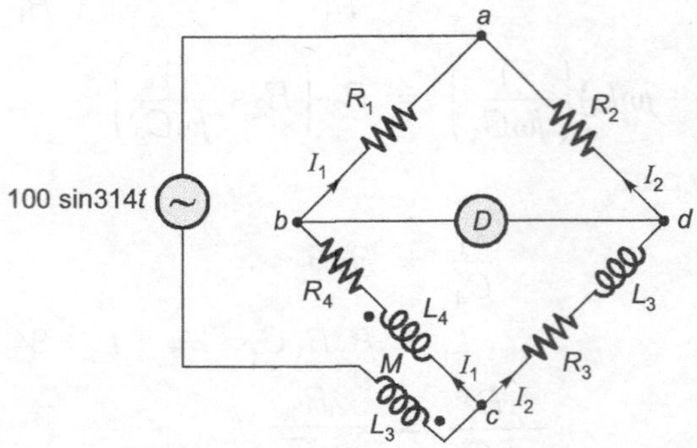

Let, M = Unknown mutual inductance

L_4 = Self inductance of secondary of mutual inductance

L_3 = Known self inductance

R_1, R_2, R_3 and R_4 = Non-inductive resistors

At balance condition voltage drop between *c* and *d* must be equal to the voltage drop between *a* and *d*. Also voltage drop across *c-b-a* must be equal to the voltage drop across *c-d-a*.

Thus, $$I_1R_1 = I_2R_2 \quad \text{...(i)}$$

and $$(I_1 + I_2)(j\omega M) + I_1(R_1 + R_4 + j\omega L_4) = I_2(R_2 + R_3 + j\omega L_3) \quad \text{...(ii)}$$

From equation (i) and (ii), we get

$$I_2\left(\frac{R_2}{R_1}+1\right)(j\omega M) + I_2\left(\frac{R_2}{R_1}\right)(R_1+R_4+j\omega L_4) = I_2(R_2 + R_3 + j\omega L_3)$$

or, $$(j\omega M)\left(1+\frac{R_2}{R_1}\right) + R_2 + \frac{R_2R_4}{R_1} + j\omega L_4\cdot\frac{R_2}{R_1} = R_3 + R_2 + j\omega L_3$$

Equating real and imaginary part, we get.

Thus, $$R_4 = \frac{R_3R_1}{R_2}$$

and $$M = \frac{L_3 - L_4R_2/R_1}{\left(1+\frac{R_2}{R_1}\right)} = \frac{R_1L_3 - R_2L_4}{R_1 + R_2}$$

It $R_1 = R_2$, $$M = \frac{L_3 - L_4}{2} \text{ and } R_3 = R_4$$

3.15 **For the bridge circuit shown determine the values of *R* and *C*. Derive the formula used.**

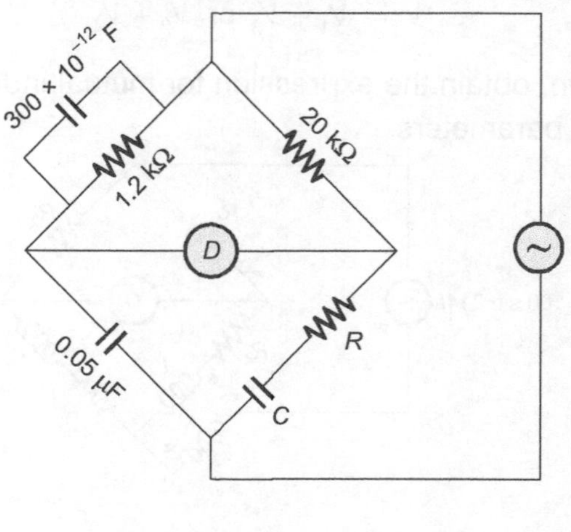

[20 marks : 2018]

Solution:

With balance of bridge,

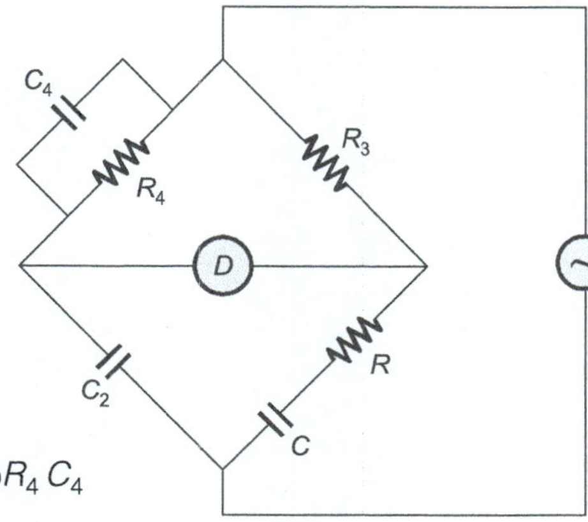

$$Z_1 Z_4 = Z_2 Z_3$$

$$\left(R+\frac{1}{j\omega C}\right)\left(\frac{R_4}{1+j\omega C_4 R_4}\right) = \frac{R_3}{j\omega C_2}$$

$$\frac{(1+j\omega RC)R_4}{(1+j\omega C_4 R_4)j\omega C} = \frac{R_3}{j\omega C_2}$$

$$C_2 R_4[1+j\omega RC] = CR_3[1+j\omega R_4 C_4]$$

Separate real and imaginary parts,

$$C_2R_4 = CR_3 \quad \Big| \quad \omega RC\cdot C_2R_4 = CR_3\,\omega R_4\,C_4$$

$$C = C_2\cdot\frac{R_4}{R_3} \quad \Big| \quad R = \frac{C_4}{C_2}\cdot R_3$$

Given,

$$C_2 = 0.05\ \mu\text{F},\ R_3 = 20\ \text{k}\Omega,$$

$$R_4 = 1.2\ \text{k}\Omega,\ C_4 = 300\ \text{pF}$$

By substituting values,

$$C = C_2\frac{R_4}{R_3} = 0.05\times10^{-6}\times\frac{1.2}{20} = 3\ \text{nF}$$

$$R = \frac{C_4}{C_2}\cdot R_3 = \frac{300\times10^{-12}}{0.05\times10^{-6}}\times20\times10^{3} = 120\ \Omega$$

3.16 The bridge network shown in the figure measures the inductance (L) and resistance (R) if the impedance between A and B is unknown.

(i) Find R and L if bridge balance is obtained for $Q = S = 1000\ \Omega$, $P = 100\ \Omega$, $r = 50\ \Omega$, $C = 1\ \mu$F.

(ii) Draw the vector diagram showing the voltage and current at every point of the network.

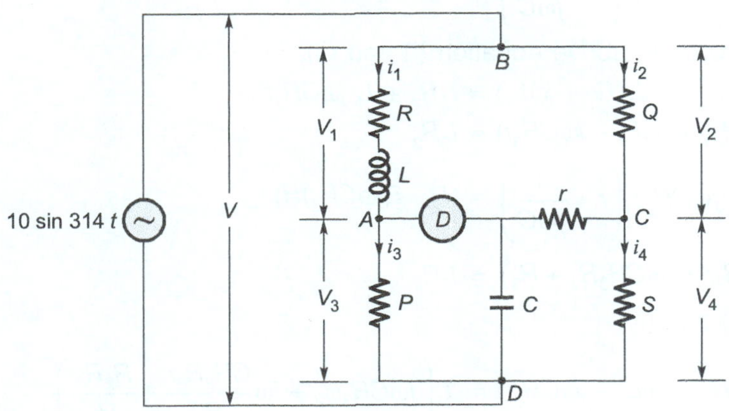

[20 marks : 2019]

Solution:

Given circuit is Anderson's bridge and can be redrawn as,

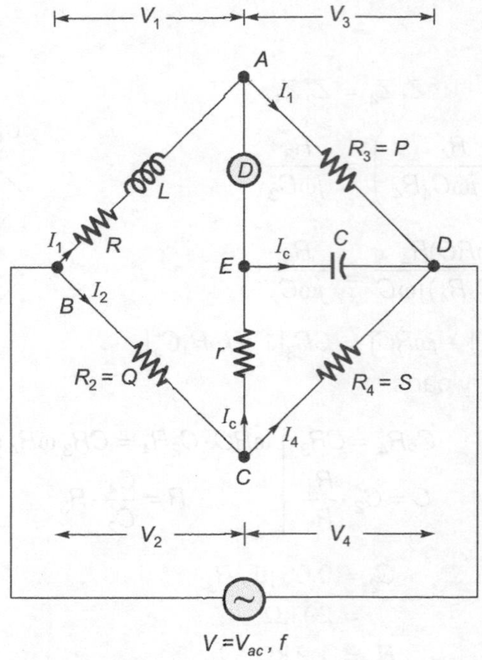

L_1 : Self inductance to be measured

R_1 : Resistance of self inductor

r, R_4, R_2, R_3 : Known non inductive resistances

C : Fixed standard capacitor

$$I_1 = I_3$$

and
$$I_2 = I_C + I_4$$

We know, by using KVL,
$$I_1 R_3 = I_C \times \frac{1}{j\omega C}$$

$$I_C = I_1 j\omega C R_3$$

Writing the balance equations for bridge,

$$I_1(R_1 + j\omega L_1) = I_2 R_2 + I_c r \quad \text{...(i)}$$

and
$$I_C\left(r + \frac{1}{j\omega C}\right) = (I_2 - I_c)R_4 \quad \text{...(ii)}$$

Substituting value of I_C in above equation (i) and (ii),

$$I_1(R_1 + j\omega L_1) = I_2 R_2 + I_1\, j\omega C R_3 r$$

or
$$I_1(R_1 + j\omega L_1 - j\omega C R_3 r) = I_2 R_2 \quad \text{...(iii)}$$

and
$$j\omega C R_3 I_1\left(r + \frac{1}{j\omega C}\right) = (I_2 - I_1 j\omega C R_3)R_4$$

$$I_1(j\omega C R_3 r + j\omega C R_3 R_4 + R_3) = I_2 R_4 \quad \text{...(iv)}$$

Using equation (iii) and (iv),

$$I_1(R_1 + j\omega L_1 - j\omega C R_3 r) = I_1\left(j\omega C R_3 R_2 + j\omega \frac{C R_2 R_3 r}{R_4} + \frac{R_2 R_3}{R_4}\right)$$

Equating real and imaginary parts,

$$R_1 = \frac{R_2 R_3}{R_4} \quad \text{...(v)}$$

$$L_1 = C\frac{R_3}{R_4}[r(R_4 + R_2 + R_2R_4)] \quad \text{...(vi)}$$

(i)
$$R = \frac{PQ}{S} = \frac{100 \times 1000}{1000} = 100\ \Omega$$

$$L = \frac{CQ}{S}[r(S+P)+PS]$$

$$= \frac{10^{-6} \times 1000}{1000}[50(1000+100)+100000)] = 0.155\ \text{H}$$

(ii) The phasor diagram can be drawn as follows:

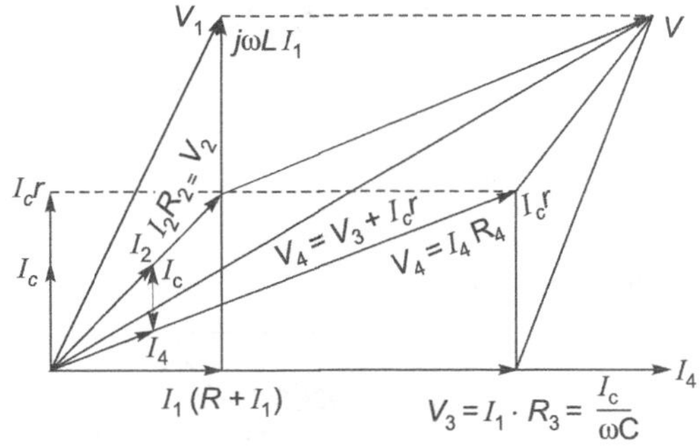

3.17 The self-capacitance or distributed capacitance of a coil is measured using Q meter. The first measurement is carried out at 2.5 MHz, when the tuning capacitor is set at 425 pF. The second measurement is carried out by increasing the frequency to 6 MHz, when the tuning capacitor is set at 60 pF. Determine the distributed capacitance of the coil.

[12 marks : 2020]

Solution:

In case of first measurement,

$$f_1 = \frac{1}{2\pi\sqrt{L(C_1 + C_d)}}$$

$$2.5 \times 10^6 = \frac{1}{2\pi\sqrt{L(425 \times 10^{-12} + C_d)}} \quad \text{...(i)}$$

In case of second measurement,

$$f_2 = \frac{1}{2\pi\sqrt{L(C_2 + C_d)}}$$

$$6 \times 10^6 = \frac{1}{2\pi\sqrt{L(60 \times 10^{-12} + C_d)}} \quad \text{...(ii)}$$

Dividing equation (ii) by equation (i), we get

$$\frac{6}{2.5} = \frac{\sqrt{L(425 + C_d)}}{\sqrt{L(60 + C_d)}}$$

$$5.76 = \frac{425 + C_d}{60 + C_d}$$

$$345.6 + 5.76\ C_d = 425 + C_d$$

$$4.76\ C_d = 79.4$$

$$C_d = 16.68\ \text{pF}$$

3.18 The following data is related to the Schering bridge shown in the figure. R_1 = 2 kΩ, C_1 = 0.5 μF, R_2 = 4 kΩ and C_3 = 0.5 μF at frequency 1 kHz. Determine the unknown resistance and capacitance of the bridge circuit and dissipation factor.

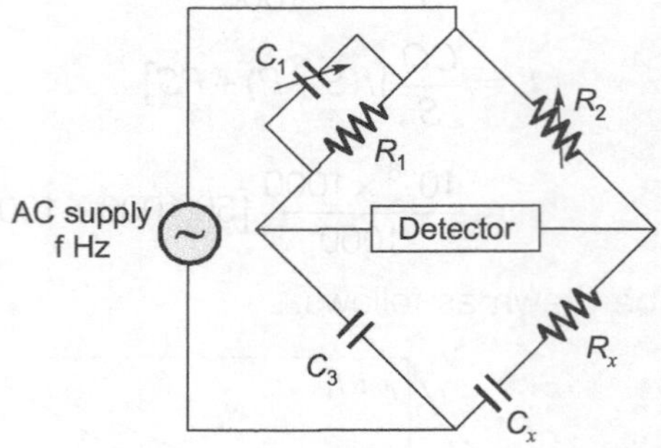

[10+10 = 20 marks : 2021]

Solution:

Bridge under balance condition :

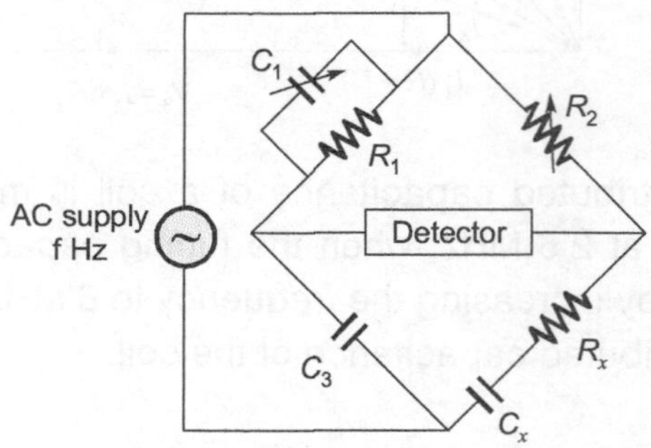

$$Z_1Z_4 = Z_2Z_3$$

$$\frac{R_1}{(1+j\omega C_1R_1)} \times \left(R_x + \frac{1}{j\omega C_x}\right) = \frac{R_2}{j\omega C_3}$$

$$\frac{R_1(1+j\omega C_xR_x)}{(1+j\omega C_1R_1)j\omega C_x} = \frac{R_2}{j\omega C_3}$$

$$R_1C_3(1+j\omega C_xR_x) = R_2(1+j\omega C_1R_1)C_x$$

Separate real and imaginary parts

$$R_1C_3 = R_2C_x \qquad \omega C_xR_xR_1C_3 = \omega C_1R_1C_xR_2$$

$$= \frac{2\times10^3}{4\times10^3}\times0.5\times10^{-6} \qquad = \frac{0.5}{0.5}\times2$$

$$C_x = 0.25\ \mu\text{F} \qquad R_x = 4\ \text{k}\Omega$$

Given : f = 1 kHz

Dissipation factor, $D = \omega C_xR_x = 2\pi\times1\times10^3\times0.25\times10^{-6}\times2\times10^3$

$= 6.28$ rad

3.19 An AC bridge with terminals A, B, C, D (consecutively marked) has in arm AB, a pure resistance; arm BC a resistance of 800 ohms in parallel with a capacitor of 0.5 μF; arm CD, a resistance of 400 ohms in series with a capacitor of 1 μF; and arm DA a resistance of 1000 ohms

(i) obtain the value of the frequency for which the bridge can be balanced by first deriving the balance equations connecting the branch impedance, and draw phasor diagram, and

(ii) calculate the value of the resistance in arm AB to produce balance.

[20 marks : 2022]

Solution:

This is Wein's bridge :

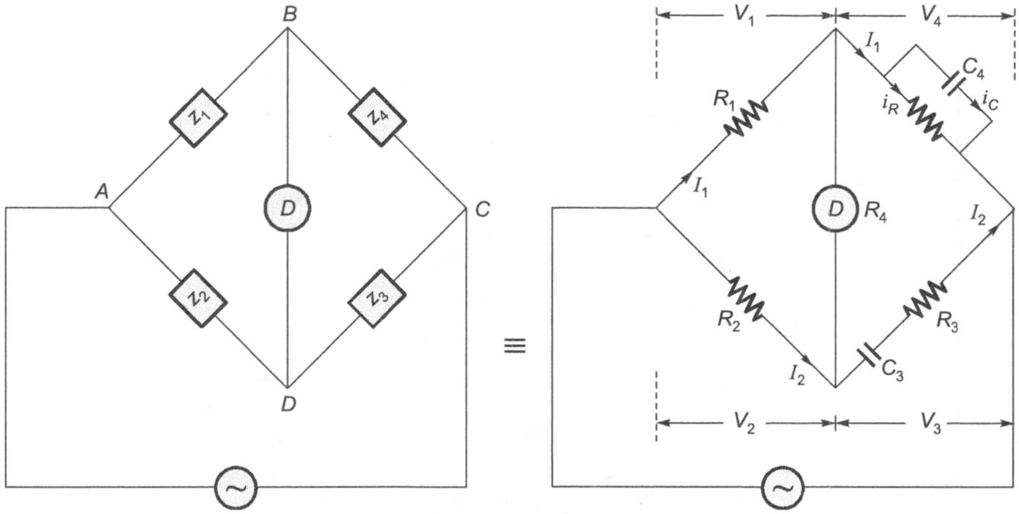

For balance equation, $Z_1Z_3 = Z_2Z_4$

$$R_1\left(R_3 + \frac{1}{j\omega C_3}\right) = R_2\left(\frac{R_4 \cdot \frac{1}{j\omega C_4}}{R_4 + \frac{1}{j\omega C_4}}\right)$$

$$R_1\left(\frac{1 + j\omega C_3R_3}{j\omega C_3}\right) = R_2\left(\frac{R_4}{1 + j\omega R_4C_4}\right)$$

$$R_1(1 + j\omega C_3R_3)(1 + j\omega R_4C_4) = R_2R_4(j\omega C_3)$$

$$R_1(1 + j\omega C_3R_3 + j\omega R_4C_4 - \omega^2C_3R_3C_4R_4) = R_2R_4(j\omega C_3)$$

By separating Img and Real on sixth side and comparing

$$R_1 - \omega^2C_3C_4R_3R_4R_1 = 0$$

$$1 - \omega^2C_3C_4R_3R_4 = 0 \Rightarrow \omega^2 = \frac{1}{R_3R_4C_3R_4}$$

Hence, $$\omega = \frac{1}{\sqrt{C_3C_4R_3R_4}} \quad ...(1)$$

By comparing Imaginary part

$$R_1[\omega C_3R_3 + \omega C_4R_4)] = \omega C_3R_2R_4$$

$$R_1[R_3C_3 + C_4R_4)] = C_3R_2R_4$$

$$\frac{R_2}{R_1} = \frac{R_3}{R_4} + \frac{C_4}{C_3} \quad ...(2)$$

Phasor :

$$V_1 = V_2 = I_1R_1 = I_2R_2$$

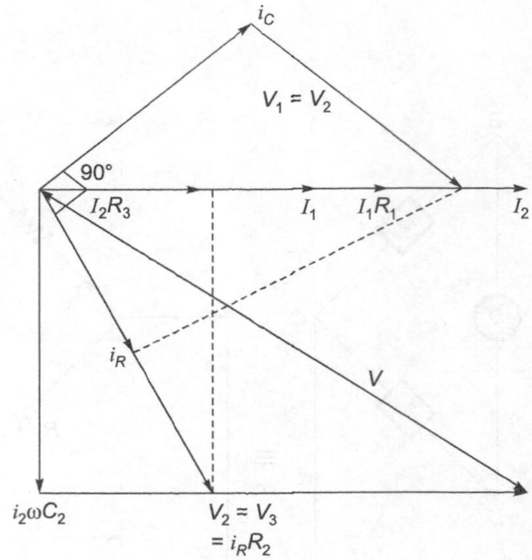

(i) $$\omega^2 = \frac{1}{R_3R_4C_3R_4} \Rightarrow \omega^2 = \frac{1}{R_3R_4C_3C_4}$$

$$R_3 = 400\ \Omega,\ R_4 = 800\ \Omega,\ C_4 = 0.5\ \mu F$$

$$C_3 = 1\ \mu F,\ R_2 = 1000\ \Omega$$

$$\omega^2 = \frac{1}{400 \times 800 \times 10^{-6} \times 0.5 \times 10^{-6}}$$

$$\omega^2 = \frac{1}{16 \times 10^{-8}} \Rightarrow \omega = \frac{1}{4} \times 10^4 \text{ rad/sec}$$

$$2\pi f = 0.25 \times 10^4 \text{ rad/sec}$$

Frequency, $$f = 398.08 \text{ Hz}$$

(ii) From eqn. (2)

$$\frac{R_2}{R_1} = \frac{R_3}{R_4} + \frac{C_4}{C_3} \quad \text{(By putting values)}$$

$$\frac{1000}{R_1} = \frac{400}{800} + \frac{0.5}{1}$$

$$\frac{1000}{R_1} = 1 \Rightarrow R_1 = 1\ k\Omega$$

3.20 (i) In low-voltage Schering bridge designed for measurement of permittivity, the branch *ab* consists of two electrodes between which the specimen under test may be inserted; arm *bc* is a non-reactive resistor R_3 in parallel with a standard capacitor C_3 and *cd* is a non-reactive resistor R_4 in parallel with a standard capacitor C_4; arm *da* is a standard air capacitor of capacitance C_2. Without the specimen between the electrodes, balance is obtained with the following values :

$$C_3 = 150 \text{ pF}$$
$$C_4 = 200 \text{ pF}$$
$$C_2 = 250 \text{ pF}$$
$$R_3 = 5 \times 10^3\ \Omega$$
$$R_4 = 10 \times 10^3\ \Omega$$

With specimen inserted, these values become

$$C_3 = 200 \text{ pF}$$
$$C_4 = 1200 \text{ pF}$$
$$C_2 = 1000 \text{ pF}$$

and R_3 and R_4 remain as previous. In each case, the frequency is $\omega = 10 \times 10^3$ rad/s. Determine the relative permittivity of the specimen.

(ii) Draw the connections and phasor diagram of Anderson's bridge along with its advantages and disadvantages.

[10 + 10 marks : 2023]

Solution:

(i) For a Schering bridge, it is given that

Arm BC : Non-reactive resistor R_3 in parallel with a standard capacitor C_3.

Arm CD : Non-reactive resistor R_4 in parallel with a standard capacitor C_4.

Arm DA : Standard air capacitor of capacitance C_2.

With the above specified parameters, the low voltage schering bridge can be drawn as below :

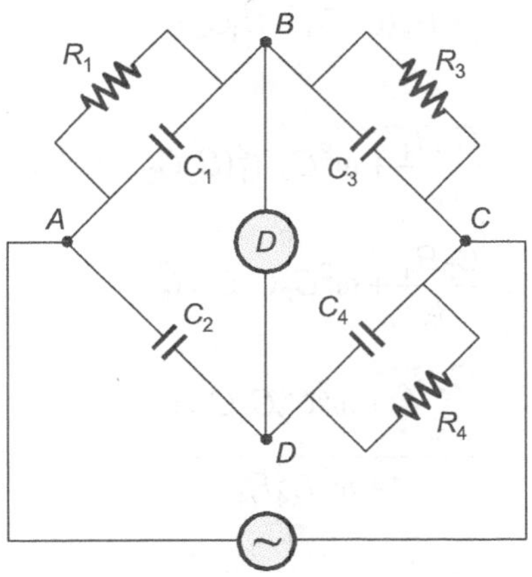

Given : Without specimen between the electrodes of capacitor C_1, the balance is obtained with following values :

$$C_3 = 150 \text{ pF}, C_4 = 200 \text{ pF and } C_2 = 250 \text{ pF}$$
$$R_3 = 5 \times 10^3 \ \Omega \text{ and } R_4 = 10 \times 10^3 \ \Omega$$

With the specimen inserted,

$$C_3 = 200 \text{ pF}, C_4 = 1200 \text{ pF and } C_2 = 1000 \text{ pF}$$
$$R_3 = 5 \times 10^3 \ \Omega \text{ and } R_4 = 10 \times 10^3 \ \Omega$$

Here,

$$Y_1 = \text{Admittance of branch } AB = \frac{1}{R_1} + j\omega C_1$$

$$Y_2 = \text{Admittance of branch } AD = j\omega C_2$$

$$Y_3 = \text{Admittance of branch } BC = \frac{1}{R_3} + j\omega C_3$$

$$Y_4 = \text{Admittance of branch } CA = \frac{1}{R_4} + j\omega C_4$$

At balance condition,

$$Y_1 Y_4 = Y_2 Y_3$$

$$\left(\frac{1}{R_1} + j\omega C_1\right)\left(\frac{1}{R_4} + j\omega C_4\right) = j\omega C_2\left(\frac{1}{R_3} + j\omega C_3\right)$$

$$\left(\frac{1}{R_1R_4} - \omega^2 C_1C_2\right) + j\omega\left(\frac{C_1}{R_4} + \frac{C_4}{R_1}\right) = -\omega^2 C_2C_3 + \frac{j\omega C_2}{R_3}$$

Separating real and imaginary parts and equating, we have

$$\frac{1}{R_1R_4} - \omega^2 C_1C_4 = -\omega^2 C_2C_3 \qquad \text{...(i)}$$

and

$$\frac{C_4}{R_1} + \frac{C_1}{R_4} = \frac{C_2}{R_3}$$

$$C_1 = \frac{C_2R_4}{R_3} - \frac{C_4R_4}{R_1} \qquad \text{...(ii)}$$

From equation (i)

$$\omega^2[-C_2C_3 + C_1C_4] = \frac{1}{R_1R_4}$$

$$R_1 = \frac{-1}{\omega^2 R_4(C_2C_3 - C_1C_4)}$$

Substituting in eqn. (ii), we get

$$C_1 = \frac{C_2R_4}{R_3} + \omega^2 C_4R_4^2(C_2C_3 - C_1C_4)$$

$$C_1[1+\omega^2 C_4^2R_4^2] = \frac{C_2R_4}{R_3} + \omega^2 C_2C_3C_4R_4^2$$

$$C_1 = \frac{\dfrac{C_2R_4}{R_3} + \omega^2 C_2C_3C_4R_4^2}{1+\omega^2 C_4^2R_4^2}$$

Here, $\omega^2 C_2C_3C_4R_4^2 \ll \dfrac{C_2R_4}{R_3}$

and $\omega^2 C_4^2R_4^2 \ll 1$

Hence, $C_1 = \dfrac{C_2R_4}{R_3}$

When there is no dielectric in the capacitor C_1, let the capacitance be C_0, we have

$$C_0 = \frac{C_2R_4}{R_3}$$

$$= \frac{250\text{ pF} \times 10 \times 10^3}{5 \times 10^3} = 500\text{ pF} \qquad \text{...(iii)}$$

When the specimen is inserted as dielectric, let its capacitance be C_s, we have

$$C_s = \frac{C_2R_4}{R_3} = \frac{1000\text{ pF} \times 10 \times 10^3}{5 \times 10^3}$$

$$= 2000\text{ pF} \qquad \text{...(iv)}$$

Without specimen between electrodes, the capacitance is given by

$$C_0 = \frac{\varepsilon_o A}{D} \qquad (\varepsilon_r = 1 \text{ for free space})$$

With specimen between electrodes, the capacitance is given by

$$C_s = \frac{\varepsilon_o \varepsilon_r A}{d}$$

where, ε_r = relative permittivity of the specimen

From above, $$\frac{C_s}{C_o} = \varepsilon_r$$

Using values of C_o and C_s from equations (iii) and (iv)

$$\varepsilon_r = \frac{C_s}{C_o} = \frac{2000}{500} = 4$$

Hence, the relative permittivity of the specimen is 4.

(ii) **Anderson's Bridge** : This bridge, in fact, is a modification of the Maxwell's inductance capacitance bridge. In this method, the self-inductance is measured in terms of a standard capacitor. This method is applicable for precise measurement of self-inductance over a very wide range of values.

Figure shows the connections and the phasor diagram of the bridge for balanced conditions.

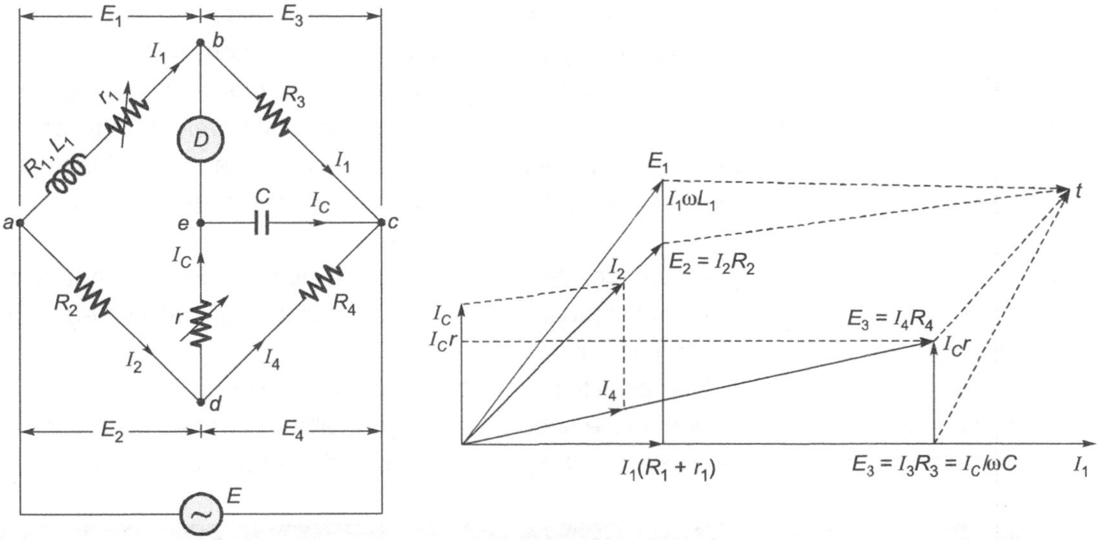

Let L_1 = self-inductance to be measured,

R_1 = resistance of self-inductor,

r_1 = resistance connected in series with self-inductor,

r, R_2, R_3, R_4 = known non-inductive resistances,

and C = fixed standard capacitor

At balance, $I_1 = I_3$ and $I_2 = I_c + I_4$

Now, $$I_1R_3 = I_C \times \frac{1}{j\omega C}$$

$\therefore$ $$I_c = I_1 j\omega CR_3$$

Writing the other balance equations

$$I_1(r_1 + R_1 + j\omega L_1) = I_2R_2 + I_cr$$

and $$I_C\left(r + \frac{1}{j\omega C}\right) = (I_2 - I_c)R_4$$

Substituting the value of I_c in the above equation, we have

$$I_1(r_1 + R_1 + j\omega L_1) = I_2R_2 + I_1 j\omega CR_3 r \text{ or } I_1(r + R_1 + j\omega L_1 - j\omega CR_3 r) = I_2R_2 \quad \text{...(i)}$$

and $$j\omega CR_3I_1\left(r + \frac{1}{j\omega C}\right) = (I_2 - I_1 j\omega CR_3)R_4$$

or $$I_1(j\omega CR_3 r + j\omega CR_3R_4 + R_3) = I_2R_4 \quad \text{...(ii)}$$

From eqn. (i) and (ii), we obtain

$$I_1(r_1 + R_1 + j\omega L_1 - j\omega CR_3r) = I_1\left(\frac{R_2R_3}{R_4} + \frac{j\omega CR_2R_3r}{R_4} + j\omega CR_3R_2\right)$$

Equating the real and the imaginary parts :

$$R_1 = \frac{R_2R_3}{R_4} - r_1$$

and
$$L_1 = C\frac{R_3}{R_4}[r(R_4 + R_2) + R_2R_4]$$

An examination of balance equations reveals that to obtain easy convergence of balance, alternate adjustments of r_1 and r should be done as they appear in only one of the two balance equations.

Advantages :

1. In case adjustments are carried out by manipulating control over r_1 and r_2 they become independent of each other. This is a marked superiority over sliding balance conditions met with low Q coils when measuring with Maxwell's bridge. A study of convergence conditions would reveal that it is much easier to obtain balance in the case of Anderson's bridge than in Maxwell's bridge for low Q-coils.
2. A fixed capacitor can be used instead of a variable capacitor as in the case of Maxwell's bridge.
3. This bridge may be used for accurate determination of capacitance in terms of induction.

Disadvantages :

1. The Anderson's bridge is more complex than its prototype Maxwell's bridge. The Anderson's bridge has more parts and is more complicated to set up and manipulate. The balance equations are not simple and in fact are much more tedious.
2. An additional junction point increases the difficulty of shielding the bridge.

Considering the above complications of the Anderson's bridge, in all the cases where a variable capacitor is permissible the more simple Maxwell's bridge is used instead of Anderson's bridge.

4. Galvanometers, Voltmeters & Ammeters

4.1 Derive the general torque equation for a moving iron instrument. The inductance of a moving iron ammeter is given by the following expression:

$$L = (20 + 10\,\theta - 2\,\theta^2)\ \mu\text{H}$$

where θ is deflection in radians. The spring constant is 24 × 10^{-6} N m/rad. Calculate the value of deflection for a current of 5 A.

[8 marks : 2001]

Solution:

General Torque Equation of Moving Iron Instruments: An expression for the torque of a moving iron instrument may be derived by considering the energy relations when there is a small increment in current supplied to the instrument. When this happens there will be a small deflection dθ and some mechanical work will be done. Let T_d be the deflecting torque.

∴ Mechanical work done = T_d . dθ

Alongside there will be a change in the energy stored in the magnetic field owing to change in inductance. Suppose the initial current is I, the instrument inductance L and the deflection θ. If the current increases by dI then the deflection changes by dθ and the inductance by dL. In order to effect an increment dI in the current there must be an increase in the applied voltage given by

$$e = \frac{d}{dt}(LI) = I\frac{dL}{dt} + L\frac{dI}{dt}$$

The electrical energy supplied $eI\,dt = I^2\,dL + IL\,dI$.

The stored energy changes from,

$$= \frac{1}{2} I^2 L \text{ to } \frac{1}{2}(I + dI)^2 (L + dL)$$

Hence the change in stored energy,

$$\frac{1}{2}[(I^2 + 2I\,dI + (dI)^2](L + dL) - \frac{1}{2} I^2 L$$

Neglecting second and higher order terms in small quantities this becomes $IL\,dI + \frac{1}{2}I^2 dL$.

From the principle of conservation of energy,

Electrical energy supplied = Increase in stored energy + Mechanical work done

$$I^2\,dL + IL\,dI = IL\,dI + \frac{1}{2} I^2\,dL + T_d\,d\theta$$

Thus, $$T_d\,d\theta = \frac{1}{2}I^2 dL$$

or deflecting torque, $$T_d = \frac{1}{2} I^2 \frac{dL}{d\theta} \quad \ldots(i)$$

T is in newton-metre, I in ampere, L in henry, and θ in radian

where, $$L = (20 + 10\,\theta - 2\,\theta^2)$$

Numerical: Rate of change of inductance with deflection

$$\frac{dL}{d\theta} = \frac{d}{d\theta}(20 + 10\theta - 2\theta^2) = (10 - 4\theta)\ \mu\text{H/rad}$$

where, $K = 24 \times 10^{-6}$ N-m/rad

At equilibrium position, $T_c = T_d$

$$\Rightarrow \quad K\theta = \frac{1}{2} I^2 \frac{dL}{d\theta}$$

$$\theta = \frac{1}{2}\frac{I^2}{K}\frac{dL}{d\theta} \quad \text{or} \quad \theta = \frac{1}{2} \times \frac{(5)^2}{24 \times 10^{-6}}(10 - 4\theta) \times 10^{-6}$$

$$\theta = 0.5208\,(10 - 4\,\theta)$$

$$\theta(1 + 2.083) = 5.208$$

$$\theta = 1.689 \text{ rad} = 96.77°$$

4.2 **Discuss the sources of error in a dynamometer instrument. Calculate the reading of an electrodynamic ammeter when a current $i(t) = 80 - 60\sqrt{2}\,\sin(\omega t + 30°)$ A is passed through it.**

[10 marks : 2002]

Solution:

Errors in Electrodynamometer Instruments: The main sources of error in a dynamometer type of instruments are due to

(a) Low torque / Weight ratio
(b) Frequency
(c) Eddy currents
(d) External magnetic fields
(e) Temperature change

(a) Torque / Weight Ratio: The magnetic field produced by air cored coils is essentially small and therefore, flux linkages per ampere in the moving coil are few. Therefore, the deflecting torque is low. Thus to produce a reasonable deflecting torque, the mmf of the moving coil must be increased. The mmf of the moving coil can be increased by increasing the current through the coil or by increasing the number of turns. The current through the moving coil cannot be increased beyond 200 mA as otherwise the springs would be heated too much. On the other hand an increase in the number of turns of the moving coil, will give rise to increased weight of moving parts resulting in a low torque / weight ratio. A low torque / weight ratio indicates a heavy moving system and therefore, the friction losses in electrodynamometer type of instrument are larger than in other types. Frictional errors tend to be high.

(b) **Frequency:** The frequency error of dynamometer type instrument is due largely as a result of variation of self-reactance of coils, with frequency. In case of voltmeters.

Deflection, $\theta = (V^2/KZ^2)\cdot\dfrac{dM}{d\theta}$

Now, impedance $Z = \sqrt{R^2 + (2\pi fL)^2}$ and therefore, the value of impedance Z increases with frequency and hence the voltmeter tend to read low as the frequency is increased.

(c) **Eddy Currents:** The effect of eddy currents is to produce a torque as a result of coupling between moving coil and the adjacent metal parts. Hence metal in the coil supports and structural part is kept to the minimum possible. Any metal that is used is kept away and is selected to have high resistivity so as to reduce eddy currents induced in it.

(d) **External Magnetic Fields:** Since the operating field in electrodynamometer instruments is weak, therefore, these instruments must be protected from the external magnetic fields.

(e) **Temperature Change:** High currents are carried by coils which produce heat. Self heating of coils produces errors. High accuracy instruments contain temperature compensating resistors which tend to neutralize the effects of temperature changes.

An electrodynamic ammeter reads rms value of current.

$$I_0 = 80,\ I_m = -60\sqrt{2}$$

$$I_{rms} = \sqrt{I_0^2 + \left(\frac{I_m}{\sqrt{2}}\right)^2} \Rightarrow I_{rms} = \sqrt{80^2 + \left(-\frac{60\sqrt{2}}{\sqrt{2}}\right)^2} = 100\text{ A}$$

4.3 **Why magnetic shielding is required in dynamometer-type instrument? Explain, how it is arranged in this instrument.**

[8 marks : 2004]

Solution:

The electrodynamometer has two fixed coils and one movable coil. Unlike PMMC where a powerful magnet produces this magnetic field B, in electrodynamometer instrument the field is produced by the current passing through the fixed coil. Hence they have a high ampere turns. In spite of this, the flux density produced is very low.

The field density is also low because of air cored coils. Because of the very low field produced even earth's magnetic field may affect the reading. Hence the movement is surrounded by a laminated steel shield to protect from the effect of strong magnetic field. The casing is of high pemeability alloy to protect the entire mechanism, shunting the external magnetic field around the instrument.

4.4 **The inductance of a moving iron instrument is given by $L = (0.01 + K_1\theta)^2\ \mu$H. Where θ is angular deflection in radians from zero position? The angular deflection in radians corresponding to currents of 1.5 and 4.5 amps are $\pi/6$ and $\pi/3$ respectively. Find the value of K_1.**

[10 marks : 2005]

Solution:

$$L = (0.01 + K_1\theta)^2\ \mu\text{H}$$

$$\frac{dL}{d\theta} = 2(0.01 + K_1\theta)\,.\,K_1$$

$$\theta = \frac{1}{2}\cdot\frac{I^2}{K}\cdot\frac{dL}{d\theta}$$

$$\frac{\pi}{6} = \frac{1}{2}\cdot\frac{(1.5)^2}{K}\cdot 2\left(0.01 + K_1 \cdot \frac{\pi}{6}\right)\cdot K_1 \quad ...(i)$$

$$\frac{\pi}{3} = \frac{1}{2}\cdot\frac{(4.5)^2}{K}\cdot 2\left(0.01 + K_1 \cdot \frac{\pi}{3}\right)\cdot K_1 \quad ...(ii)$$

From equation (*i*) and (*ii*),

$$\frac{(\pi/6)}{(\pi/3)} = \frac{(1.5)^2}{(4.5)^2}\times\frac{(0.01+K_1\cdot\pi/6)}{(0.01+K_1\cdot\pi/3)}$$

$$(0.09 - 0.02) = K_1\left(\frac{2\pi}{6} - \frac{2\pi}{3}\right);\ K_1 = -8.355 \times 10^{-3}$$

4.5 In an indicating PMMC type ammeter, the current through the coil for full-scale deflection is 5 mA. Coil resistance is 5 Ω and swamping resistance is 4 Ω. The meter is to be used for measuring current up to 1 ampere. Calculate the value of shunt resistance required. What arrangement is required to be made to compensate for temperature changes which may occur. Derive necessary expression.

[10 marks : 2006]

Solution:

The temperature error can be eliminated when the shunt and the moving-coil are made of the same material and kept at the same temperature. This method, however, is not satisfactory in practice as the temperature of the two parts are not likely to change at the same rate. Disadvantage of using copper shunt is that they are likely to be bulky as the resistivity of copper is small. Copper shunts are only occasionally used in instruments with built-in shunts.

The arrangement normally used is shown in figure. In this case, a 'swamping resistance' of manganin (which has a negligible temperature co-efficient) having a resistance 20 to 30 times the coil resistance is connected in series with the coil and a shunt of manganin is connected across this combination. Since copper forms a small fraction of the series combination, the proportion in which the currents would divide between the meter and the shunt would not change appreciably with the change in temperature.

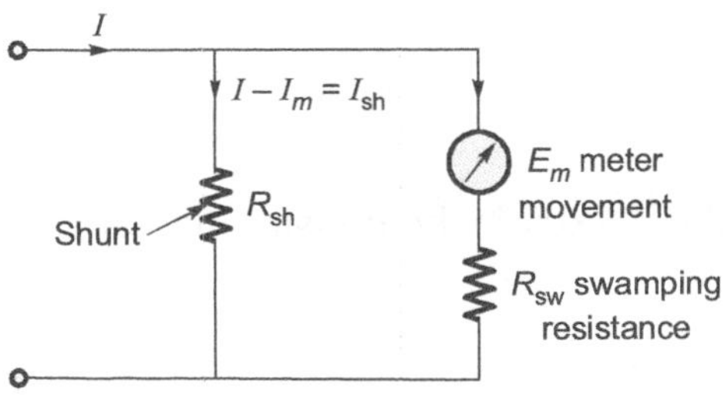

Voltage across the meter coil and swamping resistor is,

$$= 5 \times 10^{-3} \times 9 = 45 \times 10^{-3}\ \text{V}$$

$$\text{The shunt should carry} = 1 - 0.005 = 0.995\ \text{A}$$

$$\therefore \quad \text{The shunt resistance} = \frac{45\times10^{-3}}{0.995} = 0.045\ \Omega$$

4.6 A moving iron voltmeter designed to read up to 100 volts has resistance of 2000 Ω and inductance of 0.6 henry. How it can be modified to read up to 300 volts?
How the modified voltmeter can be made to read without error on both dc and 50 Hz ac?

[10 marks : 2006]

Solution:

R_v the resistance of voltmeter, L inductance and R_s the external series resistance to increase the range.

Voltmeter current, $$I = \frac{V}{\sqrt{(R_s + R_v)^2 + \omega^2 L^2}} = \frac{300}{\sqrt{(2 \times 10^3 + R_s)^2 + (2\pi \times 50 \times 0.6)^2}}$$

Voltage drop across instrument,

$$v = I\sqrt{R_v^2 + (\omega L)^2}$$

$$v = \frac{V}{\sqrt{(R_s + R_v)^2 + \omega^2 L^2}} \cdot \sqrt{R_v^2 + (\omega L)^2}$$

$$\frac{V}{v} = \frac{\sqrt{(R_s + R_v)^2 + (\omega L)^2}}{\sqrt{R_v^2 + (\omega L)^2}}$$

$$\frac{300}{100} = \frac{\sqrt{(2000 + R_s)^2 + (2\pi \times 50 \times 0.6)^2}}{\sqrt{(2000)^2 + (2\pi \times 50 \times 0.6)^2}}$$

Squaring, $9[(2000)^2 + (60\pi)^2] = (2000 + R_s)^2 + (60\pi)^2$

$$36 \times 10^6 + 9(60\pi)^2 - (60\pi)^2 = (2000 + R_s)^2$$

$$10^6 \times 89.58 = (2000 + R_s)^2$$

$$R_s = 7465\ \Omega$$

The multiplier resistance is shunted by a capacitor in order to compensate for the frequency errors introduced due to inductance of the operating coil.

$$C = 0.41 \times \frac{L}{R_s^2}$$

$$= 0.41 \times \frac{0.6}{(7465)^2} = 4.41 \times 10^{-9}\text{F} = 4.41\ \text{nF}$$

4.7 **Explain the construction of a shunt multiplier.**

[4 marks : 2009]

Solution:

The general requirements of shunts are :

(*i*) The thermal co-efficient of shunt and instrument should be low and should be as nearly as possibly the same.

(*ii*) The resistance of shunts should not vary with time.

(*iii*) They should carry the current without excessive temperature rise.

- Manganin is usually used for shunts of d.c. instruments as it gives low value of thermal emf.
- Constantan is a useful material for a.c. circuits since its comparatively high thermal emf, unidirectional is ineffective on these circuits.

4.8 **A moving coil ammeter has a fixed shunt of 0.02 Ω with a coil circuit resistance of R = 1 kΩ and needs potential difference of 0.5 V across it for full scale deflection.**

(a) Calculate the current it corresponds to.

(b) Find the value of shunt when the total current is 10 Amperes.

[12 marks : 2009]

Solution:

(a) Current through shunt $I_{sh} = \dfrac{0.5}{0.02} = 25$

Current through the meter to give full scale deflection

$$I_m = \frac{0.5}{1000} = 0.5 \times 10^{-3} \text{ A}$$

Hence, Total current $I = I_{sh} + I_m$

$$= 25 + 0.5 \times 10^{-3} = 25.0005 \text{ Amp.}$$

(b) Total current $I = I_{sh} + I_m = 10$ Amp.

As, $I_m = 0.5 \times 10^{-3}$

$$I_{sh} = I - I_m$$

$$= 10 - 0.5 \times 10^{-3} = 9.9995 \text{ Amp.}$$

$$I_{sh} = \frac{0.5}{9.9995} \approx 0.05\ \Omega$$

4.9 **Discuss the difference between the construction of PMMC and dynamometer instruments.**

[4 marks : 2009]

Solution:

Difference between the Construction of PMMC and Dynamometer Instrument:

PMMC:

(a) **Moving Coil:** The moving coil is wound with many turns of enamelled or silk covered copper wire. The coil is mounted on a rectangular aluminium former which is pivoted on jewelled bearings. The coils moves freely in the field of a permanent magnet.

(b) **Control:** The coil is supported between two jewel bearings the control torque is provided by two phosphor bronze hair springs.

(c) **Damping:** Damping torque is produced by movement of the aluminium former moving in the magnetic field of the permanent magnet.

(d) **Pointer and Scale:** The pointer is carried by the spindle and moves over a graduated scale. The pointer is of light-weight construction and, apart from those used in some inexpensive instruments, has the section over the scale twisted to form a fine blade.

Dynamometer Instrument:

(a) **Moving Coil:** A single element instrument has one moving coil. The moving coil is wound either as a self-sustaining coil or else on a non-metallic former. A metallic former cannot be used as eddy currents would be induced in it by the alternating field. Both fixed and moving coils are air cored.

(b) **Control:** The controlling torque is provided by two control springs. These springs act as leads to the moving coil.

(c) **Damping:** Air friction damping is employed for these instruments and is provided by a pair of aluminium vanes attached to the spindle at the bottom. These vanes move in sector shaped chambers.

(d) **Pointer and Scale:** The scales are hand drawn, using machine sub-dividing equipment. Diagonal lines for fine sub-division are usually drawn for main markings on the scale.

4.10 **Derive the general torque equation for a moving-iron instrument. The inductance of a moving-iron ammeter is given by the following expression. $L = (30 + 10\theta - 2\theta^2)$ μH where θ is the deflection in radians. The control spring constant is 25×10^{-6} N-m/rad. Calculate the value of deflection for a current of 5 A.**

[10 marks : 2010]

Solution:

For derivation general torque equation of moving iron instrument **(Refer Question No. 4.1).**

Rate of change of inductance with deflection,

$$\frac{dL}{d\theta} = \frac{d}{d\theta}(30+10\theta-2\theta^2) = 10-4\theta \ \mu\text{H/rad}$$

The deflection is, $\theta = \frac{1}{2}\times\left(\frac{I^2}{K}\right)\frac{dL}{d\theta}$

where, $I = 5$ A

$K = 25\times10^{-6}$ N-m/rad

$$\theta = \frac{1}{2}\times\frac{5^2}{25\times10^{-6}}\times(10-4\theta)\times10^{-6}$$

From above, we get, $\theta = 1.67$ radian $= 95.5°$

4.11 **It is required to measure the voltage across 10 kΩ resistor of the circuit shown in figure below.**

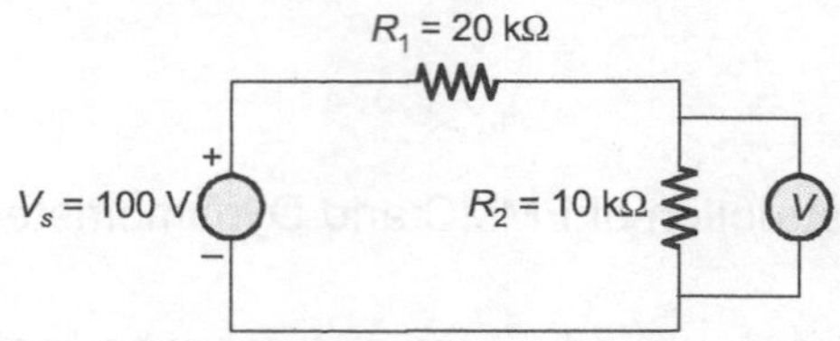

Two voltmeters are available. Voltmeter *A* is 1 mA movement and voltmeter *B* is 50 mA movement. Both use their 50 V scales. Calculate:

(i) The reading of each voltmeter.

(ii) The error from true reading. **[10 marks : 2013]**

Solution:

Actual voltage across 10 kΩ resistor, when voltmeter is not connected,

$$V_{AB} = V_s\left(\frac{R_{AB}}{R_1+R_{AB}}\right) \quad \text{(Using voltage division rule)}$$

$$= 100\times\left(\frac{10}{20+10}\right) = 33.33 \text{ V}$$

Case-1:

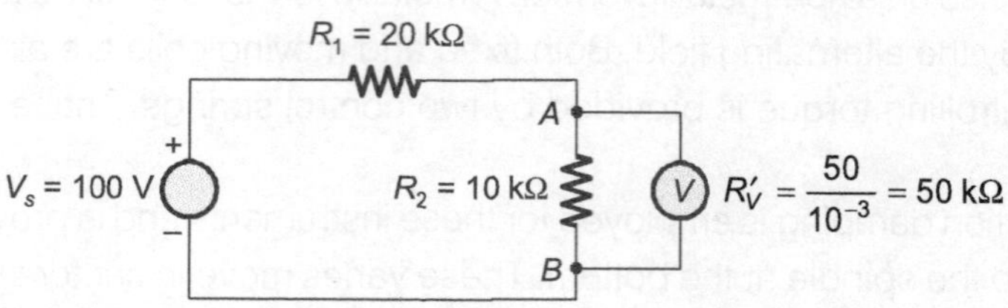

Resistance across the terminals of voltmeter,

$$R'_{AB} = \frac{10\times50}{10+50} = 8.33 \text{ kΩ}$$

$$V'_{AB} = 100\times\left(\frac{8.33}{20+8.33}\right)$$

∴ Reading of voltmeter $= 29.41$ V

$$\text{Error} = \left(\frac{\text{Measured value} - \text{True value}}{\text{True value}}\right) \times 100\%$$

$$\text{Error} = \frac{29.41 - 33.33}{33.33} \times 100 = -11.76\%$$

Case-2:

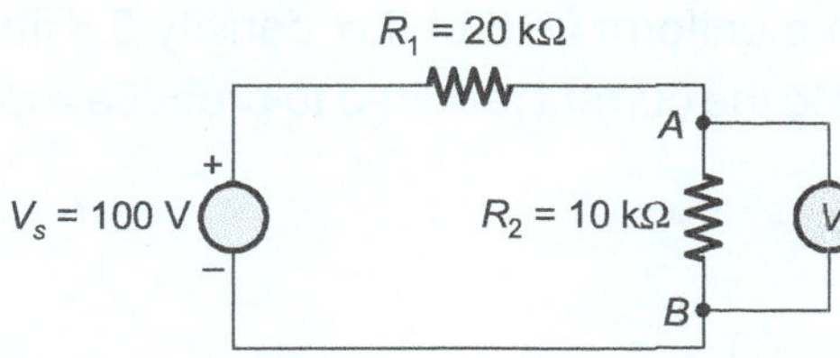

$$R_V'' = \frac{50}{50 \times 10^{-6}} = 1000\ \text{k}\Omega$$

Resistance across the terminals of voltmeter,

$$R_{AB}'' = \frac{10 \times 1000}{10 + 1000} = 9.90\ \Omega$$

$$V_{AB}'' = 100 \times \left(\frac{9.90}{20 + 9.90}\right) = 33.11\ \text{V}$$

$\therefore$ Reading of voltmeter = 33.11 V

$$\text{error} = \frac{33.11 - 33.33}{33.33} \times 100 = -0.66\%$$

4.12 A PMMC instrument has full scale deflection current of 1 mA and internal resistance of 50 Ω. How this can be converted into an 1 A ammeter and 100 V voltmeter? How the above instrument can be converted to read ac voltages?

[10 marks, 2015]

Solution:

For 1 A ammeter:

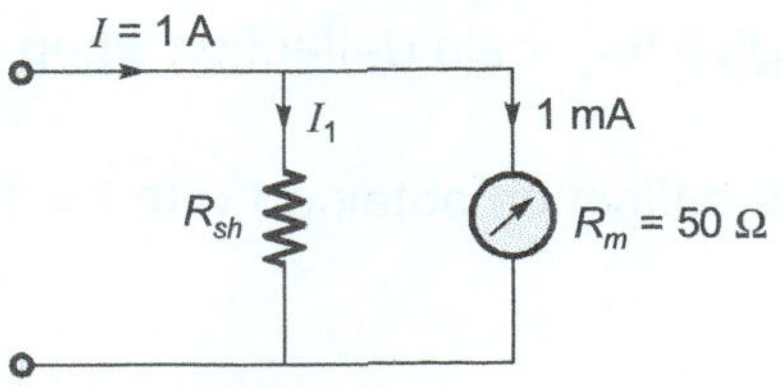

$$I_1 = 0.999\text{A}$$

and, $1\text{ m} \times 50 = 0.999 \times R_{sh}$

$$\Rightarrow \quad R_{sh} = \frac{1m \times 50}{0.999} = 0.05\ \Omega$$

For 100 V voltmeter:

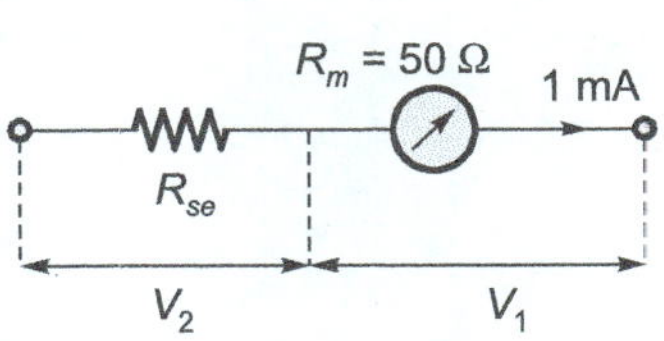

$\Rightarrow \quad V_1 + V_2 = 100\ \text{V}$

$\Rightarrow \quad 1\text{ m} \times R_{se} + 1\text{ m} \times 50 = 100$

$\Rightarrow \quad R_{se} = 99.95\ \text{k}\Omega$

4.13 A current of $-8 + 6\sqrt{2}\sin(\omega t + 30°)$ amperes is passed through a centre zero PMMC meter and moving iron meter. What would be their readings?

[10 marks, 2015]

Solution:

$$I = -8 + 6\sqrt{2}\sin(\omega t + 30°)$$

Centre zero PMMC meter, $I_{avg} = -8\text{ A}$

Moving iron meter, $I_{rms} = \sqrt{8^2 + \frac{1}{2}(6\sqrt{2})^2}$

$$= \sqrt{64+36} = 10\text{ A}$$

4.14 The coil of a PMMC instrument has 60 turns, on a former that is 18 mm wide, the effective length of the conductor being 25 mm. It moves in a uniform field of flux density 0.5 Tesla. The control spring constant is 1.5×10^{-6} Nm/degree. Calculate the current required to produce a deflection of 100 degree.

[5 marks : 2016]

Solution:

$$N = 60$$
$$W = 18\text{ mm}$$
$$l = 25\text{ mm}$$
$$B = 0.5\text{ J}$$
$$K = 1.5 \times 10^{-6}\text{ Nm/degree}$$
$$\theta = 100^\circ$$
$$I = ?$$

At balance, $T_c = T_d$

$$K\theta = NBI(l \times W)$$
$$1.5 \times 10^{-6} \times 100 = 60 \times 0.5 \times 18 \times 25 \times 10^{-6} \times I$$
$$I = 0.011\text{ A} = 11\text{ mA}$$

4.15 A moving coil ammeter has a fixed shunt of 0.02 Ω. With a coil resistance of R = 1000 Ω and a potential difference of 500 mV across it, full scale deflection is obtained.

(a) To what shunted current does it correspond?

(b) Calculate the value of R to give full scale deflection when shunted current I is
(*i*) 10 A, and (*ii*) 75 A.

(c) With what value of R, is 40% deflection obtained with I = 100?

[10 marks : 2016]

Solution:

$$R_{SH} = 0.02\ \Omega$$
$$R = 1000\ \Omega = 1\text{ k}\Omega$$
$$V_m = 500\text{ mV}$$

(a)

$$I_{sh} = \frac{V_m}{R_{sh}} = \frac{500 \times 10^{-3}}{1000} = 0.5\text{ mA}$$

$$= \frac{V_m}{R_{SH}} = \frac{500 \times 10^{-3}}{0.02\ \Omega} = 25\text{ A}$$

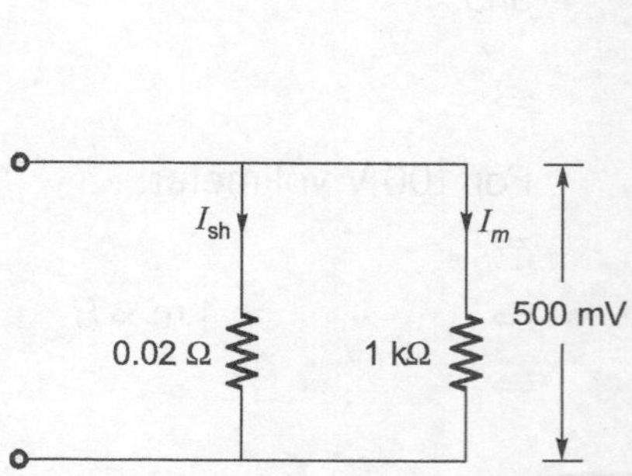

$$I_m = \frac{500 \times 10^{-3}}{1000} = 0.5\text{ mA}$$

(b) (i)

$$I_{sh} = 10\text{ A}$$
$$10 \times 0.02 = R(0.5 \times 10^{-3})$$
$$R = 400\ \Omega$$

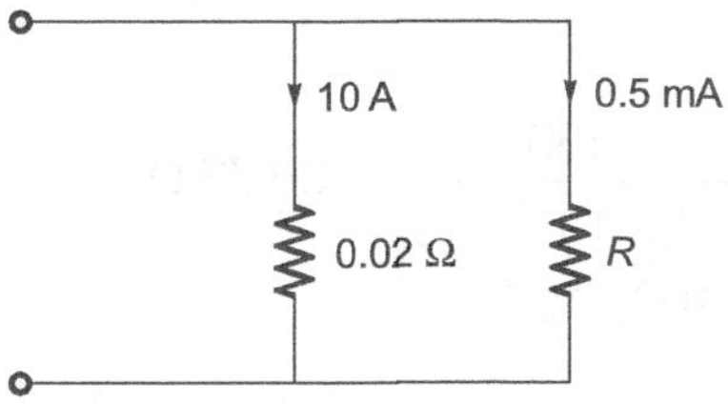

(ii)

$$I_{sh} = 75 \text{ A}$$
$$75 \times 0.02 = R(0.5 \times 10^{-3})$$
$$R = 3000 \ \Omega$$

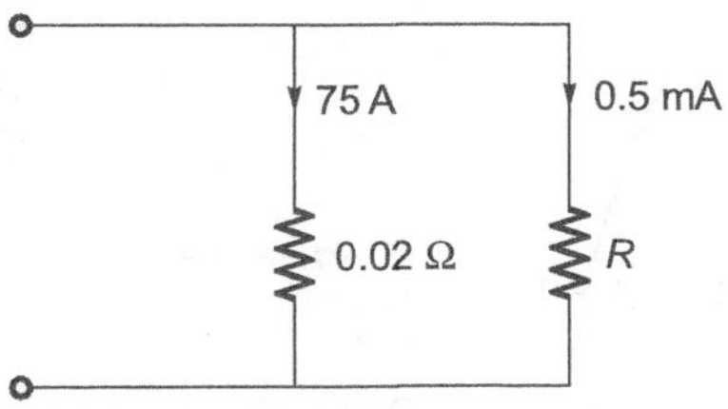

(c)

$$40\% \text{ of deflection} = 0.4 \times 0.5 = 0.2 \text{ mA}$$
$$100 \times 0.02 = 0.2 \times 10^{-3} \times R$$
$$R = 10 \text{ k}\Omega$$

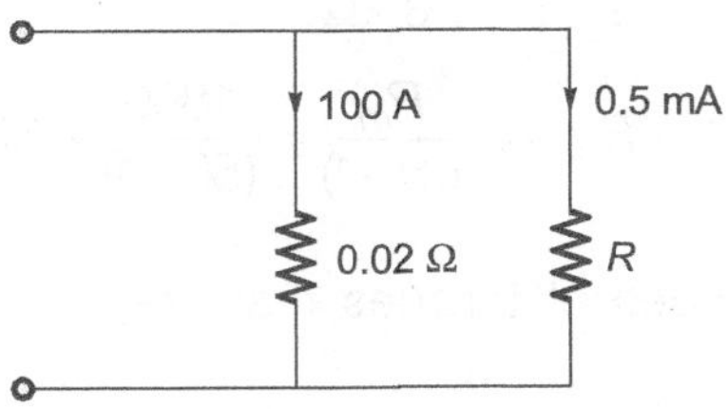

4.16 A simple shunted ammeter using a basic meter movement with an internal resistance of 1800 Ω and a full scale deflection current of 100 μA is connected in a circuit and gives reading of 3.5 mA on its 5 mA scale. The reading is checked with a recently calibrated dc ammeter which gives a reading of 4.1 mA. The implication is that the ammeter has a faulty shunt on its 5 mA range. Calculate:
(a) the actual value of faulty shunt, and
(b) the current shunt for the 5 mA range.

[10 marks : 2016]

Solution:

(a) 5 mA corresponds to 100 μA

For 3.5 mA

$$\Rightarrow \quad \frac{100 \times 10^{-6} \times 3.5}{5} = 7 \times 10^{-5} \text{ A} = I_m$$

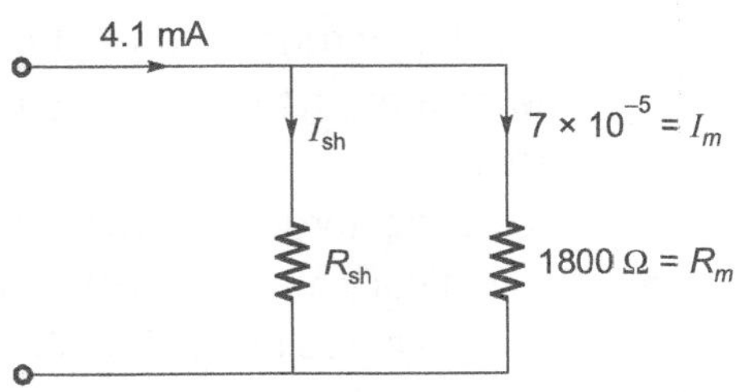

Actual current is 4.1 mA

$$I_{sh} = (4.1 \times 10^{-3} - 7 \times 10^{-5})$$
$$= 4.03 \times 10^{-3}$$
$$R_{sh} I_{sh} = R_m \cdot I_m$$
$$R_{sh} = \frac{1800 \times 7 \times 10^{-5}}{4.03 \times 10^{-3}} = 31.26 \ \Omega$$

(b) For 5 mA range:

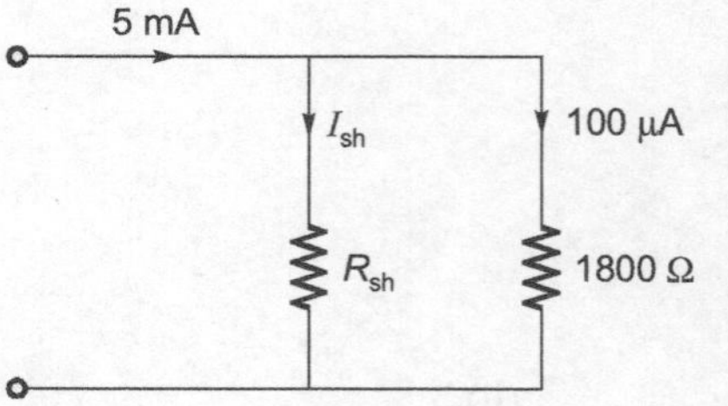

$$R_{sh} = \frac{R_m}{(m-1)} = \frac{R_m}{\left(\frac{I}{I_m}-1\right)} = \frac{180}{\left(\frac{5\times10^{-3}}{100\times10^{-6}}-1\right)} = 36.73\ \Omega$$

Now the actual value of reading is 4.1 mA.

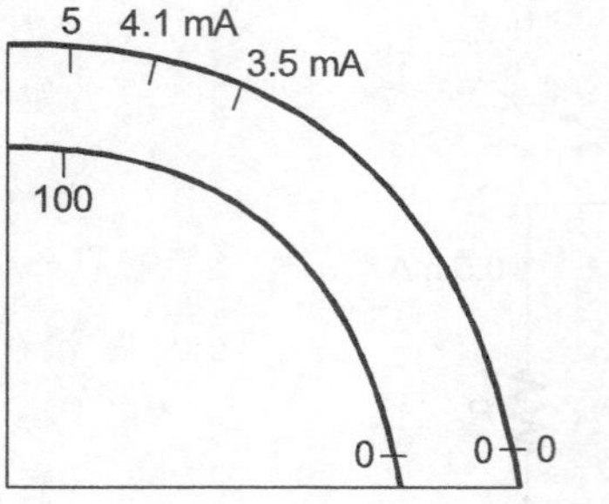

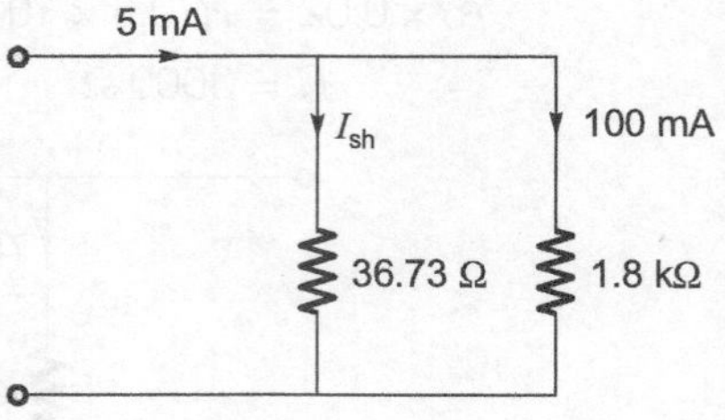

$$I_m = 100\text{ mA}, I = 5\text{ mA}$$

$$m = \frac{I}{I_m} = \frac{5\times10^{-3}}{100\times10^{-6}} = 50$$

$$R_m = 1.8\text{ k}\Omega$$

$$R_{sh} = \frac{R_m}{(m-1)} = \frac{1800}{(50-1)} = 36.73\ \Omega$$

4.17 **Explain in detail the different types of torques that are needed for satisfactory operation of any indicating instrument.**

[10 marks : 2016]

Solution:

Torques in Electromechanical Indicating Instruments:

Three types of torque are needed for the satisfactory operation of any indicating instruments which are as follows: (1) Deflecting torque (2) Controlling torque (3) Damping Torque.

Deflecting Torque (T_d):

The deflecting torque is the amount of torque needed for the pointer to move from it's zero position.

Deflecting torque (T_d) $\propto$ Measureable quantity

Controlling Torque (T_c):

- The controlling torque is in opposite to the deflecting torque. At a point where controlling torque and deflecting torque is equal, the pointer stops and shows the measured value of the reading.
- The controlling torque is also used to bring the pointer to the zero initial position if there is no deflecting torque.
- The control torque is provided by spring or gravity control.
- For vertically mounted instruments, gravity control is used.
- Spring is made up of phospher Bronze to reduce the heating effect.
- If the control spring has been failed then, the pointer moves to the maximum position of the scale except in PMMC.
- In case of PMMC instrument the pointer comes to zero initial position due to the current passing through the spring.

Damping Torque (T_D):

- Analog instruments are of 2nd order type which has a damping factor, $\delta = 0.6$ to 0.8.
- The type of damping is underdamped.
- When a deflecting force is applied to the moving system, it deflects and should come to rest at a position when controlling torque equals to and opposite to deflecting torque. But, due to inertia the pointer keeps oscillating about it's equilibrium. Therefore, a damping torque is required to damp out the oscillation of the pointer.

4.18 **Explain in detail the principle of operation of Attraction-type Moving-iron instruments.**

[10 marks : 2016]

Solution:

Moving Iron (M.I.) Instruments:

- The moving iron instruments are most common and can be constructed to measure current and voltage to an accuracy needed in most engineering works.
- Principle of operation : When the current flows through the coil a magnetic field is produced. The iron vane experiences magnetic field in one part of the vane and hence it moves (rotates) towards the minimum reluctance path and the pointer indicates the current flowing through the coil.
- Figure shows the two type of moving iron instruments namely attraction type and repulsion type.

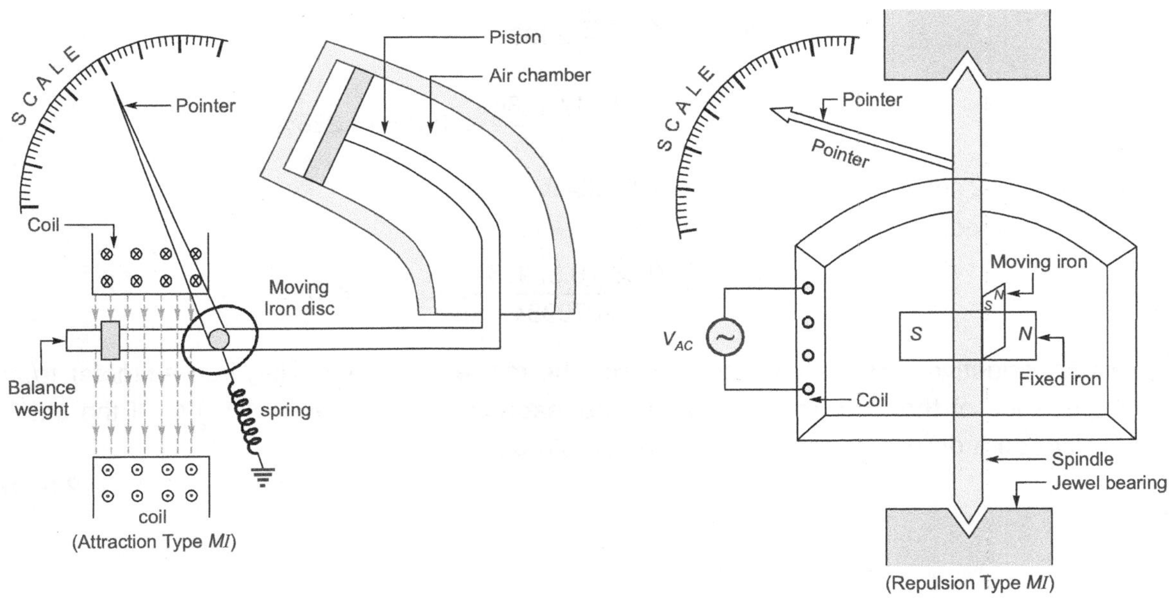

Construction of moving iron instruments

General Torque Equation of Moving Iron Instruments:

∴ Deflecting torque, $\boxed{T_d = \frac{1}{2} I^2 \frac{dL}{d\theta}}$

where,
θ = deflection of the pointer
L = Inductance of the coil
I = rms value of the current in the coil

Controlling Torque, $\boxed{T_c = k\theta}$ (k = spring constant)

At equilibrium position, $T_c = T_d$

or, $k\theta = \frac{1}{2} I^2 \frac{dL}{d\theta}$ or $\boxed{\theta = \frac{1}{2}\frac{I^2}{k}\cdot\frac{dL}{d\theta}}$ or $\boxed{\theta \propto I^2}$

Thus, the deflection is proportional to the square of the rms value of the operating current.

4.19 **A standard cell of 1.0185 V used with a simple potentiometer balances at 50 cm. Calculate:**
(i) the emf of cell which balances at 72 cm.
(ii) the percentage error in voltmeter balances at 64.5 cm when reading 1.33 V.
(iii) percentage error in an ammeter that reads 0.43 A and balance is obtained at 43.2 cm with Pd across a 2 Ω resistor in the ammeter circuit.

[12 marks : 2018]

Solution:

$$V_s = 1.0185 \text{ V}, \quad l_s = 50 \text{ cm}$$

$$1 \text{ cm} = \frac{1.0185}{50} = 0.02037 \text{ V}$$

(i) Emf of cell balanced at 72 cm = 72 × 0.02037 = 1.46664 V

(ii) For 64.5 cm, $64.5 \times 0.02037 = 1.313865 \text{ V}$

$$= \text{True voltage} = V_T$$

$$\text{Measured voltage} = V_m = 1.33 \text{ V}$$

$$\%\text{Error} = \%\in_r = \frac{V_m - V_T}{V_T} \times 100$$

$$= \frac{1.33 - 1.313865}{1.313865} \times 100 = +1.228\%$$

(iii) For 43.2 cm, $43.2 \times 0.02037 = 0.879984 \text{ V} = V_T$

Measured, $V_m = 0.43 \times 2 = 0.86 \text{ V}$

$$\%\in_r = \frac{0.86 - 0.879984}{0.879984} \times 100 = -2.27\%$$

4.20 **A non-inductive shunt is used to increase the range of 10 A moving iron ammeter to 100 A. The impedance of the instrument including the leads is (0.06 + *j*4.71 × 10^{-3}) Ω. If the combination is correct on a dc circuit, find the error on ac circuit.**

[12 marks : 2018]

Solution:

$$I_m = 10 \text{ A}$$

$$I = 100 \text{ A}$$

$$Z_m = R_m + j\omega L_m$$

$$= (0.06 + j4.71 \times 10^{-3}) \ \Omega$$

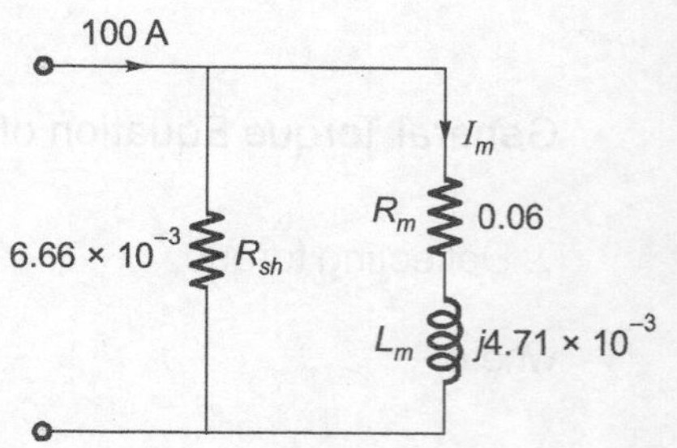

$$R_{sh} = \frac{R_m}{\left(\frac{I}{I_m} - 1\right)} = \frac{0.06}{\left(\frac{100}{10} - 1\right)}$$

$$= 6.66 \times 10^{-3} \ \Omega$$

With AC input:

$$|I_m| = \frac{I \cdot R_{sh}}{\sqrt{(R_{sh} + R_m)^2 + X_m^2}}$$

$$= \frac{100 \times 6.66 \times 10^{-3}}{\sqrt{(6.66 \times 10^{-3} + 0.06)^2 + (4.71 \times 10^{-3})^2}}$$

$$= 9.966 \text{ A} = \text{Measured value}$$

$$\%\epsilon_r = \frac{9.966 - 10}{10} \times 100 = -0.34\%$$

4.21 Calculate the reading of

(i) Moving coil voltmeter (ii) Moving iron voltmeter

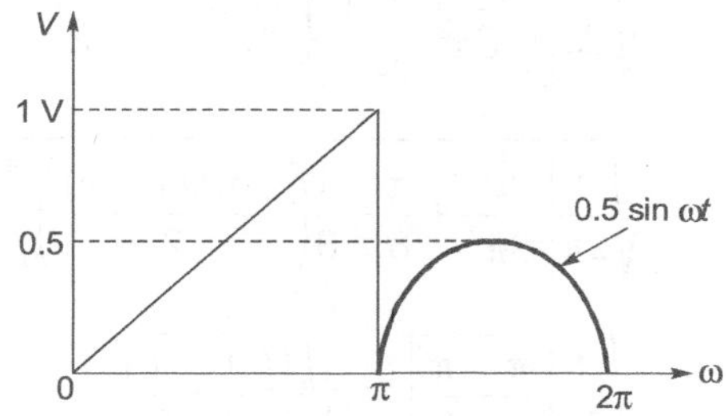

when these voltmeters are measuring the voltage of the waveform shown in the figure.

[20 marks : 2019]

Solution:

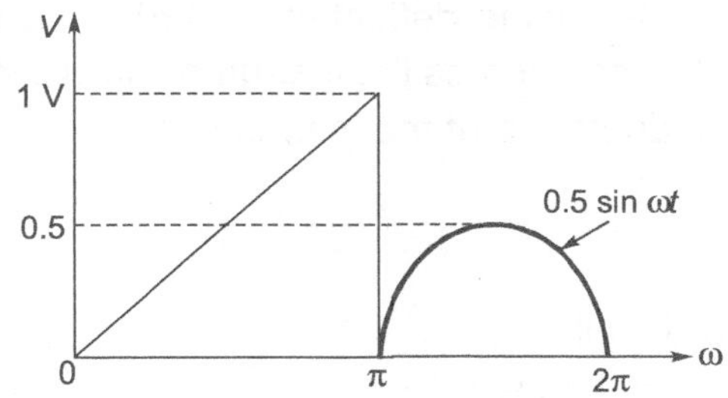

(i) Moving coil voltmeter reads average value of waveform,

$$V_{avg} = \frac{1}{2\pi} \int_0^{2\pi} V(t) \cdot d(\omega t)$$

$$= \frac{1}{2\pi} \left[\int_0^{\pi} \frac{1}{\pi} (\omega t) d(\omega t) + \int_{\pi}^{2\pi} \frac{1}{2} \sin(\omega t - \pi) d(\omega t) \right]$$

$$= \frac{1}{2\pi} \left[\frac{1}{\pi} \left(\frac{\omega t^2}{\pi} \right)_0^{\pi} + \frac{1}{2} (-\cos(\omega t - \pi))_{\pi}^{2\pi} \right]$$

$$= \frac{1}{2\pi} \left[\frac{1}{\pi} \left[\frac{\pi^2}{2} \right] + \frac{1}{2} (\cos 0 - \cos \pi) \right]$$

$$= \frac{1}{2\pi} \left[\frac{\pi}{2} + 1 \right] = \frac{2.57}{2\pi} \text{volt} = 0.41 \text{ V}$$

∴ Moving coil voltmeter reads 0.41 Volt.

(ii) Moving iron voltmeter reads RMS value of waveform,

$$V_{RMS} = \sqrt{\frac{1}{T}\int_0^T [V(t)]^2 \cdot dt}$$

$$= \sqrt{\frac{1}{2\pi}\left[\int_0^{\pi}\left(\frac{1}{\pi}\omega t\right)^2 \cdot d(\omega t) + \int_{\pi}^{2\pi}\left[\frac{1}{2}\sin(\omega t - \pi)\right]^2 d(\omega t)\right]}$$

$$= \sqrt{\frac{1}{2\pi}\left[\left(\frac{(\omega t)^3}{3\pi^2}\right)_0^{\pi} + \int_4^{2\pi}\frac{1}{4}\left(\frac{1-\cos 2(\omega t - \pi)}{2}\right)\right] \cdot d(\omega t)}$$

$$= \sqrt{\frac{1}{2\pi}\left[\frac{\pi^2}{3\pi^2} + \frac{\pi}{8} - \frac{1}{8}\left(\frac{\sin 2(\omega t - \pi)}{2}\right)_0^{2\pi}\right]}$$

$$= \sqrt{\frac{1}{2\pi}\left[\frac{\pi}{3} + \frac{\pi}{8}\right]} = \sqrt{\frac{1}{2}\left(\frac{1}{3} + \frac{1}{8}\right)}$$

$$V_{RMS} = 0.478 \text{ volt}$$

Moving iron voltmeter reads 0.478 volt.

4.22 A spring controlled, electrodynamic voltmeter has a range of 100 V, has a square low scale response, and it takes 0.08 A on d.c. for full scale deflection of 120°. The control constant is 1 x 10^{-6} N-m/degree. The true potential difference across the instrument is 100.42 V, when it reads 100 V at 50 Hz. Determine the initial mutual inductance of the instrument.

[20 marks : 2020]

Solution:

Actual voltage across the instrument

$$= 100.42 \text{ V}$$

Current through instrument when reading 100 V = 0.08 A

Impedance of the instrument

$$= \frac{100.42}{0.08} = 1255.25\ \Omega$$

$$\text{Resistance of instrument} = \frac{100}{0.08} = 1250\,\Omega$$

$$\text{Reactance of instrument} = \sqrt{(1255.25)^2 - (1250)^2} = 114.68\ \Omega$$

Mutual inductance at $\theta = 120°$ is

$$M = \frac{114.68}{2\pi \times 50} = 0.365 \text{ H}$$

The expression for deflection for a square law response voltmeter is

$$\theta = \frac{V^2}{KZ^2}\frac{dM}{d\theta}$$

For d.c. impedance, Z = resistance R and therefore,

$$\theta = \frac{V^2}{KR^2}\frac{dM}{d\theta}$$

$$\frac{V}{R} = 0.08 \text{ A},$$
$$K = 1 \times 10^{-6},$$
$$\theta = 120°$$
$$120° = \frac{(0.08)^2}{1\times10^{-6}}\frac{dM}{d\theta}$$
$$\frac{120°\times1\times10^{-6}}{0.08\times0.08} = \frac{dM}{d\theta}$$
$$\frac{dM}{d\theta} = 0.01875 \text{ H}$$
$$M = \frac{dM}{d\theta}(\theta_2-\theta_1) = 0.01875\times\left(\frac{2\pi}{3}-0\right) = 0.03927$$

Initial mutual inductance = 0.365 – 0.03927 = 0.32573 H

4.23 A dynamometer ammeter is arranged so that 1% of total current passes through the moving coil and rest through the fixed coil. The mutual inductance between the two coils varies with the angle of displacement of the moving coil from its zero position as follows :

Angle (degree)	0	15	30	60	90	105	120
Mutual inductance (μH)	–336	–275	–192	0	192	275	336

If a torque of 10.5×10^{-6} Nm is required to give the full scale deflection of 120°, calculate the current at half and full scale deflection. (Graph sheet attached).

[20 marks : 2021]

Solution:

θ	0	15	30	60	90	105	120
M	–336	–275	–192	0	192	275	336

On plotting the table on graph

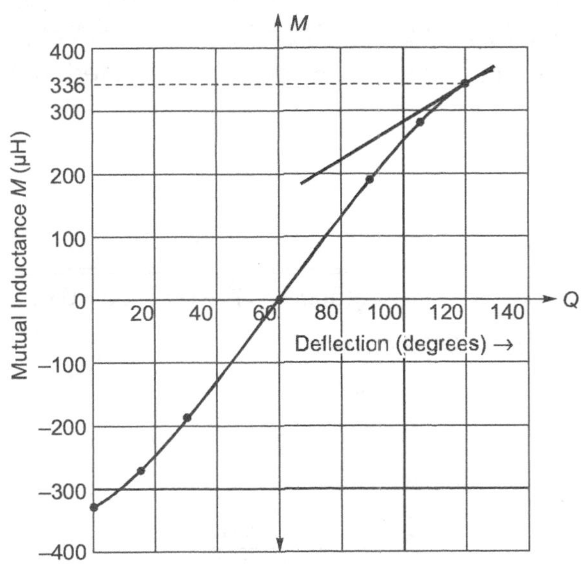

At steady state

$$|T_d| = |T_c|$$

$$K_c \cdot \theta = T_d$$

$$K_c = \frac{T_d}{\theta} = \frac{10.5\times10^{-6}}{\frac{2\pi}{3}} = 5.02\times10^{-6}\text{ N-m/rad}$$

(i) At half scale deflection, $\theta = 60° = \frac{\pi}{3}$ rad

Current through fixed coil $= I_1 = 0.99I$

Moving coil $= I_2 = 0.01I$

where, I = Total current

Scale is linear between 30° to 90°

$$\left(\frac{dM}{d\theta}\right)_{60°} = \frac{192-(-192)}{90°-30°} = 6.4\ \mu\text{H/degree}$$

$$= 366.414\ \mu\text{H/rad}$$

$$\theta = \frac{I_1\cdot I_2}{K_c}\cdot\frac{dM}{d\theta}$$

$$\frac{\pi}{3} = \frac{(0.99I)(0.01I)}{5.02\times10^{-6}}\times366.414\times10^{-6}$$

Current at half deflection, $I = 1.2$ A

(ii) At full scale deflection of $\theta = 120° = \frac{2\pi}{3}$ rad, $\frac{dM}{d\theta}$ is calculated by drawing tangent along the curve.

$$\left(\frac{dM}{d\theta}\right)_{120°} = \frac{336-200}{120-77} = 3.16\ \mu\text{H/degree} = 181.2\ \mu\text{H/rad}$$

$$\theta = \frac{I_1\cdot I_2}{K_c}\cdot\frac{dM}{d\theta}$$

$$\frac{2\pi}{3} = \frac{(0.99I)(0.01I)}{5.02\times10^{-6}}\times181.2\times10^{-6}$$

Full scale deflection current, $I = 2.42$ A

4.24 A coil of 300 V Moving Iron Voltmeter has a resistance of 500 ohms and an inductance of 0.8 H. The instrument reads correctly at 50 Hz AC supply and takes 100 mA at full scale deflection. What is the percentage error in the instrument reading, when it is connected to 200 V DC supply ?

[12 marks : 2022]

Solution:

Let series resistance is R_s.

So, $$I = 0.10 = \frac{300}{\sqrt{(500+R_s)^2+(2\pi\times50\times0.8)^2}}$$

Solving this, $R_s = 2489.45$ ohm

The current at 200 V, 50 Hz $$I_T = (0.1)\left(\frac{200}{300}\right)$$

$$I_T = 0.0667\text{ Amp}$$

The current at 200 V, DC $$I_M = \frac{V_{dc}}{R_m+R_s} = \frac{200}{500+2489.45}$$

$$I_M = 0.0669\text{ Amp}$$

The voltage reading will be proportional to the current value.
So, error in measurement

$$\therefore \quad \text{Error} = \frac{I_M - I_T}{I_T} \times 100$$

$$= \frac{0.0669 - 0.0667}{0.0667} \times 100$$

$$\text{Error} = 0.30\%$$

4.25 **A moving coil instrument is designed to give a full scale reading of 30 mA and the corresponding drop across is 72 mV. Calculate the shunt resistance required for full scale deflection corresponding to 10 Amps and the series resistance for full scale reading with 400 V.**

[10 marks : 2022]

Solution:

Given, $I_{fSD} = 30$ mA,
Corresponding voltage drop, $V = 72$ mV

Internal resistance of meter, $R_m = \frac{V}{I_{fSD}}$

$$R_m = \frac{72}{30} = 2.4\,\Omega$$

Now given, $I = 10$ A

Therefore, multiplying factor, $M = \frac{I}{I_{fSD}} = \frac{10 \times 10^3}{30} = \frac{1000}{3}$

To read the given meter $I = 10$ A at full scale deflection, the shunt resistance to be added in parallel is given as

$$R_{sh} = \frac{R_m}{m-1} = \frac{2.4}{\frac{1000}{3} - 1} = 7.2216\text{ M}\Omega$$

Similarly to read the given meter,

$$V = 400\text{ V}$$

At full scale deflection,

Multiplication factor, $M = \frac{V}{V_{fSD}} = \frac{400}{72 \times 10^{-3}} = 5555.555$

To read the given meter, $V = 400$ V at full scale deflection,
The required series multiplier resistance is given as

$$R_{se} = R_m(m-1)$$
$$= 2.4[5555.55 - 1] = 13.33\text{ k}\Omega$$

4.26 (i) **A moving-coil instrument has a resistance of 5 Ω between terminals and full-scale deflection is obtained with a current of 15 mA. This instrument is to be used with a manganin shunt to measure 100 A at full scale. Calculate the error caused by a 10°C rise in temperature :**

(1) when the internal resistance of 5 Ω is due to copper only;

(2) when a 3 Ω manganin swamping resistance is used in series with a copper coil of 2 Ω resistance.

(The resistance temperature coefficients of copper and manganin are 0.004/°C and 0.000015/°C respectively).

(ii) Draw the block diagram of ramp-type digital voltmeter and explain its functioning.

[10 + 10 marks : 2023]

Solution:

(i) Given that : $I_m = 15\text{ mA}, I = 100\text{ A}$

$$m = \frac{I}{I_m} = \frac{100}{15\times10^{-3}} = 6666.67$$

$$R_{sh} = \frac{R_m}{(m-1)} = \frac{5\ \Omega}{(6666.67-1)} = 0.75\text{ m}\Omega$$

Meter resistance for 10°C rise in temperature.

$$R_{mt} = 5(1 + 0.004 \times 10) = 5.2\ \Omega$$

Shunt resistance for 10°C rise in temperature

$$R_{sh} = 0.75\text{ m}\Omega(1 + 0.000015 \times 10) = 0.7501125\text{ m}\Omega$$

Current I_{mt} through the meter for 100 A in the main circuit for 10°C rise in temperature

$$I_{mt} = 100\times\frac{0.7501125\text{ m}\Omega}{5.2+0.7501125\text{ m}\Omega}$$

$$= 0.014423\text{ Amp} = 14.423\text{ mA}$$

$$\%\text{ Errror} = \frac{14.423\text{ mA} - 15\text{ mA}}{15\text{ mA}}\times100 = -3.845\%$$

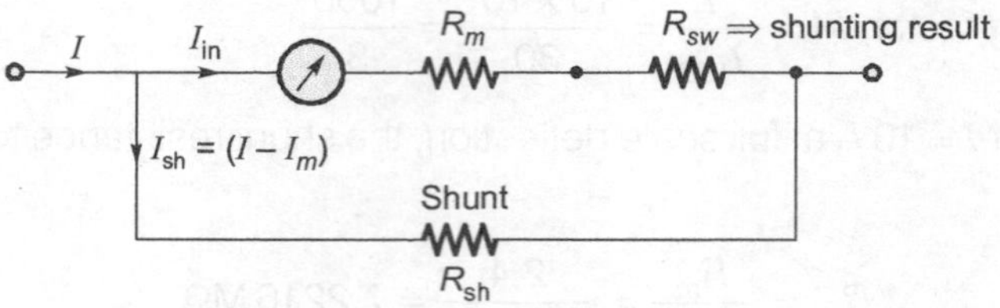

Total resistance in the meter circuit,

$$R_m + R_{SW} = 2 + 3 = 5\ \Omega$$

$$R_{sh} = \frac{R_m}{(m-1)} = \frac{5\ \Omega}{6666.67-1} = 0.75\text{ m}\Omega$$

Resistance of the meter for 10°C rise in temperature

$$R_{mt} = 2\ \Omega[1 + 0.004 \times 10] + 3\ \Omega(1 + 0.000015 \times 10]$$

$$R_{mt} = 2[1 + 0.04] + 3[1 + 0.00015] = 5.08045\ \Omega$$

Shunt resistance for 10°C rise in temperkature

$$R_{sht} = 0.75\text{ m}\Omega[1 + 0.000015 \times 10] = 0.7501125\text{ m}\Omega$$

$$I_{mt} = 100\times\frac{0.7501125\text{ m}\Omega}{5.08045+0.7501125\text{ m}\Omega}$$

$$= 0.0147625\text{ Amp}$$

$$I_{mt} = 14.7625\text{ Amp}$$

$$\%\text{ Error} = \frac{14.7625-15}{15}\times100$$

$$= -1.5833\%$$

(ii) **Ramp Type Digital Voltmeter :** The operating principle of a ramp type digital voltmeter is to measure the time that a linear ramp voltage takes to change from level of input to zero voltage (or vice versa). This time interval is measured with an electronic time interval counter and the count is displayed as a number of digits on electronic indicating tubes of the output readout of the voltmeter.

The conversion of a voltage value of a time interval is shown in the timing diagram of figure.

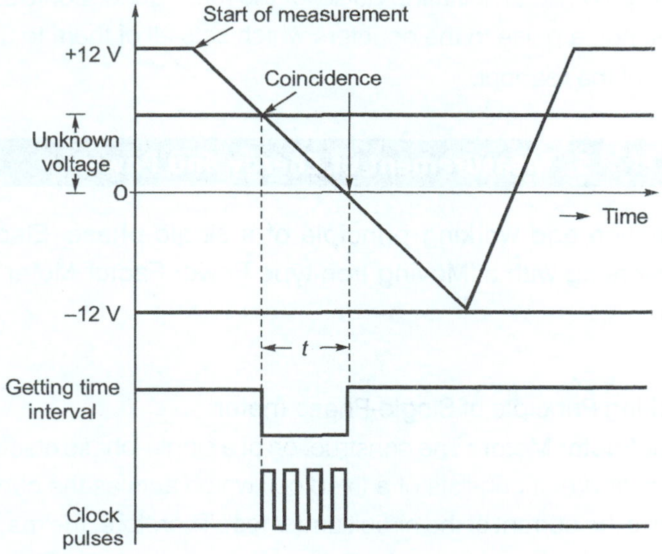

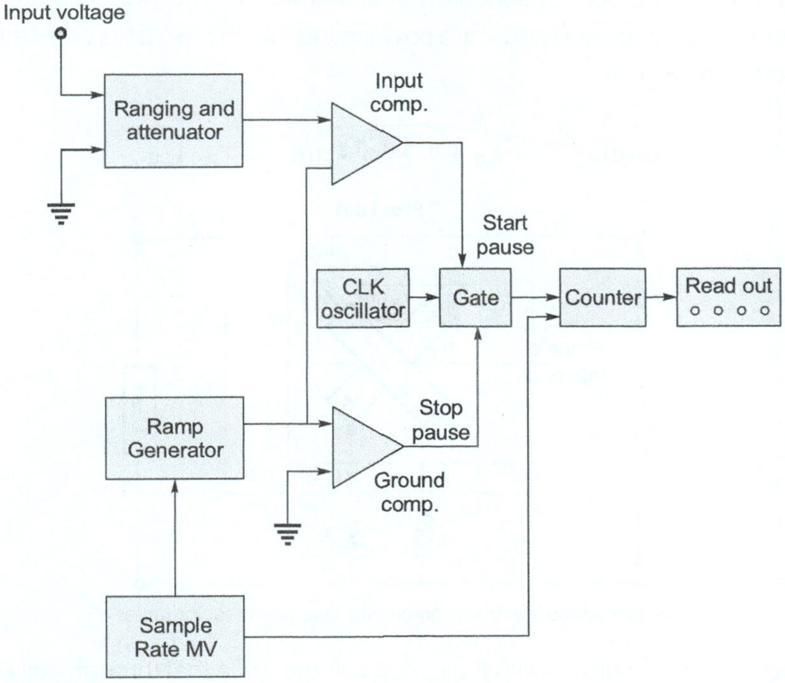

At the start of measurement a ramp voltage is initiated. A negative going ramp is shown in figure but a positive going ramp may also be used. The ramp voltage value is continuously compared with the voltage being measured (unknown voltage). At the instant the value of ramp voltage is equal to that of unknown voltage a coincidence circuit, called an input comparator, generates a pulse which opens a gate (see figure). The ramp voltage continues to decrease till it reaches ground level (zero voltage). At this instant another comparator called ground comparator generates a pulse and closes the gate. The time elapsed between opening and closing the gate is *t* as indicated in figure. During this time interval pulses from a clock pulse generator pass through the gate and are counted and displayed. The decimal number as indicated by the readout is a measure of the value of input voltage.

The sample rate multivibrator determines the rate at which the measurement cycles are initiated. The sample rate circuit provides an initiating pulse for the ramp generator to start its next ramp voltage. At the same time it sends a pulse to the counters which sets all of them to 0. This momentarily removes the digital display of the readout.

5. Measurement of Energy and Power

5.1 Describe the construction and working principle of a single-phase 'Electrodynamic Power Factor Meter'. Compare its working with a 'Moving Iron type Power Factor Meter'.

[8 marks : 2001]

Solution:

Construction and Working Principle of Single-Phase meter:

Electrodynamic Power Factor Meter : The construction of a single-phase electro dynamometer type power factor meter is shown in figure. It consists of a fixed coil which acts as the current coil. This coil is split up into two parts and carries the current of the circuit under test. Therefore, the magnetic field produced by this coil is proportional to the main current. Two identical pressure coils *A* and *B* pivoted on a spindle constitute the moving system. Pressure coil *A* has a non-inductive resistance *R* connected in series with it, and coil *B* has a highly inductive choke coil L connected in series with it. The two coils are connected across the voltage of the circuit. The values of *R* and *L* are so adjusted that the two coils carry the same value of current at normal frequency, i.e., $R = \omega L$.

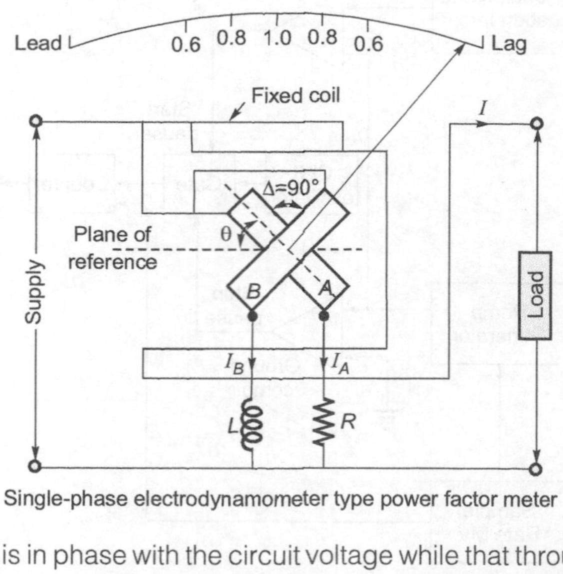

Single-phase electrodynamometer type power factor meter

The current through coil *A* is in phase with the circuit voltage while that through coil *B* lags the voltage by an angle Δ which is nearly equal to 90°. The angle between the planes of coils is also made equal to Δ. There is no controlling device. Connections to moving coils are made through thin silver or gold ligaments which are extremely flexible and thus give a minimum control effect on the moving system.

In order to simplify the problem, we assume that the current through coil *B* lags the voltage by exactly 90°. Also that the angle between planes of coils is exactly 90°, (i.e., Δ = 90°).

Now there will be two deflecting torques, one acting on coil A and the other on coil B. The coil windings are so arranged that the torques due to the two coils are opposite in direction. Therefore the pointer will take up a position where these two torques are equal.

Let us consider the case of a lagging power factor of $\cos\phi$.

Deflecting torque acting on coil A is:

$$T_A = KVI\, M_{max} \cos\phi \sin\theta$$

where, θ = angular deflection from the plane of reference.

and M_{max} = maximum value of mutual inductance between the two coils.

This torque say act in the clockwise direction.

Deflection torque acting on coil B is:

$$T_B = KVI\, M_{max} \cos(90° - \phi) \sin(90° + \theta) = KVI\, M_{max} \sin\phi \cos\theta$$

This torque acts in the anticlock wise direction. The value of M_{max} is the same in the two expressions, owing to similar constructions of the coils.

The coils will take up such a position that the two torques are equal.

Hence, at equilibrium, $T_A = T_B$

or $KVI\, M_{max} \cos\phi \sin\theta = KVI\, M_{max} \sin\phi \cos\theta$

or $\theta = \phi$

Therefore the deflection of the instrument is a measure of phase angle of the circuit. The scale of the instrument can be calibrated in directly in terms of power factor.

Comparison with the Moving Iron Power Factor Meter:

1. In electrodynamometer type P.F. meter working forces are very small as compared to moving iron type.
2. The scale of the moving iron type extends over 360°.
3. Electrodynamometer type is more accurate than the moving iron type.

5.2 Explain the phenomena of 'creeping'. If an energy meter disc makes 10 revolutions in 100 seconds when a load of 360 W is connected to it, determine the meter constant in revolutions/kWh.

[8 marks : 2002]

Solution:

In some meters a slow but continuous rotation is obtained even when there is no current flowing through the current coil and only pressure coil is energized. This is called creeping. The major cause for creeping is over-compensation for friction. If the friction compensating device is adjusted to give a driving torque to compensate for starting friction which is bigger than the running friction, there is a tendency for the disc to run even when there is no current through the current coils because the friction compensating torque is independent of the load current as the compensating device is voltage actuated.

$$\text{Load of kW} = \frac{\text{Energy consumed in kWh}}{\text{Time in hours}}$$

$$0.36 = \frac{\text{Energy consumed in kWh}}{100/3600}$$

$$\text{Energy consumed in kWh} = 0.36 \times \frac{1}{36} = 0.01$$

$$\text{Meter constant} = \frac{\text{Number of revolutions}}{\text{Energy consumed in kWh}} = \frac{10}{0.01}$$

$$= 1000 \text{ revolutions/kWh}$$

5.3 **Describe a method of power measurement for a 3-phase 3-wire unbalanced load. How can power-factor of a balanced load be determined by the method?**

[14 marks : 2002]

Solution:

In a three phase three wire system we require 3 elements but if we make the common points of the pressure coils coincide with one of the lines, then we will require only $n - 1 = 2$ elements.

Instantaneous power consumed by load

$$= V_1 I_1 + V_2 I_2 + V_3 I_3$$

Star (Wye) connection: Instantaneous reading of P_1 wattmeter,

$$P_1 = V_{13} I_1 = I_1(V_1 - V_3)$$

Instantaneous reading of P_2 wattmeter,

$$P_2 = V_{23} I_2 = I_2(V_2 - V_3)$$

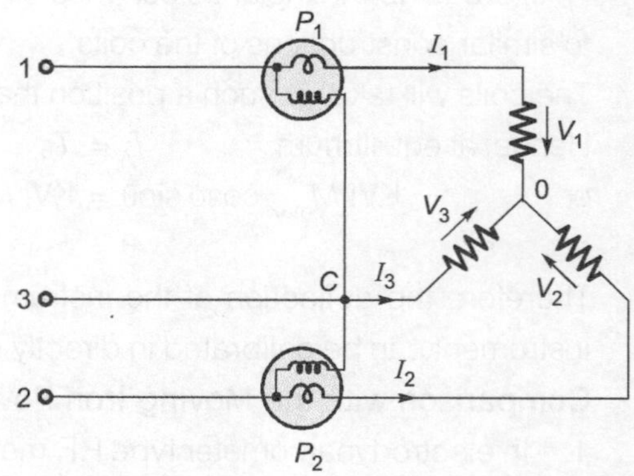

Sum of instantaneous readings of two wattmeters

$$= P_1 + P_2$$
$$= I_1(V_1 - V_3) + I_2(V_2 - V_3)$$
$$= V_1 I_1 + V_2 I_2 - V_3(I_1 + I_2)$$

By Kirchhoff's law,

$$I_1 + I_2 + I_3 = 0$$

or

$$I_3 = -(I_1 + I_2)$$

∴ Sum of instantaneous readings of two wattmeters

$$= V_1 I_1 + V_2 I_2 + V_3 I_3$$

Therefore, the sum of the two wattmeter reading is equal to the power consumed by the load.

Let V_1, V_2, V_3 be the rms values of the phase voltage and I_1, I_2, I_3 be the rms value of phase currents.

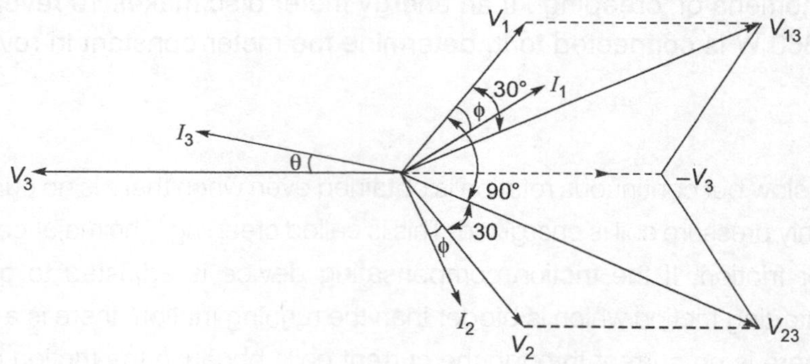

The load is balanced therefore,

Phase voltages, $V_1 = V_2 = V_3 = V$ (say)

Line voltages, $V_{13} = V_{23} = V_{12} = \sqrt{3}\, V$

Phase currents, $I_1 = I_2 = I_3 = I$ (say)

Line currents, $I_1 = I_2 = I_3 = I$

Power factor $= \cos\phi$

The phase current lag the corresponding phasor voltages by an angle ϕ.

The current through wattmeter P_1 is I_1 and voltage across its pressure coil is V_{13}. I_1 leads V_{13} by an angle $(30° - \phi)$.

∴ Reading of P_1 wattmeter,

$$P_1 = V_{13} I_1 \cos(30° - \phi) = \sqrt{3}\; VI \cos(30° - \phi)$$

The current through wattmeter P_2 is I_2 and voltage across its pressure coil is V_{23}. I_2 lags V_{23} by an angle $(30 + \phi)$.

$\therefore$ Reading of P_2 wattmeter,

$$P_2 = V_{23} I_2 \cos(30° + \phi) = \sqrt{3}\ VI \cos(30° + \phi)$$

Sum of reading of two wattmeters:

$$P_1 + P_2 = \sqrt{3}\ VI [\cos(30° - \phi) + \cos(30° + \phi)]$$
$$= 3\ VI \cos\phi$$

This is the total power consumed by load.

$\therefore$ Total power consumed by load, $P = P_1 + P_2$

Difference of readings of two wattmeters

$$P_1 - P_2 = \sqrt{3}\ VI [\cos(30° - \phi) - \cos(30° + \phi)] = \sqrt{3}\ VI \sin\phi$$

$\therefore$ $$\frac{P_1 - P_2}{P_1 + P_2} = \frac{\sqrt{3}\,VI \sin\phi}{3\,VI \cos\phi} = \frac{\tan\phi}{\sqrt{3}}$$

or $$\phi = \tan^{-1}\sqrt{3}\,\frac{P_1 - P_2}{P_1 + P_2}$$

Power factor, $$\cos\phi = \cos\left(\tan^{-1}\sqrt{3}\,\frac{P_1 - P_2}{P_1 + P_2}\right)$$

5.4 **Sketch the circuit diagram for power measurement in a 3-phase circuit using two wattmeters and show that total power is given by the algebraic sum of the wattmeters readings using vector diagrams.**
[12 marks : 2003]

Solution:

Let us consider two wattmeters connected to measure power in three phase circuit as shown below in figure.

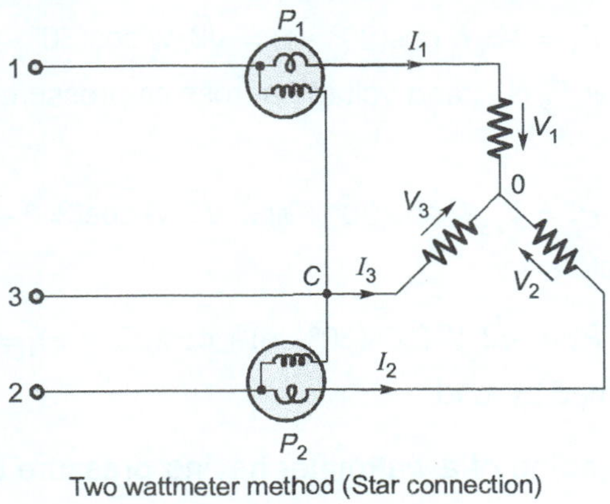

Two wattmeter method (Star connection)

Star Connection:

Instantaneous reading of P_1 wattmeter,

$$P_1 = I_1(V_1 - V_3)$$

Instantaneous reading of P_2 wattmeter,

$$P_2 = I_2(V_2 - V_3)$$

$\therefore$ Sum of instantaneous readings of two wattmeters

$$= P_1 + P_2 = I_1(V_1 - V_3) + I_2(V_2 - V_3) = V_1 I_1 + V_2 I_2 - V_3(I_1 + I_2)$$

From Kirchhoff's law:

$$I_1 + I_2 + I_3 = 0$$

or $$I_3 = -(I_1 + I_2)$$

∴ Sum of instantaneous readings of two wattmeters

$$= V_1 I_1 + V_2 I_2 + V_3 I_3$$

Therefore the sum of the two wattmeter reading is equal to the power consumed by the load.

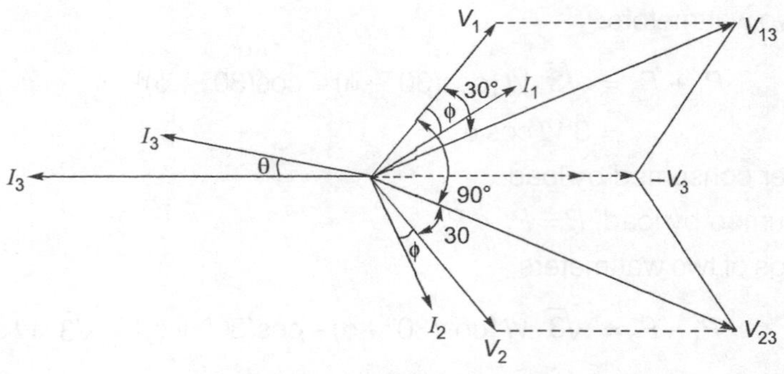

Phaser diagram for balanced star connected load

Let V_1, V_2, V_3 be the rms values of phase voltage and I_1, I_2, I_3 be the rms values of phase currents. The load is balanced therefore,

Phase voltage, $V_1 = V_2 = V_3 = V$ (say)

Line voltage, $V_{13} = V_{23} = V_{12} = \sqrt{3}\,V$

Phase currents, $I_1 = I_2 = I_3 = I$ (say)

Line currents, $I_1 = I_2 = I_3 = I$

Power factor $= \cos\phi$

The phase currents lag the corresponding phasor voltages by an angle ϕ.

The current through wattmeter P_1 is I_1 and voltage across its pressure coil is V_{13}. I_1 leads V_{13} by an angle $(30° - \phi)$.

∴ Reading of P_1 wattmeter, $P_1 = V_{13} I_1 \cos(30° - \phi) = \sqrt{3}\, VI \cos(30° - \phi)$

The current through wattmeter P_2 is I_2 and voltage across its pressure coil is V_{23}. I_2 lags V_{23} by an angle $(30° + \phi)$.

∴ Reading of P_2 wattmeter, $P_2 = V_{23} I_2 \cos(30° + \phi) = \sqrt{3}\, VI \cos(30° + \phi)$

Sum of reading of two wattmeters:

$$P_1 + P_2 = \sqrt{3}\, VI\,[\cos(30° - \phi) + \cos(30° + \phi)] = 3\, VI \cos\phi$$

This is the total power consumed by load.

5.5 Derive the expression for reading of a wattmeter having pressure coil inductance.

[10 marks : 2004]

Solution:

Let,
r_p = resistance of pressure coil,
L = inductance of pressure coil,
R = resistance in series with pressure coil,
R_p = total resistance of pressure coil circuit,
= $r_p + R$, V = voltage applied to pressure coil circuit,
I = current in the current coil circuit,

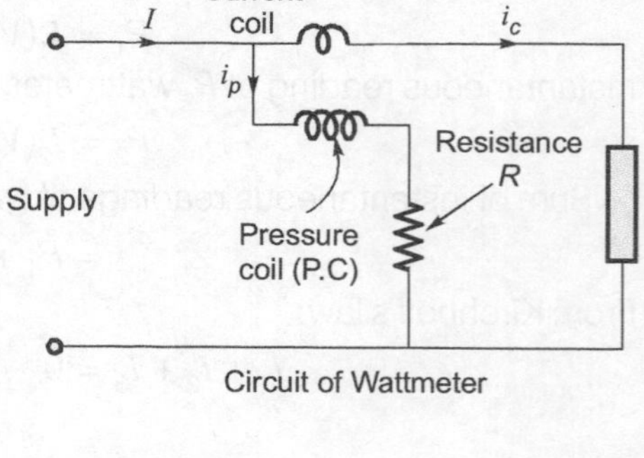

Circuit of Wattmeter

I_P = current in the pressure coil circuit,

Z_P = impedance of pressure coil circuit

$$= \sqrt{(R + r_P)^2 + (\omega L)^2}$$

If the pressure coils of the wattmeter has an inductance the current in it will lag the voltage by an angle β where:

$$\beta = \tan^{-1}\frac{\omega L}{(R + r_P)}$$

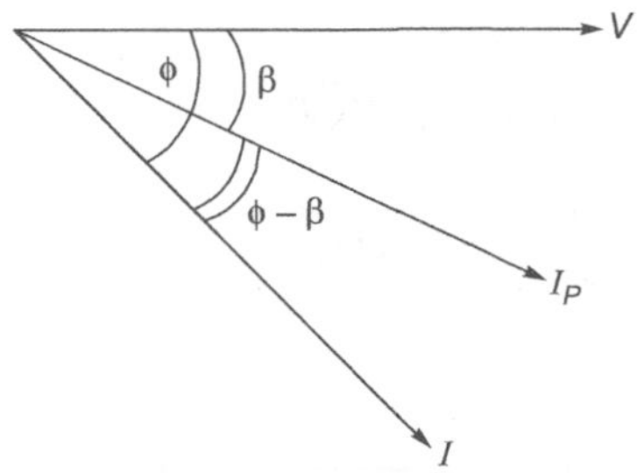

Wattmeter phaser diagram for lagging p.f.

From above figure, for lagging power factor, the angle between current in the current coil circuit and the current in the pressure coil circuit is less than ϕ, by which the load current lags the applied voltage. The angle between the pressure coil current and the current in the current coil is:

$$\phi' = \phi - \beta$$

$\therefore$ The actual wattmeter reading is:

$$= \left(\frac{I_P I}{K}\right)\cos\phi' \cdot \frac{dM}{d\theta} = \left(\frac{V}{Z_P K}\right) \cdot I \cdot \cos(\phi - \beta) \cdot \frac{dM}{d\theta}$$

Now, $$Z_P = \frac{R_P}{\cos\beta}$$

$$\text{Actual wattmeter reading} = \frac{VI\cos(\phi-\beta)}{K(R_P/\cos\beta)} \cdot \frac{dM}{d\theta} = \frac{VI}{KR_P}\cos\beta \cdot \cos(\phi - \beta) \cdot \frac{dM}{d\theta}$$

5.6 The reading of a dynamometer type wattmeter with pressure coil phase angle of 2° is 700 watts, when it is used to measure power of a single-phase inductive load supplied by 240 V single-phase a.c. source. When the wattmeter is replaced by a second wattmeter with a phase angle of 1° for the pressure coil circuit, a reading of 620 watts is obtained. Calculate the actual power and current taken by the load assuming all errors of the wattmeter, except those due to pressure coil inductance, are neglected.

[12 marks : 2004]

Solution:

$$\text{True power} = \frac{\cos\phi}{\cos\beta\cos(\phi-\beta)} \times \text{actual wattmeter reading}$$

This can be simplified as

$$\text{actual wattmeter reading} = \text{true power}(1 + \tan\phi\tan\beta)$$

$$700 = TP(1 + \tan\phi\tan 2°)$$

$$620 = TP(1 + \tan\phi\tan 1°)$$

$$\frac{700}{620} = \frac{1 + \tan 2°\tan\phi}{1 + \tan 1°\tan\phi}$$

$$700(1 + 0.0175\tan\phi) = 620(1 + 0.035\tan\phi)$$

$$700 + 12.25 \tan\phi = 620 + 21.7 \tan\phi$$
$$9.45 \tan\phi = 80$$
$$\tan\phi = \frac{80}{9.45} = 8.4656$$
$$\phi = 83.26°$$
$$\text{True power} = \frac{\text{Actual wattmeter reading}}{1 + \tan\phi \tan\beta}$$
$$= \frac{700}{1 + 0.035 \times 8.4656}$$
$$= \frac{700}{1.296} = 540 \text{ watts}$$
$$VI \cos\phi = 540$$
$$I = \frac{540}{240 \times \cos 83.26°} = 19.17 \text{ A}$$

5.7 Explain how a dynamometer type wattmeter can be used to measure power in a circuit having low power factor.

[8 marks : 2004]

Solution:

To measure power in a circuit having low power factor these features are required:

(a) **Pressure Coil Current:** The pressure coil circuit is designed to have a low value of resistance, so that the current, flowing through it, is increased to give an increased operating torque. The pressure coil current in a low power factor wattmeter may be as much as 10 times the value employed for high power factor wattmeter.

(b) **Compensation for Pressure Coil Current:** The power being measured in a low power factor circuit is small and current is high on account of low power factor gives large error in the wattmeter reading. To compensate for the pressure coil current in a low power factor wattmeter this compensation is required.

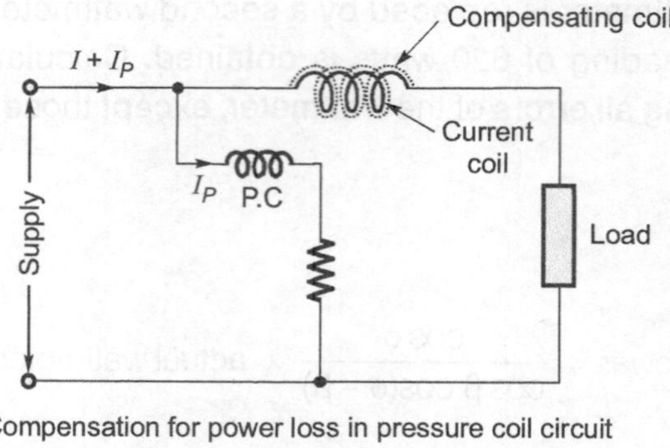

Compensation for power loss in pressure coil circuit

(c) **Compensation for Inductance of Pressure Coil:** The error caused by pressure coil inductance is $VI \sin\phi \tan\beta$. Now, with low power factor, the value of ϕ is large and therefore the error is correspondingly large. Hence in a low power factor wattmeter we must compensate for the error caused by inductance of the pressure coil. This is done by connecting a capacitor across a part of series resistance in the pressure coil circuit.

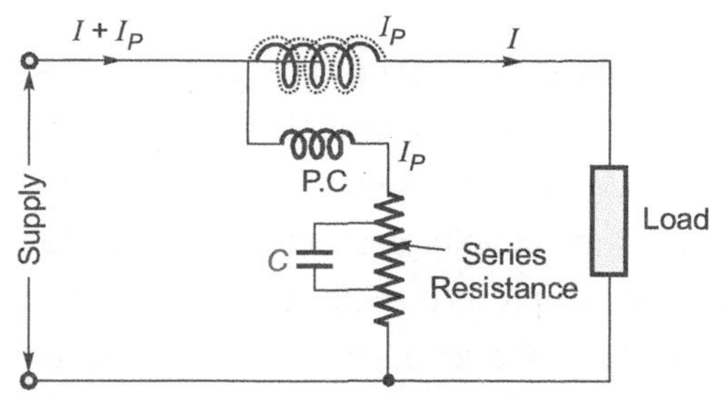

(d) **Small Control Torque:** Low power factor wattmeters are designed with to have a small control torque so that they give full scale deflection for power factors as low as 0.1.

5.8 **An electrodynamic wattmeter is employed to measure power in a single phase circuit. The load voltage is 230 V and the load current is 5 Amps. at a lagging power factor of 0.1. The wattmeter potential coil has a resistance of 10,000 Ω and an inductive reactance negligible compared to resistance. Determine percentage error in the wattmeter reading. Inductance of coil is 100 mH and pressure coil is connected on load side.**

[10 marks : 2005]

Solution:

When the pressure coil connected on the load side, the wattmeter measures power loss in the pressure coil circuit in addition to load power,

$$\text{True power} = VI\cos\phi = 230 \times 5 \times 0.1 = 115 \text{ W}$$

Let us consider only the effect of inductance here, resistance of the pressure coil circuit,

$$R_P = 10{,}000\ \Omega$$

and reactance of the pressure coil circuit = $2\pi \times 100 \times 10^{-3} \times 50 = 31.42\ \Omega$

As the phase angle of the pressure coil circuit is small,

$$\beta = \tan\beta = \frac{31.42}{10{,}000}$$

$$= 3.142 \times 10^{-3} \text{ rad} = 0.003142 \text{ rad}$$

$$\text{actual reading of the wattmeter} = \text{true power}(1 + \tan\phi\tan\beta)$$

$$= 115\,(1 + 9.95 \times 0.00314) = 118.6 \text{ Watt}$$

Power loss in pressure coil circuit,

$$= \frac{V^2}{R_P} = \frac{(230)^2}{10{,}000} = 5.29 \text{ W}$$

∴ Reading of wattmeter considering the power loss in pressure coil circuit,

$$= 118.6 + 5.29 = 123.89 \text{ Watt}$$

$$\text{Percentage Error} = \frac{123.89 - 118.60}{118.60} \times 100 = 4.46\%$$

5.9 **What is meant by phantom or fictitious loading of energy meters?**
An a.c. energy meter is tested for half hour run at a supply voltage of 230 V with a current of 12 Amps at 0.8 p.f. lag. The dial reading at the beginning of test was 58.5 and at the end was 59.5. The meter constant is 1200 revolutions per kWH. The meter revolutions registered during the test were 1150. Find out: (a) error in registration and (b) error in r.p.m. of energy meter. How error in r.p.m. can be rectified?

[12 marks : 2005]

Solution:

Phantom or fictious loading: When the current rating of a meter is high, a test with actual loading arrangements would involve a considerable waste of power. In order to avoid this 'Phantom' or 'Fictitious' loading is done. Phantom loading consists of supplying the pressure coil circuit from a circuit of required normal voltage, and the current coil circuit from a separate low voltage supply. It is possible to circulate the rated current through the current circuit with a low voltage supply as the impedance of this circuit is very low. With this arrangement the total power supplied for the test is that due to small pressure coil current at normal. Voltage, plus that due to the current circuit therefore required for testing the meter with phantom loading is comparatively very small.

$$\text{Actual kWh consumed} = VI\cos\phi \times t \times 10^{-3}$$

$$= 230 \times 12 \times 0.8 \times 1/2 \times 10^{-3} = 1.104 \text{ kWh}$$

$$\text{Registered kWh} = \frac{\text{No. of revolutions done}}{\text{Meter constant in revolution/kWh}} = \frac{1150}{1200} = 0.958 \text{ kWh}$$

$$\therefore \quad \text{Error registration} = \frac{1.104 - 0.958}{1.104} \times 100\% = 13.22\%$$

$$\text{Speed of the disc} = \text{Energy supplied per minute} \times \text{Meter constant}$$

$$= \frac{1.104}{60} \times 1200 = 22.08 \text{ rev/min.}$$

$$\text{Revolution registered/min} = \frac{1150}{30} = 38.33$$

$$\% \text{ error} = \frac{38.33 - 22.08}{22.08} \times 100 = 73.6\% \text{ (fast)}$$

It can be rectified by bringing braking magnet near to the centre of disc.

5.10 For measurement in a 3-phase load, two wattmeter method was used. With 3-phase, 440 volt system of supply and the current coils of two wattmeters are connected in *R* and *B* phases, determine the line currents, reading of each wattmeter and total input power. The load is resistive with 10 Ω, 15 Ω, 20 Ω (in star) respectively in *R*, *Y* and *B* phases.

[10 marks : 2006]

Solution:

Assume "*RYB*" sequence.

Assuming V_R as the reference,

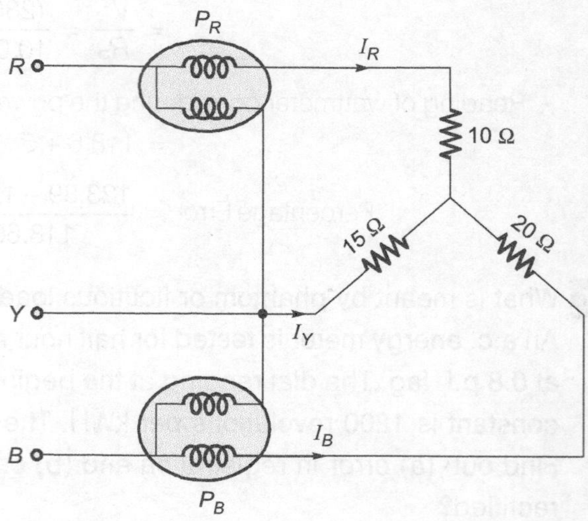

$$V_R = \frac{440}{\sqrt{3}}\angle 0^\circ = 254\angle 0^\circ \text{ V}$$

$$V_Y = 254\angle -120^\circ \text{ V}$$

$$V_B = 254\angle -240^\circ \text{ V}$$

$$I_R = \frac{254\angle 0^\circ}{10} = 25.4\angle 0^\circ \text{ Amp.}$$

$$I_Y = \frac{254\angle -120^\circ}{15} = 16.9\angle -120^\circ \text{ Amp.}$$

$$I_B = \frac{254\angle -240^\circ}{20} = 12.7\angle -240^\circ \text{ Amp.}$$

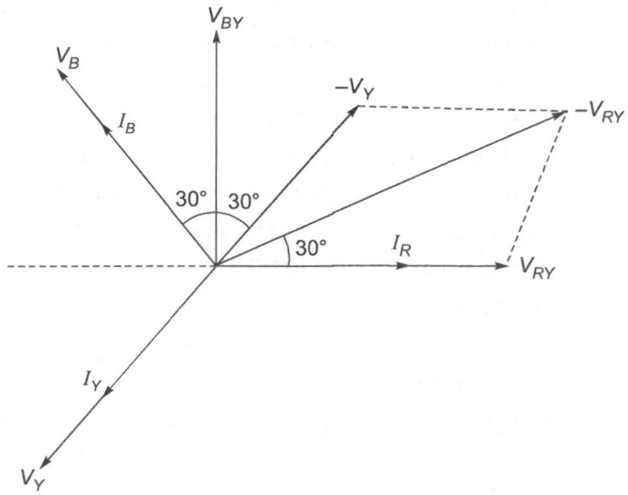

where, $V_{RY} = 440\angle 30°$ V

and $V_{BY} = 440\angle 90°$ V

$\theta_R = 30° = 30°$ {angle between V_{RY} and I_R}

and $\theta_B = 90-(-240) = 330°$ {angle between V_{BY} and I_B}

$$P_R = V_{RY} I_R (\cos\theta_R)$$
$$= 440 \times 25.4 \times \cos 30°$$
$$= 9678.7 \text{ watt}$$
$$P_B = V_{BY} I_B \cos\theta_B$$
$$= 440 \times 12.7 \times \cos 330°$$
$$= 4839.35 \text{ watt}$$
$$\text{Total power} = P_R + P_B = 9678.70 + 4839.35$$
$$= 14518 \text{ watt}$$
$$= 14.518 \text{ kW}$$

5.11 **Describe the constructional details and working of a single phase electrodynamometer type of power factor meter. Prove that the displacement of moving system is equal to the phase angle of the system.**
[12 marks : 2007]

Solution:

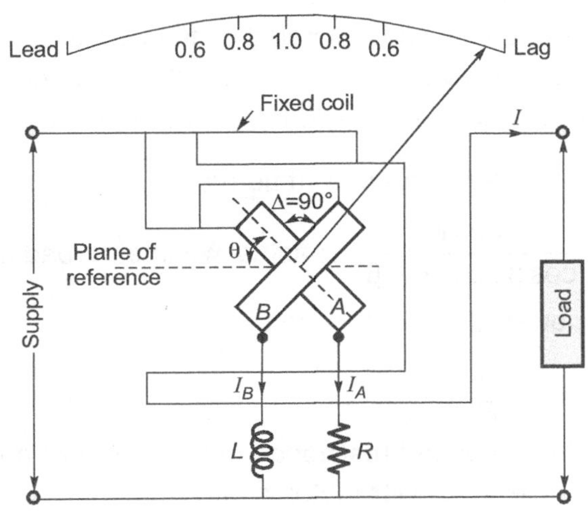

It consists of a fixed coil which acts as the current coil. This coil is split up into two parts and carries the current of the circuit under test. Therefore the magnetic field produced by this coil is proportional to the main current. Two identical pressure coils *A* and *B* pivoted on a spindle constitute the moving system. Pressure coil A has a non-inductive resistance *R* connected in series with it, and coil *B* has a highly inductive choke coil L connected in series with it. The two coils are connected across the voltage supply of the circuit.

Working: The values of *R* and *L* are so adjusted that the two coils carry the same value of current at normal frequency, i.e. $R = \omega L$. The current through coil *A* is in phase with the circuit voltage while that through coil *B* lags the voltage by angle Δ which is nearly equals to 90°.

The angle between the planes of coils is also made equal to Δ. There is no controlling device. Connections to moving coils are made through thin silver or gold ligaments which are extremely flexible and thus give a minimum control effect on the moving system.

Assume, the current through coil *B* lags the voltage by exactly 90°. Also that the angle between planes of coils is exactly 90°, (0 i.e., Δ = 90°).

Now, there will be two deflecting torques, one acting on coil *A* and the other on coil *B*. The coil winding are so arranged that the torques due to the two coils are opposite in direction. Therefore, the pointer will take up a position where these two torques are equal.

Let us consider the case of lagging power factor of cos ϕ.

Deflecting torque acting on coil *A* is:

$$T_A = KVI\, M_{max} \cos\phi \sin\theta$$

where, θ = angular deflection from the plane of reference

and M_{max} = maximum value of mutual inductance between the two coils

This torque say acts in the clockwise direction.

Deflection torque acting on coil *B* is:

$$T_B = KVI\, M_{max} \cos(90° - \phi) \sin(90° + \theta)$$
$$= KVI\, M_{max} \sin\phi \cos\theta$$

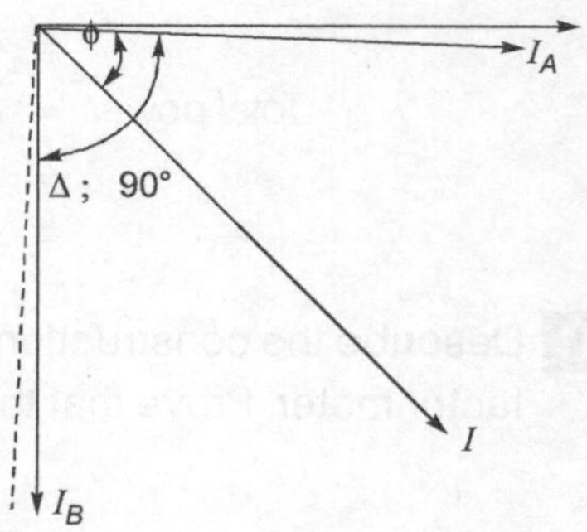

This torque acts in the anticlockwise direction. The value of M_{max} is the same in the two expressions, owing to similar constructions of the coils. The coils will take up such a position that the two torque are equal.

Hence at equilibrium $T_A = T_B$

or $KVI\, M_{max} \cos\phi \sin\theta = KVI\, M_{max} \sin\phi \cos\theta$

or $\theta = \phi$

The scale of the instrument can be calibrated in directly in terms of power factor.

The instrument must be designed for, and calibrated at, the frequency of the supply on which it is to be used. In case the meter is used for any other frequency or if the supply contains harmonics it will give rise to serious errors in the indication on account change in the value of reactance of choke coil.

5.12 Prove that for electrodynamometer type of wattmeters

$$\text{True Power} = \frac{\cos\phi}{\cos\beta \cdot \cos(\phi - \beta)} \times \text{actual wattmeter reading}$$

where, $\cos\phi$ = **power factor of the circuit**

$$\beta = \tan^{-1}\frac{\omega L}{R}$$

where L and R are the inductance and resistance of the pressure coil.

Explain why errors are large when power factor is low.

[10 marks : 2007]

Solution:

It was implied in the idealized wattmeter that the current in the pressure coil is in phase with the applied voltage. If the pressure coils of the wattmeter has an inductance the current in it will lag the voltage by an angle β where,

$$\beta = \tan^{-1}\frac{\omega L}{R}$$

It will be seen from the figure that for lagging power factor, the angle between current in the current coil circuit and current in the pressure coil circuit is less than ϕ, by which the load current lags the applied voltage. The angle between the pressure coil current and the current in the current coil is:

$$\phi' = \phi - \beta$$

∴ The actual wattmeter reading is

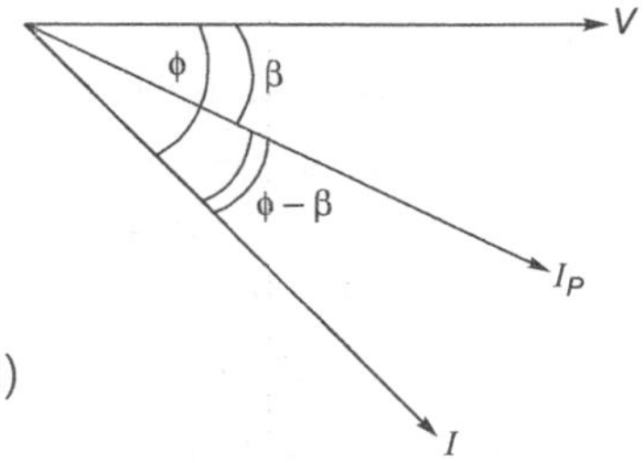

$$= \left(\frac{I_P I}{K}\right)\cdot\cos\phi'\frac{dM}{d\theta} = \left(\frac{V}{Z\cdot K}\right)\cdot I\cdot\cos(\phi-\beta)\cdot\frac{dM}{d\theta}$$

Now, $$Z = \frac{R}{\cos\beta}$$ (Z = impedance of the pressure coil)

∴ Actual wattmeter reading $$= \frac{VI\cos(\phi-\beta)}{K(R/\cos\beta)}\cdot\frac{dM}{d\theta}$$

$$= \frac{VI}{KR}\cdot\cos\beta\cdot\cos(\phi-\beta)\frac{dM}{d\theta} \quad ...(i)$$

in the absence of inductance $Z = R$ and $\beta = 0$ and therefore the wattmeter will read true power under these conditions.

∴ True power $$= \frac{I_P I}{K}\cdot\cos\phi\cdot\frac{dM}{d\theta}$$

$$= \frac{VI\cos\phi}{KR}\cdot\frac{dM}{d\theta} \quad ...(ii)$$

It is clear from the above expression that in the absence of inductance, the wattmeter reads correctly at all power factors and frequencies.

$$\text{Ratio} = \frac{\text{True power}}{\text{Actual wattmeter reading}}$$

$$= \frac{\dfrac{VI\cos\phi}{KR}\cdot\dfrac{dM}{d\theta}}{\dfrac{VI\cos(\phi-\beta)}{KR}\cdot\cos\beta\cdot\dfrac{dM}{d\theta}}$$

$$= \frac{\cos\phi}{\cos\beta\cos(\phi-\beta)}$$

Hence, True power $$= \frac{\cos\phi}{\cos\beta\cos(\phi-\beta)}\times\text{Actual watt meter reading}$$

5.13 Describe the constructional details of an electrodynamometer type wattmeter. Derive the expression for torque when the instrument is used on a.c.

[12 marks : 2008]

Solution:

Electrodynamometer Type Wattmeter Constructional Detail:

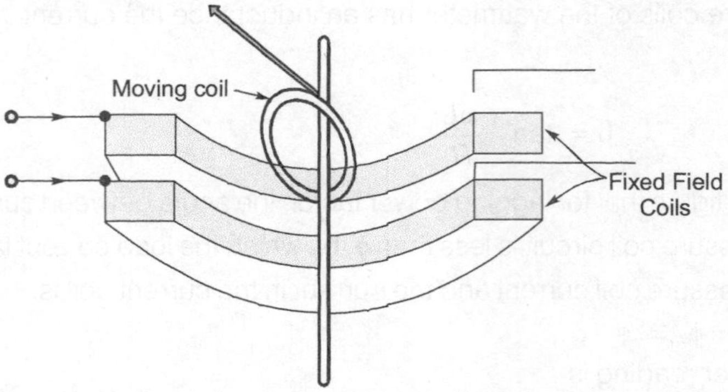

(a) Fixed Coils: The fixed coils carry the current of the circuit. They are divided into two halves. The fixed coils are wound with heavy wires.

In case of precision wattmeters, the two halves of the fixed coils are connected in series. For basic measuring range, can be connected in parallel to increase the wattmeter current range to twice of its original value.

(b) Moving Coil: The moving coil is mounted on a pivoted spindle and is entirely embraced by the fixed current coils. Spring control is used for the movement. Both fixed and moving coils are air cored.

(c) Control: Spring control is used for the instrument.

(d) Damping: Air friction damping is used. The moving system carries a light aluminium vane which moves in a sector shaped box.

(e) Scale and Pointers: They are equipped with mirror type scales and knife edge pointers to remove reading errors due to parallax.

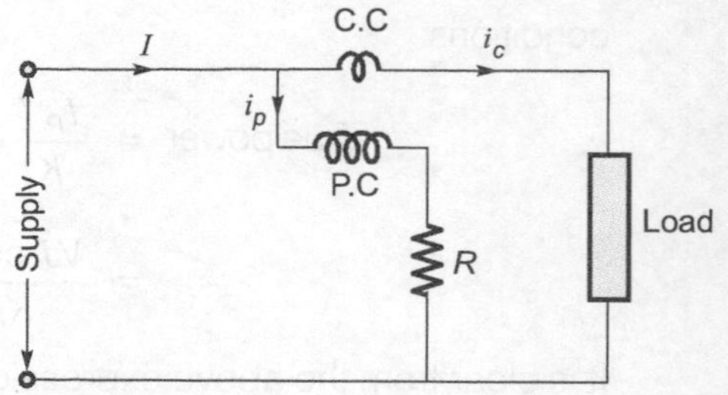

Expression for torque with a.c.:

The instantaneous torque is given by,

$$T_i = i_1 i_2 \frac{dM}{d\theta}$$

where, i_1 and i_2 = instantaneous value of current in two coils (Amp).

Let V and I be the rms values of voltage and current being measured.

$\therefore$ Instantaneous value of voltage across the pressure coil circuit

$$v = \sqrt{2}\, V \sin\omega t$$

Pressure coil can be treated as purely resistive as it has a very high resistance.

Therefore, current i_p in the P.C is in phase with the voltage and its instantaneous value is,

$$i_P = \frac{v}{R_P} = \frac{\sqrt{2}\, V \sin\omega t}{R_P} = \sqrt{2}\, I_P \sin\omega t$$

where, I_P = rms value of current in P.C circuit

If the current in the C.C lags the voltage in phase by an angle ϕ then

$$i_C = \sqrt{2}\, I_C \sin(\omega t - \phi)$$

$\therefore$ Instantaneous torque,

$$T_i = \sqrt{2} I_P \sin\omega t \times \sqrt{2} I_C \sin(\omega t - \phi) \frac{dM}{d\theta}$$

$$= 2 I_P I_C \sin\omega t \cdot \sin(\omega t - \phi) \frac{dM}{d\theta}$$

$$T_i = I_P I_C [\cos\phi - \cos(2\omega t - \phi)] \frac{dM}{d\theta}$$

Average deflecting torque, $$T_d = \frac{1}{T}\int_0^T T_i d(\omega t)$$

$$= \frac{1}{T}\int_0^T I_P I_C [\cos\phi - \cos(2\omega t - \phi)\frac{dM}{d\theta} \times d(\omega t)$$

$$= I_P I_C \cos\phi \frac{dM}{d\theta}$$

5.14 **Describe the functions of the following in a single-phase induction-type energy meter:**

(a) Shunt and Series Magnet **(b) Moving Disc**
(c) Copper-Shading bands **(d) Braking Magnet**
(e) Registering Mechanism **(f) Holes in The Disc** **[12 marks : 2008]**

Solution:

(a) **Shunt and Series Magnet:** The driving system of the meter consists of two electro-magnets. The coil of one of the electromagnets is excited by the load current. This coil is called the current coil. The coil of second electromagnet is connected across the supply and carries a current proportional to the supply voltage and is called pressure coil and these two electromagnets are called shunt and series magnet. Shunt and series magnet is the driving system of all meter.

(b) **Moving Disc:** An aluminium disc mounted a light alloy shaft. This disc is positioned in the air gap between series and shunt magnet. The upper bearing of the rotor (moving system) is a steel pin located in a hole in the bearing cap fixed to the top of the shaft.

(c) **Copper-Shading bands:** These are provided on the central limb. The position of these bands is adjustable. The function of these bands is to bring the flux produced by the shunt magnet exactly in quadrature with the applied voltage.

(d) **Braking Magnet:** A braking magnet is a permanent magnet positioned near the edge of the aluminium disc forms the braking system. The aluminium disc moves in the field of this magnet and thus provides a braking torque.

(e) **Registering Mechanism:** The function of a registering or counting mechanism is to record continuously a number which is proportional to the revolutions made by the moving system.

(f) **Holes in The Disc:** To prevent creeping two diametrically opposite holes are drilled in the disc; the disc will come to rest with one of the holes under the edge of a pole of the shunt magnet.

5.15 **The inductive reactance of the pressure-coil circuit of a dynamometer wattmeter is 0.4% of its resistance at normal frequency (50 Hz) and the capacitance is negligible.**
Calculate the percentage error and correction factor due to the reactance for load at 0.707 p.f. lagging.

[10 marks : 2009]

Solution:

The pressure coils of the wattmeter has an inductance. The current in it will lag the voltage by an angle β

where, $$\beta = \tan^{-1}\left(\frac{\text{reactance of pressure coil}}{\text{resistance}}\right)$$

$$= \tan^{-1}\left(\frac{0.4}{100}\right) = 0.229$$

Given, $$\text{p.f.} = \cos\phi = 0.707 \text{ lagging}$$

or $$\phi = 45^\circ$$

$$\text{Percentage error} = \tan\phi \cdot \tan\beta \times 100$$
$$= \tan(45°) \cdot \tan(0.229) \times 100 = 0.399\%$$

$$\text{Correction factor} = \frac{\cos\phi}{\cos\beta\cos(\phi-\beta)}$$
$$= \frac{\cos 45°}{\cos(0.229)\cos(45-0.229)} = 0.996$$

5.16 **Draw an electrodynamic type frequency meter circuit and explain its working.**

[10 marks : 2009]

Solution:

Electrodynamic Type Frequency Meter: The schematic diagram of this meter is shown in figure. The fixed coil is divided into two parts 1 and 2. The two parts of the fixed coil from two separate resonant circuits. Fixed coil 1 is in series with an inductance, L_1, and a capacitance, C_1 forming a resonant circuit of frequency, f_1 slightly below the lower end of the instrument scale.

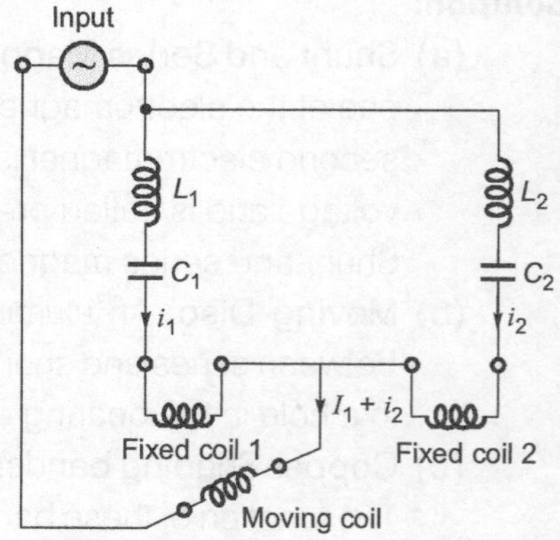

Fixed coil 2 is in series with inductance L_2 and capacitance C_2 forming a resonant circuit of frequency f_2 slightly higher than the upper end of the instrument scale. In the case of instruments for power fequency measurements, the circuits may be tuned to frequencies of 40 Hz and 60 Hz respectively with 50 Hz in the middle of the scale.

The two parts of fixed coil are arranged as shown in the diagram, their return circuits being through the movable coil. The torque on the movable element is proportional to the current through the moving coil. This current is the sum of the currents in the two parts of the fixed coil. For applied frequency, within the limits of the frequency range of the instrument, the circuit of fixed coil 1 operates above resonant frequency (as $X_{L_1} > X_{C_1}$) with current i_1 through it, lagging the applied voltage. The circuit of fixed coil 2 operated below the resonant frequency (as $X_{C_2} > X_{L_2}$) with current i_2 leading the applied voltage. One fixed coil circuit is inductive and the other is capacitive and therefore the torque produced by the two currents i_1 and i_2 act in opposition on the moving coil.

The resultant torque is a function of frequency of the applied voltage and therefore the meter scale can be calibrated in terms of frequency. The instrument scale spreads over an angle of about 90°.

The controlling torque is provided by a small iron vane mounted on the moving system.

This meter is used only for power frequency measurements and is used in power systems for monitoring the frequency.

5.17 **Describe the constructional details and working of a single phase electro-dynamometer type of power factor meter. Power that the displacement of the moving system is proportional to the phase angle of the system.**

[10 marks : 2010]

Solution:

(Refer Questions No. 5.11).

5.18 **An electric utility supplies power to 1 MW load at power factor (p.f.) 0.85 and at 11 kV. The utility wants to measure the voltage, current and power factor continuously using 250 V voltmeter, 10 A ammeter and 250 V, 10 A power factor meter. Draw circuit diagram of the scheme.**

Discuss, why the p.f. meter does not come back to zero reading like voltmeter and ammeter disconnecting the supply to it.

[10 marks : 2011]

Solution:

Assuming 1-ϕ supply, I_1 = Current drawn by load

$$= \frac{1\times10^6}{11\times10^3\times0.85} \approx 107\text{ A}$$

V_1 = Voltage across the load = 11 kV

- To measure the power factor, V_1 and I_1 are reduced to the levels of voltmeter and ammeter rating using potential transformer (PT) and current transformer.

$$\text{Ratio of PT} = \frac{11\times10^3}{250} \text{ and Ratio of CT} = \frac{107}{10}$$

Circuit diagram of the scheme is drawn below:

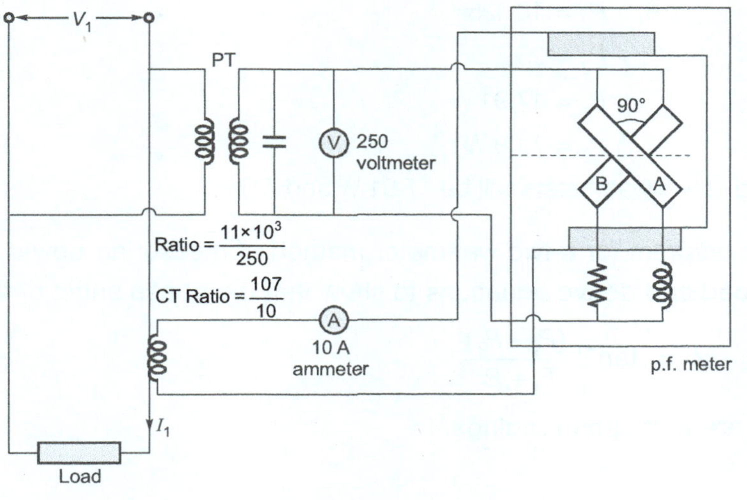

- There is no controlling system in power factor meter but present in voltmeters and ammeters. The functions of controlling system are:
  - ⇒ to produce a force equal and opposite to the force at the final steady position of pointer.
  - ⇒ to bring the moving system back to zero when the force causing the instrument moving system to deflect is removed. Therefore the p.f. meter does not come back to zero reading after disconnecting the supply to it.
- There is no controlling torque required in p.f. meter the torques produced by current coil and voltage oppose each other balance out at one point of the scale.
- In voltmeter and ammeter only one torque will be there to balance if controlling torque will be used. When current/voltage is zero then only controlling torque presents and move the needle to zero reading.

5.19 **Three single phase loads can be connected in either star or in delta to from three-phase load. Which of these connections results in higher current when connected to a three-phase supply? During the measurement of power by two-wattmeter method, the total input power to a 3-phase 440-V motor running at power factor of 0.8 was found to be 25 kW. Find the readings of the two wattmeters.**

[12 marks : 2012]

Solution:

In star connection,

$$\text{Line voltage} = \sqrt{3}\,(\text{phase voltage})$$

and Line current = phase current

In delta connection,

Line voltage = phase voltage

Line current = $\sqrt{3}$ (phase voltage)

Hence delta connections results in higher current when connected to a 3-ϕ supply.

$$\text{total power} = 25 \text{ kW} = P_1 + P_2$$

$$\phi = \tan^{-1}\left[\sqrt{3}\left(\frac{P_1 - P_2}{P_1 + P_2}\right)\right]$$

$$\frac{\sqrt{3}}{4} = \frac{P_1 - P_2}{25}$$

$$\Rightarrow \quad P_1 - P_2 = 10.825$$

$$P_1 + P_2 = 25$$

$$\Rightarrow \quad P_1 = 17.91 \text{ W}$$

$$P_2 = 7.09 \text{ W}$$

Hence the reading of two wattmeters will be 17.91 W and 7.09 W

5.20 Draw the phasor diagram for a two wattmeter method of measuring power in a 3-phase balanced star connected load and derive equations to show that the phase angle of the load

$$\phi = \tan^{-1}\frac{(P_1 - P_2)}{P_1 + P_2}$$

where P_1 and P_2 are wattmeter readings.

[10 marks : 2012]

Solution:

Let the two wattmeter P_1 and P_2 are connected to measure power in a 3-ϕ balanced star connected load as shown in figure.

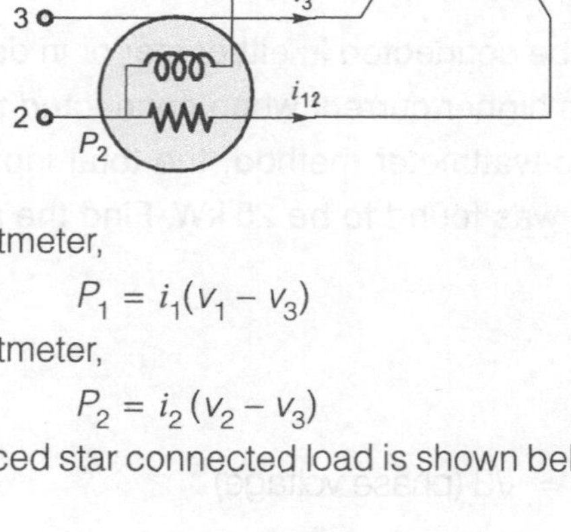

Instantaneous reading of P_1 wattmeter,

$$P_1 = i_1(v_1 - v_3)$$

Instantaneous reading of P_2 wattmeter,

$$P_2 = i_2(v_2 - v_3)$$

The phasor diagram for a balanced star connected load is shown below:

Let v_1, v_2, v_3 be the rms values of phase voltages and I_2, I_2, I_3 be the rms currents.
The load is balanced, therefore phase voltages,

$$V_1 = V_2 = V_3 = V \text{ (say)}$$

Line voltages, $$V_{12} = V_{23} = V_{31} = \sqrt{3}\text{V}$$

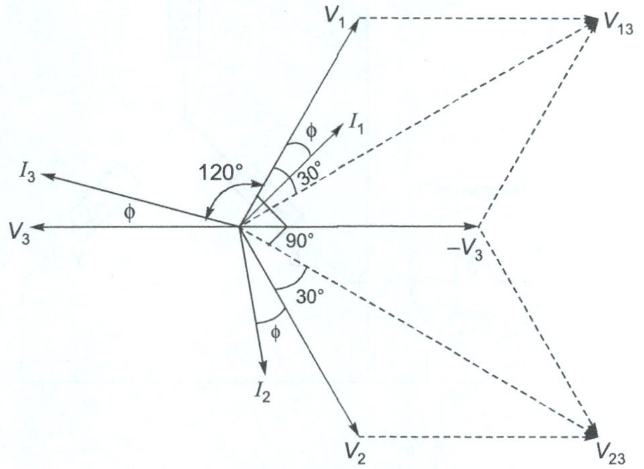

Phase currents, $I_1 = I_2 = I_3 = I$ (say)

Line currents, $I_1 = I_2 = I_3 = I$

Power factor $= \cos\phi$

The current through wattmeter P_1 is I_1 and voltage across its pressure coil is V_{13} by an angle $(30 - \phi)^\circ$.

$\therefore$ Reading of P_1 wattmeter,

$$P_1 = V_{13} I_1 \cos(30 - \phi)$$

$$= \sqrt{3} VI \cos(30 - \phi)$$

The current through wattmeter P_2 is I_2 and voltage across its pressure coil is V_{23}; I_2 lags V_{23} by an angle $(30 + \phi)^\circ$.

Therefore reading of P_2 wattmeter,

$$P_2 = V_{23} I_2 \cos(30 + \phi) = \sqrt{3} VI \cos(30 + \phi)$$

$$P_1 + P_2 = \sqrt{3} VI [\cos(30 - \phi) + \cos(30 + \phi)] = 3\, VI \cos\phi$$

and $$P_1 - P_2 = \sqrt{3} VI [\cos(30 - \phi) - \cos(30 + \phi)] = \sqrt{3} VI \sin\phi$$

$\therefore$ $$\frac{P_1 - P_2}{P_1 + P_2} = \frac{1}{\sqrt{3}} \tan\phi$$

$\Rightarrow$ $$\phi = \tan^{-1}\left[\sqrt{3} - \frac{(P_1 - P_2)}{(P_1 + P_2)}\right]$$

5.21 **The power in a balanced 3-phase, 3-wire, star-connected system is to be measured by two wattmeters. Draw neat connection and phasor diagrams. Do not derive the expressions for the power measured by each wattmeter.**

[5 marks : 2013]

Solution:

Connection diagram:

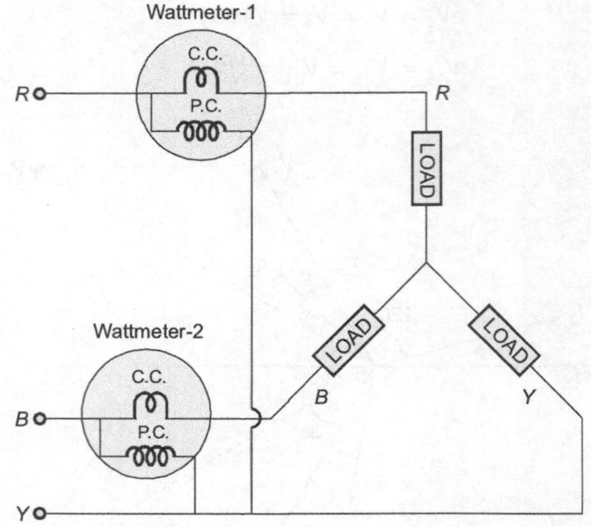

Phasor diagram:

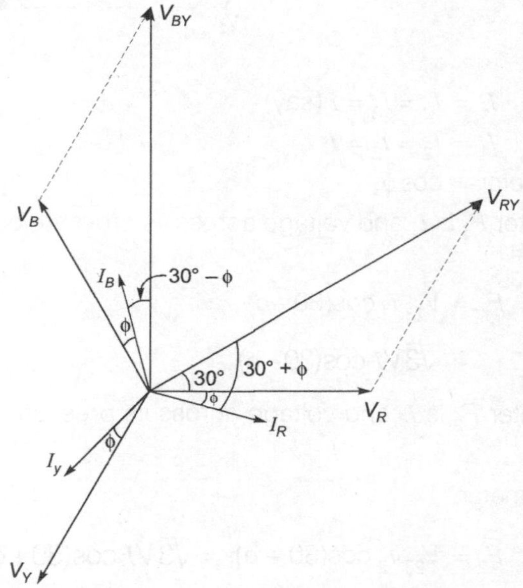

5.22 A single-phase load of 100∠–36.9° ohms is connected across terminals *RY* of a balanced 3-phase supply, 415 volts, having phase sequence *RYB*. The current coil of a wattmeter is connected in series with the load, whereas its pressure coil is connected between terminals *B* and *N* of the supply. Obtain the wattmeter reading. Assume V_{RN} as a reference phaser voltage.

[10 marks : 2014]

Solution:

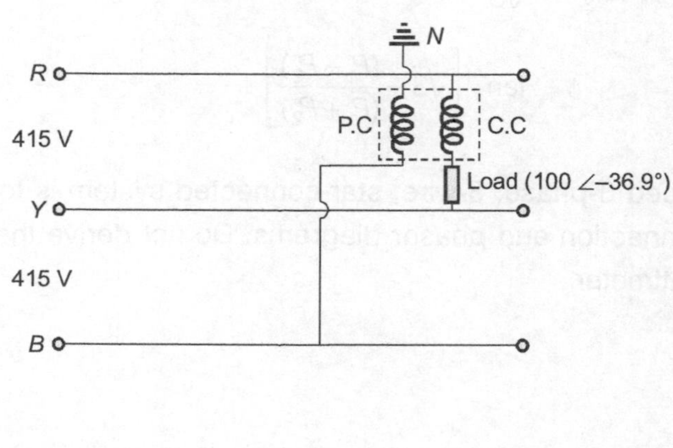

Phase sequence *RYB*,

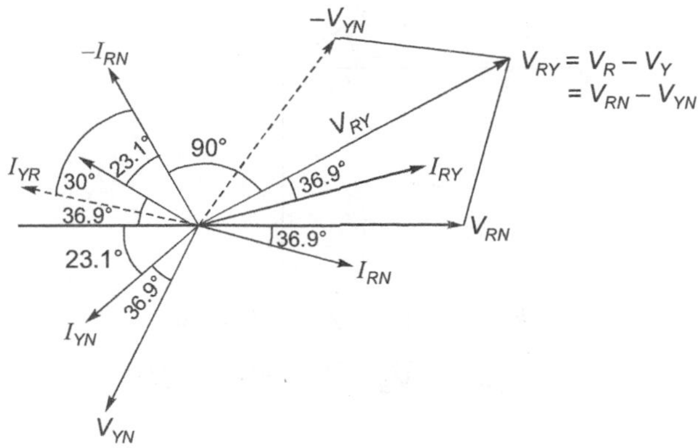

$$V_{RY} = V_R - V_Y = V_{RN} - V_{YN}$$

$$\text{Reading of wattmeter} = V_{BN} I_{RY} \cos(\angle V_{BN} - I_{RY})$$

$$= \frac{415}{\sqrt{3}} \times \frac{415}{100} \times \cos(126.9°) = -597.02 \text{ watt}$$

as this reading is negative, so change current coil terminals, i.e. current I_{RY}.

We will take as I_{YR} so that angle becomes 53.1°

$$W = \frac{415}{\sqrt{3}} \times \frac{415}{100} \times \cos(53.1°) = 597.02 \text{ watt}$$

5.23 **A wattmeter is connected as indicated in the following figure.**

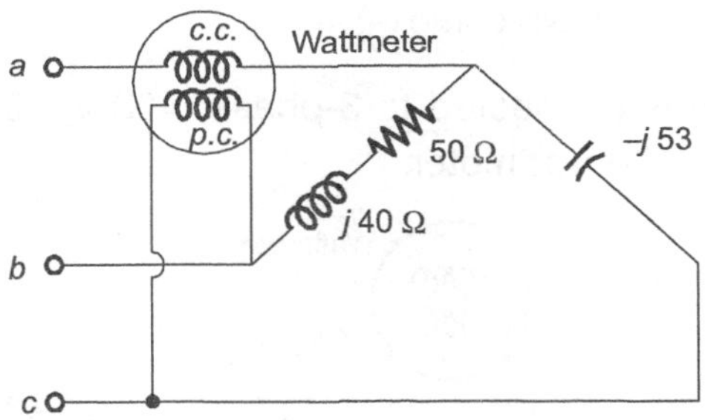

Calculate the wattmeter reading, assuming, 3-phase, 400 volts, 50 Hz balanced supply, with phase sequence abc.

[10 marks : 2015]

Solution:

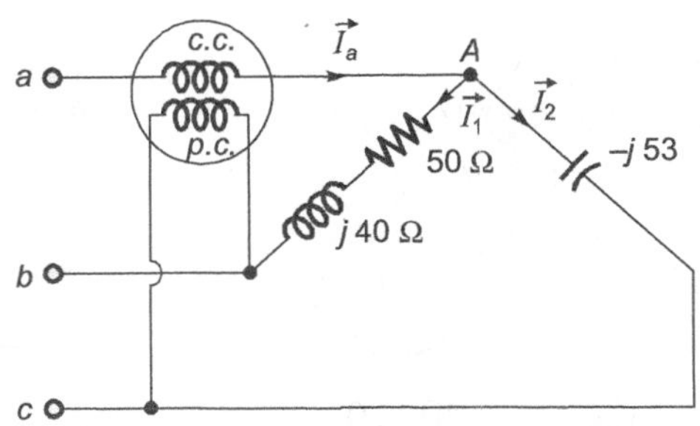

Phasor diagram is shown below,

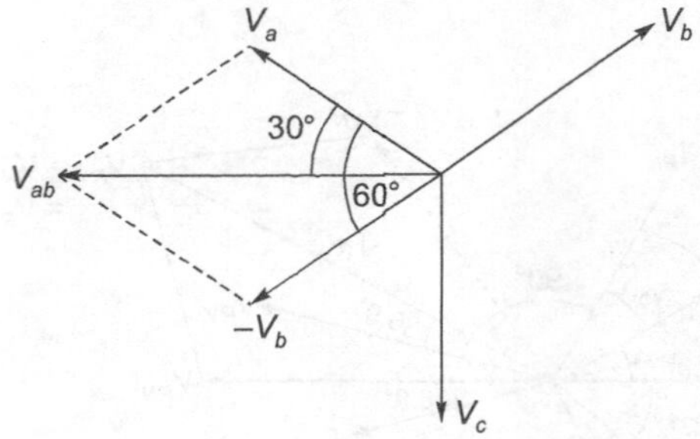

$$\vec{V}_{ab} = \vec{V}_a - \vec{V}_b = 400\angle 30°$$

$$\vec{V}_{ac} = \vec{V}_a - \vec{V}_c = 400\angle -30°$$

Applying KCL at node *A*,

$$\vec{I}_a = \vec{I}_1 + \vec{I}_2$$

$$= \frac{\vec{V}_{ab}}{(j40+50)} + \frac{\vec{V}_{ac}}{(-j53)}$$

$$= \frac{400\angle 30°}{(50+j40)} + \frac{400\angle -30°}{(-j53)} = 11.415\angle 29.353°$$

and,

$$\vec{V}_{cb} = 400\angle 90°$$

wattmeter reading,

$$W = V_{cb} \times I_a \times \cos\angle(\vec{V}_{cb} - \vec{I}_a)$$

$$= 400 \times 11.415 \times \cos(90° - 29.353°) = 2238.2 \text{ watt}$$

Note: Here $\vec{V}_{cb}$ is taken and not $\vec{V}_{bc}$, because in phasor diagram angle between $\vec{V}_{bc}$ and $\vec{I}_a$ will become more than 90° and wattmeter reading will come (–)ve.

5.24 The network shown in figure is connected to 3-phase, 400 V, 50 Hz supply with phase sequence *abc*. Calculate the reading of the wattmeter.

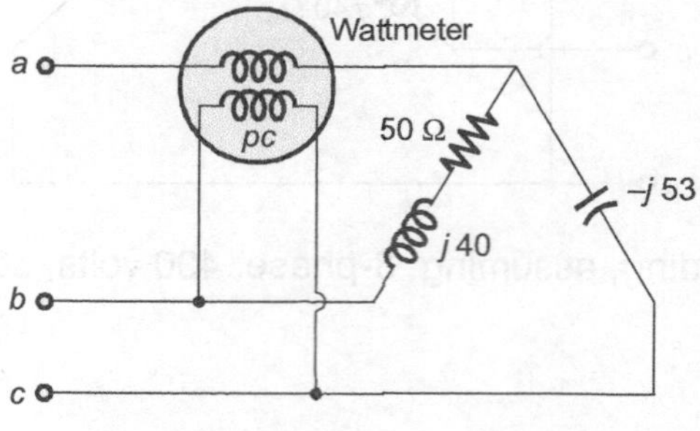

[20 marks : 2017]

Solution:

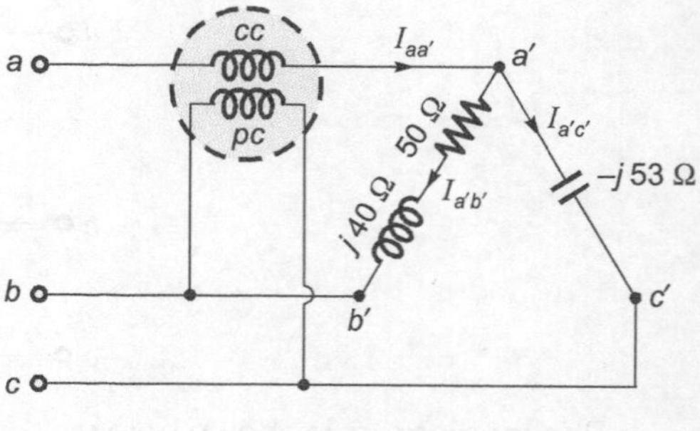

Given that,

$$V_{ab} = 400\angle 0° \text{ V}$$

$$V_{bc} = 400\angle -120° \text{ V}$$

and

$$V_{ca} = 400\angle 120° \text{ V}$$

or,

$$V_{ac} = -400\angle 120° \text{ V}$$

Also, impedance,

$$Z_{a'b'} = 50 + j40 = 64.03\angle 38.66° \ \Omega$$

and impedance,

$$Z_{a'c'} = -j53 = 53\angle -90° \ \Omega$$

Then,

$$I_{a'b'} = \frac{V_{ab}}{Z_{a'b'}} = \frac{400\angle 0°}{64.03\angle 38.66°}$$

$$= 6.247\angle -38.66° \text{ A}$$

and $$I_{a'c'} = \frac{V_{ac}}{Z_{a'c'}} = \frac{-400\angle 120°}{53\angle -90}$$

$$= -7.547\angle 210° \text{ A}$$

Then, $$I_{aa'} = I_{a'b'} + I_{a'c'}$$

$$= 6.247\angle -38.66° - 7.547\angle 210°$$

$$= 11.415\angle -0.647° \text{ A}$$

Thus, the current through current coil (cc) of the wattmeter is 11.415 A.

The potential coil of wattmeter is connected between *b* and *c*.

So, the voltage across the wattmeter is

$$V_{bc} = 400\angle -120°$$

Reading of the wattmeter, $$W = V_{bc} \times I_{aa'} \times \cos\phi$$

$$= 400 \times 11.415 \cos[-120° - (-0.647°)]$$

$$= 2238.203 \text{ W}$$

5.25 What is creeping in energy meters? State the reasons for the same and how it is avoided.

[8 marks : 2018]

Solution:

Creeping in energy meter: Creeping is a phenomenon in which the aluminium disc rotates continuously when only the voltage is supplied to pressure coil and current coil is not energized.

Reasons for creeping:

- Creeping error occurs due to extra compensation for friction in energy meter.
- The disc rotates because of the additional torque provided by the compensating vane.

Prevention of creeping:

- Drilling the hole in disc diametrically opposite can prevent creeping.
- A small piece of iron can be attached to the edge of the disc so that force of attraction occurs between pole magnets and iron piece which can prevent creeping.

5.26 The power consumed by a single-phase 11 kV load taking 100 A at 0.5 power factor lagging is measured on a dynamometer wattmeter used in conjunction with potential transformer (PT) and current transformer (CT). Determine the reading of the wattmeter considering the following data:

Nominal ratio of PT and CT are 100 : 1 and 20 : 1.

Ratio error of PT and CT are +0.8% and –0.2%.

Phase angle errors of PT and CT are +42 minutes and +90 minutes.

Phase angle of the pressure coil due to its inductance is 30 minutes.

[20 marks : 2018]

Solution:

For lagging load:

$$\phi = \alpha + \beta - \theta_p + \theta_c$$

$$\phi = \text{p.f. angle of load} = \cos^{-1} 0.5 = 60°$$

$$\beta = \text{p.c. phase angle of inductance}$$

$$= 30' = \frac{30}{60} = 0.5°$$

$$\theta_p = \text{P.T. phase angle error}$$

$$= 42' = \frac{42}{60} = 0.7°$$

$$\theta_c = \text{C.T. phase angle error}$$
$$= 30' = \frac{30}{60} = 0.5°$$
$$\alpha = \text{Phase angle between P.C. and C.C. currents of wattmeter}$$
$$= \phi + \theta_p - \theta_c - \beta$$
$$= 60 + 0.7 - 0.5 - 0.5 = 59.7°$$

Ratio error of P.T., $\sigma_P = 0.8\% = 0.008$

$$\sigma_P = \frac{K-R}{R} = \frac{K}{R} - 1$$

$$\Rightarrow \quad \frac{K}{R} = 1 + \sigma_P$$

Ratio correction factor of P.T. $= (RCF)_{PT}$

$$= \frac{R}{K} = \frac{1}{1+\sigma_P} = \frac{1}{1+0.008} = 0.9921$$

Ratio error of C.T. $= -0.2\% = \sigma_C = -2 \times 10^{-3}$

$$(RCF)_{CT} = \frac{1}{1+\sigma_C} = \frac{1}{1-2\times10^{-3}} = 1.002$$

Correction factor of wattmeter,

$$K = \frac{\cos\phi}{\cos\beta \cdot \cos\alpha}$$
$$= \frac{0.5}{\cos(0.5°)\cdot\cos(59.7°)}$$
$$= 0.9910°$$

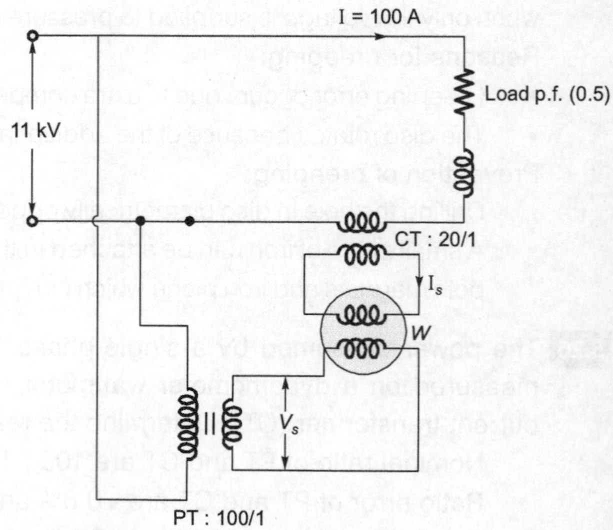

P.T. ratio $= (K_n)_{PT} = \frac{100}{1} = 100$

C.T. ratio $= (K_n)_{CT} = \frac{20}{1} = 20$

Wattmeter reading $= V_s I_s \cdot \cos\phi$

$$= \frac{11\times10^3}{100} \times \frac{100}{20} \times 0.5$$
$$= 275 \text{ W}$$

True power $= K \times (R.C.F)_{P.T} \times (R.C.F)_{C.T} \times (K_n)_{P.T} \times (K_n)_{C.T} \times \text{Wattmeter reading}$

$$= 0.9910 \times 0.9921 \times 1.002 \times 100 \times 20 \times 275$$
$$= 0.5418 \text{ MW}$$

5.27 **In a single-phase power measurement test by three-voltmeter method, the following readings were obtained. Across AC mains is 200 V; across the non-inductive resistance of 10 Ω is 110 V; across the load consisting of resistance (*R*) and inductance (*L*) is 120 V.**

(i) Calculate the power supplied to the load.

(ii) Calculate the inductive reactance (X_L) and resistance (*R*) of the load.

[20 marks : 2019]

Solution:

$$(200)^2 = (\sqrt{(110)^2 + (120)^2 + 2\times110\times120\cos\phi})^2$$

$$40000 = 12100 + 14400 + 26400\cos\phi$$

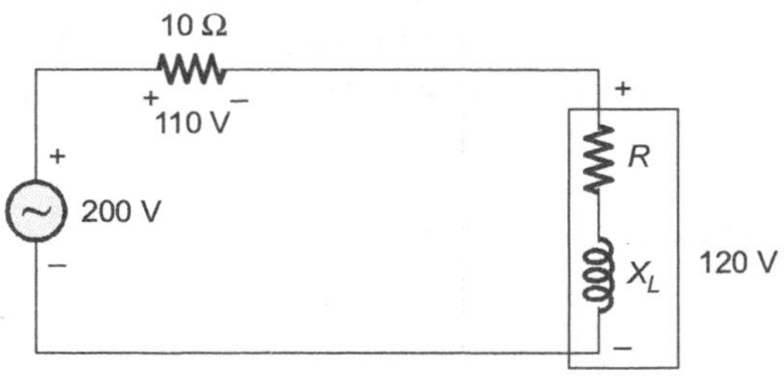

$$26400\cos\phi = 13500$$

$$\cos\phi = \frac{13500}{26400} = 0.511$$

Power supplied to the load

$$= VI\cos\phi$$

$$= 120 \times \frac{110}{10} \times 0.511 = 674.52$$

Power supplied by load = I^2R

$$(11)^2 R = 674.52$$

$$R = 5.574\ \Omega$$

$$V_R = 61.32$$

$$V_x = \sqrt{(120)^2 - (61.32)^2} = 103.15$$

$$IX_L = 103.15$$

$$X_L = 9.37\ \Omega$$

5.28 **A 230 V, single phase, watt hour meter has a constant load of 5 A passing through it for 8 hours at unity power factor. If the meter constant is 460 revolutions per kWh, how many revolutions does the meter disc make during this period? If the same meter makes 1638 revolutions when operating at 230 V and a constant load of 6 A passing through it for a certain duration at a power factor of 0.86, determine the duration of operation of the meter in hours.**

[20 marks : 2020]

Solution:

Given, $V = 230$ V, $I = 5$ A

Power factor = 1

Watt hour meter reading = 230 × 5 × 1 × 8 = 9200 Wh

Given meter constant is 460 revolutions per kWh

No. of revolutions during this period is

$$= \frac{9200 \times 4600}{1000} = 4232 \text{ revolutions}$$

Given meter constant is 460 revolutions per kWh

∴ Meter reading in 1638 revolution is

$$= \frac{1000}{460} \times 1638 = 3.56 \text{ kW}$$

Given, $V = 230$ V,

$I = 6$ A

Meter reading = 230 × 6 × 0.866 × time

230 × 6 × 0.866 × time = 3.56 kW

$$\text{time} = \frac{3.56 \times 10^3}{230 \times 6 \times 0.866} = 3 \text{ Hours}$$

5.29 A wattmeter reads 4 kW when its current coil is connected in R phase of a symmetrical 3-phase system supplying a balanced 3-phase inductive load of 25 A at 400 V. What will be the reading of the wattmeter if the connections of the current coil remain unchanged and voltage coil be connected between *B* and *Y* phases? Draw the corresponding phasor diagram.

[10+10 = 20 marks : 2021]

Solution:

$$P = VI\cos\phi$$

$$4\times 10^3 = \left(\frac{400}{\sqrt{3}}\right)\times 25\times\cos\phi$$

$$\cos\phi = 0.6928$$

$$\phi = 46.146°$$

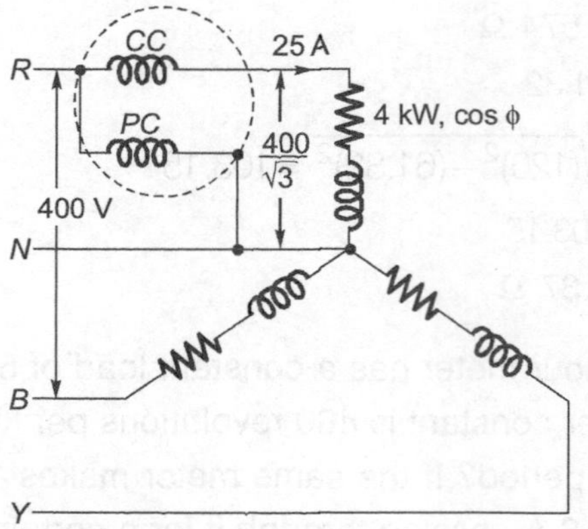

By connecting *PC* between *B* and *Y* terminals

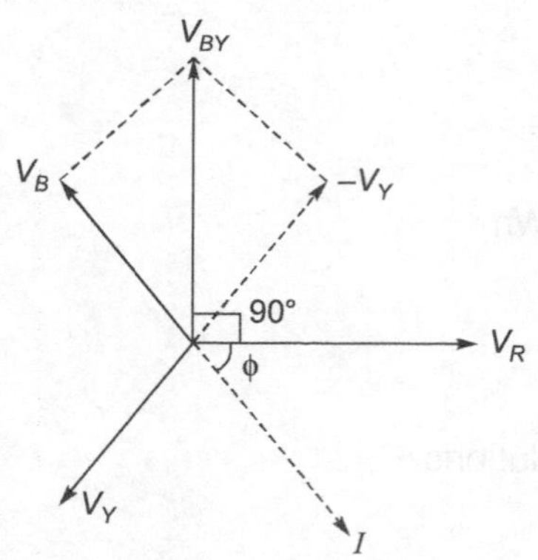

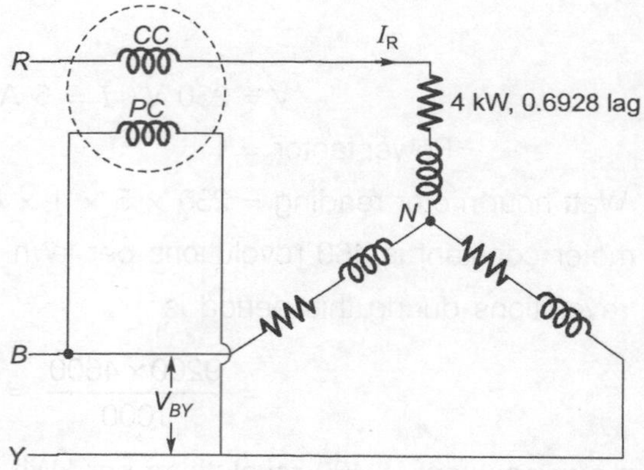

Wattmeter reading

$$P = V_{BY}I_R\cos(90° + \phi)$$

$$P = V_L I_L \sin\phi$$

$$= 400\times 25\times\sin(46.146°)$$

$$= 7.211 \text{ kVAR}$$

5.30 A 10 HP induction motor runs from a 3-phase 400 volts supply on no-load. The motor takes a line current of 4 A at a power factor of 0.208 (lag). On full load it operates at power factor of 0.88 (lag) and with an efficiency of 89%. Determine the readings on each of the two watt-meters connected to read total power (1) on no-load and (2) on full-load.

[10 marks : 2022]

Solution:

Given : $V_L = 400\text{ V}$

(1) At no load condition :

$$\phi = \cos^{-1}(0.208) = 78^\circ$$

Motor input power, $P_{in} = \sqrt{3}V_L I_L \cos\phi$

$$= \sqrt{3} \times 400 \times 4 \times 0.208$$
$$= 576.42\text{ Watt}$$

In two wattmeter method,

Power measured by wattmeter W_1 is,

$$P_1 = V_L I_L \cos(30 - \phi)$$
$$P_1 = 400 \times 4 \times \cos(30^\circ - 78^\circ)$$
$$P_1 = 1070.60\text{ Watt}$$

Similarly reading of second wattmeter (W_2)

$$P_2 = V_L I_L \cos(30^\circ + \phi)$$
$$= 400 \times 4 \times \cos(30^\circ + 78^\circ)$$
$$= -494.42\text{ Watt}$$

Therefore, $P_1 = 1070.60\text{ Watts}$

$$P_2 = -494.42\text{ Watts}$$

(2) At full load condition :

$$P_{in} = \frac{P_o}{\eta} = \frac{10 \times 746}{0.89} = 8382\text{ Watts}$$

As we know, $\sqrt{3}V_L I_L \cos\phi = P_{in}$

Line current, $I_L = \dfrac{8382}{400\sqrt{3} \times 0.88} = 13.74\text{ A}$

Full load power factor angle,

$$\phi = \cos^{-1}(0.88) = 28.35^\circ$$

Reading of first wattmeter (W_1)

$$P_1 = V_L I_L \cos(30^\circ - \phi)$$
$$= 400 \times 13.74 \times \cos(30^\circ - 28.35^\circ)$$
$$P_1 = 5493.72\text{ Watts}$$

Similarly reading of second wattmeter (W_2)

$$P_2 = V_L I_L \cos(30^\circ + \phi)$$
$$= 400 \times 13.74 \times \cos(30^\circ + 28.35^\circ)$$
$$P_2 = 2883.91\text{ Watts}$$

Therefore, $P_1 = 5493.72\text{ Watts}$

$$P_2 = 2883.91\text{ Watts}$$

5.31 A balanced 240 V, 3-phase voltage is applied to an unbalanced delta-connected load having the following phase impedances :

$$Z_{AB} = 25\angle 90^\circ\ \Omega,\ Z_{BC} = 15\angle 30^\circ\ \Omega,\ Z_{CA} = 20\angle 0^\circ\ \Omega$$

(i) Calculate the line currents.

(ii) Obtain the readings of the two wattmeters whose current coils are connected in the lines A and B, and the voltage coils are connected across the line C.

Consider ABC system for supply voltage and V_{BC} as reference.

[20 marks : 2023]

Solution:

(i) Unbalanced delta-connected load :

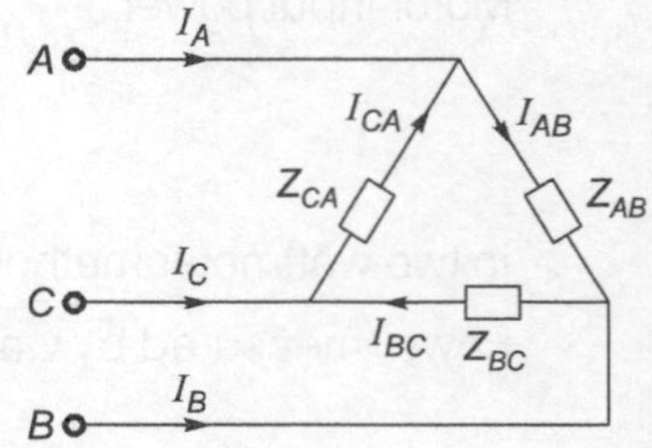

Taking V_{BC} as a reference

$$V_{BC} = 240\angle 0° \text{ V}$$
$$V_{CA} = 240\angle -120° \text{ V}$$
$$V_{AB} = 240\angle -240° \text{ V or } 240\angle 120° \text{ V}$$

Given : $Z_{AB} = 25\angle 90°\ \Omega,\ Z_{BC} = 15\angle 30°\ \Omega,\ Z_{CA} = 20\angle 0°\ \Omega$

$$I_{AB} = \frac{V_{AB}}{Z_{AB}} = \frac{240\angle 120°}{25\angle 90°} = 9.6\angle 30° \text{ A}$$

$$I_{BC} = \frac{V_{BC}}{Z_{BC}} = \frac{240\angle 0°}{15\angle 30°} = 16\angle -30° \text{ A}$$

$$I_{CA} = \frac{V_{CA}}{Z_{CA}} = \frac{240\angle -120°}{20\angle 0°} = 12\angle -120° \text{ A}$$

Line current will be :

$$I_A = I_{AB} - I_{CA}$$
$$I_A = (9.6\angle 30° - 12\angle -120°) \text{ A}$$
$$I_A = 20.87\angle 46.705 \text{ A}$$

Similarly,

$$I_B = I_{BC} - I_{AB}$$
$$= 16\angle -30° - 9.6\angle 30° \text{ A}$$
$$I_B = 13.95\angle -66.586° \text{ A}$$
$$I_C = I_{CA} - I_{BC}$$
$$= (12\angle -120° - 16\angle -30°) \text{ A}$$
$$I_C = 20\angle -173.13 \text{ A}$$

(ii)

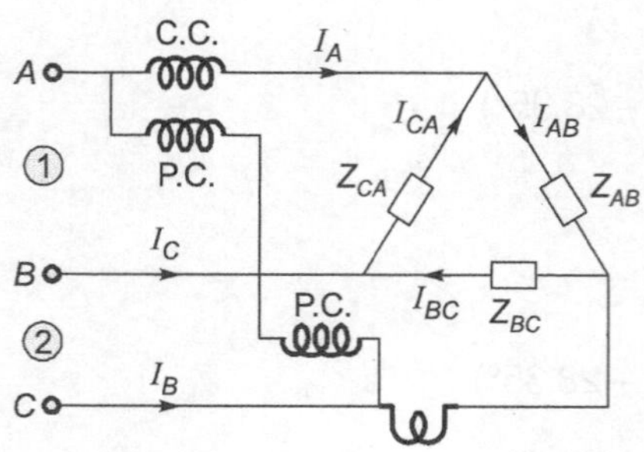

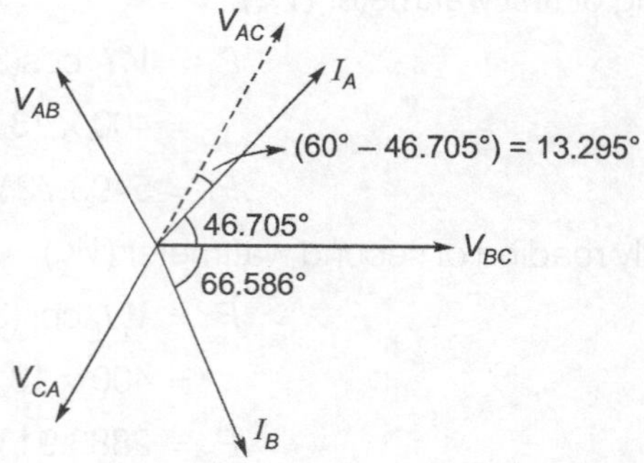

$$W_1 = V_{AC} I_A \cos(13.295)$$
$$= 240 \times 20.87 \times 0.973$$
$$= 4874.55 \text{ W}$$
$$W_2 = V_{BC} I_B \cos(66.856)$$
$$= 240 \times 13.95 \times \cos(66.586)$$
$$= 1330.40 \text{ W}$$

5.32 The power in a single-phase circuit is measured by an electrodynamometer wattmeter. The voltage across the load is 100 V and the load current is 10 A at a power factor of 0.2 lagging. The wattmeter circuit has a resistance of 3500 Ω and an inductance of 30 mH. Estimate the percentage error in the wattmeter reading when the pressure coil is connected (1) on the supply side and (2) on the load side. The current coil has a resistance of 0.1 Ω and negligible inductance. The supply frequency is 50 Hz.

[15 marks : 2023]

Solution:

Power consumed by load,

$$P = VI\cos\phi$$
$$= 100 \times 10 \times 0.2 = 200 \text{ W}$$
$$\cos\phi = 0.2$$
$$\phi = \cos^{-1} 0.2 = 78.46°$$
$$R_P = 3500\ \Omega,\ L = 30 \text{ mH},\ X_L = 2\pi \times 50 \times 30 \times 10^{-3}$$
$$= 9.42\ \Omega$$
$$\beta = \tan^{-1}\left(\frac{X_L}{R_P}\right) = \tan^{-1}\left(\frac{9.42}{3500}\right)$$
$$= 0.1542 \text{ rad} = 8.835°$$

Consider the pressure coil connected on load side :

Actual wattmeter reading $=(1 + \tan\phi \,.\, \tan\beta)P_T$

$$= [1 + \tan(78.46°).\tan 8.835°] \times 200 = 202.63 \text{ W}$$

Power loss in pressure coil $= \dfrac{V^2}{R_P} = \dfrac{(100)^2}{3500}$

$$= 2.857 \text{ W}$$

Total wattmeter reading $= 202.63 + 2.857$

$$= 205.487 \text{ W}$$

$$\% \text{ error} = \frac{P_W - P_T}{P_T} \times 100$$
$$= \frac{205.487 - 200}{200} \times 100$$
$$= 2.7435\%$$

Now, consider that pressure coil is connected on supply side,

Total power $= P_T + I^2R_C$

$$= 200 + 10^2 \times 0.1$$
$$= 210 \text{ W}$$

Impedance of load,

$$Z = \frac{V}{I} = \frac{100}{10} = 10\ \Omega$$

Resistance of load, $R_L = Z\cos\phi = 10 \times 0.2 = 2\ \Omega$

Reactance of load, $X_L = Z\sin\phi = 9.797\ \Omega$

The current coil acts as a load,

Total resistance $= 2 + 0.1 = 2.1\ \Omega$

Total reactance $= 9.797\ \Omega$

Total impedance of current coil

$$= \sqrt{(2.1)^2 + (9.797)^2}$$

$$= 10.01\ \Omega$$

Total power factor of load $= \dfrac{R_T}{Z_T} = \dfrac{2.1}{10.01}$

$$= 0.2097$$

$$\phi = \cos^{-1}(0.2097) = 77.89°$$

$$\tan\phi = 4.662$$

Reading of wattmeter, $P_W = (1 + \tan\phi \,.\, \tan\beta) \times P_T$

$$= (1 + 4.662 \times 0.1554) \times 200$$

$$= 344.89\ \text{W}$$

6. Transducers and Sensors

6.1 What are thermistors? Explain the working, construction and applications of thermistors. Compare resistance temperature characteristics of a typical thermistor and platinum.

[8 marks : 2001]

Solution:

Thermistors: Thermistor is a contraction of a term "thermal resistors". Thermistors are generally composed of semi-conductor materials. Although positive temperature co-efficient of units are available, most thermistors have a negative coefficient of temperature resistance. The negative temperature coefficient of resistance can be as large as several percent per degree celcius. This allows the thermistor circuit to detect very small changes in temperature which could not be observed with an RTD or a thermocouple.

This high sensitivity to temperature changes makes thermistors extremely useful for precision temperature measure-ments control and compensation.

Construction of Thermistors: Thermistors are composed of sintered mixture of metallic oxides such as manganese, nickel, cobalt, copper, iron. They are available in variety of sizes and shapes. The thermistors may be in the form of beads, rods, discs. Some of the commercial forms are shown in figure.

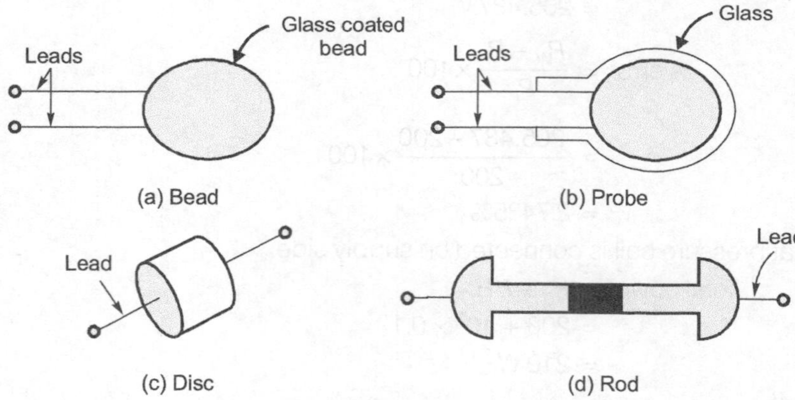

Different forms of construction of thermistors

Applications of Thermistors:

(a) **Measurement of Temperature:** A thermistor produces a large change of resistance with a small change in temperature being measured. This large sensitivity of thermistor provides good accuracy and resolution.

(b) **Control of Temperature:** Thermistor control systems are inherently sensitive, stable and fast acting and require relatively simple circuitry.

(c) **Temperature Compensation:** Because thermistors have a negative temperature coefficient of resistance-opposite to the positive coefficient of most electrical conductors and semiconductors-they are widely used to compensate for the effects of temperature on both component and circuit performance.

(d) Other Applications: The other applications of thermistors include:

(*i*) measurement of power at high frequencies.

(*ii*) measurement of thermal conductivity.

(*iii*) measurement of level, flow and pressure of liquids.

(*iv*) measurement of composition of gases.

(*v*) vacuum measurements.

(*vi*) providing time delay.

Resistance-Temperature Characteristics of a Typical Thermistor and Platinum:

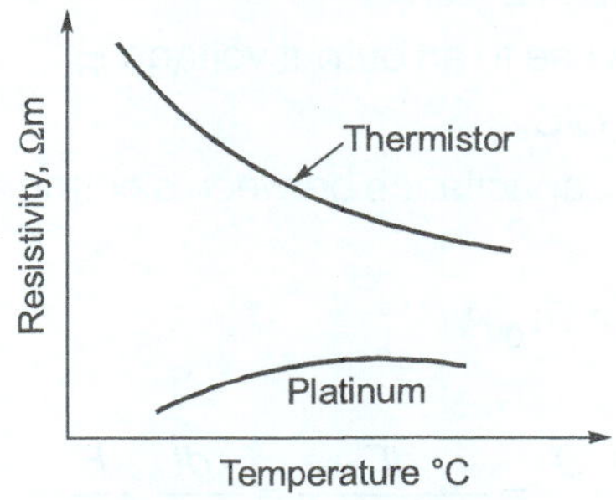

6.2 How does a 'Piezo-Electric Transducer' work? What are the common materials used for it? Derive an expression for its (a) voltage and (b) charge sensitivities. Draw an equivalent circuit for this transducer. Write uses of Piezo-Electric materials and transducers.

[12 marks : 2001]

Solution:

Piezo-Electric Transducers: A piezoelectric material is one in which an electric potential appears across certain surfaces of a crystal if the dimensions of the crystal are changed by the application of a mechanical force. This potential is produces by the displacement of charges. The effect is reversible, i.e., conversely, if a varying potential is applied to the proper axis of the crystal, it will change the dimensions of the crystal thereby deforming it. This effect is known as piezoelectric effect. Elements exhibiting piezo-electric qualities are called as electro-resistive elements.

Common piezo-electric materials include rochelle salts, ammonium dihydrogen phosphate, lithium sulphate, dipotassium tartorate, potassium dihydrogen phosphate, quartz and ceramics *A* and *B*. Except for quartz and ceramics *A* and *B*, the rest are man-made crystals grown aqueous solutions under carefully controlled conditions.

A piezo-electric element used for converting mechanical motion to electrical signals may be thought as charge generator and a capacitor. Mechanical deformation generates a charge and this charge appears as a voltage across the electrode. The voltage is $E = Q/C$.

The magnitude and polarity of the induced surface charges are proportional to the magnitude and direction of the applied force *F*. The polarity of the induced charges depends upon the direction of the applied force.

Change, $Q = d \times F$ coulomb ...(*i*)

where d = charge sensitivity of the crystal;
C/N : (it is constant for a given crystal)

and F = applied force, *N*

The force *F* causes a change in thickness of the crystal,

$$\therefore \quad F = \frac{AE}{t} \cdot \Delta t \text{ newton} \qquad ...(ii)$$

where A = area of crystal ; m^2

t = thickness of crystal ; m

and E = Young's modulus ; N/m²

Young's modulus $E = \frac{\text{stress}}{\text{strain}} = (F/A)\cdot\frac{1}{\Delta t/t} = \frac{Ft}{A\Delta t}$ N/m² ...(*iii*)

Area $A = Wl$

where W = width of crystal ; m

l = length of crystal ; m

∴ From equations (*i*) and (*ii*), we have, charge

$$Q = d\,A\,E\,(\Delta t/t) \quad ...(iv)$$

The charge at the electrodes gives rise to an output voltage E_0.

Voltage, $E_0 = Q/C_P$ ...(*v*)

where, C_P = capacitance between electrode ; F

Capacitance between electrodes

$$C_P = \epsilon_r \epsilon_0 A/t \quad ...(vi)$$

From equations (*i*), (*v*) and (*vi*)

$$E_0 = \frac{Q}{C_P} = \frac{dF}{\epsilon_r \epsilon_0 A/t} = \frac{dt}{\epsilon_r \epsilon_0}\cdot\frac{F}{A} \quad ...(vii)$$

But, $F/A = P$ = pressure or stress in N/m²

$$\therefore \quad E_0 = \frac{d}{\epsilon_r \epsilon_0} tP = gtP \quad ...(viii)$$

where $g = \frac{d}{\epsilon_r \epsilon_0}$ ...(*ix*)

'*g*' is the voltage sensitivity of the crystal. This is constant for a given crystal cut. Its units are Vm/N.

Now, $g = \frac{E_0}{tP} = \frac{E_0/t}{P}$ ...(*x*)

But, E_0/t = electric field strength, V/m

Let, $\epsilon = E_0/t$ = electric field

$$\therefore \quad g = \frac{\text{electric field}}{\text{stress}} = \frac{\epsilon}{P} \quad ...(xi)$$

Now E_0/t is the electric field intensity in the crystal and P is the pressure or the applied stress to the crystal. Therefore, crystal voltage sensitivity, g, can be defined as the ratio of electric field intensity to pressure (or stress).

From equation (*x*), charge sensitivity

$$d = \epsilon_r \epsilon_0 g \text{ ; C/N} \quad ...(xii)$$

Equivalent Circuit of Piezo-Electric Transducer: The basic equivalent circuit of a piezo-electric transducer is shown in figure.

The source is a charge generator. The value of the charge is $Q = dF$. The charge generated is across the capacitance, C_P, of the crystal and its leakage resistance R_P.

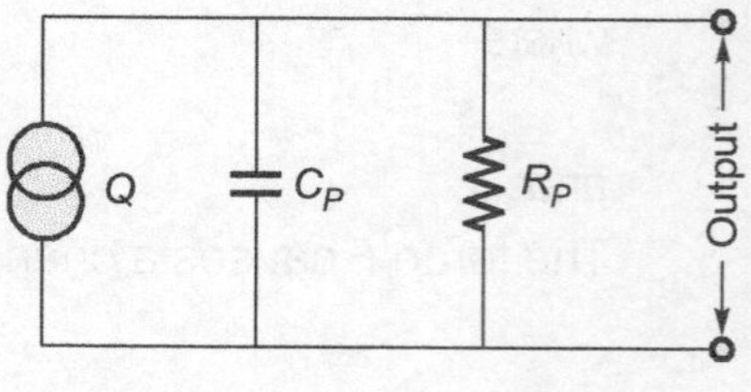

The charge generator can be replaced by an equivalent voltage source having a voltage of

$$E_0 = \frac{Q}{C_P} = \frac{dF}{C_P}$$

in series with a capacitance, C_P, and resistance, R_P as shown in figure.

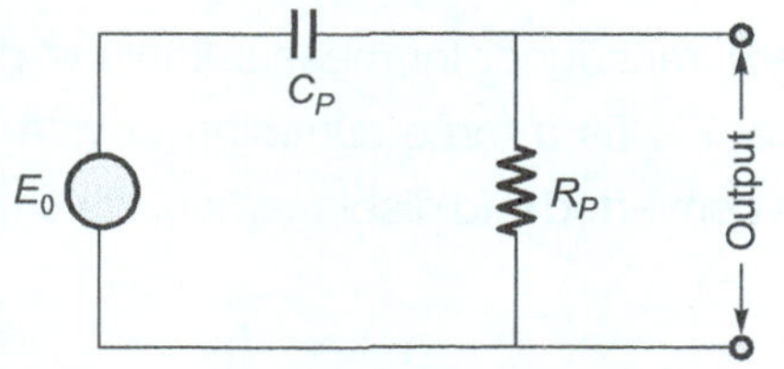

Uses of Piezoelectric Materials and Transducer: The uses of piezo-electric materials and transducers are described below:

1. The desirable properties expected out of a piezo-electric material are, stability, high output, insensitivity to temperature variations, insensitivity to variations in humidity and also the ability to be formed into a usable component. Undoubtedly the most stable material is quartz. However, its output is low. Rochelle salt, on the other hand provides, the highest output out of any piezo-electric material. But it requires protection from moisture and hence cannot be used above a temperature of 45°C.
   Because of its stability, quartz is commonly used for stabilizing electronic oscillators. The crystal is ground to proper shape and is connected in an appropriate electronic circuit whose frequency in controlled by it.
2. The use of piezo-electric transducer element is confined primarily to dynamic measurements. The voltage developed by application of strain is not held under static conditions. Hence, the elements are primarily used in the measurement of such quantities as surface roughness and in accelerometer and vibration pickups.
3. Ultrasonic generator elements also use barium titanate, a piezo-electric material. Such elements are used in industrial cleansing apparatus and also in underwater detection system known as sonar.

6.3 What is a strain gauge? A strain gauge is cemented to an iron post of length 1 m, which is subjected to a compressible force. The nominal resistance of the gauge is 200 Ω. The change in resistance of the strain gauge is measured to be 0.7 Ω. The final length of the iron post is 999 mm. Calculate the gauge factor of the strain gauge.

[12 marks : 2003]

Solution:

Strain Gauge: If a metal conductor is stretched or compressed, its resistance changes on account of the fact that both length and diameter of conductor change. Also there is a change in the value of resistivity of the conductor when it is strained and this property is called piezoresistive effect. Therefore, resistance strain gauge are also known as piezoresistive gauges. The strain gauge are used for measurement of strain and associated stress in experimental stress analysis. Secondly, many other detectors and transducers notably the load cells, torque meters, diphragm type pressure gauges, temperature sensors, accelerometers and flow meters, employ strain gauges as secondary transducers.

$$\Delta R = 0.7\ \Omega$$

$$\Delta L = L_{\text{initial}} - L_{\text{final}}$$

$$= 1000\text{ mm} - 999\text{ mm} = 1\text{ mm}$$

$$R = 200\ \Omega;\ L = 1000\text{ mm}$$

$$\text{Gauge factor} = \frac{\Delta R/R}{\Delta L/L} = \frac{0.7/200}{1/1000} = 3.5$$

6.4 Explain giving a neat sketch how an LVDT can be used to measure pressure.

[8 marks : 2003]

Solution:

The LVDT is used as a secondary transducer for measurement of pressure with Bellows or Bourdon tube acting as the primary transducer i.e. as a force summing device. The two arrangements are shown in figure (*i*) and (*ii*). The pressure is converted into displacement which is sensed by the LVDT and transduced into a voltage.

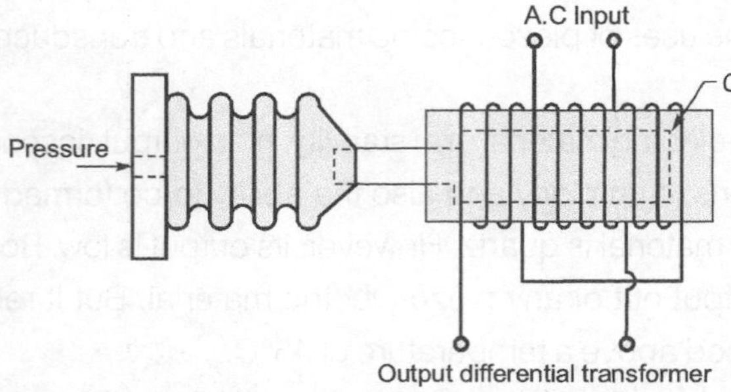

Fig. (*i*)
Measurement of pressure with Bellows and LVDT

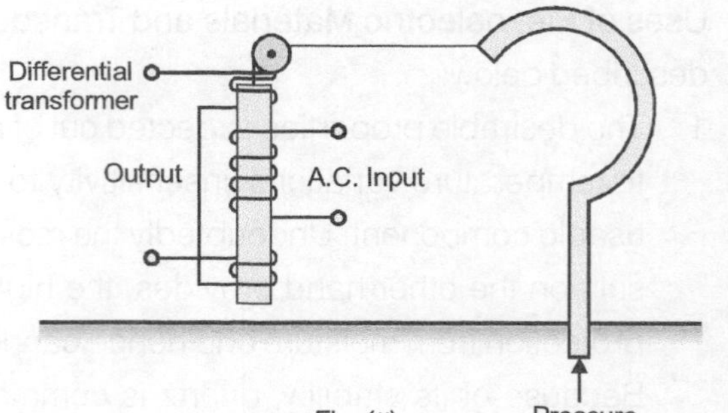

Fig. (*ii*)

6.5 **Explain how thermistor can be used for temperature measurement. The resistance at temperature *T* kelvin is given by**

$$R_T = R_0\, e^{\beta\left(\frac{1}{T} - \frac{1}{T_0}\right)}$$

where, R_0 = 1050 Ω at 27°C
the corresponding β = 3140
What is the temperature when the thermistor resistance is 2330 Ω?

[10 marks : 2003]

Solution:

A thermistor produces a large change of resistance with a small change in the temperature being measured. This large sensitivity of thermistor provides good accuracy and resolution. When this thermistor is connected in a simple series circuit consisting of a battery and micro-ammeter as shown in figure, any change in temperature causes a change in the resistance of thermistor and corresponding change in the circuit current.
The micro-ammeter may be directly calibrated in terms of temperature. The micro-ammeter may be able to give a resolution of 0.1°C.

$$R_T = R_0 e^{\beta\left(\frac{1}{T} - \frac{1}{T_0}\right)}$$

$$ln\frac{R_T}{R_0} = \beta\left(\frac{1}{T} - \frac{1}{T_0}\right)$$

$$ln\frac{2330}{1050} = 3140\left(\frac{1}{T} - \frac{1}{300}\right)$$

$$2.538 \times 10^{-4} = \frac{1}{T} - \frac{1}{300}$$

$$T = 278.76\text{ K} = 5.76°$$

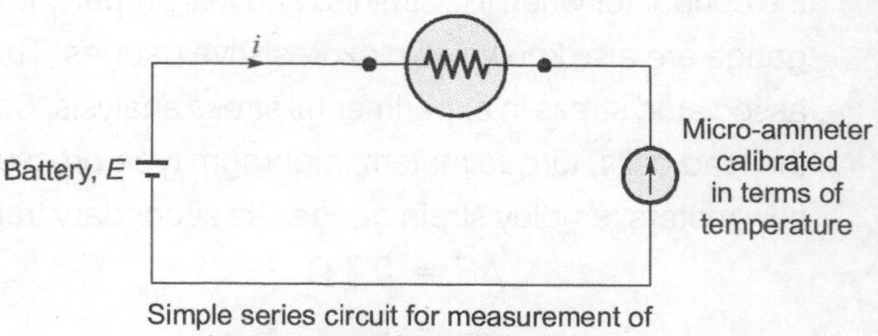

Simple series circuit for measurement of temperature using a thermistor

6.6 **What is LVDT? Explain its working with necessary diagrams and characteristics. What are its advantages and uses?**

[8 marks : 2004]

Solution:

The most widely used inductive transducer to translate the linear motion into electrical signals is the linear variable differential transformer (LVDT). The transformer consists of a single primary winding P and two secondary windings S_1 and S_2 wound on a cylindrical former. The secondary windings have equal number of turns and are identically placed on either side of the primary winding. The primary winding is connected to an alternating current source. A movable soft iron core is placed inside the former. The displacement to be measured is applied to the arm attached to the soft iron core.

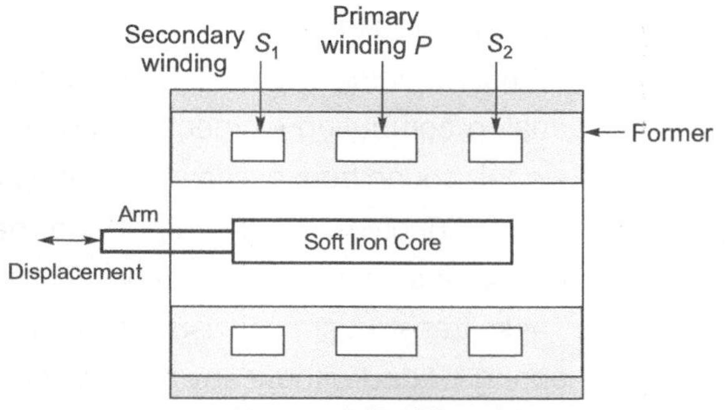

Linear variable differential transformer (L.V.D.T.)

Since the primary winding is excited by an alternating current source, it produces an alternating magnetic field which in turn induces alternating current voltages in the two secondary windings. The output voltage of secondary, S_1, is E_{S_1} and that of secondary S_2 is E_{S_2}. In order to convert the outputs from S_1 and S_2 into a single voltage signal, the two secondaries S_1 and S_2 are connected in series opposition. Thus the output voltage of the transducer is the difference of the two voltages.

Differential output voltage, $E_0 = E_{S_1} - E_{S_2}$.

When the core is at its normal (NULL) position, the flux linking with both the secondary windings is equal and hence equal emfs are induced in them. Thus at null position $E_{S_1} = E_{S_2}$. Since the output voltage of the transducer is the difference of the two voltages, the output voltage E_0 is zero at null position.

Now if the core is moved to the left of the NULL position, more flux links with winding S_1 and less with winding S_2. Accordingly output voltage E_{S_1}, of the secondary winding S_1 is more than E_{S_2}, the output voltage of secondary winding S_2. The magnitude of output voltage is, thus, $E_0 = E_{S_1} - E_{S_2}$ and the output voltage is in phase with say, the primary voltage.

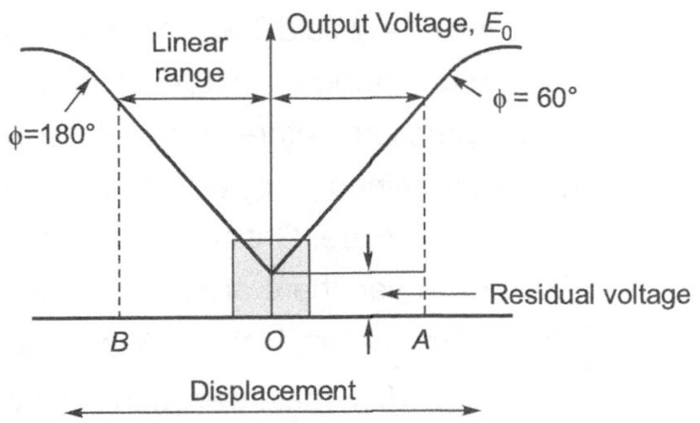

Similarly, if the core is moved to the right of the null position, the flux linking with winding S_2 becomes larger than that linking with winding S_1. This results in E_{S_2} becoming larger than E_S.

The output voltage in this case is $E_0 = E_{S_2} - E_{S_1}$ and is 180° out of phase with the primary voltae. Therefore, the two differential voltages are 180° out of phase with each other.

Advantages of LVDT:

1. The LVDTs have a very high range for measurement of displacement.
2. Ordinarily, there is no physical contact between the movable core and coil structure which means that the LVDT is a friction less device. This permits its use in critical measurements that cannot tolerate the addition of low mass core but cannot tolerate friction loading. The absence of friction between coil and core of an LVDT means that there is no wear out. This gives an LVDT essentially infinite mechanical life. This is of paramount importance in applications such as the fatigue-life testing of materials and structures. The infinite mechanical life is also important in high reliability mechanisms and systems found in aircrafts, missiles, space vehicles and critical industrial equipment.

3. The LVDT is a transformer means that there is complete isolation between excitation voltage given to the primary winding and the output produced by the secondary windings. This makes an LVDT an effective analog computing element without the need of buffer amplifiers. It also facilitates the isolation of the signal ground from excitation ground in high performance measurement and control loops.
4. The LVDT gives a high output and many a times there is no need for amplification. The transducer possesses a high sensitivity which is typically about 40 V/mm.
5. These transducers can usually tolerate high degree of shock and vibrations especially when the core is spring loaded without any adverse effects. They are simple in construction and by virtue of their being small and light in weight, they are stable and easy to align and maintain.
6. LVDT show a low hysteresis and hence repeatability is excellent under all conditions.
7. Most of LVDTs consume power which is less than 1 W.

Uses of LVDTs:

1. The LVDTs can be used in all applications where displacements ranging from fraction of a mm to a few cm have to be measured. The LVDT acting as a primary transducer converts the displacement directly into an electrical output proportional to displacement.
2. Acting as a secondary transducer it can be used as a device to measure force, weight and pressure etc. The force measurement can be done by using a load cell as the primary transducer while fluid pressure can be measured by using Bourdon tube which acts as primary transducer. The force or the pressure is converted into a voltage. In these applications the high sensitivity of LVDTs is a major attraction.

6.7 Name transducers for sensing flow rate. Explain general principle of their working. Explain working, construction, advantage and limitations of electro-magnetic flow meter.

[12 marks : 2004]

Solution:

There are number of devices for sensing the rate of fluid flow. They mainly operate on the principle of placing an obstruction in the path of fluid causing a change in fluid pressure which is dependent upon the rate of flow. Thus by measuring the difference in pressure before and after the obstruction by means of differential pressure sensor, the rate of flow may be determined. They are, in fact, mechanical means of measuring rate of liquid flow. The example of these flow meters are: Orifice plate, venturi tube and Rota meter. These are indirect means of measuring fluid flow. However, there are some direct methods of measuring fluid flow which are: Turbine meters, Electromagnetic flow meters, Hot wire Anemometers, Flow meter using Thermistors.

Electromagnetic Flow Meters: Electromagnetic flow meters are particularly suitable for the flow measurements of slurries, sludge and any electrically conducting liquid.

It consists basically of a pair of insulated electrodes buried flush in the opposite sides of a nonconducting, non-magnetic pipe carrying the liquid whose flow is to be measured.

The pipe is surrounded by an electromagnet which produces a magnetic field. The arrangement is analogous to a conductor moving across a magnetic field. Therefore, voltage is induced across the electrodes.

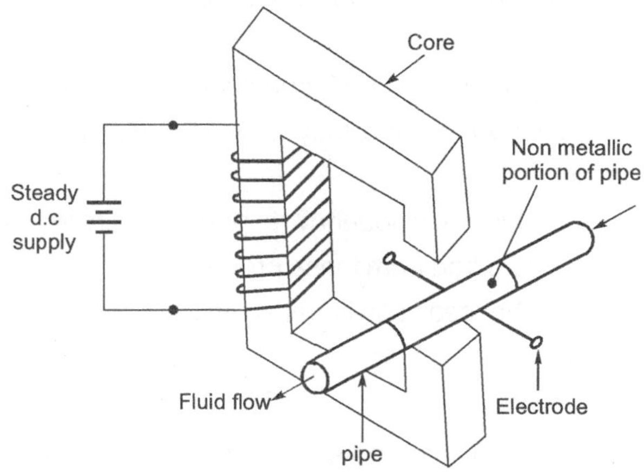

This voltage is given by:

$$E = B\,l\,v \text{ volt}$$

where, B = (flux) × (density) ; Wb/m^2,

l = length of conductor = diameter of pipe ; m

and v = velocity of conductor (flow) ; m/s

Thus assuming a constant magnetic field, the magnitude of the voltage appearing across the electrodes will be directly proportional to velocity. Non conducting pipe has to be used as the output voltage gets short circuited if metallic pipes are used. This is true when liquids of low conductivity are being measured. But when liquids of high conductivity are measured the short circuiting has no effect. Stainless steel pipes can then be used. The voltages produced are small specially at low flow rates. Therefore, the meter relies greatly on a high gain amplifier to convert the induced voltage into a usable form.

Advantages:

1. The electromagnetic flowmeters may be manufactured to measure flow in pipes of any size provided powerful magnetic field can be produced.
2. The major advantage from a fluid handling point of view is that with the electromagnetic flow meter there is no obstruction to flow that may cause pressure drops.
3. The output is linearly rated to the input.
4. The output is unaffected by changes is characteristics of liquid such as viscosity, pressure and temperature.

Limitations:

1. The operating costs are high particularly if heavy slurries are handled.
2. The conductivity of the liquid being metered should not be less than 10 μΩ/m.

6.8 **Discuss the merits and demerits of potentiometers used as transducers. A linear resistance potentiometer is 100 mm long and is uniformly wound with a wire of total resistance 10,000 Ω. Under normal conditions the slider is at the centre of the potentiometer. Determine the linear displacement when the resistance of the potentiometer as measured by wheatstone bridge is 3700 Ω. If it is possible to measure a minimum value of 5 Ω resistance with the above arrangement, determine the resolution of potentiometer in mm.**

[10 marks : 2005]

Solution:

Merits of Potentiometer:

1. They are in expensive.

2. They are simple to operate and very useful for applications where the requirements are not particularly severe.
3. They are very useful for measurement of large amplitudes of displacement.
4. Their electrical efficiency is very high and they provide sufficient output to permit control operation without further amplification.
5. It should be understood that while the frequency response of wire wound potentiometers is limited, the other types of potentiometers are free from these problems.
6. In wire wound potentiometers the resolution is limited while in cermet and metal film potentiometers, the resolution in infinite.

Demerits:

1. The chief disadvantage of using a linear potentiometer is that they require a large force to move their sliding contacts (wipers).
2. The other problems with sliding contacts are that they can be contaminated, can wear out, become misaligned and generate noise. So the life of the transducer is limited. However, recent developments have produced a roller contact wiper which, it is claimed, increases the life of the transducer by 40 times.

Numerical: The resistance of the potentiometer at its normal position

$$= \frac{10,000}{2} = 5000\ \Omega$$

Resistance of potentiometer per unit length $= \frac{10,000}{100} = 100\ \Omega/\text{mm}$

1. Change of resistance from its normal position = 5000 – 3700 = 1300 Ω
2. ∴ Displacement of wiper from its normal position $= \frac{1300}{100} = 13$ mm
3. Resolution = Minimum measurable resistance × mm/Ω

$$= 5 \times \frac{1}{100} = 0.05\ \text{mm}$$

6.9 Explain the piezoelectric phenomenon. Name two piezoelectric materials. Draw the equivalent circuit with appropriate cable connection model and charge amplifier. Obtain transfer function between amplifier output and input displacement to the piezoelectric device.

[12 marks : 2006]

Solution:

A piezo-electric material is one in which an electric potential appears across certain surfaces of a crystal if the dimensions of the crystal are changed by the application of a mechanical force. This potential is produced by the displacement of charges. The effect is reversible, i.e. conversely, if a varying potential is applied to the proper axis of the crystal, it will change the dimensions of the crystal thereby deforming it. This effect is known as piezo-electric effect. Elements exhibiting piezo-electric quantities are called as electro-resistive elements. The materials that exhibit a significant and useful piezo-electric effect are divided into two categories: (a) Natural group and (b) Synthetic group.

Quartz and Rochelle salt belong to natural group while materials like lithium sulphate, ethylene diamine tartarate belong to the synthetic group.

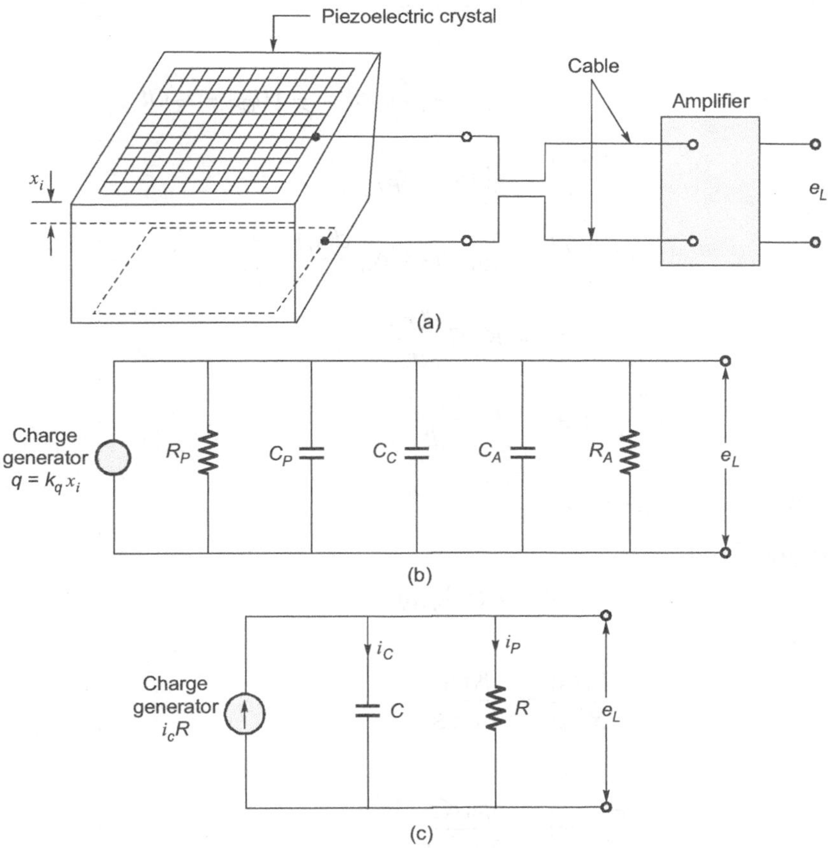

(a) (b) (c)

Consider the transducer, the connecting cable and the amplifier as a unit. The impedance of the transducer is very high and hence an amplifier with a high input impedance has to be used in order to avoid loading errors.

Charge produced, $q = K_q x_i$ coulomb

where, K_q = sensitivity; C/m

and x_i = displacement; m

Figure (b) shows the equivalent circuit:

R_P = Leakage resistance of transducer; Ω

C_P = Capacitance of transducer; F

C_C = Capacitance of cable; F

C_A = Capacitance of amplifier; F

R_A = Resistance of amplifier; Ω

The charge generator is converted into a constant current generator as shown in figure (c). The capacitance connected across the current generator is C where:

$$C = C_P + C_C + C_A$$

Resistance,
$$R = \frac{R_A R_P}{R_A + R_P}$$

Since the leakage resistance of transducer is very large (of the order of 0.1×10^{12} Ω) and therefore,

$$R = R_A$$

Converting the charge generator into a current generator

$$i_{CR} = \frac{dq}{dt} = K_q\left(\frac{dx_i}{dt}\right)$$

where i_{CR} is the current of the constant current generator.

Now,
$$i_{CR} = i_C + i_R$$

∴ Output voltage at load,
$$e_L = e_C = \frac{1}{C}\int i_C\, dt = \frac{1}{C}\int (i_{CR} - i_R)dt$$

or
$$\frac{d(e_L)}{dt} = \frac{1}{C}(i_{CR} - i_R)$$

or
$$C\frac{d(e_L)}{dt} = i_{CR} - i_R = K_q\frac{d(x_i)}{dt} - \frac{e_L}{R}$$

or
$$RC\frac{d(e_L)}{dt} + e_L = K_q R\frac{d(x_i)}{dt}$$

$$\tau\frac{d(e_L)}{dt} + e_L = K\tau\frac{d(x_i)}{dt}$$

where, $K = \text{sensitivity} = \frac{K_q}{C}$ V/m

Taking Laplace transform:

$$(\tau s + 1)\, E_L(s) = K\tau s\, X_i(s)$$

∴ Transfer function:

$$\frac{E_L(s)}{X_i(s)} = \frac{K\tau s}{1+\tau s}$$

Sinusoidal transfer function:

$$\frac{E_L}{X_i}(j\omega) = \frac{j\omega K\tau}{1+j\omega\tau}$$

6.10 **A four arm resistive bridge is used with a temperature sensor (RTD) having resistance temperature coefficient α = 0.042 Ω/°C in one of the arms. Initially the bridge is balanced with all arms having a resistance of 400 Ω. DC supply to the bridge is limited by a maximum permissible current of 30 mA through the temperature sensor. Bridge output is connected to a current responding meter having input resistance of 100 Ω and deflection sensitivity of 2°/μA. Determine the deflection in the output meter if the temperature sensor under goes a temperature rise of 30°C.**

[10 marks : 2006]

Solution:

Temperature rise of 30° increases the resistance of sensor by

$$0.042 \times 30 = 1.26\ \Omega$$

So resistance of sensor becomes 401.20 Ω.

Current through sensor is restricted is 30 mA.

So the current through arm ABC to 30 mA.

So voltage across ABC: $(400 + 401.26) \times 30 \times 10^{-3}$

$$= 801.26 \times 30 \times 10^{-3}$$
$$= 24038 \times 10^{-3} \approx 24\text{ V}$$
$$= \text{DC supply voltage}$$

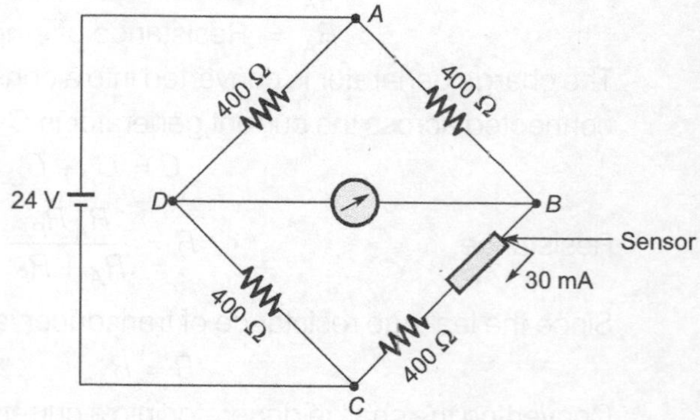

Thevenin resistance is (400)| |(400) + (400)| |(401.26)

$$= 400.315\ \Omega$$

$$V_{Th} = V_{AD} - V_{AB} = 24 \times \left(\frac{400}{800} - \frac{400}{801.26}\right)$$
$$= 24(0.5 - 0.499) = 0.0188\text{ V}$$

$\therefore$ Current I, through meter $= \dfrac{0.0188}{(400.315+100)} = 37.57\,\mu A$

Deflection of the meter $= 37.57 \times 2° = 75.14°$

6.11 Describe the use of thermocouple for the measurement of temperature.
A copper-constantan thermocouple with 0°C reference junction temperature was found to have linear calibration between 0°C and 400°C and emf at maximum temperature equal to 20.68 mV. Determine:
(a) the correction which must be made to the indicated emf if the cold junction temperature is 25°C.
(b) the temperature of the hot junction if the indicated emf is 8.92 mV with cold junction temperature of 25°C.

[10 marks : 2007]

Solution:

When two metals having different work junction are placed together, a voltage is generated at the junction which is nearly proportional to the temperature. This junction is called a thermocouple. This principle is used to convert heat energy to electrical energy at the junction of two conductors.

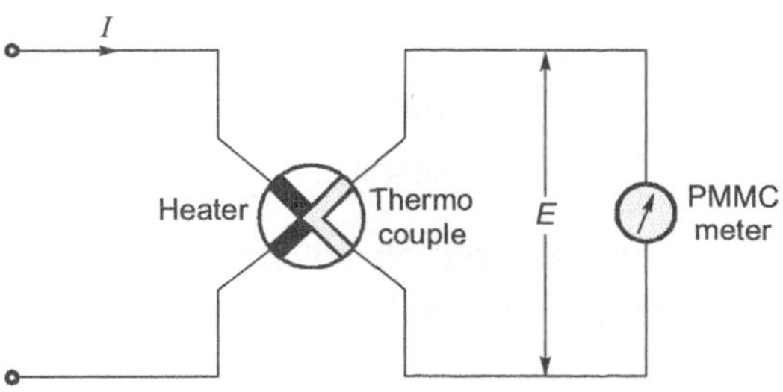

The heat at the junction is produced by the electrical current flowing in the heater element while the thermocouple produces an emf at its output terminals which can be measured with the help of a PMMC instrument. The emf produced is proportional to the temperature and hence to the rms value of the current. Therefore the scale of PMMC instrument can be calibrated to read the current passing through the heater. The thermocouple type of instruments can be used for both d.c. and a.c. applications. The most attractive feature of thermocouple instruments of current and voltages at very high frequencies. In fact, these instruments are very accurate well above a frequency of 50 MHz.

Assuming linear relationship between emf generated and temperature.

$$e = K\Delta T$$

$$K = \frac{20.68}{400}\ \text{mV/°C}$$

(a) Voltage at 25°C $= \dfrac{20.68}{400} \times 25 = 1.2925$ mV

Hence, 1.2925 mV has to be supplied to the intermediate output.

(b) Indicated emf = 8.92 mV

Corrected emf = Indicated emf + Correction

= 8.92 + 1.2925 = 10.2125 mV

as 20.68 mV corresponds to 400°C

10.2125 mV corresponds to $\dfrac{400}{20.68} \times 10.2125 = 197.53°C$

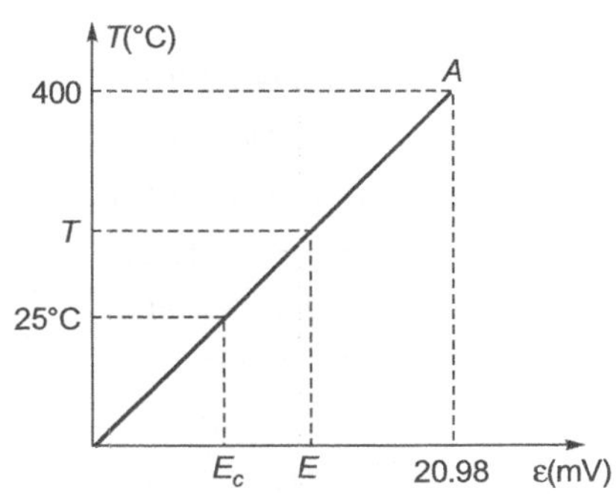

6.12 How is a differential output taken from an inductive transducer? Explain the advantages when such inductive transducers are used for measurement of displacement.

[10 marks : 2007]

Solution:

Differential output taken from an inductive transducer: Normally the change in self inductance ΔL is adequate for detection for subsequenct stages of instrumentation system. However, if the succeeding instrumentation responds to ΔL, rather then to $L + \Delta L$ the sensitivity and accuracy will be much higher. The transducer can be designed to provide two outputs one of which is an increase of self inductance. The succeeding stages of instrumentation system measure the difference between these outputs i.e. $2\Delta L$. This is known as the differential output.

The advantages of differential outputs are:

1. The sensitivity and accuracy are increased.
2. The output is less affected by external magnetic fields.
3. The effective variations due to temperature changes are reduced.
4. The effects of changes in supply voltage and frequency are reduced.

The differential arrangement consists of a coil which is divided into two parts. In response to a physical signal, which is normally a displacement, the inductance of one part increases from L to $L + \Delta L$ while that of the other part decreases from L to $(L - \Delta L)$. The change is measured as the difference of the two resulting in an output of $2\Delta L$ instead ΔL when only a single winding is used.

Example of inductive transducer:

1. LVDT
2. RVDT
3. Synchros
4. Resolver

6.13 **What are capacitive transducers? What are their advantages and disadvantages? Explain the use of capacitive transducer for the measurement of liquid level.**

[12 marks : 2008]

Solution:

Capacitive Transducers: The principle of operation of capacitive transducers is based upon the equation for capacitance of a parallel plate capacitor.

Capacitance, $$C = \frac{\epsilon A}{d} = \frac{\epsilon_0 \epsilon_r A}{d}$$

where,
A = Overlapping area of plates ; m^2
d = distance between two plates ; m
$\epsilon = \epsilon_0 \epsilon_r$ = permittivity of medium ; F/m
ϵ_r = relative permittivity
ϵ_0 = permittivity of free space ;
= 8.85×10^{-12} F/m

The capacitive transducers work on the principle of change of capacitance which may be caused by

1. Change in overlapping area A,
2. Change in distance between the plates, and
3. Change in dielectric constant.

The output impedance of capacitive transducers is high. The capacitive transducers are commonly used for measurement of linear displacement. These transducers use the following effects:

1. Change in capacitance due to change in over lapping area of plates.
2. Change in capacitance due to change in distance between the two plates.

Advantage of Capacitive Transducers:

1. They require extremely small forces to operate them and hence are very useful for use in small systems.
2. They are extremely sensitive.

3. They have good frequency response. This response is as high as 50 KHz and hence they are very useful in dynamic studies.
4. They have high input impedance and therefore the loading effects are minimum.
5. A resolution of the order of 2.5×10^{-3} mm can be obtained for these transducers.
6. The capacitive transducers can be used for applications where stray magnetic fields render the inductive transducers useless.
7. The force requirement for capacitive tranducers is very small and therefore they require small power to operate them.

Disadvantage of Capacitive Transducers:

1. The metallic parts of the capacitive transducers must be insulated from each other. To reduce the effects of stray capacitance the frames must be earthed.
2. Capacitive transducers behave non-linearly many a times due to edge effects. To overcomes this guard rings are used.
3. The output impedance of capacitive transducers tends to be high on account of their small capacitance value. This leads to loading effects.
4. The cable connecting the transducers to the measuring point is also a source of error. This results in loss of sensitivity and low frequency response (poor).
5. The capacitance of a capacitive transducer may be changed on account of presence of extraneous matter like dust particle and moisture.
6. The capacitive transducers are temperature sensitive and therefore any change in temperature adversely affects their performance.
7. The instrumentation circuitry used with these transducers is very complex.

Use of Capacitive Transducers for the Measurement of Liquid Level: Capacitive transducers using the principle of change of capacitance with change of dielectric are normally used for measurement of liquid levels. The electrodes are two concentric cylinders and the non conducting liquid acts as dielectric. At lower end of the outer cylinder there are holes which allow passage of liquid. In case these holes are small they provide mechanical damping of the surface variation.

The value of capacitance for this capacitor is,

$$C = 2\pi\epsilon_0 \times \frac{\epsilon_1 h_1 + \epsilon_2 h_2}{ln(r_2 / r_1)}$$

where, h_1 = height of liquid; m
h_2 = height of cylinder above liquid; m
ϵ_1 = relative permittivity of liquid
ϵ_2 = relative permittivity of vapour above liquid
r_2 = inside radius of outer cylinder ; m
r_1 = outside radius of inner cylinder ; m
ϵ_0 = permittivity of free space ; F/m

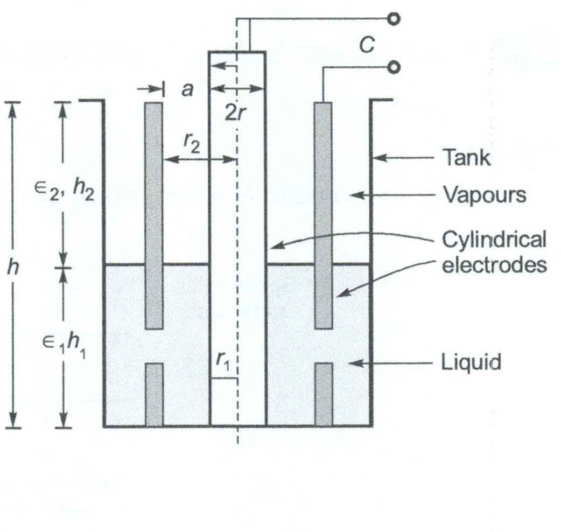

if $h >> r_2$ and $r_2 >> r_2 - r_1 >> a$

Then, $r_2 = r + a$ and $r_1 = r$

$$\epsilon = 2\pi\epsilon_0 \frac{\epsilon_1 h_1 + \epsilon_2 h_2}{ln\left(1 + \frac{a}{r}\right)}$$

6.14 **Describe the construction of a foil-type strain gauge and explain the advantages over wire-wound strain gauge. A strain gauge has a resistance of unstrained gauge of 120 Ω and the gauge factor of −12. What is the resistance value of the gauge when it is subjected to a strain of 1%?**

[10 marks : 2008]

Solution:

Bonded Metal Foil Strain Gauges: This class of strain gauges is only an extension of the bonded metal wire strain gauges. Metal foil strain gauges use identical or similar materials to wire strain gauges and are used today for most general purpose stress analysis applications and for many transducers.

The sensing elements of foil gauges are formed from sheets less than 0.005 mm thick by photo etching processes.

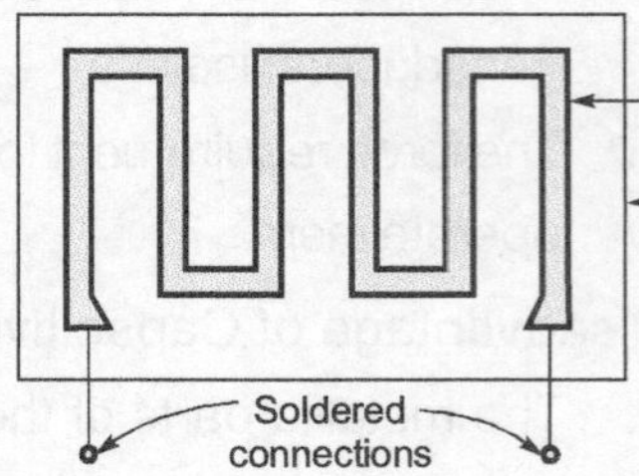

Advantage Over Wire Wound Strain Gauge:

1. Foil type gauges have a much greater heat dissipation capacity as compared with wire wound strain gauges on account of their greater surface area for the same volume. For this reason, they can be used for higher operating temperature range. Also the large surface area leads to better bonding.
2. The sensing elements of foil gauges are formed from sheets less than 0.005 mm thick by photo-etching processes, which allow greater flexibility with regard to shape.
3. Local increase in area reduces the transverse sensitivity which is a spurious input since the gauge is designed to measure the strain component along the length of grid elements.
4. For foil type strain gauges, the manufacturing process also easily provides convenient soldering tabs, which are integral to the sensing grid.
5. Foil type of gauges can be applied to curved surface, the minimum safe bend radius can be as small as 1.5 mm in some strain gauge.

$$R = 120\ \Omega$$

$$G_f = -12$$

$$\text{Strain} = \frac{\Delta L}{L} = \frac{1}{100}$$

$$G_f = \frac{\frac{\Delta R}{R}}{\frac{\Delta L}{L}} = \frac{\frac{\Delta R}{120}}{\frac{1}{100}} = \frac{-12 \times 120}{100} = \Delta R = -14.4$$

$$\text{Resistance value of gauge} = R + \Delta R$$

$$= 120 - 14.4 = 105.6\ \Omega$$

6.15 **Is tachometer a transducer? Give a circuit to measure d.c. output from a.c. tachometer.**

[10 marks : 2009]

Solution:

Yes, tachometer is a transducer which convert speed in term of voltage.

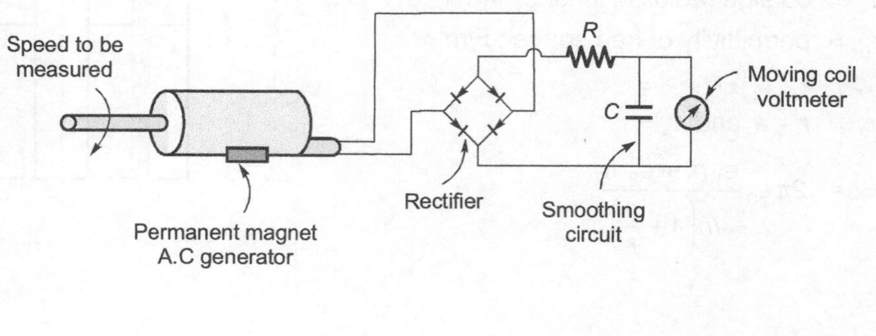

The tachometer generator has rotating magnet which may be either a permanent magnet or an electromagnet. The coil is wound on the stator and therefore the problems associated with commutator are absent.
The rotation of the magnet causes an emf to be induced in the stator coil. The amplitude and frequency of this emf are both proportional to the speed of rotation. Thus either amplitude or frequency of induced voltage may be used as a measure of rotational speed.
When amplitude of induced voltage is used as a measure of speed, the circuit shown above is used. The output voltage of a.c. tachometer generator is rectified and is measured with a permanent magnet moving coil instrument.

6.16 Calculate the gauge factor of a strain gauge if a 1.5 mm-diameter conductor that is 24 mm long changes length by 1 mm and diameter by 0.02 mm under a compression force.

[10 marks : 2009]

Solution:

The gauge factor can be written as

$$= \underbrace{1}_{\substack{\text{Resistance}\\ \text{change due to}\\ \text{change of length}}} + \underbrace{2\nu}_{\substack{\text{Resistance}\\ \text{change due to}\\ \text{change in area}}} + \underbrace{\frac{\Delta P/P}{\in}}_{\substack{\text{Resistance}\\ \text{change due to}\\ \text{piezo resistive effect}}}$$

If the change in the value of resistivity of a material when strained is neglected, the gauge factor is

$$G_f = 1 + 2\nu$$

$$\text{Poission's Ratio} = \nu = \frac{\text{lateral strain}}{\text{longitudinal strain}} = \frac{\Delta D/D}{\Delta L/L}$$

where,

$$\Delta D = 0.02 \text{ mm}, D = 1.5 \text{ mm}$$

$$\Delta L = 1 \text{ mm}, L = 24 \text{ mm}$$

$$\nu = \frac{\frac{0.02}{1.5}}{\frac{1}{24}} = 0.32$$

$$\therefore \quad \text{Gauge factor} = 1 + 2\nu = 1 + 2 \times 0.32 = 1.64$$

6.17 (i) Why is it not possible to measure the temperature of a body instantaneously?

(ii) Two strips of metal A and B with thermal expansion coefficients α_A and α_B ($\alpha_A > \alpha_B$) are bonded together at temperature T_1 to form a bimetallic strip. Draw the profile of the bonded strips when $T_2 > T_1$ and $T_2 < T_1$. On what factors does the bending of the bimetallic strip depend? What are the practical applications of bimetallic strips?

[4 + 8 marks : 2010]

Solution:

(i) These are various kind of thermometer based on material properties and such thermometers relay on the relation between pressure and volume and temperature of their thermometric material. When a quantity of heat enters or leaves a body of the material, the material must expand or contract to its final volume or reach its final pressure and must reach its final temperature with practically no delay, some of the heat that enters can be considered to change the latent heat of expansion at constant temperature and the rest of it can be considered to change the temperature of the body at constant volume and is called the specific heat at constant volume.

Some materials do not have this property and take some time to distribute the heat between temperature and volume change.

(ii) • All metals expand or contract with change in temperature and the thermal coefficient of expansion is not the same for all metals and therefor their rates of expansion or contraction are different. The difference in thermal expansion rate is used to produce deflections proportional to temperature changes.

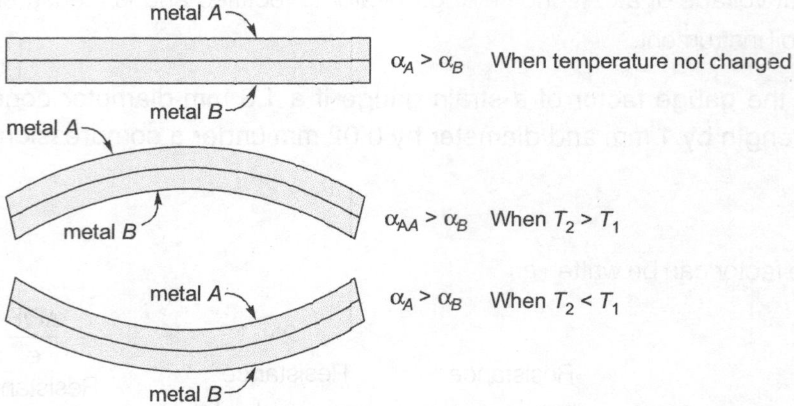

- Since all metals try to change their physical dimensions at different rates when subjected to same change in temperature, these two metallic strip change their lengths at different rates. The differential change of expansion of two metals result in bending of the bimetallic strip with change in temperature.
- The bimetallic strip bends towards the side whose metal has a lower thermal expansion coefficient when there is increase in temperature and several happens when there is decrease in temperature.
- Bimettalic strips are used in process industries for focal temperature measurement. Example in refineries, oil burners, fire vulcanizers, not wire heaters, tempering tanks etc.

6.18 **(i) How can strain gauges be used to transduce pressure?**

(ii) Explain with a diagram, the placement of gauges, explaining which of the gauges are in tension and compression respectively.

(iii) Give the connection diagram of the wheat stone's bridge to the used.

[4 + 4 + 4 marks : 2010]

Solution:

In strain gauge type pressure transducers, 4 strain gauges are connected in a wheat stone bridge. Which is connected via a rod to a diaphragm.

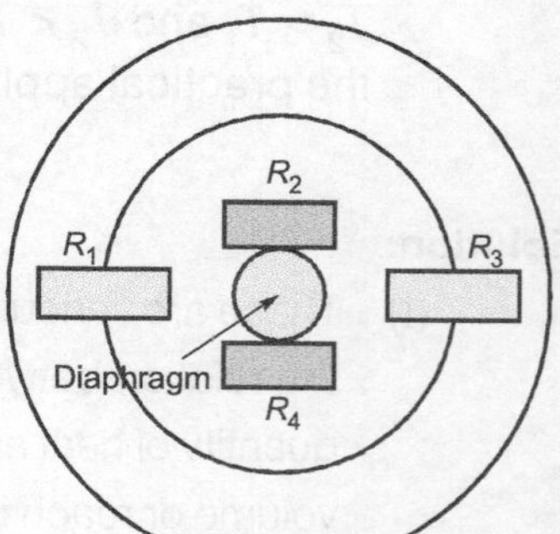

- Application of pressure produces a small displacement, this displacement increases tension in two wires and decreases in the other two, thereby increasing the resistance of two wires which are in tension and decreasing the resistance of the remaining two wires. This causes an unbalance of the bridge producing an output voltage which is proportional to the input displacement and hence to the applied pressure.
- R_2 and R_4 will be stretched along their width and R_1 and R_3 will be stretched along their length.
- All four strain gauges are mounted on to the surface of a circular diaphragm in a radial configuration 90° apart.

- The two strain gauges R_2 and R_4 that are pointing tangentially will be stretched cross the width, hence they will be in compression reducing their resistance.
- The two strain gauges R_4 and R_3 that are pointing perpendicularly will be stretched along their length hence they will be in tension increasing their resistance.

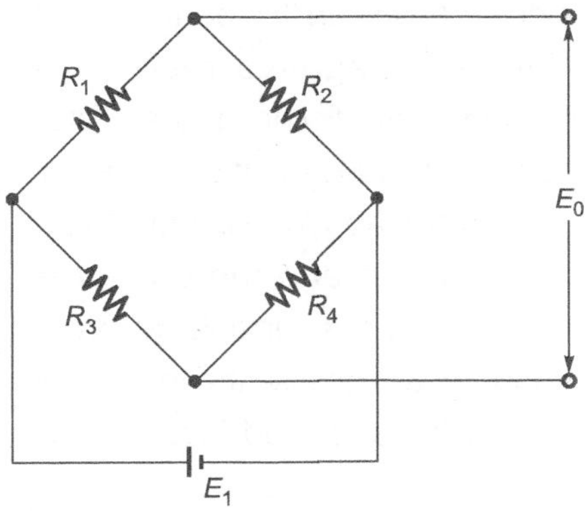

6.19 Suggest a negative temperature coefficient device for the measurement of temperature (0°C – 150°C). Describe the resistance temperature characteristic, voltage-current characteristic and current time characteristic of thermistor. Draw a circuit for measurement of temperature using thermistor. The output relation for a thermistor transducer is given by

$$R = R_0\, e^{\beta\left(\frac{1}{T}-\frac{1}{T_0}\right)}$$

For T_0 = 300 k, β = 3420, R_0 = 1 kΩ and R = 2 kΩ calculate T.

[10 marks : 2011]

Solution:

Thermistors are generally composed of semi-conductor materials. Most thermistors have a negative coefficients of temperature resistance. The negative temperature coefficients of resistance can be as large as several percent per degree celcius. This allows the thermistor circuits to detect very small changes in temperature which could not be observed with an RTD or thermocouple.

1. **Resistance-Temperature Characteristic:**

   The characteristic is shown in figure (a). Between –100°C and 40°C, the thermistor changes its resistivity from 10^6 to 10^0 Ω-cm, a factor of 10^6. This explains the high sensitivity of thermistor for measurement of temperature.

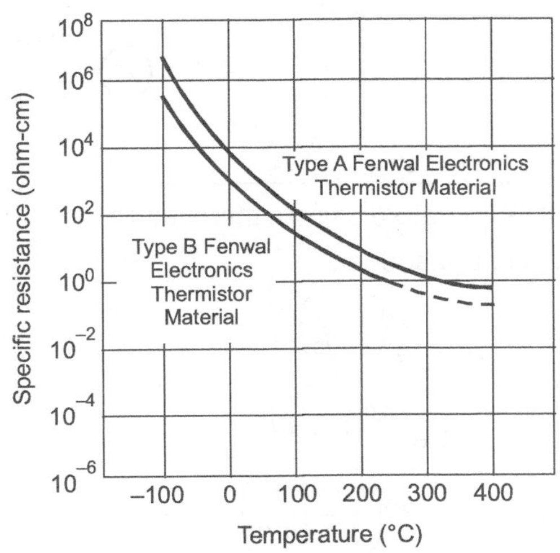

(a) Typical thermistor temperature resistance

2. **Voltage-Current Characteristic:**
These characteristics are shown in figure (b). It shows that the voltage drop across a thermistor increases with increasing current until it reaches a peak value beyond which the voltage drop decreases as the current increases. In this portion of the curve, the thermistor exhibits a negative resistance characteristic. If a very small voltage is applied to the thermistor, the resulting small current does not produce sufficient heat to raise the temperature of the thermistor above ambient. Under this condition, Ohm's law is followed and the current is proportional to the applied voltage. Larger currents, at larger applied voltages, produce enough heat to raise the thermistor temperature above the ambient temperature and its resistance then decreases.

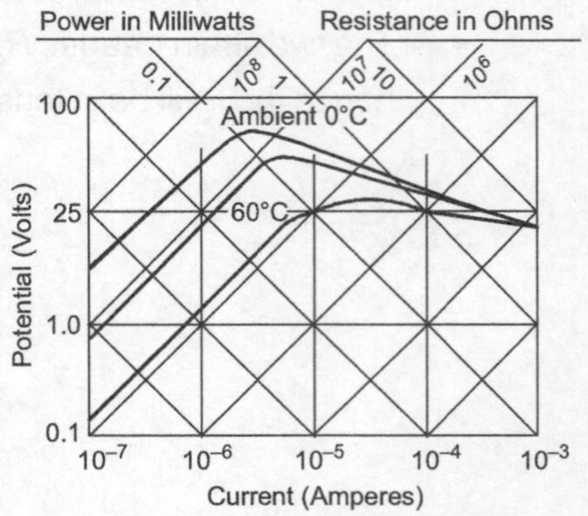

(b) Voltage current characteristic

As a result more current is then drawn and the resistance decreases further. The current continues to increase until the heat dissipation of the thermistor equals the power supplied to it.

3. **Current-Time Characteristic:**
These characteristics are shown in figure (c). It indicates the time decay to each maximum current as a function of the applied voltage when the heating effect occurs in a thermistor, a certain finite time is required to heat and the current to build upto a maximum steady state value.

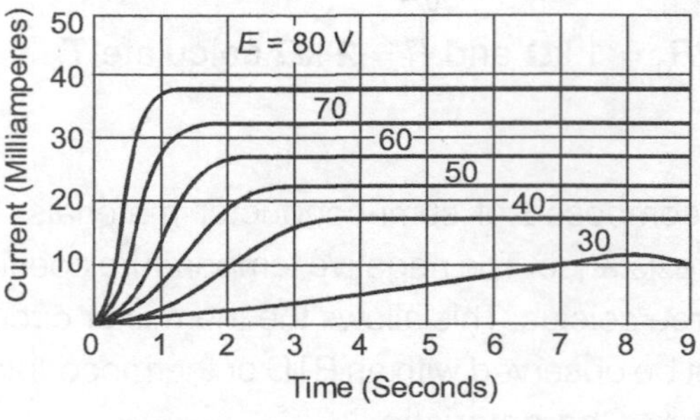

(c) Current-time characteristics

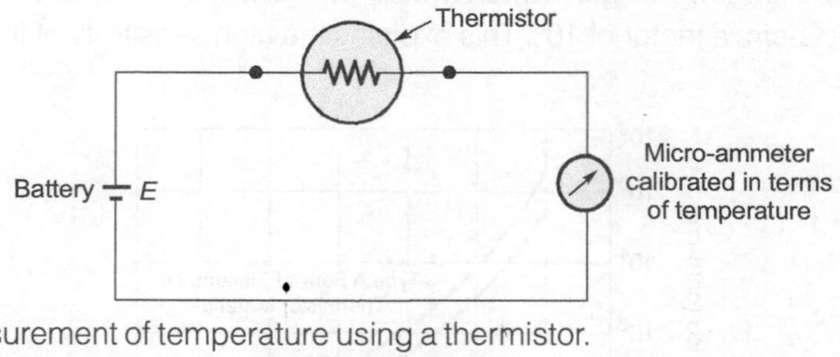

Circuit for measurement of temperature using a thermistor.

Output relation for a thermistor transducer,

$$R = R_0 e^{\beta\left(\frac{1}{T} - \frac{1}{T_0}\right)}$$

$$\Rightarrow \quad 2 = 1\, e^{3420\left(\frac{1}{T} - \frac{1}{300}\right)}$$

$$ln\, 2 = 3420\left(\frac{1}{T} - \frac{1}{300}\right)$$

$$\Rightarrow \quad T = 282.2 \text{ k}$$

6.20 **Explain the principle of operation and construction details of a Resistance Temperature Device (RTD).**

[12 marks : 2012]

Solution:

Resistance thermometer, also called resistance temperature detectors (RTDs), are used to measure temperature by correlating the resistance of the RTD element with temperature.

The resistance of a conductor changes when its temperature is changed. This property is utilised for measurement of temperature.

The variation of resistance R with temperature T can be represented by the following relationship for most of the metals.

$$R = R_o(1 + \alpha_1 T + \alpha_2 T^2 + \cdots)$$

Where, R_o = Resistance of at $T = 0$

$\alpha_1, \alpha_2, \cdots \alpha_n$ = are constants

The requirement of a conductor material to be used in RTD's are:

- The charge in resistance of material per unit charge in temperature should be as large as possible.
- The material should have a high value of resistivity so that minimum volume of material is used for the construction of RTD.
- The resistance of materials should have a continuous and stable relationship with temperature.

Platinum is especially suited for RTD's as it can withstand high temperatures while maintaining excellent stability.

Construction:

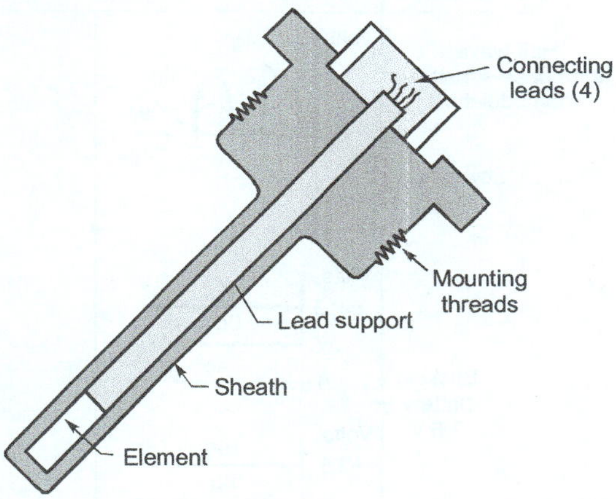

6.21 **Draw a sketch of a variable reluctance accelerometer and explain its working.**

[10 marks : 2012]

Solution:

The figure shows an accelerometer using variable—reluctance principle. Since the force required to accelerate a mass is proportional to the acceleration, the springs supporting the mass deflect in proportion to the acceleration; thus a displacement measurement allows an acceleration measurement. The mass is of iron and thus serves as both an inertial element for transducing acceleration to force and a magnetic circuit element for transducing motion to reluctance.

The instrument is constructed in such a way that the iron core is halfway between the two E frames when the acceleration is zero, and thus giving zero output voltage for zero acceleration. However to detect motion on both sides of zero (corresponding to plus or minus accelerations), the following adjustments are done. With zero-acceleration input, the iron core and springs were adjusted so that the core is offset to one side by an amount equal to the spring deflection corresponding to the maximum plus scale (say + 4g). Thus, with no acceleration, the output voltage is some specific value instead of zero (2.5 V in this case). Then, when maximum scale of acceleration (+4g) is applied, the output voltage will be 5.0 V and the minimum scale of acceleration (–4g) is applied, the output voltage will be zero.

The primary coils set up a flux dependent on the reluctance of the magnetic path. The main reluctance is the air gap. When the core is in the neutral position, the flux is same for both halves of the secondary coil; and since they are connected in series opposition, the net output voltage is zero. A motion of the core increases the reluctance (air gap) on one side and decreases it on the other, causing more voltage to be induced into one half of the secondary coil than the other and thus a net output voltage. Motion in the other direction causes the reverse action, with a 180° phase shift occurring at null. The output voltage is half wave, non-phase-sensitive rectified (demodulated) and filtered to produce an output of the same form as the acceleration input. If the 2.5 V output for zero — acceleration is objectionable, it can be bucked out with a 2.5 V battery of opposite polarity connected externally to the accelerometer. The actual full-scale motion of the mass in this particular instrument is just a few thousandths of a cm, which gives a displacement sensitivity for the variable — reluctance element of almost 500 V/cm.

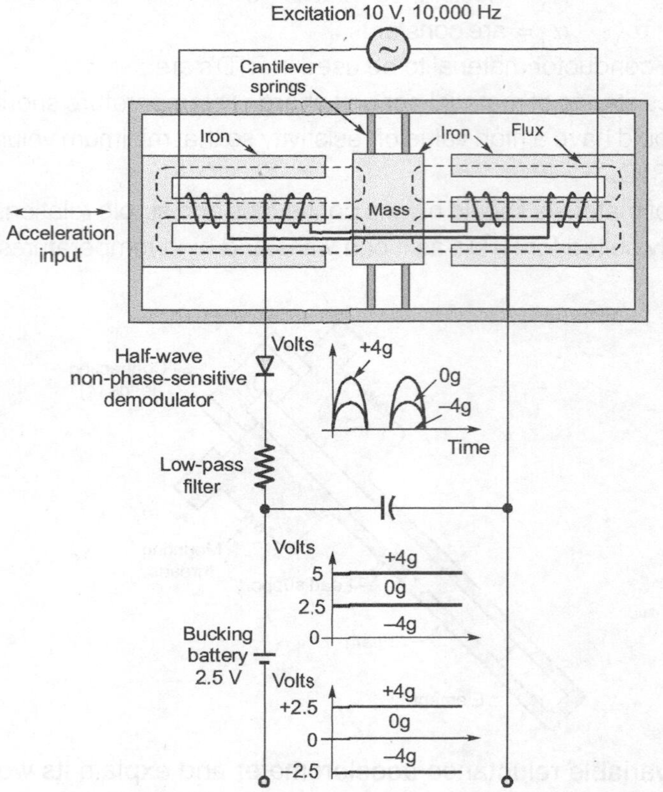

6.22 **Explain basic principle of operation of a thermocouple.** **[5 marks : 2013]**

Solution:

Thermocouple Instruments: When two metals having different work functions are placed together, a voltage is generated at the junction which is nearly proportional to the temperature. This junction is called a Thermocouple. This principle is used to convert heat energy to electrical energy at the junction of two conductors as shown in figure.

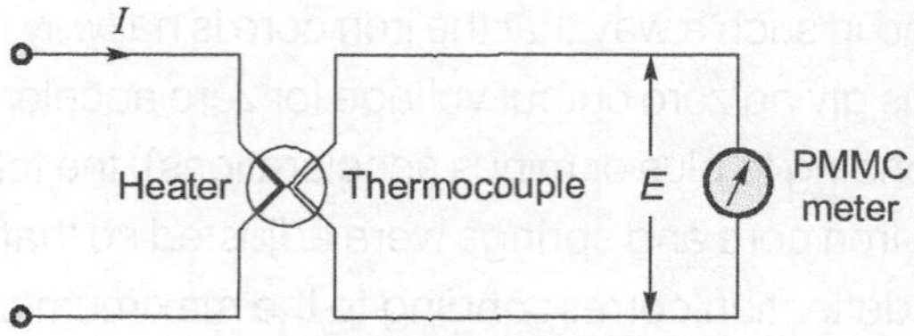

Circuit of Thermocouple

The heat at the junction is produced by the electrical current flowing in the heater element while the thermocouple produces an emf at its output terminals which can be measured with the help of a PMMC instrument. The emf produced is proportional to the temperature and hence to the rms value of the current. Therefore the scale of PMMC instrument can be calibrated to read the current passing through the heater. The thermocouple type of instruments can be used for both d.c. as well as a.c. applications. The most attractive feature of thermocouple instruments is that they can be used for measurements of current and voltages at very high frequencies. In fact, these instruments are very accurate well above a frequency of 50 MHz.

The thermal emf developed in a circuit composed of two dissimilar metals with junctions kept at absolute temperatures T_1 and T_2 (with $T_1 > T_2$) may by approximately written as:

$$E = a(T_1 - T_2) + b(T_1 - T_2)^2$$

where a and b are constants whose value depend upon the metals used.

6.23 Explain the principle operation of the thermocouple. List the types of the thermocouple and discuss how the temperature i.e. 0°C to 256°C and –256°C to 0°C can be sensed using thermocouple.

[8 marks : 2014]

Solution:

Thermocouple instrument:

Principle: When two metals having different work functions are placed together a voltage is generated at the junction which is nearly proportional to the temperature. This junction is called thermocouple.

Types of thermocouple:

Type *E*: Choromel(Nickel-10% chromium) positive leg.
Constantan (Nickel-45% copper) negative leg.
Temperature range → –330 to 1600° F.

Type *J*: Iron as positive leg, constantan as negative leg.
Temperature range → 32 to 1400°F.

Type *K*: Chromel as positive leg.
Alumel (Nickel-5% aluminium and silicon) as negative leg.
They find widest use at temperatures above 1000°F.

Type *N*: Nicrosil (Nickel-14% chromium-15% silicon) as positive leg.
Nilsil (Nickel-4.5%, silicon-1% magnesium) as negative leg.
Temperature range is 32-2300°F.

Type *T*: Copper as positive leg.
Constantan as negative leg.
Temperature range is –330 to 700°F.

Seeback effect:

According to this effect a voltage is induced between junction of two different metals when they are held at different temperature. The voltage induced in a thermocouple can be given by

$$E = a(4\theta) + b(4\theta)^2$$

Where, 4θ is temperature different between the junctions (°C).

a and *b* are constants.

a >> *b* usually so, equation may be written as

$$E = a(4\theta)$$

$$4\theta \simeq \left(\frac{E}{a}\right)$$

The output of thermocouple in the range of mV which can be detected with PMMC instrument or dc potentiometer after amplifying it.

6.24 **The inductance of a certain moving iron ammeter is expressed as $L = \left(10 + 3\theta - \frac{\theta^2}{4}\right)$ μH, where θ is deflection in radians from the zero position. The control spring torque is 25 × 10^{-6} N-m/rad. What would be the deflection of a pointer in radian when the meter carries a current of 5 A?**

[8 marks : 2014]

Solution:

$$L = \left(10 + 3\theta - \frac{\theta^2}{4}\right) \mu H$$

As we know that,

At equation, $$T_C = k\theta = T_D = \frac{I^2}{2} \cdot \frac{dL}{d\theta}$$

$$\Rightarrow \quad 25 \times 10^{-6}\,\theta = \frac{(5)^2}{2}\left(3 - \frac{\theta}{2}\right) \times 10^{-6}$$

$$25\,\theta = \frac{25}{2}\left(3 - \frac{\theta}{2}\right)$$

$$2\theta = \left(3 - \frac{\theta}{2}\right)$$

$$4\theta = 6 - \theta$$

$$5\theta = 6$$

$$\theta = \left(\frac{6}{5}\right) \text{ radian}$$

6.25 **What are different types of Electrical resistance strain gauges? Strain gauge with gauge factor of 2 is fastened to a metallic member subjected to a stress of 1000 kg/cm^2. The modulus of elasticity of metal is 2 × 10^6 kg/cm^2. Calculate percentage change in resistance of the strain gauge. What is the value of Poisson's ratio?**

[10 marks : 2015]

Solution:

Types of resistance strain gauges,

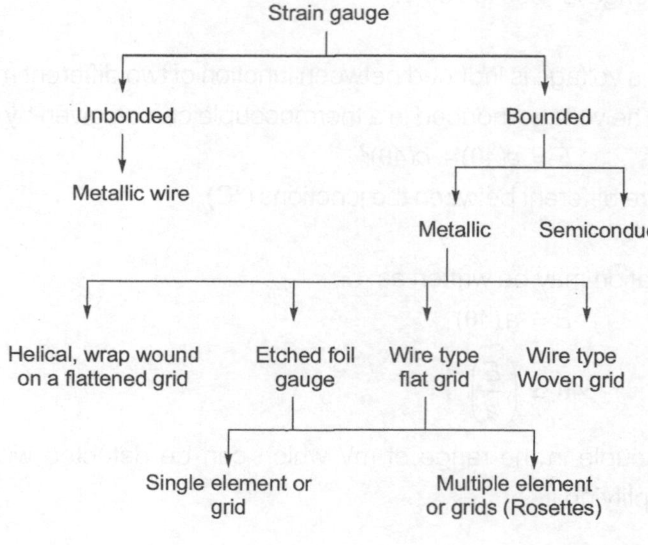

$$G.F. = 2$$
$$\text{stress} = 1000 \text{ kg/cm}^2$$
$$\text{modulus of elasticity} = 2 \times 10^6 \text{ kg/cm}^2$$
$$\text{modulus of elasticity} = 1000/(\Delta l/l)$$

$$\Rightarrow \quad \frac{\Delta l}{l} = \text{strain} = \frac{1000}{2\times10^6} = 5\times10^{-4}$$

and
$$G.F. = \frac{\Delta R/R}{\Delta l/l}$$

$$\Rightarrow \quad \frac{\Delta R}{R} = 2 \times 5 \times 10^{-4} = 1 \times 10^{-3}$$

$$\% \text{ change} = 10^{-3} \times 100 = 0.1$$

or
$$\text{change} = 0.1\%$$

Also,
$$G.F. = 1 + 2 \times \text{Poisson's ratio}$$

$$\text{Poisson's ratio} = \frac{G.F.-1}{2} = \frac{2-1}{2} = \frac{1}{2}$$

6.26 Differentiate between the following:

(i) Transducers and Inverse Transducers

(ii) Analog and Digital Transducers

(iii) Active and Passive Transducers

[12 marks : 2017]

Solution:

Transducers	Inverse Transducers
1. A transducer is defined as a device which converts a non-electrical quantity into an electrical quantity.	1. An inverse transducer is defined as a device which converts an electrical quantity into a non-electrical quantity. It is a precision actuator which has an electrical input and a low power non-electrical output.
2. e.g.: LVDT, Capacitive Transducers, Strain gauge, etc.	2. e.g.: Piezoelectric Crystal
Analog Transducers	**Digital Transducers**
1. These transducers convert input quantity into an analog output which is a continuous function of time.	1. These transducers convert the input quantity into an electrical output which is in the forms of pulses.
2. e.g.: LVDT, Strain gauge, Thermocouple etc.	2. e.g.: Shaft encoder, Digital tachometers and Limit switches etc.
Active Transducers	**Passive Transducers**
1. These transducers do not require an auxiliary power source to produce their output. They are also known as "self generating" type since they develop their own voltage or current output.	1. These transducers derive the power required for transduction from an auxiliary power source. They are also known as "externally powered transducers". They also derive part of the power required for conversion from the physical quantity under measurement.
2. e.g.: Photovoltaic cells, Piezoelectric Crystal, Thermocouples etc.	2. e.g.: Resistive, Inductive and Capacitive Transducers etc.

6.27 A parallel plate capacitive transucer uses plates of area 450 mm^2 which are separated by a distance 0.3 mm and having air as dielectric. Determine :

(i) The change in capacitance if the transducer is subjected to a linear displacement which reduces distance between plates to 0.27 mm.

(ii) The ratio of per unit change of capacitance to per unit change of displacement.

(iii) If a mica sheet of 0.02 mm is inserted in the gap, determine the value of actual capacitance and the change in capacitance for the same displacement and the ratio of per unit change in capacitance to per unit change in displacement. Assume the dielectric constant of mica to be 8.

[20 marks : 2021]

Solution:

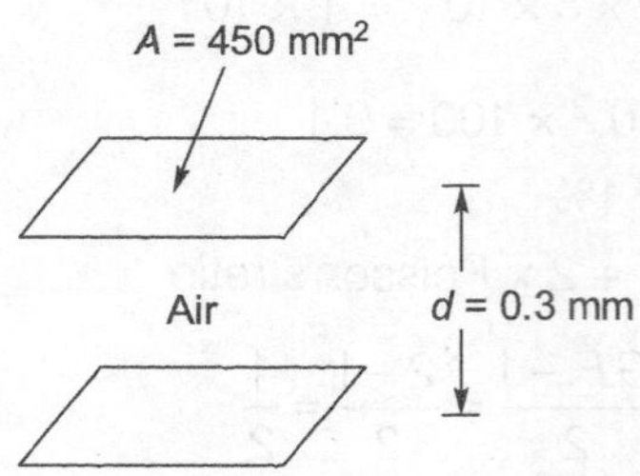

$$C = \frac{\varepsilon_o \varepsilon_r A}{d} = \frac{8.854\times10^{-12}\times450\times10^{-6}\times1}{0.3\times10^{-3}}$$

$$= 13.281 \text{ pF}$$

(i) Due to displacement new distance between plates become 0.27 mm

$$C_{\text{new}} = \frac{8.854\times10^{-12}\times450\times10^{-6}\times1}{0.27\times10^{-3}} = 14.756 \text{ pF}$$

Charge in capacitance

$$\Delta C = 14.756 - 13.281$$

$$= 1.475 \text{ pF}$$

(ii) Ratio of per unit change in capacitance to per unit change in displacement

$$\frac{\frac{\Delta C}{C}}{\frac{\Delta x}{x}} = \frac{\Delta C}{\Delta x}\times\frac{x}{C} = \frac{1.475}{(0.3-0.27)}\times\frac{0.3}{13.281} = 1.110$$

(iii) Assume Mica is kept at center of plates.

(a) With dielectric of mica : $\epsilon_r = 8$, $t = 0.02$ mm

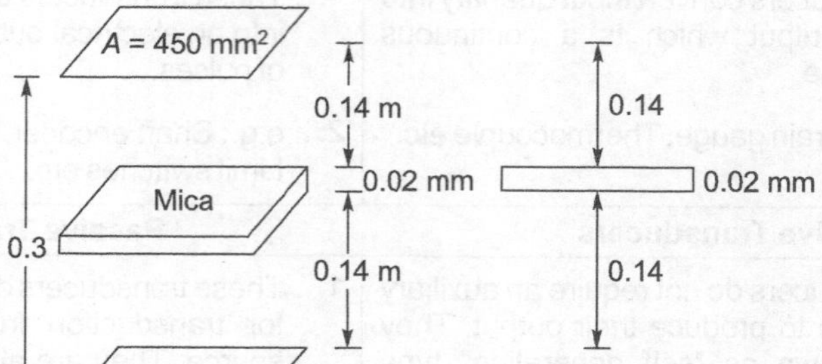

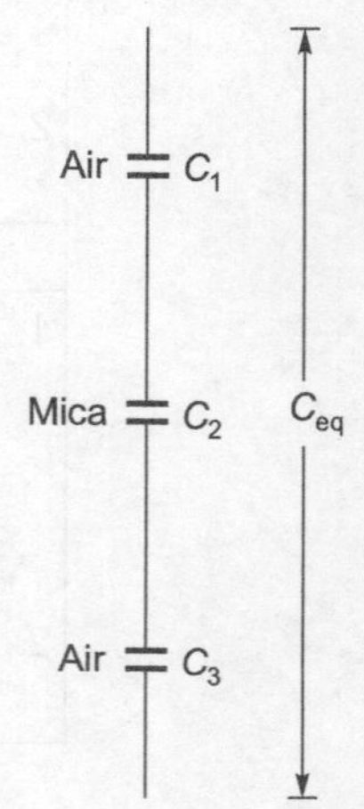

$$C_1 = C_3 = \frac{8.854\times10^{-12}\times450\times10^{-6}\times1}{0.14\times10^{-3}}$$

$$= 28.459 \text{ pF}$$

$$C_2 = \frac{8.854\times10^{-12}\times450\times10^{-6}\times8}{0.02\times10^{-3}}$$

$$= 1.594 \text{ nF}$$

$$\frac{1}{C_{eq}} = \frac{1}{C_1}+\frac{1}{C_2}+\frac{1}{C_3}$$

$$= \frac{2}{28.459\times10^{-12}}+\frac{1}{1.594\times10^{-9}}$$

Equivalent capacitance, $C_{eq} = 14.103 \text{ pF}$

(b) With change in linear displacement

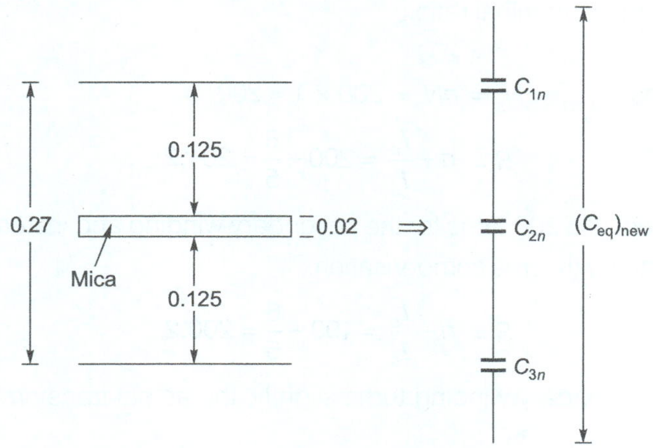

$$C_{1n} = C_{3n} = \frac{8.854\times10^{-12}\times450\times10^{-6}\times1}{0.125\times10^{-3}} = 31.87 \text{ pF}$$

$$C_{2n} = 1.594 \text{ nF}$$

$$\frac{1}{(C_{eq})_{new}} = \frac{2}{31.87\times10^{-12}}+\frac{1}{1.594\times10^{-9}}$$

$$(C_{eq})_{new} = 15.77 \text{ pF}$$

$$\Delta C = 15.77 - 14.103 = 1.667 \text{ pF}$$

$$\Delta x = 0.3 - 0.27 = 0.03 \text{ mm}$$

Ratio of per unit change in capacitance to per unit change in displacement.

$$\frac{\frac{\Delta C}{C}}{\frac{\Delta x}{x}} = \frac{\Delta C}{\Delta x}\times\frac{x}{C} = \frac{1.667}{0.03}\times\frac{0.3}{14.103} = 1.182$$

7. Instrument Transformers

7.1 State only the methods to minimise the ratio error of a current transformer.

A 1000/5 A, 50 Hz current transformer has a bar primary and has total secondary burden of 1 Ω (noninductive). Calculate the primary current and phase angle error for a rated current of 5 amperes. The magnetising current is 100 A and excitation loss is negligible. Assume there is no turns compensation.

[20 marks : 2017]

Solution:

Turn compensation: Actual transformation ratio, $R = n + \frac{I_e}{I_s}$.

Thus, if nominal ratio equal to the turn ratio, the actual transformation ratio becomes more than the nominal ratio. Now if reduce the turns ratio and keep the nominal ratio equal to the earlier value, the actual transformation ratio will be reduced. This would make actual transformation ratio nearly equal to the nominal ratio. Let us consider a 1000/5 A C.T. with loss component equal to 0.6 percent of primary winding current.

Nominal ratio, $K_n = \frac{1000}{5} = 200$

Loss component, $I_e = \frac{0.6}{100} \times 1000 = 6 \text{ A}$

Let the number of primary turns, $N_P = 1$

If the turn ratio is equal to the nominal ratio.

i.e., $n = 200$

Secondary winding turns, $N_s = nN_p = 200 \times 1 = 200$

Actual ratio, $R = n + \frac{I_e}{I_s} = 200 + \frac{6}{5} = 201.2$

Now suppose if we do not use 200 turns for the secondary winding and instead use 199 turns.

Actual transformation ratio with turns compensation,

$$R = n + \frac{I_e}{I_s} = 199 + \frac{6}{5} = 200.2$$

Thus, by reducing the secondary winding turns slightly, the actual transformation ratio is made equal to nominal ratio.

Usually the best number of secondary winding turns is one or two less than the number which would make actual transformation ratio equal to nominal ratio of the current transformer.

The phase angle error is almost not effected or very little by a change of one or two turns in the secondary winding.

Given that:

A bar primary, $N_P = 1$

Secondary winding current, $I_s = 5 \text{ A}$

Magnetizing current, $I_m = 100 \text{ A}$

Excitation loss is negligible i.e., $I_e \approx 0$

Turn ratio, $n = \frac{1000}{5} = 200$

Now primary winding current, $I_P = \sqrt{(nI_s)^2 + (I_m)^2}$

$$= \sqrt{(200 \times 5)^2 + (100)^2} = 1004.987 \text{ A}$$

As the secondary burden is non-inductive i.e. purely resistive, the secondary current is in phase with secondary induced voltage and the secondary power factor is unity or $\delta = 0$ and $I_e \approx 0$ (excitation loss is negligible)

So, $I_0 = \sqrt{I_m^2 + I_e^2} = I_m$ $(\because I_e = 0)$

Phase angle, $\theta = \frac{180}{\pi}\left[\frac{I_m \cos\delta - I_e \sin\delta}{nI_s}\right]$

$$= \frac{180}{\pi}\left[\frac{100\cos 0° - 0}{200 \times 5}\right]$$

$$= \frac{18}{\pi} = 5.7296°$$

7.2 A 1000/5 A, 50 Hz current transformer at its rated load of 50 VA has an iron loss of 0.5 W and a magnetizing current of 8 A. Calculate the ratio error and phase angle when rated output is supplied to a meter whose resistance is 0.4 Ω and inductance is 0.7 mH.

[12 marks : 2019]

Solution:

Nominal ratio, $k_n = \frac{1000}{5} = 200$

In the absence of any other data. The turns ratio is taken equal to the nominal ratio $n = 200$. Neglecting the burden of the secondary winding, the total burden of the secondary circuit is equal to the burden of the meter.

∴ Burden of secondary circuit = 50 VA

Voltage across primary winding E_P

$$E_P = \frac{VA}{I_P} = \frac{50}{I_P} = \frac{50}{1000} = 0.05 \text{ V}$$

Loss component of current, $I_e = \frac{\text{Iron loss}}{E_P} = \frac{0.5}{0.05} = 10 \text{ A}$

Magnetizing current, $I_m = 8 \text{ A}$

Secondary circuit phase angle,

$$\delta = \tan^{-1}\frac{X}{R} = \tan^{-1}\left[\frac{2\pi\times 0.7\times 10^{-3}\times 50}{0.4}\right] = 28.8°$$

$$\cos\delta = \cos(28.8) = 0.876$$

$$\sin\delta = \sin(28.8) = 0.481$$

Actual ratio, $R = n + \frac{I_e\cos\delta + I_m\sin\delta}{I_s} = 200 + \frac{10\times 0.876 + 8\times 0.481}{5}$

$= 202.52$

$$\text{Ratio error} = \frac{k_n - R}{R}\times 100 = \frac{200 - 202.52}{202.52}\times 100 = -1.244\%$$

Phase angle, $\theta = \frac{180}{\pi}\left[\frac{8\times 0.876 - 10\times 0.481}{200\times 5}\right] = 0.12°$

7.3 (i) Distinguish between a power transformer and current transformer (CT) and discuss the effect of secondary CT open circuit, while primary winding is energized.

(ii) A 1000/5, 50 Hz CT has a bar primary and a resistive burden of 12.5 VA at rated current. The secondary winding has a leakage inductance of 0.96 mH. The core requires a magnetising mmf of 16 AT and a loss component of 12 AT, when rated current flows in the secondary. Calculate the secondary turns at which the ratio error is eliminated.

[10 + 10 marks : 2022]

Solution:

(i) Differences between a Power Transformer and Current Transformer :

Power Transformer	Current Transformer
Power transformer is used to convert voltage using the number of windings in the two coils as ratio of turns in both winding.	Current transformer have only one winding placed around a wire to measure current in that winding.
More emphasis is on efficiency of transformer then accuracy.	There should be minimum phase and ratio error hence measurement can be accurate.
Power transformer used for both purposes, i.e, stepdown and step up voltage and current.	Current transformer used for only step down current hence it can be measured by ordinary measuring instruments.
Power transformer can be used for open and close circuit of secondary both.	Current transformer can not be used at secondary open circuit while primary is energised.
Resistance of coil of both primary and secondary is higher than current transformer is winding.	Resistance of coil with which current transformer is secondary winding is connected is small.
Power transformer used for high power rating.	While current transformer is used for low power (VA) rating.
Eddy and hysteresis losses are higher than current transformer.	Eddy and hysteresis losses are lower than power transformer.

Effect of Current Transformers secondary open circuit while primary is energised.

If the secondary is open circuit while primary is energised, the primary winding mmf remains same. Because current in secondary winding producing counter mmf that reducing primary mmf. If secondary is O.C., then there is no secondary current, hence, no counter mmf. Hence, there will be very large primary mmf which saturated core flux hence very high voltage induced on secondary side. This high voltage causing failure of insulation on secondary side which will dangerous to the person working nearer to current transformer.

(ii) Given : $\frac{I_P}{I_S} = \frac{1000}{5} = 200 = n$

Since primary has only one turn.

So, $I_m = 16$ Amp

$I_e = 12$ Amp

$I_o = \sqrt{16^2 + 12^2} = 20$ Amp

$$\alpha = \tan^{-1}\left(\frac{I_e}{I_m}\right) = \tan^{-1}\left(\frac{12}{16}\right) = 36.87^\circ$$

The secondary is purely resistive so

$$(R_l + R_s) = \frac{12.5}{5^2} = 0.5 \text{ ohm}$$

and $X_l + X_s = 2\pi \times 50 \times 0.96 \times 10^{-3} = 0.301$ ohm

So, $$\delta = \tan^{-1}\left(\frac{X_s + X_l}{R_s + R_l}\right) = \tan^{-1}\left(\frac{0.301}{0.5}\right) = 31.05^\circ$$

So, to eliminate ratio error, the actual ratio must be as nominal ratio.

$$R = n + \frac{I_o \sin(\alpha + \delta)}{I_s}$$

$$200 = \frac{N_s}{1} + \frac{20\sin(36.87° + 31.05°)}{5}$$

$$= N_s + 3.71$$

$$N_s = 196.29 \simeq 196$$

So, secondary turn should be equal to 196 turns, in order to eliminate the ratio error.

7.4 **A current transformer having a bar primary is rated at 500/5 A, 50 Hz with an output of 20 VA. At rated load with non-inductive burden, the inphase and quadrature components (referred to the flux) of the exciting mmf are 8 A and 10 A respectively. The number of turns in the secondary winding is 98 and the impedance of the secondary winding is (0.4 + *j*0.3) Ω. Calculate the ratio and phase angle errors.**

[12 marks : 2023]

Solution:

Primary winding turns, $N_P = 1$

Secondary winding turns, $N_S = 98$

∴ Turns ratio, $n = 98$

Nominal ratio, $K_n = \frac{500}{5} = 100$

Magnetizing current components is in phase while the loss component is in quadrature with the flux.

∴ Magnetizing mmf = 8 A

Mmf equivalent ot loss = 10 A

Magnetizing current, $I_m = \frac{\text{Magnetizing mmf}}{\text{Primary winding turns}}$

$$= \frac{8}{1} = 8 \text{ A}$$

Loss component, $I_e = \frac{\text{Loss mmf}}{\text{Primary winding turns}}$

$$= \frac{10}{1} = 10 \text{ A}$$

Output volt-ampere, $VA = 20$

Impedance of secondary load burden

$$= \frac{20}{(5)^2} = \frac{20}{25} = 0.8\ \Omega$$

It is given that the external burden is purely resistive.

∴ Resistance of external burden = 0.8 Ω. Reactance of external burden = 0 Ω.

Resistance of total secondary circuit burden

$$= 0.8\ \Omega + 0.4\ \Omega$$

$$= 1.2\ \Omega$$

Reactance of total secondary circuit burden

$$= 0 + 0.3\ \Omega = 0.3\ \Omega$$

Secondary phase angle,

$$\delta = \tan^{-1}\frac{0.3}{1.2}$$

$$= 14.036°$$

So, $\cos\delta = 0.970$ and $\sin\delta = 0.242$

Actual ratio,
$$R = \frac{I_e\cos\delta + I_m\sin\delta}{I_s} + n$$
$$= 98 + \frac{10\times 0.97 + 8\times 0.242}{5}$$
$$= 98 + \frac{9.7 + 1.936}{5}$$
$$= 100.327$$

Ratio (current) error
$$= \frac{100 - 100.327}{100.327}\times 100$$
$$= -0.326\%$$

Phase angle,
$$\theta = \frac{180}{\pi}\left[\frac{I_m\cos\delta - I_e\sin\delta}{nI_s}\right]$$
$$= \frac{180}{\pi}\left[\frac{8\times 0.97 - 10\times 0.242}{98\times 5}\right]$$
$$= 0.624^\circ$$

8. Digital Meter and Electronic Instruments

8.1 What are the advantages of a 'digital' voltmeter over 'analog' type? What are its types? With block diagram, explain working of an integrating type. Compare its performance with other types.

[12 marks : 2001]

Solution:

Advantages of Digital voltmeter over Analog type:

1. **Stability:** 0.008% of the reading for 6 month period
2. **Accuracy:** ±0.005% of the reading, i.e., they have greater accuracy
3. Instant read off
4. More durability and capacity to read true rms values
5. **High sensitivity:** 1 μV/1 V
6. Calibration can be done independent of measuring circuit
7. Output can be printed. BCD output can be had for digital processing or recording

The Four Major Types are:

1. Counter or ramp
2. Successive approximation
3. Dual slope or dual ramp integrating
4. Flash or parallel

Integrating Type: The integrating type has four major components

1. a voltage integrator
2. an analog voltage comparator,
3. logic circuits and
4. reference voltage

Working: The circuit is essentially voltage to time converter in which the time required to convert the analog input voltage V_{in} is compared with the conversion time of a precision reference voltage, V_{ref}. Initially the counter is reset to 0 and the capacitor is fully discharged. A start of conversion pulse sent by control logic connects the switch to the input analog voltage V_{in} and the capacitor starts getting charged by the steady current V_{in}/R. The operational amplifier is connected as an integration.

$$V_{01} = -\int_0^{t_1} I\,dt = -\frac{V_{in}}{RC}\int_0^{t_1} dt = -V_{in}\frac{t_1}{RC} = -\frac{V_{in}T_1}{RC} \quad \text{... } (i)$$

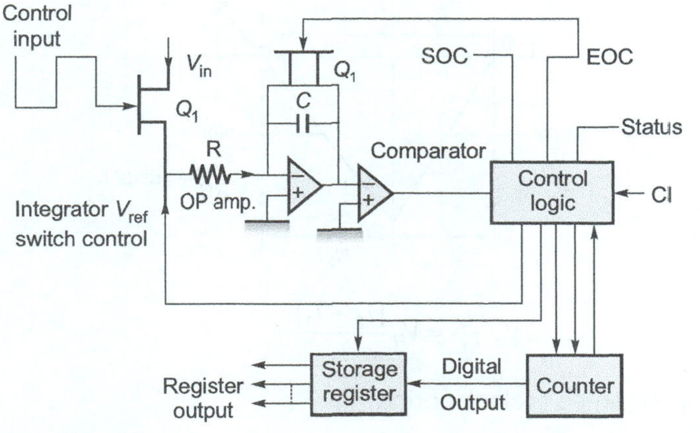

i.e., the unknown voltage V_{in} is proportional to $1/t$ or frequency of the output voltage.

So, the steady current has been converted into a ramp voltage of slope V_{in}/RC.

As soon as voltage starts ramping down, the comparator threshold of 0 voltage (+ earthed) is exceeded and the control logic causes the counter to start counting up until it reaches where all bits are 0 except MSB of 1 i.e., 10 .... 00. For the comparator, $V_{01} \le 0$ is 1 and $V_0 > 0$ as 0. The 1 in MSB of the counter changes the switch to the reference voltage V_{ref}. V_{ref} is opposite in polarity to that of input voltage and here as V_{in} is taken positive V_{ref} is negative.

When the switch is changed to V_{ref}, the counter is reset again and starts counting up from zero. The capacitor starts getting charged in the opposite direction. When the positive going ramp reaches zero voltage at t_2, the comparator changes state to 0, inhibits the clock and stops the counter and its output transferred to storage register. C is discharged during the time $t_1 - t_2$ and output at time t_2 is

$$V_{02} = V_{01} - \left(\frac{1}{RC}\int_{t_1}^{t_2} V_R\,dt\right) = V_{01} - \frac{V_R}{RC}(t_2 - t_1)$$

As V_{02} is 0 at t_2,

$$V_{01} - \frac{V_R}{RC}(t_2 - t_1) = 0$$

$$-V_{in}\frac{t_1}{RC} = -\frac{V_R}{RC}(t_2 - t_1)$$

$$V_{in} = V_R\frac{t_2 - t_1}{t_1} = V_2\frac{T_2}{T_1} \quad \text{...}(ii)$$

T_1 is a constant since it is the time for the counter to reach 10 .... 00. Thus for an n-bit counter

$$T_1 = \frac{2^n}{f} = 2^n T_C$$

The time interval $t_2 - t_1 = N/f = N\,T_c$ where N is the number of counts during the period $t_2 - t_1$, f is the frequency of clock pulses, $1/f = T_c$, clock period.

$$\therefore \quad V_{in} = V_R\frac{t_2 - t_1}{T_1} = V_R \times \frac{N}{f} \times \frac{f}{2^n}$$

$$V_{in} = N\frac{V_R}{2^n}$$

or $$N = \frac{V_{in}}{V_R}2^n \quad ...(iii)$$

If the number of counts during t_1 is N_1 and during $t_2 - t_1$ is N_2, the above equation can also be written as,

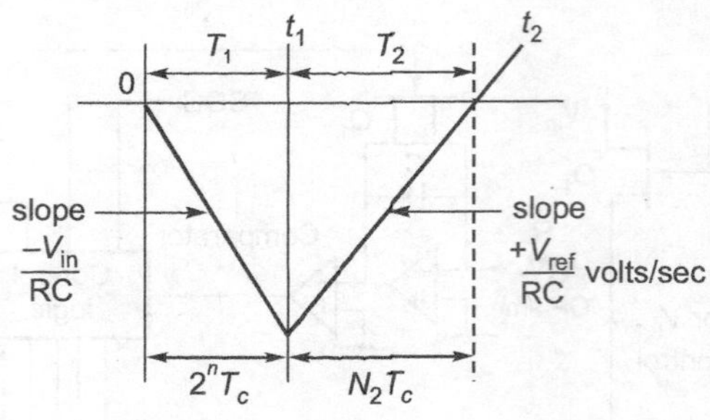

$$V_{in} = V_R\frac{N_2}{N_1} = V_R\frac{[T_2 - T_1]}{T_1} \quad ...(iv)$$

Comparison of conversion times τ (clock period)
Counter type $(2^n - 1)\,\tau$
Successive Approximation $n\tau$, Flash 1τ.

8.2 What are the advantages of digital voltmeters and limitations compared to analog indicating type voltmeters?

[10 marks : 2006]

Solution:

Advantages:

1. A digital voltmeter (DVM) displays the value of a.c. or d.c. voltage being measured directly as discrete numerals in the decimal number system. Numerical readout of DVMs is advantageous since it eliminates observational errors committed by operators.
2. The errors on account of parallex and approximations are entirely eliminated.
3. The use of digital voltmeters increases the speed with which readings can be taken.
4. The output of digital voltmetres can be fed to memory devices for storage and future computations.
5. A digital voltmeter is a versatile and accurate voltmeter which has many laboratory applications. On account of developments in the integrated circuit (IC) technology, it has been possible to reduce the size, power requirements and cost of digital voltmeters. For the same accuracy, a digital voltmeter is less costly than its analog counter part.
6. High resolution.
7. Transfer and processing of digital signals without loss of accuracy.

Limitation: Digital instruments are relatively complex and consist of large number of parts which individually will react to change in temperature and humidity.

8.3 Explain with the help of a circuit diagram, the working of a digital voltmeter. Why is it considered to be accurate?

[10 marks : 2009]

Solution:

Digital Voltmeters (DVMs) are normally used for measuring voltages, current, flux densities etc., and they provide an instant read out. A general block diagram of a DVM is

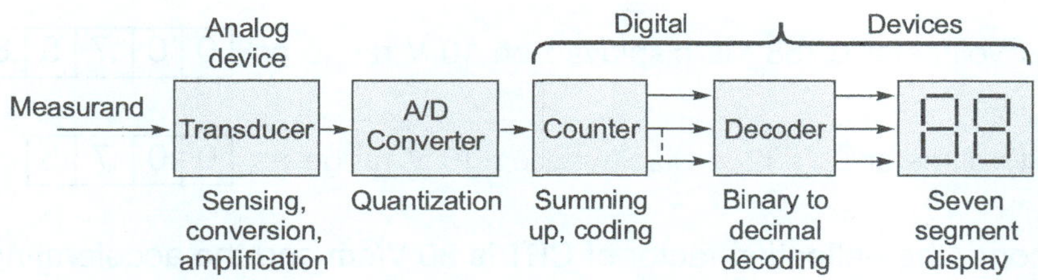

Resolution of a decade digital voltmeter = $1/10^n$, n-number of bits sensitivity = minimum full-scale × resolution. As may be seen from the block diagram, the AD converter is the heart of the DVM. The AD converters used is mostly dual slope integrator. The conversions are done repeatedly on a cyclic basis and not a start-stop command. Generally the conversion rate is 50 to 60 conversions per second, the same as supply frequency of 50-60 Hz.

Digital voltmeter is considered to be accurate because a digital voltmeter (DVM) displays the value of a.c. or d.c. voltage being measured directly as discrete numerals in the decimal number system. Numerical readout of DVMs is advantageous since it eliminates observational errors committed by operators. The errors on account of parallax and approximations are entirely eliminated. The use of digital voltmeters increases the speed with which readings can be taken.

8.4 **Explain the advantages of electronic voltmeters over conventional voltmeters. Describe the working of a differential amplifier type electronic voltmeter using two FETs.**

[10 marks : 2010]

Solution:

(Refer Question No. 8.1)

8.5 **A $4\frac{1}{2}$ digit and a $3\frac{1}{2}$ digit voltmeter on 10 V range are used for voltage measurements.**

(i) Find the resolution of each meter.

(ii) How would the reading 0.7582 be displayed on these two meters?

[12 marks : 2019]

Solution:

(i) $4\frac{1}{2}$ digit voltmeter is on 10 V range.

1	0	.0	0	0

In 10 V range the minimum voltage can be measured by $4\frac{1}{2}$ digit voltmeter is 00.001 V i.e. 1 mV.

∴ Resolution = 1 mV for $4\frac{1}{2}$ digit voltmeter.

$3\frac{1}{2}$ digit voltmeter, is on 10 V range.

1	0	.0	0

In 10 V range, the minimum voltage can be measured by $3\frac{1}{2}$ digit voltmeter is 00.01 V i.e. 10 mV.

∴ Resolution = 10 mV for $3\frac{1}{2}$ digit voltmeter.

(ii) In $4\frac{1}{2}$ digit voltmeter 0.7582 is displayed on 10 V range as | 0 | 0 | .7 | 5 | 8 |.

In $3\frac{1}{2}$ digit voltmeter 0.7582 is displayed on 10 V range as | 0 | 0 | .7 | 5 |.

8.6 **In an oscilloscope, the deflection factor of CRT is 80 V/cm and the accelerating voltage is 2500 V. What is the minimum distance required from center of deflection plates to screen that allows full deflection of 4 cm on the oscilloscope screen?**

[20 marks : 2020]

Solution:

Consider basic layout of CRT shown in figure :

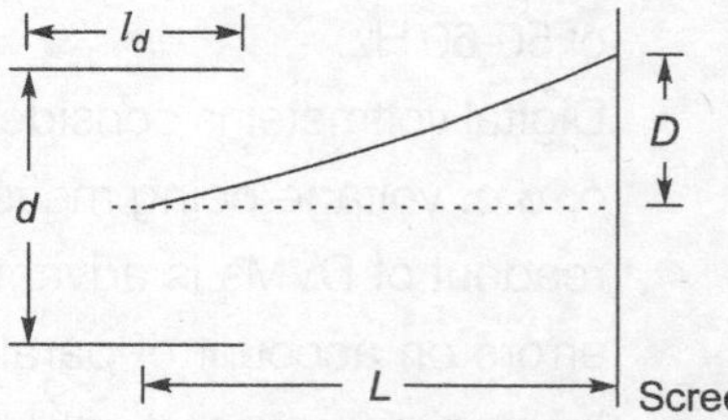

Given :

Deflection Factor = 80 Volt/cm

Accelerating Voltage, V_a = 2500 volts, D = 4 cm

$$\text{Deflection, } D = \frac{L \cdot l_d \cdot V_d}{2V_a \cdot d}$$

$$\text{Sensitivity} = S' = \frac{D}{V_d} = \frac{L \cdot l_d}{2V_a \cdot d}$$

$$\text{Deflection factor} = D.F = \frac{1}{S'} = \frac{V_d}{D}$$

$$\Rightarrow \quad V_d = 80 \text{ Volt/cm} \times 4 \text{ cm} = 320 \text{ Volt}$$

$$L = \frac{2V_a \cdot dD}{l_d \cdot V_d} = \frac{2D \cdot V_a}{V_d} \times \left(\frac{d}{l_d}\right) = \frac{2 \times 4 \text{ cm} \times 2500 \text{ Volt}}{320 \text{ Volt}} \times K$$

where, $$K = \frac{d}{l_d} = \text{Constant}$$

$$L = 62.5 \text{ cm} \times K$$

Minimum value of L depends on value of K

Taking $K = 1$

$$L = 62.5 \text{ cm}$$

8.7 **Draw the circuit diagram and explain the operation of a practical Q-meter and write its applications.**

[20 marks : 2022]

Solution:

Circuit diagram:

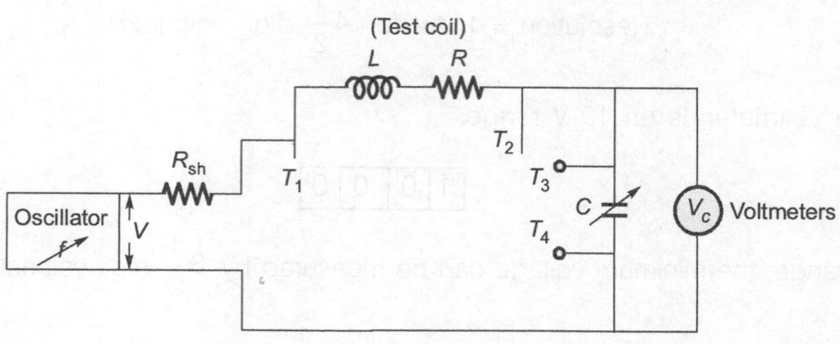

Working principle:

This meter is based on the principle of series resonance. Under series resonance,

Inductive reactance, X_L = capacitive reactance, X_C

$$I = \frac{V}{R}$$

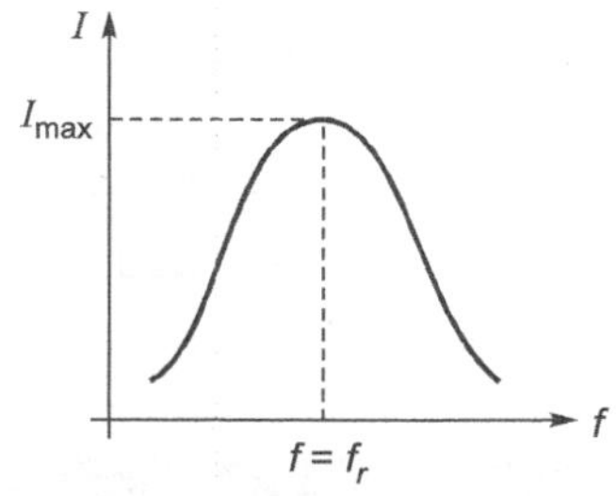

Quality factor, $Q = \frac{X_L}{R} = \frac{X_C}{R}$

Voltage across capacitor, $V_C = IX_C$

$$= \frac{V}{R}X_C = V \cdot \frac{X_C}{R} = V \cdot Q$$

$$V_C = V \cdot Q$$

$$V_C \propto Q$$

Voltage across capacitor is directly proportional to quality factor of the meter. Hence this meter is also called voltage magnifier. By varying the capacitance 'C' the circuit is resonated and the voltage across capacitor is measured by voltmeter which is calibrated in terms of quality factor of the coil.

Application:

1. **Measurement of quality factor Q of test coil :**

   Test coil is connected between T_1 and T_2 and by varying 'C' circuit is resonated. For reducing error, R_{sh} must be of low value (in mΩ range)

$$Q_T \text{(true value)} = \frac{\omega L}{R}$$

$$Q_m \text{(measured value)} = \frac{\omega L}{R + R_{sh}} = \frac{Q_T}{\left(1 + \frac{R_{sh}}{R}\right)}$$

$$\because \quad Q_T = Q_m\left(1 + \frac{R_{sh}}{R}\right)$$

$\because$ For $Q_T \approx Q_m \Rightarrow R_{sh}$ should be low

2. **Measurement of unknown capacitance :**

   Test capacitance, C_T is connected between terminals T_3 and T_4. By varying C to value C_1, the circuit is resonated.

   Again varying C to a new value C_2, new value of $V_C = V_1$ is calculated

For $C = C_1$, $f = \frac{1}{2\pi\sqrt{L(C_1 + C_T)}}$

For $C = C_2$, $f = \frac{1}{2\pi\sqrt{LC_2}}$

Both frequency are same,

$$\frac{1}{2\pi\sqrt{L(C_1 + C_T)}} = \frac{1}{2\pi\sqrt{LC_2}}$$

$$\because \quad C_2 = C_1 + C_T$$

$$C_T = C_2 - C_1$$

$$\because \quad C_1 = \text{with } C_T$$

$$C_2 = \text{without } C_T$$

$$C_T = \text{test capacitance}$$

3. Q-meter is used to calculated self or distributed capacitance of the coil

Distributed capacitance, $C_d = \dfrac{C_1 - n^2C_2}{(n^2 - 1)}$

$\because$ where, $n = \dfrac{f_2}{f_1}$

$f_2 \rightarrow$ frequency calculated when $C \rightarrow C_2$

$f_1 \rightarrow$ frequency calculated with $C \rightarrow C_1$

4. Q-meter is used in communication application and self inductance of the coil.

9. Data Acquisition System, Recorders

9.1 What are Electronic Counters? How can these be used to measure frequency? Explain the working of a Decade Counter.

[12 marks : 2001]

Solution:

Electronic Counters: Electronic counters are capable of making many measurements involving frequency, time, phase angle, radiation events and totalizing electric events.

The electronic counter normally employs a frequency divider circuit known as a scaler. A scaler produces a single pulse for every set of number of input pulses.

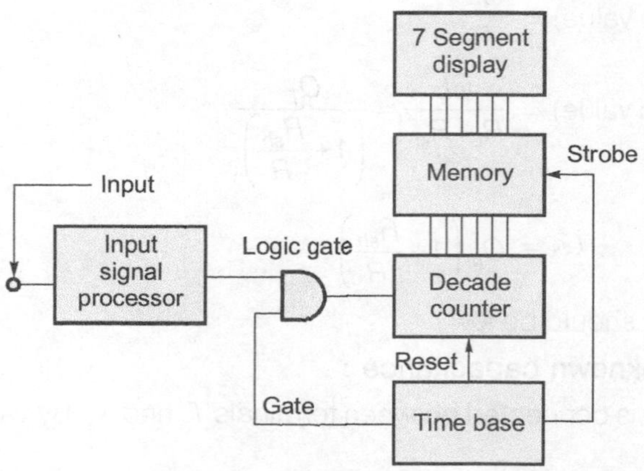

Block diagram of a Frequency Counter

Digital Frequency Counter: The measurement of frequency and time has been greatly simplified with the advent of integrated circuits, logic gates and seven segment displays.

The block diagram of a simple frequency counter is shown in figure.

The signal whose frequency is to be measured is fed into the input signal processor. This consists of an amplifier to boost the weak signal which may be of millivolts. Besides it has an alternator and comparator to adjust for slow rise time of the input signal.

If f_c is the clock (crystal) frequency and f_x the frequency of the signal to be measured, in period measurement, number of pulses counted is

$$N_P = \frac{f_c}{f_x}$$

In frequency measurement, for 1 sec gate time,

$$N_f = f_x$$

The cross over frequency (f_0) at which, $N_P = N_f$ is

$$\frac{f_c}{f_x} = f_x$$

or $$f_x^2 = f_c \text{ or } f_x = \sqrt{f_c}$$

Signals lower than f_0 are measured in 'period' mode and higher than f_0 in the 'frequency' mode.

Decade Counter: Decade counter is a circuit of flip-flops in cascade, which counts in the base 10 (decimal number system). This means that there is a sequence of ten distinct counts in increasing order 4 F/Fs in cascade progress through 16 distinct states. Hence to get a count of 10, a minimum of 4 F/Fs are required, resetting the output of each F/Fs to 0 after the desired 10 counts. Figure shows the decade counter using 4 negative edged triggered F/Fs in cascade.

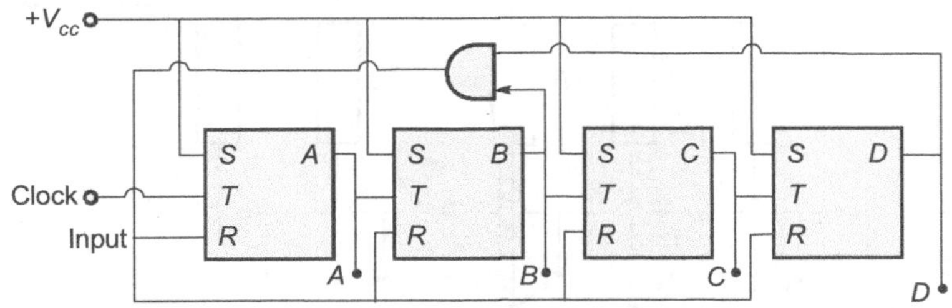

Decade Counter

The outputs of the flip-flop *B* and *D* are high after 10 pulses applied to the circuit. Therefore, the output signal of the decade counter is 1010. This output has to be reset on the very next pulse which is done by the use of an AND gate that resets all F/Fs to 0, when the outputs of *B* and *D* are 1.

At the beginning all the F/Fs are reset to 0000. The clock pulse is applied to the trigger input *T* of the F/F. Since this is an negative edged triggerd F/F, at the negative edge or falling edge of the trigger input the F/F A will toggle, and hence the output of F/F A changes to level 1; all other F/Fs undergoes no change. The outputs from the F/Fs will be 0001. At the next clock pulse the F/F A will toggle back to 0, and the output of F/F *A* falls from 1 to 0 and is applied to the *T* input of the next F/F *B*, toggling it. The output of the F/F *B* changes to 1 and the output of the decade counter goes to 0010. Similarly, as the clock progresses, the F/F toggles in a straight sequence upto 10 (binary 1010). On the very next pulse and AND gate is enabled, which makes the reset input F/F *B* high. As soon as the clock pulse arrives, all F/Fs all reset to 0 and the output of the decade counter goes back to 0000. Therefore by using the AND gate the counter is reset at the tenth pulse. An F/F divides its input frequency by two and acts as a frequency divider.

9.2 What are digital to Analog converters? Describe, with necessary diagrams, working of a 'binary weighted resistance D/A converter'.

[12 marks : 2001]

Solution:

Digital to Analog Converter: In D/A conversion, the input to the converter is in the form of bits, 0 or 1 (binary digit) or as BCD and is converted to a voltage or current proportional to the digital value, A general block diagram is,

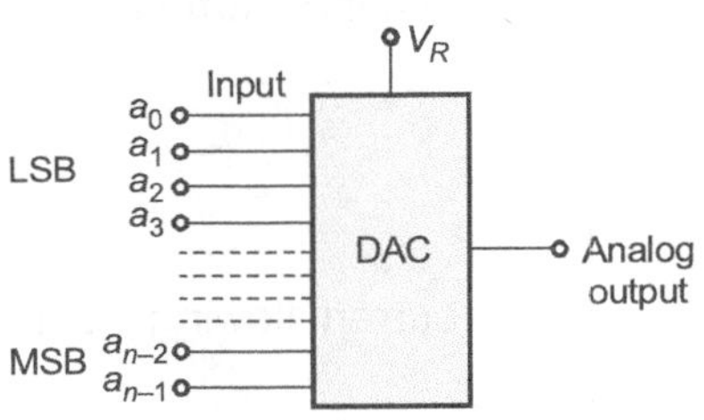

$a_0, a_1, a_2, \ldots, a_{n-2}, a_{n-1}$ is the word length for *n* bits. V_R is the reference voltage and it is the maximum voltage possible at the analog output.

In general, analog output = $K \times$ digital input

where, K is the proportionality factor and a constant for given DAC.

Binary-Weighted resistance D/A Converter: A general structure of a digital to analog converter is

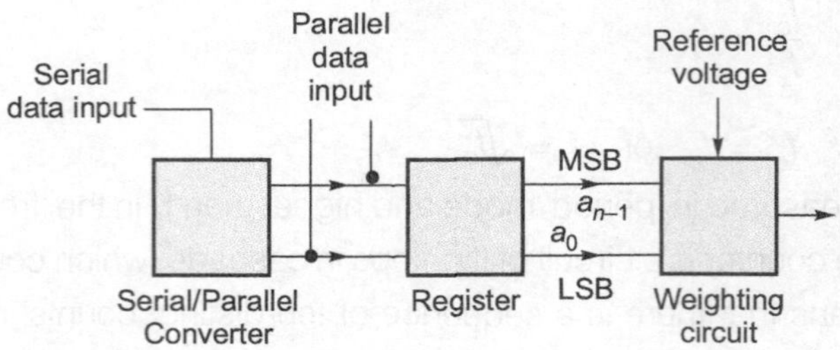

We will consider the weighting circuit only and is for a 4 bit as follows:

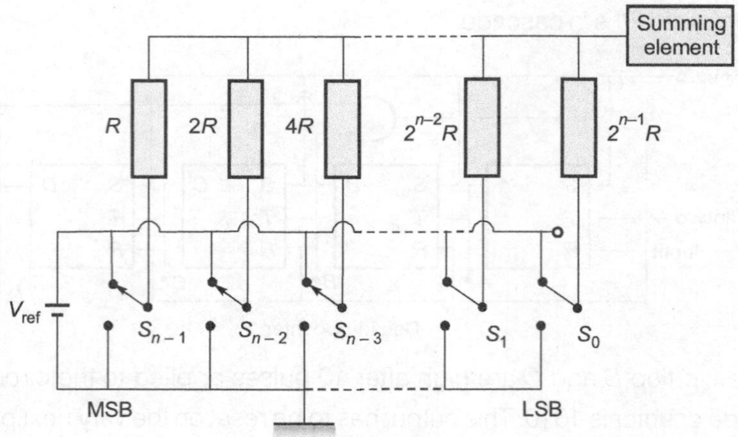

The major components of the binary weighted resistance DAC are

1. a weighted resistor network, R to $2^{n-1}R$.
2. n switches one for each bit applied to input.
3. reference voltage V_{ref}.
4. Summing element which sums up the currents flowing in the resistors to give an output proportional to input, usually an op-amp.

When a particular switch is closed, current flows through the connected resistor and the circuit is complete through the summing element. So is the LSB switch and when it is closed current is

$$I(\text{LSB}) = \frac{V_{ref}}{2^{n-1}R}$$

This current is the least.

When the MSB switch S_{n-1} is closed.

$$\text{Current } I(\text{MSB}) = \frac{V_{ref}}{R}$$

and is the highest. So the currents are in multiples of 2, 2^{n-1}. When the switch is connected to supply the bit $a_i = 1$ and when grounded $a_i = 0$. The resistors need not be in multiples of 2. It can be in submultiples of 2, $2^{-(n-1)}$. If all the switches are closed, all the resistors come in parallel. If, say, two switches are closed these two resistors are in parallel while the grounded resistors are in parallel with the load (input resistance of summing element). A binary number N of n-bits is

$$N = a_0 2^0 + a_1 2^1 + a_2 2^2 + \ldots + a_{n-2} 2^{n-2} + a_{n-1} 2^{n-1}$$

$$N = \sum_{i=0}^{n-1} a_i 2^i \qquad \ldots(v)$$

The current flowing into the summing element assuming to be shorted is

$$I_{SC} = S_{n-1}\frac{V_R}{R} + S_{n-2}\frac{V_R}{2R} + \ldots + \frac{S_1 V_R}{2^{n-2}R} + \frac{S_0 V_R}{2^{n-1}R} \qquad \ldots(vi)$$

If each switch position corresponds to one bit

$S_{n-1} = a_{n-1}$; $S_{n-2} = a_{n-2}$; ... $S_1 = a_1$; $S_0 = a_0$

then equation (vi) can be written as

$$I_{SC} = \frac{V_R}{2^{n-1}R}(a_{n-1}2^{n-1} + a_{n-2}2^{n-2} + \ldots + a_1 2^1 + a_0 2^0)$$

$$I_{SC} = \frac{V_R}{2^{n-1}R}\sum_{i=0}^{n-1} a_i 2^i \quad \ldots(vii)$$

The output voltage $I_{SC}R$ of DAC is proportional to the bit corresponding to switches connected to V_R i.e., $a_i = 1$.

Maximum output current will flow when all switches are closed i.e., $a_i = 1$.

$$I_{max} = \frac{V_R}{R}\cdot\frac{2^n - 1}{2^{n-1}} \quad \ldots(viii)$$

Similarly, when all switches are closed, all the resistors come in parallel so that total resistance

$$R_T = \frac{V_R}{I_{max}} = \frac{2^{n-1}R}{2^n - 1} \quad \ldots(ix)$$

I_{min} will be only when one resistor is through,

$$I_{min} = \frac{V_R}{R}\cdot\frac{1}{2^{n-1}}$$

9.3 **Draw neat diagram for 4-bit R-2R ladder and Weighted-Resistor types *D* to *A* converters and compare them.**

[10 marks : 2002]

Solution:

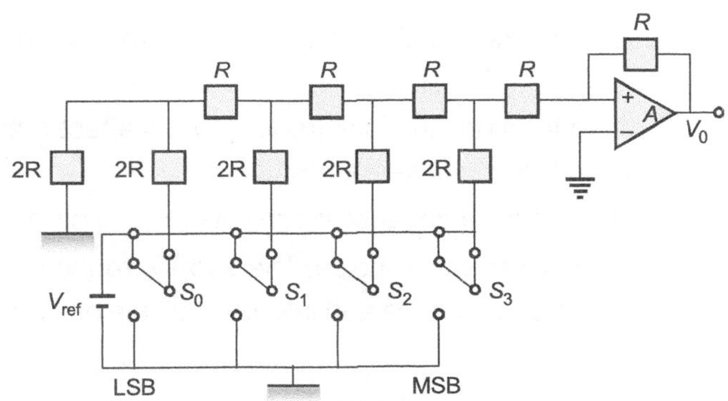

R-2R Ladder type

$$V_0 = V_R\left(\frac{S_0}{2^4} + \frac{S_1}{2^3} + \frac{S_2}{2^2} + \frac{S_3}{2^1}\right)$$

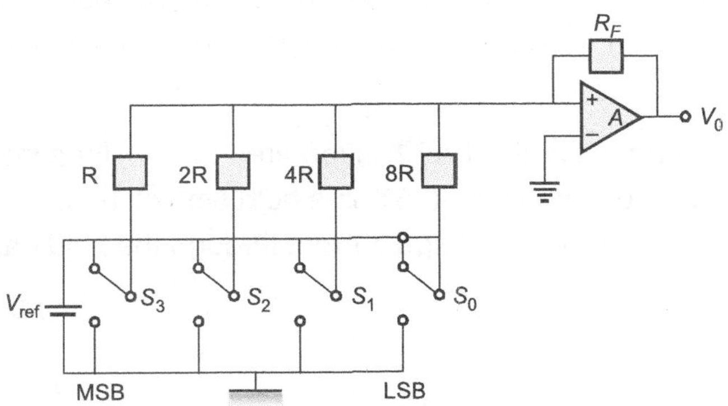

Weighted Resistor Type

$$V_{out} = -\frac{V_{ref}}{R}(S_3 2^0 + S_2 2^{-1} + S_1 2^{-2} + S_0 2^{-3})$$

In weighted resistor type *DA* convertor, the resistor values increase in multiples of 2. If the value for MSB is 2 kΩ, then for a 10 bits the value for LSB is $2^9 \times 2$ k = 1.024 MΩ. It is difficult to obtain high precision at very high values. It is also difficult to fabricate high value resistance by integrated circuit methods.

In the R-2R ladder network only two values of resistors are used though each bit requires now two resistors instead of one.

9.4 Discuss the advantages and disadvantages of an integrating type *A* to *D* converter under the following heads:

(a) Speed of operation (b) Change in R-C values

(c) Presence of periodic noise (d) Offset voltage of the operational amplifier.

[10 marks : 2002]

Solution:

Advantages:

1. Accuracy depends only on the stability of the reference voltage V_{ref} and the clock period over a short interval of time.
2. Errors due to propagation delay times in the electronic circuitry are eliminated.
3. *C* and *R* need not have high precision. Nor temperature effects on them have any effect as *CR* does not figure in equation. The *RC* time constant does not affect the input voltage measurement.
4. The comparator offset voltage and current are also automatically compensated as the input signal crosses zero voltage twice.
5. Very good conversion accuracies are obtained in the double ramp method. Hence this is used in digital voltmeters.
6. As $(V_{in}/V_{ref}) = (t_2 - t_1)/t_1$, long term drift in clock frequency will affect t_2 and t_1 in the same proportion and so insensitive to slow drifts in clock frequency.
7. It is relatively simple and inexpensive because it does not require precision components.
8. A major advantage is its low sensitivity to noise. The supply frequency interference can be almost eliminated by choosing the integrating time and integral multiple of supply frequency.

Disadvantages:

1. It has a very slow conversion time 10 to 100 ms. Hence if is mostly used in DVMs where fast conversion times are not required.
2. This type of circuit where a capacitor is charged and discharged is known as an integrate and dump integrator. In this the response is relatively flat for frequencies below $f_1(1/t_1)$ and so integrating A/D convertors are used only for dc and slowly varying signals if the integrity of the signals is to be presented.

9.5 A S/H circuit has a hold-capacitor of 0.1 nF. It is required to work for a maximum input of 5 V and the hold value should not drop by more than 0.5% in a hold time of 1 μs.

Calculate the maximum permissible leakage current through the hold capacitor.

[10 marks : 2002]

Solution:

For S/H circuit, $$C\frac{dV_0}{dt} = I$$

where, $\Delta V = \dfrac{0.5}{100}$

and $\Delta t = 1 \times 10^{-6}$ sec

$$0.1 \times 10^{-9} \times \frac{0.5}{100} \times \frac{1}{10^{-6}} = I$$

Leakage current, $I = 0.05 \times 10^{-5} = 0.5\ \mu A$

9.6 **What are the requirements of an ideal S/H circuit? How are they realized in practical S/H circuit?**

[8 marks : 2003]

Solution:

The circuit of a S/H circuit is

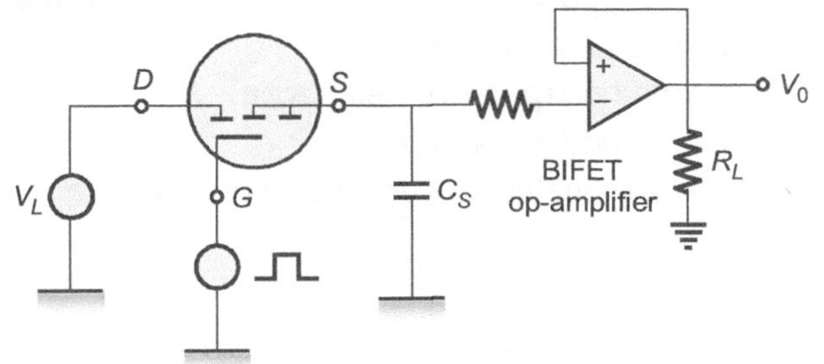

The E-MOSFET switch is on when the pulse to the gate is positive and charges the capacitor C_s equal to the input voltage for a specified period called sampling period or acquisition time. This time should be as small as possible about 0.1 μ sec.

So the input to this MOSFET switch shown as V_L is from a unity gain op-amp as its output resistance is very low-making CR_s the time constant very low.

The charged capacitor should retain the charge until the arrival of the next pulse called hold time t_H. This should be high for which the capacitor is connected to the input of a op-amp as its input resistance is very high, $10^5\ \Omega$. The loss of charge of capacitor is called droop rate which should be small as possible, 1 mV/ms.

Aperture time is the delay (reaction time) for the FET switch to turn off by removal of constant signal on gate. This is of the order of 10 ns and less. Because of the delay the sampled voltage may be a little higher than what it should be.

9.7 **Mention the various types of analog to digital converters in the increasing order of speed of operation. Why is a dual slope ADC preferred in a digital voltmeter? An 8-bit successive approximation type of ADC uses a clock frequency of 1 MHz. Calculate the time of conversion.**

[12 marks : 2003]

Solution:

Increasing Order of Speed of Operation of ADC Converter: Counter or ramp < Dual slope or Dual ramp integrating < Successive Approximation < Flash or Parallel.

Very good conversion accuracies are obtained in the dual slope ADC. Hence this is used in digital voltmeter.

Time of conversion for successive approximation-ADC is $n\tau$.

where, n = number of bits

τ = clock period

$$\tau = \frac{1}{1 \times 10^6} = 1\ \mu \text{ sec}, \quad n = 8$$

Conversion time = 8 μ sec

9.8 **Differentiate clearly between D to A and A to D conversion. Name different methods of accomplishing each. With necessary diagram, explain working of simultaneous A/D converter. What are its advantages and disadvantages?**

[12 marks : 2004]

Solution:

In D to A conversion, the input to the converter is in the form of bits, 0 or 1 or as BCD and is converted to a voltage or current proportional to the digital value.

An A to D, converter converts the analog is signal into corresponding digital word usually binary.

ADCs are more complicated than DACs and hence more extensive. An ADC consists of a comparator, a number of registers, control logic and a DAC. While conversion by a DAC is instantaneous.

Analog to Digital (A/D) Conversion Techniques: Analog to digital (A/D) conversion is slightly more complex than digital-to analog (D/A) conversion. A number of methods may be used for this purpose. The following four methods are most commonly used:

(a) Successive approximation method (potentiometric type)
(b) Voltage to-time conversion method (ramp type)
(c) Voltage to frequency conversion method (integration type)
(d) Dual slope integration method
(e) Re-circulating remainder system

Digital to Analog Conversion Techniques:

(a) Binary weighted resistor method
(b) R-2R ladder method

Simultaneous A/D Conversion or Flash A/D Converter: This type uses an array of comparators connected in parallel, each comparator switching on in a definite ratio of the reference voltage. The comparator compares the input analog voltage at a particular ratio of the reference voltage. When the input signal is less than the reference input voltage, comparator output is 0. When the input signal is greater than the reference signal, the comparator output is 1.

It may be seen that $2^n - 1$ comparators are required for n bit output. A combinational circuit encodes the states of $2^n - 1$ comparators into n parallel bits. The difference in voltage between each comparator is 1 LSB.

Let us take a 3 bit comparator.

Number of comparators $= 2^n - 1 = 7$ and there will be 8 resistors of R Ω each.

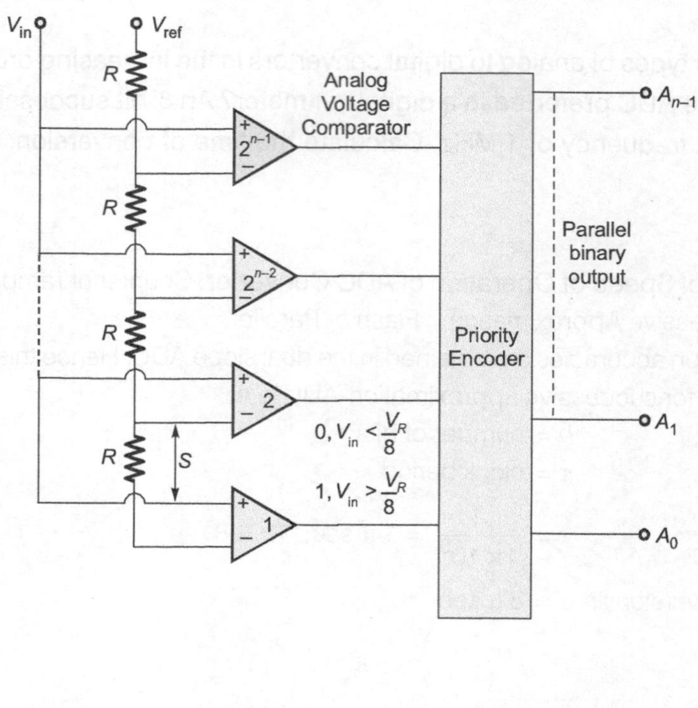

$$\text{Reference voltage of 1st comparator} = V_R \times \frac{R}{8R} = \frac{V_R}{8}\ V$$

$$\text{Reference voltage of 2nd comparator} = V_R \times \frac{2R}{8R} = \frac{V_R}{4}\ V$$

$$\text{Reference voltage of 7th comparator} = V_R \times \frac{7R}{8R} = \frac{7}{8} V_R\ V$$

$$\text{Resolution} = 1\ \text{LSB} = \frac{V_R}{8} = \frac{V_R}{2^n}$$

Advantage: Its chief advantage is the fast conversion time. In fact, its conversion time is the fastest of the ADCs as the comparisons of the input voltage are carried out simultaneously. Its conversion time is independent of the number of bits. The conversion is asynchronous, its speed being determined by the propagation delays of the comparators and the encoding logic. Its conversion time is about 20 ns for 3 bit.

Disadvantage: The number of comparators is $2^n - 1$ and so as bits increase the number of comparators become enormous. It is not used for more than 3 or 4 bit word. As the no. of comparators increases cost becomes high. Its resolution is also relatively poor.

9.9 Explain the working of binary ladder digital to analog converter using R-2R ladder network with four input voltages.

If in the above DAC a reference voltage of 12 V is used, what will be its resolution and full scale output?

[10 marks : 2005]

Solution:

R-2R Ladder Network:

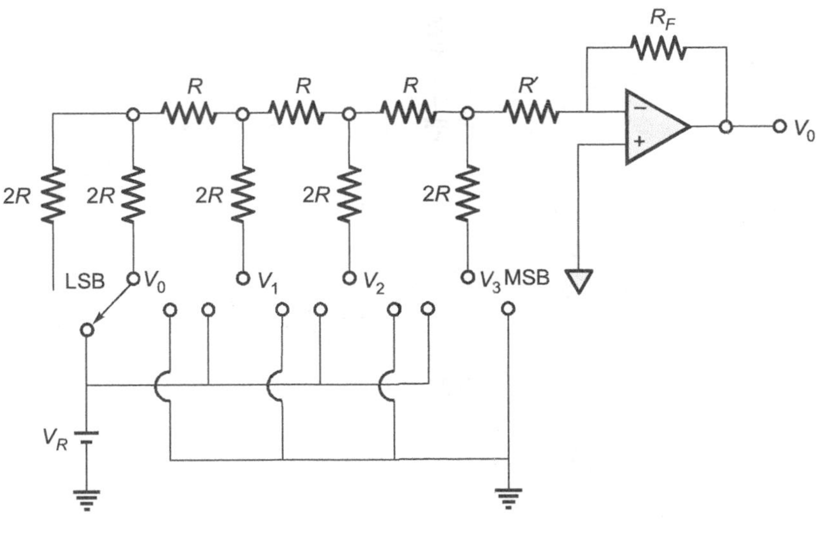

$$V_n = b_n V_R$$
$$n = 0, 1, 2$$
$$b_n = 1 \text{ when switch is connected to } V_R$$
$$= 0 \text{ when switch is connected to ground.}$$

Output of above circuit can be obtained by applying super-position theorem.

Case-I: Taking V_0 only

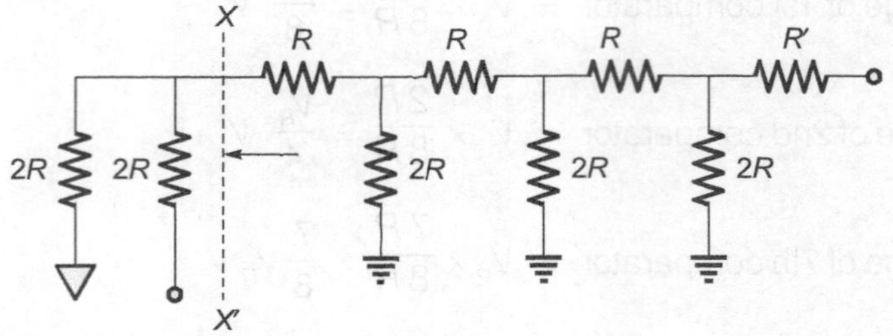

Thevenin's to the left of XX':

$$V_{TH1} = \frac{V_0}{2}, \quad R_{TH1} = R$$

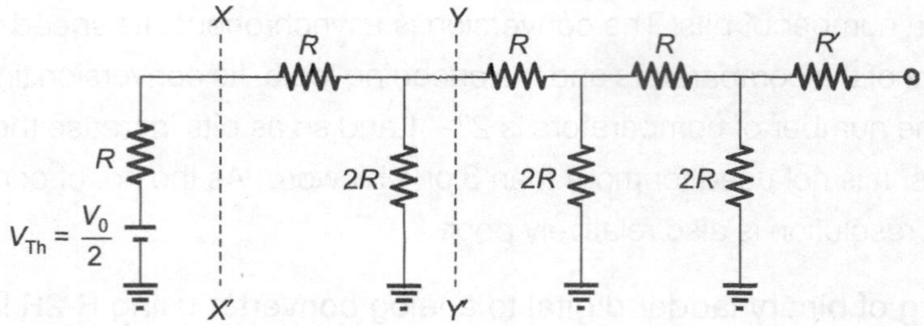

Thevenin's equivalent to the left of YY':

$$V_{TH2} = \frac{V_0}{4}, \qquad R_{TH2} = R$$

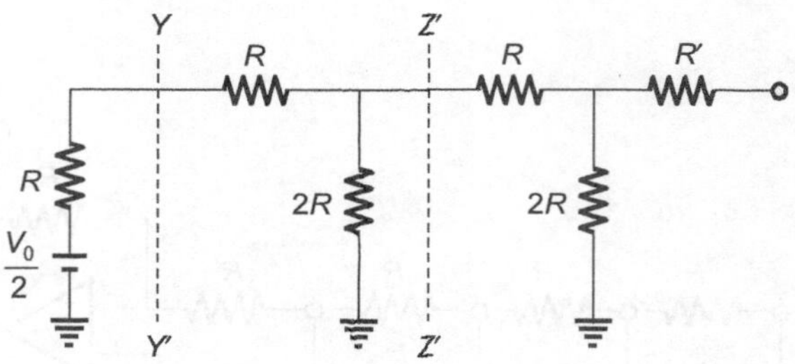

Thevenin's equivalent to the left of ZZ':

$$V_{TH_3} = \frac{V_0}{8}, \quad R_{TH_3} = R$$

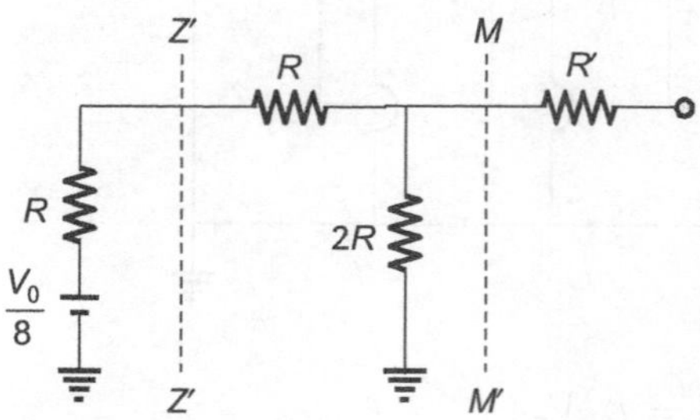

Thevenin's equivalent to the left of MM':

$$V_{TH_4} = \frac{V_0}{16}, \quad R_{TH_4} = R$$

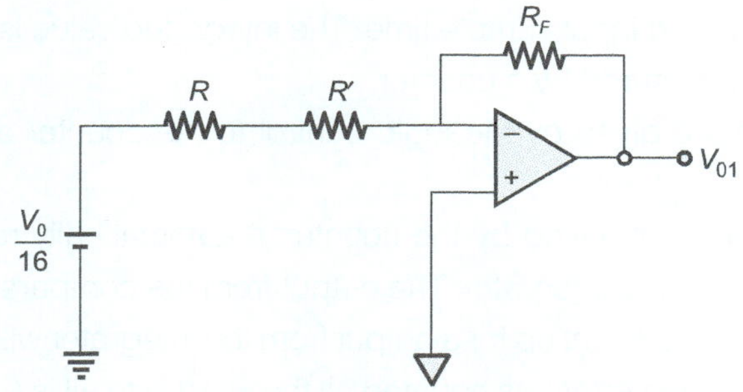

$$V_{01} = \frac{-R_F}{R+R'} \cdot \frac{V_0}{16}$$

Similarly, $$V_{02} = \frac{-R_F}{R+R'} \cdot \frac{V_1}{8};\quad V_{03} = \frac{-R_F}{R+R'} \cdot \frac{V_2}{4};\quad V_{04} = \frac{-R_F}{R+R'} \cdot \frac{V_3}{2}$$

Applying superposition theorem = $V_0 = V_{01} + V_{02} + V_{03} + V_{04}$

$$V_0 = \frac{-R_F}{(R+R')} \cdot \frac{1}{16} [2^3 V_3 + 2^2 V_2 + 2^1 V_1 + 2^0 V_0]$$

or $$V_0 = \frac{-R_F}{R+R'} \cdot \frac{1}{2^4} [2^3 . b_3 V_R + 2^2 . b_2 V_R + 2^1 . b_1 V_R + 2^0 . b_0 V_R]$$

$$V_0 = \frac{-R_F}{R+R'} \cdot \frac{V_R}{2^4} [2^3 . b_3 + 2^2 . b_2 + 2^1 . b_1 + 2^0 . b_0]$$

Resolution for, $$V_R = 12\text{ V} = \frac{12}{2^4 - 1} = 0.8$$

$\therefore$ Full-scale output = 12 V

9.10 Explain the principle of operation of a dual slope integrator type of analog to digital converter with the help of block and timing diagram.
An analog voltage signal with highest significant frequency of 1 kHz is to be digitally coded, with a resolution of 0.1% covering the range of 0-10 volts. To avoid loss of information determine. The minimum sampling rate, minimum number of bits in the digital code, analog value of least significant bit and the dynamic range of A/D converter in dB.

[10 marks : 2006]

Solution:

The block diagram of a dual slope integration A/D converter is shown in figure.

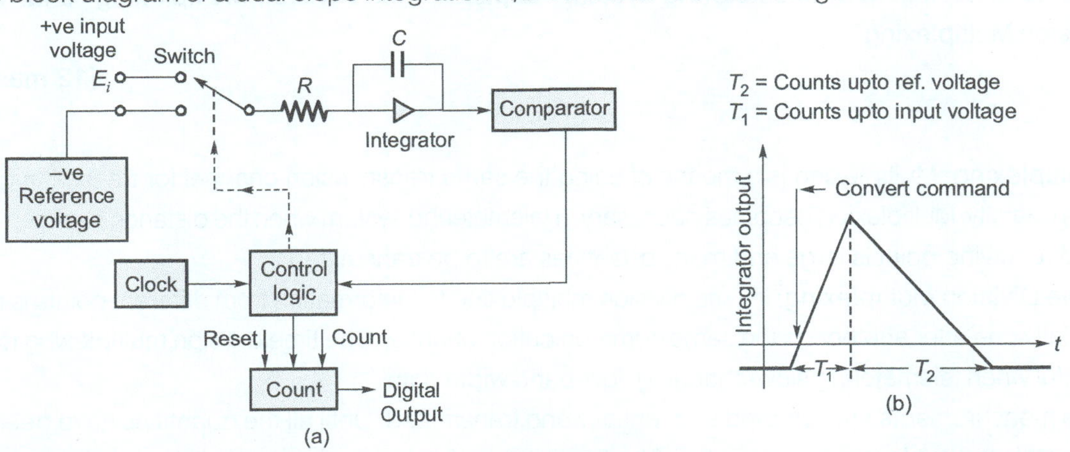

The reference voltage and the input analog voltage are sequentially connected to the integrator with the help of a switch. The reference voltage and the input analog voltage must be of opposite polarity. The input voltage is integrated for a fixed input sample time. The integrated value is then discharged at a fixed rate and the time to do this is measured by a counter.

Two types of commands are given by the logic control to the counter and they are: (a) Reset and (b) Convert.

When a convert command is received by the counter, it automatically reset to all zeros and the switch connects the input voltage to the integrator. The output from the comparator is designed such that at this time it will permit the counter to count up for a output from the integrator will be steadily increasing in value. On the next count after the converter has counted all the way up to all is (i.e. the next count will cause it to go all zeros and start over again), the switch changes position. The input voltage is dis-connected and the switch connects the reference voltage to the integrator. Therefore the integrator now integrates the opposite polarity voltage, which causes the output to decrease towards zero voltage. Mean while, the counter is counting up from zero again. When the output of the integrator goes to zero, it causes the comparator to switch its output, thereby stopping the counter via the control logic. The binary number in the counter at this time is proportional to the amount of time that it took the integrator to integrate down from its starting point to zero. Therefore the binary count is proportional to input voltage. Figure (b) illustrates the output from the integrator during a conversion cycle. The difference in the slopes of the curve is due to difference in voltage level of input voltage E_i and reference voltage E_r.

Suppose, T_1 = Counts when the input voltage is applied

T_2 = Counts when the reference voltage is applied

Input voltage is, $$E_i = \left(\frac{T_2}{T_1}\right) \times E_r$$

1. $$\text{Resolution} = 0.1\% = \frac{0.1}{100} = \frac{1}{1000} = \frac{1}{2^n}$$

   Minimum number of bits $n = 10$

2. Analog value of LSB $= \frac{1}{2^n} \times 10 = \frac{1}{1024} \times 10 \text{ V} = 9.76 \times 10^{-3}$ V

3. In order that the signal may not be lost, the sampling frequency should be at least 5 times the highest frequency in the signal.

   ∴ Minimum sampling frequency $f_s = 5\, f_h = 5 \times 1 = 5$ kHz

4. Dynamic range of converter = 6 × number of bits = 6 × 10 = 60 db

9.11 What is multiplexing in telemetering system? Explain "Time Division Multiplexing" and "Frequency Division Multiplexing".

[12 marks : 2007]

Solution:

Multiplexing: Multiplexing is a means of using the same transmission channel for transmitting more than one quantity. Multiplexing becomes necessary in telemetering system when the distance between treatmitting and receiving point is large and many quantities are to be transmitted.

Time Division Multiplexing: In time division multiplexing the information from different points is transmitted serially one after another on the same communication channel. The time division multiplexing is especially useful when telemetering slow-changing, low-band width data.

The measurements are sampled sequentially and transmitted. Until all the quantities have been sampled. After this the cycle is again repeated. The time to complete transmission of all the measurements may be sufficiently small so that one quantity cannot change considerably during the interval between sampling.

The order of transmission is controlled either cyclically by means of pulse generators or by command signals.

Frequency Division Multiplexing: In frequency division multiplexing, several information can be simultaneously transmitted using different carrier frequencies and employing modulation techniques.

By suitable choice of the magnitudes of the different carrier frequencies, several data may simultaneously be transmitted over a common communication line. These can be recovered separately at the receiving end by using demodulation and filters.

9.12 Explain with the help of a functional block diagram, the principle of operation of a digital frequency meter.

[10 marks : 2008]

Solution:

Functional Block diagram of Digital Frequency Meter:

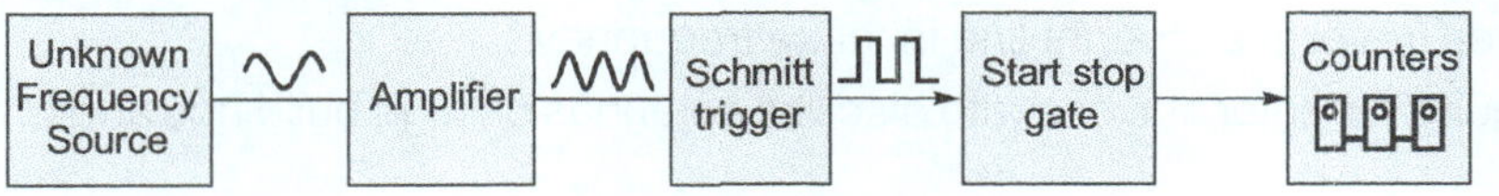

Principle of Operation:

1. The unknown frequency signal is fed to the schmitt trigger. The signal may be amplified before being applied to schmitt trigger.
2. In a schmitt trigger, the signal is converted into a square wave with very fast rise and fall times, then differentiated and clipped.
3. As a result, the output from a schmitt trigger is a train of pulses, one pulse, for each cycle of the signal.
4. The output pulses from the schmitt trigger are fed to start stop gate when this gate opens (start), the input pulses pass through this gate and are fed to an electronic counter which starts registering the input pulses.
5. When the gate is closed (stop), the input of pulses to counter ceases and it stops counting.
6. The counter displays the number of pulses that have passed through it in the time interval between start and stop. If this interval is known, the pulse rate and hence the frequency of the input signal can be known. Frequency of unknown signal,

$$f = \frac{N}{t}$$

where, f = frequency of unknown signal
N = number of counts displayed by counter
t = time interval between start and stop of gate

9.13 Define the term 'Telemetry' and explain why it is necessary to use it in a instrumentation system. Name different types of telemetry systems. Give salient features of RF telemetry.

[10 marks : 2008]

Solution:

Telemetry: Telemetry is the technology which enables a user to collect data from several measurement points at inaccessible or inconvenient locations, transmit that data to a convenient location, and present the several individual measurements in a usable form.

Its Necessity in an Instrumentation System: In modern measurement systems, the various components comprising the system are usually located at a distance from each other. It, therefore, becomes necessary to transmit the data or an information between them through some form of communication channels.

The transmission of a measured variable to a remote point is an important function in instrumentation system because of the size and complexity of modern industrial plants. The most common variables encountered in industrial plants are temperature, pressure and flow and transmission of data of these systems take great length from the place of measurement to the place of data recording or display. This will result into excessive measurement lags. Hence there is a need for fast transmission of data.

Different Types of Telemetry Systems: Two types of telemetry systems are used:

1. Land line telemetry
2. R.F. (radio frequency) telemetry

Salient Features of R.F. Telemetry:

1. There is no physical link between the transmitting and receiving stations.
2. The link between the transmission station and the receiving station can only be established through radio links.
3. R.F. telemetry is usually more suitable if data is to be transmitted over distances greater than 1 km.
4. Certain parts of the R.F. spectrum allocated for telemetry and microwave links above about 4 MHz, and radio waves travel in a straight line in these frequency.
5. These require repeater stations with disc like antennas on high buildings and towers every 30 to 60 km.

9.14 Describe with the help of a block diagram a multi-channel data acquisition system (DAS).

[6 marks : 2009]

Solution:

Data-acquisition systems are used to measure and record analog signals in basically two different ways:

(a) Signals which originate from direct measurement of electrical quantities. These signals may be d.c. or a.c. voltages, frequency or resistance, etc.

(b) Signals which originate from use of transducers.

Digital Data Acquisition System:

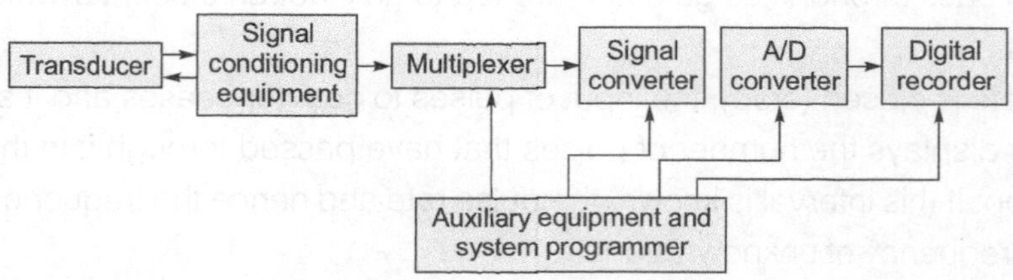

(a) **Transducers:** They convert a physical quantity into an electrical signal which is acceptable by the data acquisition system.

(b) **Signal Conditioning Equipment:** Signal conditioning equipment includes any equipment that assists in transforming the output of transducer to the desired magnitude or form required by the next state of the data acquisition system.

(c) **Multiplexer:** A multiplexer accepts multiple inputs and connects them sequentially to one measuring instrument.

(d) **Signal Converter:** A signal converter translates the analog signal to a form acceptable by the analog to digital (A/D) converter.

(e) **Analog to Digital Converter:** An A/D converter converts the analog voltage to its equivalent digital form.

(f) **Auxiliary Equipment:** This contains devices for system programming functions and digital data processing.

(g) **Digital Recorders:** Records of information in digital form may be had on punched cards, perforated paper tapes, type written pages, floppy, discs, magnetic tape or a combination of these systems.

(h) **Digital Printers:** After all the tests have been completed and the data generated, it becomes necessary to record the numbers and in some cases reduce the data to a more meaningful form. A digital printer can be specified to interface with an electronic instrumentation system in order to perform this work, and thus provide a high quality hard copy for records and minimizing the labour of the operating staff.

9.15 A 2-bit D/A converter using R-2 R adder network was constructed using 2 kΩ and 8 kΩ resistors. If the bit length or the converter is to be increased by two, using resistors of the same values, what would be the total number of resistors required in the augmented version? Give the circuit diagram of the augmented converter.

[12 marks : 2010]

Solution:

In a n-bit D/A converter, resistors of two values R and $2R$ are required. For each bit two resistors are used R and $2R$ resistors.

Since the available resistors are 2 kΩ and 8 kΩ the above condition can be fulfilled by connecting two 2 kΩ resistors in series to make equivalent resistor of 4 kΩ.

So in the augmented version, for each bit three resistors are used, two resistors of 2 kΩ and one resistor of 8 kΩ.

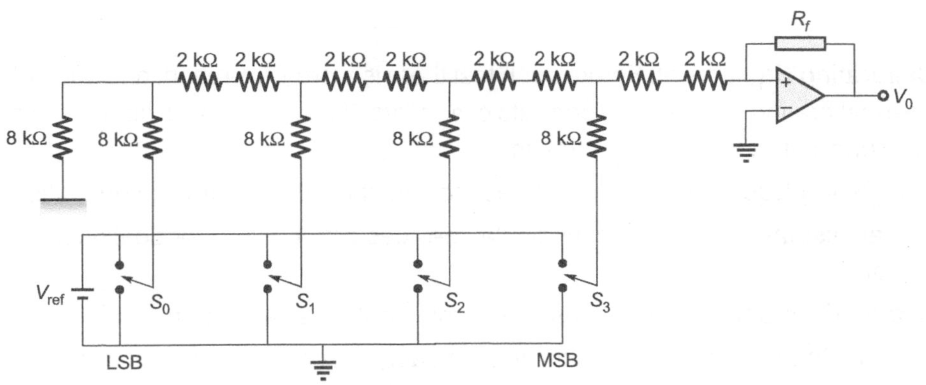

Number of bits in augmented version = 4

Hence total number of resistors required = 4 × 3 = 12

9.16 Draw the line diagram of data acquisition system from process plant to computer system. Explain the function of each component. What do you understand by smart transducer? Discuss it in brief.

[10 marks : 2011]

Solution:

Data-acquisition systems are used to measure and record analog signals in basically two different ways:

(a) Signals which originate from direct measurement of electrical quantities. These signals may be d.c. or a.c. voltages, frequency or resistance, etc.

(b) Signals which originate from use of transducers.

Types of Instrumentation Systems: The instrumentation systems can be classified into two distinct categories:

(a) **Analog System:** These systems deal with information in analog form. An analog signal may be defined as continuous function, such as a plot of voltage versus time or displacement versus force.

(b) **Digital System:** A digital quantity may consist of a number of discrete or discontinuous pulses whose time relationship contains information about the magnitude and the nature of the quantity under measurement.

Components of an Analog Data-Acquisition System:

(a) **Transducers:** It is desirable that an emf obtained from the transducer proportional to the quantity being measured is used as an input to the data acquisition system.

(b) **Signal Conditioning Equipment:** Signal conditioning equipment includes any equipment that assists in transforming the output of transducer to the desired magnitude or form required by the next stage of the data acquisition system. It also produces the required conditions in the transducers so that they work properly.

(c) **Multiplexer:** Multiplexing is the process of sharing a single channel with more than one output. Thus a multiplexer accepts multiple analog inputs and connects them sequentially to one measuring input.

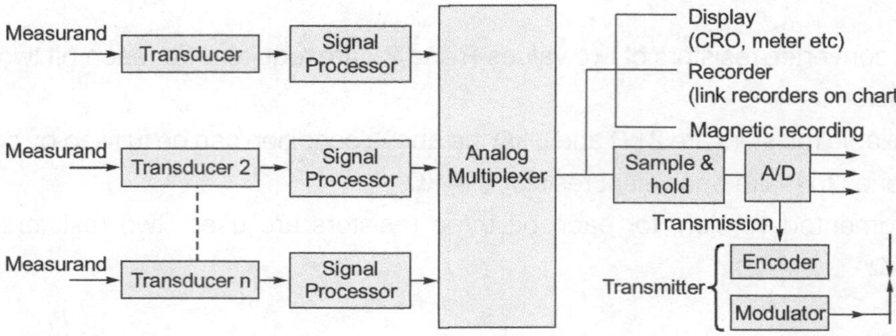

(d) **Calibrating Equipment:** Before each test there is a pre-calibration, and often after each test there is a post-calibration. This usually consists of a millivolt calibration of all input circuits and shunt calibration of all bridge-type transducer circuits.

(e) **Integrating Equipment:** It is often desirable to know the integral or summation of a quantity.

(f) **Visual Display Devices:** Visual display devices are required for continuous monitoring of the input signals.

(g) **Analog Recoders:** The methods of recording data in analog form.

(h) **Analog Computers:** The function of a data acquisition system is not only to record data acquired by the transducers and the sensors but also to reduce this data to the desired form. An analog computer may be used as a data reduction device. The output voltage of an analog computer can either be recorded in analog form or be converted to a digital form for recording and further computation.

(i) **High Speed Cameras and TV Equipment:** In many industrial processes, engine testing, and aerodynamic testing it is not possible for the test operator to have a view of the equipment being tested. Therefore, closed circuit TV is used to enable the operator to make visual observations of the test. Also high speed cameras are employed to obtain a complete visual record of the process for further analysis.

Smart Transducer:

As sensors and actuators becomes more complex they provide support for various modes of operation and interfacing. Some applications require additionally fault-tolerance and distributed computing. Such high full functionality can be achieved by adding an embedded micro-controller to the classical sensor actuator, which increases the ability to cope with complexity at a fair price. This integration of an analog or digital sensor/actuator, a processing unit and a communication interface is reference to as smart transducer.

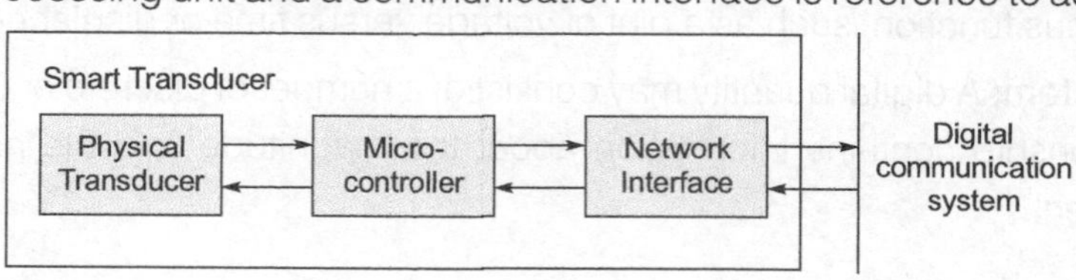

9.17 **Describe the principle of frequency measurement using digital technical. Draw its block diagram. The unknown input signal of 2 V square wave is 3.5 kHz. Determine the display indication if the gate enable time is (i) 0.1 second (ii) 1 second and (iii) 10 seconds.**

[10 marks : 2011]

Solution:

Functional Block diagram of Digital Frequency Meter:

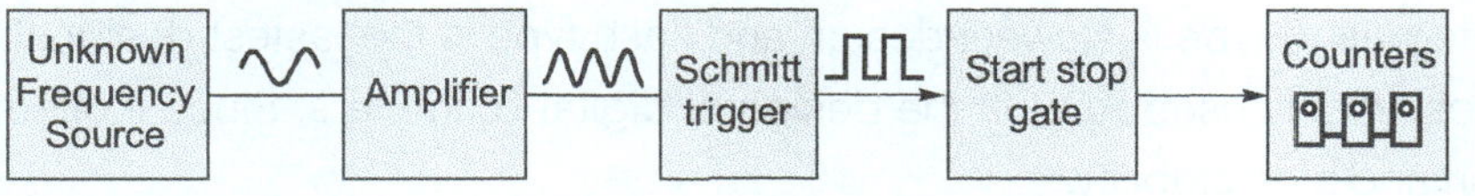

Principle of Operation:

1. The unknown frequency signal is fed to the schmitt trigger. The signal may be amplified before being applied to schmitt trigger.
2. In a schmitt trigger, the signal is converted into a square wave with very fast rise and fall times, then differentiated and clipped.
3. As a result, the output from a schmitt trigger is a train of pulses, one pulse, for each cycle of the signal.
4. The output pulses from the schmitt trigger are fed to start stop gate when this gate opens (start), the input pulses pass through this gate and are fed to an electronic counter which starts registering the input pulses.
5. When the gate is closed (stop), the input of pulses to counter ceases and it stops counting.
6. The counter displays the number of pulses that have passed through it in the time interval between start and stop. If this interval is known, the pulse rate and hence the frequency of the input signal can be known. Frequency of unknown signal,

$$f = \frac{N}{t}$$

where, f = frequency of unknown signal

N = number of counts displayed by counter

t = time interval between start and stop of gate

f = frequency of the signal = 3.5 kHz = 3500 Hz

Assuming 5 digit display

(i) $t = 0.1$ second

$$f = \frac{N}{t}$$

$$N = ft = 3500 \times 0.1 = 350$$

display = 00350

(ii) $t = 1$ second,

$$f = \frac{N}{t}$$

$$N = ft = 3500 \times 1 = 3500$$

display = 03500

(iii) $t = 10$ second,

$$N = ft = 3500 \times 10 = 35000$$

display = 35000

9.18 **Mention various types of analog to digital converter in increasing order of speed of operation. Why is dual slope ADC preferred in digital voltmeter? An 8 bit successive approximation type ADC uses a clock frequency of 1 MHz. Calculate the time of conversion.**

[10 marks : 2014]

Solution:

Various types of digital voltmeters are:

(*i*) Dual slope integrating type

(*ii*) Ramp type

(*iii*) Counter type

(*iv*) Successive approximation type

(*v*) Flash type

Dual slope integrating type is slowest design and flash type is the fastest design. Dual slope integrating type AC is most widely used ADC in the design of digital voltmeters. Since it offers high accuracy, high noise rejection and high stability.

$$\text{Clock frequency} = 1\text{ MHz}$$

$$\text{Time of conversion} = ntpd$$

$$= 8\times\frac{1}{1\times10^6} = 8\ \mu\text{sec.}$$

9.19 With a block schematic, explain the essential functional operations of a generalised data acquisition system.

[12 marks : 2021]

Solution:

Data acquisition systems are being used in ever increasing large and wide fields in a variety of industrial and scientific areas including the aerospace biomedical and telemetry industries.

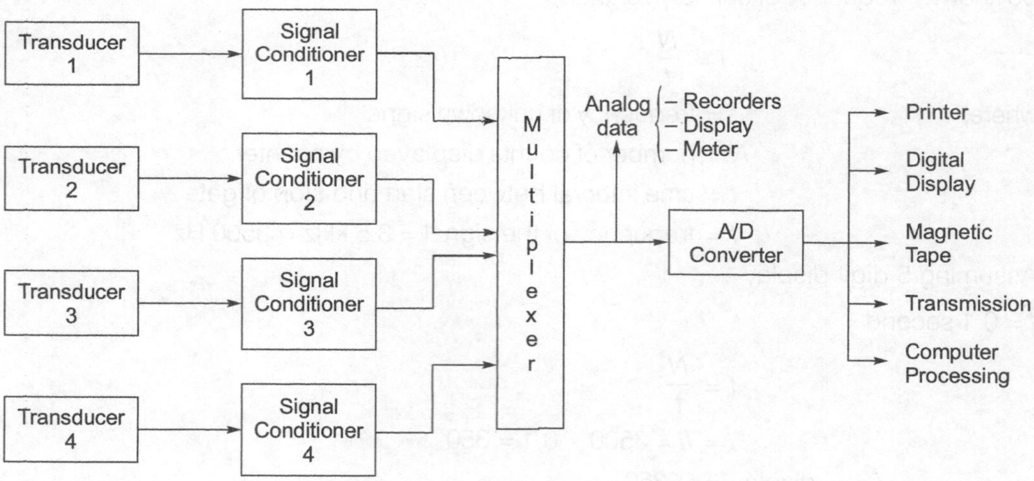

The various components and their functions are described below :

1. **Transducers :** They convert a physical quantity into an electrical signal which is acceptable by the data acquisition system.
2. **Signal conditioning equipment :** They amplify and convert signal into easy to read and compatible form.
3. **Multiplexer :** Multiplexing is the process of sharing a single channel with more than one input. Thus a **multiplexer** accepts multiple analog inputs and connects them sequentially to one measuring instrument. Another name for a multiplexer is **scanner**.
4. **Analog to digital converter (A/D converter) :** An A/D converter converts the analog voltage to its equivalent digital form. The output of the A/D converter may be fed to digital display devices for visual

display or may be fed to digital recorders for recording. It may be fed to a digital computer for data reduction and further processing.

Essential functional operations of a generalized data acquisition system are :

1. It must acquire the necessary data, at correct speed and at the correct time.
2. Use of all data efficiently to inform the operator about the state of the plant.
3. It must monitor the complete plant operation to maintain on-line optimum and safe operations.
4. It must provide an effective human communication system and be able to identify problem areas, thereby minimising unit availability and maximising unit through point at minimum cost.
5. It must be able to collect, summarise and store data for diagnosis of operation and record purpose.
6. It must be able to compute unit performance indices using on-line, real-time data.
7. It must be flexible and capable of being expanded for future requirements.
8. It must be reliable, and not have a down time greater than 0.1%.

■■■■

5 Computer Fundamentals

Revised Syllabus of ESE: *Number systems, Boolean algebra (Part of Digital Electronics), arithmetic functions, Basic Architecture, Central Processing Unit, I/O and Memory Organisation; peripheral devices, data represenation and programming, basics of Operating system and networking, virtual memory, file systems; Elements of programming languages, typical examples.*

1. Number System

1.1 What is the dimension of a quantity? Derive dimensions of capacitance, resistance and inductance The binary representation of 100 110 is numerically equivalent to:

(a) The decimal representation of 46
(b) The octal representation of 46
(c) The octal representation of 26
(d) The decimal representation of 26

[2 marks : 2003]

Solution: (b)

$$(\underline{100}\ \underline{110})_2 = (4\,6)_8$$

100 110 is equivalent to octal representation of 46.

1.2 Perform the following binary substraction and give the result in decimal value also:
11001 – 1100

[4 marks : 2017]

Solution:

Given: 11001 – 1100

First number in decimal $(11001)_2 = (27)_{10}$

Second number in decimal $(1100)_2 = (12)_{10}$

Their difference is, $27 - 12 = (15)_{10}$

Result in binary $(15)_{10} = (1111)_2$

1.3 (i) How is division exactly done by ALU?
(ii) Prove with illustration that NAND is a "Universal GATE".
(iii) How will you implement a two-way switch using minimum number of logic gates?

[7 + 6 + 7 = 20 marks : 2018]

Solution:

(i) Division:

$$\text{Dividendend} = \underset{(Q)}{\text{Quotient}} * \text{Divisor} + \underset{(R)}{\text{Remainder}}$$

e.g.:

$$7 \div 2 = \begin{Bmatrix} Q = 3 \\ R = 1 \end{Bmatrix}$$

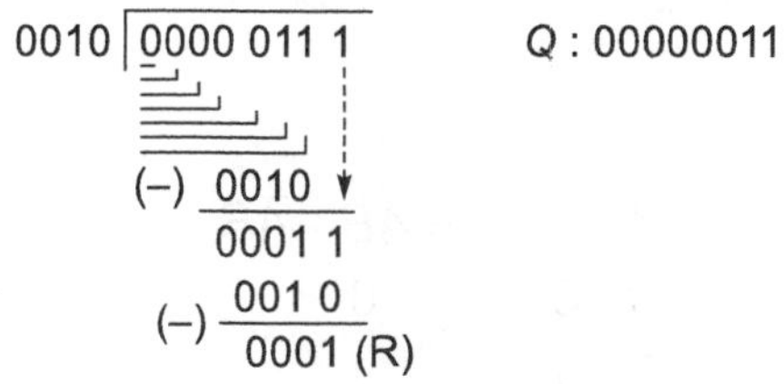

Dividend scanned from MSB to LSB in bitwise. If it is less than or equal to divisor then put 1' in quotient and subtract divisor from dividend (or) put '0' in quotient move to next bit scan of dividend.

32-bit Division: $\left[\dfrac{2n \text{ bit}}{n \text{ bit}}\right]$

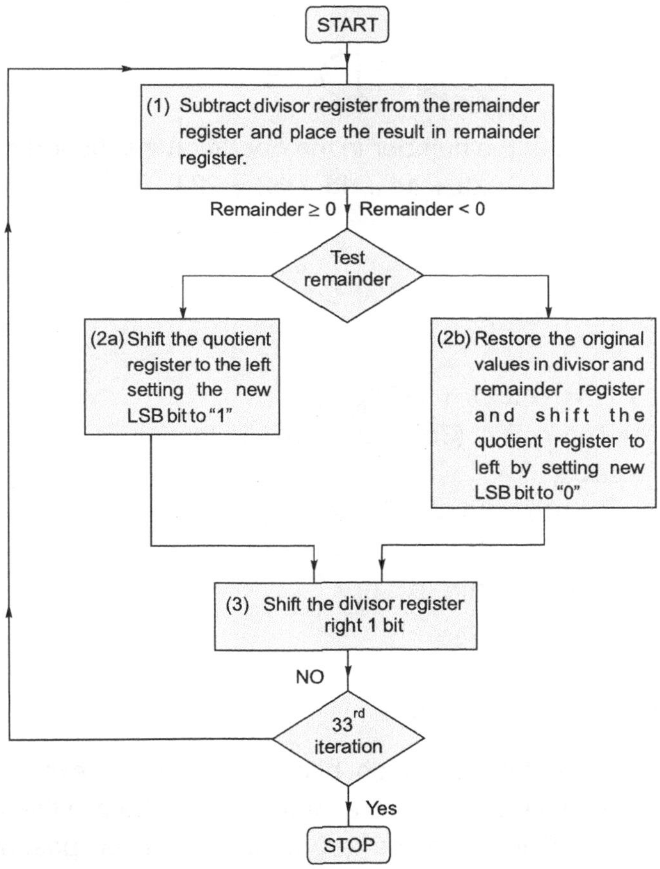

(ii) NAND gate is called as universal gate as it is possible to implement any logic function using it.

(a) Realization of NOT gate using NAND gate

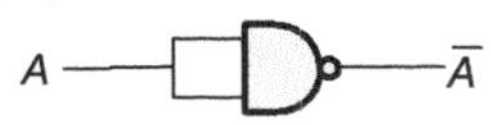

(b) Realization of AND gate using NAND gate

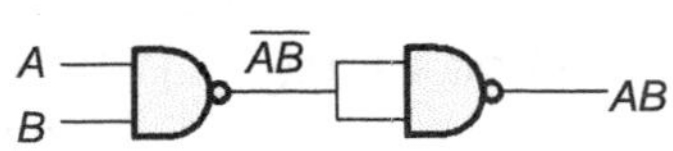

(c) Realization of OR Gate using NAND gate

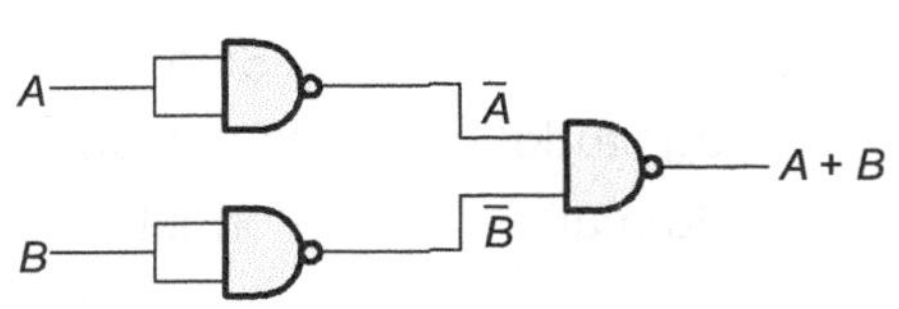

(iii) Two way switch using minimum number of logic gates.
$A = 0$ and $B = 1$ (or) $A = 1$ and $B = 0$
Therefore,

A	B	$X = A\bar{B} + \bar{A}B$
0	0	0
0	1	1
1	0	1
1	1	0

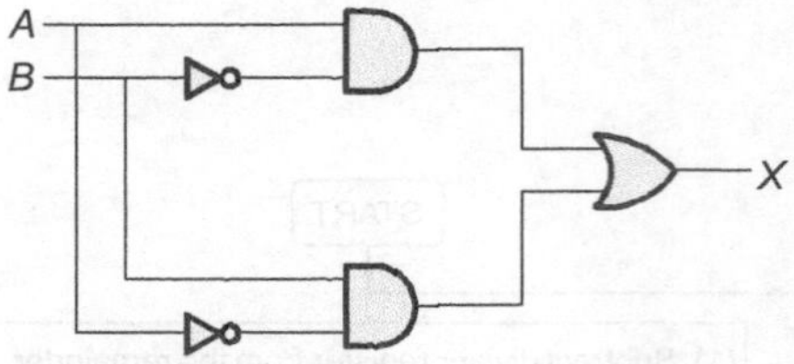

1.4 **Determine the possible base of the number in the operation mentioned below:**

$$23 + 44 + 14 + 32 = 223$$

[4 marks : 2019]

Solution:

Assuming the base of numbers be: b
We can write,

$$(23)_b + (44)_b + (14)_b + (32)_b = (223)_b$$

$$(2b^1 + 3) + (4b^1 + 4) + (b^1 + 4) + (3b^1 + 2) = 2b^2 + 2b + 3$$

On solving above expression,

$$10b + 13 = 2b^2 + 2b + 3$$

or $$2b^2 - 10b + 2b - 10 = 0$$

$$b(2b - 10) + 1\,(2b - 10) = 0$$

$$b = -1$$

$$b = 5$$

$\therefore$ Acceptable value of base, $b = 5$

1.5 **A 36-bit floating point binary number has eight bits plus sign for the exponent and twenty-six bits plus sign for mantissa. The mantissa is a normalized fraction. Numbers in the mantissa and exponent are in signed-magnitude form. Determine the largest and smallest positive quantities that can be represented excluding zero.**

[10 marks : 2021]

Solution:

- 36-bit format :

Exponent	Mantissa
[1 bit + 8 bit]	[1 bit + 26 bit]
9 bit	27 bit
↓	↓
Sign magnitude format	Sign magnitude format

- Exponent range : 9-bit sign matnitude range ⇒ {–255 to +255}
- Mantissa Range (Normalized mantissa), i.e., 1.M

$$\text{Min 'M'} = 1.000.....(26)0\text{'s} = 1$$

$$\text{Max 'm'} = 1.1111.....(26)1\text{'s} = (2 - 2^{-26})$$

In the floating point data minimum and maximum exponents are reserved for special value (0, +∞ and –∞) i.e., (–255 and +255).

∴ Representable data exponents are {–254 to +254}.

Smallest representable positive data is $+(1)*2^{-254}$.

Largest represental positive data is $+(2 - 2^{-26})*2^{+254}$.

1.6 Construct full :

(i) conjunctive normal form for the statement $P \rightarrow Q$;

(ii) disjunctive normal form for the statement $(P \rightarrow \{Q \vee R)\} \wedge (P \vee Q)$.

[6 + 6 = 12 marks : 2023]

Solution:

To construct the full Conjunctive Normal Form (CNF) for the implication P → Q, we need to convert it into a series of conjunctions. Here's the process:

Step 1: Negate P and rewrite the implication as ¬P ? Q.

Step 2: Convert ¬P ? Q into CNF form.

To convert ¬P ? Q into CNF form, we can apply the distributive law:

(¬P ? Q) is equivalent to (¬P ? Q) ? (P ? Q).

Therefore, the CNF form for P -> Q is:

(¬P ? Q) ? (P ? Q).

To construct the full disjunctive normal form (DNF) for the given formulas, let's first break down each formula into its constituent parts:

Formula 1: (P -> (Q v R))

Formula 2: (P v Q)

Now, let's convert each formula into its DNF form separately:

DNF for Formula 1:

Step 1: Apply the implication equivalence (P -> Q) = (~P v Q) to the main implication:

(~P v (Q v R))

DNF for Formula 2:

This formula is already in disjunctive form.

Now, we need to combine the two formulas to obtain the full DNF.

To combine them, we'll distribute Formula 2 over Formula 1:

(P v Q) ? (~P v (Q v R))

Now, we can expand the conjunction to get all the possible disjunctive clauses:

(P ? ~P) v (P ? (Q v R)) v (Q ? ~P) v (Q ? (Q v R))

Simplifying further:

False v (P ? (Q v R)) v (Q ? ~P) v (Q ? (Q v R))

Since False v anything = anything, we can simplify the expression:

(P ? (Q v R)) v (Q ? ~P) v (Q ? (Q v R))

This is the full disjunctive normal form for the given formulas.

2. Boolean Algebra

2.1 The Boolean expression $X = B + A \cdot \bar{B} + A \cdot B$ is equivalent to

(a) $A + B$ (b) $\bar{A} \cdot B$
(c) $\overline{A + B}$ (d) $A \times B$

[2 marks : 2005]

Solution: (a)

$$X = B + A \cdot \bar{B} + A \cdot B$$

$$= B + A(\bar{B} + B) = A + B$$

2.2 The Boolean expression $ABC + \bar{A}\bar{B}C + AB\bar{C} + \bar{A}\bar{B}\bar{C}$ is equivalent to

(a) OR GATE (b) AND GATE
(c) EX-NOR GATE (d) EX-OR GATE

[2 marks : 2007]

Solution: (c)

From K-map,

A \ BC	00	01	11	10
0	1	1		
1			1	1

$$Y = \bar{A}\bar{B} + AB \equiv \text{EX-NOR Gate}$$

2.3 The Boolean expression $ABCD + A\bar{B}CD + ABC\bar{D} + A\bar{B}C\bar{D}$ is equivalent to

(a) A (b) AC
(c) ABC (d) 1

[2 marks : 2008]

Solution: (b)

K-map of Boolean expression,

AB \ CD	00	01	11	10
00				
01				
11			1	1
10			1	1

The equivalent of Boolean expression,

$$Y = AC$$

2.4 Simplify the Boolean function: $F = xy + x'y + xy'$. [4 marks : 2017]

Solution:

Given: $F = xy + x'y + xy'$

to simplify we can use K-map,

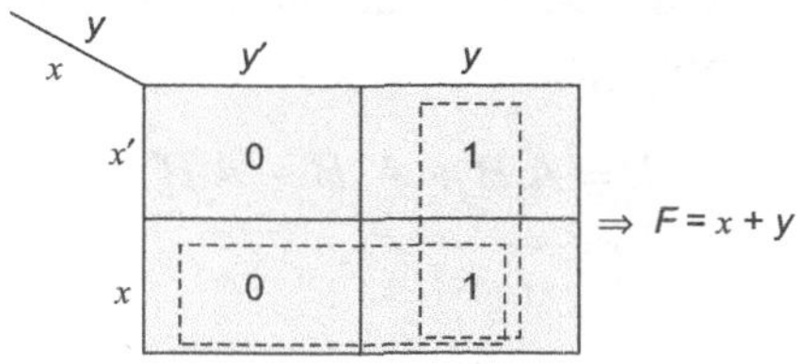

Therefore simplified Boolean expression will be : $F = x + y$

2.5 **Find the minimal sum of products (SOP) expression of the following boolean function $f(a, b, c, d)$**

$$f(a, b, c, d) = abc + abd + a'bc' + cd + bd'$$

Where a', b', c' and d' are complements of variables a, b, c and d respectively.

[4 marks : 2019]

Solution:

For SOP expression of given boolean function,

$$\begin{aligned} f(a, b, c, d) &= abc + abd + a'bc' + cd + bd' \\ &= abc(d + d') + abd(c + c') + a'bc'(d + d') + cd(a + a')(b + b') + bd'(a + a')(c + c') \\ &= abcd + abcd' + abc'd + a'bc'd + a'bc'd' + ab'cd + a'bcd + a'b'cd + abcd' + abc'd' \\ &\quad + a'bcd' + a'bc'd' \end{aligned}$$

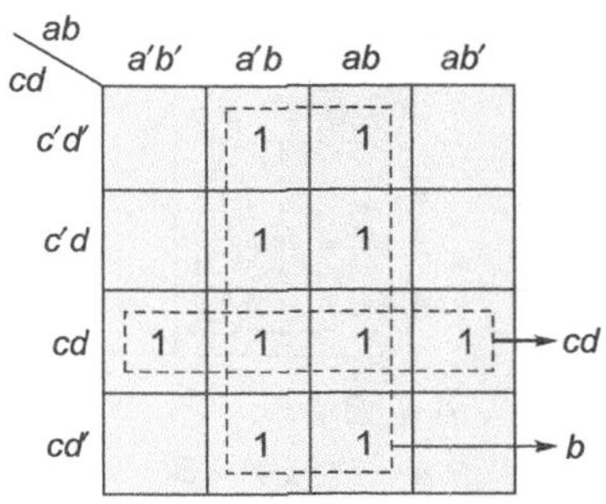

$$f = b + cd$$

2.6 **Find minimal disjunctive normal form (DNF) for the expressions $\bar{x}yz + xyz + xy\bar{z}$ and $x + yz + z\bar{x}y + \bar{y}xz$.**

[6 marks : 2020]

Solution:

Given, $f = \bar{x}yz + xyz + xy\bar{z}$

Corresponding K-map can be represented as shown below :

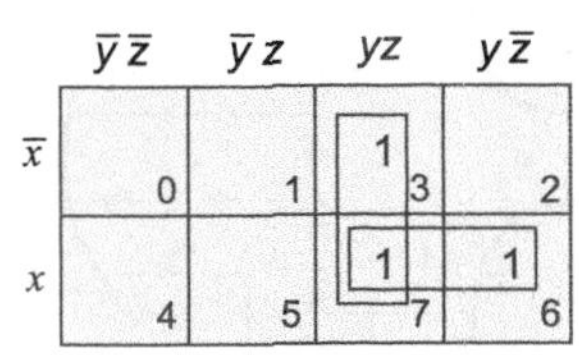

$f = xy + yz = (x \cap y) \vee (y \cap z) \quad \Rightarrow \quad f = x + yz + z\bar{x}y + y\bar{x}z$

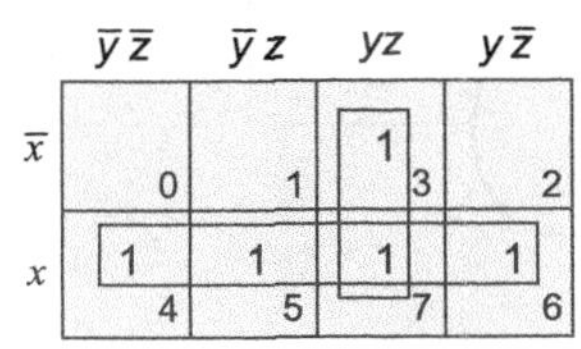

$f = x + yz = x \cup (y \cap z)$

2.7 Construct a switching circuit for the Boolean function using logic gates. Simplify the function and draw the simplified circuit using logic gates

$$L = A.B + A'.B + A.B'$$

[10 marks : 2021]

Solution:

Given : $L = A{\cdot}B + A'{\cdot}B + A{\cdot}B'$

Switching circuit for Boolean function is shown below.

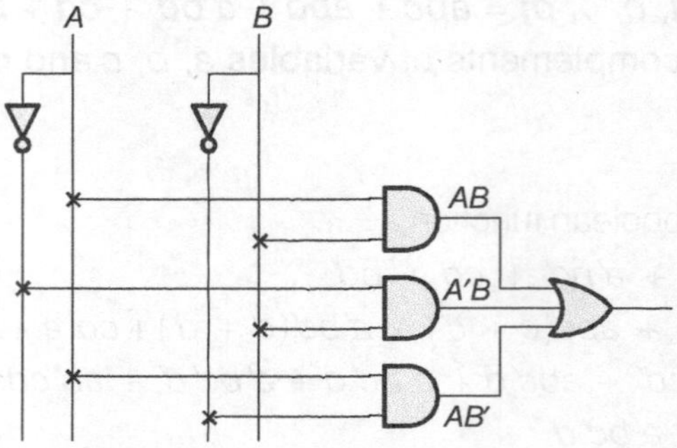

Simplified expression :

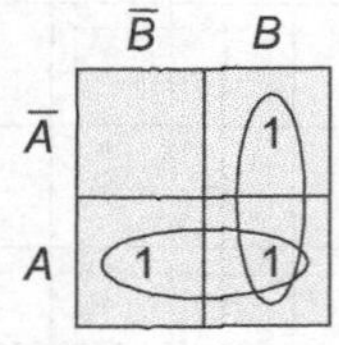

$f = A + B$

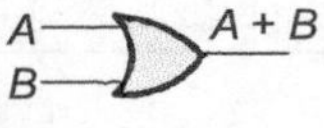

2.8 Minimize the combinational logic shown in figure.

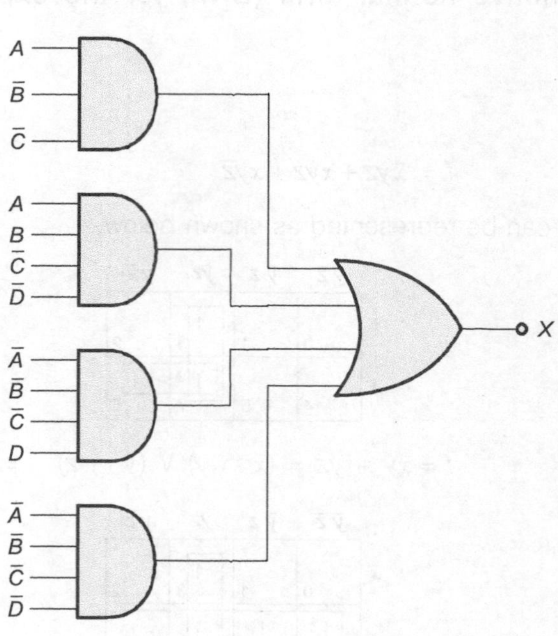

[10 marks : 2022]

Solution:

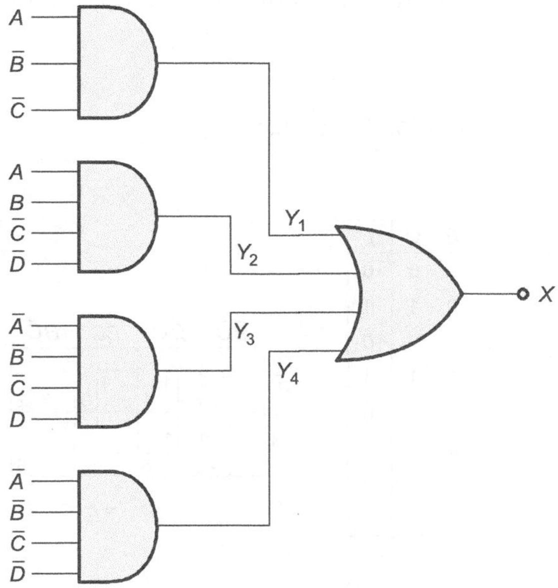

For AND gate

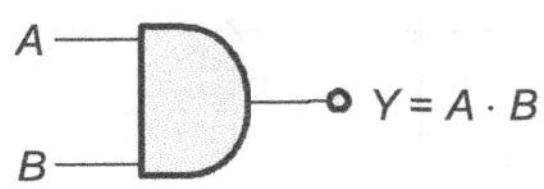

For OR gate

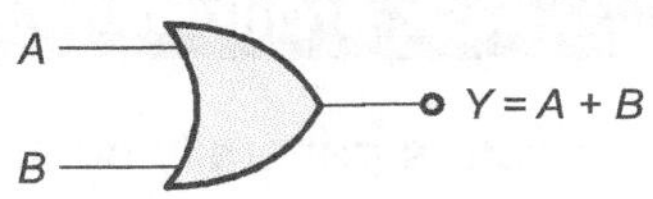

For diagram,

$$Y_1 = A\bar{B}\bar{C},$$

$$Y_2 = AB\bar{C}\bar{D},$$

$$Y_3 = \bar{A}\bar{B}\bar{C}D$$

$$Y_4 = \bar{A}\bar{B}\bar{C}\bar{D}$$

$$X = Y_1 + Y_2 + Y_3 + Y_4$$

$$X = A\bar{B}\bar{C} + AB\bar{C}\bar{D} + \bar{A}\bar{B}\bar{C}D + \bar{A}\bar{B}\bar{C}\bar{D}$$

Realizing K-map

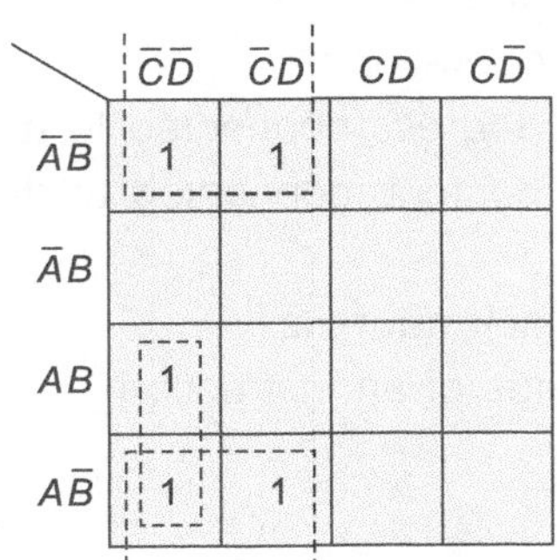

$$X = \bar{B}\bar{C} + A\bar{C}\bar{D}$$

2.9 **Design a circuit that accepts a 3-bit number and gives an output 0, if input represents even decimal number and gives an output 1, if input represents an odd decimal number.**

[12 marks : 2023]

Solution:

For given data we can draw K-map for the circuit.

Truth Table of the circuit :

	A	B	C	f
0	0	0	0	0
1	0	0	1	1
2	0	1	0	0
3	0	1	1	1
4	1	0	0	0
5	1	0	1	1
6	1	1	0	0
7	1	1	1	1

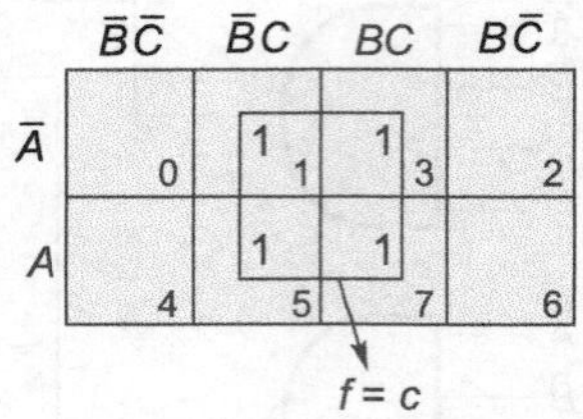

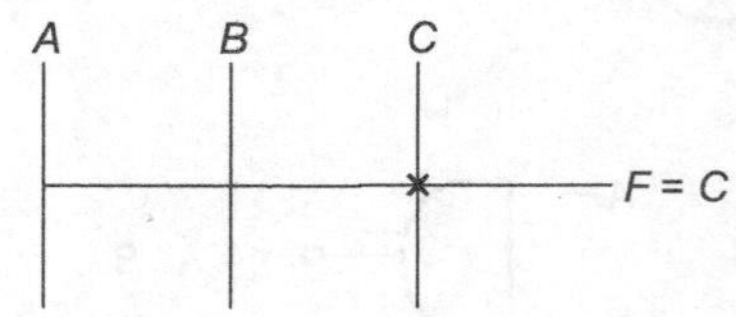

3. Basic Architecture

3.1 **(i) What are the different types of registers generally contained in the central processing unit? Explain functions of each register used in computer systems.**

(ii) List the steps involved in interrupt drivers I/O with a suitable pseudo code/flowchart from the view of in I/O module.

[10+10 = 20 marks : 2019]

Solution:

(i) Registers are one of the three components of central processing unit which include: Registers, ALU and control unit. For execution registers is a internal storage elements of the CPU, which may include following essential kinds:

1. PC : Holds the instruction address
2. IR : Holds the opcode
3. ACC : Holds the ALU I/P and ALU O/P.
4. TR : Holds the ALU 2nd operation
5. MAR : Holds the memory address connected to address lines of a system bus.
6. MBR : Holds the memory content, connected to data lines of a system bus.
   MDR
7. SP : Holds the top of stack address
8. Flag register : Holds the status of an instruction
9. GPR : Holds the data
   (General purpose register)

(ii) Following steps are followed in interrupt driver I/O when interrupt goes high.

(a) The microprocessor completes the execution of the instruction OUT.

(b) It senses the line INTR high and that interrupt is enabled.

(c) The microprocessor disables the interrupt, stops execution and sends out a control signal $\overline{\text{INTA}}$ (interrupt ac knowledge).

(d) The $\overline{\text{INTA}}$ (active low) enables the tristate buffer and the instruction EFH is placed on the data bus.

(e) The microprocessor sent the address of next instruction to the stack from program counter and program sequence will be to the interrupt service routine (ISR) by loading the vector location of the ISR into PC.

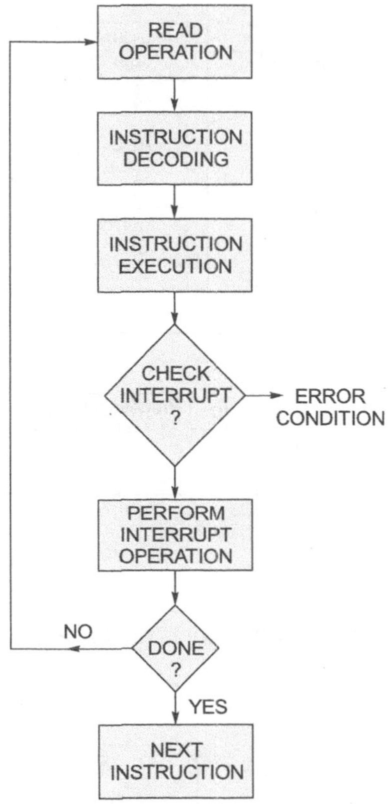

4. CPU Organization

4.1 Draw the basic organization of a computer system and indicate the flow of instruction and data.

[10 marks : 2017]

Solution:

Computer: Computer is a computation device which is used to process the data under the control of a program.

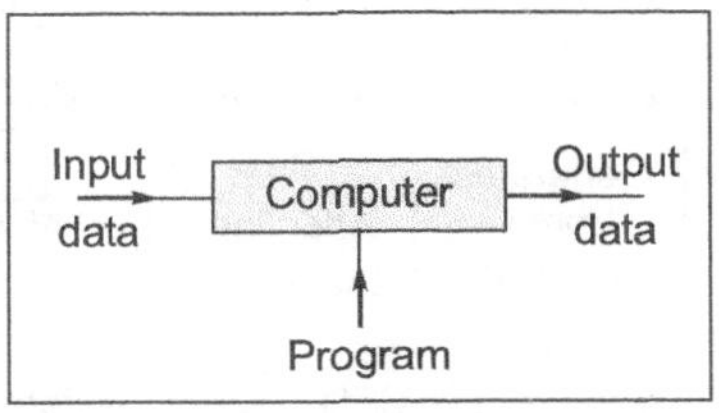

- Computer architecture follows the Von Neumann architecture principle i.e. stored program concept.

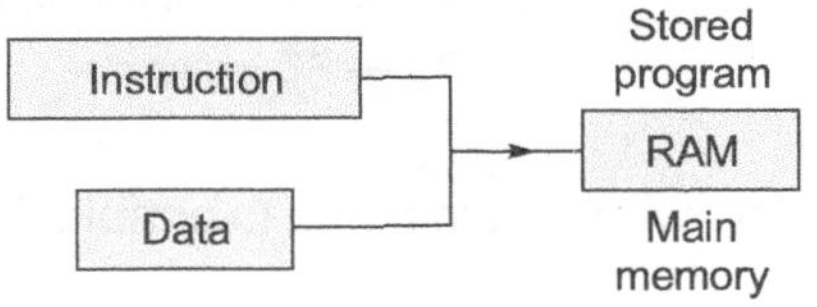

- To execute the user application program which is present in the main memory. Computer system contain 3 basic components.
  1. CPU [Processing unit]
  2. Memory [Storage unit]
  3. Input/Output [External communication]

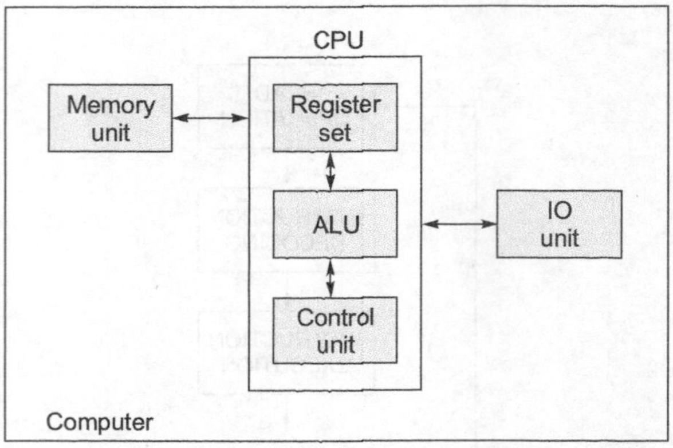

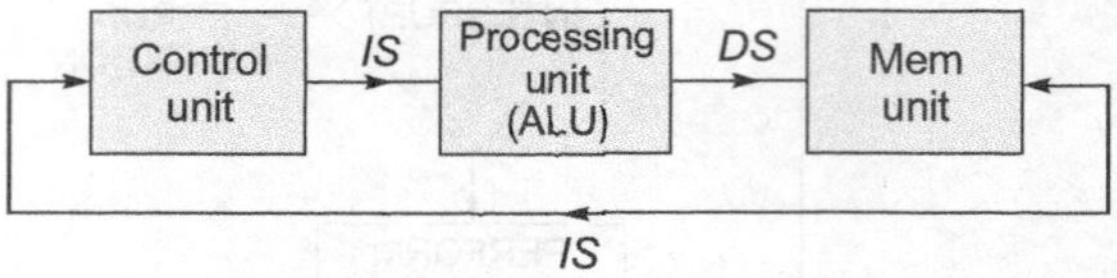

IS : Instruction stream
DS : Data stream

- Instruction execution flow in the computer is described using instruction cycle. It contain two subcycles.
  1. Fetch cycle
  2. Execution cycle

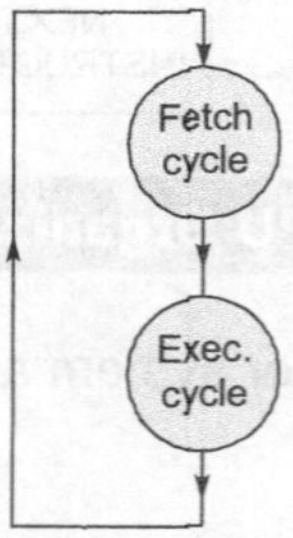

Fetch cycle: Read the instruction from the main memory based on the program counter called as instruction fetch. At the end of this process PC will be incremented to next sequential instruction address.

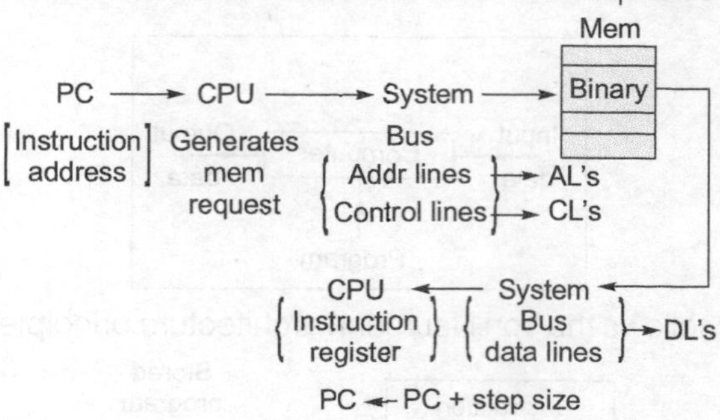

Execution cycle:

- In this cycle instruction decoding will be take place to enable the hardware to perform the operation.

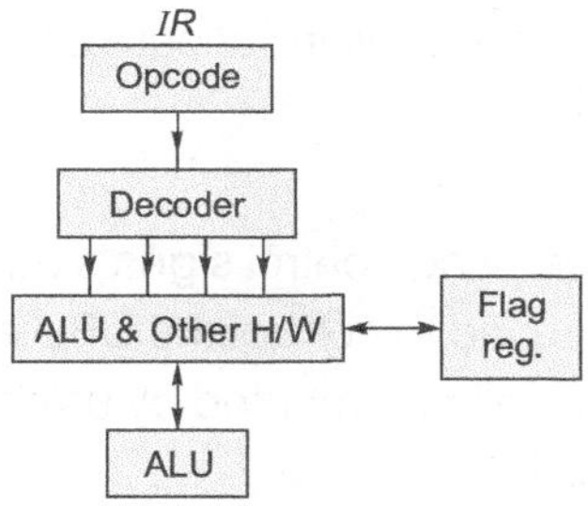

- During the operand fetch, CPU reads the data using the addressing mode.
  e.g.: Direct addressing mode

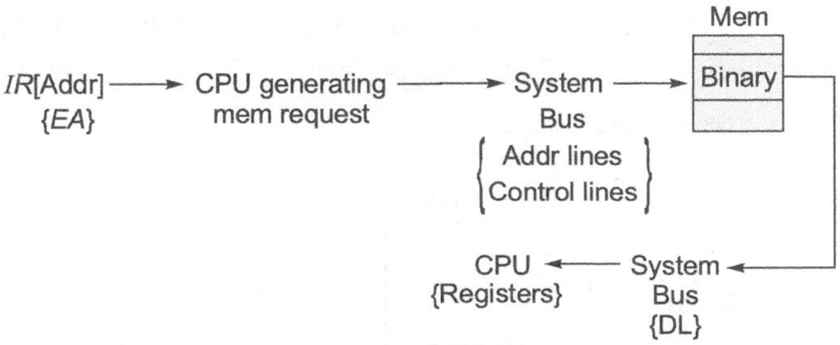

- When the data is available then CPU process the data based on the enabled H/W (Opcode) later result will be stored into a destination using addressing mode.

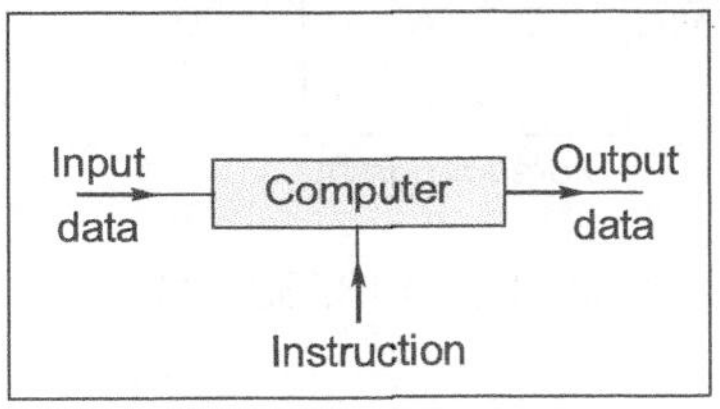

4.2 **Explain in brief the functions of each of the units shown in (Q.No.4.1).** **[10 marks : 2017]**

Solution:

Computer components:

1. CPU
2. Memory
3. IO

CPU: CPU is a processing unit which contain 3 internal components used to process the instructions.

1. Register
2. ALU
3. Control unit

Register: It is a internal storage element in the CPU used to hold the information temporarily. Mandatary registers in the CPU is,

PC
IR
Accumulator
Temp Register
MAR
MBR
SP
GPR
PSW

ALU: It is a hardware circuit which is connected with various hardware elements used to perform the arithmetic and logical operations.

Control unit:

- Control unit is responsible for generating control signals during the execution of an instructions.
- Control unit is a brain of the CPU, used to control the CPU actions.
- In the control unit, control signals are implemented by using the following approaches.
  1. Hardwired CU [SOP form of CS]
  2. μprogrammed CU[Encoded binary form of CS]

Memory:

- It is a storage component in the CPU. It is used to store the user application program.
- Memory chip is organized into cells. Every cell is addressable cell. Memory chip recognizes the control signals to indicate the type of operation.

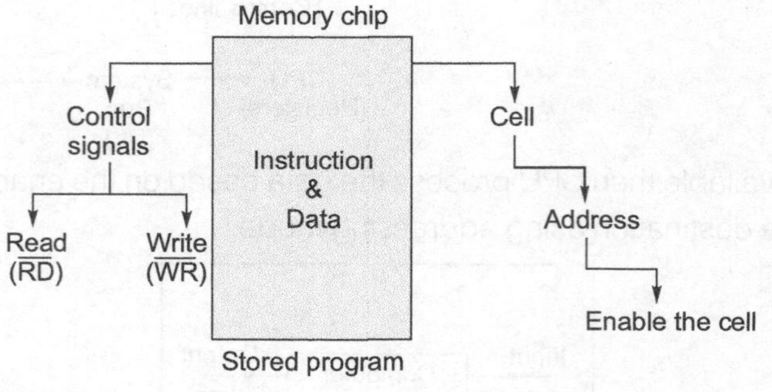

e.g.: Memory chip configurations 64 kB, 1 MB, 32 GB, etc.

IO:

- It is a electro-magnetic device used to perform the IO operations like $\overline{IORD}$ and $\overline{IOWR}$. It is used to provide the communication between the user and H/W.
- IO devices are categorized into 2 types:
  1. Input devices
  2. Output devices
- IO devices are electro-magnetic components and CPU is a electronic component. So there is a difference existing in terms of
  1. Operating modes
  2. Word formats
  3. Data transfer rate

  Therefore, to synchronize the IO speed with processor speed, IO interface logic is used. So all the IO devices are connected to system bus via IO interface logic.

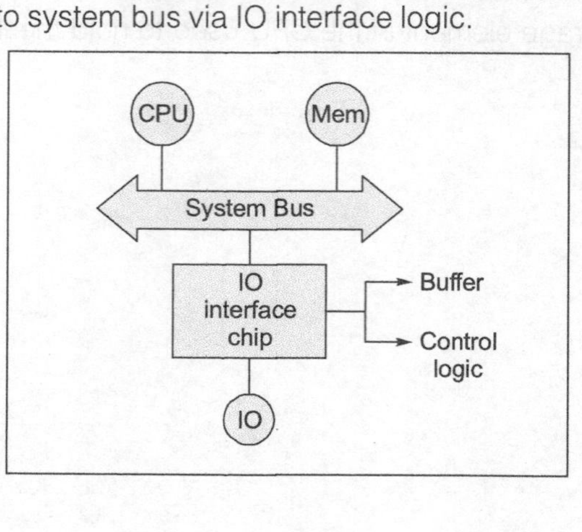

4.3 Explain in brief the following and differentiate between

I. Stack and Queue

II. Sort and Search

[12 marks : 2020]

Solution:

I.

Stack	Queue
1. It is an ordered list where the new item is added and existed element is deleted from only one end called as the top of the stack (TOS). All the insertions and deletions in a stack is done from the TOS. So, the last element added will be the first to be removed from the stack. ∴ Stack is called as last in first out (LIFO) type of list. 2. Stack follows LIFO mechanism. 3. In a stack, same end is used to insert and delete the elements. 4. Stack is using only one pointer to refer the TOS. 5. It performs two operations known as push and POP. 6. Stack implemation is easy. 7. Stack does not have variants.	1. It is a collection of similar type of elements. It contains. It contain two ends named as real and front. The addition of new elements takes place at one end called rear end. Deletion of he existing elements takes place at the other end called front end. ∴ Queue is called as first in first out (FIFO) type of list. 2. Queue follows FIFO mechanism. 3. In a Queue two different ends are used to insert and delete the elements. 4. Queue uses two pointers to refer the front and the rear ends of a queue. 5. It performs two operations known as enqueue and dequeue. 6. Queue implemation is tricky. 7. Queue has variants like circular queue, priority queue, etc.

II.

Search	Sort
1. Searching is method for finding a target value within a list. It sequentially checks each element of the list for the target value until a match is found (or) until all the elements have been searched.	1. Sorting is the method of arranging the element in particular order mostly either in ascending (or) descending order.
2. Popular types of searching are: λ linear search λ Binary search	2. Types of sorting techniques are: λ Bubble sort λ Insertion sort λ Selection sort λ Quick sort λ Merge sort λ Heap sort

4.4 A machine has 32 bit Architecture, with 1 word long instruction. It has 64 registers each of which is 32 bits long. It needs to support 48 instructions which have an immediate operand in addition to two register operands. If the immediate operand is an unsigned integer, find the maximum value of the immediate operand.

[10 marks : 2022]

Solution:

$$\text{Word size} = 32 \text{ bit}$$

$$\text{Instruction size} = 1\text{ W} = 32 \text{ bit}$$

Instruction format

32-bit

Opcode	Reg	Reg	Immediate
$\log_2 48$	$\log_2 64$	$\log_2 64$	14 bit
↓ 6 bit	↓ 6 bit	↓ 6 bit	

$$\text{Immediate field size} = 14 \text{ bit}$$

So unsigned data range is

$\{0 \text{ to } (2^{14} - 1)\}$

$\{0 \text{ to } 16383\}$

So, max value is 16383

5. Memory Organization

5.1 **In view of important parameters, give the comparison between RAM and ROM in tabular form.**

[4 marks : 2008]

Solution:

Comparison between RAM and ROM

Parameter	RAM	ROM
1. Data stroage	Temporary. Vanishes when power is switched off.	Permanent. Power failure does not affect data stored.
2. Data entry	Data can be entered fast. No special program-writers are required.	Special program-writers are required. So, data entry is normally slow. But this is being rectified.
3. Data read out	Data will be read-out at the same time it is entered.	Data read-out only after it is entered.
4. Repeatability of read out	Once the data is erased, it cannot be recovered.	Data is not erased; hence it can be read out any number of times.
5. Erasability	Easily erasable, very fast operation.	Not easily erasable, slower operation.
6. Packing density	Currently large.	Currently low.
7. Static/dynamic operations	Static and dynamic operation exist.	Only static ROMs exist.
8. Use of memory	Data entered in any computer is first stored in RAM. If the data is to be stored permanently it will be saved in the hard disk.	This memory stores data of a permanent nature such as operating-system commands, look-up tables, etc., which are not stored in hard disks.

5.2 **Bring out the concept of virtual memory and briefly point out its operation.**

[10 marks : 2017]

Solution:

Virtual Memory:

- Virtual memory is a memory management capability of an OS that uses hardware and software to allow a computer to compensate for physical memory shortages by temporarily transferring data from RAM to disk storage.
- Virtual address space is increased using active memory in RAM and inactive memory in hard disk drives to form contiguous addresses that hold both the application and its data.
- Virtual memory concept is implemented using:
  (i) Paging
  (ii) Segmentation

Paging:

- In this technique, main memory and secondary memory both are divided into a equal sized pages based on the page size.
- After the organization memory cells are grouped into frames in main memory and pages in secondary memory.

$$\text{\# Frames in main memory (MM)} = \frac{\text{MM size}}{\text{Page size}}$$

$$\text{\# Pages in secondary memory (SM)} = \frac{\text{MM size}}{\text{Page size}}$$

- Consider 8 kB secondary space and 4 kB main memory space organized into a 1 kB pages.

$$\text{\# Frames} = \frac{4\text{K}}{1\text{K}} \Rightarrow 4$$

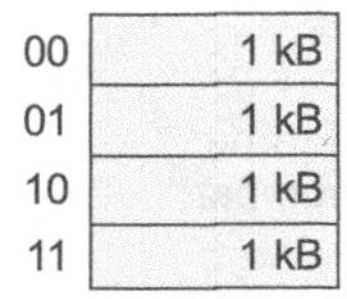

$$\text{\# Pages} = \frac{8\text{K}}{1\text{K}} \Rightarrow 8$$

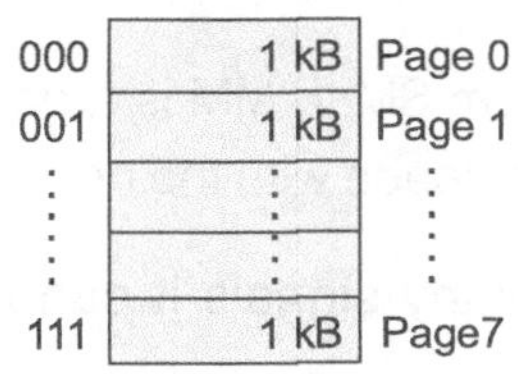

- After organization, address format is changed, i.e.;

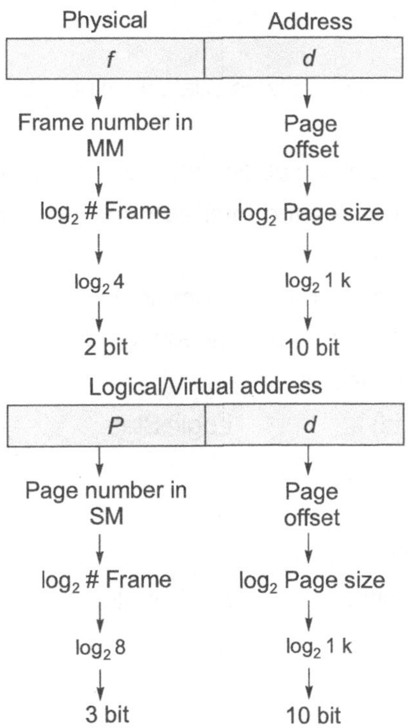

- MM contain less number of frames and SM contain more number of the pages.
- During the mapping process OS takes the responsibility to copy the data from SM to MM.
- OS maintains the page table in main memory to specify the MM information i.e. which frame contain which page.
- Structure of the page table is

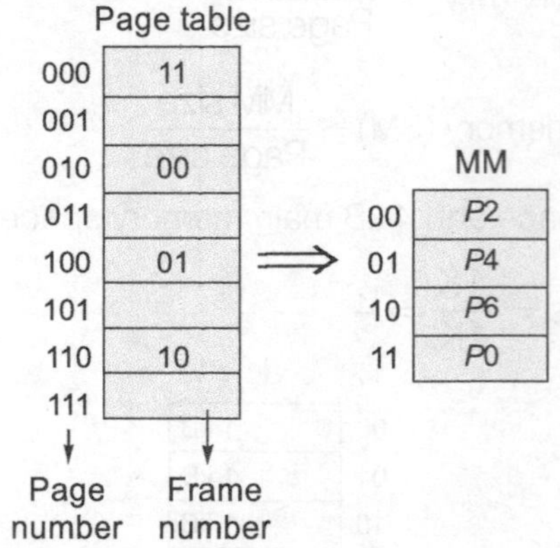

- During the program execution, CPU refers the page table in MM. When the required page is present in the MM the CPU directly access the data from MM and executes the program, otherwise generates the trap to OS then OS transfers the data from SM to MM and page table is updated correspondingly later signal to CPU to execute the code
- Like this data will be transferred from SM to MM is a pagewise. Therefore, we can execute the huge programs on a small amount of MM space with the help of **virtual memory**.

5.3 What is an RS-232 interface? How many signals it can handle? How many wires are sufficient for operation?

[6 marks : 2018]

Solution:

(ii)
- RS-232 defines the connection between data terminal equipment (DTE) and data circuit terminating equipment (DCE).
- DTE is an end user device, such as a computer that can be used to send data over a network.
- DCE is a device that provides the interface between the DTE and the network and is often a modem or terminal adapting.
- An RS-232 compatable interface was commonly used for computer serial communication ports which were originally interneded for connecting the computer to a modem.
- RS-232 signal line voltage levels

Signal voltage level (Volts)	Logic State
–3 to –25	1
+3 to +25	0

- RS-232 control line voltage levels

Control voltage levels (volts)	Logic state
–3 to –25	OFF
+3 to +25	ON

RS – 232 signal

(i) **TXD :** Transmitted data (Data transmitted from the DTE to the DCE)

(ii) **RXD :** Received Data (Data Transmitted from the DCE to the DTE)

(iii) **RTS :** Request to send (set to '0' by the DTE to prepare the DCE to receive data)

(iv) **CTS :** Clear to send (set to '0' by the DCE to acknowledge RTS and allow the DTE to transmit)

(v) **DTR :** Data terminal ready set to '0' by the DTE to indicate that it is ready to be connected.

(vi) **DSR :** Data set ready (set to '0' by the DCE to indicate an active connection)

(vii) **DCD :** Data carrier detect (set to '0' by the DCE when a connection has been established with a remote device.

(viii) **RI :** Ring indicator (set to '0' by the DCE when it detects a ring signal from the telephone line.

- There are four type of lines defined in the RS-232 specification:

(i) Data (ii) Control

(iii) Timing (iv) Ground

5.4 (i) Can we used semiconductor memory for secondary storage? Justify your answer.

(ii) What exactly is the role of Control Unit in CPU? How is it different from that of an Arithmetic Logic Unit?

(iii) How exactly is an array stored in main memory? Illustrate with the help of storage of a matrix. What will be the exact address of element *A*(3, 7) of an 8 × 9 array *A* assuming that location of the first element is *d*.

[5+6+9 = 20 marks : 2018]

Solution:

(i)
- Semiconductor memory is a digital electronic data storage device name as primary memory. They are different types of implementations using various technologies ie RAM & ROM.
- Most types of semiconductor memories have the property of random Access, which means that it takes the same amount of time to access any memory location. So data can be efficiently accessed in any random order.

  This "contrast" with data storage media such as hard disk and CD's which read and write data consecutively, therefore data can only be accessed in the same sequence it was written.
- Semiconductor memory access time is less than other types of data storage, Usually nano seconds. But hard disk is in the range of milliseconds.

  So, semiconductor memories are not used for secondary memory storage device. Secondary memory devices are magnetic and optical memories.
- Secondary memory is always non-volatile.
- Secondary memory is capable to store huge amount of data.
- Secondary memory is slow.
- Secondary memory is an external memory.
- Secondary memory devices are connected to the computer through cables.

(ii) Control Unit:
- Control unit is a brain of the CPU
- Control unit generates the sequence of control signals during the execution of an instruction.
- All the CPU actions are controlled by the control unit.
- Control unit is designed by the designer and fixed in the CPU

  (Therefore control unit, is a mandatary component in the CPU)
- Control unit is a mandatory component in the CPU.
- Control unit is implemented in two ways

  (i) Hard wired control unit

  (ii) μprogrammed control unit

- In hard wired CU design, control signal is represented as a SOP expression, implemented using AND & OR gates.
- In μprogrammed CU design, all the μprograms are programmed into a control memory permanently i.e. ROM.

Arithmetical Logical Unit (ALU):

- It is a computational unit, used to perform the arithmetic and logical operations.
- It contain different logic circuits to perform the various operations.
- All the logic circuits are controlled by the control unit during the instruction execution.
- ALU is under the control of a control unit.
- ALU operations are performed on a accumulator and temporary register and stores the result into accumulator.

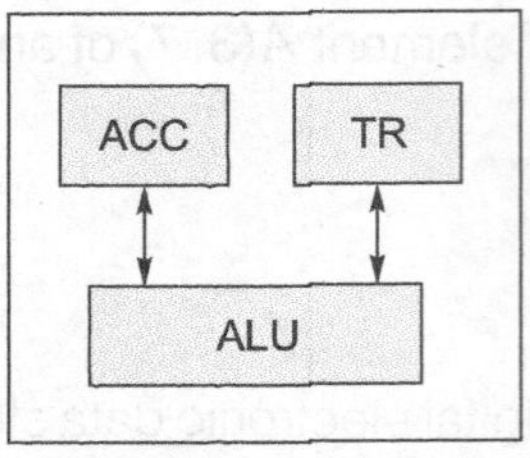

- This total unit operations are controlled by the control unit.

(iii)
- Array is an ordered set of homogeneous elements.
- Array elements are stored in the sequential memory cells in a row-wise sequence by default.

a[u] [u]; array "a" contain 4 rows and 4 columns.

$$\begin{bmatrix} a[0]\,[0] & a[0]\,[1] & a[0]\,[2] & a[0]\,[3] \\ a[1]\,[0] & a[1]\,[1] & a[1]\,[2] & a[1]\,[3] \\ a[1]\,[0] & a[2]\,[1] & a[2]\,[2] & a[2]\,[3] \\ a[3]\,[0] & a[3]\,[1] & a[3]\,[2] & a[3]\,[3] \end{bmatrix}$$

Memory storage:

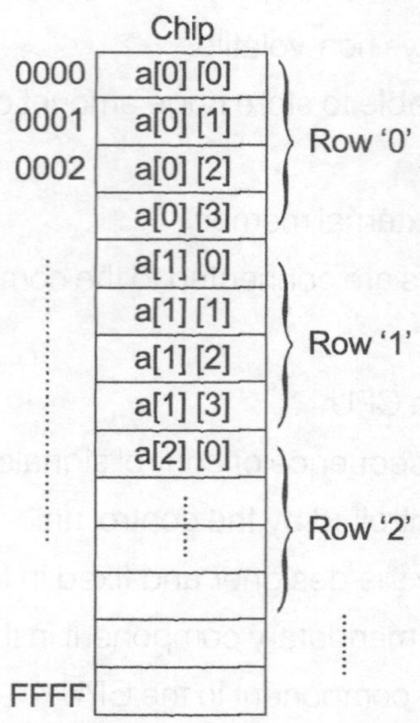

- Array elements are accessed by the memory in 2 ways.
  1. Row: Major indexing (Row-wise access)
  2. Column: Major indexing (Col-wise access)

- Array:

  a[8] [9]; contain 8 rows and 9 columns.

$$\begin{bmatrix} a[0][0] & a[0][1] & L & a[0][8] \\ a[1][0] & & & \\ a[2][0] & & & \\ a[3][0] & a[3][7] & & \\ a[4][0] & & & \\ M & & & \\ a[7][0] & & & a[7][8] \end{bmatrix}$$

Address is in decimal:

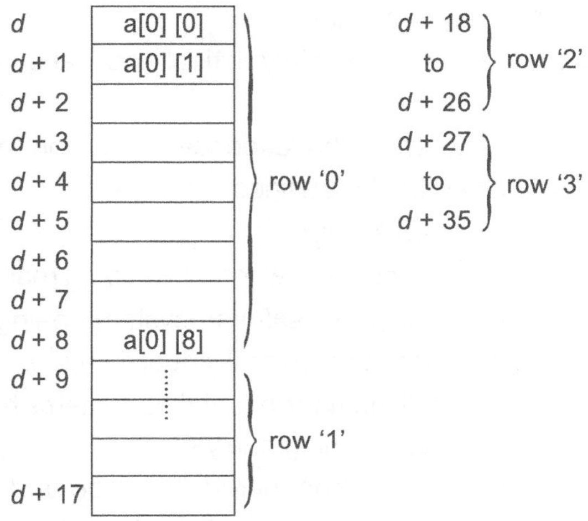

in this a[3] [7] Address is [d + 34]

If address given in hexa decimal.

Then,

row 0 : d to 15 cells
row 1 : 16 to 1E cells
row 2 : 1F to 27 cells
row 3 : 28 to 30 cells

So, a[3] [7] element address is "2F".

5.5 **(i) Is the use of virtual memory advisable for real time systems? Justify briefly your answer.**
(ii) Compare briefly and precisely the LINUX OS with Windows-NT OS.
(iii) For File management how does a DBMS help?

[7+5+8 = 20 marks : 2018]

Solution:

(i)
- Real time system is any information processing system which has to respond to externally generated input stimuli within a finite and specified period. Correctness depends not only on the logical result but also, the time it was delivered.
- System whose correctness depends on their temporal aspects as well as their functional aspects called as real-time systems.

Performance Measure:

- Timeliness on timing constrains (dead lines).
- Speed/average case performance are less significant.
- Key property of a real-time system is predictability on timing constraints.

For example:

- Real time monitoring systems.
- Signal processing systems (Radar)
- Online transaction systems.
- Embedded control systems (Automotives, Robots, Aircrafts, etc.)
- Time constraints in the real-time systems, two kinds
  (i) Hard time constraints
  (ii) Soft time constraints
- A hard real time system is one where the response time is specified as an absolute value. This time is normally dictated by environment.
  For example Safety critical applications.
  [A failure i.e. missing deadlines in a safety critical application can lead to loss of human life (or) severe economical damage].
- A soft real time system is one where the response time is normally specified as an average value. This time is normally dictated by the business or market.
  For example: Air line reservation system.
- Memory management is a major concern when developing real-time and embedded applications.
- Predictability issues have resulted in real-time systems being most of the times strictly static, Avoiding dynamic allocation/deallocation and virtual memory.
- Conventionally, the use of virtual memory in real-time systems has been avoided, the main reason being the difficulties it provides to timing analysis.
- However, as these systems are getting increasing large and complex, there is now a need to escape from this strictly static memory management.
- There is a trend towards systems where different functions are implemented by concurrent processes. Such systems need spatial separation between processes, which can be easily implemented via use of the MMU (Memory management unit) or commercial processors.
- Some systems have a limited amount of physical memory available. So far attempts to provide realtime address spaces have focused on the predictability of virtual to physical address translation and not implemented demand paging.
- virtual memory raises predictability issues at two levels.
  (i) Level of address translation (Duration of address translation is hard to predict)
  (ii) Level of paging activity (knowing) whether (or) not a reference to a virtual page will result in a pagefault is hard to predict).
- Demand paging is carefully avoided, all physical pages are voluntarily created in memory at process load time. (or) wired in memory, to avoid unpredictability due to page faults.
- Finally, both dynamic and virtual memory management are widely used concepts in desktop and server environments. Despite the benefits of these concepts, their use in the context of embedded real-time system has been limited so far.
- Existing virtual memory management systems have runtimes that depend on the size of the requested memory.
- For this reason embedded real time systems retrain from using virtual memory, because these mechanisms are viewed as a main obstacle to right estimates of worst-case execution times and to an efficient scheduling of processes with deadlines.
- In many cases, fixed (contiguous) memory regions are assigned to each process and there is a static memory management strategy for each process. This typically leads to a large overestimation of the memory resources which are really needed.

- Due to a page in and pageout times are unavoidable and significant in the virtual memory implementation. So this may leads to system failure. So most of the applications doesn't uses virtual memory demand paging implementation in real time systems.
- This is a research area to reduce page in and pageout times and cost-performance metrices are the challenges.
- In the hard real time systems dead lines are very critical and time constraints are very strict. So virtual memory concept is not fruitful in such applications.
- In the soft real time systems dead lines are nominal.
  For example: Banking systems, Airline reservation system, etc.
  In such application client server environments present. So virtual memory concept is used in such application to minimize hardware requirements (mean-space).

(ii)

Linux	Windows
1. Full access to the source code	1. No access to the source code
2. Licensing freedom	2. Licensing restrictions
3. Online peer support	3. Paid help desk support
4. Full H/w support	4. Partial H/w support
5. Command line	5. No command line
6. Centralized application installation	6. Noncentralized application installation
7. Flexibility	7. Regidity
8. Automated removable media	8. Nonautomated removable media
9. Multilayered run levels	9. Single layered run levels

(iii)
- File management system (FMS) is a data base management system that allows access to single file and table at a time.
- FMS accommodate flate files that have no relation to other files.
- In FMS, physical details required to access the file i.e. file name, file type, file size etc.
- DBMS allows access to multiple files (or) tables at a time.

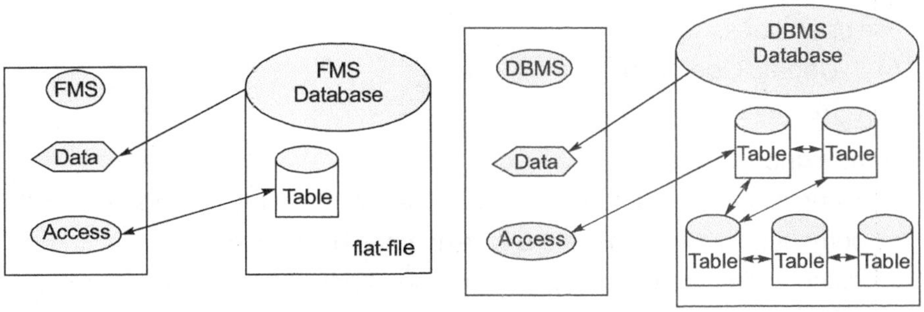

Relations between the files

- DBMS is a collection of interrelated data and a set of programs to access those data. Some of the well known DBMS are Microsoft Access, Microsoft SQL server, Oracle, SAP, etc.
- DBMS does not remove redundency completely, it controls. The amount of redundancy in the database because in database approach, a single repository of data is maintained that is defined once and then accessed by many users.
- Relative data and information is stored collectively in file formats.
- A file is a sequence of records stored in binary format.
- A disk drive is formatted into several blocks that can store records. File records are mapped onto those disk blocks.

- File organization defines how file records are mapped onto disk blocks.
- We have four types of file organization to organize file records.

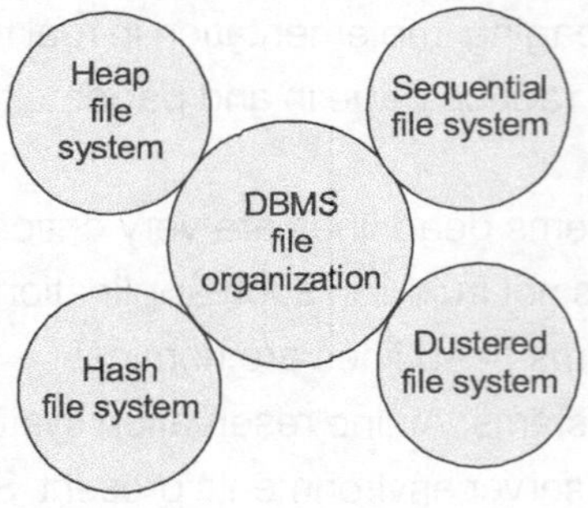

Heap file Organization:

- When a file is created using heap file organization, the operating system allocates memory area to that file without any further accounting details. File records can be placed anywhere in that memory area. It is the responsibility of the software to manage the records.
- Heap file does not support any ordering, sequencing (or) indexing on its own.

Sequential file organization:

- Every file record contains a data field (attribute) to uniquely identify that record.
- In this organization records are placed in the file in some, sequential order based on the unique key field (or) search key.

Hash file organization:

- Hash file organization uses hash function computation on some fields of the records.
- The output of the hash function determines the location of disk block where the records are to be placed.

Clustered file organization:

- This organization is not considered good for large databases. In this mechanism, related records from one or more relations are kept in the same disk block. i.e. the ordering of records is not based on primary key or search key.

  File operations:

  (i) Updata Operations

  (ii) Retrieval Operations

DBMS-Indexing:

- Indexing is a data structure technique to efficiently retrieve records from the database files based on some attributes on which the indexing has been done.
- Indexing is defined based on its indexing attributes. Different types of indexing

  (i) Primary Index

  (ii) Secondary Index

  (iii) Clustering Index

DBMS-Hashing:

- For a huge database structure, it can be almost next to impossible to search all the index values through all its level and then reach the destination data.
- Hashing is an effective technique to calculate the direct location of a data record on the disk without using index structure.
- Hashing uses hash functions with search keys as parameters to generate the address of a data record.

5.6 In virtual memory based system, suppose we have an average of one page fault after every 10,000,000 instructions. A normal instruction takes 4 ns (4 nano seconds), and a page fault causes the instructions to take an additional 10 milli seconds. What is the average instruction time, taking page faults into account?

[10 marks : 2019]

Solution:

For given virtual memory system,

Instruction execution time without fault = 4 nsec

$$\text{Page fault rate for instruction} = \frac{1}{10{,}000{,}000}$$

Extra delay time due fault of instruction = 10 msecond

= 10^{-2} second

Average instruction time,

$$4 \times 10^{-9} + \frac{1}{10{,}000{,}000} \times 10^{-2} \text{ sec} = 5 \text{ nsec}$$

5.7 Execution of a sequence of instructions of a program involves 200 instruction fetch operations, 100 memory operand read operations, and 80 memory operand write operations. Find the average memory access time in executing this sequence of instructions if the memory access time is 2 ns for a read operation with a hit in cache, 5 ns for a read operation with a miss in cache, 3 ns for a write operation with a hit in cache and 10 ns for a write operation with a miss in cache. The cache hit ratio is 0.9. Consider the time taken for fetch operation to be equal to that of read operation.

[10 marks : 2020]

Solution:

#instruction fetch = 200

#memory read operation = 100

#memory write operation = 80

HC = 0.9

$$T_{avg\ r} = H_r T_c + (1 - H_r)(T_M + T_L) \text{ (given in question)}$$

$$= (0.9 * 2 \text{ ns}) + (1 - 0.9)\, 5 \text{ ns} = 2.3 \text{ nsec}$$

$$T_{avg\ w} = H_w T_C + (1 - H_w)(T_M + T_C)$$

$$= (0.9 * 3 \text{ ns}) + (1 - 0.9)\, 10 \text{ ns}$$

$$= 3.7 \text{ ns}$$

$$T_{avg} = \frac{(f_r * T_{avg\,r}) + (f_\omega * T_{avg\,\omega})}{(f_r + f_\omega)}$$

$$= \frac{300 \times 2.3 + 80 \times 3.7}{380} = 2.594 \text{ ns}$$

5.8 Distinguish between

(i) High level language and low level language.

(ii) Machine cycle and instruction cycle.

(iii) Memory mapped Input/Output and Input/Output mapped Input/Output.

[4 × 3 = 12 marks : 2022]

Solution:

(i) High level language and low level language,

The difference between high level and low level languages:

S. No.	High Level language	Low Level language
1.	It is programmer friendly language.	It is machine friendly language.
2.	High level language is less memory efficient.	Low level language is high memory efficient.
3.	It is easy to understand.	It is tough to understand.
4.	It is simple to debug.	It is complex to debug.
5.	It is simple to maintain	It is complex to maintain comparatively.
6.	It is portable.	It is non-portable.
7.	It can run on any platform.	It is machine dependent.
8.	Compiler or interpreter for translation.	It need's assembler for translation.
9.	It is used widely for programming.	It is not commonly used now a days in programming.

(ii) Machine cycle and instruction cycle

The difference between machine cycle and instruction cycle is as follows:

Machine Cycle	Instruction Cycle
The steps that get performed by the processor getting employed in a device and all the instruction's that get implemented.	A process by which a computer takes an instruction given by a program then understands it and executes if from memory.
Processes	
Fetch, decode, execute and store	Fetch, decode, execute and run
Component's	
Memory unit and central processing unit.	Arithmetic logic unit, registers, data and memory.
Value	
The steps required by CPU to fetch and execute an instruction is called an instruction cycle.	The time required by the microprocessor to complete the operation of accessing memory or I/O device is called machine cycle.

(iii) Difference between memory mapped I/O and I/O mapped I/O.

The microprocessor cannot do anything by itself therefore, it needs to be linked with memory, extra peripherals or I/O device. This linking is called interfacing.

The interfacing of the I/O device in 8085 can don in two mass:

1. Memory mapped I/O interfacing : In this kind of interfacing, we assign a memory address that can be used in the same manner as we use a normal memory location.
2. I/O mapped I/O interfacing : A kind of interfacing in which we assign an 8-bit address value of the input/output devices which can be access using IN and OUT interaction is called I/O mapped IO interfacing.

Difference between memory-mapped I/O interfacing and I/O mapped I/O interfacing.

	Features	Memory mapped I/O	I/O mapped I/O
1.	Addressing	I/O devices are accessed like any other memory location.	They cannot be accessed like any other memory location.
2.	Address size	16-bit	8-bit
3.	Instruction used	The instruction used are LDA and STA etc.	The instruction used are IN and OUT.
4.	Cycles	Cycles involved during operation are memory read, memory write.	Cycles involved during operation are IO read and IO write in the case of IO mapped IO.
5.	Registers communicating	Any register can communicate with I/O device.	Only accumulator can communicate with IO devices.
6.	Space involved	2^{16} I/O parts are possible to be used for interfacing.	any 2^8 i.e. 256 IO parts are available.
7.	$IO/\overline{M}$ signal	During writing for read cycle ($IO/\overline{M} = 0$).	During writing or read cycle ($IO/\overline{M} = 1$).
8.	Control signal	No separate control signal required since we have unified memory space.	Special control signal's are used.

5.9 What is Random Access Memory? Explain Static Random Access Memory and Dynamic Random Access Memory.

[10 marks : 2023]

Solution:

RAM (Random Access Memory) :

- The information stored in the RAM is lost when the power supply to the PC (or) laptop is switched off.
- RAM is a volatile memory. It is generally known as main memory.
- RAM is used to read and write the data into it which is accessed by the CPU randomly.
- RAM is used to store the data that is currently processed by the CPU.
- Most of the programs and data that are modifiable are stored in the RAM.
- Integrated RAM chips are available in two forms :
  (a) SRAM (Static RAM);
  (b) DRAM (Dynamic RAM)

SRAM	DRAM
1. In this memory, data is stored using the state of a six transistor memory cell.	1. In this RAM, each bit of a data is stored in a separate capacitor within a specific integrated circuit.
2. SRAM has less access time and faster than DRAM.	2. DRAM has higher access time and slower than SRAM.
3. It is costlier than DRAM.	3. DRAM cost is less compared to SRAM.
4. It requires constant power supply. So, it consumes more power.	4. Less power consumption because information is stored in the capacitor.
5. SRAM size is less than DRAM size.	5. DRAM size is large so more storage capacity present.

6. Basic Concepts of Operating Systems

6.1 What is a page-fault; how is it managed? [10 marks : 2017]

Solution:

Page fault:

- When the CPU executing the program, initially refers the page table in main memory. If the required page is not present in the memory then operation becomes miss. This situation is called as page fault.
- When page fault occurs then CPU generate the TRAP signal to OS. So, OS takes the responsibility to copy the data from SM to MM. After the copying updates the page table later signals to CPU to execute the program.
- During the mapping if no space in MM (MM is full) then page replacement policies are used to replace the MM data.
- Different page replacement policies used in the OS to replace data is,
  1. FIFO
  2. LRU
  3. Optimal
- In FIFO, replace the page which is having longest time stamp.
- In LRU, replace the page which is present in the MM longest time without reference.
- In optimal replacement, replace the page which is not required in the future or required after a long time. It is not in the use.

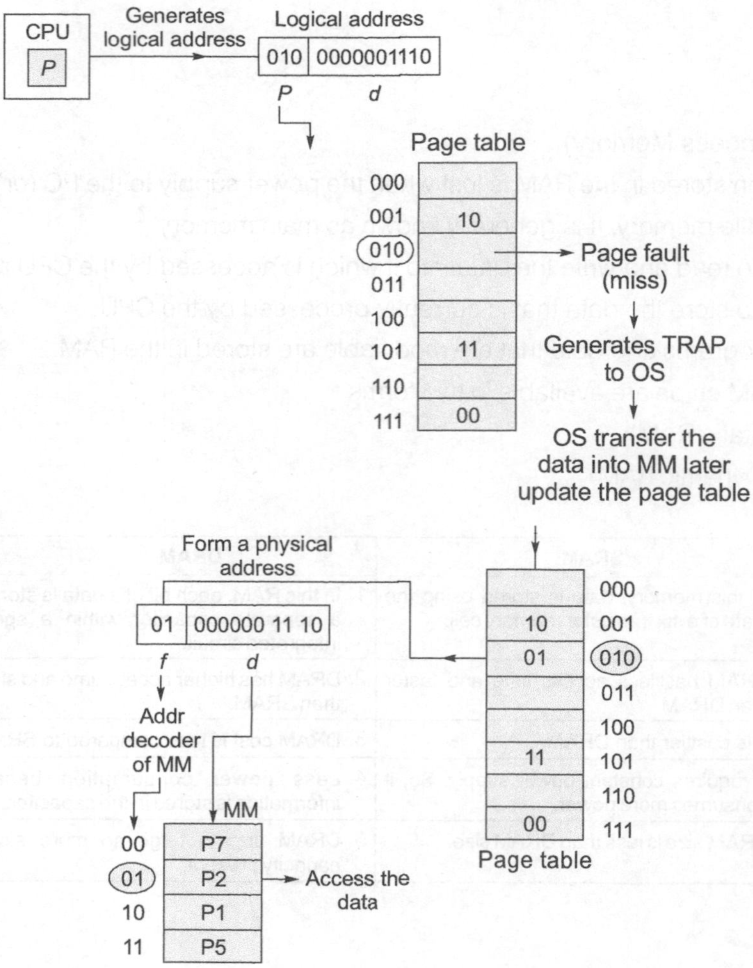

7. Networking Fundamentals

7.1 **What is a computer network? Distinguish between LAN and WAN.**

[10 marks : 2017]

Solution:

A computer network is a group of computer systems and other computing hardware devices that are linked together through communication channels to facilitate communication and resource sharing among a wide range of users.

LAN: Local area network is a computer network which is limited to a small office, single building, multiple buildings inside a campus. LAN is a private network owned and maintained by a single organization.

WAN: A wide area network spans over multiple geographic locations which is composed of multiple LAN's. It is impossible for a small to medium organisation to pull network cables between their two offices in two different countries located thousands of kilometres away. Network service provider provide the connectivity solutions for wide area networks.

LAN:

1. It is a private computer network that connects computer in a small physical areas.
2. LAN has higher bandwidth rates. Current LAN's runs on bandwidth of 100 Mbps, 1 Gbps or 10 Gbps.
3. LAN bandwidth rates are almost constant.
4. LAN use ethernet as the LAN standard fast ethernet 100 Mbps or Gigabit ethernet 1/10 Gbps.

WAN

1. WAN is a type of computer network to connect offices which are located in different geographical locations.
2. WAN has lower bandwidth rates compared with LAN. Current WAN runs on bandwidths of 4 Mbps, 8 Mbps, 20 Mbps, 50 Mbps, etc.
3. WAN connectivity solutions are dependent on internet service providers (ISP's)
4. WAN uses technologies like VPN (Vertical Private Network) over internet, frame relay, leased lines as WAN connectivity solutions.

7.2 **Explain in brief what is network topology by illustrating four major topology schemes.**

[10 marks : 2017]

Solution:

Computer networks can be broken down historically into topologies, which is a technique of connecting computers.

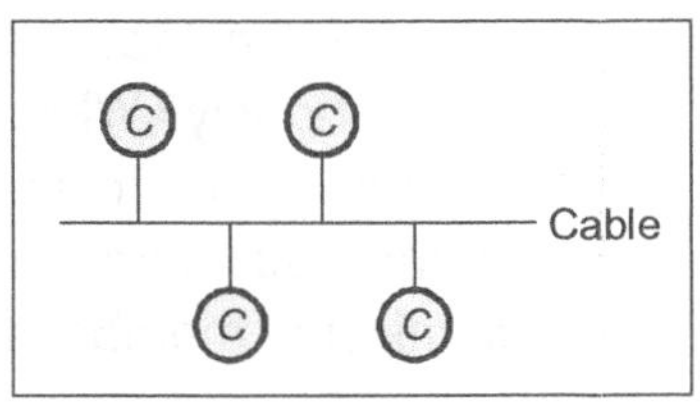

Bus Topology:

- In this design, single cable connects all computers and the information intended for the last node on the network must run through each connected computer.
- If a cable broken, all computers connected down the line cannot reach the network.
- The benefit of a bus topology is a minimal use of cabling.

Star Topology:

- In this design of a network, a central node extends a cable to each computer on the network.
- On a star network computers are connected independently to the center of the network.
- If a cable is broken the other computer can operate without problems.
- Star topology requires alot of cabling.

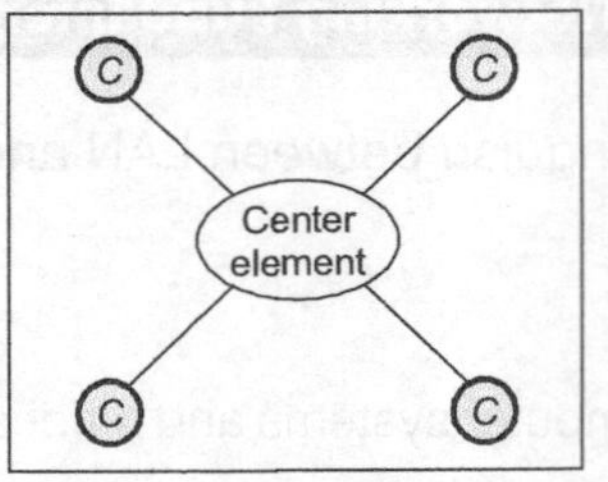

Ring Topology:

- In this design, computers are connected via a single cable, but the end nodes also are connected to each other.
- In this design, the signal circulates through the network until it finds the intended recipient.
- If a network node is not configured properly or it is down temporarily for another reason, the signal will make a number of attempts to find its destination.

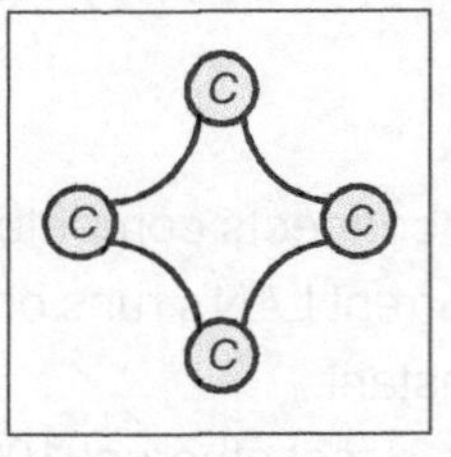

Mesh Topology:

- In this design there exists direct and dedicated link between each pair of hosts.
- It is used when high performance and high redundancy is required for a small number of hosts.
- If a network contain 'n' hosts, each host must have $(n-1)$ physical interfaces.

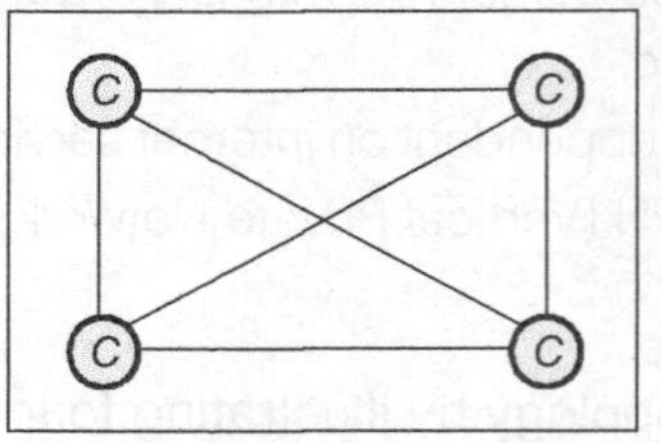

7.3 A company wishes to transmit numerous 10,000 bytes files between two computer systems connected through a computer network. Each byte consists of 8 data bits, no parity bits used in this application. Error is not a factor for transmission. What is the overhead percentage if a single file is sent asynchronously? What is the overhead percentage if the file is sent synchronously using blocks with 1000 data bytes in each block and each block is accompanied by 15 special bytes? Assume in asynchronous communication 1 start bit and 2 stop bits sent with each data byte. Comment, which transmission method is most efficient.

[10 marks : 2019]

Solution:

Given:

Number of files to be transferred = 10000 bytes

Each byte = 8 bits

For asynchronous file transfer:

1 + start bit + 2 stop bit extra for each data byte

$\therefore$ Total overhead for asynchronous = (1 + 2) × (10000) = 30000

$\therefore$ Total size in bits for asynchronous transfer including overhead

$$= 80 \text{ k} + 30 \text{ k}$$
$$= 110 \text{ kbits}$$

$$\text{Overhead percentage} = \frac{30}{110} \times 100 = 27.27\%$$

$$\text{Transfer efficiency} = \frac{80}{110} \times 100 = 72.72\%$$

In case of synchronous file transfer,
15 special bytes + 1000 data bytes

$$= (15 \times 8) \text{ special bits} + 1000 \times 8 \text{ data bits}$$
$$= 120 \text{ spec} + 8000 \text{ data}$$
$$\text{total size} = 8120 \text{ bits}$$

$$\text{Overhead percentage} = \frac{120}{8120} \times 100 = 1.478\%$$

$$\text{Transfer efficiency} = \frac{8000}{8120} \times 100 = 98.52\%$$

7.4 Name the layers of Open Systems Interconnection (OSI) model created by the International Organization for Standardization for different computer systems to communicate with each other using standard protocols. Mention the important functions of each of these layers in brief.
[20 marks : 2020]

Solution:

OSI reference model describes how information from a software application in one computer moves through a physical medium to the software application in another computer.

OSI consists of 7 layers and each layer performs a particular network function, named as

Responsibility of the host {
- Application layer — layers-7
- Presentation layer — layers-6
- Session layer
- Transport layer

Responsibility of the network {
- Network layer
- Datalink layer
- Physical layer — layers-1

OSI model is divided into two layers : upper layers and lower layers.

Upper layers of the OSI model mainly deals with the application related issues and lower layers of the OSI model delays with the data transport issues.

Functions of physical layer:

- line configuration
- data transmission
- Topology
- Signals

Functions of a data link layer:

- Framing
- Physical addressing
- Flow control
- Error control
- Access control

Functions of a Network layer :

- Internetworking
- Addressing
- Routing
- Packetizing

Functions of a transport layer :

Two protocols are used in this layer

1. Transmission control protocol (TCP)
2. User datagram protocol (UDP)
   - Service point addressing
   - Segmentation and reassembly
   - Connection control
   - Flow control
   - Error control

Functions of a session layer :

- Dialog control
- Synchronization

Functions of presentation layer :

- Translation
- Encryption
- Compression

Functions of application layer :

- File transfer, across and management
- Mail services
- Directly services

7.5 What are the functions of Data link layer and Network layer according to ISO-OSI-7 layer Network Architecture.

[12 marks : 2022]

Solution:

- The ISO-OSI is a 7 layer architecture for implementing the network system.
- The 7 layer are application, presentation, session, transport, network, data link, physical.

The functions of data link layer and network layer are as follows:

1. **Data link layer:**
   - ensures **node to node** delivery of messages
   - **error** correction
   - division of data **into frames** on sending side and recombination of frames on receiving side.
   - **access control** through MAC ID identification of device.
   - **Physical addressing** after creating frames, data link layer adds physical address to header.
2. **Network layer:**
   - **Routing**: determines the route for data from sender to receiver.
   - Logical addressing : in order to identify various devices on the networks.
   - Ensures transmission of data across networks.
   - Sender and receivers IP address added to header by network layer.

7.6 If the signal speed in the cable is 2×10^5 km/sec find the maximum length of cable required to transmit the data at a rate of 600 Mbps in an Ethernet LAN with frame size of 12000 bits.

[8 marks : 2022]

Solution:

For ethernet cable if *V* is velocity of signal, *L* is length of frame and BW is bandwidth and *d* is length of cable. They are related as follows :

$$d \leq \frac{L \times V}{2 \times BW}$$

So,
$$d \leq \frac{12000 \times 2 \times 10^5 \text{ km/s}}{2 \times 600 \times 10^6} \leq 2 \text{ km}$$

So maximum cable length can be 2 km.

8. Programming Languages Methodology

8.1 Give the output of the following program:

```
int main ( )
{
char str [ ] = "India\0\BIX\0";
print f ("% s\n", str);
return 0;
}
```

[12 marks : 2017]

Solution:

India 0 BIX
{It returns the string constant}

8.2 (i) Write down algorithm in Pseudocode for sorting an array in descending order. Specify the name of the algorithm you have used.

(ii) Write a program segment in any higher level language for linear search problem. (Specify which language you are using).

[6 + 6 = 12 marks : 2018]

Solution:

(i) **Bubble sort:** In bubble sort, in every step it compare two adjacent elements and if the lower value is on the left side of the higher value. Then bubble sort swaps them. Continue the same process until the lower value move to the end of an array.

- Repeat the same logic to the next element and so on.
- After one iteration the lowest value is located at the end of the array.
- So, $(n-1)$ iterations are required to sort the array.

(3 2 8 7 6) initial array

Pass-1: 3 > 2; True; No swap

2 > 8; False; Swap

(3 8 2 7 6)

2 > 7; False; Swap

(3 8 7 2 6)

2 > 6; False; Swap

(3 8 7 6 (2))

At the end of pass 1, smallest element adjusted at the end of array.

Pass-2: 3 > 8; False; Swap
8 3 7 6 (2)
3 > 7; False; Swap
8 7 3 6 (2)
3 > 6; False; Swap
8 7 6 (3) (2)

Pass-3: 8 > 7; True; No swap
7 > 6; True; No swap
When # swaps "0" in the pass then terminate the process.

∴ Array is in sorted order.

Pseudo code:

```
Function bubble sort (array a)
for i in 1 → a.length – 1 do
for j in 1 → (a.length – i) – 1 do
if a[j] < a[j + 1]
swap (a[j], a[j + 1]);
```

(ii) Linear search (sequential search)
(c-code)

```
    # include <conio.h>
    # include < stdio.h>
void Main ()
    {
        int a[50], n, i, item;
        printf("enter number of elements \n");
        scanf("%d", &n);
        printf(" enter %d elements", n);
        for (i = 0; i < n; i++)
        scanf("%d", &a[i]);
        printf("enter item to search:\n");
        scanf("%d", & item);
        for (i = 0; i < (n – 1); i++)
        {
            If (item = = a[i])
            {
            printf("item found at location %d :\n", i + 1);
            break;
            }
    If (i > n – 1)
        printf(" item not in array:\n");
}
    getch();
}
```

8.3 What computation is performed by the given flowchart? Perform dry run and write down problem statement.

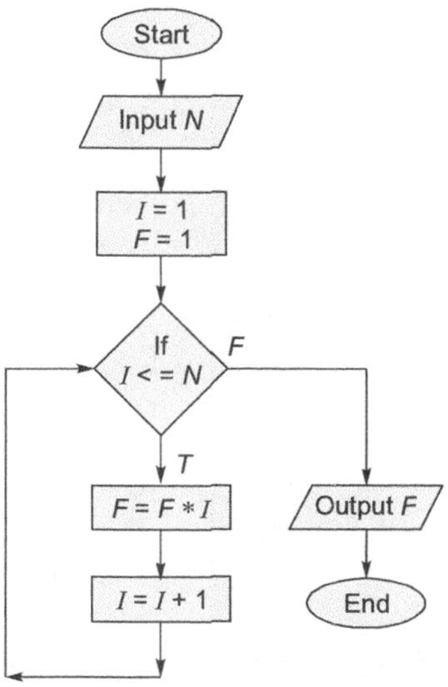

[4 marks : 2019]

Solution:

Given flow chart corresponds to computation of factorial of any given number N.

Step-1 : Initialize input variable N, I and F given values 1 to I and F both.

Step-2 : Initialize loop by checking whether I value is less than equal to N. If yes go to step-3 else to step-6.

Step-3 : Perform multiplying operation $F = F \times I$.

Step-4 : Perform increment in value of I by 1.

Step-5 : Go back to loop at step-2.

Step-6 : Output value

Assuming a value $N = 3$

Step-1 : $I = 1, F = 1$

Step-2 : $(1 < = 3)$ True

Step-3 : $F = 1 \times 1 = 1$

Step-4 : $I = 1 + 1 = 1$

Step-2 : $(2 < = 3)$ True

Step-3 : $F = 2 \times 1 = 2$

Step-4 : $I = 2 + 1 = 3$

Step-2 : $(3 < = 3) =$ True

Step-3 : $F = 2 \times 3 = 6$

Step-4 : $I = 3 + 1 = 4$

Step-2 : $(4 < = 3) =$ False

Step-6 : Output : 6

8.4 Write a program in any programming language to find highest common factor (HCF) of two positive integer numbers.

[10 marks : 2019]

Solution:

C program to find highest common factor (HCF) of two positive integers.

```
# include <stdio.h>
        int main {
        int a, b, x, y, t, hcf ;
        printf ("Enter two integer\n");
        scanf ("%d %d", & x, &y);
        a = x;
        b = y;
        while (b! = 0) {
        t = b;
        b = a % b;
        a = t;
}
        hcf = a;
print f ("Highest common factor of %d and %d is %d \n", x, y, hcf);
return 0;
}
```

8.5 **Write a program in any programming language of sum of the following series up to *N* terms:**

$$\lfloor\underline{1} + \lfloor\underline{2} + \lfloor\underline{3} + \lfloor\underline{4} + \lfloor\underline{5} + ... + \lfloor\underline{N}$$

where $\lfloor\underline{N}$ = factorial of *N*.

[10 marks : 2019]

Solution:

Program to find sum of series of $\lfloor\underline{1} + \lfloor\underline{2} .... \lfloor\underline{N}$

```
# include <stdio.h>
int main ( )
{
int num, i, j, fact, sum = 0; //variables defining
printf ("enter the last number of series:\n");
scanf ("%d", & num); // last number of series
for (i = 1; i< = num, i++) // loop for finding factorial and sum
{
    fact = 1 ;
    if (i! = num)
    printf ("%d! + ", i);
    else
    printf ("%d! = ", i);
for (j = 1, j < = i, j++)
    fact = fact * j;
    sum = sum + fact ;
}
    print f ("%d", sum);
    return 0;
}
```

8.6 **Using an iterative method, write *C* program segment to generate first $n(n \geq 8)$ Fibonacci numbers.**
[6 marks : 2020]

Solution:

In the fibonacci series, except for the first two terms of the sequence every other is the sum of the previous two.

Shown below terms in the series with respect to (n = 8) is

$$\underset{\text{First}}{\underset{\downarrow}{0}} \quad \underset{\text{Second}}{\underset{\downarrow}{1}} \quad 1 \quad 2 \quad 3 \quad 5 \quad 8 \quad 13$$

Code:

```
#include <stdio.h>
int main ()
{
int n, first = 0, second = 1, next, i ;
printf ("enter the number of terms\n") ;
scanf ("%d", %n);
printf ("first %d terms of Fibonacci series are :\n", n);
for (i = 0; i < n; i++)
{
if (i < = 1)
next = i;
else
{
next = first + second;
first = second;
second = next;
}
printf ("%d\n", next);
}
return 0;
}
```

8.7 **Write a pseudo code or in any standard programming language for interchanging the values of two variables :**

I. Using a third variable.

II. Not using any extra variable.

[8 marks : 2020]

Solution:

Swap two numbers with third variable

Assign variable-1 value to a temp variable

Assign variable-2 value to variable-1

Assign temp variable value to variable-2

Code :

step - 1 : Start
step - 2 : enter x, y, temp
step - 3 ; temp = x
step - 4 : $x = y$
step - 5 : y = temp
step - 6 : print x, y
step - 7 : stop

Swap two numbers without third variable

Code :

step - 1 : Start
step - 2 : Enter x, y
step - 3 : print x, y
step - 4 : $x = x + y$
step - 5 : $y = x - y$
step - 6 : $x = x - y$
step - 7 : print x, y
step - 8 : stop

8.8 Derive the expression for the number of comparisons required in the worst case for sorting an array of '*n*' elements using Bubble sort. Calculate it for an array of 100 elements. When will such maximum number comparisons occur in Bubble sort ?

[10 marks : 2020]

Solution:

Bubble sort is a simple algorithm which is used to sort a given set of '*n*' elements provided in form of an array with '*n*' number of elements.

Bubble sort compares all the elements one by one and sort them based on their values.

```
#include <stdio.h>
void bubblesort (int arr[ ], int n]
{
int i, j, temp;
for (i = 0; i < n; i ++)
{
for (j = 0; j < n – (i – 1); j ++)
{
if (arr [ j ] > arr [j + 1];
{
        temp = arr [ j ];
    arr [ j ] = arr [j + 1];
      arr [j + 1] = temp;
                    }
                    }
}
printf ("sorted array: ")
```

```
for (i = 0; i < n; i++)
{
      printf ("sorted array:");
         for (i = 0; i < n; i++)
                              {
         printf ("%d", arr[i]) ;
                              }
}
int main ()
{
int arr [100], i, n, step, temp;
printf ("enter the number of elements to be sorted:");
scanf ("%d", % n);
for (i = 0; i < n; i++)
{
printf ("enter element no %d,", i + 1);
scan f("%d", & arr [i]);
}
bubble sort (arr, n);
return 0;
}
```

In bubble sort, (n – 1) comparisons will be done in the 1st pass, (n – 2) in 2nd pass, (n – 3) in 3rd pass and so on.

So the total # of comparisons will be

(n – 1) + (n – 2) + (n – 3) + ... + 3 + 2 + 1

$$\text{Sum} = \frac{n(n-1)}{2}$$

i.e., $O(n^2)$

For the arr[100] elements

$$\text{Sum} = \frac{100(100-1)}{2} = 4950 \text{ comparisons required in the worst case.}$$

8.9 Write a C-program to check whether a number is a perfect number or not.

[12 marks : 2021]

Solution:

Perfect number means the sum of factors of given number is equal to given number.

For example : 4
factors 1, 2
(1 + 2) = 3
(3 == 4)
False. So, 4 is not a perfect number.

For example : 6
factors 1, 2, 3
(1 + 2 + 3) = 6
(6 == 6)
True. So, 6 is a perfect number.

```
#include <stdio.h>
#include <conio.h>
main()
```

```
{
    int i, n, sum=0;
    clrscr();
    printf("Enter a number");
    scanf("/d", &n);
    i=1;
    while (i < n)
    {
    If (n%i==0)
    Sum=Sum + i;
    i++;
}
if (sum==n)
    printf("In %d is a perfect number");
else
printf("In%d is not a perfect number");
}
```

8.10 (i) What are trigraph characters? How are they useful? Translate the following trigraph sequences into their equivalent symbols :

(I) ?? = (II) ??) (III) ??< (IV) ??/ (V) ??– (VI) ??!

(ii) From the keywords and identifiers given below, segregate the keywords and identifiers in two separate groups :

(I) break (II) ifloat (III) unsigned (IV) size (V) vshort (VI) goto (VII) sizeof (VIII) reverse (IX) enum (X) register (XI) iff (XII) continue

[6+6 = 12 marks : 2021]

Solution:

(i) Trigraph character means use the combination of 3 charactes to produce a single character. A trigraph is a sequence of three characters, the first two of which are question marks.

The symbols [,], {, }, ^, \, I, ~, # are frequently used in C-program. But in the late 1980s, there were code sets in use where the ASCII character positions for these were used for national language variant charactes.

This means that it was hard to write C-code on machines that used these sets. So, trigraph symbols are used to solve this problem.

??= → #

??) →]

??< → {

??/ → \

??– → ~

??! → I

(ii) • Keywords are pre-defined reserved words used in programming that have special meanings to the compiler.

• Identifier is a user defined word.

Keyword :	Identifier :
Break	ifloat
Unsigned	size

goto vshort
size of reverse
enum iff
register
continue

8.11 **Using any high level language, write a computer program to compute the area enclosed by a curve $f(x) = x^2 + 1$, between $x = A$ and $x = B$, by using trapezoidal approximation.**

[20 marks : 2021]

Solution:

The basic working principle of the trapezoidal method C program is that, this method splits the area under the curve into a number of trapezims. Then, area of each trapezium is calculated and all the individual areas are summed up to give the total area under the curve which is the value of difinite integral.

```
#include <stdio.h>
float   start_point,
        end_point,
        total_area;
int  numtraps;
main()
{
    void input (void);
    float find_area (float a, float b, int n);
    printf("Area under a curve");
    input ();
    total_area = find_area (start_point, end_point, numtraps);
    printf("Total area = %f", total_area),
    }
    void input(void)
    {
    printf("||n Enter lower limit:");
    scanf(:%f", & start_point);
    printf("Enter upper limit :");
    scanf("%f", &end_point);
    printf("Enter number of trapezoids:")
    scanf("%d", &numtraps);
    }
    float find_area (float a, float b, int n)
    {
    float base, lower, h1, h2;
    float function_x(float x);
    float trap_area(float h1, float h2, float base);
base = (b – 1)/n;
    lower = a;
    for (lower = a; lower <=(b – base); lower = lower + base)
    {
    h1 = function_x(lower);
    h1 = function_x(lower + base);
```

```
total_area += trap_area(h1, h2, base);
}
return (total_area);
float trap_area(float height_1, float height_2, float base)
{
float area;
area = 0.5*(height_1 + height_2)*base;
return(area):
}
float function_x(float x)
{
    return(x * x + 1);
}
```

8.12 **Write a program in any programming language to compute the money accumulated in a cumulative fixed deposit of a bank for the principal *P* after *n* years. The rate of interest is *r* percent per year compounded annually. The program should read the values of *P*, *r* and *n*. It should display the accumulated money as *F*.**

[12 marks : 2022]

Solution:

The bank compounds the money anually, i.e., the fixed deposit principal amount, i.e., P will be compounded at a interest rate *r* for *n* years to accumulate the amount *F*.

Hence, the formula used, $$F = P\left(1 + \frac{r}{100}\right)^n$$

The C-program to calculate *F* is given below :

```
# include <stdio.h>
# include <math.h> // for using pow, i.e., power function
void main ( )
{
float P, r, n;
float F,
    printf {"\n Enter the principal amount");
    scanf("%f", &P)
    printf("\n Enter the interest rate");
    scanf("%f", %r);
    printf("\n Eter the number of years for which the amount will be compounded");
    scanf("%f", &n);
    F = P*C pow((1 + r/100), n))
    printf ("\n The accumulated moey is % f", F);
}
```

8.13 **Write a program in C to find whether the given number is even or odd and if it is odd, find whether it is prime or not.**

[10 marks : 2023]

Solution:

```
# include <stdio.h>
main ( )
{
    int n, i, count=0;
    printf("Enter any number");
    scanf("%d",&n);
    if (n%2==0)
        {
            printf("even number");
        }
        else
        {
            printf("odd number");
        for (i=1; i<=n; i++)
        {
            if (n%i==0)
            {
            count ++;
            }
        }
        if (count==2)
        printf("Prime Number");
    else
    printf("Not a Prime Number");
}
```

■■■■

6 Basic Electronics Engineering

Revised Syllabus of ESE: *Basics of Semiconductor diodes and transistors and characteristics, Junction and field effect transistors (BJT, FET and MOSFETS), different types of transistor amplifiers, equivalent circuits and frequency response; oscillators and other circuits, feedback amplifiers.*

1. Semiconductor Physics

1.1 The Fermi function for hole is expressed as

(a) $\dfrac{1}{1+\exp\left(\dfrac{E-E_F}{kT}\right)}$

(b) $\dfrac{1}{1+\exp\left[-\dfrac{(E-E_F)}{kT}\right]}$

(c) $\exp\left[-\left(\dfrac{E-E_F}{kT}\right)\right]$

(d) None

[2 marks : 2001]

Solution: (b)

Fermi dirac function for hole i.e. probability that a state E is filled (occupied) by hole

$$= 1 - f(E)$$

$$= 1-\frac{1}{1+e^{(E-E_F)/K_T}} \qquad f(E) - \text{Fermi dirac function for } e^-.$$

$$= \frac{e^{(E-E_F)/K_T}}{1+e^{(E-E_F)/K_T}} = \frac{1}{1+e^{(-(E-E_F)/kT)}}$$

2. Semiconductor Diode, Characteristics and Circuits

2.1 A voltage source $V_{AB} = 4\sin\omega t$, is applied across the terminals A and B of the circuit. The diodes are assumed to be ideal. The impedance offered by the circuit across the terminals A and B in kilo ohm is

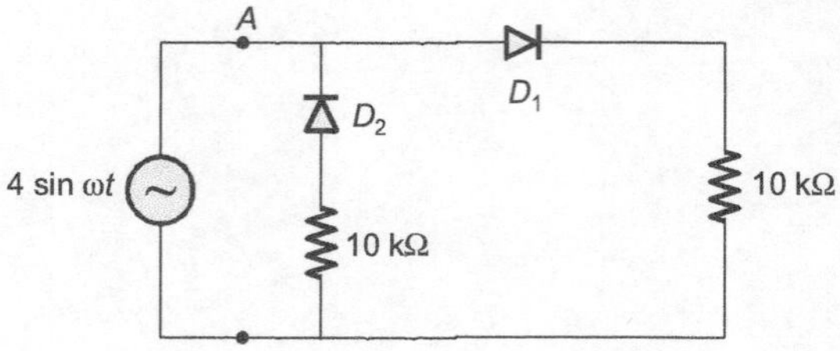

(a) 5
(b) 20
(c) 10
(d) 100

[2 marks : 2003]

Solution: (c)

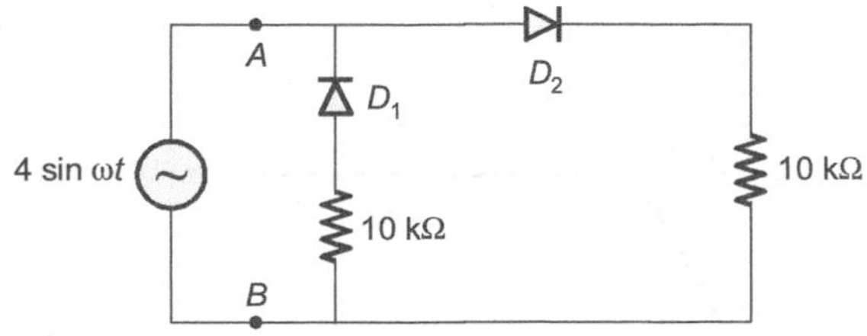

In +ve half cycle D_1 – off (R.B.)
D_2 – on (F.B.)
∴ Equivalent circuit will be

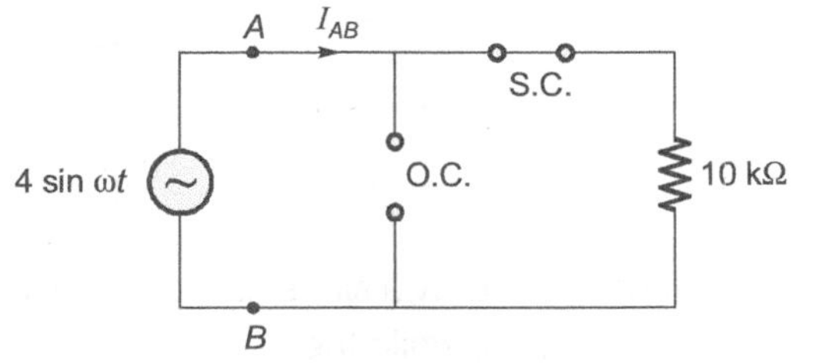

∴ $$V_{AB} = 4\sin\omega t$$

$$I_{AB} = \frac{V_{AB}}{10\text{ k}\Omega}$$

∴ $$R_i = \frac{V_{AB}}{I_{AB}} = 10\text{k}\Omega$$

For –ve half cycle,
D_1 on, D_2 off
Equivalent circuit,

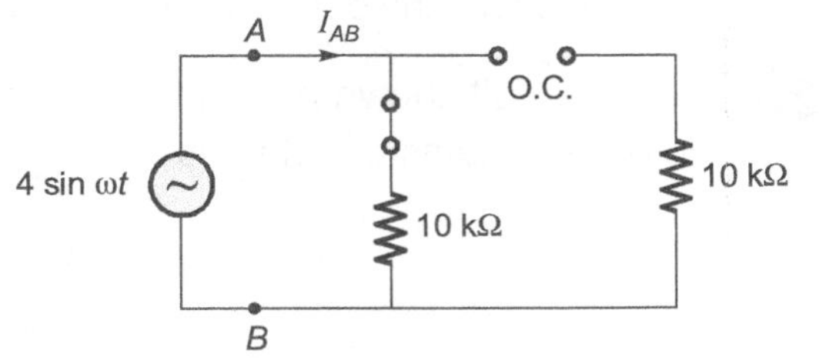

$$V_{AB} = 4\sin\omega t$$

$$I_{AB} = \frac{4\sin\omega t}{10\text{ k}\Omega}$$

∴ $$\frac{V_{AB}}{I_{AB}} = R_i = 10\text{ k}\Omega$$

2.2 The peak voltage rating of a diode in a bridge rectifier is 'X' times larger than that of a full-wave rectifier feeding the same dc output voltage. where the value of 'X' is

(a) 0.5 (b) 1.0

(c) $\sqrt{2}$ (d) 2.0

[2 marks : 2004]

Solution: (a)

Peak inverse voltage of bridge rectifier = V_m
Peak inverse voltage of full wave rectifier = $2V_m$

2.3 Draw the neat circuit of a voltage doubler. Explain its operation. Draw the waveforms for the voltages across the two capacitors.

[10 marks : 2004]

Solution:

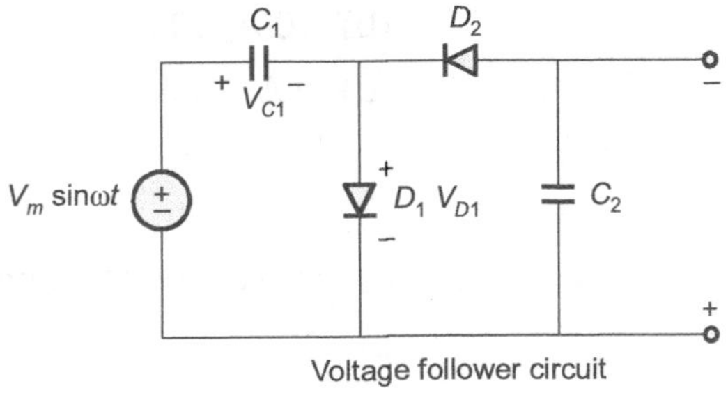

Voltage follower circuit

The figure shows a circuit-composed of two sections in cascade, a clamp circuit formed by C_1 and D_1 and peak rectifier formed by D_2 and C_2.

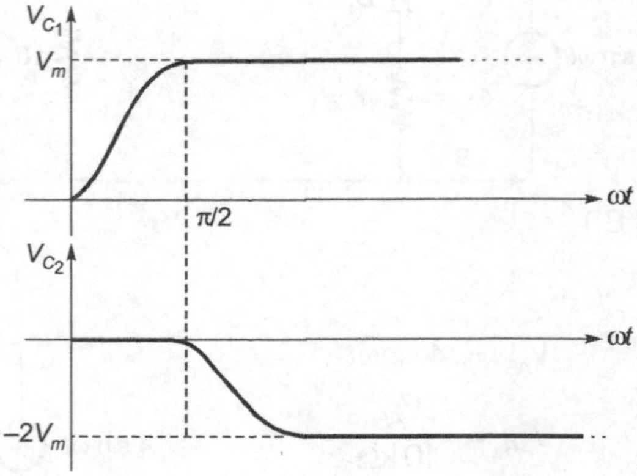

When excited by a sinusoidal of amplitude V_m the clamping section provides the waveform shown. Assuming ideal diodes, while the positive peaks are damped to 0 V, the negative peak reaches $-2V_m$. In response to this wave form, the peak-detector section provides, across capacitor C_2 a negative dc voltage of magnitude $2V_m$. Because the output voltage is double the input peak, the circuit is known as a voltage doubler.

2.4 In the circuit shown below, the zener diode is ideal. The minimum value of R_L up to which the output voltage remains constant is

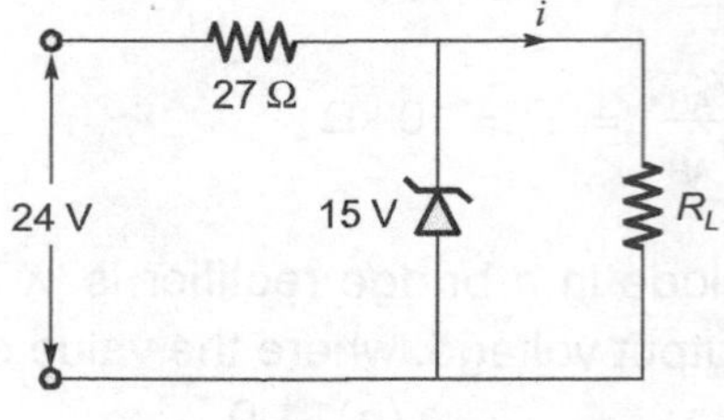

(a) 27 Ω (b) 45 Ω
(c) 15 Ω (d) 24 Ω

[2 marks : 2005]

Solution: (b)

The minimum value of 'R_L' for the zener diode to be a voltage regulator.

$$R_L = \frac{15}{i_{max}}, \qquad \text{where,} \quad i_{max.} = \frac{24-15}{27}$$

$$R_{L(min)} = 45\ \Omega$$

2.5 A single phase Full-Bridge diode rectifier delivers a constant load current of 10 A. Average and RMS values of source current, are respectively

(a) 5 A, 10 A (b) 10 A, 10 A
(c) 5 A, 5 A (d) 0 A, 10 A

[2 marks : 2007]

Solution: (b)

For a constant value of load current, average and r.m.s. value of the current are equal.

2.6 A Zener regulator circuit is shown below in which the nominal supply voltage is 10 V but can vary by ± 1 V. The ratings of Zener are; $V_z = 6.8$ V, $I_z = 5$ mA, $r_z = 20\ \Omega$ and $I_{zk} = 0.2$ mA.

Determine:

(i) The output voltage for no load when supply is 10 V.

(ii) Change in output voltage at no load due to supply variation of ± 1 V.

(iii) Change in output voltage when $R_L = 2$ kΩ.

(iv) Output voltage at nominal supply when $R_L = 0.5$ k.

(v) Minimum value of R_L for which Zener operates in breakdown region.

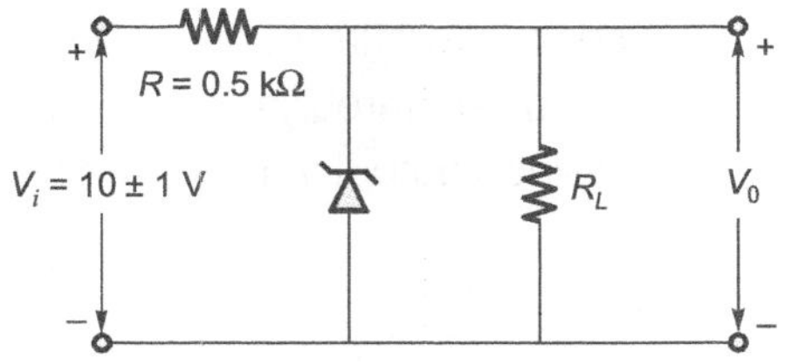

[15 marks : 2007]

Solution:

(i)

$$V_0 = 6.8 = V_{z0} + I_z \times r_z$$
$$= V_{z0} + 5 \times 10^{-3} \times 20$$
$$\therefore \quad V_{zo} = 6.8 - 5 \times 10^{-3} \times 20 = 6.7 \text{ V}$$

$$v_0\,(\text{no-load}) = 6.7 + \frac{(10-6.7)}{520} \times 20 = 6.827 \text{ V} \qquad (\therefore R + r_z = 0.5 \times 10^3 + 20 = 520)$$

(ii) Change in output voltage,

$$\Delta V_0 = \frac{\Delta V_i \times 20}{520} = \pm 38.46 \text{ mv/V} \qquad (\therefore \Delta V_i = 1 \text{ V})$$

Line regulation = 38.46 mv/V

(iii)

$$R_L = 2 \text{ k}\Omega,\ I_L = 6.8/2\text{K}$$
$$= 3.4 \text{ mA (approx)}$$
$$\Delta I_z = -3.4 \text{ mA}$$
$$\Delta V_z = -3.4 \times 10^{-3} \times 20 = -68 \text{ mV}$$

(iv) If, $R = R_L = 0.5$ kΩ

Then, $I_L = 6.8/0.5\text{ K} = 13.6$ mA

This is not possible, because the current (I) supplied through R is only 6.4 mA. Therefore zener must be cut-off.

$$V_0 = V_i \times \frac{R_L}{R + R_L} = 10 \times \frac{0.5\text{ K}}{0.5\text{ K} + 0.5\text{ K}} = 5 \text{ V}$$

(v) Minimum resistance required for diode in break down region.

$$I_z = I_{zk} = 0.2 \text{ mA}$$
$$V_z = V_{zk} = 6.7 \text{ V}$$

Lowest current supplied through $R = \dfrac{9-6.7}{0.5\text{K}} = 4.6$ mA

The load current = 4.4 mA

The corresponding, $R_L = \dfrac{6.7\text{ V}}{4.4\text{ mA}} = 1.5$ kΩ

2.7 An input of $E = \pm 50$ V is applied to the clipping circuit. The output current from the circuit is to be $I_L = 20$ mA and the negative output voltage does not exceed –0.5 V. Calculate the value of R_1. Specify the diode in terms of forward current, power dissipation and peak reverse voltage. Assume the reverse saturation current I_S is 5 μA and forward voltage of diode, $V_F = 0.7$ V.

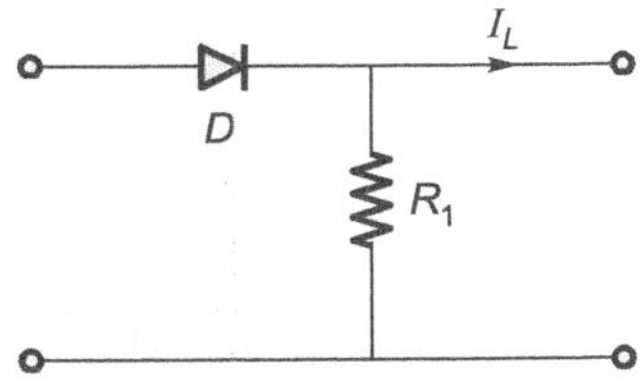

[12 marks : 2011]

Solution:

When $E = -50$ V

The diode D is reverse biased and current through the diode is reverse saturation current $I_s = 5\ \mu$A.

Also, a load current of $I_L = 20$ mA flows out of the circuit,

$\therefore$ Current through $R_1 = I_s + I_L$

But since output voltage can't exceed –0.5 V,

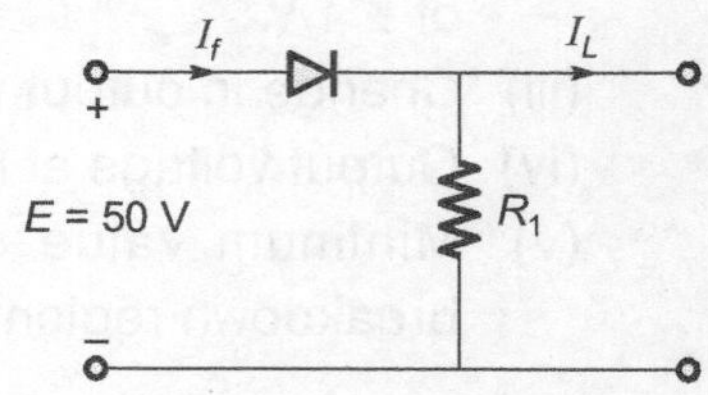

$$\therefore \qquad R_1 = \left|\frac{0.5}{5\times10^{-6} + 20\times10^{-3}}\right| \simeq 25\ \text{k}\Omega$$

Now, in forward bias,

i.e. when $E = 50$ V

Current through R_1, $\quad I_f - I_L = \dfrac{50-0.7}{R_1}$

$$\Rightarrow \qquad I_f = I_L + \frac{49.3}{R_1} \Rightarrow I_f = 21.97\ \text{mA}$$

Power dissipation, $\quad P_D = V_f \times I_f = 15.38$ mW

Peak inverse voltage, $\quad V_P = (50 - 0.5)$ V $= 49.5$ V

(50 V is maximum negative value of E and taking $R_1 = 100\ \text{k}\Omega$ voltage drop across R_1 is equal to 0.5 V so rest voltage will be across diode).

2.8 **Answer the following with respect to the circuit shown:**

(i) Why is it called a 'shunt regulator'?

(ii) Which is the regulating element?

(iii) Draw the *v-i* characteristic of the regulating element.

(iv) Mark the portion of the curve used for regulation.

(v) Show the range of current over which regulator will operate satisfactorily.

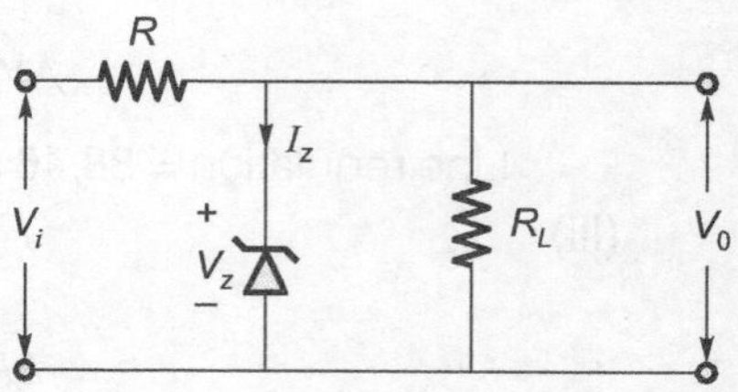

[4 marks : 2012]

Solution:

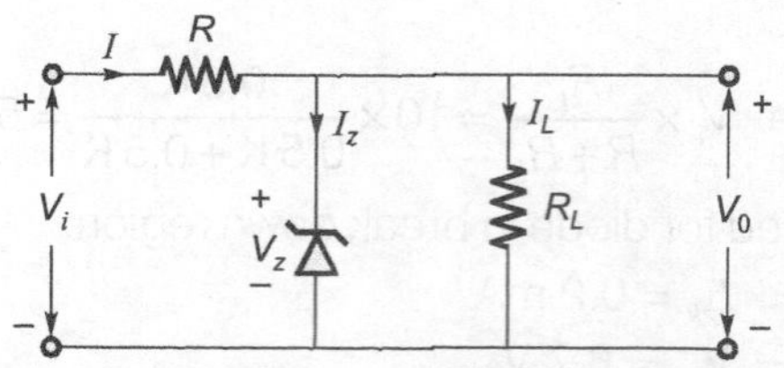

- The control element (Zener diode) is in parallel with the load that's why it is called a shunt regulator.
- The regulating element is Zener diode.
- For fixed values of R_L, the voltage V_i must be sufficiently large to turn the Zener diode on.

The minimum turn-on voltage,

$$V_{i,\min} = \frac{(R+R_L)V_Z}{R_L}$$

$$\therefore \qquad I_{\max} = I_Z + I_L; \quad \text{where, } I_L = \frac{V_Z}{R_L}$$

The maximum value of V_i

$$V_{i\max} = I_{\max} R + V_Z$$

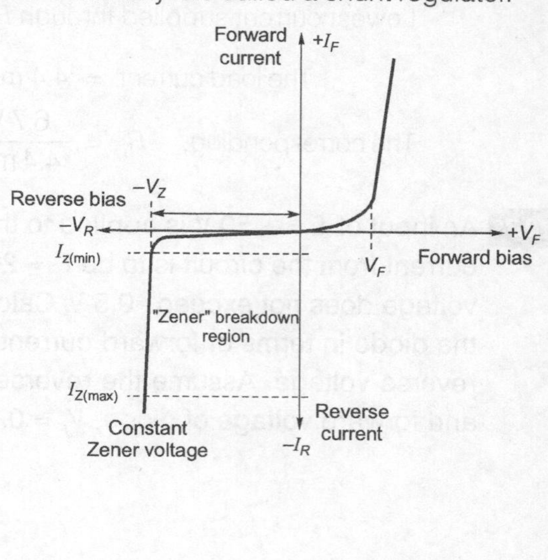

2.9 In the circuit shown, calculate and sketch the current $i(t)$ over the period $0 \le \theta \le 2\pi$. Assume the diodes to be ideal.

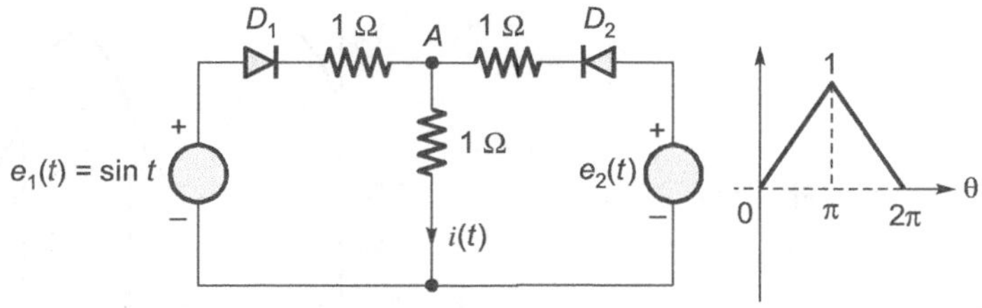

[10 marks : 2012]

Solution:

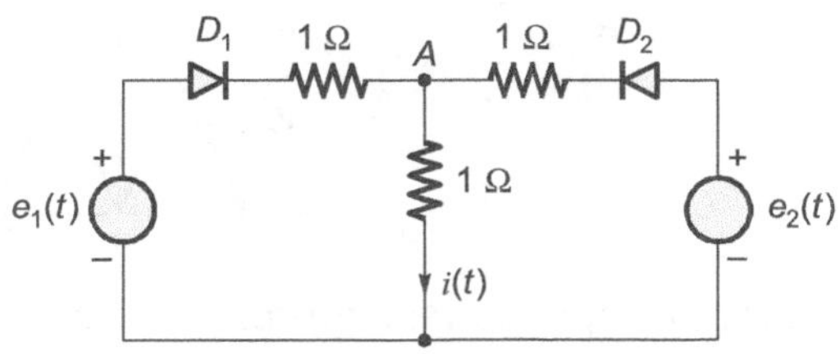

Between the duration 0 and π both diode will be ON

$$e_1(t) = 2i_1(t) + i_2(t)$$
$$e_2(t) = i_1(t) + 2i_2(t)$$
$$i_1(t) + i_2(t) = \frac{1}{3}\left(\sin t + \frac{1}{\pi}t\right)$$

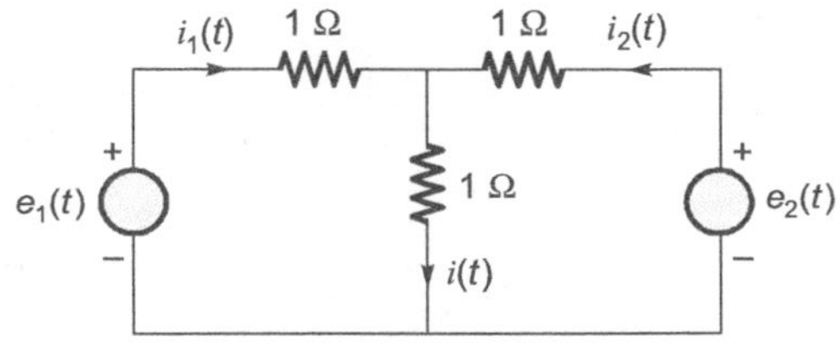

Between the duration π and 2π only diode D_2 will be ON

$$i(t) = \frac{e_2(t)}{2} = 1 - \frac{t}{2\pi}$$

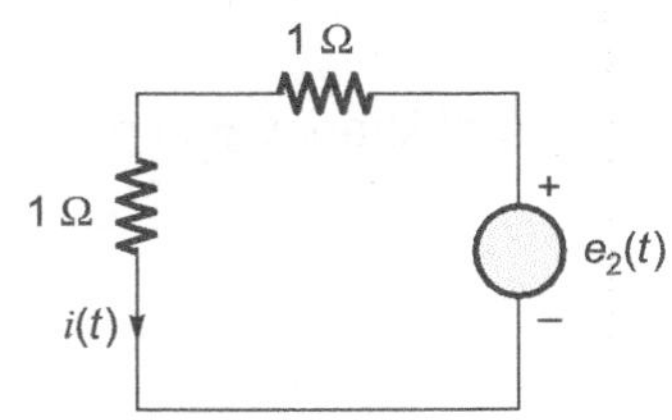

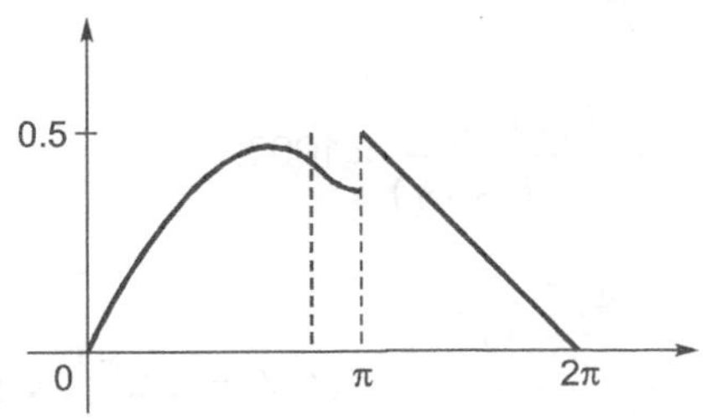

2.10 The efficiency of rectification η_r is defined as the ratio of the dc output power $P_{dc} = V_{dc} I_{dc}$ to the input power $\left(\frac{1}{2\pi}\right)\int_0^{2\pi} v_i \, i d\alpha$; show that for the full wave-rectifier circuit

$$\eta_r = \frac{81.2}{1 + \frac{R_F}{R_L}} \%$$

[4 marks : 2013]

Solution:

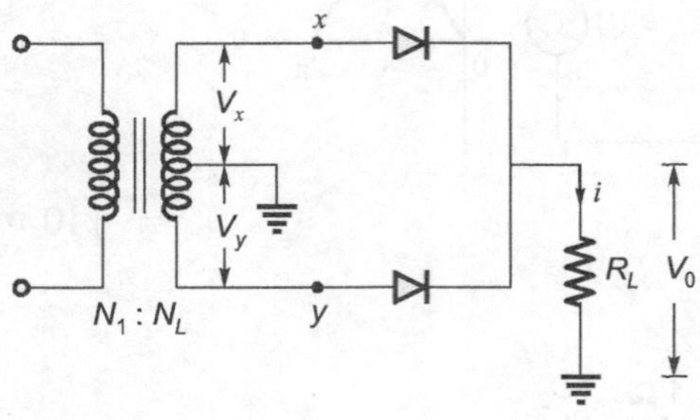

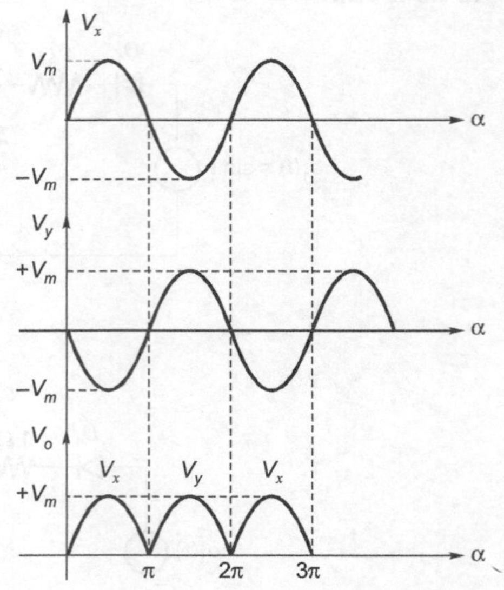

Input Power: For $0 < \alpha < \pi$, D_1 will draw input power from upper half of the winding and for $\pi < \alpha < 2\pi$, D_2 will draw power from lower half of the winding.

$$P_i = \frac{1}{2\pi}\left[\int_0^{\pi} i(R_F + R_L)\, i \cdot d\alpha + \int_{\pi}^{2\pi} i(R_F + R_L)\, i \cdot d\alpha\right]$$

$$P_i = (R_F + R_L) \times \frac{1}{2\pi}\int_0^{2\pi} i^2\, d\alpha$$

$$P_i = I_{rms}^2 (R_F + R_L) \text{ and } P_{DC} = I_{DC}^2 \times R_L$$

and efficiency η is given by

$$\%\eta = \frac{P_{DC}}{P_i} \times 100\%$$

$$\%\eta = \frac{I_{DC}^2 \times R_L}{I_{rms}^2 (R_F + R_L)} \times 100\%$$

$$\%\eta = \frac{\frac{4I_m^2}{\pi^2} \times R_L}{\frac{I_m^2}{2}(R_F + R_L)} \times 100\% = \frac{0.81}{1 + \frac{R_F}{R_L}} \times 100\%$$

2.11 **Assuming forward voltage drop across diodes to be 0.7 V, draw the transfer characteristic of the clipper circuit shown in figure.**

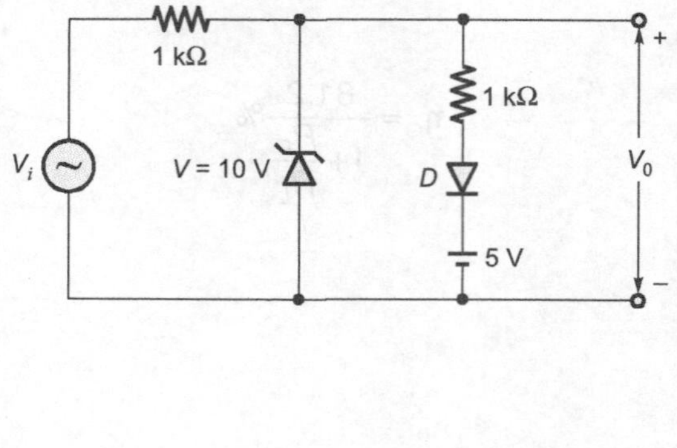

[4 marks : 2014]

Solution:

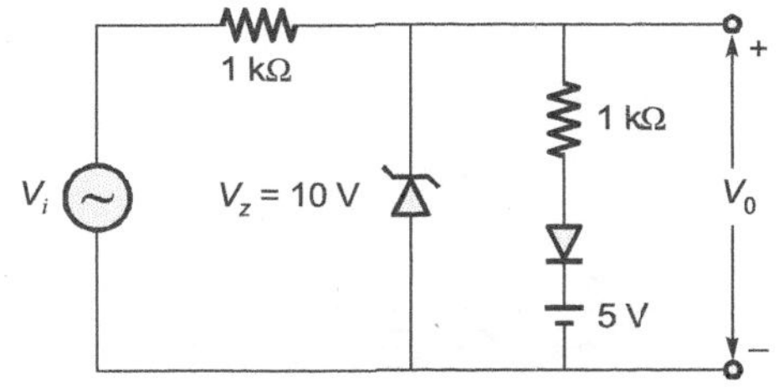

When V_i is +ve and $V_i > 5.7$ V
Assume diode is conducting,

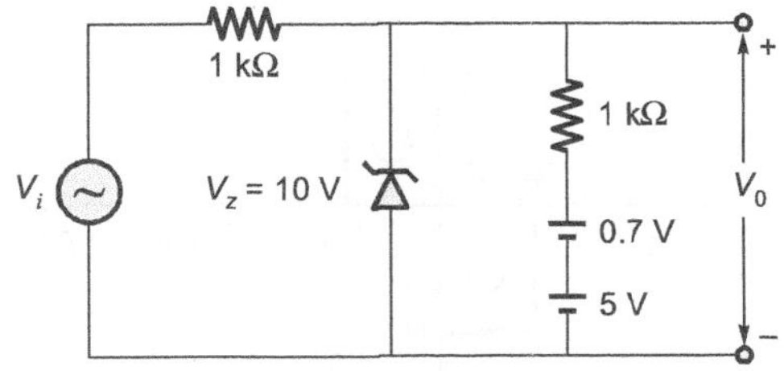

Assume the zener diode is reverse biased and having voltage less than breakdown voltage

$$V_i - I(1\text{ k}\Omega) - I(1\text{ k}\Omega) - 0.7 - 5 = 0$$

$$V_i - I(2\text{ k}\Omega) - 5.7 = 0$$

$$I(2\text{ k}\Omega) = V_i - 5.7$$

$$I = \left(\frac{V_i - 5.7}{2\text{ k}\Omega}\right)$$

$$V_0 = 5.7 + 1 \times I$$

$$= 5.7 + \left(\frac{V_i - 5.7}{2}\right) = \left(\frac{5.7 + V_i}{2}\right)$$

When, $V_i < 5.7$ V

$V_0 = V_i$ (∵ Diode is in off state)

If, $5.7 < V_i < 14.3$

$$V_0 = \left(\frac{5.7 + V_i}{2}\right)$$

Diode conducts and zener diode is reverse biased,

When, $V_i > 14.3$ V

⇒ $V_0 = 10$ volts

Breakdown occurs in zener diode.

When, $V_i < 0$ V

$V_0 = -0.7$ V

Transfer characteristics:

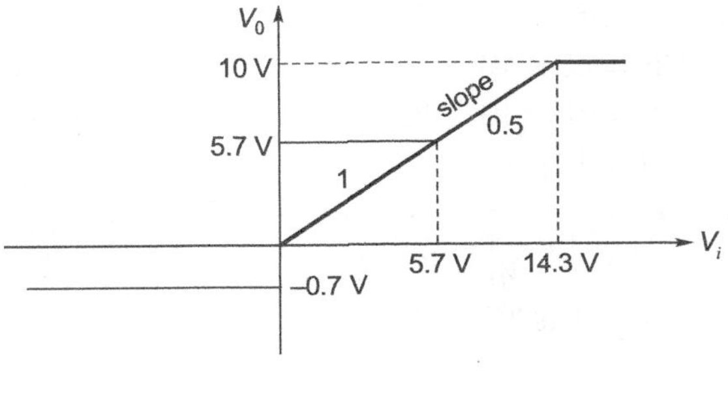

2.12 Draw the variation of space charge, electric field and electrostatic potential across a semiconductor p-n junction.

[4 marks : 2017]

Solution:

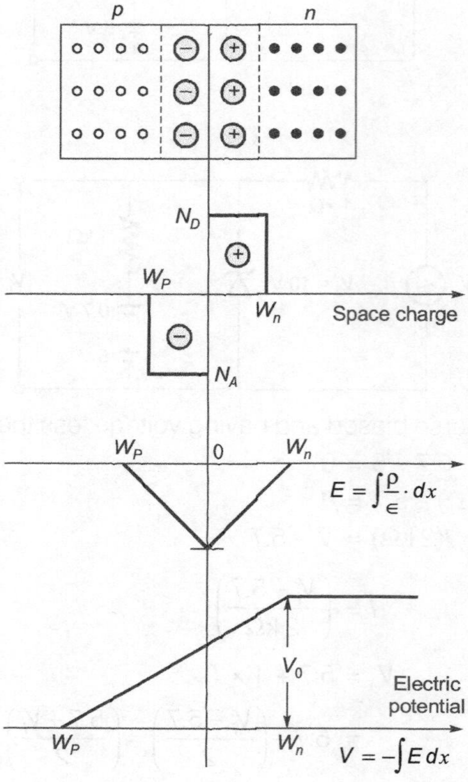

2.13 A Silicon diode that has an ohmic resistance of 0.5 W with reverse saturation current $I_0 = 10^{-12}$ A and $\eta = 2.0$, consumes 50×10^{-12} W extra power compared to ideal diode. Diode operating temperature is 350° K.

(i) Determine the fraction of the applied voltage that falls across the ohmic resistance.

(ii) Solve part (i), when diode current is 1 A.

(iii) Compare results of part (i), (ii) and draw conclusion.

[8 + 2 + 2 = 12 marks : 2018]

Solution:

Silicon diode can be represented as,

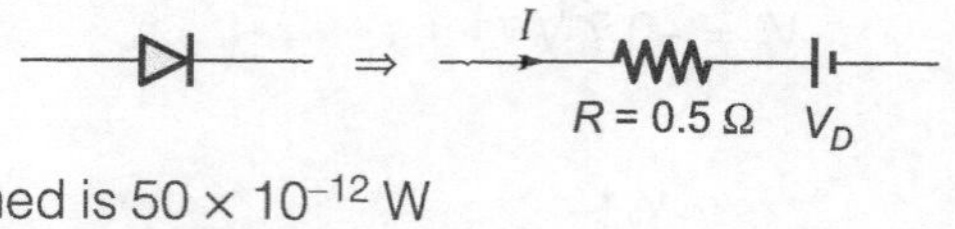

(i) Given that power consumed is 50×10^{-12} W

$$\therefore \quad I^2R = 50 \times 10^{-12} \text{ W}$$

$$I^2 \times 0.5 = 50 \times 10^{-12} \text{ W}$$

$$I = 10\ \mu\text{A}$$

Now, using diode equation,

$$I = I_0\left(e^{\frac{V_D}{\eta V_T}} - 1\right)$$

When diode is forward biased '1' can be ignored

$$\therefore \quad I = I_0 \cdot e^{\frac{V_D}{\eta V_T}}$$

$$I_0 = 10^{-12}\text{ A},$$

$$\eta = 2,$$

$$V_T = \frac{KT}{q} = \frac{T}{11600} = \frac{350}{11600} = 30.17 \text{ m volts}$$

$$10\,\mu\text{A} = 10^{-12} \cdot e^{\frac{V_D}{2\times30.17\times10^{-3}}}$$

$$V_D = 0.9725\text{ V}$$

$$\text{Voltage across resistance} = 0.5 \times 10\,\mu\text{A} = 5\,\mu\text{V}$$

Hence, fraction of voltage across resistance,

$$= \frac{5\times10^{-6}}{5\times10^{-6}+0.9725} = 5.14\times10^{-6}\text{ V}$$

(iii) When, $I = 1\text{ A}$

$$1\text{ A} = 10^{-12} \cdot e^{\frac{V_D}{2\times30.17\times10^{-3}}}$$

$$V_D = 1.66\text{ Volts}$$

Voltage across resistance, $V_R = 0.5 \times 1 = 0.5\text{ V}$

Fraction of voltage across resistance $= \dfrac{0.5}{0.5+1.66} = 0.23\text{ V}$

(iii) **Conclusion:** From (i) and (ii) it can be seen that as forward current through diode increases the fraction of voltage across resistance increases rapidly.

2.14 (i) **For the circuit shown in below figure using ideal diodes, find the values of the voltage V and current I indicated in the figure.**

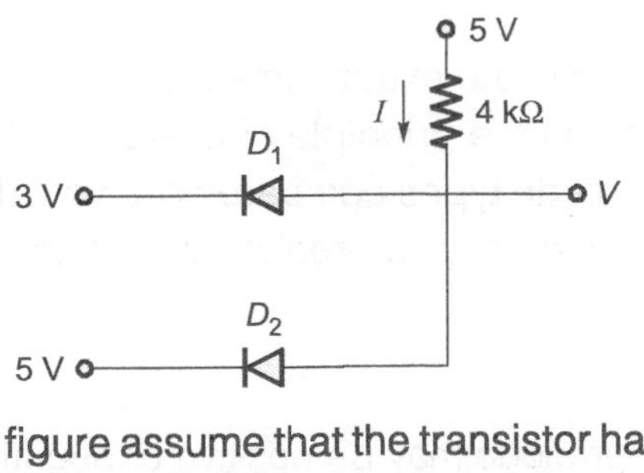

(ii) **For the circuits in the below figure assume that the transistor have very large β. Some measurements have been made on these circuits. The results are indicated in the figure. Find the values of the voltage V_2.**

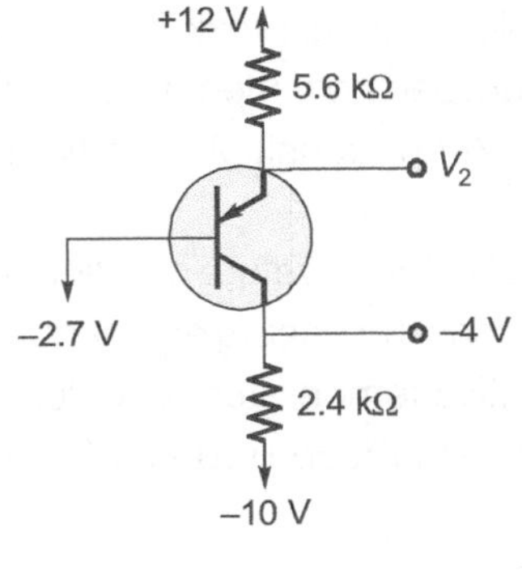

[6 + 6 = 12 marks : 2019]

Solution:

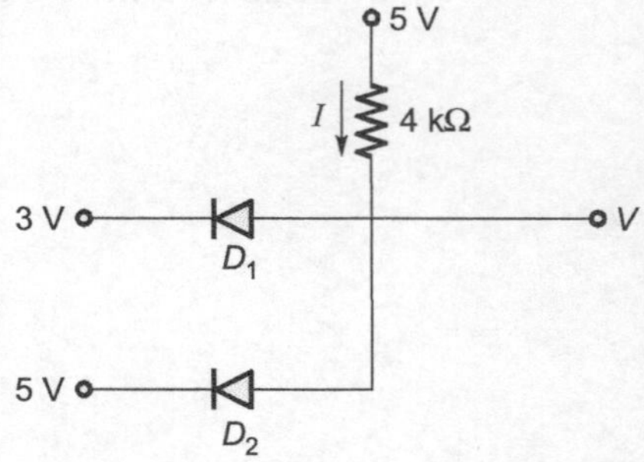

(i) For the given circuit as all diodes are ideal.

Current through diode D_2

As potential is same at both ends, So $I_{D2} = 0$.

Current through diode D_1

$$I_{D1} = \frac{5-3}{4\,k\Omega} = \frac{2}{4\,k\Omega} = 0.5\,\text{mA}$$

when D_1 conducts circuit becomes,

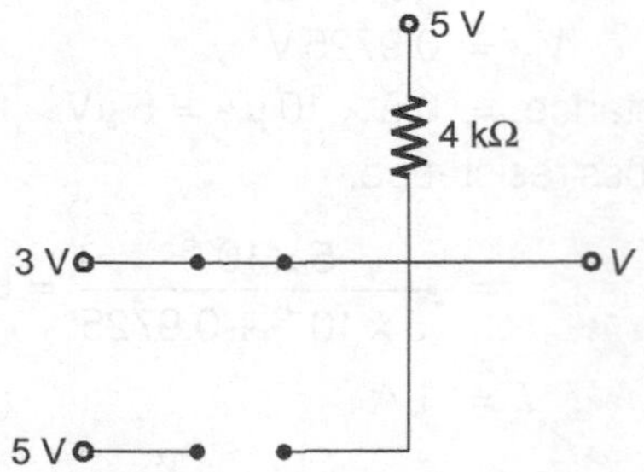

$\therefore$ Voltage at given terminal = 3 V

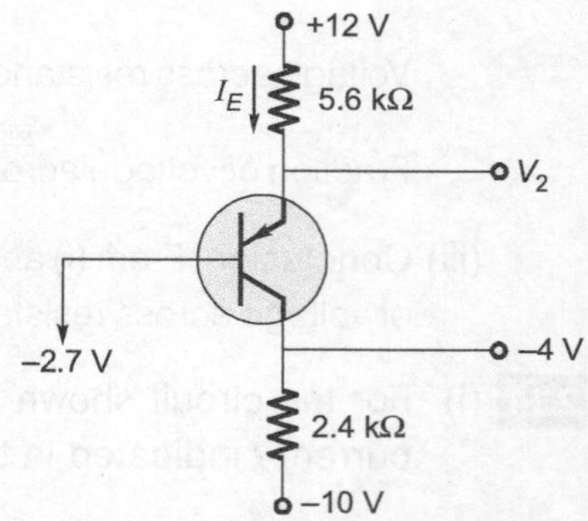

(ii) For the given p-n-p transistor. β is very large.

$$I_E \simeq I_C$$

Collector current is given by

$$I_C = \frac{-4-(-10)}{2.4\,k\Omega} = 2.5\,\text{mA} \simeq I_E$$

Apply KVL,

$$V_2 = -5.6\,kI_E + 12\,\text{V}$$
$$= 12 - (5.6 \times 2.5)$$
$$= -2\,\text{V}$$

2.15 Not all "zener" diodes breakdown in the exact same manner. Some operate on the principle of zener breakdown, while other operate on the principle of avalanche breakdown. How do the temperature coefficients of these two zener diode types compare? Are you able to discern whether a zener diode uses one principle or the other just from its breakdown voltage rating? Justify your answer.

[10 marks : 2019]

Solution:

Temperature coefficient of a zener diode may be positive or negative depending upon the breakdown or zener voltage, v_z.

For lower breakdown voltages (less than 5 V), breakdown takes place due to tunneling effect and the phenomenon is Zener Breakdown. While for higher breakdown voltages (exceeding 8 V), Avalanche Breakdown occurs due to impact ionization. In between both mechanism contribute to breakdown.

Despite the fact that Avalanche and Zener constitute two types of diodes, the name Zener is commonly applied to both types of diodes.

Low voltage Zener diodes have negative temperature coefficient since Zener breakdown takes place more easily with increase in temperature (tunneling occurs earlier due to increase in thermal energy).

High voltage Zener diodes have positive temperature coefficient because breakdown takes place due to Avalanche effect and increase in thermal vibration reduces kinetic energy gain, thus increasing breakdown voltage.

2.16 Calculate the output voltage V_0 at 7 ms in the figure shown below if *a*, ±10 V square wave of 250 Hz source is applied to R = 10 Ω, C = 20 μF. The diode is ideal and capacitor is initially uncharged.

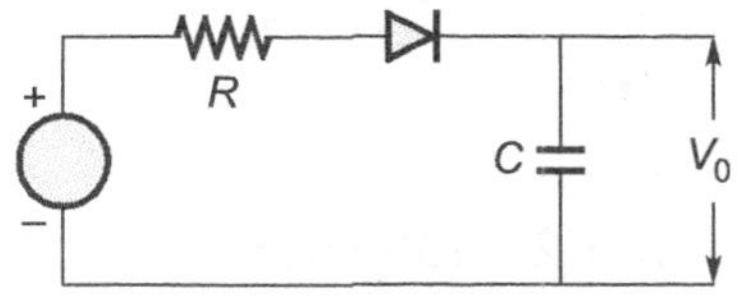

[12 marks : 2020]

Solution:

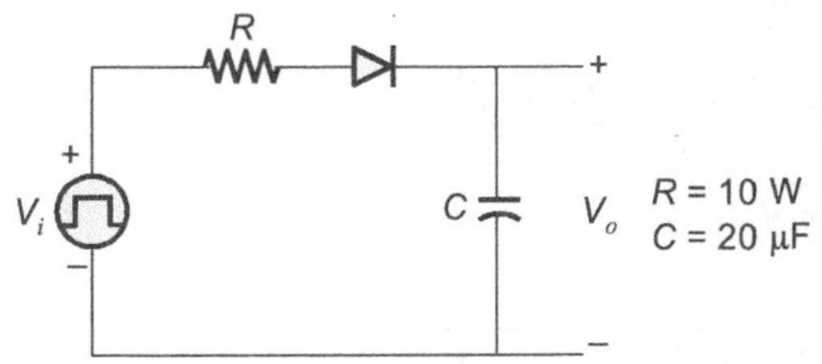

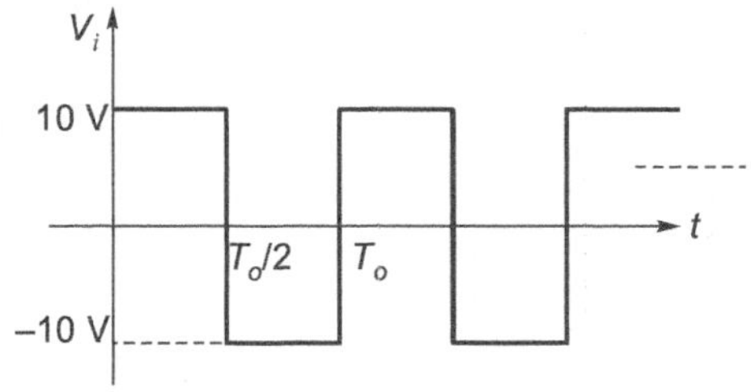

$$T_o = \frac{1}{f_o} = \frac{1}{250} = 4 \text{ m sec}$$

$$\frac{T_o}{2} = 2 \text{ m sec}$$

For the positive half cycle of V_i, diode is forward bias and capacitor will start to change through resistance R.

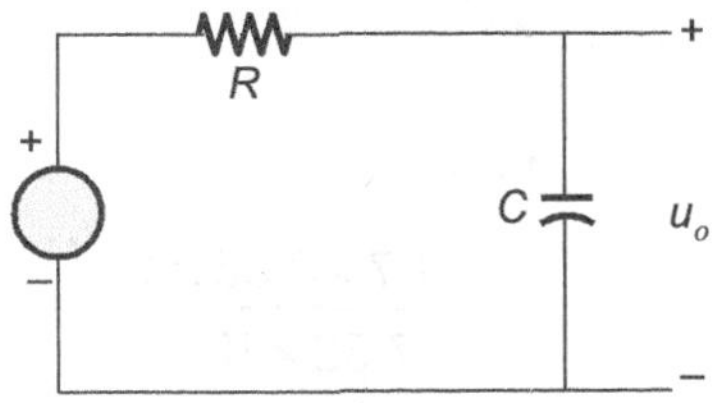

Charging time constant (τ) = RC

$$\tau = 10 \times 10^{-6} \text{ sec} = 0.01 \text{ m sec}$$

So, $\quad 5\tau = 5 \times 0.01 \text{ m sec} = 0.05 \text{ m sec}$

Hence, $\quad 5\tau < \frac{T_o}{2}$

So, capacitor will fully charge upto peak value of V_i, during first positive cycle only.

So, $\quad V_o = 10 \text{ V}$

For next negative half cycle of V_i, diode is reverse biased. So, circuit will be open.

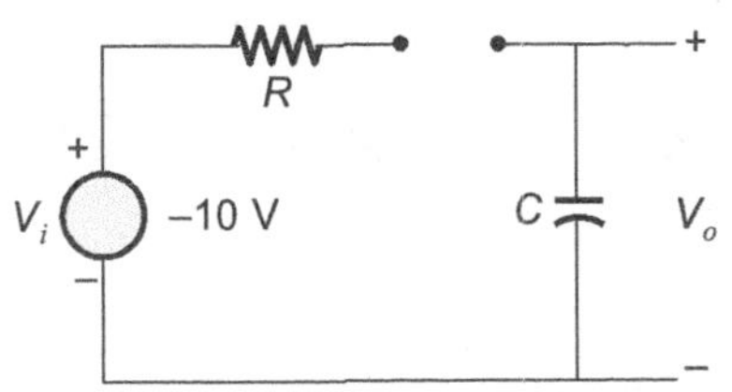

Here in this case, there is no discharge path available for capacitor to discharge. So, capacitor will always retain its voltage at constant 10 V for t > 2 m sec.

$$V_0\big|_{t=7 \text{ msec}} = 10 \text{ volts}$$

2.17 A pn-junction is doped with an acceptor ion concentration $N_A = 2 \times 10^{16}$ cm^{-3} and a donor ion concentration $N_D = 9 \times 10^{15}$ cm^{-3}. Determine the capacitance of the device with $V_R = 2$ V. Dielectric constant of silicon $\varepsilon_{Si} = 11.7 \times 8.85 \times 10^{-14}$ F/cm.

[10 marks : 2021]

Solution:

Given :

$$N_A = 2 \times 10^{16} \text{ cm}^{-3}$$

$$N_D = 9 \times 10^{15} \text{ cm}^{-3}$$

For silicon,

$$n_i = 1.5 \times 10^{10} \text{ cm}^{-3}$$

Built-in potential :

$$V_{bi} = V_T \ln \frac{N_A N_D}{n_i^2}$$

$$V_{bi} = 0.026 \ln \frac{2 \times 10^{16} \times 9 \times 10^{15}}{2.25 \times 10^{20}}$$

$$= 0.7126 \text{ V}$$

Depletion width in reverse bias :

$$W = \sqrt{\frac{2\varepsilon(V_{bi} + V_R)}{e} \times \left(\frac{1}{N_A} + \frac{1}{N_D}\right)}$$

$$W = \sqrt{\frac{2 \times 11.7 \times 8.85 \times 10^{-14} \times 2.7126}{1.6 \times 10^{-19}}\left(\frac{1}{2 \times 10^{16}} + \frac{1}{9 \times 10^{15}}\right)}$$

$$W = 7.52 \times 10^{-5} \text{ cm}$$

In reverse bias, diode has transition capacitance.

$$C_T = \frac{A\varepsilon}{W}$$

Assume, junction area

$$A = 1 \text{ cm}^2$$

$$C_T = \frac{1 \times 11.7 \times 8.85 \times 10^{-14}}{7.52 \times 10^{-5}}$$

$$C_T = 13.77 \text{ nF}$$

2.18 The circuit shown below, shows two diodes with reverse saturation currents of I_{S1} and I_{S2} placed in parallel.

(i) Prove that the parallel combination operates as a diode.

(ii) If the total current is I_T, then determine the current carried by each diode.

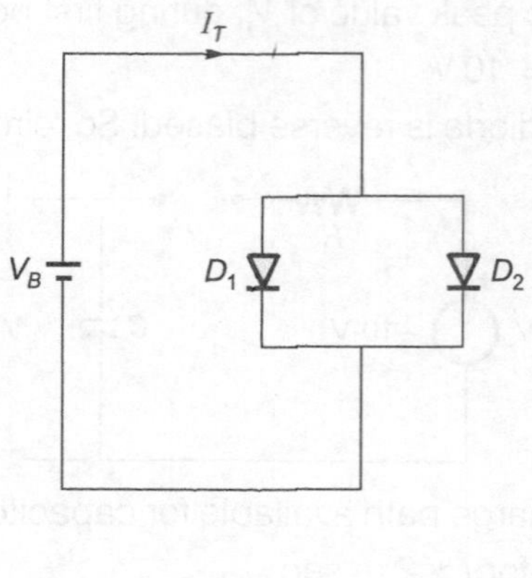

[6+6 = 12 marks : 2021]

Solution:

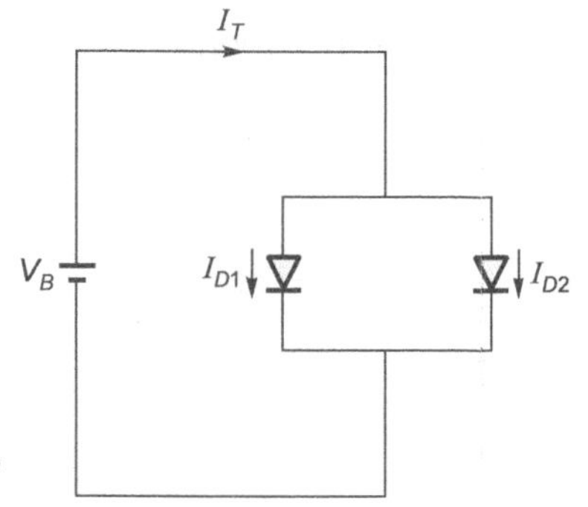

(i)

$$I_{D1} = I_{S1}(e^{V_B/\eta V_T} - 1)$$

$$I_{D2} = I_{S2}(e^{V_B/\eta V_T} - 1)$$

$$I_T = I_{D1} + I_{D2} = (I_{S1} + I_{S2})(e^{V_B/\eta V_T} - 1)$$

$$= I_S(e^{V_B/\eta V_T} - 1),\ I_S = I_{S1} + I_{S2}$$

Thus, parallel connection of two diodes will act as a diode with reverse saturation current $I_S = I_{S1} + I_{S2}$.

(ii)

$$\frac{I_{D2}}{I_{D1}} = \frac{I_{S2}(e^{V_B/\eta V_T} - 1)}{I_{S1}(e^{V_B/\eta V_T} - 1)} = \frac{I_{S2}}{I_{S1}}$$

KCL :

$$I_T = I_{D1} + I_{D2}$$

$$I_T = I_{D1} + \frac{I_{S2}}{I_{S1}} \times I_{D1}$$

$$I_T = I_{D1}\left(1 + \frac{I_{S2}}{I_{S1}}\right) \Rightarrow I_{D1} = \frac{I_T}{1 + \frac{I_{S2}}{I_{S1}}}$$

$$I_T = I_{D1} + I_{D2}$$

$$I_T = \frac{I_{S1}}{I_{S2}} \times I_{D2} + I_{D2}$$

$$I_T = \left(1 + \frac{I_{S1}}{I_{S2}}\right) I_{D2} \Rightarrow I_{D2} = \frac{I_T}{1 + \frac{I_{S1}}{I_{S2}}}$$

2.19 For the diode-resistor-battery circuit shown below, the diode D_1 is assumed to be ideal. Sketch the time-average of V_{out}, as the battery voltage V_B is varied from $-\infty$ to $+\infty$, if the input voltage is a sinusoidal voltage with an amplitude equal to V_P.

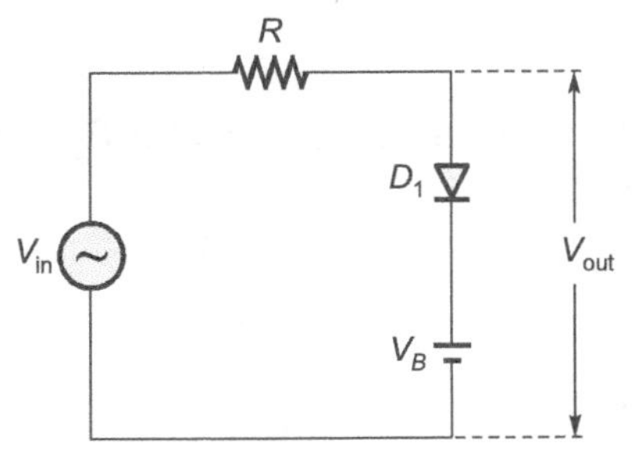

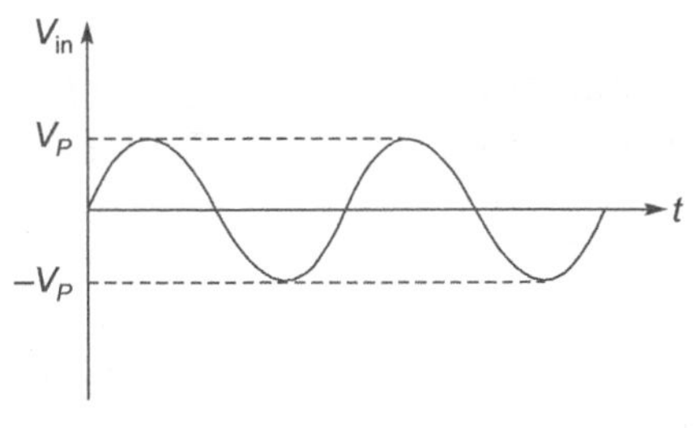

[20 marks : 2021]

Solution:

Given configuration :

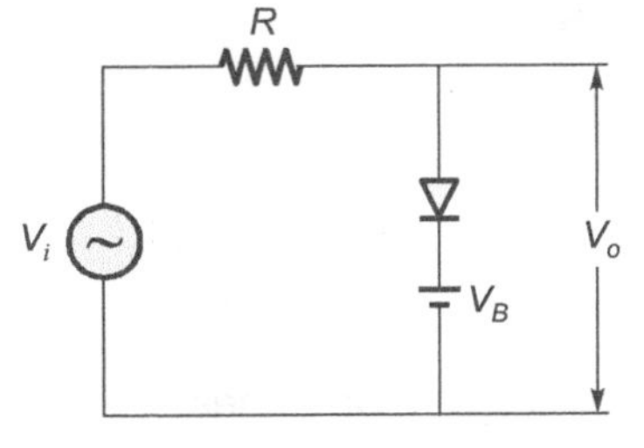

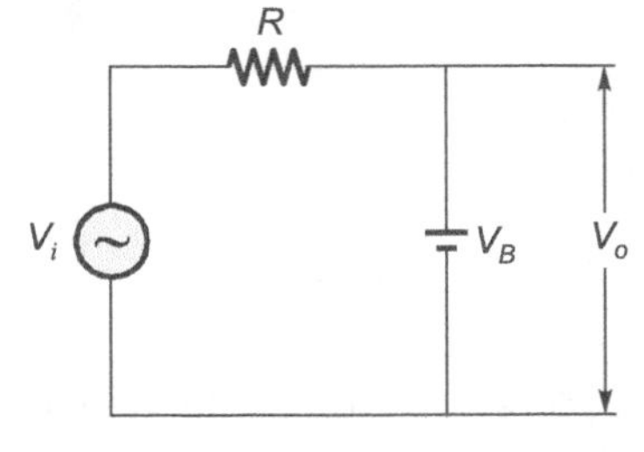

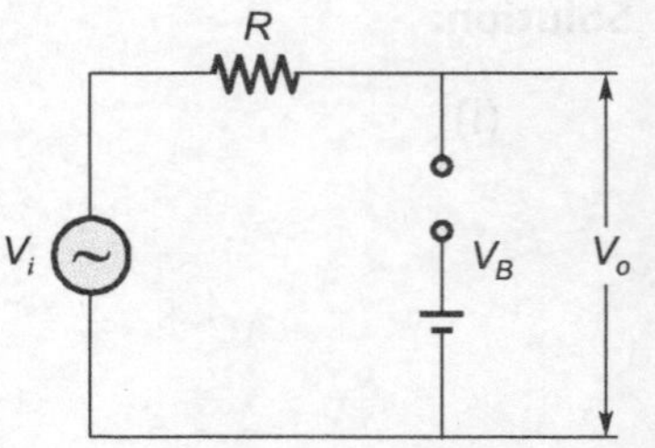

Case 1 : When $V_i > V_B \Rightarrow$ Diode is ON

$V_o = V_B$

Case 2 : When $V_i < V_B \Rightarrow$ Diode is OFF

$V_o = V_i$

Take $V_B = 0$:

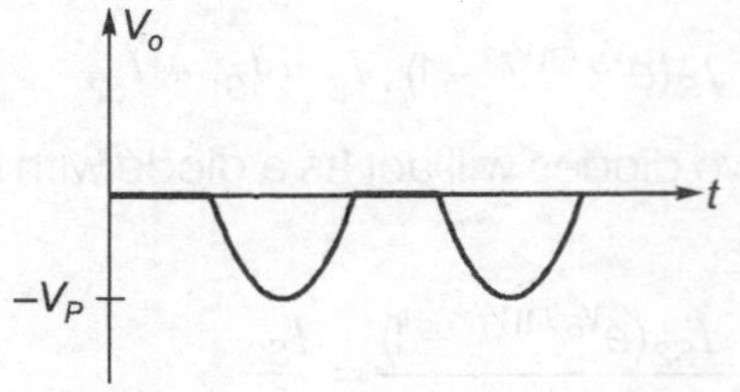

Average Output Voltage : $V_o = -\frac{V_P}{\pi}$

Take $V_B \geq V_P$: Diode is always OFF.

$\Rightarrow$ $V_o = V_i \Rightarrow$ Average Output Voltage = 0

Take $V_B \leq -V_P$: Diode is always ON

$\Rightarrow$ $V_o = V_B \Rightarrow$ Average Output Voltage = V_B

Take $-V_P \leq V_B \leq V_P$: Average Vo will vary non-linearly.

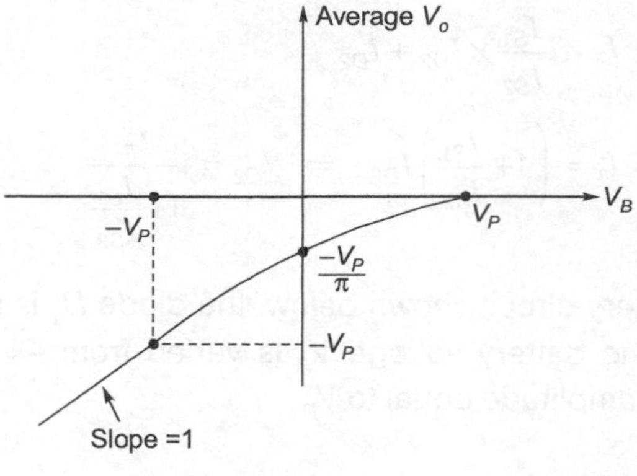

2.20 **Find the currents through diodes D_1 and D_2 and the voltage across resistor R_4 in the figure shown. Assume forward voltage drop of 0.7 V for each diode.**

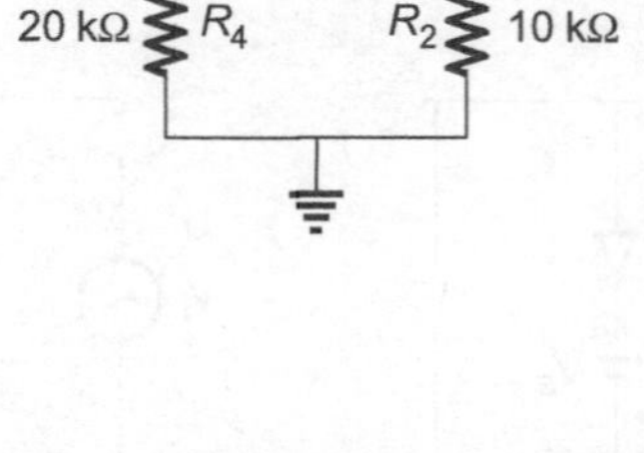

[12 marks : 2022]

Solution:

Consider diode D_1 on and diode D_2 off, hence the circuit diagram becomes,

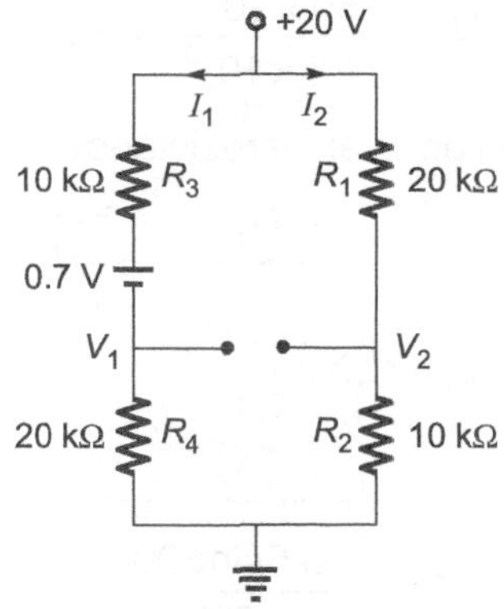

Current I_1 flows through diode D_1, equation becomes,

$$-20 + 10k \times I_1 + 0.7 + 20k \times I_1 = 0$$

$$I_1 = 0.64 \text{ mA} \quad \text{...(i)}$$

Current I_2 flows through R_1 and R_2

$$-20 + I_2 \times (20k + 10k) = 0$$

$$I_2 = \frac{20}{30k} = 0.67 \text{ mA} \quad \text{..(iii)}$$

Voltage, $V_1 = I_1 \times 20\text{ k}$

$= 12.87$ V

Voltage, $V_2 = I_2 \times 10\text{ k} = 6.67\ V_1$

As $V_1 > V_2$, hence our assumption is correct, D_1 is on and D_2 is off

Current through D_1, $I_1 = 0.64$ mA

Current through D_2, $= 0$

Voltage drop across resistor $R_4 = I_1 \times 20\text{ k}$

$= 12.87$ V

2.21 Find the circulating current in the circuit shown below operated at a room temperature of 27° C, if the source voltage V_S = 1.2 + 0.5 sin 500t Volts. Assume ηV_T = 50 mV and V_{ON} = 0.7 volts. Neglect contact resistance.

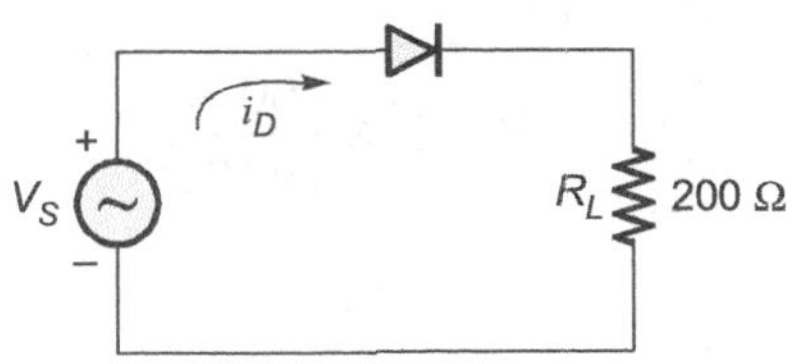

[8 marks : 2022]

Solution:

Given,

$$V_s = \underbrace{1.2}_{\text{DC component}} + \underbrace{0.5\sin 500t}_{\text{AC component of supply}} \text{ Volts}$$

For DC component,

$$V_1 = 1.2 \text{ Volt}$$

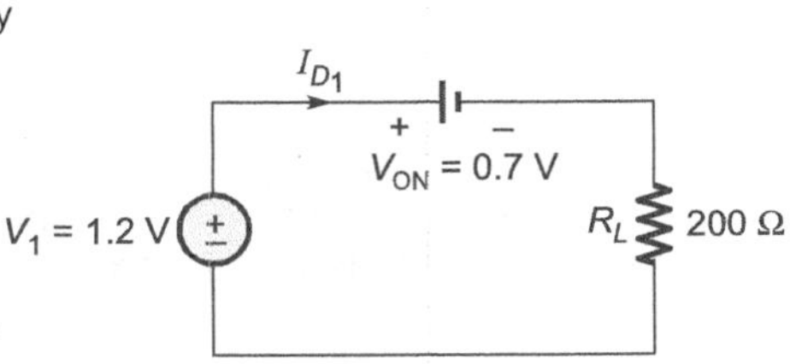

Current due to DC component of supply,

$$I_{D1} = \frac{V_1 - V_{ON}}{R_L} = \frac{1.2 - 0.7}{200} = 2.5 \text{ mA}$$

For AC component of supply,

$$V_2 = 0.5 \sin 500t \text{ Volt}$$

Dynamic AC resistance of diode is given as,

$$r_d = \frac{nV_T}{I_{D_1}} = \frac{50\times10^{-3}}{2.5\times10^{-3}} = 20\ \Omega$$

Here, we will replace the diode with its equivalent resistance for AC supply

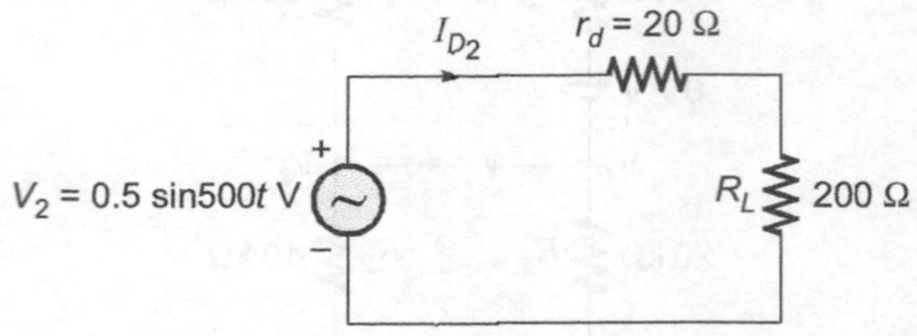

$$I_{D_2} = \frac{V_2}{r_d + R_L} = \frac{0.5\sin 500t}{220} = 2.27 \sin 500t \text{ mA}$$

Therefore, circulating current i_D is given as

$$i_D = I_{D_1} + I_{D_2} = (2.5 + 2.27\sin 500t)\text{ mA}$$

2.22 **(i) The reverse recovery time t_{rr} of a diode is 2 μs. In the conducting mode to reverse blocking mode operation, the diode needs the rate of fall of forward current of 50 amperes/μs. Determine the stroage charge and the peak reverse current.**

(ii) A diode with 500 mW power dissipation at 25°C has 5 mW/°C derating factor. If the forward voltage drop remains constant at 0.7 V, calculate the maximum forward current at 50°C.

[6 + 6 marks : 2023]

Solution:

(i) Given reverse recovery time,

$$t_{rr} = 2\ \mu\text{s}$$

$$\frac{di}{dt} = \frac{I_F}{T} = 50 \text{ amp/μs}$$

Storage charge, $Q_S = ?$

Peak reverse current, $I_{RR} = ?$

$$Q_S = \frac{1}{2}\times t_{rr}^2\frac{di}{dt}$$

$$= \frac{1}{2}\times(2\times10^{-6})^2\times\frac{50\text{ A}}{1\times10^{-6}}$$

$$Q_S = \frac{1}{2}\times4\times10^{-6}\times50$$

$$Q_S = 100\ \mu\text{C}$$

Peak reverse current I_{RR}

$$Q_S = \frac{1}{2}\times I_{RR}\times t_{rr}$$

$$I_{RR} = \frac{Q_S\times2}{t_{rr}} = \frac{100\mu\text{C}\times2}{2\ \mu\text{S}}$$

$$I_{RR} = 100\text{ A}$$

(ii) At 25° C, derating factor

$$D = 5\text{ mW/°C}$$

$$V_D = 0.7\text{ V}$$

I_F at 50° C = ?

$$\Delta P_D = D(T_j - T_A)$$

$$= 5\ \text{mW/°C}\ (50° - 25°\ \text{C}) = 125\ \text{mW}$$

$$P_D(50°\ \text{C}) = 500\ \text{mW} - 125\ \text{mW} = 375\ \text{mW}$$

$$P_D(50°\ \text{C}) = V_D I_F$$

$$\Rightarrow \qquad I_F = \frac{P_D}{V_D} = \frac{375\ \text{mW}}{0.7\ \text{V}}$$

$$I_F = 535.71\ \text{mA}$$

3. BJT-Characteristics, Biasing and Thermal Stabilization

3.1 A JFET has channel length, l = 10 μm, channel width W = 100 mm, channel height without any depletion = 2.5 μm. The channel is n-type with donor level of $N_D = 10^{22}/\text{m}^3$ and $m_n = 0.1500\ \text{m}^2/\text{V-sec}$. The depletion width from each side of the gate junction is 0.25 μm. How does this semiconductor bar behave without forward biasing of the gate junction between two ends of the channel length?

[4 marks : 2001]

Solution :

$$\text{ON drain resistance} = \frac{V_{DS}}{I_D} = r_{DS}(\text{ON}) = \frac{1}{TqN_D\,\mu_N\left(\dfrac{W}{L}\right)}$$

$$= \frac{10\times10^{-6}}{2.5\times10^{-6}\times1.6\times10^{-19}\times0.15\times10^{22}\times100\times10^{-6}} = 166.67\ \Omega$$

When ($V_{GS} = 0$), that is gate and source are shorted, the channel is fully opened. The drain current increase linearly with a small drain voltage and the bar behaves like a simple semi-conductor resistor.

3.2 Explain how you obtain h_{fe} parameter of a transistor graphically.

[18 marks : 2002]

Solution:

The hybrid h-model of a transistor in *CE* configuration can be obtained by first writing the terminal equation using the graphical symbol

$$V_{be} = h_{ie} I_b + h_{re} V_{ce}$$
$$I_c = h_{fe} I_b + h_{oe} V_{ce}$$

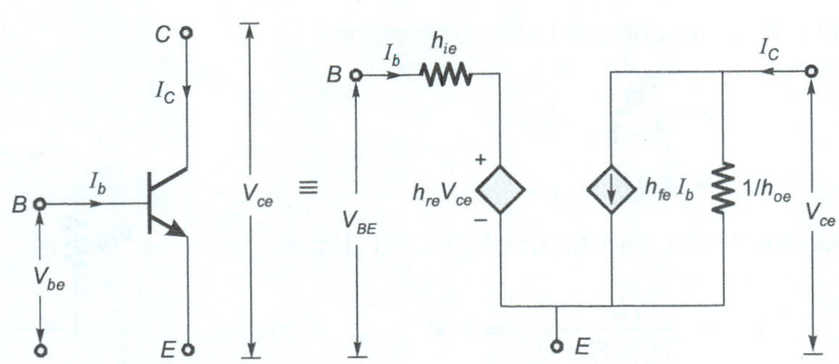

Graphic symbol h-parameter model

It is possible to calculate h-parameters for a given Q-point from the output and input characteristics of transistor by using definition of the parameters.

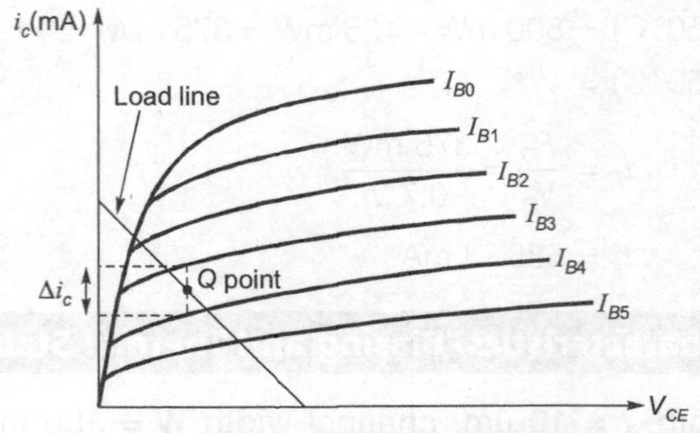

Calculate h_{fe} from CE output configuration

$$h_{fe} = \left.\frac{\Delta i_C}{\Delta i_B}\right|_{V_{CE}} = \frac{\Delta i_C}{I_{B3} - I_{B4}}$$

3.3 Consider the circuit shown below. Let V_{CC} = 5 V, R_C = 1 kΩ, $V_{BE(ON)}$ = 0.7 and β = 120. Choose R_E and determine R_1 and R_2 such that the circuit is bias stable and that V_{CEQ} = 3 V.

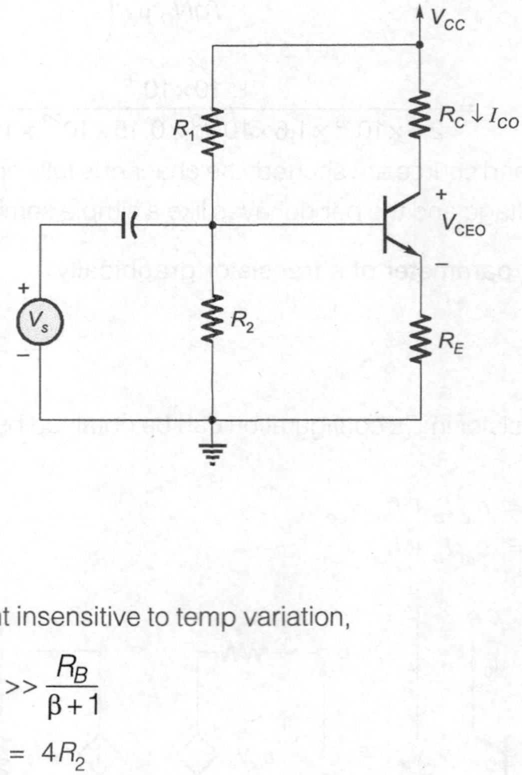

[10 marks : 2006]

Solution:

To make the operating point insensitive to temp variation,

$$R_E >> \frac{R_B}{\beta+1}$$

Let, $\quad R_1 = 4R_2$

Now, Thevenine equivalent circuit will for base-ground circuit,

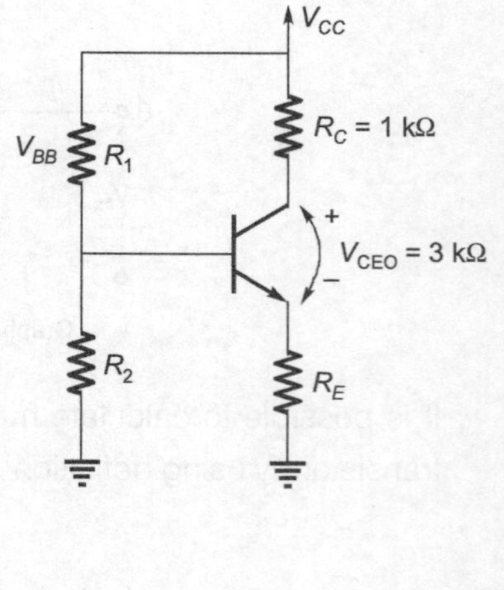

$$V_{th} = \frac{R_2}{R_1 + R_2} \cdot V_{CC} = 1\text{ V}$$

$$R_{th} = R_1 || R_2 = 4R_2 || R_2 = \frac{4}{5}R_2$$

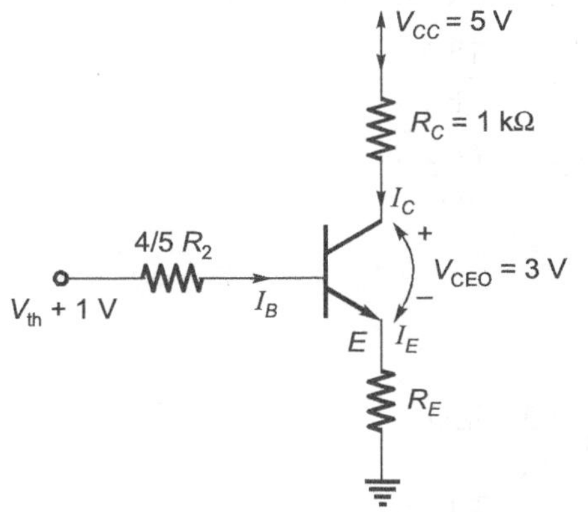

So KVL in CE region,

$$V_{CC} - R_C I_C - V_{CEQ} - I_E R_E = 0$$

$$5 - I_C - 3 - V_E = 0$$

$$2 = I_C + V_E$$

$\because$ β is very high.

Assuming, $\quad I_B \simeq 0$

$\therefore \quad V_E = V_{th} - V_{BE\ on} = 0.3\text{ V}$

$\therefore \quad I_C = 1.7\text{ mA} \simeq I_E$

$\therefore \quad R_E = \dfrac{V_E}{I_E} = \dfrac{0.3}{1.7\text{mA}} = 0.17647 = 0.177\text{ k}\Omega$

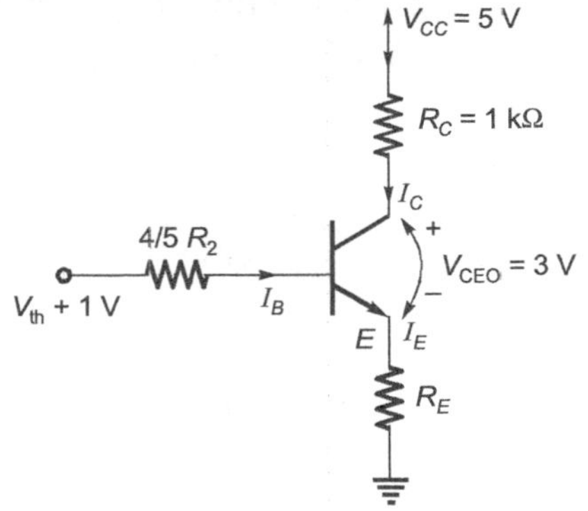

$\because \quad R_E >> \dfrac{R_B}{\beta+1}; \qquad R_B = \dfrac{4}{5}R_2 << (\beta+1)R_E$

$$R_2 << \frac{5}{4} \times 121 \times 0.177\text{k}\Omega$$

$\Rightarrow \quad R_2 << 26.77\text{ k}\Omega$

Let, $\quad R_2 = 1\text{ k}\Omega$

Then, $\quad R_1 = 4R_2 = 4\text{ k}\Omega$

3.4 The bipolar transistor in figure is specified to have β_F in the range of 8 to 40. The load resistance is $R_C = 11\ \Omega$. The d.c. supply voltage is $V_{CC} = 200$ V and the input voltage to the base circuit is $V_B = 10$ V. If $V_{CE(sat)} = 1.0$ V and $V_{BE(sat)} = 1.5$ V, find

(a) the value of R_B that results in saturation with an ODF of 5,

(b) the β_{forced}, and

(c) the power loss P_T in the transistor.

[6 marks : 2006]

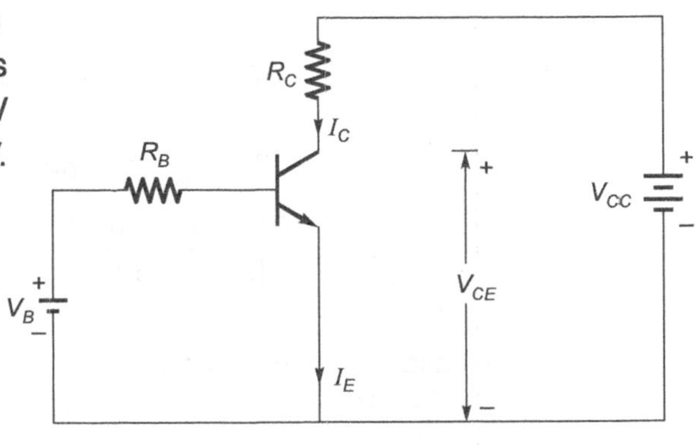

Solution:

When the transistor is in saturation, the collector current will be

$$V_c = V_{CE}(\text{sat}) = 1\text{ V}$$

$\therefore \quad \text{Collector current} = \dfrac{200-1}{11} \cong 18.1\text{ A}$

To saturate the transistor with lowest 'β' we need to provide a base of atleast

$$I_B = \frac{I_c(sat)}{\beta_{min}} = \frac{18.1}{8} = 2.26 \text{ A}$$

With overdrive factor 5, $I_B = 5 \times 2.26 = 11.3$ A

(a) Thus required value of R_B

$$\frac{10-1.5}{R_B} = 11.3$$

$$R_B = \frac{8.5}{11.3} = 0.752\Omega$$

(b) $$\beta_{forced} = \frac{I_{csat}}{I_B} = \frac{18.1}{11.3} = 1.60$$

(c) Power loss in transistor, $P_T = V_{CE} \times I_{CE} = (V_{CE\,sat}) \times (I_{CE\,sat})$

$$P_T = 1 \times 18.1$$
$$= 18.1 \text{ Watt}$$

3.5 The circuit shown uses a silicon transistor with β = 50, V_{BE} = 0.6 V, V_{CC} = 22.5 V and R_C = 5.6 kΩ. Find the values of the resistors R_E, R_1 and R_2 so that Q point is set at V_{CE} = 12 V and I_C = 1.5 mA. The stability factor, S must be ≤ 3.

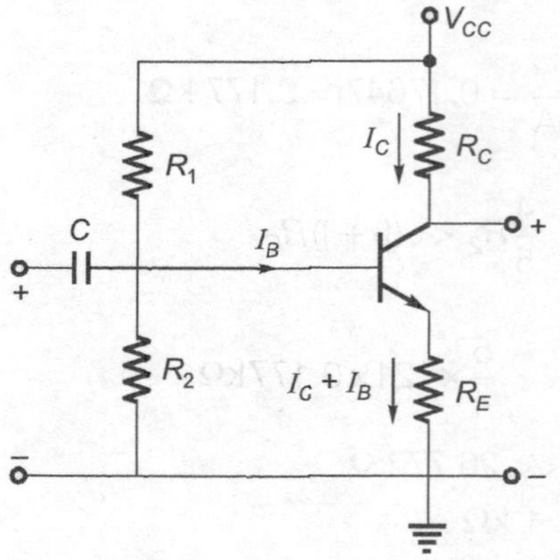

[20 marks : 2011]

Solution:

Given that, $\beta = 50$, $V_{BE} = 0.6$ V, $V_{CC} = 22.5$ V

$$R_C = 5.6 \text{ k}\Omega$$

$$V_{CE} = 12 \text{ V}, I_C = 1.5 \text{ mA} = 1.5 \times 10^{-3} \text{ A}$$

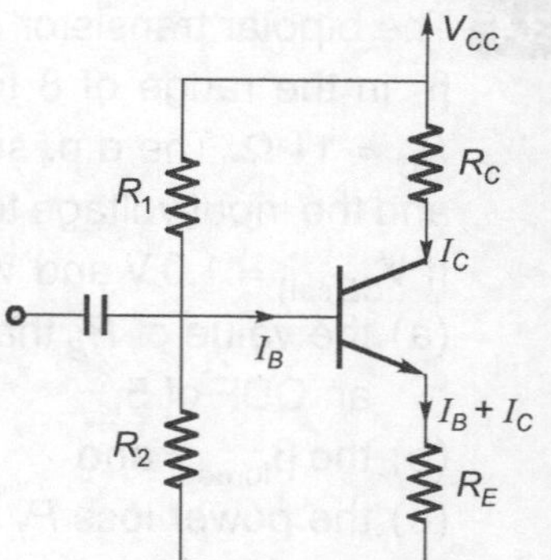

$$\therefore \quad I_B = \frac{I_c}{\beta} = \frac{1.5\times10^{-3}}{50}$$

$$I_B = 3 \times 10^{-5} \text{ A}$$

∴ Now from collector emitter loop we have

$$V_{CC} = I_C R_C + V_{CE} + (I_B + I_C) R_E$$

$$22.5 = (1.5 \times 10^{-3}) \times 5.6 \times 10^3 + 12 + (1.5 \times 10^{-3} + 3 \times 10^{-5}) R_E$$

$$2.1 = (1.5 + 0.03) \times 10^{-3} R_E$$

$$\Rightarrow \quad R_E = \frac{2.1}{1.53\times10^{-3}}$$

$$R_E = 1.3725 \times 10^3 \ \Omega = 1.373 \text{ k}\Omega$$

Given, $S \leq 3$

we know, $S = \dfrac{(\beta+1)(R_{th}+R_E)}{R_{th}+(\beta+1)R_E}$ where $R_{th} = (R_1 \| R_2)$

$$\frac{(\beta+1)(R_{th}+R_E)}{R_{th}+(\beta+1)R_E} \leq 3$$

$$(\beta + 1)(R_{th} + R_E) \leq 3[R_{th} + (\beta + 1R_E)]$$

$$51(R_{th} + R_E) \leq 3R_{th} + 3 \times 51\ R_E$$

$$48R_{th} \leq 102\ R_E$$

$$R_{th} \leq \frac{102}{48} R_E$$

$$R_{th} \leq 2.125\ R_E$$

$$R_{th} \leq 2.916\ k\Omega \quad ...(i)$$

Now from the equivalent circuit,

$$V_{TH} = \frac{R_2}{(R_1+R_2)} V_{CC}$$

$\therefore$ From base – Emitter loop

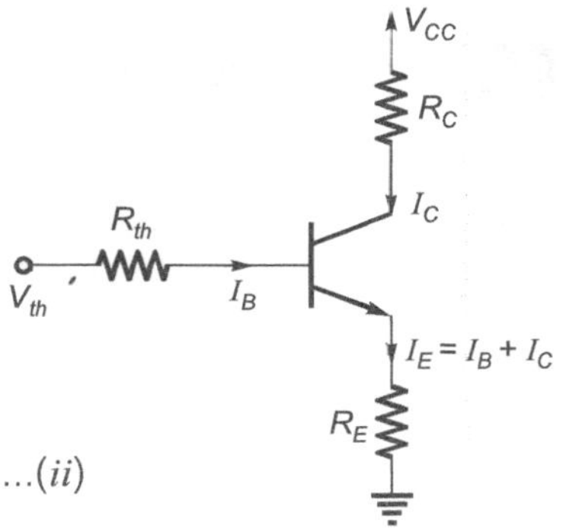

$$V_{th} = R_{th} I_B + V_{BE} + I_E R_E$$

$$V_{th} = 0.03 \times 10^{-3}\ R_{th} + 0.6 + 2.1$$

$$\frac{R_2}{R_1+R_2} \times 22.5 = 0.03 \times 10^{-3}\ R_{th} + 2.7$$

$$\frac{R_2}{R_1+R_2} \times 22.5 = 0.03 \times 10^{-3} \times \frac{R_1 R_2}{R_1+R_2} + 2.7$$

$$22.5\ R_2 = 0.03 \times 10^{-3}\ R_1 R_2 + 2.7(R_1 + R_2) \quad ...(ii)$$

From equation (*i*), $R_{th} \leq 2.916 \times 10^3\ \Omega$

$$\frac{R_1 R_2}{R_1+R_2} \leq 2.916 \times 10^3\ \Omega$$

$$R_1 R_2 \leq (2.916 \times 10^3)(R_1 + R_2)\ \Omega^2$$

Taking, $R_1 R_2 = 2.916 \times 10^3\ (R_1 + R_2)\ \Omega^2 \quad ...(iii)$

From equation (*ii*) and (*iii*) we have

$$22.5\ R_2 = 0.03 \times 10^{-3} \times 2.916 \times 10^3\ (R_1 + R_2) + 2.7\ (R_1 + R_2)$$

$$22.5\ R_2 = (2.78748)\ (R_1 + R_2)$$

$$R_2 = 0.1414\ R_1$$

Let, $R_2 = 5\ k\Omega$

Then, $R_1 = 35.359\ k\Omega$

3.6 Explain briefly the early effect phenomenon in BJT. [5 marks : 2013]

Solution:

Early Effect: The transition region at a junction is the region of uncovered charges on both sides of the junction at the positions occupied by the impurity atoms. As the voltage applied across the CB junction increases, the transition region penetrates deeper into the collector and base. Since the doping in the base is ordinarily substantially smaller than that of the collector, the penetration of the transition region into the base is much larger than into the collector. As a result the electrical base width of the transistor is reduced in comparison to metallurgical base width. This modulation of the effective base width by the collector voltage is known as the **Early effect** after J.M. Early, who first interpreted it.

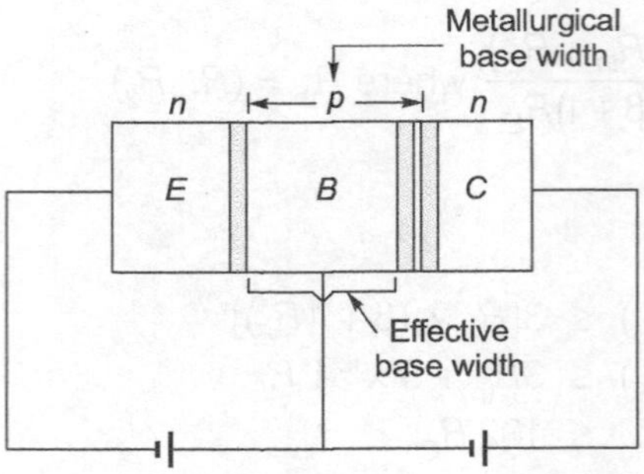

The decrease in base width with increasing reverse collector voltage has three consequences:

- There is less chance for recombination within the base region. Hence α increases with increasing $|V_{CB}|$.
- The concentration gradient of minority carriers is increased within the base. Since the hole current injected across the emitter is proportional to the gradient of minority carriers at emitter junction, then I_E increases with increasing reverse collector voltage.
- For extremely large voltages, base width may be reduced to zero causing voltage breakdown in the transistor. This phenomenon is known as **punch through** or **reach through.**

3.7 For the circuit shown below:

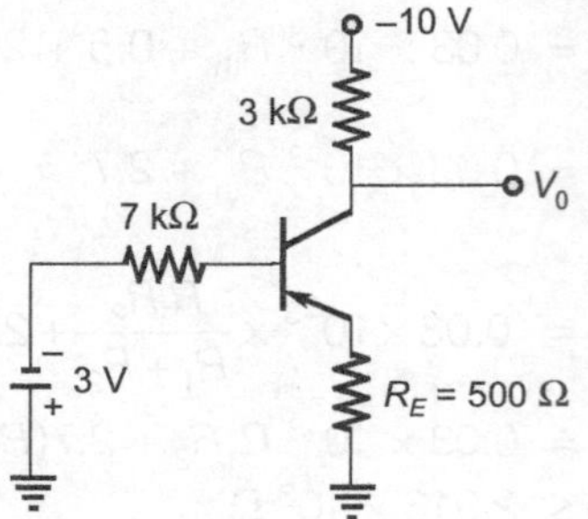

$\beta = h_{FE} = 100$

(i) Verify if the transistor is in cut-off, saturation or in the active region.

(ii) Find V_0.

[15 marks : 2013]

Solution:

(i) Assuming the transistor in active mode, applying KVL,

$-500\, I_E - V_{EB} - (7\text{ k}\Omega)I_B + 3 = 0$

$$I_B = \frac{I_E}{\beta+1} = \frac{I_E}{101}$$

$$3 - 0.7 = \frac{7 \times I_E}{101} + 0.5 I_E$$

$$I_E = 4.04 \text{ mA}$$

$$I_C = 0.99 \times 4.04 = 4 \text{ mA}$$

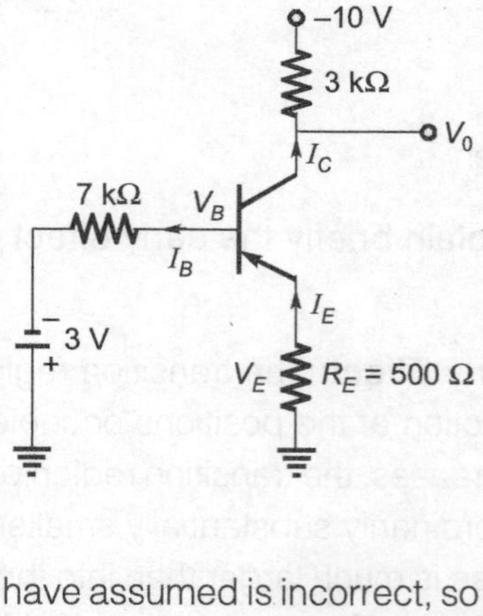

To find V_B : $V_B = R_e I_E - V_{EB}$

$$V_B = -(0.5)(4.04) - 0.7 = -2.72 \text{ V}$$

$$V_0 = -10 + R_C I_C$$

$$V_0 = -10 + 3 \times 4 = 2 \text{ V}$$

So, collector base diode is also forward biased. Hence, what we have assumed is incorrect, so transistor is in saturation mode.

(ii) $V_0 = 2$ V

3.8 The following circuit is used as a CE amplifier employing potential divider bias. If h_{fe} = 50 and V_{BE} = 0.6 V, determine the values of R_C, R_E, R_{B_1} and R_{B_2} for quiescent operating point I_C = 1 mA and V_{CE} = 7.5 V. (Assume $R_B = R_{B_1} \| R_{B_2} = 10\,R_E$)

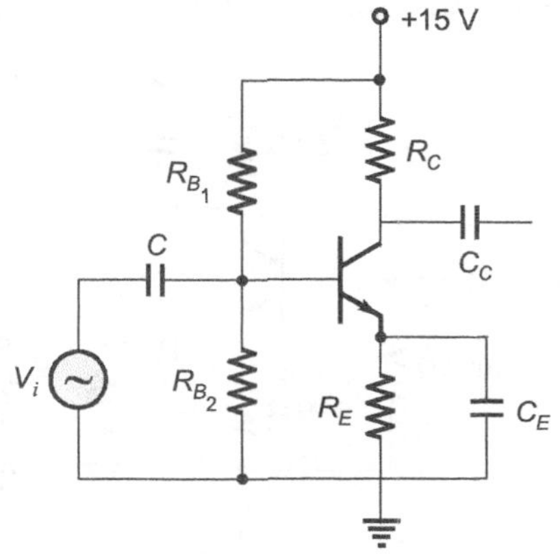

[15 marks : 2015]

Solution:

Given:

$$h_{fe} = \beta = 50$$
$$V_{BE} = 0.6\text{ V}$$
$$I_C = 1\text{ mA}$$
$$V_{CE} = 7.5\text{ V}$$

Now from the given circuit,

$$V_{CC} = R_C I_C + I_E R_E + V_{CE}$$

Since, $\beta = 50$ so $\alpha = 0.98$

and $$I_E = \frac{I_C}{\alpha} = 1.02\text{ mA}$$

and $$I_B = \frac{I_C}{\beta} = 0.02\text{ mA}$$

So, $$15 = 1 \times 10^{-3} \times R_C + R_E \times 1.02 \times 10^{-3} + 7.5$$

$\Rightarrow$ $$7.5 = (R_C \times 10^{-3}) + (R_E \times 1.02 \times 10^{-3}) \quad ...(1)$$

Now let, $R_C = R_E$

so, from equation (1) we get, $R_C = R_E = 3.7\text{ k}\Omega$

From the assumption, $R_B = R_{B1} \| R_{B2} = 10\,R_E = 37\text{ k}\Omega$

Consider the circuit shown in figure,

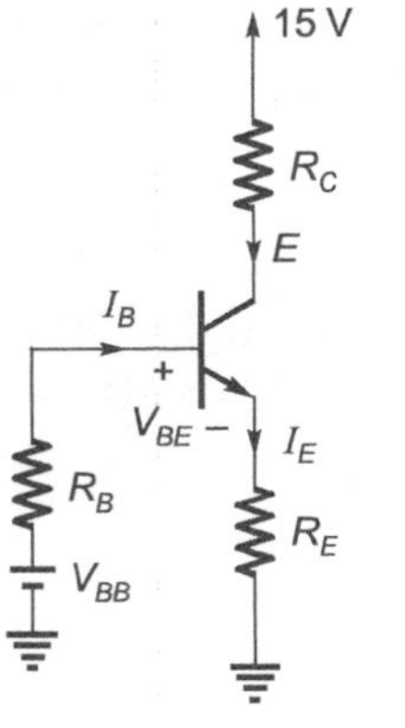

Now, applying kVL in the base-emitter loop,

$$V_{BB} = R_B I_B + V_{BE} + I_E R_E$$

so, $$V_{BB} = 5.114$$

Since, $$V_{BB} = \frac{15 \times R_{B2}}{R_{B1} + R_{B2}} = 5.114\text{ V}$$

Also, $$R_B = \frac{R_{B1} R_{B2}}{R_{B1} + R_{B2}} = 37\text{ k}\Omega$$

$\Rightarrow$ $$R_{B1} = 108.5\text{ k}\Omega \text{ and } R_{B2} = 56.1\text{ k}\Omega$$

Thus, $R_{B1} = 108.5\ \text{k}\Omega \quad R_{B2} = 56.1\ \text{k}\Omega$

$R_E = 3.7\ \text{k}\Omega$

$R_C = 3.7\ \text{k}\Omega$

3.9 Find the Q-point and the maximum symmetrical V_{out} for the transistor amplifier shown in the following figure. Show both dc and ac load lines. Assume standard design rules.

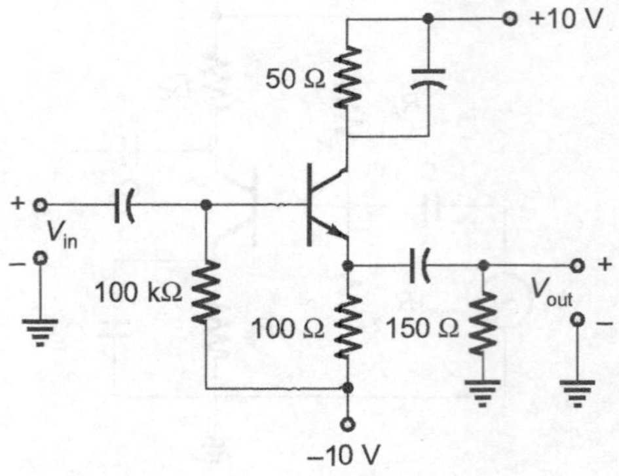

[10 marks : 2016]

Solution:

The given circuit is

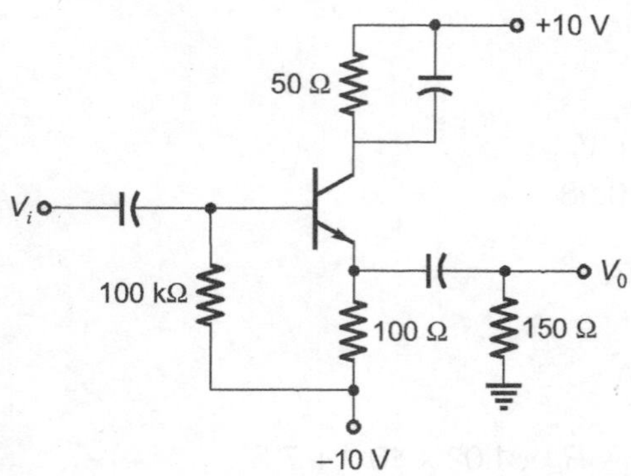

DC Model:

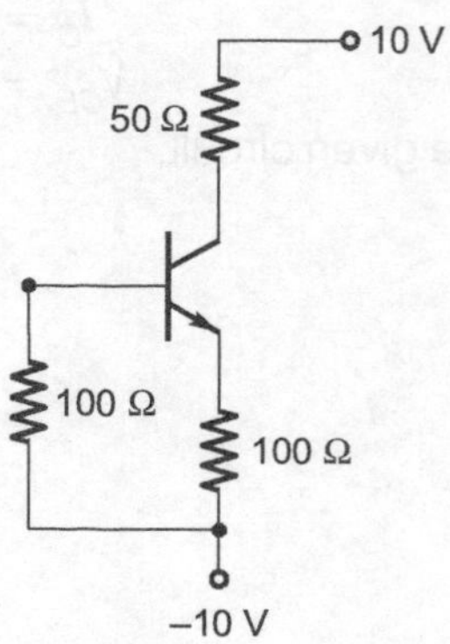

The operating point $Q((I_C)_Q, (V_{CE})_Q)$

DC load line:

$$\frac{2V_{CC}}{R_C + R_E} = \frac{2 \times 10}{100 + 50} = 133.3\ \text{mA}$$

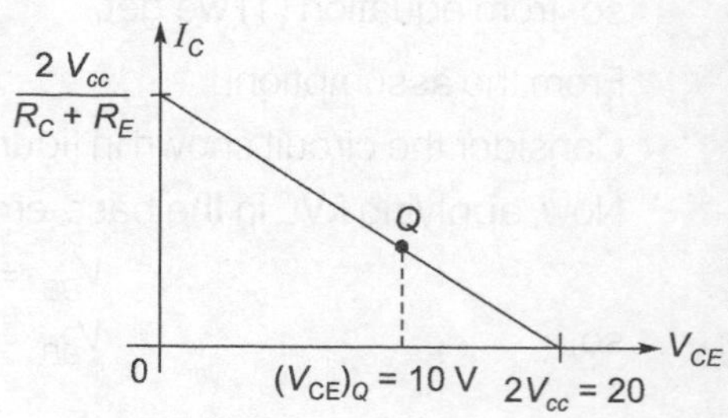

To get the maximum symmetrical V_{out} for the transistor amplifier the Q point should be centre of dc load line.

$$\therefore \quad (V_{CE})_Q = \frac{2V_{CC}}{2} = 10\ \text{V}$$

To find $(I_C)_Q$ apply kVL to dc model at output loop

$$10 - (I_c)_Q \times 50\ \Omega - (V_{CE})_Q - 100 \times (I_C)_Q - (-10) = 0$$

$$(I_C)_Q = \frac{20 - 10}{150} = \frac{10}{150} = 66.6\ \text{mA}$$

AC model:

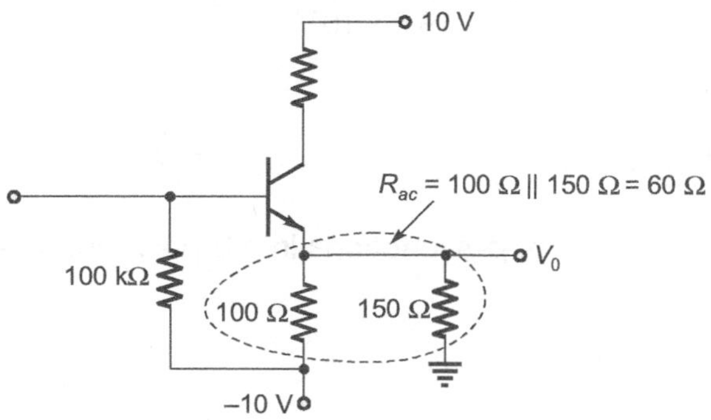

AC load line:

$$(I_C)_Q + \frac{(V_{CE})_Q}{R_{ac}} = 66.6\text{mA} + \frac{10}{60\Omega}$$
$$= 66.6 \text{ mA} + 166.6 \text{ mA}$$
$$= 233.2 \text{ mA}$$
$$(V_{CE})_Q + (I_C)_Q R_{ac} = 10 + 66.6 \text{ mA} \times 60\ \Omega$$
$$= 13.99 \text{ V}$$

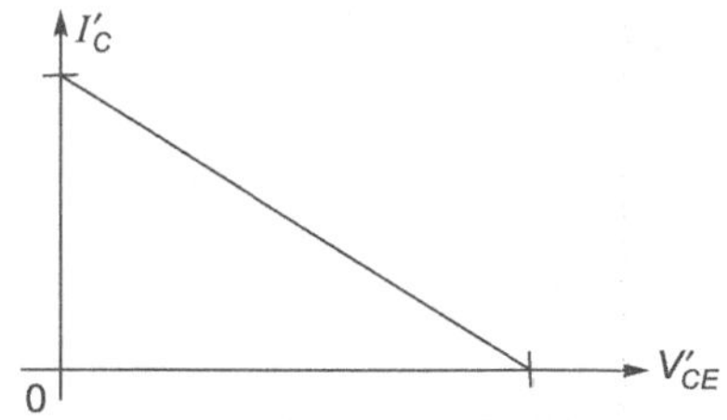

3.10 Find R such that at dc, $V_{out} = 0$ for the circuit shown in the following figure. Also find I_{CQ1} and I_{CQ2}. Assume that $V_{BE} = 0.7$ V and $\beta = 100$ for both transistors.

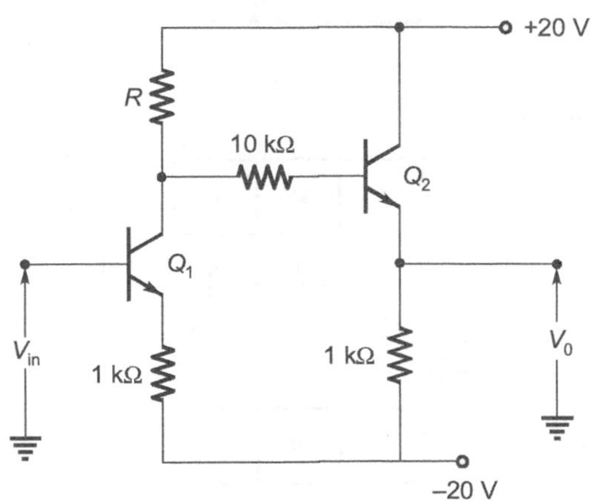

[10 marks : 2016]

Solution:

$$I_{E_{Q_1}} = \frac{-0.7-(-20)}{1\text{k}\Omega} = 19.3 \text{ mA}$$

$$I_{E_{Q_1}} \approx I_{C_{Q_1}}$$

$$I_{B_{Q_2}} = \frac{I_{C_{Q_2}}}{\beta}$$

$$I_{E_{Q_2}} = \frac{0-(-20)}{1\text{k}} = 20 \text{ mA}$$

$$I_{B_{Q_2}} = \frac{I_{C_{Q_2}}}{\beta} = \frac{20 \times 10^{-3}}{100} = 200\ \mu\text{A}$$

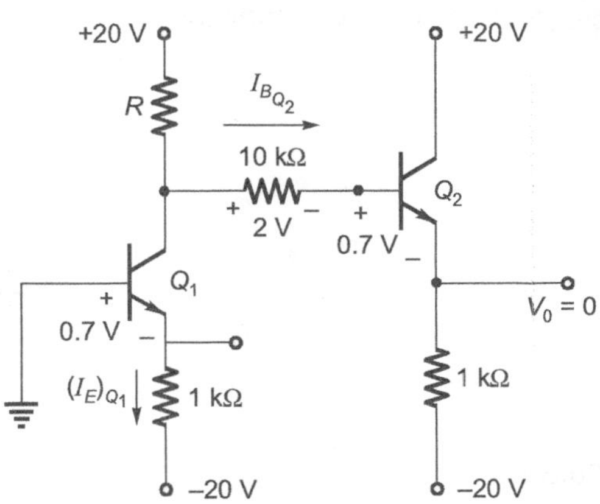

$$\therefore \qquad R = \frac{20 - V_c}{I_{C_{Q_1}}} = \frac{20 - 2.7}{19.3 \times 10^{-3}} = 896\ \Omega$$

$$I_{C_{Q_1}} = 19.3\text{ mA}$$

$$I_{C_{Q2}} = 20\text{ mA} \qquad (I_{C_{Q_2}} \simeq I_{E_{Q_2}})$$

3.11 **Discuss the biasing conditions of *p-n* junctions in a Bipolar Junction transistor under saturation and cut-off conditions.**

[4 marks : 2017]

Solution:

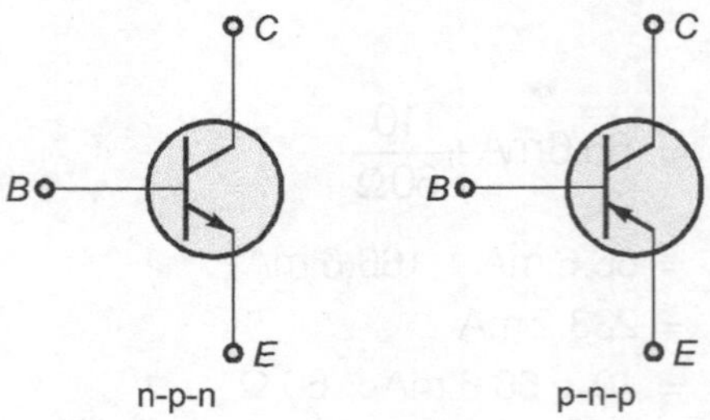

For example:

Block diagram of p-n-p transistor is considered,

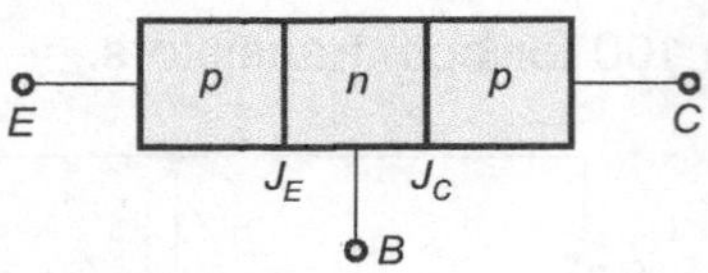

At saturation J_E is *FB*, J_C is forward bias,

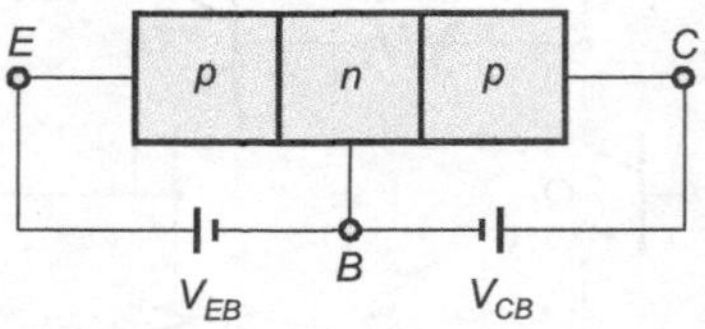

Cut-off J_E is reverse bias, J_C is reverse bias,

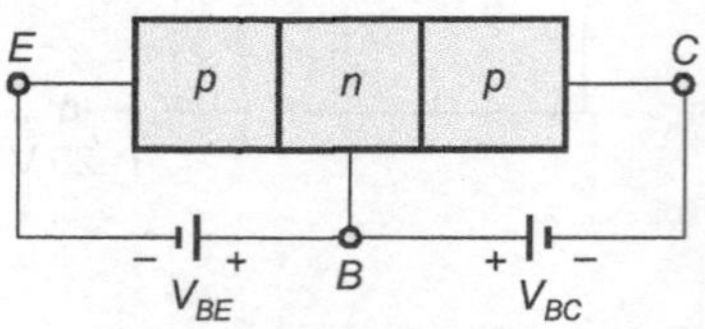

3.12 **Determine whether the transistor circuit shown can be termed as being saturated.**

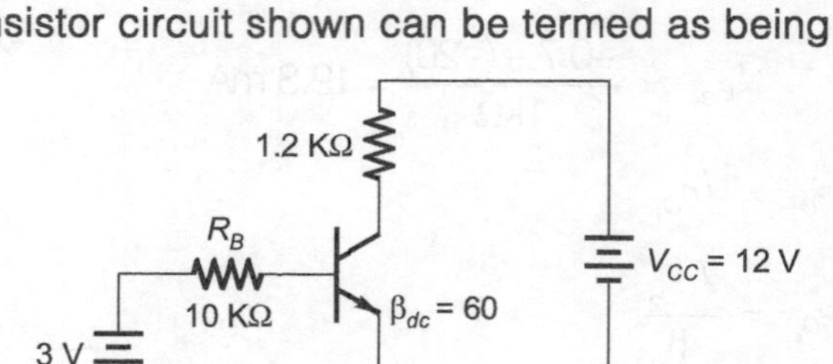

[4 marks : 2017]

Solution:

Assume *Q* is in active,

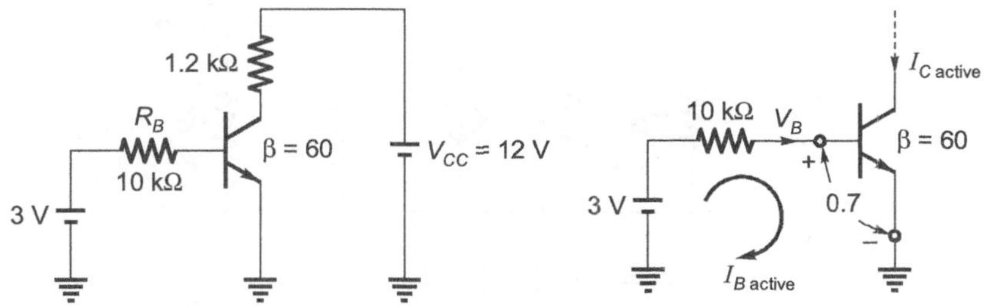

$$V_{BE\,\text{active}} = 0.7\text{ V}$$

$$I_{B\,\text{active}} = \frac{3-0.7}{10\text{ k}} = 0.23\text{ mA}$$

$$I_{C\,\text{active}} = \beta I_{B\,\text{active}}$$

$$= 0.23 \times 10^{-3} \times 60 = 13.8\text{ mA}$$

$$V_C = V_{CC} - I_C R_C$$

$$= 12 - 1.2\text{ k} \times 13.8 \times 10^{-3} = -4.56\text{ V}$$

where, $V_B = 0.7$ V

Our assumption is wrong. Q is in saturation.

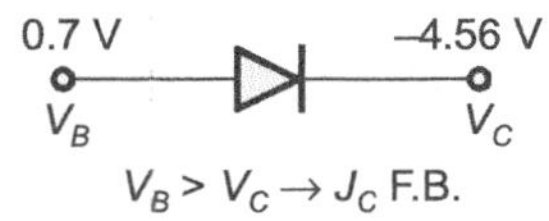

3.13 **Obtain the expression for voltage gain to given circuit,**

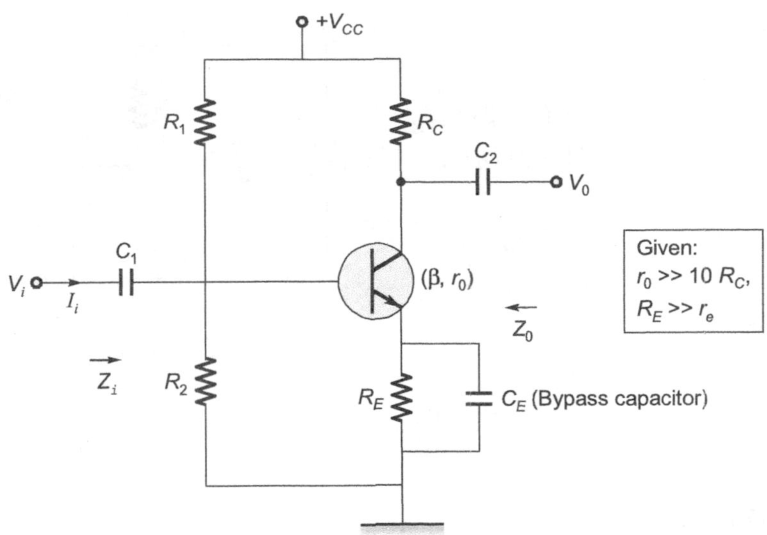

(i) With bypass capacitor.

(ii) Without bypass capacitor.

(iii) Calculate voltage gain in part (i) and (ii), if R_1 = 90 kΩ, R_2 = 10 kΩ, R_E = 0.8 kΩ, β = 200, r_0 = 50 kΩ, R_C = 2.2 kΩ, and V_{CC} = +15 V. Compare both voltage gains and write conclusion in short.

[5+5+10 = 20 marks : 2018]

Solution:

(i) With bypass capacitor:

While drawing small signal approximate equivalent circuit, capacitor will get short circuit,

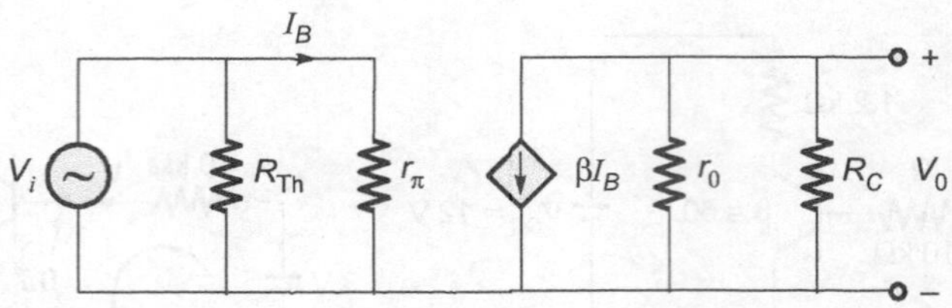

At output: $V_0 = -\beta I_B (R_C \| r_0)$

As $r_0 >> 10\,R_C$

$\therefore$ $R_C \| r_0 = R_C$

$\therefore$ $V_0 = -\beta I_B \cdot R_C$

At input: $V_{in} = r_\pi \times I_B$

$$\therefore \quad \frac{V_0}{V_{in}} = -\frac{\beta R_C}{r_\pi}$$

$$\therefore \quad \text{Voltage gain } A_v = -\frac{\beta R_C}{r_\pi}$$

where, $r_\pi = (1+\beta) r_e$

and $r_e = \dfrac{V_T}{I_C}$

(ii) Without bypass capacitor:

As $r_0 >> 10\,R_C$ it can be ignored

At output: $V_0 = -\beta I_B \times R_C$

At input, $V_{in} = I_B \cdot r_\pi + (1+\beta) R_E I_B$

$\therefore$ $V_{in} = (r_\pi + (1+\beta) R_E) I_B$

As $R_E >> r_e$

$\therefore$ $V_{in} \approx (1+\beta) R_E I_B$

$V_{in} \approx \beta R_E I_B$

$$\therefore \quad \frac{V_0}{V_{in}} = \frac{-\beta I_B R_C}{\beta I_B R_E}$$

$$\therefore \quad A_V = \text{voltage gain} = -\frac{R_C}{R_E}$$

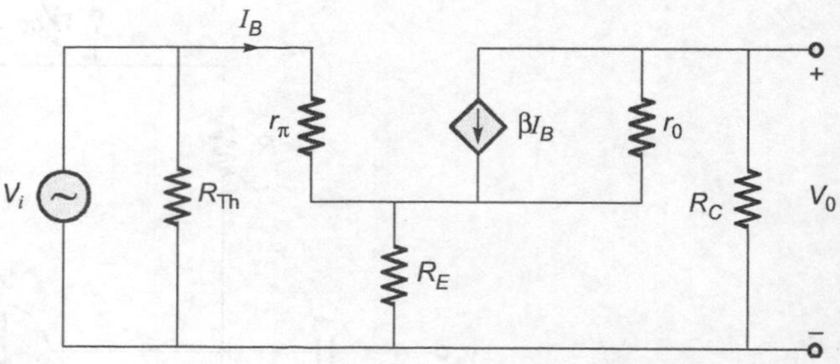

(iii) Doing DC analysis for part (i):

In loop (A),

$$1.5\text{ V} = I_B \cdot 9\text{ k}\Omega + 0.7 + (1+\beta) R_E \times I_B$$

$$I_B = \frac{1.5 - 0.7}{(9 + 201 \times 0.8)\text{ k}\Omega} = 4.7\,\mu\text{A}$$

$$I_E = (1+\beta) I_\beta = 0.947\text{ mA},$$

$$I_C = \beta I_B = 0.942\text{ mA}$$

$$r_\pi = (1+\beta) \cdot \frac{V_T}{I_C} = 5.54\text{ k}\Omega$$

$\therefore$ Voltage gain with bypass capacitor,

$$A_V = \frac{-\beta R_C}{r_\pi} = -79.42$$

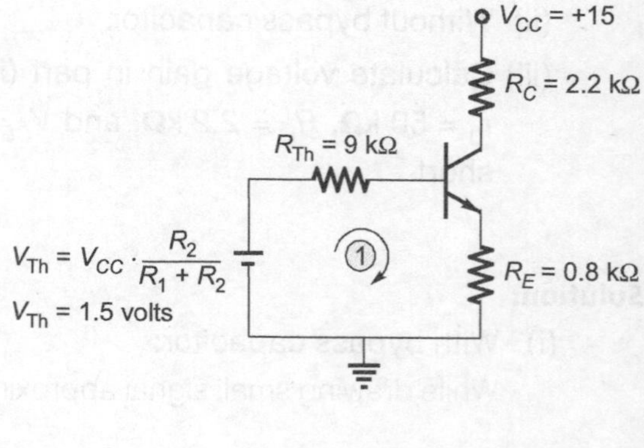

and, voltage gain without bypass capacitor,

$$A_V = -\frac{R_C}{R_E} = -2.75$$

Conclusion: Voltage gain with bypass capacitor is larger than that of without bypass capacitor, this is because bypass capacitor removes negative feedback of 'R_E' and improve voltage gain.

3.14 **Bipolar junction transistors (BJTs) are considered "normally off" devices, because their natural state with no signal applied to the base is no conduction between emitter and collector, like an open switch. Are junction field effect transistors (JFETs) considered the same? Why or why not? Justify your answer.**

[6 marks : 2019]

Solution:

Normally there is a channel between drain and source, JFET operates as a ON switch during normal condition (absence of gate signal). When a reverse bias voltage is applied to the gate terminal, then the width of drain to source voltage starts decreasing and at a particular voltage pinch off region is created and JFET stops conducting. In BJT, if no current is flowing through base terminal than there will be no current between emitter and collector and it works as a normally off device. BJT will starts operating when there is a base current is the base terminal.

3.15 **Find R_B in the figure shown below for silicon transistor with β = 100 and negligible leakage current, if V_{CC} = 20 V, V_{EE} = 5 V, R_E = 100 Ω, R_C = 2 kΩ and I_C = 6 mA.**

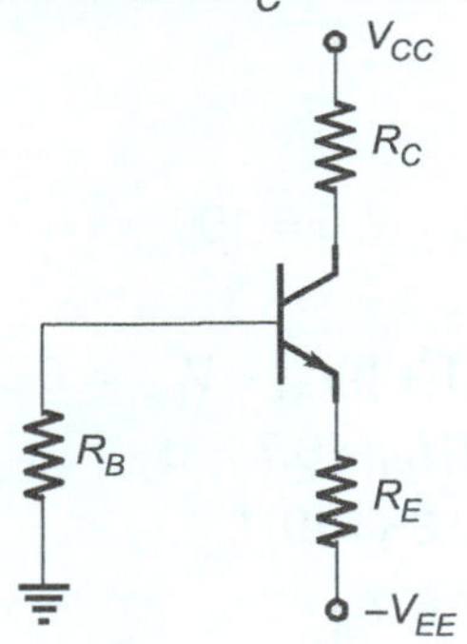

[10 marks : 2020]

Solution:

Given BJT Circuit can be represented as shown below :

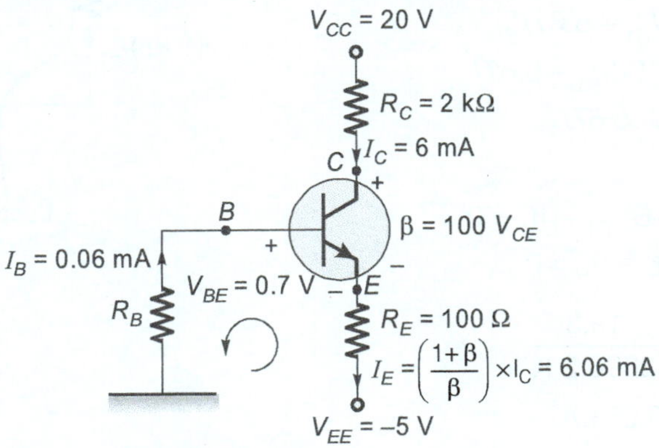

$$I_B = \frac{I_C}{\beta} = \frac{6}{100} = 0.6 \text{ mA}$$

$$I_E = I_B + I_C = 6.06 \text{ mA}$$

Since given transistor is npn type transistor.

So,
$$I_B + I_C = I_E$$
$$I_B = I_E - I_C = 6.06 - 6 = 0.06 \text{ mA}$$

Apply KVL in input loop

$$I_B R_B + V_{BE} + I_E R_E - V_{EE} = 0$$

$$I_B R_B = V_{EE} - I_E R_E - V_{BE}$$

$$= 5 - 0.1 \times 6.06 - 0.7 = 3.694 \text{ Volts}$$

$$R_B = \frac{3.694}{0.06}$$

$$= 61.56 \text{ k}\Omega$$

3.16 Find I_{C1} in the circuit shown in the figure below. Assume that the two transistors are matched and $V_{BE} = 0.7$ V, $\beta = 100$.

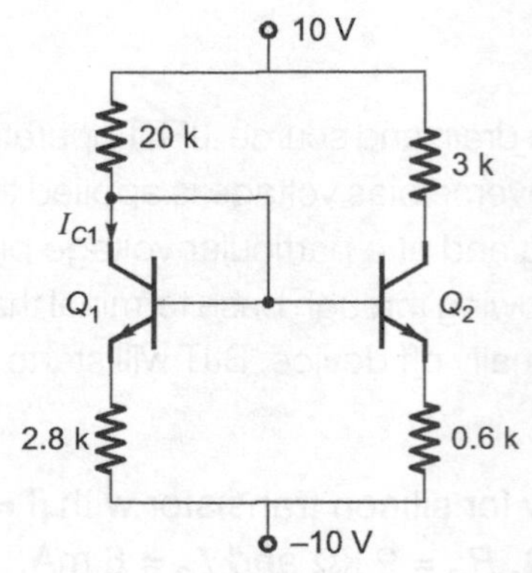

[12 marks : 2020]

Solution:

Given : $V_{BE} = 0.7$ V, $\beta = 100$

Applying KVL in loop (1)

$$V_{BE2} + (1+\beta)I_{B2} \times 0.6K - 2.8K(1+\beta)I_{B1} - V_{BE1} = 0$$

$$0.7 + 101 I_{B2} \times 0.6K - 2.8K \times 101 I_{B1} - 0.7 = 0$$

$$101 I_{B2} \times 0.6K = 2.8K \times 101 I_{B1}$$

$$I_{B2} = 4.67 I_{B1}$$

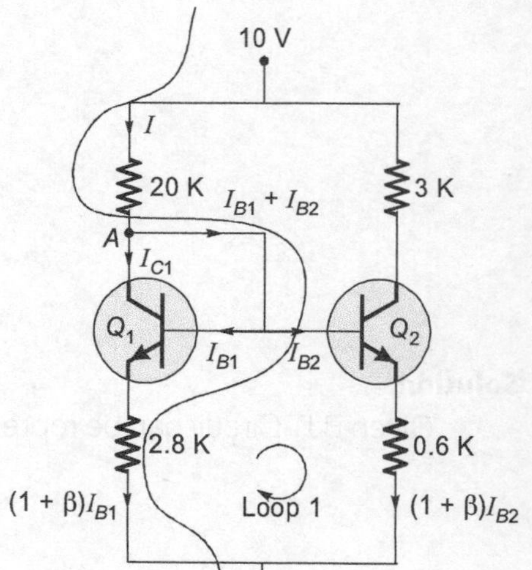

On applying KCL on node (A)

$$I = I_{C1} + I_{B1} + I_{B2}$$

$$I_R = I_{C1} + 5.67 I_{B1}$$

$$I_R = 100 I_{B1} + 5.67 I_{B1}$$

$$= 105.67 I_{B1}$$

On applying KVL in loop (2)

$$-10 + 20K I_R + V_{BE1} + 2.8K(1+B)I_{B1} - 10 = 0$$

$$20K \times 105.67 I_{B1} + 101 I_{B1} \times 2.8K = 20 - 0.7$$

$$I_{B1} = \frac{19.3}{2396.2 \text{ K}}$$

$$I_{B1} = 8.05 \ \mu\text{A}$$

and

$$I_{C1} = \beta I_{B1}$$

$$I_{C1} = 0.8054 \text{ mA}$$

4. Field Effect Transistors

4.1 $r_{d(ON)}$ for $T2$ = 100 kΩ ; $r_{d(OFF)}$ for $T2 = 10^5$ MΩ
$T_{d(ON)}$ for $T1$ = 1 kΩ ; $r_{d(OFF)}$ for $T1 = 10^4$ MΩ
The value of V_0 for V_1 in logic 1 for the shown figure is
(a) 5 V (b) 4.8 V
(c) 0.05 V (d) 0 V

[2 marks : 2001]

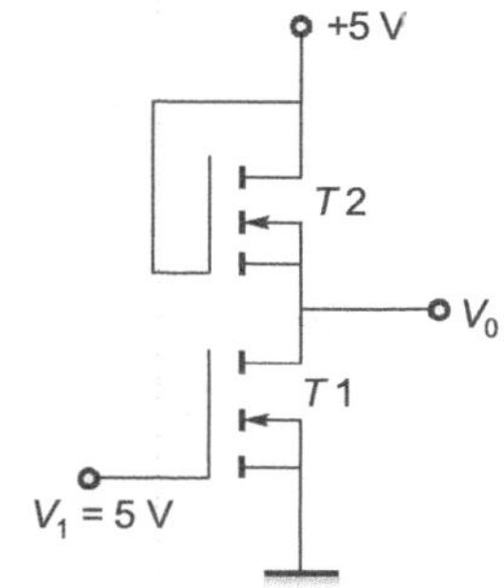

Solution: (c)

Both are n-type mosfets

V_E is +ve, $V_{DG} = 0 = V_{GD}$ for T_2

$\therefore$ $V_{DG} \geq V_{GS} - V_t$

$V_{DG} \geq -V_t$

$0 \geq -V_t$ true

So T_2 is always on

For $V_1 \rightarrow$ high i.e. 1

T_1 is also on so circuit will be

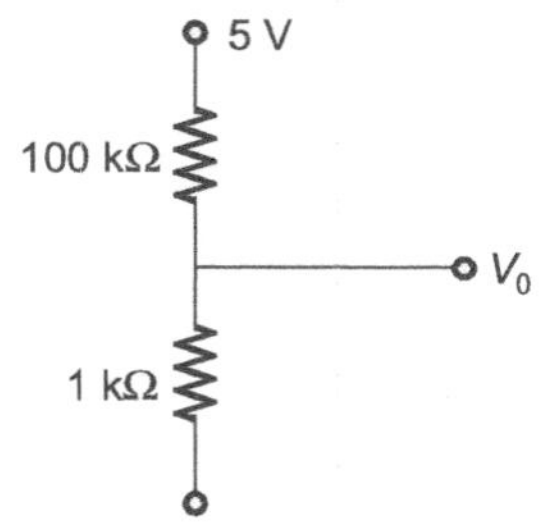

$$V_0 = \frac{1}{101} \times 5 = 0.0495 = 0.05 \text{ V}$$

4.2 Define pinch-off voltage for a JFET. [4 marks : 2002]

Solution:

For a JFET, when 'V_{DS}' is increased 'I_D' also increases, same time depletion region also increases, channel width becomes small, the condition at which the channel is nearly dropped, is called pinch off. After the occurrence of pinch off, further increase of 'V_{DS}' does not increase (I_D). The voltage of 'V_{DS}' at which pinch off occurs, at voltage (V_{GS} = 0) is called pinch off voltage.

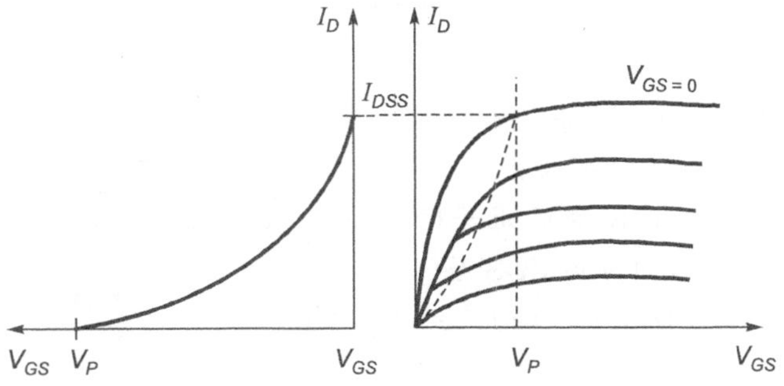

Another explanation:

The voltage of 'V_{GS}' beyond which drain current reduces for zero in saturation region.

From definition second, pinch off voltage is negative for n-channel JFET.

4.3 (i) Draw, with proper labelling, the steady state switching models (circuit diagrams and equivalent circuit) of a MOSFET. How many different types of MOSFETS are available?
Name them. How do they respond to different gate (or input) voltage? Explain with necessary diagram, how the turn-on time of a MOSFET can be reduced?

(ii) Two MOSFETS which are connected in parallel similar to the below figure, carry a total current I_T = 25 A. The drain to source voltage of M_1 is V_{DS1} = 2.0 V and that of M_2 is V_{DS2} = 3.5 V.

Determine the drain current of each transistor and difference in current sharing if the current sharing series resistances are (a) R_{S1} = 0.35 Ω, R_{S2} = 0.2 Ω, and (b) R_{S1} = R_{S2} = 0.5 Ω.

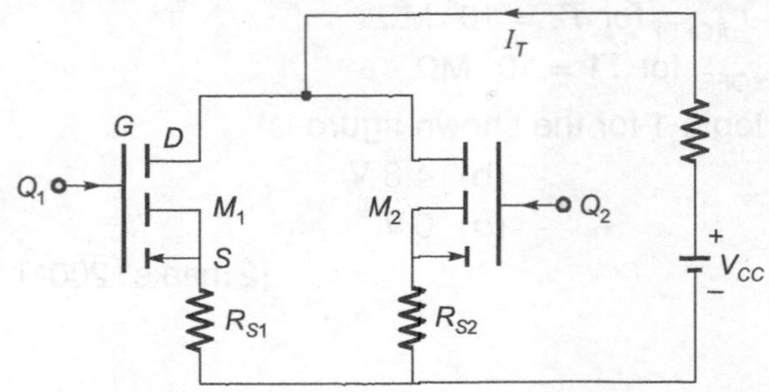

[10 marks : 2005]

Solution:

(i)

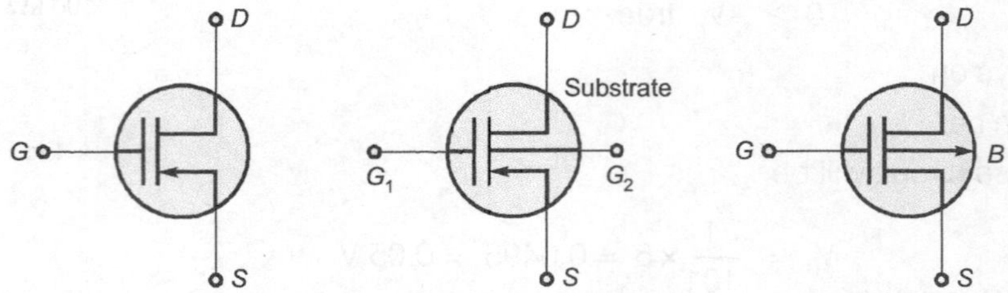

p-channel MOSFET

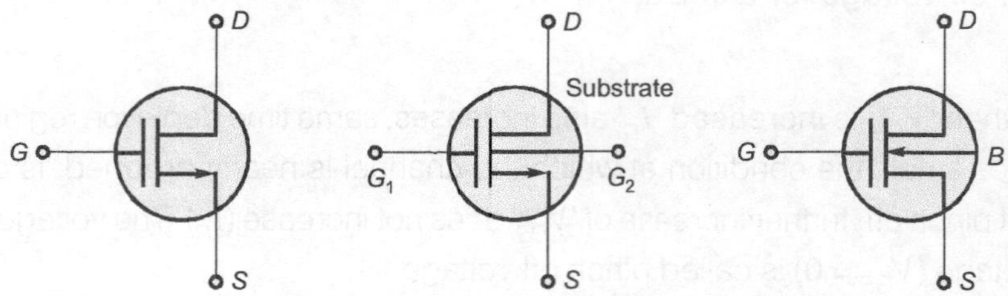

n-channel MOSFET

Different types of MOSFET are as follow:

(a) n-channel depletion type MOSFET

(b) p-channel depletion type MOSFET

(c) n-channel enhancement type MOSFET

(d) p-channel enhancement type MOSFET

For getting turn on time of MOSFET reduced, the gate voltage should be a fast rise time step signal.

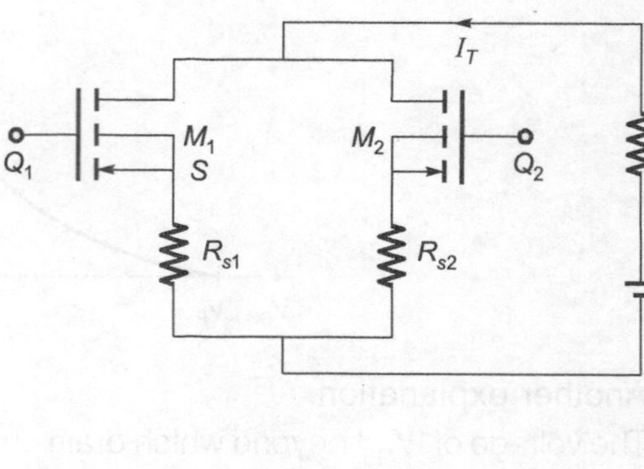

(ii) For (a) first-MOSFET

$$V_D - 2 - 0.35\, I_{D1} = 0$$

For second MOSFET,

$$V_D - 3.5 - 0.2\, I_{D2} = 0$$

$$\therefore \quad I_{D2} - 1.75\, I_{D1} + 7.5 = 0$$

and
$$I_{D1} + I_{D2} = 25 \text{ A}$$

$$\therefore \quad I_{D1} = 11.8 \text{ A and } I_{D2} = 13.18 \text{ A}$$

(b)
$$R_{s1} = R_{s2} = 0.5\ \Omega$$

$$2 + 0.5\, I_{D1} = 3.5 + 0.5\, I_{D2}$$

$$I_{D1} - I_{D2} = 3$$
$$I_{D1} + I_{D2} = 25$$
$$\therefore \quad I_{D1} = 14 \text{ A and } I_{D2} = 11 \text{ A}$$

4.4 Explain the transfer characteristics of *n*-channel and *p*-channel MOSFETS along with the output characteristics of an n-channel enhancement MOSFET.

[8 marks : 2006]

Solution:

The n-channel enhancement MOSFET operates in the triode region when V_{GS} is greater than V_t and the drain voltage is lower than the gate by at least V_t volts. In the triode region,

$$i_D = k_n' \frac{\omega}{L}\left[(V_{GS} - V_t)V_{DS} - \frac{1}{2}V_{DS}^2\right]$$

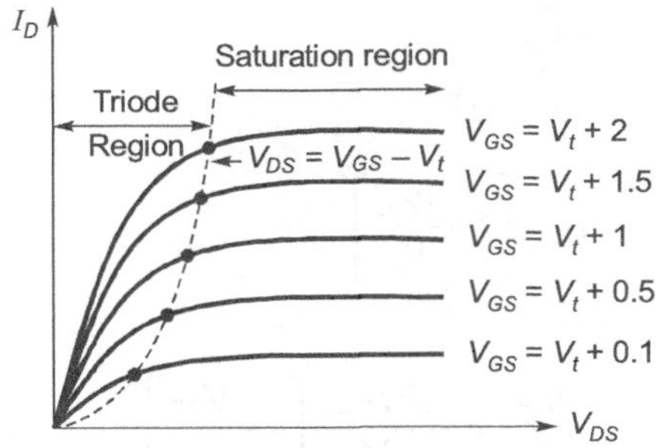

Output characteristics of n-channel MOSFET

The *n*-channel enhancement MOSFET operates in the saturation region when V_{GS} is greater than 'V_t' and the drain voltage does not fall below the gate voltage by more than 'V_t' volts.

$$i_D = \frac{1}{2}k'_n \frac{\omega}{L}(V_{GS} - V_t)^2$$

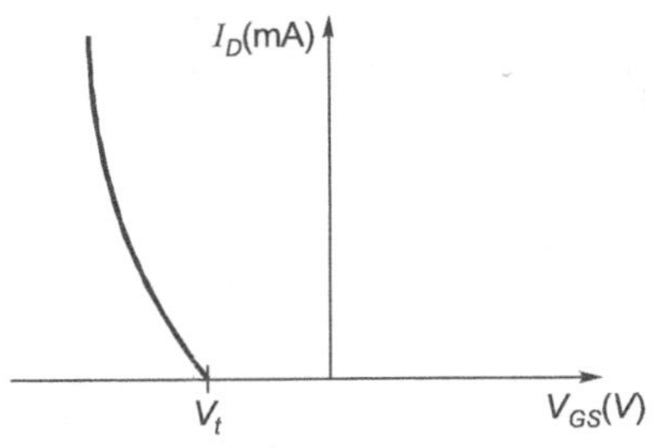

Transfer characteristic of p-channel MOSFET

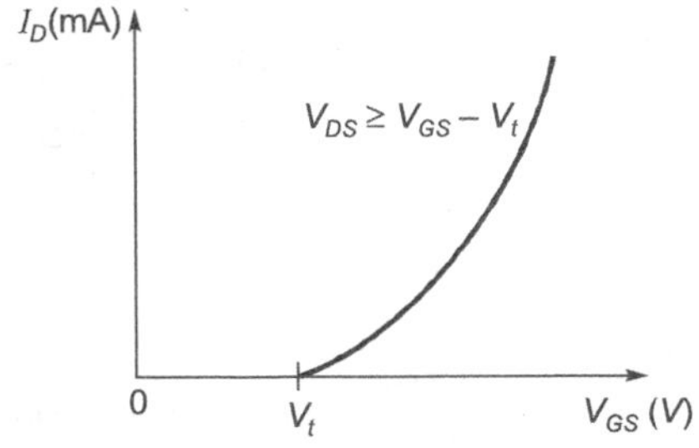

Transfer characteristics of n-channel MOSFET

Transfer characteristic shows that there will be no current until and unless

$V_{GS} \geq V_t$ for (n-MOS) and $V_{GS} \leq V_t$ for (p-MOS)

4.5 (i) What is meant by the term pinch off voltage of a JFET? Draw the drain characteristics of a JFET for V_{GS} = 0 V and show the pinch off voltage in the diagram.

(ii) Define the terms transconductance and amplification factor of a JFET.

(iii) In an FET, as V_{GS}, the gate-to-source voltage is varied from –1.5 V to –2 V, keeping V_{DS} constant, I_D of an FET decreases from 7.5 mA to 5.5 mA. What is the trans-conductance of the FET? If the a.c. drain resistance is 210 kilo ohms, find the amplification factor.

[4 + 4 + 4 = 12 marks : 2017]

Solution:

(i) **Pinch-off voltage V_P:**

It is the minimum V_{DS} voltage where drain current I_D will be saturated,

$$V_P = \frac{a^2 q V_A}{2\epsilon_0 \epsilon_r} \quad \text{(p-channel)}$$

$$V_P = \frac{a^2 q V_D}{2\epsilon_0 \epsilon_r} \quad \text{(n-channel)}$$

where a is half-channel height.

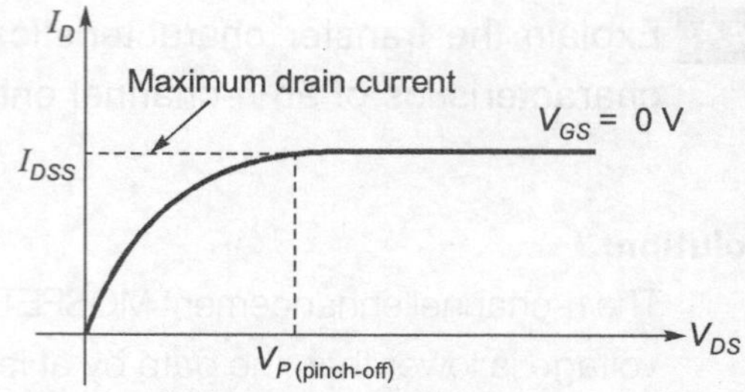

Drain characteristics of JFET at $V_{GS} = 0$ V

(ii) Transconductance g_m is defined as change in drain current I_D per unit change in input voltage V_{GS}.

$$g_m = \frac{\partial I_D}{\partial V_{GS}}$$

Transfer characteristic equation of JFET

$$I_D = I_{DSS}\left[1 - \frac{V_{GS}}{V_P}\right]^2$$

Differentiating with respect to V_{GS},

$$g_m = \frac{\partial I_D}{\partial V_{GS}} = -\frac{2I_{DSS}}{V_P}\left[1 - \frac{V_{GS}}{V_P}\right]$$

Maximum trans-conductance is,

$$g_{m0} = -\frac{2I_{DSS}}{V_P} \qquad [V_{GS} = 0\text{ V}]$$

Amplification factor μ is defined as change in drain source voltage V_{DS} with change in input gate source voltage V_{GS}.

$$\mu = \frac{\Delta V_{DS}}{\Delta V_{GS}} = \frac{\Delta V_{DS}}{\Delta I_D} \cdot \frac{\Delta I_D}{\Delta V_{GS}}$$

$$\frac{\Delta V_{DS}}{\Delta I_D} = r_d \quad \text{(ac drain resistance)}$$

$$\frac{\Delta I_D}{\Delta V_{GS}} = g_m \quad \text{(Transconductance)}$$

$$\mu = g_m r_d$$

(iii) V_{GS} is varied from –1.5 V to –2 V

I_D decreases from 7.5 mA to 5.5 mA.

Then,
$$g_m = \frac{\Delta I_D}{\Delta V_{GS}} = \frac{7.5\text{ mA} - 5.5\text{ mA}}{-1.5-(-2)} = \frac{2\text{ mA}}{0.5\text{ V}} = 4\text{ mS}$$

If ac drain resistance, $r_d = 210\text{ k}\Omega$

(μ) amplification factor = ?

$$\mu = g_m r_d$$

$$\mu = 4 \times 10^{-3} \times 210 \times 10^3$$

$$= 840$$

4.6 How an *n*-channel enhancement mode MOSFET can be used to switch a motor on and off? Justify your answer.

[6 marks : 2019]

Solution:

If the absence of signal, MOSFET normally operates as off switch. When we apply the signal to the gate terminal, a channel is formed between drain and source and MOSFET becomes ON and motor will starts operating.

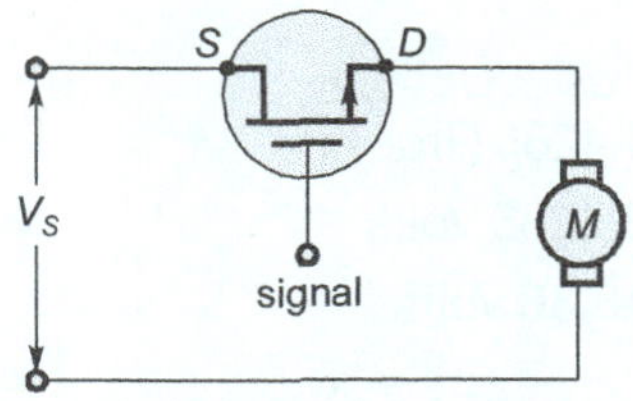

4.7 Determine V_{GS} and V_{DS} for the E-MOSFET circuit shown in the figure below. The minimum values of $I_{D(ON)}$ = 200 mA at V_{GS} = 4 V and Gate to Source threshold voltage $V_{GS(Th)}$ = 2 V.a

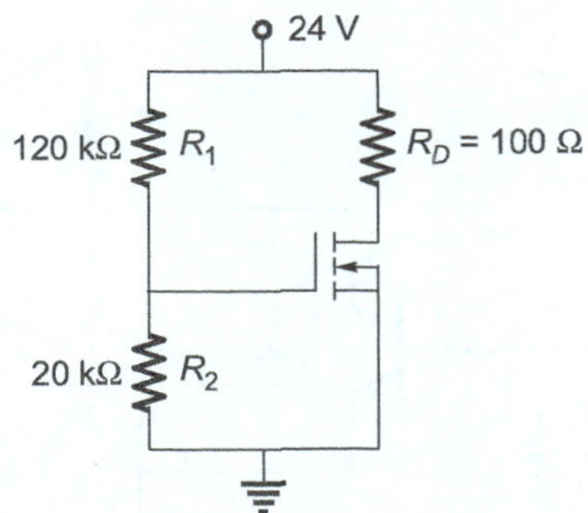

[12 marks : 2020]

Solution:

Taking Thevenin equivalent on gate terminal

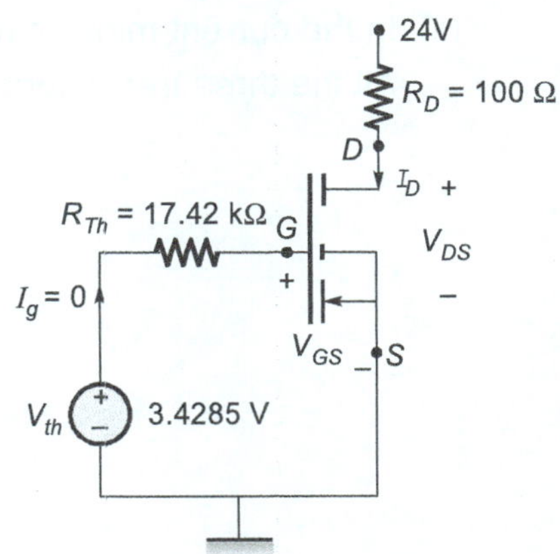

$$V_{Th} = \frac{R_2}{R_1 + R_2} \times V_{DD}$$

$$V_{Th} = \frac{20}{120+20} \times 24 = 3.4285 \text{ volts}$$

$$R_{Th} = R_1 \,//\, R_2 = (120 \,//\, 20) = 17.42 \text{ k}\Omega$$

Equivalent circuit for figure given in problem is

Since $I_g = 0$, so $V_G = 3.4285$ volts

$V_S = 0$ Volts

$V_{GS} = 3.4285$ Volts

Since given MOSFET is N-channel. So, drain current is given by

$$I_{DS} = \text{Kn}(V_{GS} - V_{Th})^2$$

Assuming MOSFET operating in saturation region.

Given that $V_{GS} = 4$ V, $V_{GS(Th)} = 2$ and $I_{D(ON)} = 200$ mA

$$0.2 = K_n(4-2)^2$$

$$K_n = \frac{0.2}{4} = 0.05 \text{ A/V}^2$$

Now, $V_{GS} = 3.4285$ V and $V_{GS(Th)} = 2$

$$I_{DS} = K_n(V_{GS} - V_{Th})^2$$
$$I_{DS} = 0.05(3.4285 - 2)^2$$
$$I_{DS} = 0.10203 \text{ A}$$

and $V_{DS} \geq V_{GS} - V_{th}$ [TRUE],

so, MOSFET is operating in saturation region.

$$I_{DS} = V_{DD} - I_{DS} \cdot R_D$$
$$V_{DS} = 24 - 100 \times 0.10203 = 13.80 \text{ Volts}$$

For saturation region

$$V_{DS} \geq V_{GS} - V_{Th}$$
$$13.80 \geq 1.4285 \text{ (True)}$$

So, $V_{GS} = 3.4285$ Volts

$V_{DS} = 13.80$ Volts

4.8 **(i) Calculate the DC voltages V_o and V_i, and the bias currents for the feedback pair of transistors shown in the figure below, given that β_1 = 100, β_2 = 150, V_{CC} = 15 volts, R_C = 200 Ω and R_B = 1 MΩ :**

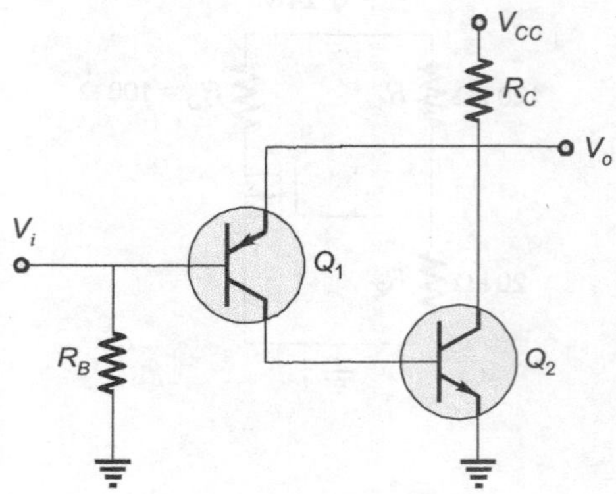

(ii) In the current mirror circuit shown in the figure below, the current is mirrored in two transistors. All the three transistors are identical. Calculate the load current I assuming β = 100 :

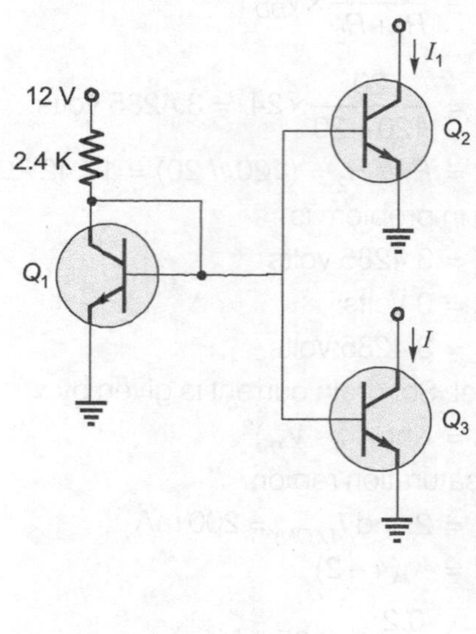

[10 + 10 marks : 2023]

Solution:

(i) Given circuit can be drawn as :

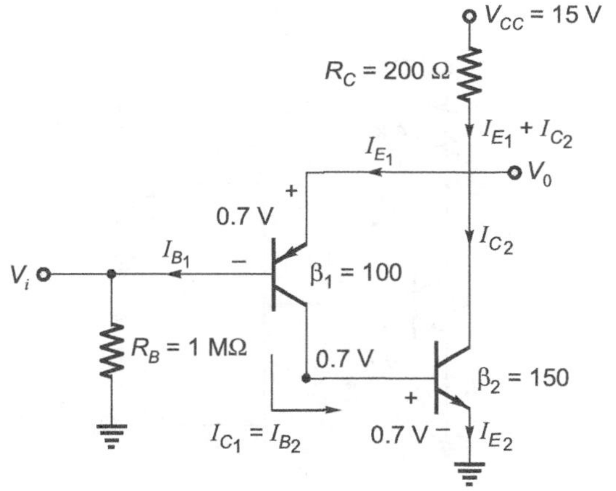

$$V_o = 0.7 + V_{in}$$
$$V_{in} = I_{B1} \times R_B$$
$$= 10^6 I_{B1}$$
$$= 10^6 \frac{I_{E1}}{(1+\beta_1)}$$
$$= 10^6 \times \frac{I_{E1}}{100}$$
$$= 10^4 I_{E1} \quad \Rightarrow \quad V_o = 0.7 + 10^4 I_{E1} \qquad ...(1)$$
$$V_o = V_{CC} - (I_{E1} + I_{C2})R_C$$
$$= V_{CC} - (I_{E1} + \beta_2 I_{B2})R_C$$
$$= V_{CC} - (I_{E1} + \beta_2 I_{C1})R_C$$
$$= V_{CC} - (I_{E1} + \beta_2 I_{E1})R_C$$

where, $I_{C1} = I_{E1}$ (β is large)

$$= V_{CC} - (1 + \beta_2) I_{E1} R_C \qquad (\beta_2 >> 1)$$
$$= V_{CC} - 150 I_{E1} \times 200\ \Omega$$
$$= 15 - 3I_{E1} \times 10^4 \qquad ...(2)$$

Equating equation (1) and (2)

$$0.7 + 10^4 I_{E1} = 15 - 3I_{E1} \times 10^4$$
$$4I_{E1} \times 10^4 = 15 - 0.7$$
$$4I_{E1} \times 10^4 = 14.3$$
$$I_{E1} = \frac{14.3}{4 \times 10^4} = 0.35 \text{ mA}$$
$$V_i = 10^4 I_{E1}$$
$$= 10^4 \times 0.35 \text{ mA}$$
$$= 10 \times 0.35 = 3.5 \text{ V}$$
$$V_i = 3.5 \text{ V}$$
$$V_o = 15 - 3I_{E1} \times 10^4$$
$$= 15 - 0.3 \times 0.35 \text{ mA} \times 10^4$$
$$= 15 - 1.05 \times 10$$
$$= 15 - 10.5 = 4.5 \text{ V}$$
$$V_o = 4.5 \text{ V}$$

(ii)

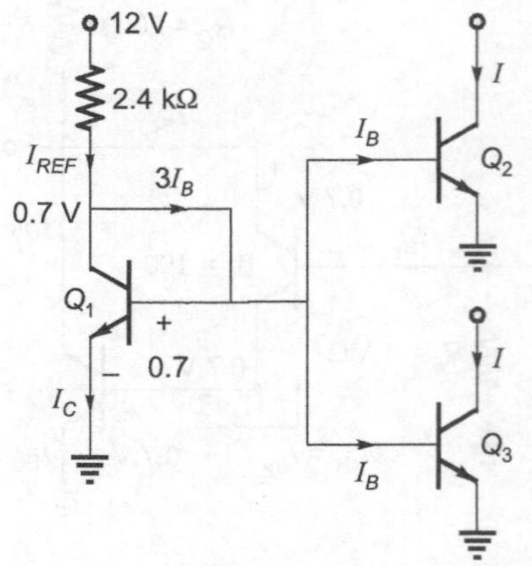

Given : Q_1, Q_2, Q_3 are identical.

$$I_{REF} \text{ in } Q_1 = \frac{12-0.7}{2.4K}$$

$$= 4.7 \text{ mA}$$

$$I_{REF} = I_C + 3I_B$$

$$= I_C + 3\frac{I_C}{\beta}$$

$$= I_C\left(1+\frac{3}{\beta}\right)$$

Here, $I_C = I$ (current mirror)

$$= I\left(1+\frac{3}{\beta}\right)$$

$$= I_C + \frac{1}{\beta} = I_C\left(1+\frac{3}{\beta}\right)$$

Here, $I_C = I$ (current mirror)

$$= I\left(1+\frac{3}{\beta}\right)$$

$$I = \frac{I_{REF}}{1+\frac{3}{\beta}} = \frac{4.7 \text{ mA}}{1+\frac{3}{100}}$$

$$= \frac{4.7 \text{ mA}}{1.03} = 4.56 \text{ mA}$$

$$I = 4.56 \text{ mA}$$

5. BJT and FET Amplifiers

5.1 For the Darlington amplifier shown in figure below, obtain its Z-matrix for identical T_1 and T_2 having $h_{fe} = 100$, $R_E = h_{ie} = 1$ k.

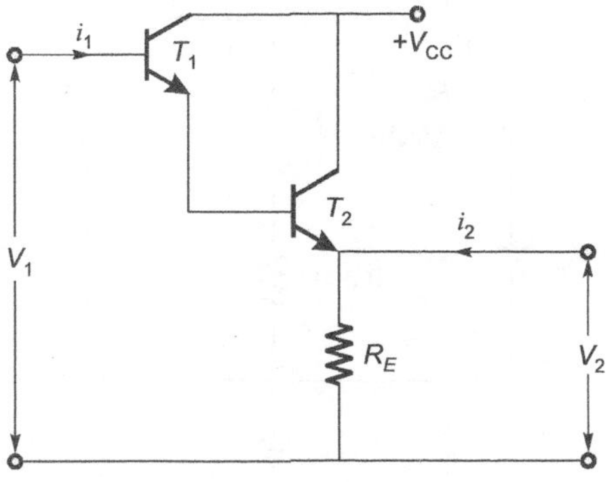

[10 marks : 2001]

Solution:

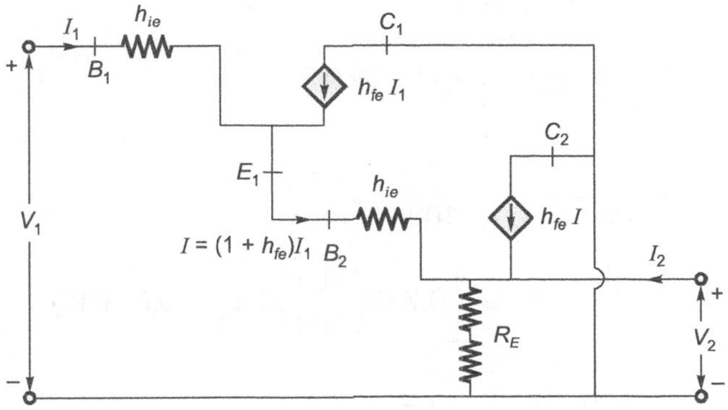

By applying KVL, $$V_1 = h_{ie} \times I_1 + (1 + h_{fe})\, I_1 \times h_{ie} + (1 + h_{fe})\, I\, R_E + I_2\, R_E$$

$$= \left[h_{ie} + (1 + h_{fe})h_{ie} + (1 + h_{fe})^2\, R_E\right] I_1 + R_E \times I_2$$

KVL in output loop, $$V_2 = I_2\, R_E + (1 + h_{fe})^2\, R_E I_1 = (1 + h_{fe})^2\, R_E I_1 + R_E I_2$$

Thus, $$[z] = \begin{bmatrix} h_{ie} + (1 + h_{fe})h_{ie} + (1 + h_{fe})^2 R_E & R_E \\ (1 + h_{fe})^2\, R_E & R_E \end{bmatrix} = \begin{bmatrix} 10.303\text{ k}\Omega & 1\text{k}\Omega \\ 10.201\text{k}\Omega & 1\text{k}\Omega \end{bmatrix}$$

5.2 The BJT amplifier of figure below has $h_{fe} = 100$, $V_{BE} = 0.7$ V, $I_{co} = 0$. Calculate the values of R_1 and R_c such that its $I_c = 1$ mA and $V_{CE} = 2.5$ V.

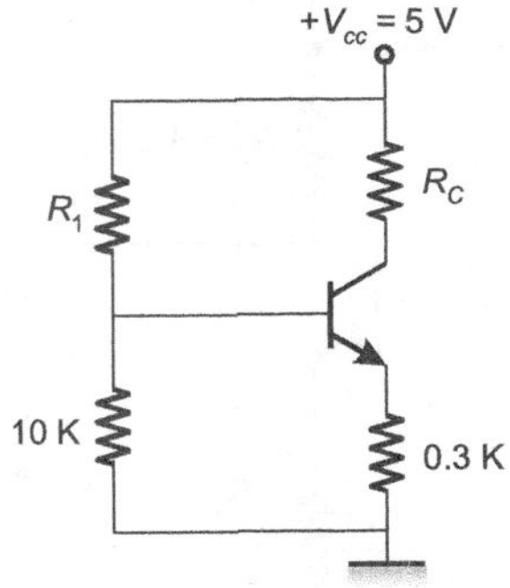

[10 marks : 2001]

Solution:

Thevenin equivalent of Base-ground terminal,

$$\therefore \quad V_{\text{base}} = \frac{10}{R_1 + 10} \times 5 \text{ and } R_{\text{base}} = \frac{10\, R_1}{10 + R_1}\text{ k}\Omega$$

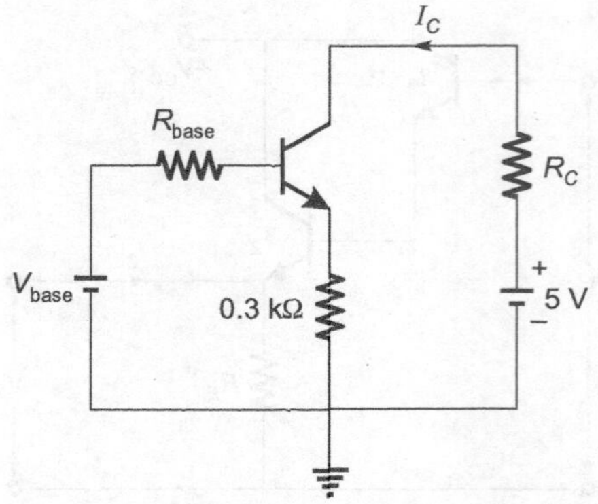

Let us consider BJT is in active region by applying.
KVL around the base-emitter loop, then we get,

$$V_b - I_B R_b - V_{BE} - (I_B + I_C)\,0.3 = 0 \quad \text{...}(i)$$

$$\frac{50}{R_1+10} - I_B \times \left(\frac{10R_1}{10+R_1}\right) - 0.7 = I_B(101)\times 0.3 \quad (\because I_C = \beta I_B)$$

$$\frac{50}{R_1+10} - \left(\frac{R_1\times 10}{10+R_1}\right)\times\frac{1}{100} - 0.7 = \frac{1}{100}\,(101)\times 0.3$$

$$\therefore \quad 500 - R_1 = \left[0.7 + 1\left(\frac{101}{100}\right)0.3\right]\times 10(R_1+10)$$

$$\therefore \quad 500 - R_1 \cong 10\,R_1 + 100 \quad \left(\because 0.7 + \frac{101}{100}\times 0.3 \cong 1\right)$$

$$11\,R_1 = 400$$

$$\therefore \quad R_1 = 36.36\ \text{K}\Omega$$

Applying KVL in collector-emitter loop,

$$5 - I_C R_C - V_{CE} - (I_B + I_C)\,0.3 = 0$$

$$\therefore \quad R_C = \frac{5 - 2.5 - 0.303}{1} = 2.194\ \text{K}\Omega$$

5.3 **For a single-stage transistor amplifier shown in the circuit below, find the voltage gain A_v, input impedance Z_{in} and V_{CE}. Take β = 200 and r_e = 30 MV/I_E.**

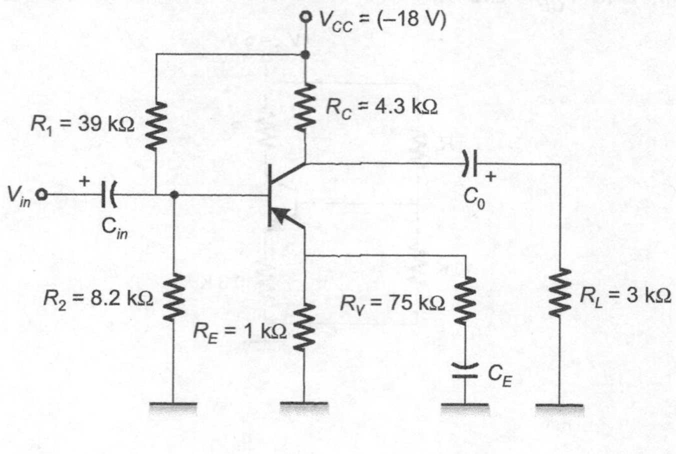

[18 marks : 2002]

Solution:

For dc analysis all the capacitors will be open and equivalent circuit will be

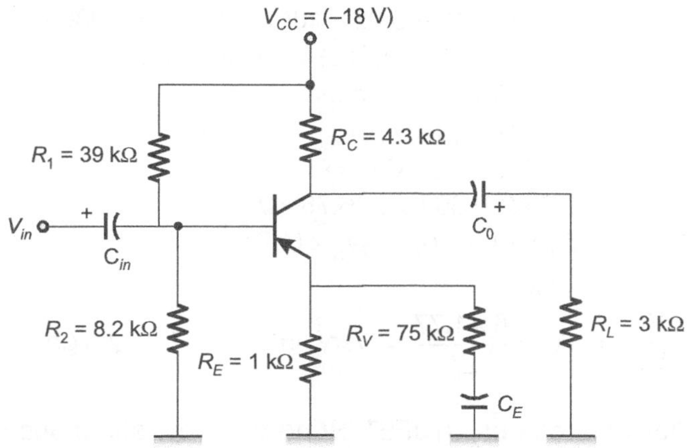

∴ Thevenin equivalent circuit is

$$V_{th} = \frac{8.2}{39+8.2} \times (-18) = -3.127 \text{ V} = -3.13 \text{ V}$$

$$R_{th} = 39 \parallel 8.2 \text{ k}\Omega = 6.775 \text{ k}\Omega = 6.78 \text{ k}\Omega$$

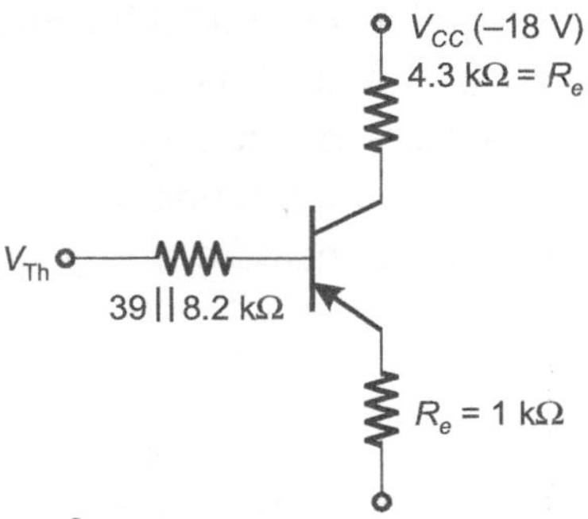

$$\therefore \quad 0 - I_E R_E - 0.7 - 6.78\, I_B - V_{th} = 0$$

$$-I_E - 0.7 - 6.78\frac{I_E}{(\beta+1)} - (-3.13) = 0$$

$$I_E + \frac{6.78 I_E}{201} = 2.43$$

$$\therefore \quad I_E = 2.35 \text{ mA}$$

$$\therefore \quad r_e = 30 \text{ MV}/I_E = 30 \times 10^6 \text{ V}/2.35 \times 10^{-3} \text{ A}$$

$$= 12.765 \times 10^9 \ \Omega = 12765.95 \text{ M}\Omega$$

$$V_{CE} = -V_{EC} = (18 - I_E R_E - I_C R_C)$$

$$= -(18 - 2.35 \times 1 - 2.35 \times 4.3)$$

$$V_{CE} = -5.81 \text{ V} \qquad \therefore I_E \simeq I_C$$

Now ac equivalent circuit will be

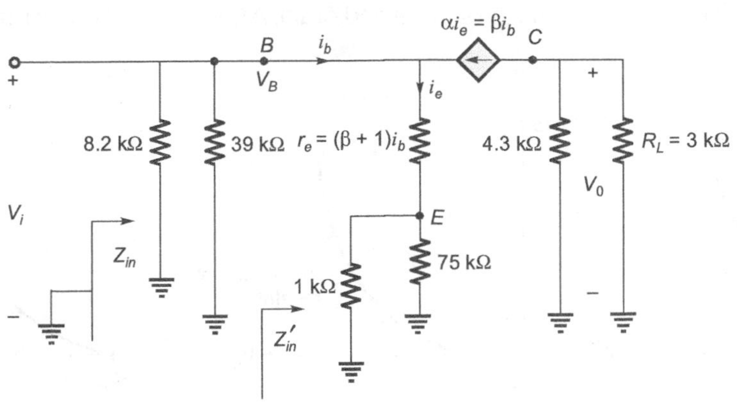

$$Z_{in}' = V_B / i_b$$

Now,

$$V_B = i_e r_e + i_e (1 \parallel 75 \text{ k}\Omega)$$

$$V_B = (\beta + 1)\, r_e i_b + (\beta + 1)\, i_b (0.9868 \text{ k}\Omega)$$

$$\frac{V_B}{i_b} = (\beta+1)r_e + (\beta+1)0.9868$$

$$Z_{in}' = (\beta + 1) \times 12765.95 \times 10^3 \text{ k}\Omega + 201 \times 0.9868$$
$$Z_{in}' = 2565955.95 \times 10^3 \text{ k}\Omega + 198.3468 \text{ k}\Omega$$
Z_{in}' – very high can take as ∞.
$$\therefore \quad Z_{in} = 8.2 \text{ k}\Omega \,||\, 39 \text{ k}\Omega \,||\, Z_{in}$$
$$Z_{in} = 8.2 \text{ k}\Omega \,||\, 39 \text{ k}\Omega \,||\, \infty$$
$$Z_{in} = 8.2 \text{ k}\Omega \,||\, 39 \text{ k}\Omega = 6.78 \text{ k}\Omega$$
$$V_0 = \beta i_b \times (4.3 \,||\, R_L) = \beta i_b \times (1.77)$$
$$V_{in} = i_b Z_{in}'$$
$$\therefore \quad \frac{V_0}{V_{in}} = A_v = \frac{\beta \times 1.77}{Z_{in}'} = \text{very small} \because Z_{in}' \text{ very high.}$$

5.4 **Develop a voltage follower circuit using JFET. Bring out the distinctions between the constructional and operational features of a JFET and MOSFET.**

[10 marks : 2003]

Solution:

Junction field effect transistor (JFET): For a JFET construction, Gate is heavily doped and channel is lightly doped. There is no constructional difference between drain and source in JFET, voltage between drain and source is increased 'I_D' also increases. Same time, depletion region also increases so that channel width will decrease, the condition at which channel is nearly closed is called pinch off voltage. After occurrence of pinch off, further increase in 'V_{DS}' does not cause increase of (I_D) and pinch off voltage can be controlled by (V_{GS}) voltage.

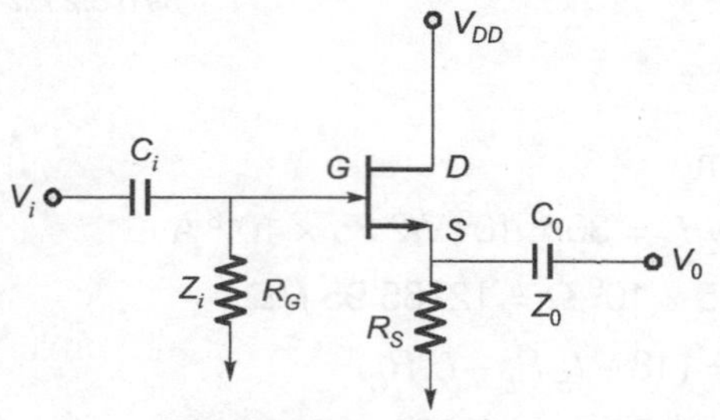

Voltage follower circuit

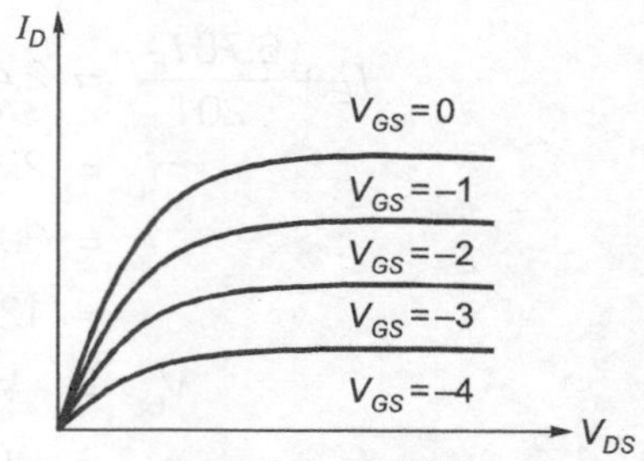

For n-channel JFET

MOSFET:

An n-channel MOSFET consists of a lightly doped p-type, substrate into which two highly doped n-regions are diffused and these two sections act as the source and drain. A thin layer of SiO_2 is made over the surface of the structure.

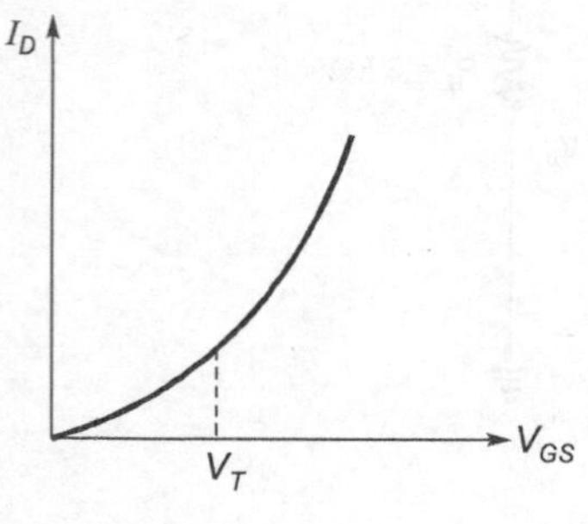

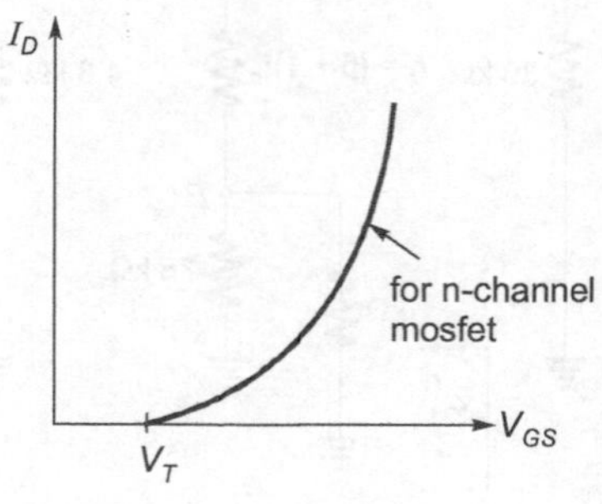

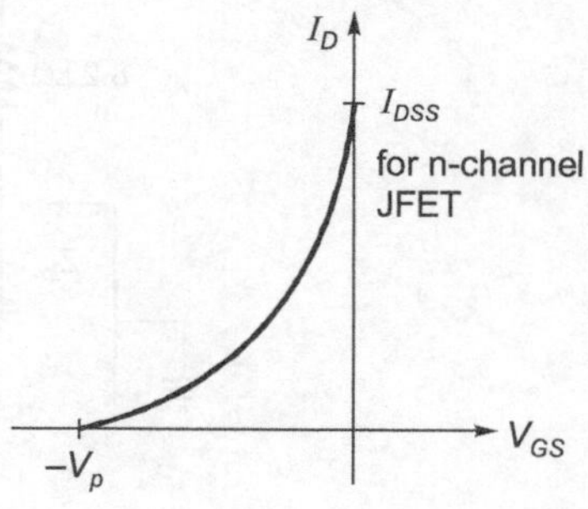

Upon application of positive voltage at the gate (for n-type) negative charges are induced on the semi conductor site. As the magnitude of the gate voltage increases, more negative charges accumulate which leads to the higher conductivity and current flows from source to drain through the induced channel.

JFET can only work in depletion mode but MOSFET can work in both depletion mode and enhancement mode.

5.5 Consider the idealized push-pull Class *B* power amplifier as shown below:

Given that: $R_2 = 0$, $V_{CC} = 20$ V, $N_2 = 2N_1$, $R_L = 20\ \Omega$ and the transistors have $h_{FE} = 20$. The input is a sinusoid. For the maximum output signal at $V_m = V_{CC}$, determine:

(i) the output signal power.

(ii) the collector dissipation in each transistor.

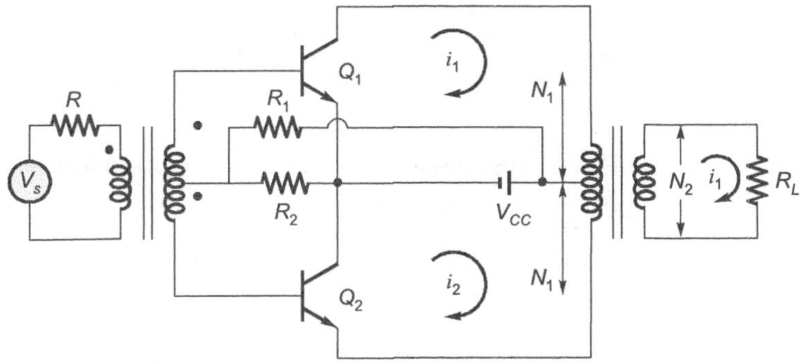

[12 marks : 2006]

Solution:

The power output $P_{ac} = 1/2\, I_m V_m$ where $V_m = V_{CC} - V_{min}$

The DC current under load is the average of the half sine wave $I_{dc} = I_m/\pi$

P_{dc} from supply for two transistor = $2 \times V_{CC} \times I_m/\pi$

Ideally, $$\eta = \frac{\pi}{4} \times 100 = 78.5\%$$

(i) $$P_o(ac) = \frac{V_{CC}^2}{2R_L'} = \frac{20 \times 20}{2 \times 5} = 40 \text{ watts}$$

where, $$R_L' = \frac{R_L}{K^2} = \frac{20}{4} = 5\ \Omega \text{ and } I_{max} = \frac{V_{CC}}{R_L'}$$

(ii) $$P_{in}(\text{dc max}) = \frac{2V_{CC}^2}{\pi R_L'} = \frac{2 \times 20^2}{\pi \times 5} = 50.9 \text{ W}$$

$\therefore$ Power dissipation by transistor,

$$P_D = P_{in} - P_0 = 10.9 \text{ Watts}$$

5.6 What is the basic advantage of push-pull operation as compared to parallel operation of transistors? Draw the circuit diagram of a transformer-coupled push-pull power amplifier using npn transistors and explain its operation. Also show DC and AC load lines.

[12 marks : 2008]

Solution:

Consider transformer coupled push-pull amplifier given in figure below. When $V_s = 0$, transistor (T_1 and T_2) are in cut-off. When 'V_s' is non-zero. R_L' is satisfied to (i_1 and i_2). The combined action of i_1 and i_2 is used for getting full waveform.

$$\text{Load resistance} = R_L' = R_L\left[\frac{N_1/2}{N_2}\right]^2 = \left(\frac{R_L}{4}\right)\left(\frac{N_1}{N_2}\right)^2$$

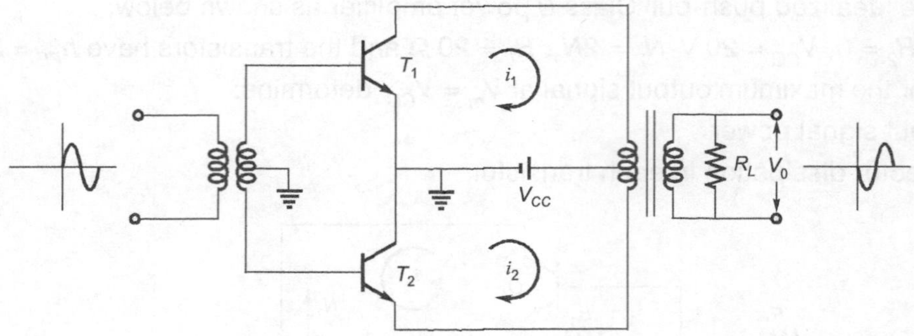

A transformer push-pull amplifier lets us assume the d.c. resistance of transformer winding is negligible and the d.c. load line is almost vertical as indicated in figure below.

D.C. power output $P(\text{d.c.}) = V_{cc} I_{dc}$

$$I_{dc} = I_1(\text{d.c.}) + I_2(\text{d.c.}) \approx 2I_m/\pi$$

where $I_m \rightarrow$ peak value of current

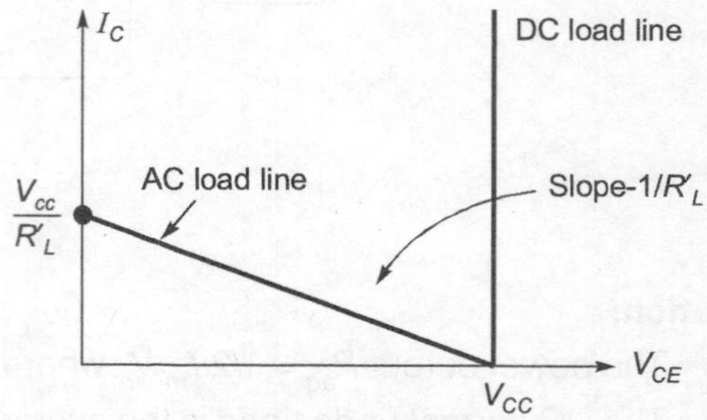

Load line for push-pull amplifier

$$P(\text{d.c.}) = \frac{2V_{CC} I_m}{\pi}$$

AC output power, $$P_0(\text{a.c.}) = V_{rms} I_{rms} = \frac{I_{rms}^2}{R_L'}$$

$$P(a.c.) = I_m^2 \left(\frac{R_L'}{2}\right)$$

where, $$I_{rms} = \left(\frac{I_m}{\sqrt{2}}\right)$$

$$P(\text{d.c.})(\text{max.}) = \frac{2V_{CC}^2}{\pi R_L'}$$

$$P_0(\text{a.c.})(\text{max.}) = \frac{V_{CC}^2}{2R_L'}$$

$\therefore$ $$\eta(\text{efficiency}) = \frac{P_0(a.c.)}{P(d.c.)} = \frac{\pi I_m R_L'}{4V_{CC}}$$

$$\eta_{max} = \frac{\pi}{4} \text{ or } 78.54\%$$

Then we can conclude that maximum efficiency is deserved when (I) become maximum.

Power dissipated by transistor:

$$P_d = P_{in} - P_{out}$$

For getting maximum power, $\frac{dP_D}{dI_m} = 0$

$$\left(\frac{2V_{CC}}{\pi}\right) - I_m R_L' = 0$$

$\therefore$ $$I_m = \frac{2V_{CC}}{R_L' \times \pi}$$

$$\therefore \qquad P_D(\text{max}) = \left(\frac{2V_{cc}}{\pi}\right)\left(\frac{2V_{cc}}{R'_L \times \pi}\right) - \left[\frac{(2V_{cc})^2}{\left(\frac{\pi}{R'_L}\right)}\right]\left(\frac{R'_L}{2}\right) = \frac{2V_{cc}^2}{\pi^2 R'_L}$$

From this, maximum power dissipated by the transistor is not at the maximum current. The basic advantage of push-pull amplifier is that one half of the circuit pushes the signal on the high on the other hand other half pulls the signal on the low.

5.7 Design a common emitter amplifier using the circuit shown in figure to obtain the maximum symmetrical output voltage swing. The design must be dc bias stable.
(Use $V_{BE} = -0.6$ V, and $\beta = 200$).
Determine:

(*i*) I_{CQ} and V_{CEQ}
(*ii*) R_1 and R_2
(*iii*) Maximum symmetrical output voltage swing
(*iv*) Power rating of the transistor
(*v*) Amplifier ac power output

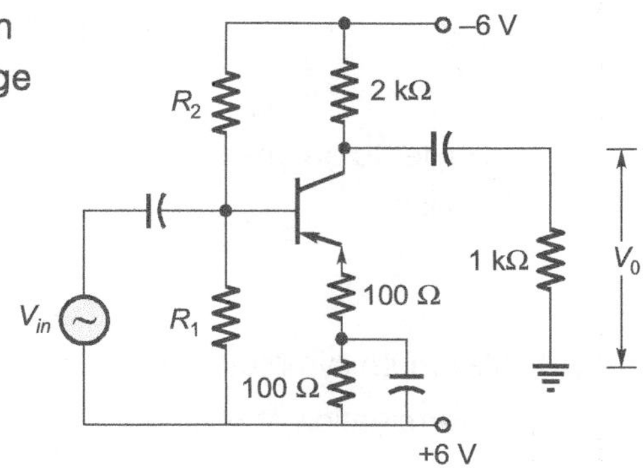

[15 marks : 2014]

Solution:

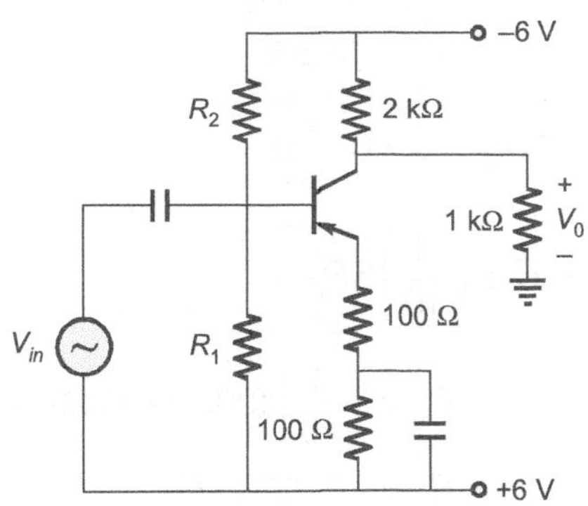

(*i*) Getting the operating point:

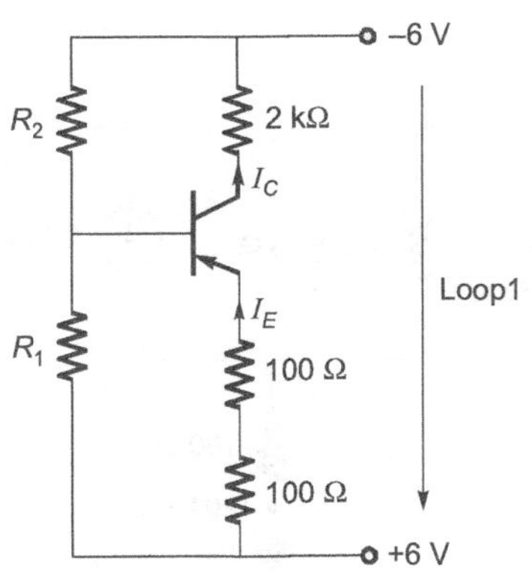

Applying KVL to Loop 1:

$$-6 + (2\text{ k}\Omega)\, I_C + V_{CE} + (200)\, I_E - 6 = 0$$

$$12 = (2200)\, I_C + V_{CE}$$

Drawing the load line:

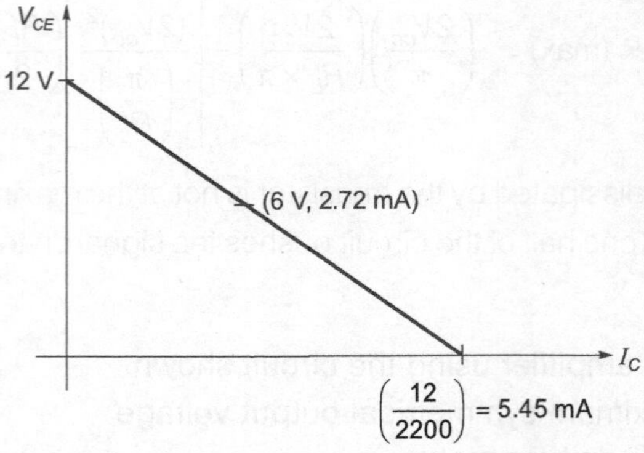

In order to get maximum symmetrical output voltage swing the operating must be at the mid of the dc load line.

i.e.,
$$V_{ca} = 6\text{ V}$$
$$I_{CQ} = 2.7272\text{ mA}$$

(ii) Maximum symmetrical output voltage swing = 6 V

(iii) Determination of R_1 and R_2:

$$V_{th} = \frac{-6R_1}{R_1 + R_2}$$

$$R_{th} = \frac{R_1 R_2}{R_1 + R_2}$$

$$I_B = 13.636\ \mu\text{A}$$

$$-\left(-\frac{6R_1}{R_1 + R_2}\right) + I_B R_{th} + 0.6 + 200I_E - 6 = 0$$

For the transistor to be active $R_1 \le \left(\frac{1}{10}\right)\beta R_E$

$$R_1 = \left(\frac{1}{10}\right)(200)\,(200\ \Omega) = 4\text{ k}\Omega$$

Similarly,
$$R_2 = 40.3334\text{ k}\Omega$$

(iv) Power rating of the transistor = $V_{CQ} \cdot I_{CQ} = 16.3632$ mW

(v) Small signal model:

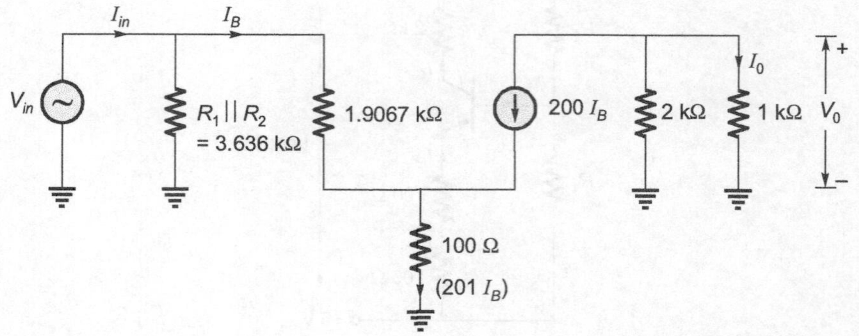

Voltage gain: (A_V),
$$I_0 = \frac{(201I_B) \times 2\text{ k}\Omega}{3\text{ k}\Omega} = 134\ I_B$$

⇒
$$V_0 = (134)\,(1\text{ k}\Omega)\,I_B$$

$$V_{in} = (1.9067\text{ k}\Omega)\, I_B + (100)(201)\, I_B$$

$$V_{in} = (22\text{ k}\Omega)\, I_B$$

$$\Rightarrow \qquad I_B = \left(\frac{V_{in}}{22\text{ k}\Omega}\right)$$

$$V_0 = \frac{(134\text{ k}\Omega)V_{in}}{22\text{ k}\Omega}$$

$$\left(\frac{V_0}{V_{in}}\right) = 6.1$$

Current gain (A_I):

$$I_0 = 134\, I_B = \frac{I_{in}(3.636)}{22 + 3.636} = I_{in}\left(\frac{3.636}{25.636}\right)$$

$$I_0 = 134\, I_{in}\left(\frac{3.636}{25.636}\right)$$

$$\left(\frac{I_0}{I_{in}}\right) = \frac{134 \times 3.636}{25.636} = 19.005$$

Power gain:

$$A_P = A_V A_I$$

$$= 6.1 \times 19.005 = 116$$

5.8 **The parameters of FET used in the amplifier given below are g_m = 4 m℧ and r_d = 30 kW. Assuming C to be short circuit for the given frequency, determine the small signal voltage gain.**

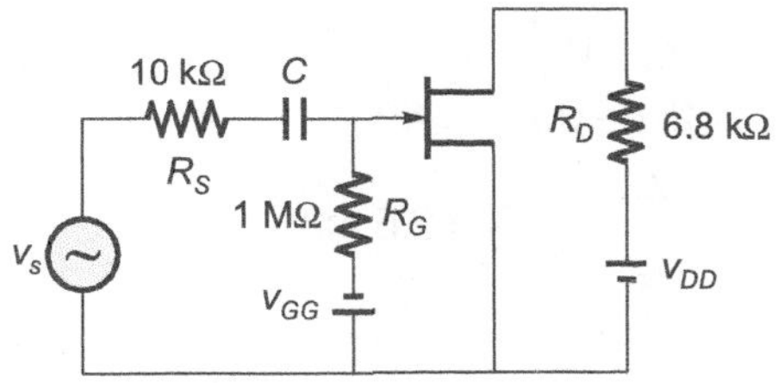

[4 marks : 2015]

Solution:

Given:

$$g_m = 4\text{ m℧}$$

$$r_d = 30\text{ k}\Omega$$

Assuming the capacitance C is short circuited at given frequency.

Now, Drawing the small signal model

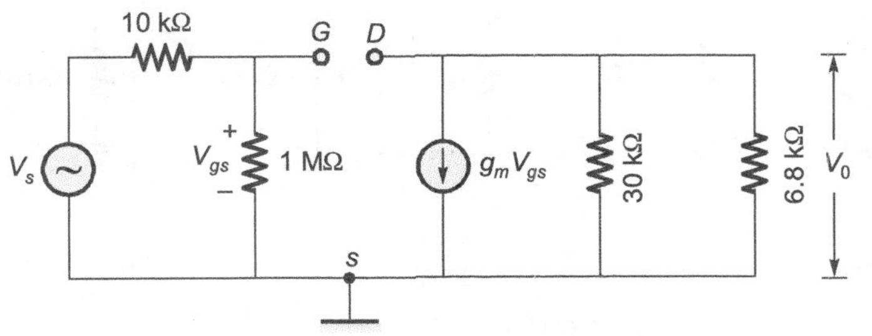

We need to find the small signal voltage gain (V_o/V_s)

So applying voltage division rule we get

$$V_{gs} = V_s\left(\frac{10^6}{10^6 + 10\times10^3}\right) = 0.99\,V_s$$

and

$$V_o = -g_m\,V_{gs}(30\text{ k}\Omega \parallel 6.8\text{ k}\Omega)$$

$$\frac{V_o}{V_s} = -21.95$$

So voltage gain is –21.95.

5.9 (i) For the common emitter amplifier shown, find the values of the collector current I_C and V_{CE} if β_{dc} = 150.

(ii) Draw the a.c. equivalent circuit of the base of the transistor amplifier. Find the value of the input resistance and the voltage gain. Discuss the effect of the capacitor C_3 on the voltage gain when V_{in} = 10 mV. (Please refer to amplifier circuit above)

[8 + 12 = 20 marks : 2017]

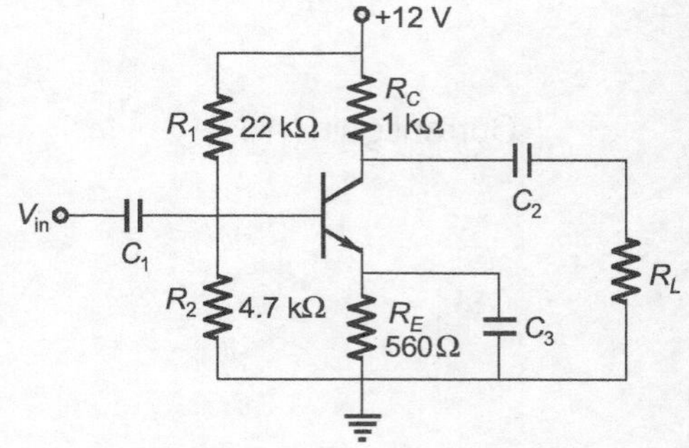

Solution:

(i) We have to calculate I_C and V_{CE}

As β is very high I_B can be neglected,

$$V_B = \frac{V_{CC}\times R_2}{R_1 + R_2}$$

$$= 12\times\frac{4.7\text{ k}}{22\text{ k}+4.7\text{ k}} = 12\times\frac{4.7\text{ k}}{26.7\text{ k}} = 2.1\text{ V}$$

$$V_{BE} = 0.7\text{ V}$$

$$V_E = V_B - V_{BE}$$

$$= 2.1 - 0.7 = 1.4\text{ V}$$

$$I_E = \frac{V_E}{R_E} = \frac{1.4}{560} = 2.5\text{ mA}$$

$$I_C \approx I_E = 2.5\text{ mA}$$

$$V_C = V_{CC} - I_C R_C$$

$$= 12 - 1\text{ k}\times 2.5\text{ mA} = 9.5\text{ V}$$

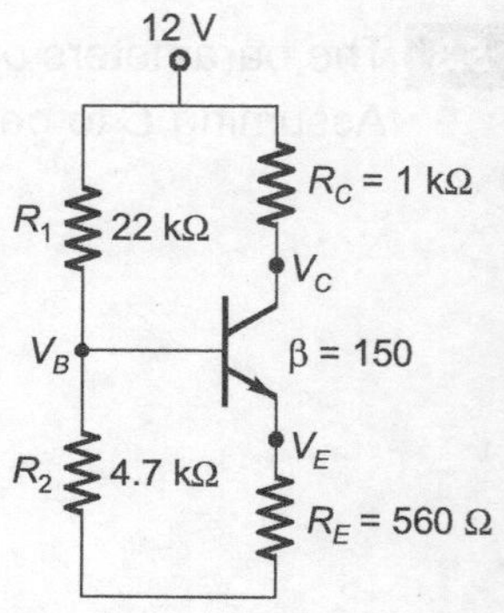

Then, $$V_{CE} = V_C - V_E = 9.5\text{ V} - 1.4\text{ V} = 8.1\text{ V}$$

Therefore, $$I_C = 2.5\text{ mA}$$

$$V_{CE} = 8.1\text{ V}$$

(ii) **Case-I:** When capacitor C_3 is short-circuited.

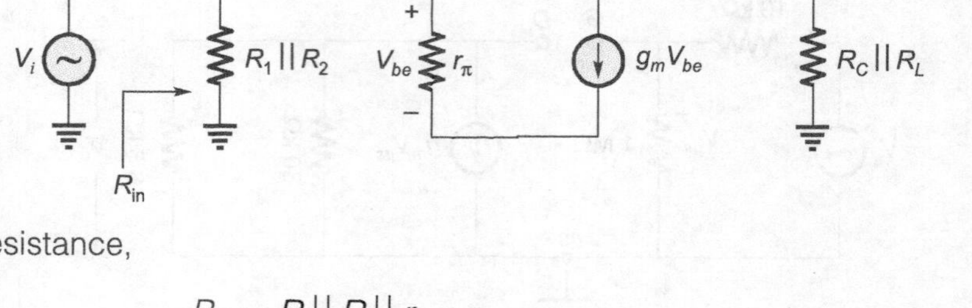

1. Input resistance,

$$R_{in} = R_1 \parallel R_2 \parallel r_\pi$$

Now, $r_\pi = \dfrac{V_T}{I_B} = \dfrac{\beta \cdot V_T}{I_C} = \dfrac{150 \times 25 \times 10^{-3}}{2.5 \times 10^{-3}} = 1.5\ \text{k}\Omega$

and $R_1 = 22\ \text{k}\Omega$ and $R_2 = 4.7\ \text{k}\Omega$

$\therefore$ $R_{in} = 1.08$ kW

2. Voltage gain,

$$V_0 = -g_m(R_C || R_L)\, V_{be}$$

Now, $V_{be} = V_{in}$

$$\therefore \quad \frac{V_0}{V_{in}} = -g_m(R_C || R_L)$$

$$g_m = \frac{\beta}{r_\pi} = \frac{150}{1.5 \times 10^3} = 100\ \text{mA/V}$$

$$\frac{V_0}{V_{in}} = -100 \times 10^{-3} \times (1 \times 10^3 || R_L)$$

Assume R_L to be 1 kΩ we get,

$$\frac{V_0}{V_{in}} = -100 \times 10^{-3} \times (1 \times 10^3 || 1 \times 10^3)$$

$$= -100 \times 10^{-3} \times 500 = -50$$

$$\therefore \quad V_0 = -50\ V_{in} = -50 \times 10 \times 10^{-3}\ \text{V} = -0.5\ \text{V}$$

Case-II: When the capacitor C_3 is removed, we get,

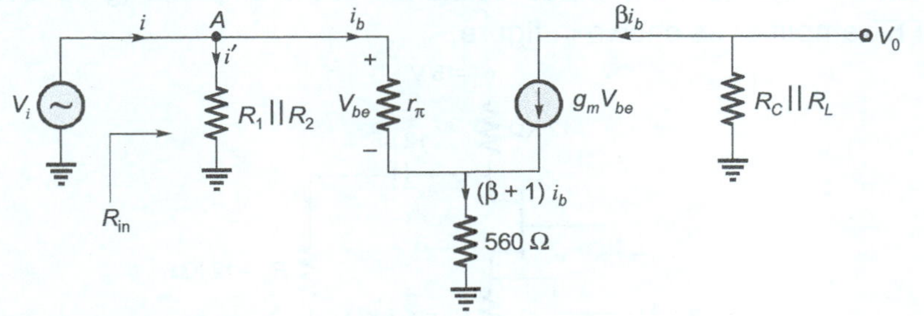

Now, applying KCL at node A we get,

$$i = i' + i_b$$

$$\therefore \quad i = \frac{V_{in}}{R_1 || R_2} + \frac{V_{in}}{r_\pi + (\beta+1)R_E}$$

$$\frac{V_{in}}{i} = R_{in} = R_1 || R_2 || r_\pi + (\beta + 1)R_E$$

$$\frac{V_{in}}{i} = 22\ \text{k}\Omega || 4.7\ \text{k}\Omega || (1.5\ \text{k}\Omega + 151 \times 560)$$

$$= 22\ \text{k}\Omega || 4.7\ \text{k}\Omega || 86.06\ \text{k}\Omega$$

$$R_{in} = 3.7\ \text{k}\Omega$$

Now, to calculate the voltage gain,

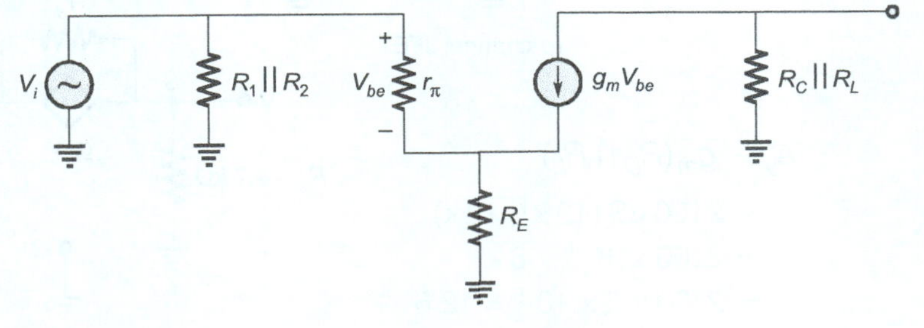

$$V_{in} = V_{be} + g_m R_E V_{be}$$
$$= (1 + g_m R_E)\,V_{be}$$

Also,
$$V_0 = -g_m(R_C || R_L)\,V_{be}$$
$$V_0 = -\frac{g_m(R_C||R_L)\cdot V_{in}}{1+g_m R_E}$$
$$V_0 = -\frac{100\times10^{-3}(1\text{k}\Omega||1\text{k}\Omega)}{(1+100\times10^{-3}\times560)}\cdot(10\times10^{-3})$$
$$= -8.77\times10^{-3}\text{ V} = -8.77\text{ mV}$$

Conclusion:

When C_3 was short-circuit then,

$$R_i = 1.08\text{ k}\Omega$$
$$V_0 = -0.5\text{ V}$$

and gain = −50

But when C_3 was removed then,

$$R_i = 3.7\text{ k}\Omega$$
$$V_0 = -8.77\text{ mV}$$

and gain = −0.877

Thus, adding C_3 reduces the input resistance of the amplifier while increasing the small signal gain of the amplifier.

5.10 **Find the minimum voltage gain and input resistance of the amplifier (g_m = 2100 μ Siemens) for the circuit of the JFET amplifier as shown in figure.**

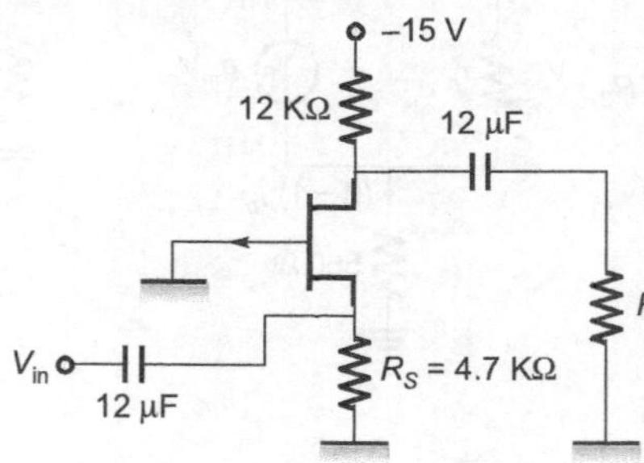

[10 marks : 2017]

Solution:

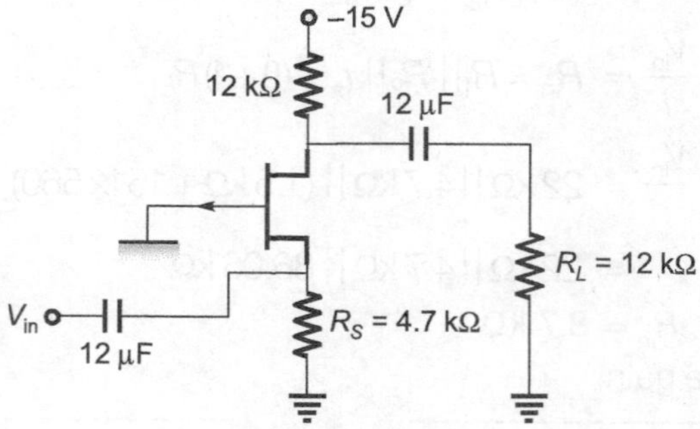

(p-channel JFET)

AC model:

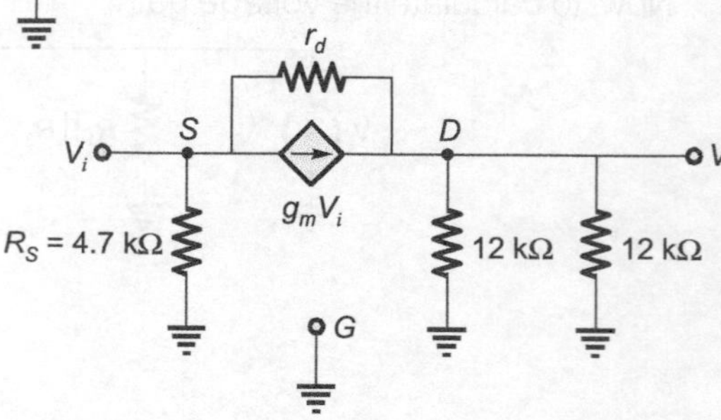

$$A_V = g_m(R_D || R_L)$$
$$= 2100\,\mu\text{S}\,(12\text{ k}||12\text{ k})$$
$$= 2100\times10^{-6}\times6\text{ k}$$
$$= 2100\times6\times10^{-3} = 12.6$$

$$Z_i = R_S || \frac{1}{g_m} = 4.7\text{k} || 476.19$$
$$= 4.7 \text{ k} || 476.19 = 432\ \Omega$$

5.11 Write the equation related to the drain current (I_D) and Gate to source voltage V_{GS} explaining all the parameters for

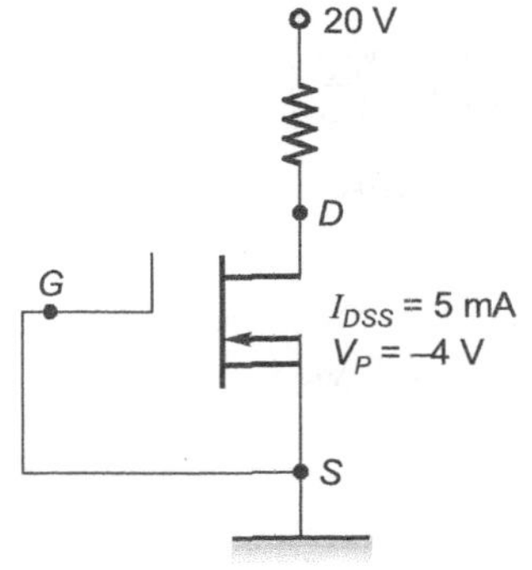

(i) Depletion type MOSFET
(ii) Enhancement type MOSFET
(iii) Determine V_{DS} for the circuit shown in the figure.

[4 + 4 + 4 = 12 marks : 2018]

Solution:

(i) Depletion type MOSFET:

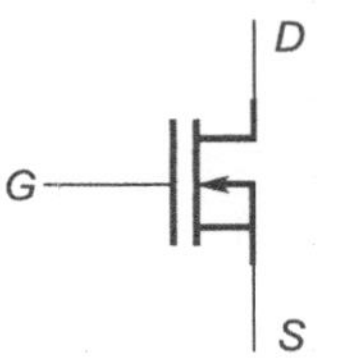

$$I_D = I_{DSS}\left(1 - \frac{V_{GS}}{V_t}\right)^2$$

Where, $$I_{DSS} = \frac{1}{2}K_n' \frac{W}{L} \cdot V_t^2$$

and $$K_n' = \mu_n C_{ox} = \text{process transconductance}$$

$$V_t = \text{Threshold voltage}$$

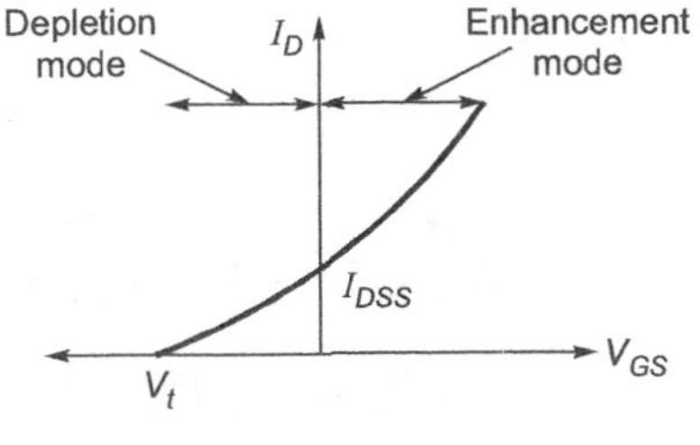

(ii) Enhancement MOSFET:

(a) Triode region:

$$I_D = K_n\left[(V_{GS} - V_T)V_{DS} - \frac{V_{DS}^2}{2}\right]$$

where, $$K_n = K_n' \frac{W}{L}$$ ($K_n', \frac{W}{L}$ as defined earlier)

$$V_{DS} < V_{GS} - V_t \rightarrow \text{Triode region}$$

$$V_{OD} = V_{GS} - V_T = \text{overdrive voltage}$$

(b) Saturation region:

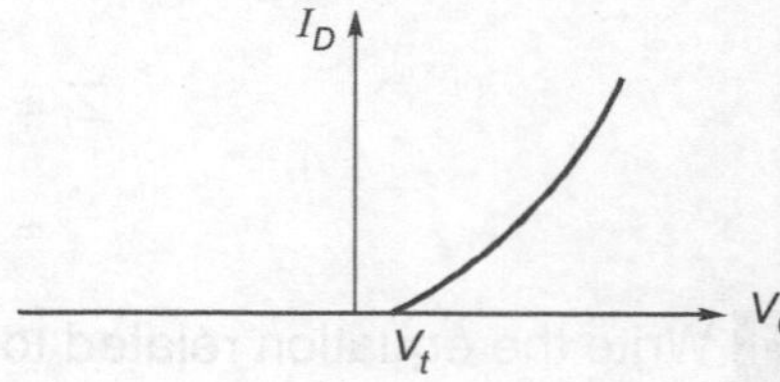

$$I_D = \frac{1}{2}K_n(V_{GS} - V_t)^2$$

$V_{DS} > V_{GS} - V_t \rightarrow$ Saturation region

(iii) $V_{GS} = 0$

$V_{GS} > V_P$ hence JFET is in saturation region

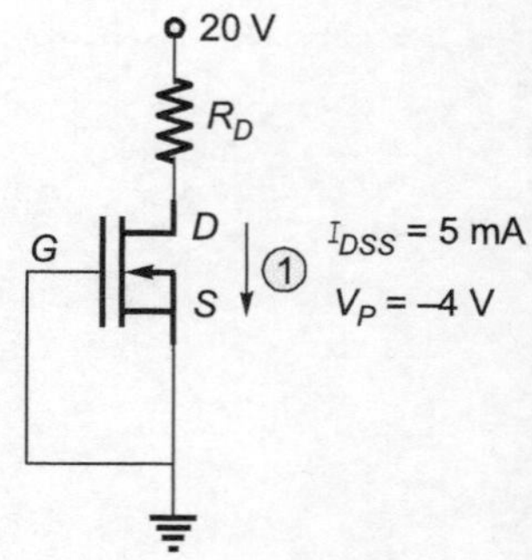

$$I_D = I_{DSS}\left(1 - \frac{V_{GS}}{V_P}\right)^2$$

$\therefore$ $I_D = I_{DSS} = 5$ mA

KVL in loop (i),

$$-20 + I_D R_D + V_{DS} = 0$$

$\therefore$ $V_{DS} = 20 - 5 \text{ mA } X R_D$

Let, $R_D = 1$ kΩ

$\therefore$ $V_{DS} = 15$ V

5.12 For the common-emitter amplifier shown in the figure, let $V_{CC} = 9$ V, $R_1 = 27$ kΩ, $R_2 = 15$ kΩ, $R_E = 1.2$ kΩ and $R_C = 2.2$ kΩ. The transistor has β = 100 and $V_A = 100$ V (V_A = Early voltage). Calculate the dc bias current I_E. If the amplifier operates between a source for which $R_s = 10$ kΩ and a load of 2 kΩ (R_L), replace the transistor with its hybrid-p model, and find the values of R_i and voltage gain v_o/v_s. Assume $V_{BE} = 0.7$ V, V_T (Thermal voltage) = 25 mV.

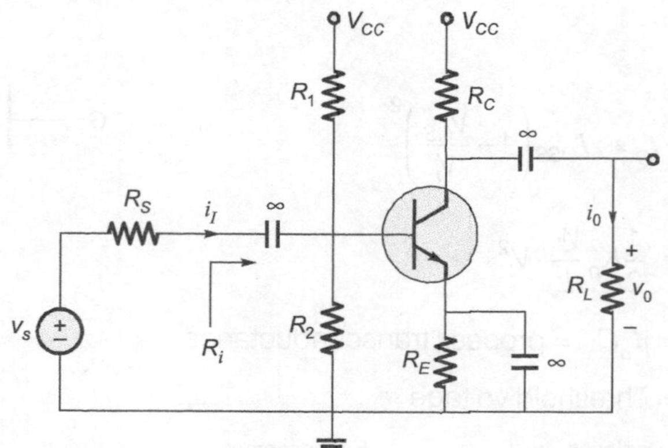

[20 marks : 2019]

Solution:

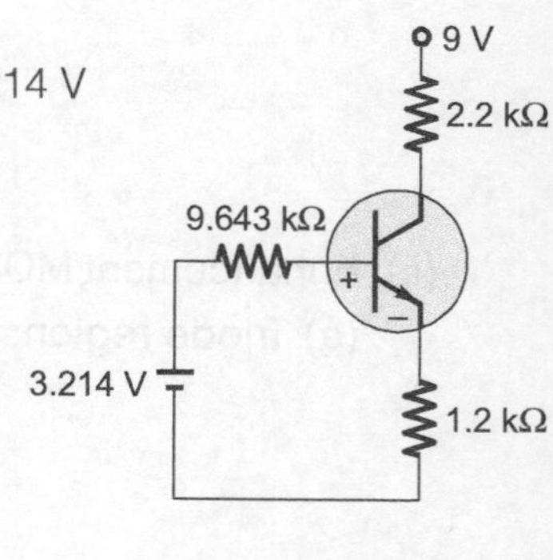

$$V_{th} = \frac{R_2}{R_1 + R_2} \times V_{CC} = \frac{15}{15 + 27} \times 9 = 3.214 \text{ V}$$

$$R_{th} = R_1 \parallel R_2 = 27 \parallel 15 = 9.643 \text{ k}\Omega$$

$$-3.215 + 9.643 I_B + 0.7 + 1.2 I_E = 0$$

$$I_B(9.643 + 1.2 \times 101) = 2.514$$

$$I_B = 19.214 \text{ μA}$$

$$I_E = 1.941 \text{ μA}$$

$$I_C = 1.921 \text{ μA}$$

$$r_0 = \frac{V_A}{I_C} = \frac{100}{1.921} = 52.06 \text{ k}\Omega$$

$$r_e = \frac{V_T}{I_E} = \frac{25}{1.941}$$

$$= 12.88\ \Omega$$

$$r_\pi = \beta r_e = 1.288\ \text{k}\Omega$$

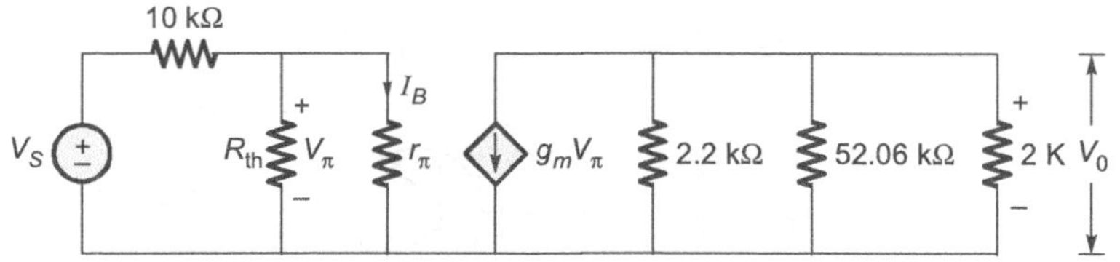

$$V_0 = -g_m V_\pi (2.2 \parallel 52.06 \parallel 2)$$

$$= \frac{-1 \times 1.0269}{12.88 \times 10^{-3}} \times V_\pi = -79.732\ V_\pi$$

$$V_0 = -79.732 \times \left(\frac{R_i}{R_i + R_s}\right) V_s$$

$$R_i = R_{th} \parallel r_\pi = 9.643 \parallel 1.288$$

$$R_i = 1.1362\ \text{k}\Omega$$

$$V_0 = -79.732 \times \frac{1.136}{1.136 + 10} \times V_S$$

$$\frac{V_0}{V_S} = -8.134$$

$$A_V = -8.134$$

5.13 In the given JFET as amplifier shown in the figure, the drain current changes from 4 mA to 6 mA when the gate voltage is changed from –3.8 V to –3.5 V in the amplifier circuit. Calculate the voltage gain of the amplifier.

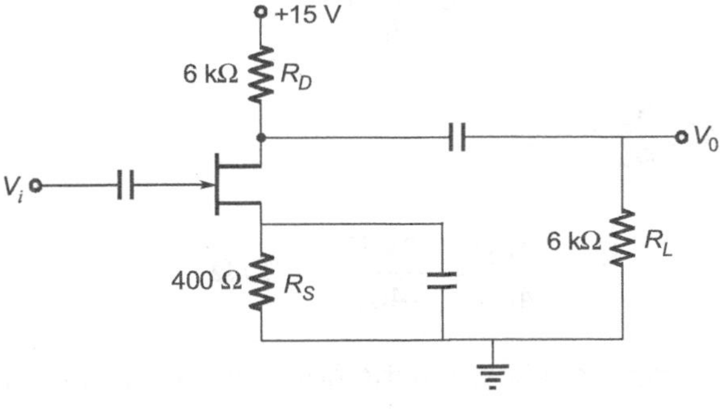

[12 Marks : 2019]

Solution:

$$\text{Transe conductance} = g_m = \frac{\Delta I_D}{\Delta V_{gs}} = \frac{6\ \text{mA} - 4\ \text{mA}}{-3.5 + 3.8} = \frac{1}{150}\text{S}$$

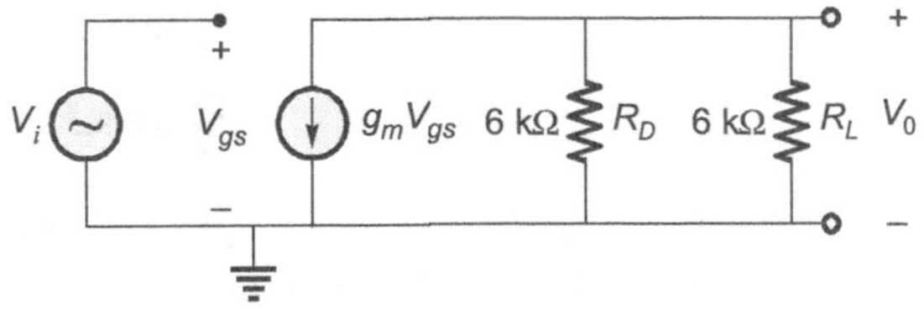

R_S is short circuited through by pass capacitor

Voltage gain, $$A_V = -g_m \left[R_D \,||\, R_L\right] = -\frac{1}{150}[6\text{ k} \,||\, 6\text{ k}] = -20$$

5.14 Common emitter (CE) amplifier shown in the figure has voltage gain 400 when $R_E = 0$. Stability is brought through negative feedback by adding resistor R_E. Find the value of resistor R_E using feedback concepts so that final voltage gain is equal to 200.

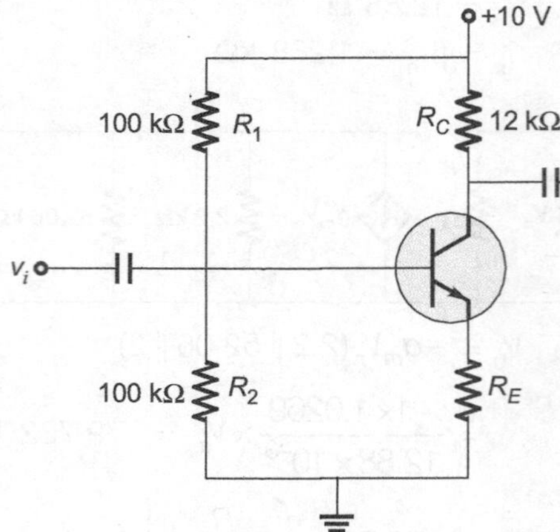

[10 marks : 2019]

Solution:

Feedback factor,
$$\beta = \frac{V_f}{V_0} = -\frac{R_E}{R_C}$$

$$A_{Vf} = \frac{A}{1 + A\beta}$$

$$A = -400$$

$$A_{Vf} = -200$$

$$-200 = \frac{-400}{1 + \left(400\frac{R_E}{R_C}\right)}$$

$$1 + \left(400\frac{R_E}{R_C}\right) = 2$$

$$R_E = \frac{R_C}{400} = \frac{12 \times 10^3}{400} = 30\ \Omega$$

5.15 **Find the voltage gain of JFET amplifier shown in the figure below for the drain to source current with gate shorted, I_{DSS} = 10 mA, cut-off voltage $V_{GS(OFF)}$ = –4 V and I_D = 2 mA. If a load resistance of 4.7 kΩ is a.c. coupled to the output of this amplifier, calculate the percentage change in voltage gain.**

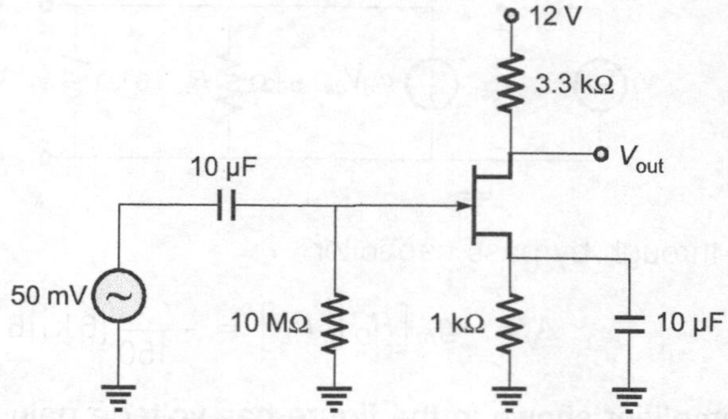

[10 marks : 2020]

Solution:

Given :

$$I_{DSS} = 10 \text{ mA}$$
$$I_D = 2 \text{ mA}$$
$$V_{GS(OFF)} = -4 \text{ V}$$

$$I_D = I_{DSS}\left[1 - \frac{V_{GS}}{V_{GS(OFF)}}\right]^2 \quad ...(1) \quad \text{Shockley equation}$$

Transconductance g_m is given by

$$g_m = \frac{\partial I_D}{\partial V_{GS}}$$

On performing partial differentiation of eqn. (1)

$$\frac{\partial I_D}{\partial V_{GS}} = \frac{-2I_{DSS}}{V_{GS(OFF)}}\left[1 - \frac{V_{GS}}{V_{GS(OFF)}}\right]$$

$$g_m = \frac{-2I_{DSS}}{V_{GS(OFF)}} \times \sqrt{\frac{I_D}{I_{DSS}}}$$

On putting values

$$g_m = \frac{-2\times 10}{-4} \times \sqrt{\frac{2}{10}} = 5\sqrt{\frac{2}{10}}$$
$$= 2.33 \text{ mA/V} = 2.23 \text{ mS}$$

⇒ Equivalent small signal modal of given problem figure.

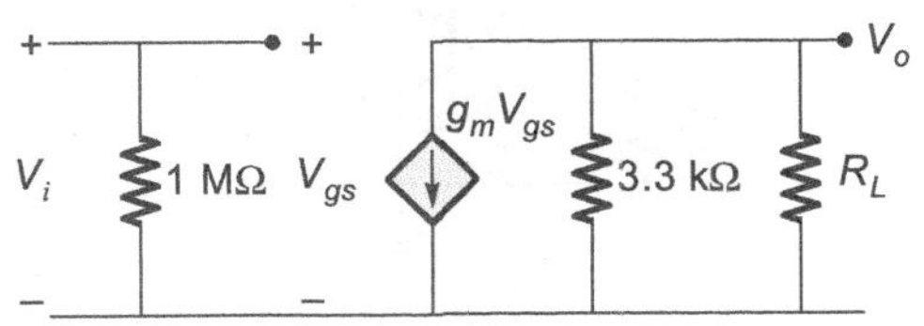

Case 1 : When $R_L = \infty$ (absent)

$$A_V = \frac{V_o}{V_i} = \frac{V_o}{V_{gs}} = \frac{-g_m \cdot V_{gs}(3.3 \text{ k}\Omega)}{V_{gs}} = -g_m\,(3.3 \text{ k}\Omega)$$

$$A_v = -2.23 \times 3.3 = -7.359$$

Case 2 : When $R_L = 4.7 \text{ k}\Omega$

$$A_V' = \frac{V_o}{V_i} = \frac{V_o}{V_{gs}} = \frac{-g_mV_{gs}(3.3 \text{ k}\Omega \parallel 4.7\text{k}\Omega)}{V_{gs}}$$
$$= -2.23 \text{ ms} \times 1.9387 \text{ k}\Omega$$
$$A_V' = -4.3234$$

$$\% \text{ change in voltage gain} = \frac{|A_V| - |A_V'|}{|A_v|} \times 100$$
$$= 41.25\%$$

5.16 **Assuming the Early Voltage $V_A = \infty$, compute the voltage gain and input impedance of the circuit shown below. I_{DC} is an ideal current source while V_{Bias} is an ideal voltage source used to bias Q_1 and Q_2 in the active mode.**

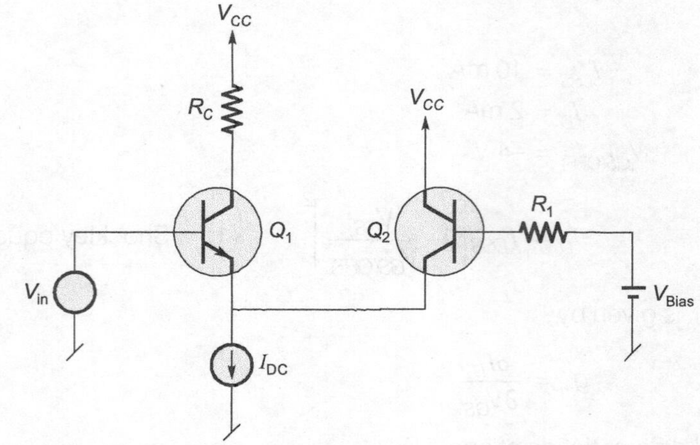

[20 marks : 2021]

Solution:

AC equivalent circuit is,

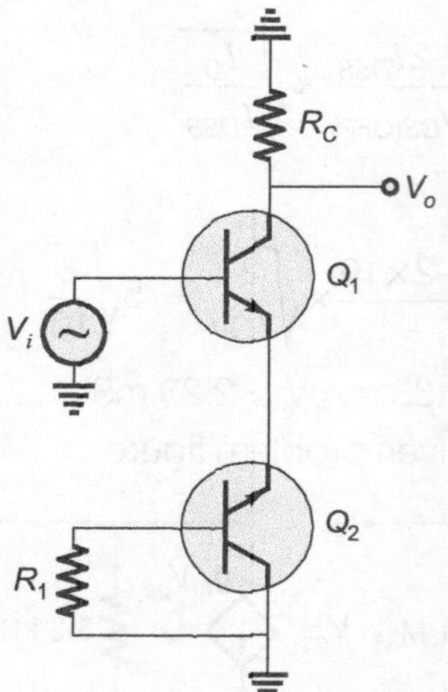

To find R_{eq} of Q_2 :

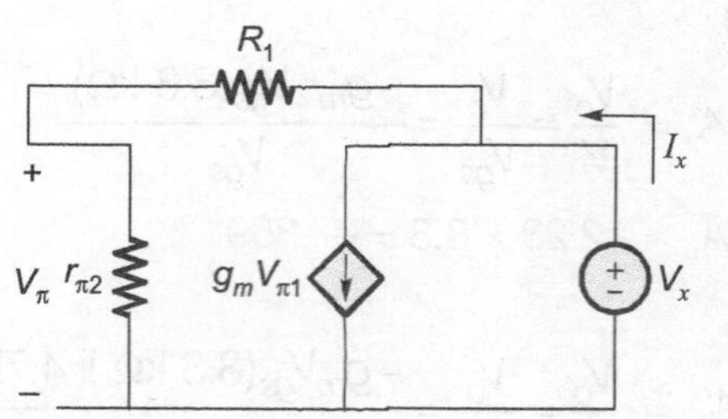

$$V_\pi = \frac{V_x r_{\pi 2}}{R_1 + r_{\pi 2}}$$

On applying KCL :

$$I_x = g_{m2} V_\pi + \frac{V_x}{R_1 + r_{\pi 2}}$$

$$I_x = \frac{g_{m2} V_x r_{\pi 2}}{R_1 + r_{\pi 2}} + \frac{V_x}{R_1 + r_{\pi 2}}$$

$$\frac{I_x}{V_x} = \frac{\beta_2}{R_1 + r_{\pi 2}} + \frac{1}{R_1 + r_{\pi 2}} = \frac{1 + \beta_2}{R_1 + r_{\pi 2}}$$

$$R_{eq} = \frac{V_x}{I_x} = \frac{R_1 + r_{\pi 2}}{1 + \beta_2}$$

Replace Q_1 with π-model :

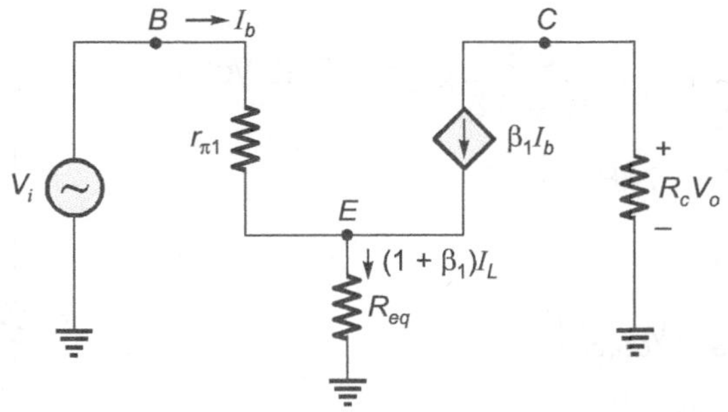

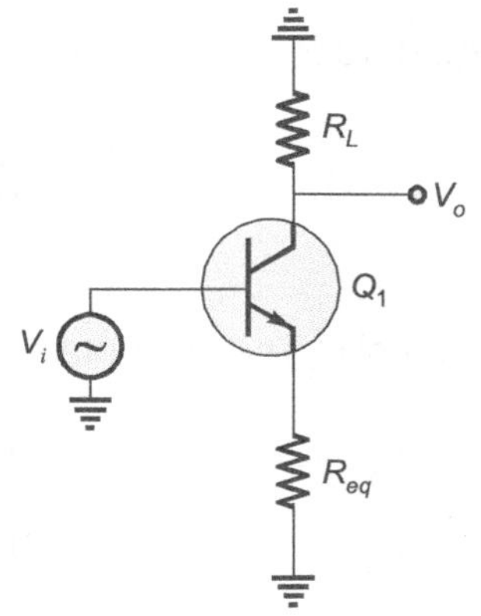

Voltage gain,
$$\frac{V_o}{V_i} = \frac{-\beta_1 I_b R_c}{I_b r_{\pi 1} + (1+\beta_1) I_b R_{eq}} = \frac{-\beta_1 R_c}{r_{\pi 1} + (1+\beta_1) R_{eq}}$$

$$A_V = \frac{-\beta_1 R_c}{r_{\pi 1} + \frac{(1+\beta_1)}{(1+\beta_2)}(R_1 + r_{\pi 2})}$$

KVL in input loop :

$$V_i = I_b r_{\pi 1} + (1 + \beta_1) I_b R_{eq}$$

Input Resistance :
$$\frac{V_i}{I_b} = R_i = r_{\pi 1} + \frac{(1+\beta_1)}{1+\beta_2} \times (R_1 + r_{\pi 2})$$

5.17 Find V_{GS}, I_D, V_{DS} and the AC output voltage of the E-MOSFET amplifier shown in the figure for an input of 25 m volts. The device parameters are I_{DON} = 18 mA at V_{GS} = 10 Volts, $V_{GS(th)}$ = 2.5 Volts and g_m = 23 mS.

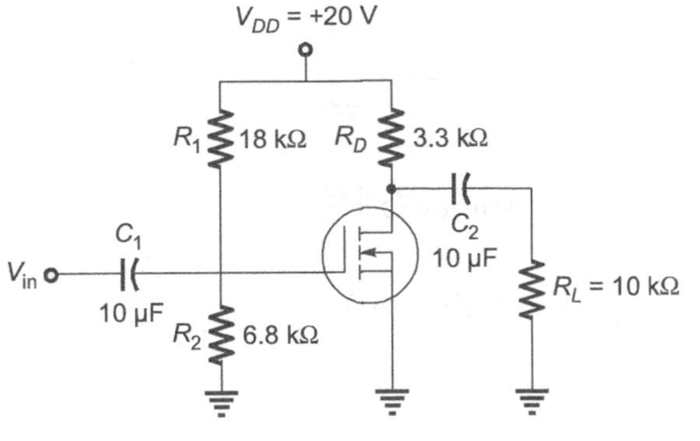

[12 marks : 2022]

Solution:

From the given conditions:

$$I_{DON} = 18 \text{ mA at } V_{GS} = 10 \text{ V and } V_{GS(th)} = 2.5 \text{ V}$$

As we know,

$$I_{DON} = K_n\left[V_{GS} - V_{GS(th)}\right]^2, \qquad \text{When } K_n = \frac{\mu_n C_{ox}}{2}\left(\frac{W}{L}\right)$$

$$18 = K_n[10 - 2.5]^2$$

$$K_n = 0.25 \text{ mA/V}$$

Now, DC equivalent circuit of given amplifier

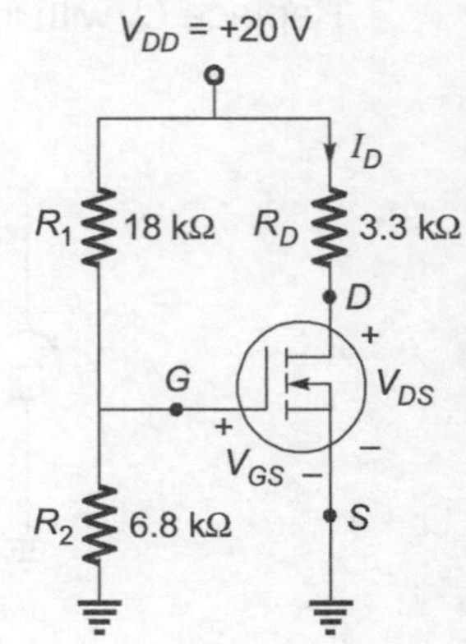

Apply voltage division at gate terminal

$$V_G = \frac{R_2}{R_1 + R_2} V_{DD} = \frac{6.8}{18 + 6.8} \times 20 = 5.48 \text{ Volt}$$

Therefore, $$V_{GS} = V_G - V_S = 5.48 - 0 = 5.48 \text{ Volt}$$

Now, drain current (I_D) can be obtained as

$$I_D = K_n\left[V_{GS} - V_{GS(\text{th})}\right]^2$$

$$I_D = 0.32\,[5.48 - 2.5]^2 = 2.84 \text{ mA}$$

Therefore, $$V_{DS} = V_{DD} - I_D R_D$$

$$V_{DS} = 20 - 2.84 \times 3.3 = 10.628 \text{ Volt}$$

Now, AC analysis:

Replacing the MOSFET by its AC equivalent circuit,

Capacitance → Short circuit

DC supply → Disabled

Since, $$V_A = \infty$$

⇒ $$r_0 = \infty$$

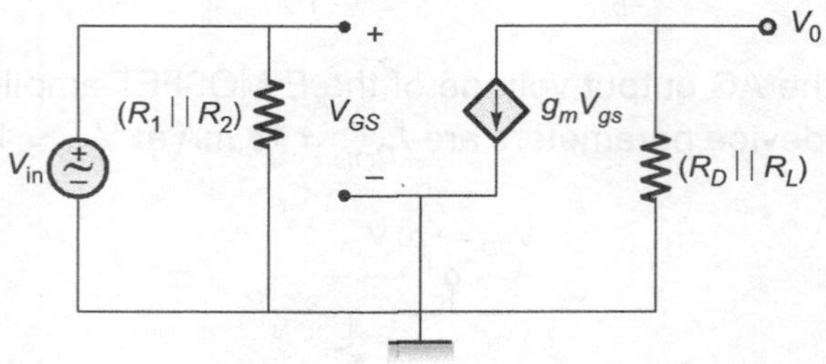

$$V_0 = -g_m V_{gs}(R_D \parallel R_L)$$

$$\frac{V_0}{V_i} = -g_m(R_D \parallel R_L) \qquad [\because V_{in} = V_{gs}]$$

Therefore AC output voltage,

$$V_0 = -23 \times 10^{-3} \times [3.3 \parallel 10] \times 10^3 \times 25 \times 10^{-3}$$

$$= -1.426 \text{ Volt}$$

5.18 The JFET circuit shown in the figure is operating in the mid frequency range. Identify the type of feedback.

(i) Draw the mid frequency small signal model indicating the output, input, feedback points and the closed loop feedback block diagram indicating the forward gain and feedback factor.

(ii) Write the equation for gain without feedback and with feedback.

(iii) Calculate the amplifier gain with and without feedback if R_1 = 80 kΩ, R_2 = 20 kΩ, R_0 = 25 kΩ and R_D = 5 kΩ. Assume FET g_m = 5.5 mS.

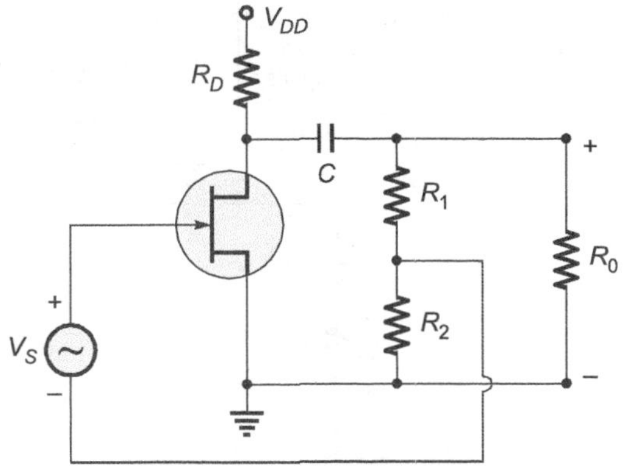

[2 + 6 + 6 + 6 = 20 marks : 2022]

Solution:

Voltage divider network of R_1 and R_2 result in feedback

If feedback network is connected to O/P node then consider Voltage sampling

If it is not connected to I/P node then consider Series mixing

∴ Voltage series feedback is present

(i)

$$-V_f = \frac{V_0 R_2}{R_1 + R_2}$$

$$\frac{V_f}{V_0} = \beta = \frac{-R_2}{R_1 + R_2}$$

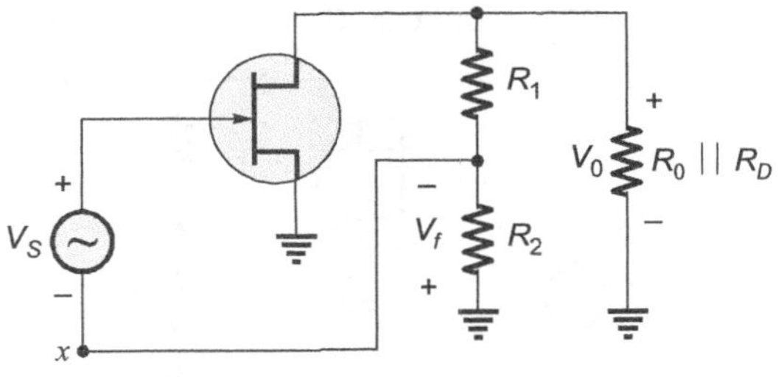

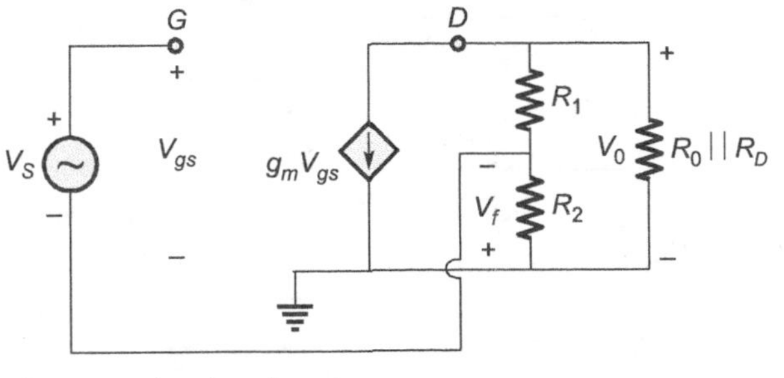

Block diagram for voltage series feedback,

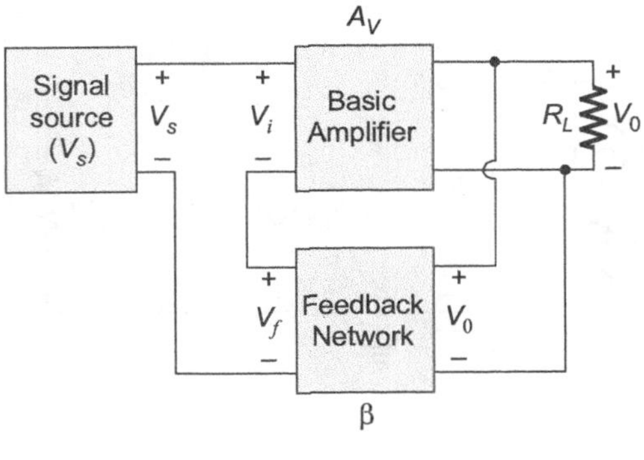

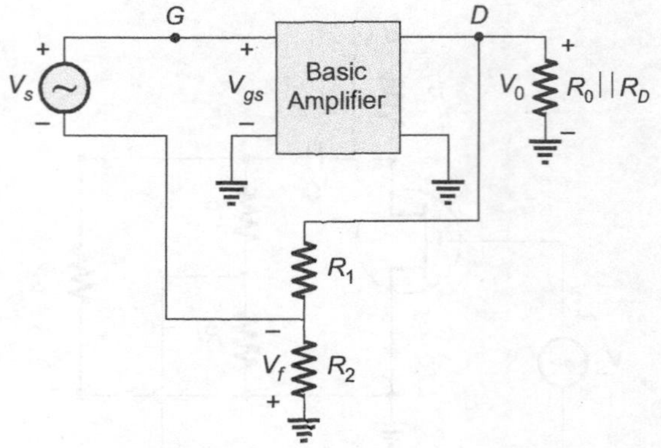

(ii) Gain without feedback

Drawing circuit without feedback

(*a*) To draw input circuit, ground O/P node

$\Rightarrow R_1$ and R_2 appear in parallel between node x and ground

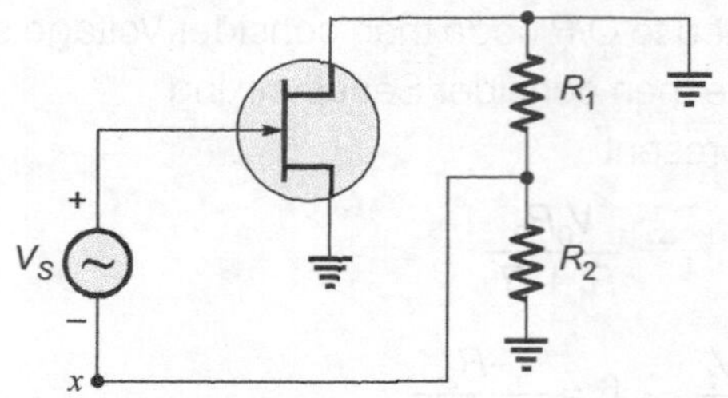

(*b*) To draw output circuit, break input loop

$\Rightarrow R_1$ and R_2 appear in series between drain and ground

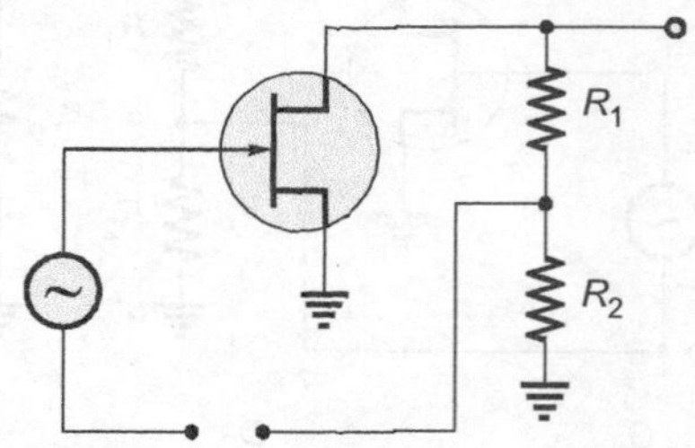

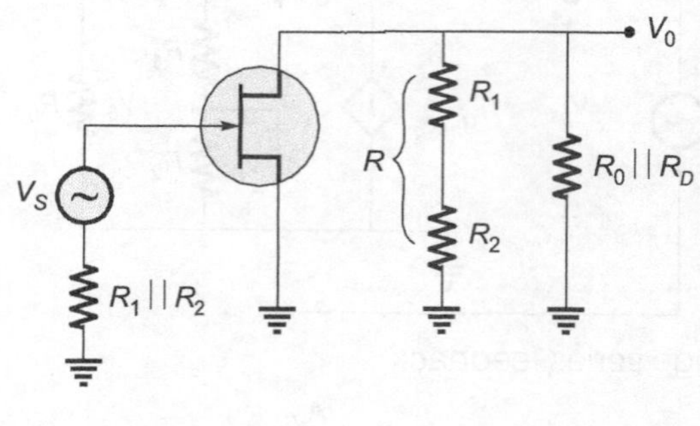

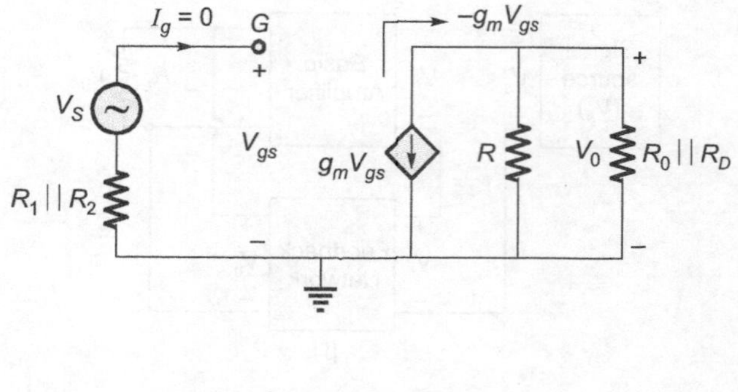

Gain without feedback, $A_v = \frac{V_0}{V_s} = \frac{-g_m V_{gs}(R \| R_0 \| R_D)}{V_{gs}}$

$$A_v = -g_m V_{gs}(R \| R_0 \| R_D),$$
$$R = R_1 + R_2$$

Voltage gain with feedback,

$$A_{vf} = \frac{A_v}{1+\beta A_v}$$

$$\beta = \frac{-R_2}{R_1 + R_2}$$

(iii) Gain with feedback, $\beta = \frac{-R_2}{R_1 + R_2} = \frac{-20}{80+20} = -0.2$

$$A_v = -g_m(R \| R_0 \| R_D),$$
$$R = R_1 + R_2 = 100 \text{ k}\Omega$$
$$A_v = -5.5\,(100 \,||\, 25 \,||\, 5)$$
$$A_v = -22$$
$$A_{vf} = \frac{A_v}{1+\beta A_v} = \frac{-22}{1+(-0.2)(-22)}$$
$$A_{vf} = -4.07$$

and gain without feedback,

$$A_v = -22$$

5.19 For the JFET amplifier circuit shown in the figure below, g_m = 2 mS, r_d = 200 kΩ, C_{gs} = 10 pF, C_{gd} = 2 pF, R_S = 1 kΩ, R_1 = 10 MΩ, R_2 = 100 kΩ, R_D = 5 kΩ, C_{C1} = C_{C2} = 0.1 μF. Assume output capacitor C_o = 10 pF, C_S and R_L to be very large. Find :

(i) mid-frequency gain;

(ii) lower cut-off frequency;

(iii) higher cut-off frequency;

(iv) gain-bandwidth product

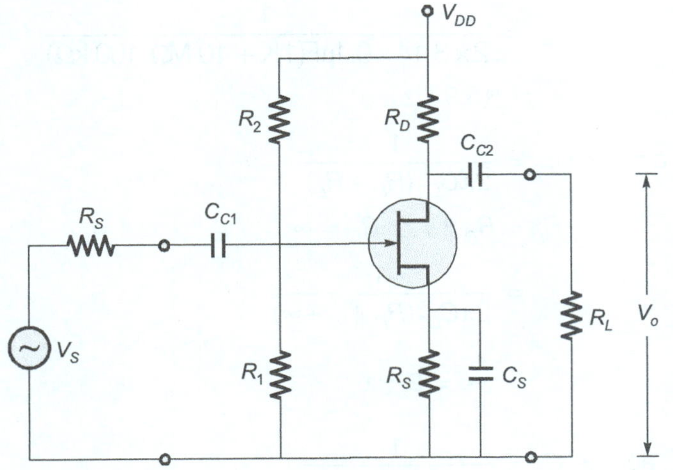

[20 marks : 2023]

Solution:

Given :

$$g_m = 2\text{ mS},\quad R_D = 5\text{ k}\Omega$$
$$r_d = 200\text{ k}\Omega,\quad C_{C1} = C_{C2} = 0.1\ \mu\text{F}$$
$$C_{gs} = 10\text{ pF},\quad C_o = 10\text{ pF}$$
$$C_{gd} = 2\text{ pF},\quad R_s = 1\text{ k}\Omega$$
$$R_1 = 10\text{ M}\Omega,\quad R_2 = 100\text{ k}\Omega$$

AC Model for the given circuit

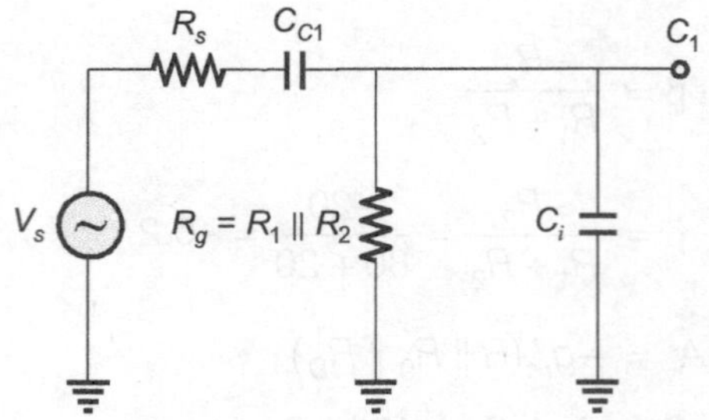

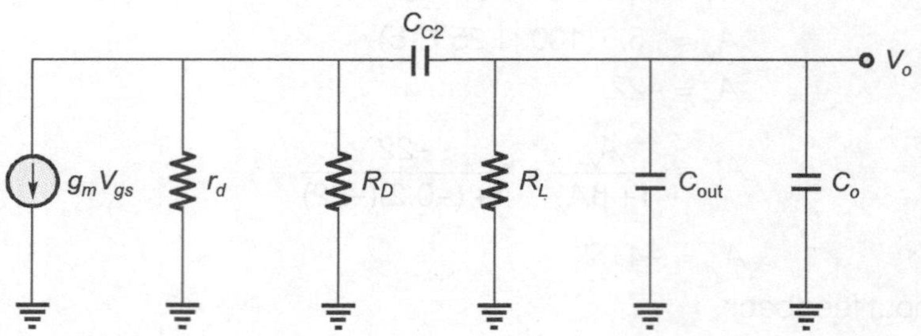

(i) Mid Frequency Gain :

$$A_v \text{ (bypass)} = -g_m R_D \parallel r_d \parallel R_L$$
$$= -2\text{ mS } 5\text{ K} \parallel 200\text{ K} \parallel \infty$$
$$= -2\text{ mS } 5\text{ K} \parallel 200\text{ K} \qquad \text{(Assume } R_L = \infty)$$
$$= -2\text{ mS} \times 4.87\text{ K} = -9.74$$

(ii) Lower cut-off frequency

$$f_L \text{ across } C_{C1} = \frac{1}{2\pi C_{C1}(R_s + R_g)}$$
$$= \frac{1}{2\times 3.14\times 0.1\mu\text{F}(1\text{K} + 10\text{ M}\Omega\ 100\text{ k}\Omega)}$$
$$= 15.9\text{ Hz}$$

$$f_L \text{ across } C_{C2} = \frac{1}{2\pi C_{C2}(R_D + R_L)}$$

where $R_D = R_D \parallel r_d;\quad R_L = \infty$

$$= \frac{1}{2\pi C_{C2}(R_D \parallel r_d + \infty)}$$
$$= \frac{1}{\infty} = 0\text{ Hz}$$

$$f_L \text{ across } C_S = \frac{1}{2\pi C_S\left(R_s \parallel \dfrac{1}{g_m}\right)}$$

$$= \frac{1}{2\pi \times \infty \left(R_s \parallel \frac{1}{g_m} \right)} = 0 \text{ Hz} \qquad (C_S \text{ is large})$$

So, lower cut off frequency,

$$f_L = 15.9 \text{ Hz}$$

(iii) Higher cut-off frequency

$$f_{H1} = \frac{1}{2\pi R_s \parallel R_g C_{in}} = \frac{1}{2\pi R_s \parallel R_g \{C_{gs} + C_{gd}(1 + A_V)\}}$$

$$= \frac{1}{2 \times 3.14 \times 1\text{ K} \parallel 99\text{ K}\{10 \text{ pf} + (1 + 9.74)2 \text{ pf}\}} = 5.10 \text{ MHz}$$

$$f_{H2} = \frac{1}{2\pi r_d \parallel R_D \left\{ C_o + C_{gd}\left(1 + \frac{1}{A_v}\right) \right\}}$$

$$= 2.67 \text{ MHz}$$

Net f_H:

$$\frac{1}{f_H} = 1.1 \times \sqrt{\frac{1}{f_{H1}^2} + \frac{1}{f_{H2}^2}}$$

$$f_H = 2.15 \text{ MHz}$$

(iv) Gain bandwidth product

$$f_T = \frac{g_m}{2\pi(C_{gs} + C_{gd})} = \frac{2 \text{ mS}}{2 \times 3.14(10 \text{ pF} + 2 \text{ pF})} = 26.53 \text{ MHz}$$

5.20 A dual input, balanced output differential amplifier is configured using silicon transistors which are identical having h_{ie} = 2.8 kΩ as shown in the figure below:

(i) Calculate the differential gain, common mode gain and CMRR.

(ii) What is the peak-to-peak output voltage Vo, if V_{S1} is 50 mV peak-to-peak at 2 kHz and V_{S2} is 30 mV peak-to-peak at 2 kHz?

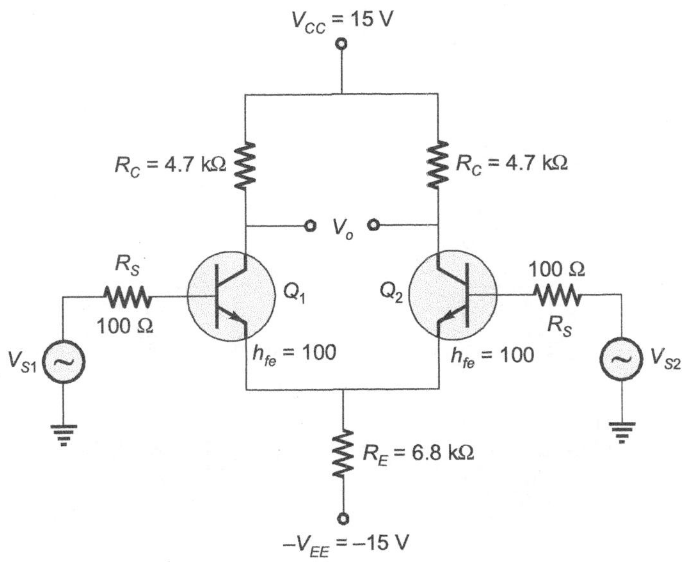

[20 marks : 2023]

Solution:

Given dual input balanced output differential amplifier Q_1 and Q_2 are identical.

$$h_{ie} = 2.8\ \text{k}\Omega$$

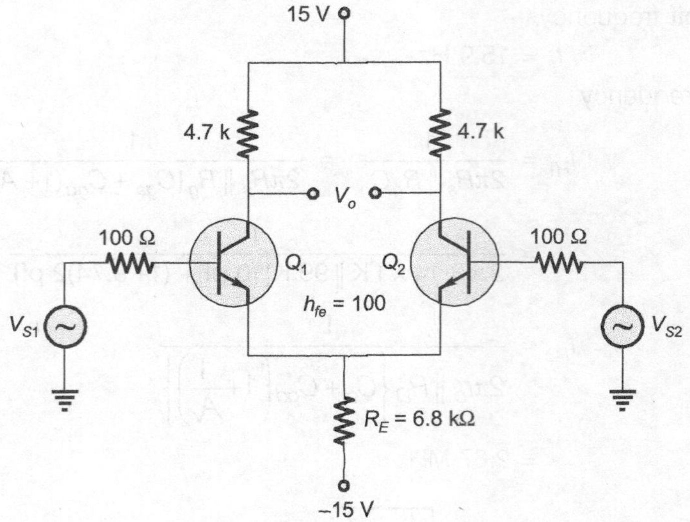

(i) Differential gain A_d :

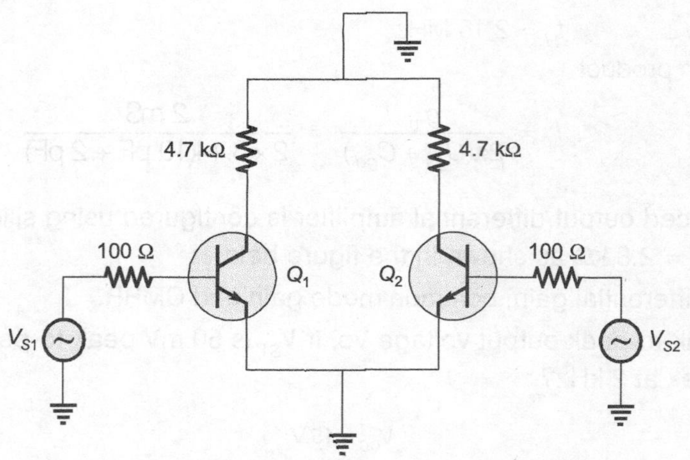

For differential mode :

$$V_{s1} = \frac{V_d}{2}$$

$$V_{s2} = -\frac{V_d}{2}$$

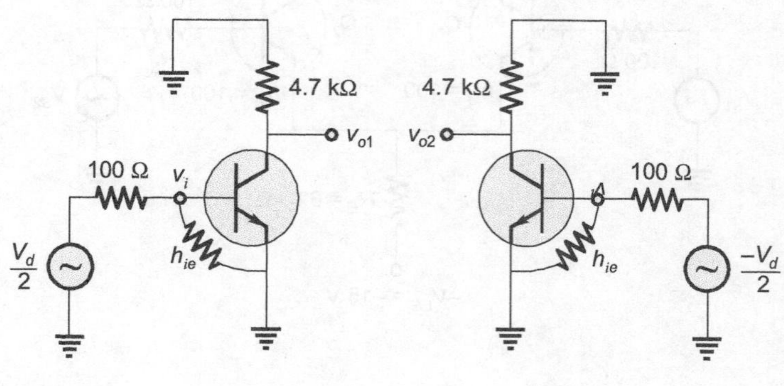

Here,

$$v_d = 50 \text{ mV} - 30 \text{ mV}$$
$$= 20 \text{ mV (P-P)}$$

$$\frac{V_{o1}}{V_i} = \frac{-h_{ge}R_L}{h_{ie}} = \frac{-100 \times 4.7k}{2.8k}$$

$$\frac{V_{o1}}{V_i} = -167.85$$

$$\frac{v_i}{v_d/2} = \frac{h_{ie}}{h_{ie}+R_s} = \frac{2.8 \text{ K}}{2.8 \text{ K} + 100 \ \Omega}$$
$$= 0.965$$

$$\frac{v_i}{v_d} = \frac{0.965}{2} = 0.482$$

$$A_{d1} = \frac{V_{o1}}{v_d} = -167.85 \times 0.482 \quad \text{(Unbalanced gain)}$$
$$= -80.90$$

Similarly,

$$A_{d2} = \frac{V_{o2}}{v_d} = 80.90 \quad \text{(Unbalanced gain)}$$

$$A_d = A_{d2} - A_{d1}$$
$$= 80.90 - (-80.90)$$
$$= 161.8 \text{ (balanced gain)}$$

A_d (differential gain) = 161.8

Common mode gain AC

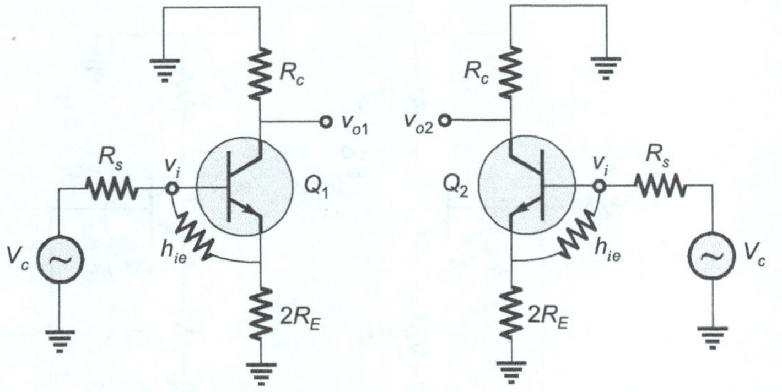

Q_1 is a unbypass circuit and Q_2 is similar circuit.

$$\frac{V_{o1}}{v_i} = \frac{-R_C}{r_e + 2R_E}$$

$$h_{ie} = \beta r_e$$

$$r_e = \frac{2.8k}{100} = 28 \ \Omega$$

$$\frac{V_{o1}}{V_i} = \frac{-4.7k}{28 \ \Omega + 2 \times 6.8k}$$
$$= -0.344$$

$$\frac{V_i}{V_c} = \frac{h_{ie} + 2(1+\beta)R_E}{R_s + h_{ie} + 2(1+\beta)R_E} = 0.99$$

$$A_c = \frac{V_{o1}}{V_c} = -0.344 \times 0.99$$

$$= -0.34$$

$$|A_c| = 0.34 \text{ (unbalanced)}$$

$$|A_c| = 0 \text{ (balanced)}$$

$$\text{CMRR} = \frac{|A_d|}{|A_c|_{\text{balanced}}}$$

$$= \infty \text{ (Theoretical)}$$

$$\text{CMRR} = \frac{A_d \text{ balanced}}{A_c \text{ Unbalanced}}$$

$$= \frac{161.8}{0.34}$$

$$= 475.88 \text{ (balanced)}$$

$$\text{CMRR} = \frac{A_d \text{ unbalanced}}{A_c \text{ unbalanced}}$$

$$= \frac{80.90}{0.34} = 237.94 \text{ (unbalanced)}$$

Additional Information : If R_{c1} and R_{c2} are not equal.
To calculate common mode gain

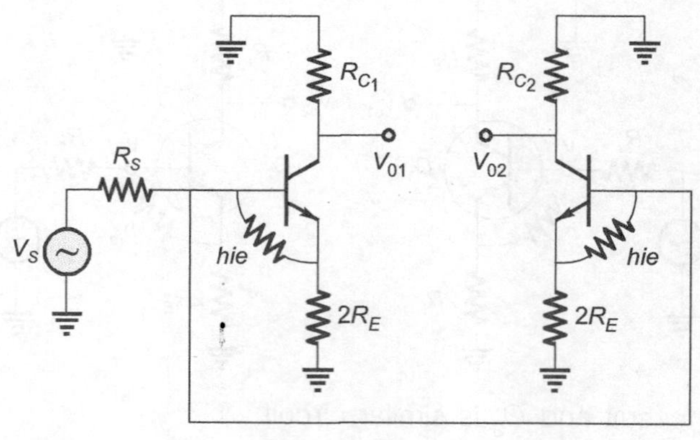

$$\frac{V_{o1}}{V_s} = \frac{-h_{fe}I_bR_c}{2I_bR_s + h_{ie}I_b + (1+h_{fe})I_b \times 2R_E}$$

$$\frac{V_{o1}}{V_s} = \frac{-h_{fe}R_{c1}}{2R_s + h_{ie} + 2(1+h_{fe})R_E}$$

$$\frac{V_{o2}}{V_s} = \frac{-h_{fe}R_{c2}}{2R_s + h_{ie} + 2(1+h_{fe})R_E}$$

For balanced output

$$A_c = \frac{V_{o1} - V_{o2}}{V_s}$$

$$A_c = \frac{h_{fe}(R_{c2} - R_{c1})}{2R_s + h_{ie} + 2(1+h_{fc})R_E}$$

If $R_{c1} = R_{c2}$, then $A_c = 0$

If $R_{c1} \neq R_{c2}$, then $A_c \neq 0$

Assume $R_{c1} = R_c + 1\% = 1.01R_c$

$R_{c2} = R_c - 1\% = 0.99R_c$

$$A_c = \frac{h_{fe}(0.99R_c - 1.01R_c)}{2R_s + h_{ie} + 2(1+h_{fe})R_E} = -6.828 \times 10^{-3}$$

$$\text{CMRR} = \left|\frac{A_D}{A_C}\right| = \frac{161.8}{6.828 \times 10^{-3}} = 23.69 \times 10^3$$

6. Frequency Response

6.1 When two amplifiers each of bandwidth, f_H = 10 kHz are cascaded, the overall bandwidth becomes

(a) 10 kHz
(b) 6.4 kHz
(c) 5 kHz
(d) 20 kHz

[2 marks : 2006]

Solution: (b)

f_H = 10 kHz for one amplifier

B.W = upper cutoff frequency f_H

For n-stage amplifier (cascaded)

Upper cut-off frequency is given by

$$f_H^* = f_H\sqrt{2^{1/n} - 1}$$

$f_H \rightarrow$ for single stage

∴ For $n = 2$

$$\therefore \quad f_H^* = f_H\sqrt{2^{1/2} - 1}$$

$= 0.64 \times 10$ kHz

$= 6.4$ kHz

6.2 Find the value of the lower cut-off frequency for the MOSFET amplifier depicted in figure. Assume $C_1 = 0.001\ \mu F$, $R_G = 10\ M\Omega$, $R_D = 4.7\ k\Omega$, $I_{GSS} = 25$ nA@ $V_{GS} = -10$ V.

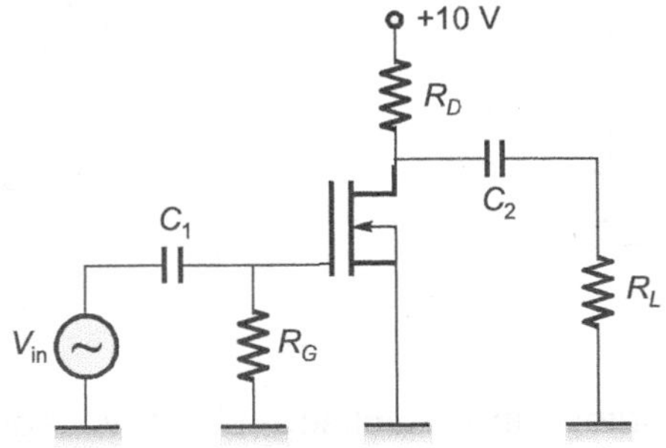

[10 marks : 2017]

Solution:

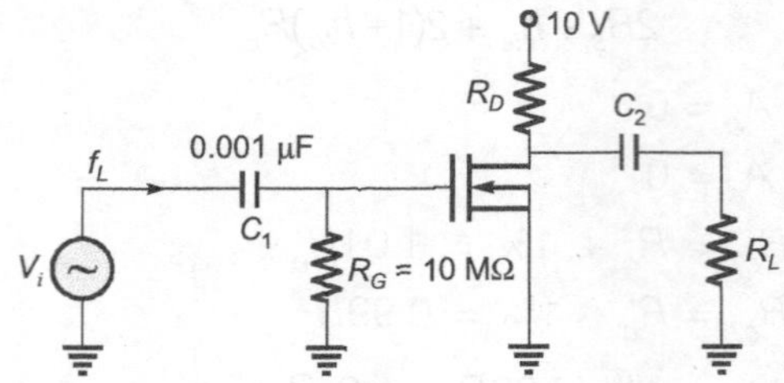

Lower cut-off frequency,

$$f_L = C_1 \times R_G$$
$$= 0.001\ \mu F \times 10\ M\Omega$$
$$= 0.001 \times 10^{-6} \times 10 \times 10^{6}$$
$$= 0.01\ Hz$$

7. Feedback Amplifiers

7.1 Make a distinction between the "voltage" feedback and "current" feedback in amplifier circuits. Discuss the merits in each case and derive the expression for the net output impedance in each case.

[15 marks : 2003]

Solution:

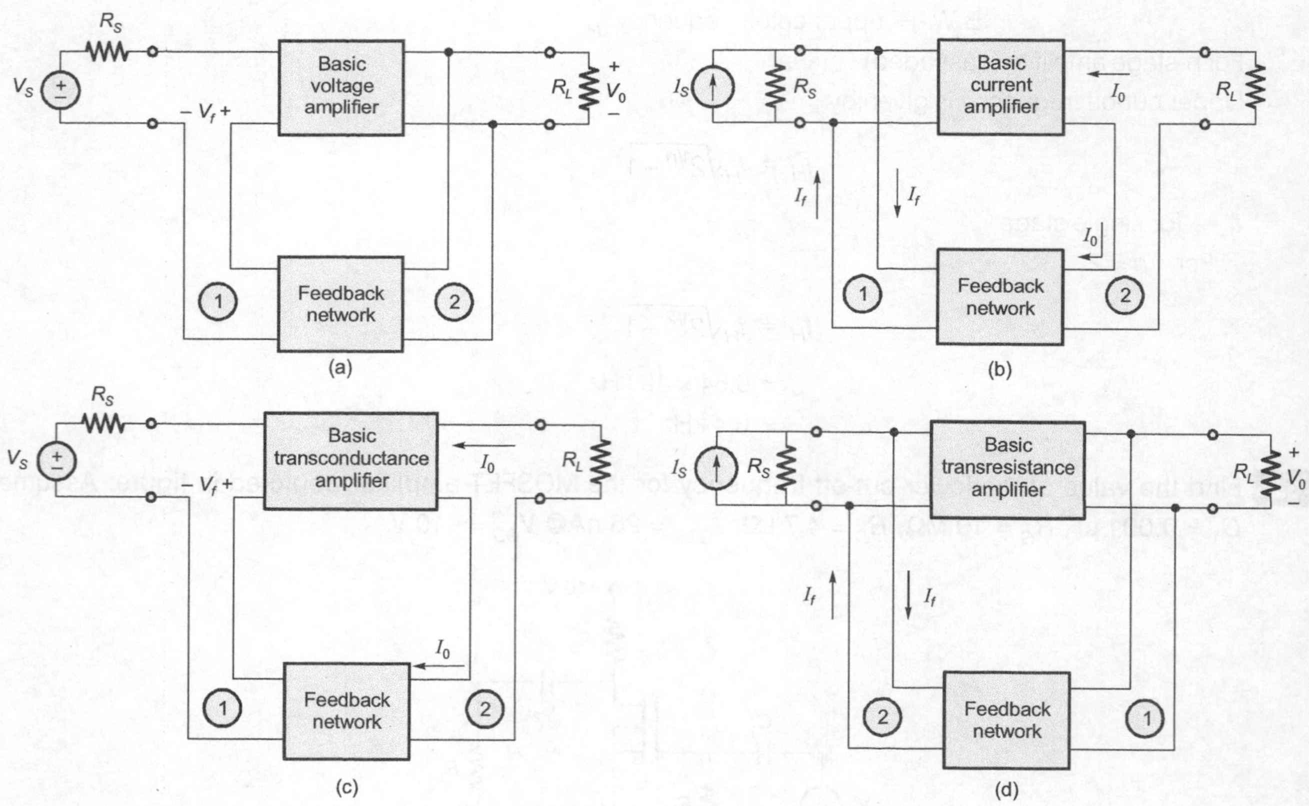

The four basic feedback topologies: (a) voltage-mixing voltage-sampling (series-shunt) topology; (b) current-mixing current-sampling (shunt-series) topology; (c) voltage-mixing current sampling (series-series) topology; (d) current-mixing voltage-sampling (shunt-shunt) topology.

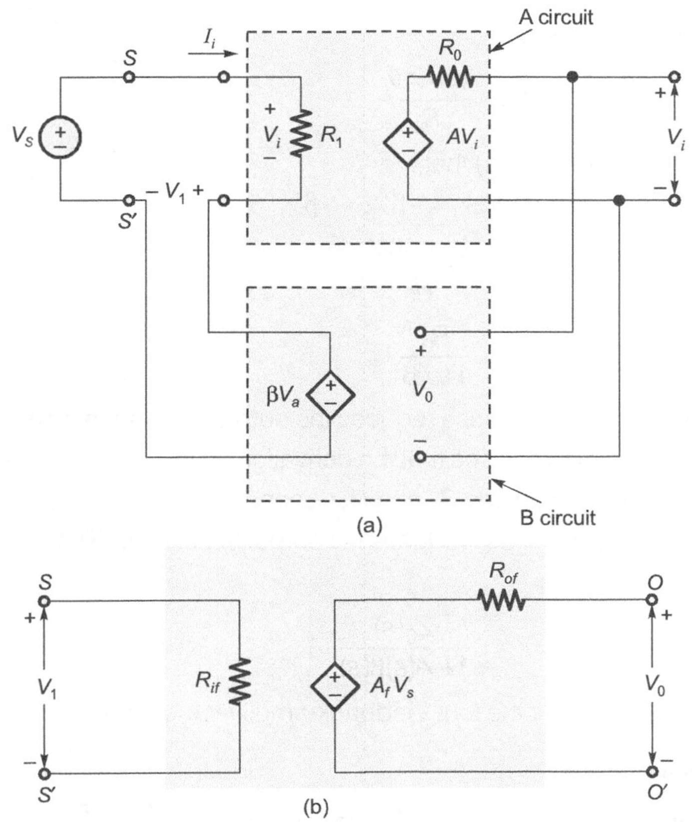

The series-shunt feedback amplifier: (a) ideal structure and (b) equivalent circuit.

Thus, $$R_{if} = R_i(1 + A\beta) \qquad \text{...}(i)$$

That is, in this case the negative feedback increases the input resistance by a factor equal to the amount of feedback. Since the derivation above does not depend on the method of sampling (shunt or series), it follows that the relationship between R_{if} and R_i is a function only of the method of mixing.

Since the feed back voltage V_f subtracts from V_S, the voltage that appears across R_i – that is, V_i – becomes quite small $[V_i = V_s/(1 + A\beta)]$. Thus the input current I_i becomes correspondingly small and the resistance seen by V_s becomes large. Finally, it should be pointed out the equation (i) can be generalized to the form.

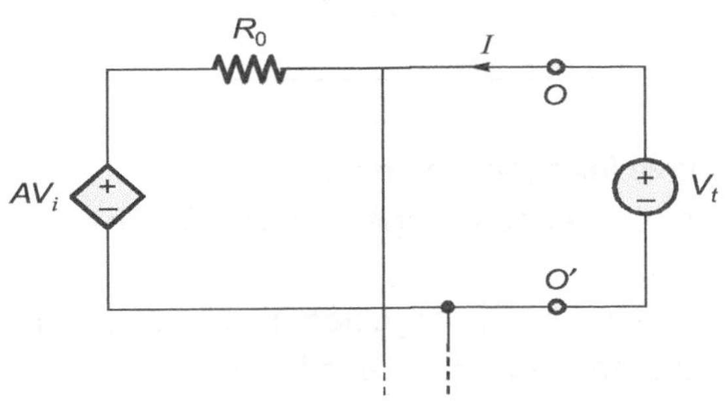

Measuring the output resistance of the feedback amplifier

$$Z_{if}(s) = Z_i(s)\,[1 + A(s)\,\beta(s)]$$

To find the output resistance, R_{of} of the feedback amplifier in figure we reduce V_S to zero and apply a test voltage V_t at the output as shown in figure (2),

$$R_{of} = \frac{V_t}{I}$$

From the above figure, we can write,

$$I = \frac{V_t - AV_i}{R_0}$$

and since $V_s = 0$ it follows from Fig. (a) that

$$V_i = V_f = -\beta V_0 = -\beta V_t$$

Thus,
$$I = \frac{V_t + A\beta V_t}{R_0}$$

Leading to
$$R_{of} = \frac{R_0}{1 + A\beta} \quad ...(ii)$$

That is, the negative feedback in this case reduces the output resistance by a factor equal to the amount of feedback. With a little though one can see that the derivation of equation (*ii*) does not depend on the method of mixing. Thus the relationship between R_{of} and R_0 depends only on the method of sampling. This, in effect, means that voltage-sampling feedback reduces the output resistance. Finally, we note that equation (*i*) can be generalized to

$$Z_{of}(s) = \frac{Z_0(s)}{1 + A(s)\beta(s)} \quad ...(iii)$$

Table: Summary of relationship for the four feedback-amplifier topologies.

Feedback Amplifier	x_i	x_0	x_f	x_s	A	β	Z_{if}	Z_{of}
Series-shunt (voltage amplifier)	V_i	V_0	V_f	V_s	$\frac{V_0}{V_i}$	$\frac{V_f}{V_0}$	$Z_i(1+A\beta)$	$\frac{Z_0}{1+A\beta}$
Shunt-series (current amplifier)	I_i	I_0	I_f	I_s	$\frac{I_0}{I_i}$	$\frac{I_f}{I_0}$	$\frac{Z_i}{1+A\beta}$	$Z_0(1+A\beta)$
Series-series (transconductance amplifier)	V_i	I_0	V_f	V_s	$\frac{I_0}{V_i}$	$\frac{V_f}{I_0}$	$Z_i(1+A\beta)$	$Z_0(1+A\beta)$
Shunt-shunt (transresistance amplifier)	I_i	V_0	I_f	I_s	$\frac{V_0}{I_i}$	$\frac{I_f}{V_0}$	$\frac{Z_i}{1+A\beta}$	$\frac{Z_0}{1+A\beta}$

7.2 **(i) State the effects of negative feedback in an amplifier.**
(ii) Draw the circuit diagram of a series voltage negative feedback and derive an expression of its input impedance.
(iii) Find the input impedance of a series voltage negative feedback if the gain of the amplifier without feedback is 100,000 and β, the feedback ratio is 0.02.

[4 + 8 + 2 = 14 marks : 2017]

Solution:

(i) Effects of negative feedback in an amplifier:

- Stability of ac gain
- Increase in input impedance by series
- Decrease in output impedance by shunt feedback
- Linearity

- Noise reduction
- Gain decreases and bandwidth increases

(ii) Series voltage negative feedback,

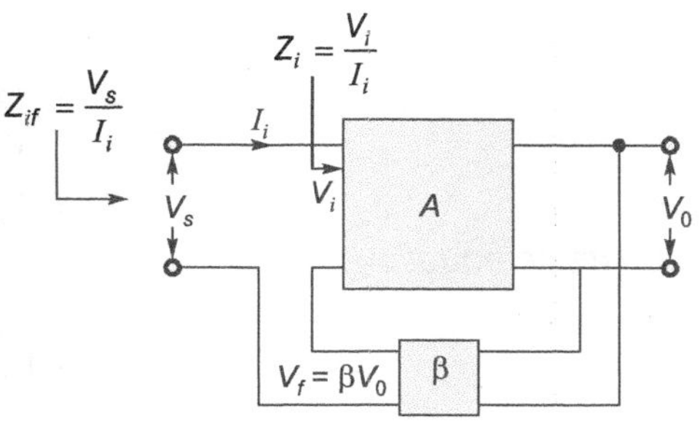

$$V_i = V_s - V_f$$
$$= V_s - \beta V_0$$
$$= V_s - A\beta V_i$$
$$V_i(1 + A\beta) = V_s$$

Divide by I_i on both sides,

$$\frac{V_i}{I_i}(1+A\beta) = \frac{V_s}{I_i}$$

(iii) Voltage series,

$$Z_{if} = Z_i(1 + A\beta)$$

Z_{if} = Input impedance with feedback

Z_i = Input impedance without feedback

$$A = 100{,}000$$
$$\beta = 0.02$$
$$Z_{if} = Z_i(1 + 100{,}000 \times 0.02)$$
$$= Z_i(1 + 2000) \qquad \text{Neglect 1 (one)}$$
$$Z_{if} = Z_i \times 2000$$

7.3 **(i) Explain Miller effect capacitance in brief.**

(ii) For the given circuit, prove that,

(a) Miller effect input capacitance, $C_{M_i} = (1 - A_v)\,C_f$

(b) Miller effect output capacitance, $C_{M_0} = \left(1 - \frac{1}{A_v}\right)C_f$

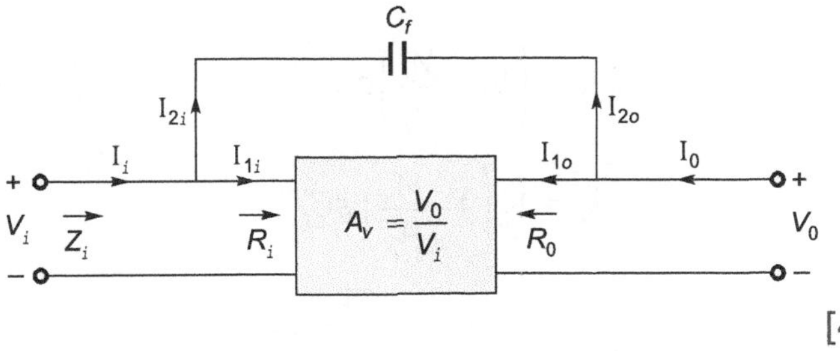

[4 + 16 = 20 marks : 2018]

Solution:

(i) **Miller Effect Capacitance:** The miller effect accounts for the increase in the equivalent input capacitance of an amplifier due to amplification of the effect of capacitance between the input and output terminals. The miller capacitance due to parasitic capacitance between the output and input of active devices is a major factor limiting their gain at high frequencies.

(ii) Given, circuit is as follows:

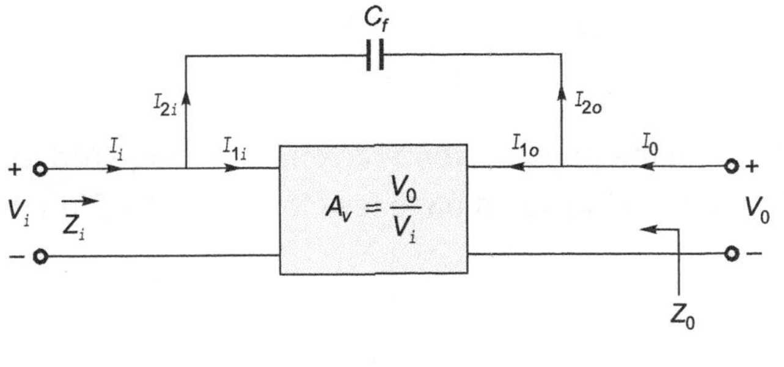

At input terminals:

$$I_i = \frac{V_i - V_0}{1/sC_f} \quad \text{...(i)}$$

$$A_v = \frac{V_0}{V_i}$$

$$\Rightarrow \quad V_0 = A_v V_i \quad \text{...(ii)}$$

Substitute equation (ii) in (i),

$$I_i = (V_i - A_v V_i)sC_f$$

$$\Rightarrow \quad I_i = V_i(1 - A_v)sC_f$$

$$\Rightarrow \quad \frac{V_i}{I_i} = Z_i = \frac{1}{sC_f(1 - A_v)}$$

Now,

$$Zi = \frac{1}{sC_{M_i}}$$

$$\therefore \quad \frac{1}{sC_{M_i}} = \frac{1}{sC_f(1 - A_v)}$$

$$\therefore \quad C_{M_i} = C_f(1 - Av)$$

At output terminals:

$$I_0 = \frac{V_0 - V_i}{1/sC_f} \quad \text{...(iii)}$$

$$V_0 = A_v V_i$$

$$\Rightarrow \quad V_i = \frac{V_0}{A_v} \quad \text{... (iv)}$$

Putting equation (iv) in (iii),

$$\therefore \quad I_0 = \left(V_0 - \frac{V_0}{A_v}\right)sC_f$$

$$I_0 = V_0\left(1 - \frac{1}{A_V}\right).sC_f$$

$$\Rightarrow \frac{V_0}{I_0} = Z_0 = \frac{1}{sC_f\left(1 - \frac{1}{Av}\right)}$$

Now,

$$Z_0 = \frac{1}{sC_{M_0}}$$

$$\therefore \quad \frac{1}{sC_{M_0}} = \frac{1}{sC_f\left(1 - \frac{1}{A_v}\right)}$$

$$\therefore \quad C_{M_0} = C_f\left(1 - \frac{1}{A_v}\right)$$

7.4 **Find the feedback factor β of the negative feedback network required for an amplifier with open loop gain A_0 = 2000 ± 200 to reduce the variation to less than ± 0.2%. Find the overall gain of the system with feedback.**

[8 marks : 2020]

Solution:

Given : Open loop gain, $A_o = 2000 \pm 200$

$$\frac{dA_f}{A_f} = \pm 0.2\% = \pm\frac{0.2}{100}$$

We have
$$\frac{dA_f}{A_f} = \frac{dA_o}{A_o} \times \frac{1}{(1+A_o\beta)}$$

where
A_f = Gain with feedback
A_o = Open loop gain

$$\pm\frac{0.2}{100} = \frac{\pm 200}{2000} \times \frac{1}{(1+A_o\beta)}$$

$$(1+A_o\beta) = \frac{200}{20\times 0.2} = 50$$

$$A_o\beta = 49 \Rightarrow \beta = \frac{49}{2000} = 0.0245$$

Gain with feedback

$$A_f = \frac{A_1}{1+A_o\beta} = \frac{2000}{50} = 40$$

$$A_f = 40$$

7.5 Draw the schematic diagram of a series-series feedback amplifier and obtain expression for :
(i) closed loop gain
(ii) input impedance
(iii) output impedance
The forward amplifier is characterised by finite input impedance R_i and finite output impedance R_o.
[3+3×3 = 12 marks : 2021]

Solution:

Series-Series Feedback :

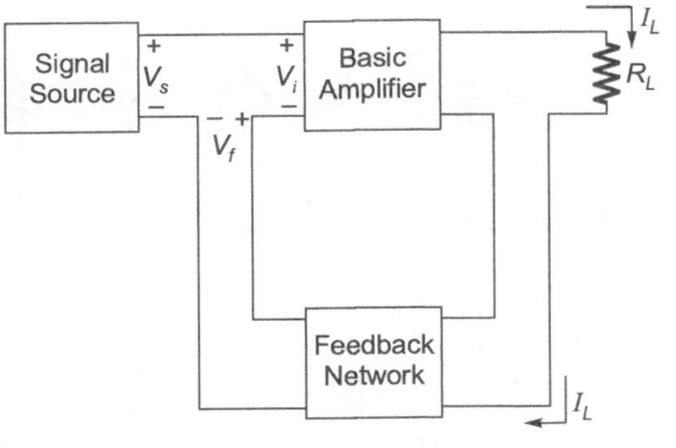

It is transconductance amplifier.

(i) Open-loop transconductance gain is,

$$G_M = \frac{I_L}{V_i} \quad ...(1)$$

KVL :
$$V_s = V_i + V_f \text{ where } V_f = \beta I_L$$
$$V_s = V_i + \beta I_L, \text{ where } I_L = G_M V_i$$
$$V_s = V_i + \beta G_M V_i$$

$$V_s = V_i(1 + \beta G_M) \quad \text{...(2)}$$

Closed-loop transconductance gain is

$$G_{Mf} = \frac{I_L}{V_s} = \frac{I_L}{V_i(1+\beta G_M)}$$

$$G_{Mf} = \frac{G_M}{1+\beta G_M}$$

(ii)

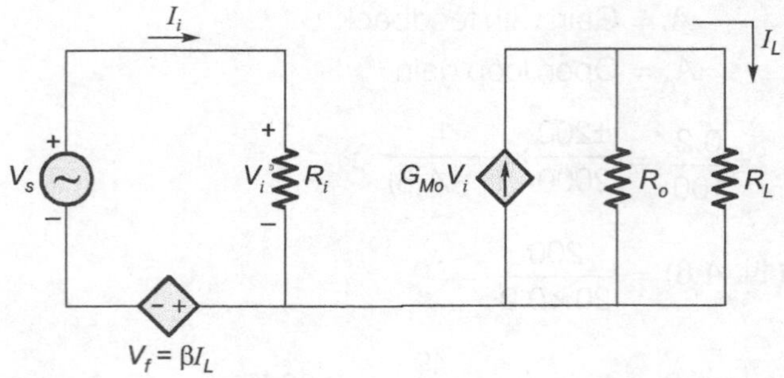

$$I_L = \frac{G_{Mo}V_i \times R_o}{R_o + R_L}$$

$$I_L = G_M V_i$$

where $$G_M = \frac{G_{Mo}R_o}{R_o + R_L}$$

G_{Mo} : Short-circuite transconductance gain

KVL : $$V_s = V_i + V_f = V_i + \beta I_L$$

$$V_s = V_i + \beta G_M V_i = V_i(1 + \beta G_M)$$

$$R_{if} = \frac{V_s}{I_i} = \frac{V_i(1+\beta G_M)}{I_i} \Rightarrow R_{if} = R_i(1 + \beta G_M)$$

(iii) To find R_{of} :

(a) Disable V_s ; (b) Disconnect R_L ; (c) Apply V_x across output port

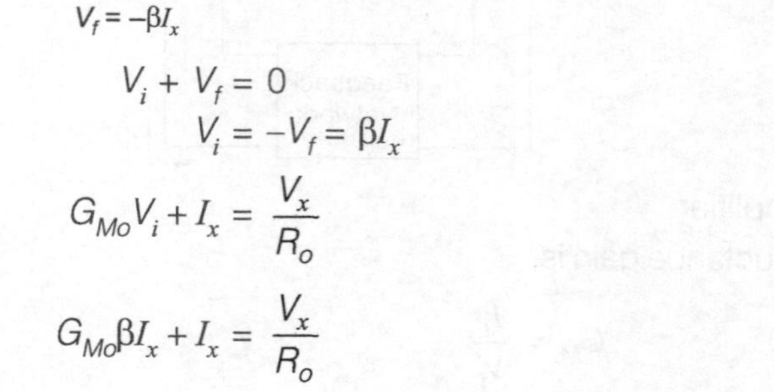

KVL : $$V_i + V_f = 0$$

$$V_i = -V_f = \beta I_x \quad \text{...(1)}$$

KCL : $$G_{Mo}V_i + I_x = \frac{V_x}{R_o}$$

$$G_{Mo}\beta I_x + I_x = \frac{V_x}{R_o}$$

$$\frac{V_x}{I_x} = R_{of} = R_o(1 + \beta G_{Mo})$$

8. Oscillators

8.1 **The simplified circuit of radio frequency oscillator is shown in figure below. Obtain the condition of oscillation. Select the suitable components in place of Z_1, Z_2 and Z_3 to yield a Colpitts oscillator.**

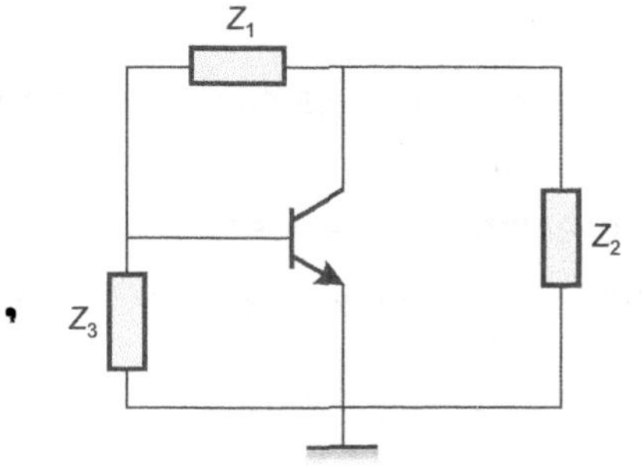

[20 marks : 2001]

Solution:

The amplifying device works into a load Z_L formed by Z_2 in parallel with Z_1, Z_3 in series.

Gain without feedback, $A = \dfrac{-\mu Z_L}{Z_L + r_d}$, $\beta = \dfrac{Z_1}{Z_1 + Z_3}$

r_d – output resistance of amplifier
μ – gain of amplifier without feedback

and
$$Z_L = Z_2 \parallel (Z_1 + Z_3)$$

$$A\beta = \frac{-\mu Z_1 Z_2 (Z_1 + Z_3)}{Z_2(Z_1 + Z_3)^2 + r_d(Z_1 + Z_2 + Z_3)(Z_1 + Z_3)}$$

$$= \frac{-\mu Z_1 Z_2}{Z_2(Z_1 + Z_3) + r_d(Z_1 + Z_2 + Z_3)}$$

Here (Z_1, Z_2, Z_3) are reactances either inductance or capacitance ωL or $\dfrac{-1}{\omega C}$ so $Z_1 = jx_1$ and $Z_2 = jx_2$ and $Z_3 = jx_3$

Substituting,
$$A\beta = \frac{\mu x_1 x_2}{jr_d(x_1 + x_2 + x_3) - x_2(x_1 + x_3)}$$

For the phase shift to be zero, the j term should vanish. So $x_1 + x_2 + x_3 = 0$ and

$$A\beta = \frac{\mu x_1 x_2}{-x_2(x_1 + x_3)} = \frac{-\mu x_1}{x_1 + x_3}$$

The system oscillates at the resonant frequency given by,

$$A\beta = \frac{\mu x_3}{x_2} (\text{as } x_1 + x_3 = -x_2)$$

If,
$$A\beta = 1 \text{ then } \mu = A_v = \frac{x_2}{x_3}$$

As per Bark hausen criterion, Aβ should not be less than 1 and positive so that x_2 and x_3 should be of the same kind, x_1 should be of the opposite kind.

For colpitt oscillator,

$$x_2, x_3 = C \text{ and } x_1 = L$$

8.2 **Draw the electrical equivalent circuit diagram of a crystal oscillator. Derive expression for the resonant frequencies of the oscillator and shown the variation of reactance of the crystal with frequency.**

[6 marks : 2017]

Solution:

Crystal oscillator:

- A piezoelectric crystal, such as quartz, exhibits electromechanical resonance characteristics in response to a voltage applied across the crystal.
- The oscillations are very stable over time and temperature.
- The oscillation frequency is determined by the crystal dimensions. This means that crystal oscillators are fixed frequency devices.

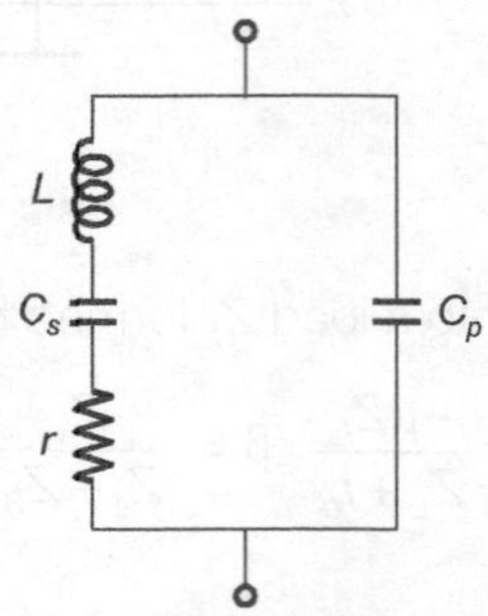

Equivalent circuit

Here, Q-factor is very high

$$Q = \frac{\omega L}{r} \quad \text{(where } r \text{ is very small)}$$

The crystal impedance is calculated as,

$$z(s) = z_1 \,||\, z_2$$

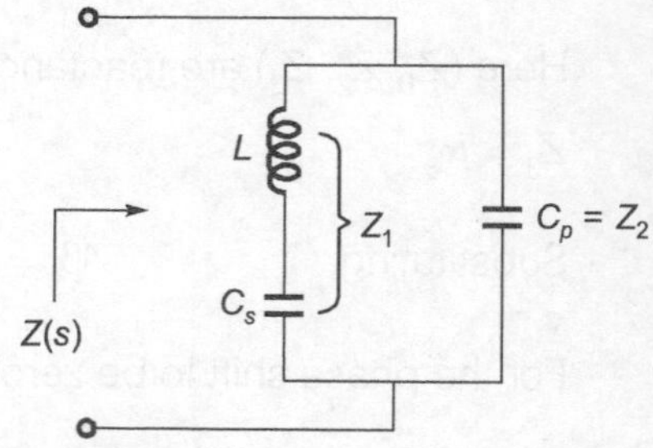

$$= \left(sL + \frac{1}{sC_s}\right) || \frac{1}{sC_P} = \frac{\frac{1}{sC_p}\left[sL + \frac{1}{sC_s}\right]}{sL + \frac{1}{sC_s} + \frac{1}{sC_p}}$$

$$= \frac{\frac{1}{sC_p}\left[\frac{s^2LC_s + 1}{sC_s}\right]}{\frac{s^2LC_sC_p + C_p + C_s}{sC_sC_p}} = \left[\frac{s^2LC_s + 1}{s^2LC_sC_p + C_p + C_s}\right]\frac{1}{s}$$

$$= \frac{\left[1 + \frac{1}{s^2LC_s}\right]}{\left[1 + \frac{C_p + C_s}{s^2LC_s C_p}\right]} \cdot \frac{1}{s} = \frac{1}{sC_p} \frac{\left[s^2 + \frac{1}{LC_s}\right]}{\left[s^2 + \frac{C_p + C_s}{LC_s C_p}\right]}$$

From the above equation, crystal has two resonance frequencies ω_s (series resonance)

$$\omega_s = \frac{1}{\sqrt{LC_s}}$$

Parallel resonance, $$\omega_p = \frac{1}{\sqrt{L\left[\dfrac{C_s C_p}{C_s + C_p}\right]}}$$

Crystal are available with resonance frequency in the range of few kilo hertzs to hundreds of mega hertzs.

Crystal colpitis oscillator:

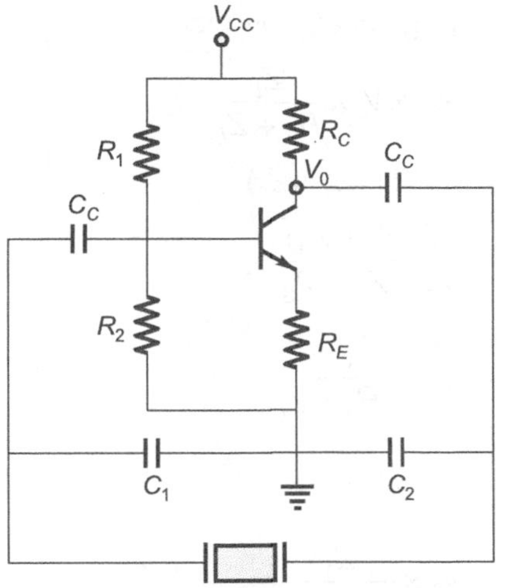

Crystal (Replacement of inductor)

8.3 (i) State whether the given statement is true or false for practical oscillators with reasons. "Loop gain is generally made slightly larger than unity."

(ii) In a general form of oscillator circuit given in figure.

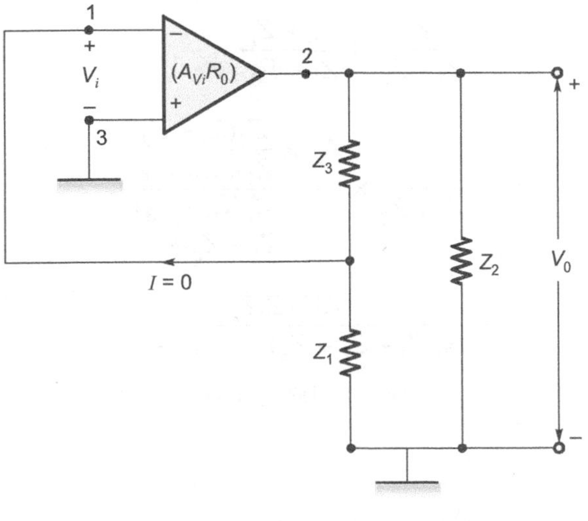

Given impedances Z_1, Z_2 and Z_3 are purely reactive. Prove that if Z_1 and Z_2 are capacitors then Z_3 must be an inductor and vice-versa.

[5 + 15 = 20 marks : 2018]

Solution:

(i) The given statement is true.Mathematically, the Barkhausen criteria guarantees sustained oscillations, but the parameters of any physical system can not be maintained constant for any length of time. If we make $A\beta = 1$ initially, then if $A\beta$ decreases then oscillations will not sustain. We therefore need a mechanism for forcing Ab to remain unity. This is accomplished by keeping $A\beta$ greater than unity, and using a non linear gain control circuit. Thus as the power supply is turned on oscillations will grow in amplitude. When the amplitude reaches the desired level, the non-linear network comes into action and causes loop gain to reduce to unity so that oscillations sustain.

(ii) Drawing small signal model of opamp.

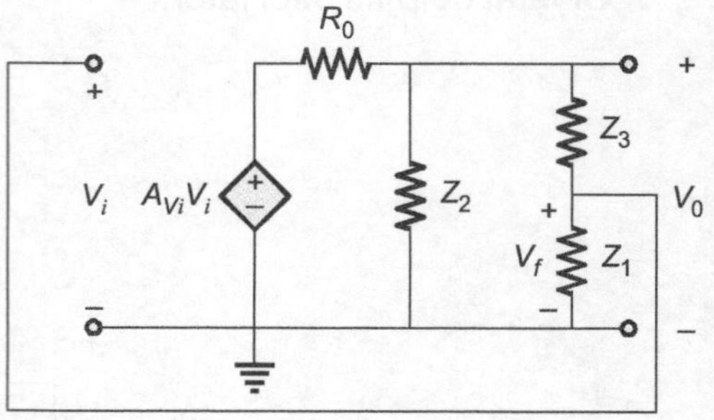

Let gain without feedback is 'A'

Hence, without feedback, applying voltage division,

$$V_0 = A_{vi} \times V_i \times \frac{Z_L}{R_0 + Z_L}$$

where,

$$Z_L = (Z_2) \,||\, (Z_1 + Z_3)$$

$$= \frac{Z_2(Z_1 + Z_3)}{Z_1 + Z_2 + Z_3}$$

$$\therefore \quad A = \frac{V_0}{V_i} = \frac{A_{vi} \cdot Z_L}{R_0 + Z_L} \quad \text{...(i)}$$

Gain with feedback:

When feedback is applied,

$$V_f = V_0 \times \frac{Z_1}{Z_1 + Z_3}$$

Feedback factor,

$$\beta = \frac{V_f}{V_0} = \frac{Z_1}{Z_1 + Z_3} \quad \text{...(ii)}$$

From equation (i) and (ii),

Using Barkhausen criteria, for sustained oscillation

$$\angle A\beta = 0 \text{ or } 360°$$

$$A\beta = \frac{Z_1}{Z_1 + Z_3} \times \frac{A_{Vi} \times Z_L}{Z_L + R_0}$$

$$A\beta = \frac{Z_1}{Z_1 + Z_3} \times \frac{A_{Vi} \times \dfrac{Z_2(Z_1 + Z_3)}{Z_1 + Z_2 + Z_3}}{R_0 + \dfrac{Z_2(Z_1 + Z_3)}{Z_1 + Z_2 + Z_3}}$$

$$A\beta = \frac{AV_i \cdot Z_1 Z_2}{R_0(Z_1 + Z_2 + Z_3) + Z_2(Z_1 + Z_3)}$$

As Z_1, Z_2 and Z_3 are purely reactive hence they are of the form

$$Z = jX$$

Let,

$$Z_1 = jX_1$$

$$Z_2 = jX_2$$

$$Z_3 = jX_3$$

Where X is either ωL or $-\dfrac{1}{\omega C}$

$$A\beta = \frac{-A_{vi} \cdot X_1 X_2}{jR_0(X_1 + X_2 + X_3) - X_2(X_1 + X_3)}$$

For

$$\angle A\beta = 360° \text{ or } 0$$

$$X_1 + X_2 + X_3 = 0$$

or, $$X_1 + X_2 = -X_3$$

which implies that sign of one of the reactances is opposite of other two.

Hence if Z_1, Z_2 are capacitor then Z_3 must be an inductor and vice-versa.

8.4 Explain the Barkhausen criterion for an oscillator circuit. How will the oscillator circuit's performance be affected if the Barkhausen criterion falls below 1, or goes much above 1?

[8 marks : 2019]

Solution:

An oscillator is a type of feedback amplifier in which part of the output is feedback to the input via a feedback circuit. If the signal feedback is of proper magnitude and phase, the circuit produces alternating currents or voltages. To visualize the requirements of an oscillator, consider the block diagram of figure. This diagram looks identical to that of the feedback amplifiers. However, here the input voltage is zero ($v_{in} = 0$). Also the feedback is positive because most oscillators use positive feedback.

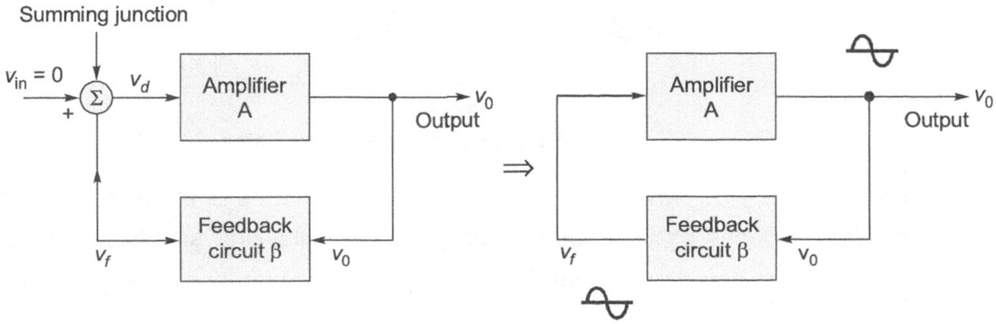

In the block diagram of figure,

$$v_d = v_f + v_{in}$$
$$v_0 = A\,v_d$$
$$v_f = \beta v_0$$

Using these relationships, the following equation is obtained:

$$\frac{v_0}{v_{in}} = \frac{A}{1 - A\beta}$$

However, $v_{in} = 0$ and $v_0 \neq 0$ implies that,

$$A\beta = 1 \quad \text{...(i)}$$

Expressed in polar form,

$$\boxed{A\beta = 1\angle 0^\circ \text{ or } 360^\circ} \quad \text{...(ii)}$$

Equation (ii) gives two requirements for oscillation:

- the magnitude of the loop gain Aβ must be at least 1, and
- the total phase shift of the loop gain Aβ must be equal to 0° or 360°.

The condition given by equation (2) is known as ***Barkhausen criterion.***

In figure if the amplifier causes a phase shift of 180°, the feedback circuit must provide an additional phase shift of 180° so that the total phase shift around the loop is 360°. The type of waveform generated by an oscillator depends on the components in the circuit hence may be sinusoidal, square or triangular. In addition the frequency of oscillation is determined by the components in the feedback circuit.

For perfect oscillation $|A\beta| = 1$. But due to non-linearity present in the circuit $|A\beta|$ becomes less than one. So, value of $|A\beta|$ in set slightly larger then unity and with time is comes very close to unity. If the value of $|A\beta|$ is set very higher than unity then the amplitude increases with time and system becomes unstable.

8.5 Derive the expression for the frequency and the condition for starting of sustained oscillation in a transistorised R-C phase shift oscillator. Neglect h_{0e} and h_{re}. Assume $R >> h_{ie}$ and a load resistance R_L is a.c. coupled to the oscillator.

[20 marks : 2020]

Solution:

RC Phase shift oscillator using BJT basic layout is shown below.

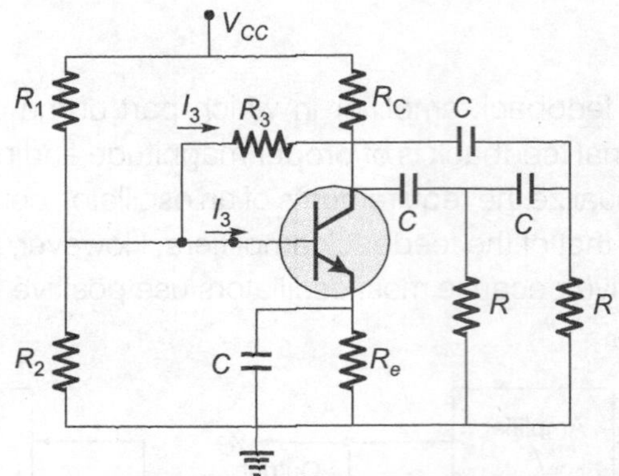

(a) Transistor phase shift oscillator

- The resistor $R_3 = R - R_i$, where $R_i \simeq h_{ie}$ is the input resistance of the transistor. This choice makes the three R-C sections of the phase shifting network alike and simplifies the calculations.
- We assume that the biasing resistors R_1, R_2 and R_e have no effect on the signal operation and neglect these in the following analysis.

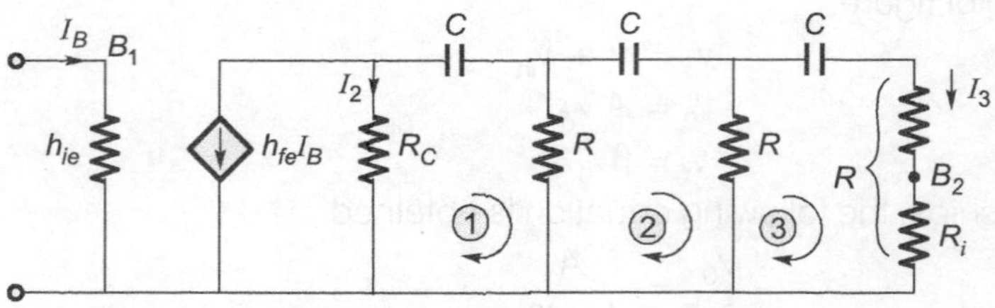

(b) Equivalent circuit for this oscillator

- Hence, we imagine the loop broken at the base between B_1 and B_2, but in order not to change the loading on the feedback network, we place R_i from B_2 to ground. If we assume a current I_B to enter the base at B_1, the loop current gain equals $\frac{I_3}{I_B}$ and is found by writing KVL for three meshes.

KVL in Loop 1 :

$$I_1X_C + R(I_1 - I_2) + R_C(I_1 + h_{fe}I_B) = 0$$

$$\Rightarrow \quad (R_C + R + X_C)I_1 - RI_2 + R_Ch_{fe}I_B = 0$$

$$\Rightarrow \quad (R + R_C + X_C)I_1 - RI_2 = -R_Ch_{fe}I_B \qquad \text{(i)}$$

KVL in Loop 2 :

$$I_2X_C + R(I_2 - I_3) + R(I_2 + I_1) = 0$$

$$\Rightarrow \quad -RI_1 + (2R + X_C)I_2 - RI_3 = 0 \qquad \text{(ii)}$$

KVL in Loop 3 :

$$I_3(R + X_C) + R(I_3 - I_2) = 0$$

$$\Rightarrow \quad RI_2 + (2R + X_C)I_3 = 0 \qquad \text{(iii)}$$

Writing equations in matrix form

$$\begin{bmatrix} (R+R_C+X_C) & -R & 0 \\ -R & (2R+X_C) & -R \\ 0 & -R & (2R+X_C) \end{bmatrix} \begin{bmatrix} I_1 \\ I_2 \\ I_3 \end{bmatrix} = \begin{bmatrix} -R_C h_{fe} I_B \\ 0 \\ 0 \end{bmatrix}$$

$$\begin{aligned} \Delta &= (R + R_C + X_C)[(2R + X_C)2 - R^2] - (-R)[(-R)(2R + X_C)] + 0 \\ &= (R + R_C + X_C)[4R^2 + X_C^2 + 4RX_C - R^2] + R(-2R^2 - RX_C) + 0 \\ &= (R + R_C + X_C)[3R^2 + X_C^2 + 4RX_C] - 2R^3 - R^2X_C \\ &= 3R^3 + RX_C^2 + 4R^2XC + 3R^2R_C + R_CX_C^2 + 4RR_CX_C + 3R^2X_C + X_C^3 \\ &\quad + 4RX_C^2 - 2R^3 - R^2X_C \end{aligned}$$

$$\Delta = R^3 + X_C^3 + 3R^2R_C + 6R^2X_C + 5RX_C^2 + R_CX_C^2 + 4RR_CX_C$$

$$\Delta_3 = \begin{vmatrix} R+R_C+X_C & -R & -R_Ch_{fe}I_B \\ -R & 2R+X_C & 0 \\ 0 & -R & 0 \end{vmatrix}$$

$$= -R_Ch_{fe}I_B[R^2] = -R_Ch_{fe}R^2I_B$$

$$I_3 = \frac{\Delta_3}{\Delta} = \frac{-R_CR^2h_{fe}I_B}{R^3 + X_C^3 + 3R^2R_C + 6R^2X_C + 5RX_C^2 + R_CX_C^2 + 4RR_CX_C}$$

$$\Rightarrow \quad \frac{I_3}{I_B} = \frac{-\frac{R_C}{R}h_{fe}}{1+\left(\frac{X_C}{R}\right)^3 + 3\frac{R_C}{R} + 6\frac{X_C}{R} + 5\left(\frac{X_C}{R}\right)^2 + \frac{R_C}{R}\left(\frac{X_C}{R}\right)^2 + 4\frac{R_C}{R}\cdot\frac{X_C}{R}}$$

Put $\quad \frac{R_C}{R} = K$

$$\Rightarrow \quad \frac{I_3}{I_B} = \frac{-Kh_{fe}}{1+\left(\frac{X_C}{R}\right)^3 + 3K + \frac{6X_C}{R} + 5\left(\frac{X_C}{R}\right)^2 + K\left(\frac{X_C}{R}\right)^2 + 4K\frac{X_C}{R}}$$

Put $\quad X_C = \frac{1}{j\omega C}$

$$\Rightarrow \quad \frac{I_3}{I_6} = \frac{-Kh_{fe}}{1+\left(\frac{1}{j\omega RC}\right)^3 + 3K + 6\frac{1}{j\omega RC} + 5\left(\frac{1}{j\omega RC}\right)^2 + K\left(\frac{1}{j\omega RC}\right)^2 + \frac{4K}{j\omega RC}}$$

So, $\quad 4 - \frac{29}{K^2} = 0$

$$\Rightarrow \quad \frac{29}{K^2} = 4$$

$$\Rightarrow \quad K = \sqrt{\frac{29}{4}} = 2.7$$

So, minimum h_{fe} required to sustain oscillations is

$$h_{fe} \geq (4 \times 2.7) + 23 + \frac{29}{2.7}$$

$$\Rightarrow \quad h_{fe} \geq 44.5$$

A transistor with a small-signal common emitter short-circuit current gain less than 44.5 cannot be used in this phase shift oscillator.

8.6 For the RC phase shift oscillator circuit shown below, determine the condition of oscillation and the frequency of oscillation. What will be the frequency of oscillation (in Hz) if $R_1 = R_2 = 10\ \text{k}\Omega$ and $C_1 = 1.0$ nF? Also calculate the minimum value of R_3 required to sustain sinusoidal oscillations.

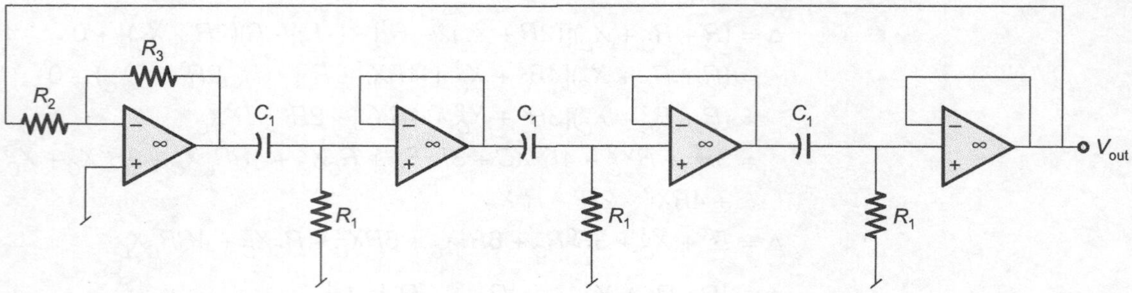

[20 marks : 2021]

Solution:

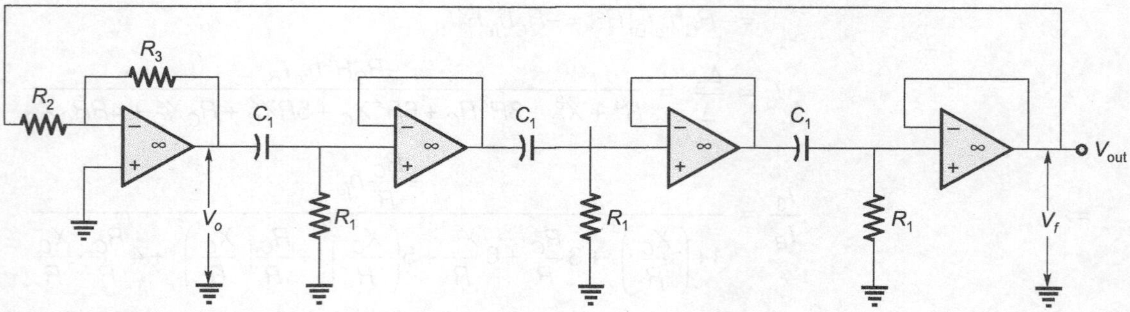

Consider single RC N/W :

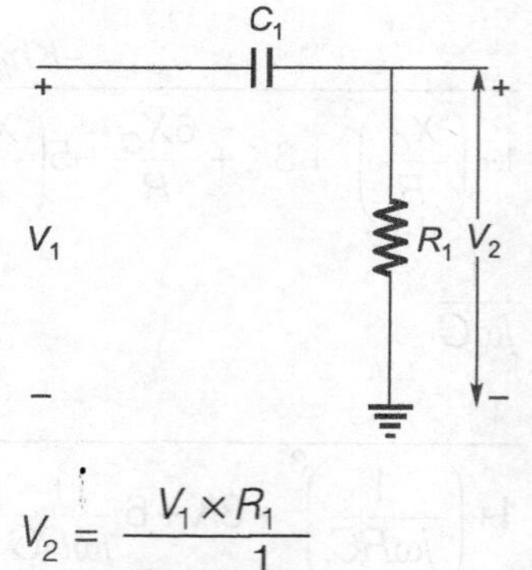

$$V_2 = \frac{V_1 \times R_1}{R_1 + \frac{1}{j\omega C_1}}$$

$$\frac{V_2}{V_1} = \frac{1}{1+\frac{1}{j\omega R_1 C_1}} = \frac{1}{1-j\alpha}, \quad \alpha = \frac{1}{\omega R_1 C_1}$$

Now, Feedback factor $= \beta = \frac{V_f}{V_o} = \left(\frac{1}{1-j\alpha}\right)^3$

$$\beta = \frac{1}{1-(j\alpha)^3 - 3j\alpha + 3(j\alpha)^2}$$

$$\beta = \frac{1}{1+j\alpha^3 - j3\alpha - 3\alpha^2} = \frac{1}{(1-3\alpha^2)+j(\alpha^3-3\alpha)}$$

$$\text{Loop gain} = A \times \beta, \text{ where } A = \frac{-R_3}{R_2}$$

$$= \frac{A}{(1-3\alpha^2) + j(\alpha^3 - 3\alpha)} \quad ...(1)$$

At $\omega = \omega_o$: Phase of loop gain = 360°

$$\Rightarrow \quad \alpha^3 - 3\alpha = 0$$

$$\alpha = \sqrt{3}$$

$$\frac{1}{\omega_o R_1 C_1} = \sqrt{3} \Rightarrow \omega_o = \frac{1}{R_1 C_1 \sqrt{3}} \Rightarrow f_o = \frac{1}{2\pi R_1 C_1 \sqrt{3}}$$

At $\omega = \omega_o$, $$\text{Loop gain} = \frac{A}{(1-3\times 3) + j\times 0} = \frac{-A}{8}$$

For sustained oscillations : |Loop gain| ≥ 1

$$\frac{|A|}{8} \geq 1 \Rightarrow |A| \geq 8$$

Given : $$R_1 = R_2 = 10\text{ k}\Omega,\ C_1 = 1\text{ nF}$$

$$f_o = \frac{1}{2\pi R_1 C_1 \sqrt{3}} = 9188.8\text{ Hz}$$

$$|A| \geq 8 \Rightarrow \frac{R_3}{R_2} \geq 8$$

$$\Rightarrow \quad R_3 \geq 8R_2$$

$$R_3 \geq 80\text{ k}\Omega$$

■■■■

7 Engineering Mathematics

Revised Syllabus of ESE: *Matrix theory, Eigen values & Eigen vectors, system of linear equations, Numerical methods for solution of non-linear algebraic equations and differential equations, integral calculus, partial derivatives, maxima and minima, Line, Surface and Volume Integrals. Fourier series, linear, nonlinear and partial differential equations, initial and boundary value problems, complex variables, Taylor's and Laurent's series, residue theorem, probability and statistics fundamentals, Sampling theorem, random variables, Normal and Poisson distributions, correlation and regression analysis.*

1. Linear Algebra

1.1 Find the eigen values and eigen vectors of the matrix $\begin{bmatrix} -2 & 2 & -1 \\ 2 & 1 & -2 \\ -3 & -6 & 0 \end{bmatrix}$. [20 marks : 2017]

Solution:

The characteristic matrix is, $A-\lambda I = \begin{bmatrix} -2-\lambda & 2 & -1 \\ 2 & 1-\lambda & -2 \\ -3 & -6 & 0-\lambda \end{bmatrix}$.

Characteristic equation is $|A-\lambda I| = 0$.

$(-2-\lambda)((1-\lambda)(-\lambda)-12)-2(-2\lambda-6)-1(-12+3-3\lambda) = 0$

$(-2-\lambda)(-\lambda+\lambda^2-12)+4\lambda+12+12-3+3\lambda = 0$

$2\lambda-2\lambda^2+24+\lambda^2-\lambda^3+12\lambda+7\lambda+21 = 0$

$$-\lambda^3-\lambda^2+21\lambda+45 = 0$$

$$\lambda^3+\lambda^2-21\lambda-45 = 0$$

$$(\lambda-5)(\lambda^2+6\lambda+9) = 0$$

$$\lambda = 5, -3, -3$$

For $\lambda = 5$, consider $(A-5I)x = 0$

$$\begin{bmatrix} -7 & 2 & -1 \\ 2 & -4 & -2 \\ -3 & -6 & -5 \end{bmatrix}\begin{bmatrix} x_1 \\ x_2 \\ x_3 \end{bmatrix} = \begin{bmatrix} 0 \\ 0 \\ 0 \end{bmatrix}$$

$R_2 \to R_2 + 2\frac{R_1}{7}, \quad R_3 \to R_3 - \frac{3}{7}R_1$

$$\begin{bmatrix} -7 & 2 & -1 \\ 0 & -\frac{24}{7} & -\frac{16}{7} \\ 0 & -\frac{48}{7} & -\frac{32}{7} \end{bmatrix} \begin{bmatrix} x_1 \\ x_2 \\ x_3 \end{bmatrix} = \begin{bmatrix} 0 \\ 0 \\ 0 \end{bmatrix}$$

$R_3 \rightarrow R_3 - 2R_2$

$$\begin{bmatrix} -7 & 2 & -1 \\ 0 & -\frac{24}{7} & -\frac{16}{7} \\ 0 & 0 & 0 \end{bmatrix} \begin{bmatrix} x_1 \\ x_2 \\ x_3 \end{bmatrix} = \begin{bmatrix} 0 \\ 0 \\ 0 \end{bmatrix}$$

rank = 2 No. of variables = 3

Free variables = 3 – 2 = 1

The equation is $-7x_1 + 2x_2 - x_3 = 0$ ...(1)

$$-\frac{24}{7}x_2 - \frac{16}{7}x_3 = 0 \quad ...(2)$$

Let $x_3 = k$, $-\frac{24}{7}x_2 - \frac{16}{7}k = 0 \Rightarrow x_2 = -\frac{2}{3}k$

Substitute $x_3 = k$, $x_2 = -\frac{2}{3}k$ in equation (1), then $x_1 = -\frac{k}{3}$

Put, $k = 3$, $x_1 = -1, \quad x_2 = -2, \quad x_3 = 3$

For $\lambda = -3$, consider $(A + 3I)x = 0$

$$\begin{bmatrix} 1 & 2 & -1 \\ 2 & 4 & -2 \\ -3 & -6 & 3 \end{bmatrix} \begin{bmatrix} x_1 \\ x_2 \\ x_3 \end{bmatrix} = \begin{bmatrix} 0 \\ 0 \\ 0 \end{bmatrix}$$

$R_2 \rightarrow R_2 - 2R_1, \quad R_3 \rightarrow R_3 + 3R_1$

$$\begin{bmatrix} 1 & 2 & -1 \\ 0 & 0 & 0 \\ 0 & 0 & 0 \end{bmatrix} \begin{bmatrix} x_1 \\ x_2 \\ x_3 \end{bmatrix} = \begin{bmatrix} 0 \\ 0 \\ 0 \end{bmatrix}$$

rank = 1 No. of variables = 3

Free variables = 3 – 1 = 2

The equation is, $x_1 + 2x_2 - x_3 = 0$

Put, $x_1 = k_1, \quad x_2 = k_2$

$$k_1 + 2k_2 - x_3 = 0$$

$\therefore$ $x_3 = k_1 + 2k_2$

$\therefore$
$$\begin{bmatrix} x_1 \\ x_2 \\ x_3 \end{bmatrix} = \begin{bmatrix} k_1 \\ k_2 \\ k_1 + 2k_2 \end{bmatrix} = k_1 \begin{bmatrix} 1 \\ 0 \\ 1 \end{bmatrix} + k_2 \begin{bmatrix} 0 \\ 1 \\ 2 \end{bmatrix}$$

Eigen vectors are $\begin{bmatrix} 1 \\ 0 \\ 1 \end{bmatrix} \begin{bmatrix} 0 \\ 1 \\ 2 \end{bmatrix}$.

The eigen values are 5, –3, –3.

$\lambda_1 = 5,\qquad X_1 = \begin{bmatrix} -1 \\ -2 \\ 3 \end{bmatrix}$

$\lambda_2 = -3,\qquad X_2 = \begin{bmatrix} 1 \\ 0 \\ 1 \end{bmatrix}$

$\lambda_3 = -3,\qquad X_3 = \begin{bmatrix} 0 \\ 1 \\ 2 \end{bmatrix}$

1.2 Let $A = \begin{bmatrix} 5 & -2 & 0 \\ -2 & 6 & 2 \\ 0 & 2 & 7 \end{bmatrix}$ and $B = A^3 - 2A^2 - 5A + 6I$, where I is the identity matrix, then calculate the determinant of B.

[12 marks : 2018]

Solution:

Given, $$A = \begin{bmatrix} 5 & -2 & 0 \\ -2 & 6 & 2 \\ 0 & 2 & 7 \end{bmatrix}$$

Characteristic equation of A is $|A - \lambda I| = 0$

∴ It becomes,

$$\begin{vmatrix} 5-\lambda & -2 & 0 \\ -2 & 6-\lambda & 2 \\ 0 & 2 & 7-\lambda \end{vmatrix} = 0$$

on expanding the determinant,

$$\lambda^3 - 18\lambda^2 + 99\lambda - 162 = 0$$

$$\Rightarrow \quad (\lambda - 3)(\lambda - 9)(\lambda - 6) = 0$$

∴ Eigen values of A are 3, 6, 9.

Eigen value of B can be found out as below:

$$B = A^3 - 2A^2 - 5A + 6I$$

Eigen value correspondign to 3, $= 3^3 - 2 \times 3^2 - 5 \times 3 + 6 = 0$

Eigen value corresponding to 6 , $= 6^3 - 2 \times 6^2 - 5 \times 6 + 6 = 120$

Eigen value corresponding to 9 , $= 9^3 - 2 \times 9^2 - 5 \times 9 + 6 = 528$

Determenant of B = product of eigen values

$$\therefore \quad |B| = 0 \times 120 \times 528 = 0$$

1.3 Determine the eigen vectors of the matrix $\begin{bmatrix} -2 & 2 & -3 \\ 2 & 1 & -6 \\ -1 & -2 & 0 \end{bmatrix}$.

Show that those eigen vectors are linearly independent.

[12 marks : 2019]

Solution:

Characteristic equation of A is

$$\begin{vmatrix} -2-\lambda & 2 & -3 \\ 2 & 1-\lambda & -6 \\ -1 & -2 & -\lambda \end{vmatrix} = 0$$

$$(-2-\lambda)[-\lambda+\lambda^2-12]-2(-2\lambda-6)-3(-4+1-\lambda)=0 \quad \ldots (i)$$

$$\lambda^3+\lambda^2-21\lambda-45=0$$

By trial: If $\lambda = -3$, then $-27+9+63-45=0$

So, $(\lambda+3)$ is one factor of equation (i).

The remaining factors are obtained on dividing (i) by $\lambda + 3$.

−3	1	1	−21	−45
	0	−3	6	45
	1	−2	−15	0

$$(\lambda+3)(\lambda^2-2\lambda-15)=0$$

$$(\lambda+3)(\lambda-5)(\lambda+3)=0$$

$$\therefore \quad \lambda = 5, -3, -3$$

To find the eigen vectors for corresponding eigen values, we will consider the matrix equation:

$$(A-\lambda I)X=0$$

$$\begin{bmatrix} -2-\lambda & 2 & -3 \\ 2 & 1-\lambda & -6 \\ -1 & -2 & -\lambda \end{bmatrix}\begin{bmatrix} x \\ y \\ z \end{bmatrix} = \begin{bmatrix} 0 \\ 0 \\ 0 \end{bmatrix} \quad \ldots (ii)$$

On putting $\lambda = 5$ in equation (ii),

$$\begin{bmatrix} -7 & 2 & -3 \\ 2 & -4 & -6 \\ -1 & -2 & -5 \end{bmatrix}\begin{bmatrix} x \\ y \\ z \end{bmatrix} = \begin{bmatrix} 0 \\ 0 \\ 0 \end{bmatrix}$$

We have,

$$-7x+2y-3z=0$$

$$2x-4y-6z=0$$

$$\frac{x}{-12-12} = \frac{y}{-6-42} = \frac{z}{28-4} = k$$

$$\frac{x}{-24} = \frac{y}{-48} = \frac{z}{24} = k$$

$$\frac{x}{1} = \frac{y}{2} = \frac{z}{-1} = k$$

$$x = k,\ y = 2k;\ z = -k$$

Hence, the eigen vector,

$$x_1 = \begin{bmatrix} k \\ 2k \\ -k \end{bmatrix} = k\begin{bmatrix} 1 \\ 2 \\ -1 \end{bmatrix}$$

Put $\lambda = -3$ in equation (ii), it becomes

$$\begin{bmatrix} 1 & 2 & -3 \\ 2 & 4 & -6 \\ -1 & -2 & 3 \end{bmatrix}\begin{bmatrix} x \\ y \\ z \end{bmatrix} = \begin{bmatrix} 0 \\ 0 \\ 0 \end{bmatrix}$$

We have,

$$x + 2y - 3z = 0$$
$$2x + 4y - 6z = 0$$
$$-x - 2y + 3z = 0$$

Here first, second and third equations are same.

Let, $x = k_1;\ y = k_2;\ z = \frac{1}{3}(k_1 + 2k_2)$

Hence, the eigen vector is $\begin{bmatrix} k_1 \\ k_2 \\ \frac{1}{3}(k_1 + 2k_2) \end{bmatrix}$

Let $k_1 = 0$; $k_2 = 3$; Hence $X_2 = \begin{bmatrix} 0 \\ 3 \\ 2 \end{bmatrix}$

Since the matrix is non-symmetric. The corresponding eigen vectors X_2 and X_3 must be linearly independent.

This can be done by choosing $k_1 = 3$, $k_2 = 0$ and hence $X_3 = \begin{bmatrix} 3 \\ 0 \\ 1 \end{bmatrix}$.

Hence,

$$X_1 = \begin{bmatrix} 1 \\ 2 \\ -1 \end{bmatrix};\quad X_2 = \begin{bmatrix} 0 \\ 3 \\ 2 \end{bmatrix};\quad X_3 = \begin{bmatrix} 3 \\ 0 \\ 1 \end{bmatrix}$$

$$|X_1\, X_2\, X_3| = \begin{vmatrix} 1 & 0 & 3 \\ 2 & 3 & 0 \\ -1 & 2 & 1 \end{vmatrix} = 1(3) + 3(4 + 3) = 3 + 21 = 24$$

As determinant of vectors is not equal to zero. Therefore, these eigen vectors are linearly independent.

1.4 **(i) If a square matrix A of order n with entries in field F has n distinct eigenvalues, then prove that matrix A is similar to a diagonal matrix.**

(ii) Find the matrix P which diagonalizes the matrix associated with the quadratic form

$$3x^2 + 5y^2 + 3z^2 - 2yz + 2zx - 2xy$$

[10 + 10 = 20 marks : 2020]

Solution:

(i) Let A be matrix order $n \times n$ and let A has n distinct eigen values, i.e., all its eigen values are distinct.

$\Rightarrow$ All the eigen vectors of $A_{n \times n}$ are linearly independent.

($\because$ No. of distinct eigen values $\Rightarrow$ No. of linearly independent eigen vectors)

Then the modal matrix $P = [X_1 \quad X_2 \, ... \, X_n]$

(where $X_1, X_2, ... X_n$ are eigenvectors of A) is a non-singular matrix.

$$\Rightarrow \qquad |P| \neq 0$$

$\Rightarrow P^{-1}$ exists.

$\therefore$ By diagonalization :

$$D = P^{-1}AP \text{ is possible.}$$

Hence, A and diagonal matrix (D) are similar matrices whose determinants and eigen values are same.

Hence Proved.

(ii) Given quadratic form : $3x^2 + 5y^2 + 3z^2 - 2yz + 2xz - 2xy$.

The symmetric matrix can be written as

$$A = \begin{matrix} & x & y & z \\ x & 3 & -1 & 1 \\ y & -1 & 5 & -1 \\ z & 1 & -1 & 3 \end{matrix}$$

For Eigen Values (λ) :

By solving characteristic equation $|A - \lambda I| = 0$

$$|A - \lambda I| = \begin{vmatrix} 3-\lambda & -1 & 1 \\ -1 & 5-\lambda & -1 \\ 1 & -1 & 3-\lambda \end{vmatrix} = 0$$

$$= (3-\lambda)[(5-\lambda)(3-\lambda) - 1] + 1[(\lambda - 3) + 1] + 1[(1 - (5 - \lambda)] = 0$$

$$= (3-\lambda)[15 - 8\lambda + \lambda^2 - 1] + \lambda - 2 + \lambda - 4 = 0$$

$$= -\lambda^3 + 11\lambda^2 - 36\lambda + 36 = 0$$

Solving we get $\qquad \lambda = 2, 3, 6$

For $\lambda_1 = 2$ in $(A - \lambda_1 I)X_1 = 0$

We get equations,

$$x - y + z = 0 \qquad ...(1)$$

$$-x + 3y - z = 0 \qquad ...(2)$$

$$x - y + z = 0 \qquad ...(3)$$

Solving (1) and (2)

$$\begin{matrix} x & y & z & \\ -1 & 1 & 1 & -1 \\ 3 & -1 & -1 & 3 \end{matrix}$$

$$\Rightarrow \qquad \frac{x}{1-3} = \frac{y}{-1+7} = \frac{z}{3-1}$$

$$\Rightarrow \qquad \frac{x}{-2} = \frac{y}{0} = \frac{z}{2} = K \text{ (say)}$$

$$\Rightarrow \qquad X_1 = \begin{bmatrix} x \\ y \\ z \end{bmatrix} = K \begin{bmatrix} -1 \\ 0 \\ 1 \end{bmatrix}$$

2. For $\lambda = 3$ in $(A - \lambda_2 I)X_2 = 0$, we get

$$0x - y + z = 0 \quad \text{...(4)}$$
$$-x + 2y - z = 0 \quad \text{...(5)}$$
$$x - y + 3z = 0 \quad \text{...(6)}$$

Solving (4), (5)

x	y	z	
−1	1	0	−1
2	−1	−1	2

$$\Rightarrow \quad \frac{x}{-1} = \frac{y}{-1} = \frac{z}{-1} = K \text{ (say)}$$

$$\Rightarrow \quad X_2 = \begin{bmatrix} x \\ y \\ z \end{bmatrix} = K\begin{bmatrix} 1 \\ 1 \\ 1 \end{bmatrix}$$

For $\lambda = 6$ in $(A - \lambda_3 I)X_3 = 0$

We get equations

$$-3x - y + z = 0 \quad \text{...(7)}$$
$$-x - y - z = 0 \quad \text{...(8)}$$
$$x - y - 3z = 0 \quad \text{...(9)}$$

Solving (7), (8)

x	y	z	
−1	1	−3	−1
−1	−1	−1	−1

$$\frac{x}{2} = \frac{y}{-4} = \frac{z}{2} = K \text{ (say)}$$

$$X_3 = \begin{bmatrix} x \\ y \\ z \end{bmatrix} = K\begin{bmatrix} 1 \\ -2 \\ 1 \end{bmatrix}$$

Hence, the required modal matrix for the given quadratic form is

$$P = \begin{bmatrix} \frac{X_1}{\|X_1\|} & \frac{X_2}{\|X_2\|} & \frac{X_3}{\|X_3\|} \end{bmatrix}_{3\times 3} = \begin{bmatrix} -\frac{1}{\sqrt{2}} & \frac{1}{\sqrt{3}} & \frac{1}{\sqrt{6}} \\ 0 & \frac{1}{\sqrt{3}} & \frac{-2}{\sqrt{6}} \\ \frac{1}{\sqrt{2}} & \frac{1}{\sqrt{3}} & \frac{1}{\sqrt{6}} \end{bmatrix}$$

1.5 **An elastic membrane in the x_1x_2-plane with boundary circle $x_1^2 + x_2^2 = 1$ is stretched so that a point $P(x_1, x_2)$ goes over into the point $Q(y_1, y_2)$ given by**

$$y = \begin{bmatrix} y_1 \\ y_2 \end{bmatrix} = AX = \begin{bmatrix} 5 & 3 \\ 3 & 5 \end{bmatrix}\begin{bmatrix} x_1 \\ x_2 \end{bmatrix}$$

Find the principal directions, i.e., the directions of the position vector X of P for which the direction of the position vector Y of Q is same or exactly opposite. What shape does the boundary circle take under this deformation?

[10 marks : 2021]

Solution:

We are looking for the vector X such that

$$Y = \lambda X$$

$$\Rightarrow \qquad AX = \lambda X$$

Characteristic equation of A is $|A - \lambda I| = 0$

$$\begin{vmatrix} 5-\lambda & 3 \\ 3 & 5-\lambda \end{vmatrix} = (5-\lambda)^2 - 3^2 = 0$$

$$\Rightarrow \qquad \lambda_1 = 8, 2$$

The eigen vectors for $\lambda_1 = 8$ are given by

$$(A - 8I)X = 0$$

$$\Rightarrow \qquad -3x_1 + 3x_2 = 0$$

$$3x_1 - 3x_2 = 0$$

On solving, we get $X = \begin{bmatrix} x_1 \\ x_2 \end{bmatrix} = \begin{bmatrix} 1 \\ 1 \end{bmatrix}$

For $\lambda_2 = 2$: Eigen vectors given by $(A - 2I)X = 0$

$$\Rightarrow \qquad 3x_1 + 3x_2 = 0$$

$$3x_1 + 3x_2 = 0$$

On solving, we get $X = \begin{bmatrix} x_1 \\ x_2 \end{bmatrix} = \begin{bmatrix} 1 \\ -1 \end{bmatrix}$

Here, $X = \begin{bmatrix} 1 \\ 1 \end{bmatrix}$ makes an angle 45° with positive x-direction with stretching factor $\lambda = 8$ and $X = \begin{bmatrix} 1 \\ -1 \end{bmatrix}$ makes an angle 135° with positive x_1-direction with stretching factor $\lambda = 2$.

The eigen vectors shows that the principal directions of the membrane is stretched by factors 8 and 2 respectively.

Choosing the principal directions of a new u_1u_2-coordinate system, say

$$u_1 = r\cos\phi, \; u_2 = r\sin\phi$$

Here, we notice that unstretched circular membrane boundary at point has coordinates ($\cos\phi$, $\sin\phi$) in new system.

Hence, after stretch it will be

$$u_1 = 8\cos\phi, \; u_2 = 2\sin\phi \qquad \left(\because \sin\frac{3\pi}{4} = \sin\frac{\pi}{4}\right)$$

Hence, before stretch it is

$$\cos^2\phi + \sin^2\phi = 1$$

but deformed boundary after stretch is an ellipse.

i.e., $$\frac{u_1^2}{8^2} + \frac{u_2^2}{2^2} = 1$$

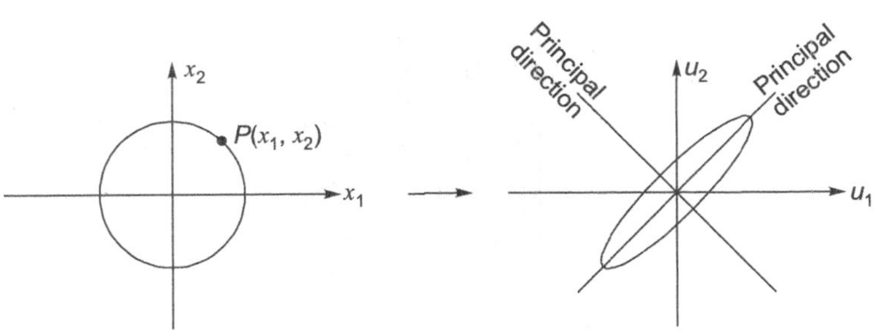

1.6 Form the homogeneous system of equation to balance the chemical equation

$$C_2H_6 + O_2 \rightarrow CO_2 + H_2O$$

Solve it by Gauss-Jordan elimination method to obtain the balanced chemical equation.

[12 marks : 2021]

Solution:

To balance the given reaction, assign variable x_1, x_2, x_3, x_4 to each chemical compounds of given reaction as

$$x_1C_2H_6 + x_2O_2 \rightarrow x_3CO_2 + x_4H_2O \quad ...(1)$$

Comparing the number of atoms of carbon (C), hydrogen (H), oxygen (O) in reactant side with the number of atoms of product side. The linear equations from the above reaction (1) as follows :

$$\begin{aligned} C &: 2x_1 = x_3 && \Rightarrow && 2x_1 - x_3 = 0 \\ H &: 6x_1 = 2x_4 && \Rightarrow && 6x_1 - 2x_4 = 0 \\ O &: 2x_2 = 2x_3 + x_4 && \Rightarrow && 2x_2 - 2x_3 - x_4 = 0 \end{aligned}$$

These linear equations can be written in matrix form $AX = 0$.

$$\begin{bmatrix} 2 & 0 & -1 & 0 \\ 6 & 0 & 0 & -2 \\ 0 & 2 & -2 & -1 \end{bmatrix} \begin{bmatrix} x_1 \\ x_2 \\ x_3 \\ x_4 \end{bmatrix} = \begin{bmatrix} 0 \\ 0 \\ 0 \\ 0 \end{bmatrix}$$

Reduce A into Echelon form by Gauss elimination

$$R_1 \rightarrow \frac{1}{2}R_1,\ R_2 \rightarrow \frac{1}{2}R_2,\ R_3 \rightarrow \frac{1}{2}R_3$$

$$\begin{bmatrix} 1 & 0 & -\frac{1}{2} & 0 \\ 3 & 0 & 0 & -1 \\ 0 & 1 & -1 & -\frac{1}{2} \end{bmatrix}$$

$$R_2 \rightarrow R_2 - 3R_1 \qquad \cong \begin{bmatrix} 1 & 0 & -\frac{1}{2} & 0 \\ 0 & 0 & \frac{3}{2} & -1 \\ 0 & 1 & -1 & -\frac{1}{2} \end{bmatrix}$$

$$R_2 \leftrightarrow R_3 \qquad \cong \begin{bmatrix} 1 & 0 & -\frac{1}{2} & 0 \\ 0 & 1 & -1 & -\frac{1}{2} \\ 0 & 0 & \frac{3}{2} & -1 \end{bmatrix}$$

$$R_2 \rightarrow \frac{2}{3}R_3 \qquad \cong \begin{bmatrix} 1 & 0 & -\frac{1}{2} & 0 \\ 0 & 1 & -1 & -\frac{1}{2} \\ 0 & 0 & 1 & -\frac{2}{3} \end{bmatrix}$$

It is in Echelon form ⇒ $\rho(A) = 3 <$ number of unknown

$\rho(A) = 3 <$ Number of unknown $= 4$

System has infinitely many solutions.

$$\text{No. of free variables} = \text{No. of unknowns} - \rho(A) = 4 - 3 = 1$$

Let $x_4 = K$

Writing the equations from Echelon form :

$$x_1 - \frac{1}{2}x_3 = 0 \quad \text{...(1)}$$

$$x_2 - x_3 - \frac{1}{2}x_4 = 0 \quad \text{...(2)}$$

$$x_3 - \frac{2}{3}x_4 = 0$$

Put $x_4 = K$ in (3) ⇒ $x_3 = \frac{2}{3}K$

Substitute in (2) ⇒ $x_2 = \frac{2}{3}K + \frac{1}{2}K = K\left[\frac{7}{6}\right]$

Substitute in (1) ⇒ $x_1 = \frac{K}{3} = \left[\frac{1}{3}\right]K$

$$\therefore \quad X = \begin{bmatrix} x_1 \\ x_2 \\ x_3 \\ x_4 \end{bmatrix} = K\begin{bmatrix} \frac{1}{3} \\ \frac{7}{6} \\ \frac{2}{3} \\ 1 \end{bmatrix}$$

Let $K = 6$: $X = \begin{bmatrix} 2 \\ 7 \\ 4 \\ 6 \end{bmatrix}$

Hence, balanced equation is

$$2C_2H_6 + 7O_2 \rightarrow 4CO_2 + 6H_2O$$

1.7 **Find the eigen values and the corresponding eigen vectors of the matrix $\begin{bmatrix} 3 & -1 & -2 \\ -1 & 3 & 2 \\ -2 & 2 & 6 \end{bmatrix}$. Is this matrix diagonalizable? Justify your answer.**

[12 marks : 2022]

Solution:

Let $A = \begin{bmatrix} 3 & -1 & -2 \\ -1 & 3 & 2 \\ -2 & 2 & 6 \end{bmatrix}$

The characteristic equation is $|A - \lambda I| = 0$

$$\begin{vmatrix} 3-\lambda & -1 & -2 \\ -1 & 3-\lambda & 2 \\ -2 & 2 & 6-\lambda \end{vmatrix} = 0$$

$$(3-\lambda)[(3-\lambda)(6-\lambda)-4]-(-1)[-1(6-\lambda)-2\times(-2)]-2[-1\times 2-(-2)\times(3-\lambda)] = 0$$

$$(3-\lambda)[18-3\lambda-6\lambda+\lambda^2-4]+[\lambda-6+4]-2[-2+6-2\lambda] = 0$$

$$(3-\lambda)[\lambda^2-9\lambda+14]+\lambda-2+4\lambda-8 = 0$$

$$3\lambda^2-27\lambda+42-\lambda^3+9\lambda^2-14\lambda+5\lambda-10 = 0$$

$$-\lambda^3+12\lambda^2-36\lambda+32 = 0$$

$$\lambda^3-12\lambda^2+36\lambda-32 = 0$$

$$\lambda = 8, 2, 2$$

Hence, eigen values are 8, 2, 2.

If X, Y, Z be the components of an eigen vector corresponding to the eigen value λ, we have

$$[A-\lambda I]\begin{bmatrix} X \\ Y \\ Z \end{bmatrix} = 0$$

For $\lambda = 8$,

$$\begin{bmatrix} 3-8 & -1 & -2 \\ -1 & 3-8 & 2 \\ -2 & 2 & 6-8 \end{bmatrix}\begin{bmatrix} X \\ Y \\ Z \end{bmatrix} = 0$$

$$\begin{bmatrix} -5 & -1 & -2 \\ -1 & -5 & 2 \\ -2 & 2 & -2 \end{bmatrix}\begin{bmatrix} X \\ Y \\ Z \end{bmatrix} = 0$$

$$-5X-Y-2Z = 0$$

$$-2X+2Y-2Z = 0$$

$$\frac{X}{2+4} = \frac{Y}{-6} = \frac{Z}{-12} = K$$

$$\frac{X}{6} = \frac{Y}{-6} = \frac{Z}{-12} = K$$

$$\frac{X}{1} = \frac{Y}{-1} = \frac{Z}{-2} = K$$

$$\text{Eigen vector} \equiv \begin{bmatrix} 1 \\ -1 \\ -2 \end{bmatrix}$$

For $\lambda = 2$

$$\begin{bmatrix} 3-2 & -1 & -2 \\ -1 & 3-2 & 2 \\ -2 & 2 & 6-2 \end{bmatrix}\begin{bmatrix} X \\ Y \\ Z \end{bmatrix} = 0$$

$$\begin{bmatrix} 1 & -1 & -2 \\ -1 & 1 & 2 \\ -2 & 2 & 4 \end{bmatrix}\begin{bmatrix} X \\ Y \\ Z \end{bmatrix} = 0$$

$$X - Y - 2Z = 0$$
$$-X + Y + 2Z = 0$$
$$X - Y - 2Z = 0 \Rightarrow \text{Only one independent equation}$$

Let $Y = K_1$ and $Z = K_2$

$$\text{Vector} = \begin{bmatrix} X \\ Y \\ Z \end{bmatrix} = \begin{bmatrix} K_1 + 2K_2 \\ K_1 \\ K_2 \end{bmatrix}$$

$$= K_1 \begin{bmatrix} 1 \\ 1 \\ 0 \end{bmatrix} + K_2 \begin{bmatrix} 2 \\ 0 \\ 1 \end{bmatrix}$$

Hence, $$\text{Eigen vector} \equiv \begin{bmatrix} 1 \\ -1 \\ -2 \end{bmatrix}, \begin{bmatrix} 1 \\ 1 \\ 0 \end{bmatrix}, \begin{bmatrix} 2 \\ 0 \\ 1 \end{bmatrix}$$

Necessary and sufficient condition for square matrix A to be diagonalized is that all the eigen vectors of A must be linearly independent.

Since, $$\begin{vmatrix} 1 & -1 & -2 \\ 1 & 1 & 0 \\ 2 & 0 & 1 \end{vmatrix} \neq 0$$

Here, all eigen vectors are linear independent. Hence, matrix are diagonalized.

1.8 Suppose A is a 3 × 3 diagonalizable matrix. Then

(i) show that that each eigen value of A is 0 or 1, if $A^2 = A$;

(ii) find the trace of the matrix $B = A + A^3 + A^{-1}$, if the eigen values of A are 2, 3, –2.

[12 marks : 2023]

Solution:

(i) As the matrix A is diagonalizable, there exists a non-singular matrix P, such that $P^{-1}.A.P = D$, where D is a diagonal matrix whose diagonal elements are the eigen values of A.

Thus, $$D = \text{diagonal}\{d_1 \, ...d_n\}$$

As eigen values are 0 or 1 only.

We see that $$(d_i)^2 = (d_i); \text{ for every } i$$

That implies that

$$D^2 = \text{diagonal}\{(d_1)^2, ... (d_n)^2\}$$
$$= \text{diagonal}\{(d_1), ... (d_n)\}$$
$$= D$$

Hence, $$A^2 = [PDP^{-1}]^2 = PDP^{-1} \,.\, PDP^{-1}$$
$$= PDIDP^{-1} = PD^2P^{-1} = PDP^{-1} = A$$

Therefore, $$A^2 = A$$

(ii) Given eigen values of A are 2, 3, –2.

So, eigen values of B corresponding to 2.

$$B = A + A^3 + A^{-1}$$
$$= 2 + 2^3 + \frac{1}{2}$$

$$= 2+8+\frac{1}{2} = \frac{21}{2}$$

Eigen values of B corresponding 3.

$$B = A + A^3 + A^{-1}$$

$$= 3+3^3+\frac{1}{3}$$

$$= 3+27+\frac{1}{3} = \frac{9+81+1}{3}$$

$$= \frac{91}{3}$$

Eigen values of B corresponding-2

$$B = A + A^3 + A^{-1}$$

$$= -2+(-2)^3+\frac{1}{-2}$$

$$= -2-8-\frac{1}{2}$$

$$= -\frac{21}{2}$$

Trace of the matrix B = Sum of diagonal elements

Sum of diagonal elements = Sum of eigen values

$\therefore$ $$\text{Trace }[B] = \frac{22}{2}+\frac{91}{3}-\frac{21}{2} = 30.33$$

2. Calculus

2.1 Find the absolute maximum and minimum values of the function $f(x, y) = x^2 + 3y^2 - y$ over the region $x^2 + 2y^2 \le 1$.

[20 marks : 2017]

Solution:

$$f(x, y) = x^2 + 3y^2 - y$$

$$\frac{\partial f}{\partial x} = 2x \qquad \frac{\partial f}{\partial y} = 6y-1$$

$$r = \frac{\partial^2 f}{\partial x^2} = 2 \qquad s = \frac{\partial^2 f}{\partial x \partial y} = 0, \quad t = \frac{\partial^2 f}{\partial y^2} = 6$$

$$\frac{\partial f}{\partial x} = 0; \qquad \frac{\partial f}{\partial y} = 0$$

$$2x = 0; \qquad 6y-1 = 0$$

$$x = 0; \qquad y = \frac{1}{6}$$

Stationary point is (0, 1/6).

At (0, 1/6), $r = 2, \quad s = 0, \quad t = 6$

$$rt - s^2 = 12 - 0 = 12 > 0$$

$$r = 2 > 0$$

$\therefore$ $f(x, y)$ has minimum value at (0, 1/6)

$\therefore$ The minimum value of the function is $f(0, 1/6)$

$$= 0+\frac{3}{36}-\frac{1}{6}=-\frac{3}{36}=-\frac{1}{12}$$

Given that, boundary of the region,

$$x^2 + 2y^2 \leq 1$$

The boundary of the region is,

$$x^2 + 2y^2 = 1$$

$$x^2 = 1 - 2y^2$$

The given function is, $f(x, y) = x^2 + 3y^2 - y$

$$f(y) = 1 - 2y^2 + 3y^2 - y$$

$$f(y) = 1 - y + y^2$$

$$f'(y) = -1 + 2y, \qquad f''(y) = 2$$

$$f'(y) = 0, \qquad \text{At } y = 1/2$$

$$-1 + 2y = 0, \qquad f''(y) = 2 > 0 \quad \text{(Relative minima)}$$

Since, $x^2 = 1 - 2y^2$

For $y=\frac{1}{2}$, $x^2 = 1-2\left(\frac{1}{2}\right)^2=\frac{1}{2}$

$\therefore$ $x = \pm\frac{1}{\sqrt{2}}$

Hence, minimum points are $\left(\frac{-1}{\sqrt{2}},\frac{1}{2}\right)\left(\frac{1}{\sqrt{2}},\frac{1}{2}\right)$.

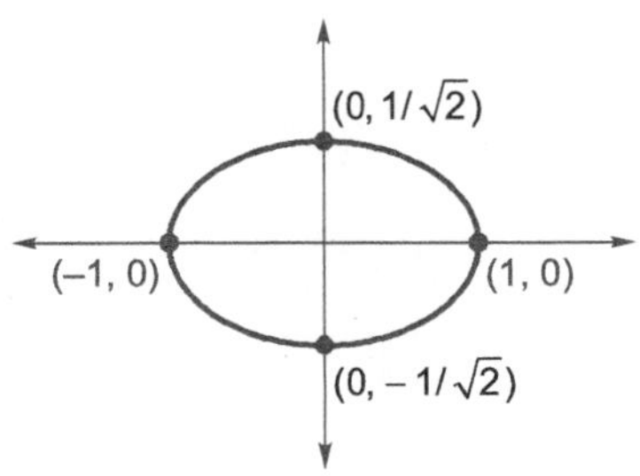

Minimum value of $f(x, y) = f\left(-\frac{1}{\sqrt{2}},\frac{1}{2}\right)=\frac{3}{4}$

$$f(x, y) = f\left(\frac{1}{\sqrt{2}},\frac{1}{2}\right)=\frac{3}{4}$$

Since the equation of the boundary is $x^2 + 2y^2 = 1$,

At vertices of the ellipse, of the function, $\frac{x^2}{1}+\frac{y^2}{1/2}=1$

$$\frac{x^2}{1^2}+\frac{y^2}{\left(\frac{1}{\sqrt{2}}\right)^2}=1$$

Values are, $f(1, 0) = 1, \qquad f(-1, 0) = 1$

$$f\left(0,\frac{1}{\sqrt{2}}\right) = 0+\frac{3}{2}-\frac{1}{\sqrt{2}}=\frac{3\sqrt{2}-2}{2\sqrt{2}}$$

$$f\left(0,-\frac{1}{\sqrt{2}}\right) = 0+\frac{3}{2}+\frac{1}{\sqrt{2}}=\frac{3\sqrt{2}+2}{2\sqrt{2}}$$

Absolute minimum value of $f(x, y) = -\frac{1}{12}$ At $\left(0,\frac{1}{6}\right)$

Absolute maximum value of $f(x, y) =\frac{3+\sqrt{2}}{2}$ At $\left(0,-\frac{1}{\sqrt{2}}\right)$

2.2 Find the volume enclosed between the surfaces.

$$x^2 + y^2 + z^2 = 16 \text{ and } x^2 + y^2 - 4x = 0$$

[20 marks : 2017]

Solution:

Required volume is $\iiint_D dV$.

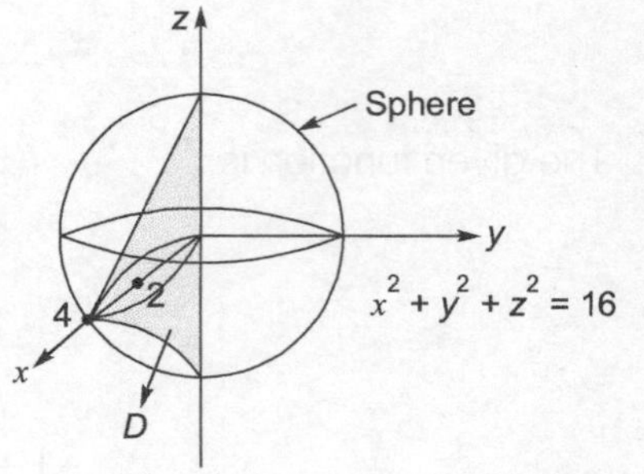

Where D can be described by

$D(r,\theta,z)$; where, $r = 4\cos\theta, 0 \le \theta \le \pi; 0 \le z \le \sqrt{16-r^2}$

Since, $x^2 + y^2 - 4x = 0,$

$$x^2 + y^2 = 4x$$
$$r^2 = 4r\cos\theta$$
$$r = 4\cos\theta$$

and $x^2 + y^2 + z^2 = 16$

$$r^2 + z^2 = 16$$
$$z^2 = 16 - r^2$$
$$z = \sqrt{16-r^2}$$

$\therefore$

$$V = \iiint_D dV = \int_0^{\pi} \int_{r=0}^{4\cos\theta} \int_{z=0}^{\sqrt{16-r^2}} r\,dr\,d\theta\,dz$$

$$= \int_0^{\pi} \int_0^{4\cos\theta} (z)\Big|_0^{\sqrt{16-r^2}} r\,dr\,d\theta$$

$$= \int_0^{\pi} \left(\int_0^{4\cos\theta} \sqrt{16-r^2} \cdot r\,dr \right) d\theta$$

Let, $16 - r^2 = t^2$

$$-2r\,dr = 2t\,dt$$
$$-r\,dr = t\,dt$$

Now, $r = 0, \quad t = 4$

$r = 4\cos\theta, \quad t = 4\sin\theta$

$$V = \int_0^{\pi} \left(\int_4^{4\sin\theta} t(-t\,dt) \right) d\theta = \int_0^{\pi} \left(\int_4^{4\sin\theta} (-t^2)dt \right) d\theta = \int_0^{\pi} \left(-\frac{t^3}{3}\Bigg|_4^{4\sin\theta} \right) d\theta$$

$$= \int_0^{\pi} -\frac{1}{3}(64\sin^3\theta - 64)\,d\theta$$

$$= \frac{-(64)}{3} \int_0^{\pi} (\sin^3\theta - 1)\,d\theta = \frac{(64)}{3} \int_0^{\pi} (1 - \sin^3\theta)\,d\theta$$

$$= \frac{64}{3} \times 2 \int_0^{\pi/2} (1 - \sin^3\theta)\,d\theta = \frac{128}{3}\left[\frac{\pi}{2} - \frac{2}{3}\right] = \frac{128\pi}{6} - \frac{256}{9}$$

2.3 Find the difference between the values of $\int_C \phi d\vec{r}, \phi(x, y) = x^2y + 2y$ from (1, 1, 0) to (2, 4, 0) along the curve $y = x^2$, $z = 0$ and along the straight line joining (1, 1, 0) and (2, 4, 0). Hence evaluate $\int_C (\nabla \cdot \vec{f})\, d\vec{r}$, where $\vec{f} = \frac{1}{4}x^4 y\hat{i} + y^2\hat{j} + xy\hat{k}$ along the curve which is a parabola $y = x^2$, $z = 0$ from (1, 1, 0) to (2, 4, 0).

[20 marks : 2019]

Solution:

Given that, $\phi(x, y) = x^3y + 2y$

Along the curve $y = x^2$:

$$= \int_{(1,1,0)}^{(2,4,0)} (x^3y + 2y)(dx\hat{i} + dy\hat{j})$$

As, $y = x^2$, $dy = 2xdx$

$$= \int_1^2 (x^3(x^2) + 2(x^2))\,(dx\hat{i} + 2xdx\hat{j})$$

$$= \int_1^2 (x^5 + 2x^2)\,(dx\hat{i} + 2xdx\hat{j}) = \int_1^2 (x^5 + 2x^2)\,dx\hat{i} + (2x^6 + 4x^3)\,\hat{j}$$

$$= \left(\frac{x^6}{6} + \frac{2x^3}{3}\right)_1^2 \hat{i} + \left(\frac{2x^7}{7} + x^4\right)_1^2 \hat{j}$$

$$= \left[\frac{16}{6} + \frac{16}{3} - \frac{1}{6} - \frac{2}{3}\right]\hat{i} + \left[\frac{256}{7} + 16 - \frac{2}{7} - 1\right]\hat{j} = 15.16\hat{i} + 51.28\hat{j}$$

Along the straight line from (1, 1, 0) to (2, 4, 0)

$$\text{Slope of line} = \frac{4-1}{2-1} = \frac{3}{1} = 3$$

$$(y - 1) = 3(x - 1)$$

$$y - 1 = 3x - 3$$

$$y = 3x - 2$$

$$\therefore \quad dy = 3dx$$

$$= \int_{(1,1,0)}^{(2,4,0)} (x^3y \times 2y) \times (dx\hat{i} + dy\hat{j})$$

$$= \int_1^2 [x^3(3x - 2) + 2(3x - 2)]\,[dx\hat{i} + 3dx\hat{j}]$$

$$= \int_1^2 [3x^4 - 2x^3 + 6x - 4]\,dx[\hat{i} + 3\hat{j}]$$

$$= \int_1^2 \left[\frac{3x^5}{5} - \frac{2x^4}{4} + \frac{6x^2}{2} - 4x\right]_1^2 [\hat{i} + 3\hat{j}]$$

$$= \left[\left(\frac{3\times 32}{5}-\frac{32}{4}+\frac{6\times 4}{2}-8\right)-\left(\frac{3}{5}-\frac{2}{4}+\frac{6}{2}-4\right)\right][\hat{i}+3\hat{j}]$$

$$= \left[\left(\frac{96}{5}-\frac{32}{4}+12-8\right)-\left(\frac{3}{5}-\frac{2}{4}-1\right)\right][\hat{i}+3\hat{j}]$$

$$= 16.1\hat{i}+48.3\hat{j}$$

The difference between the values

$$= \left|15.61\hat{i}+51.28\hat{j}\right|-\left|16.1\hat{i}+48.3\hat{j}\right|$$

$$= 53.60 - 50.91 = 2.69$$

$$\int(\nabla\cdot\vec{f})\,d\vec{r} = \int(x^3y+2y)\times(dx\hat{i}+dy\hat{j})$$

Along the curve $y = x^2$ from (1, 1, 0) to (2, 4, 0)

$$= 15.16\hat{i}+51.28\hat{j}$$

2.4 By converting into a line integral, evaluate $\iint_S (\nabla\times\bar{F})\cdot\hat{n}\,dS$, where $\bar{F}=(x^2+y-4)\hat{i}+3xy\hat{j}+(2xy+z^2)\hat{k}$ and S is the surface of paraboloid $x^2 + y^2 + z = 4$ above xy-plane.

[10 marks : 2020]

Solution:

$$\vec{F} = (x^2+y-4)\hat{i}+3xy\hat{j}+(2xy+z^2)\hat{k}$$

$\int_C \vec{F}\cdot d\vec{r}$ where C is the boundary of paraboloid $x^2 + y^2 + z = 4$

For surface of paraboloid over xy-plane

$$\partial\vec{r} = i\,dx + j\,dy$$

$$\int_C\left[(x^2+y-4)\hat{i}+3xy\hat{j}+(2xy+z^2)\hat{k}\right]\left[idx+\hat{j}dy\right]$$

$$\int_C (x^2+y-4)dx+3xy\,dy$$

Now put

$$x = 2\cos\theta$$
$$y = 2\sin\theta$$
$$dx = -2\sin\theta\,d\theta$$
$$dy = 2\cos\theta$$

$$\int_0^{2\pi}\left(-4\cos^2\theta+2\sin\theta-4\right)(-2\sin\theta)d\theta+24\sin\theta\cos^2 d\theta$$

$$I = 4\int_0^{2\pi}(4\sin\theta\cos^2\theta-\sin^2\theta+\sin\theta)d\theta$$

$$\int_0^{2\pi}\sin\theta\,d\theta = 0$$

$$\int_0^{2\pi} 4\sin\theta\cos^2\theta = 0 \qquad \because \int_0^{2\pi} \cos^n\theta\sin\theta = 0$$

$$4\int_0^{2\pi} -\sin^2\theta d\theta = -16\int_0^{\pi/2} \sin^2\theta d\theta = -16\times\frac{1}{2}\times\frac{\pi}{2} = -4\pi$$

2.5 Show that

(i) $\int_C (yz-1)dx + (z+xz+z^2)dy + (y+xy+2yz)dz$ is independent of path of integration from (1, 2, 3) to (4, 5, 6). Hence, evaluate the integral.

(ii) Let $\vec{F} = z\hat{j} + z\hat{k}$ represent the flow of a liquid. Find the flux of $\vec{F}$ through the surface S given by that portion of the plane $z = 6 - 3x - 2y$ in the first octant oriented upward.

[10+10 = 20 marks : 2021]

Solution:

(i)

$$I = \int_C (yz-1)dx + (z+xz+z^2)dy + (y+xy+2yz)dz$$

We know that

$$I = \int_C Pdx + Qdy + Rdz \text{ is independent of path}$$

iff

$$\frac{\partial P}{\partial y} = \frac{\partial Q}{\partial x};\ \frac{\partial Q}{\partial z} = \frac{\partial R}{\partial y},\ \frac{\partial P}{\partial z} = \frac{\partial R}{\partial x}$$

Given :

$$P = yz - 1,\ Q = z + xz + z^2,\ R = y + xy + 2yz$$

$$\frac{\partial P}{\partial y} = z = \frac{\partial Q}{\partial x},\ \frac{\partial Q}{\partial z} = 1 + x + 2z = \frac{\partial R}{\partial y},\ \frac{\partial P}{\partial z} = y = \frac{\partial R}{\partial x}$$

$\Rightarrow I$ is independent of path.

Evaluate : $\int_A^B Pdx + Qdy + Rdz$

$$\int_A^B (yz-1)dx + (z+xz+z^2)dy + (y+xy+2yz)dz$$

$$= \int_A^B (zdy + ydz) + (yzdx + xzdy + xydz) + (z^2dy + 2yzdz) - dx$$

$$= \int_A^B d(yz) + \int_A^B d(xyz) + \int_A^B d(yz^2) - \int_A^B dx$$

$$= (yz)_{(2,3)}^{(5,6)} + (xyz)_{(1,2,3)}^{(4,5,6)} + (yz^2)_{(2,3)}^{(5,6)} - (x)_1^4$$

$$= (30-6) + (120-6) + (180-18) - (4-1)$$

$$= 24 + 114 + 162 - 3$$

$$= 135 + 162$$

$$= 297$$

(ii) $$\vec{F} = z\hat{j} + z\hat{k}$$

$$\text{Flux of } \vec{F} = \iint_S \vec{F}\cdot\hat{n}dS = \iint_S \vec{F}\cdot\hat{n}\frac{dxdy}{\hat{n}\cdot\hat{k}} \quad \text{(by taking projection of } S \text{ on } xy\text{-plane)}$$

where $\hat{n}$ = Outward unit normal vector to plane

$$\hat{n} = \frac{\nabla\phi}{|\nabla\phi|} = \frac{3i+2j+k}{\sqrt{14}}$$

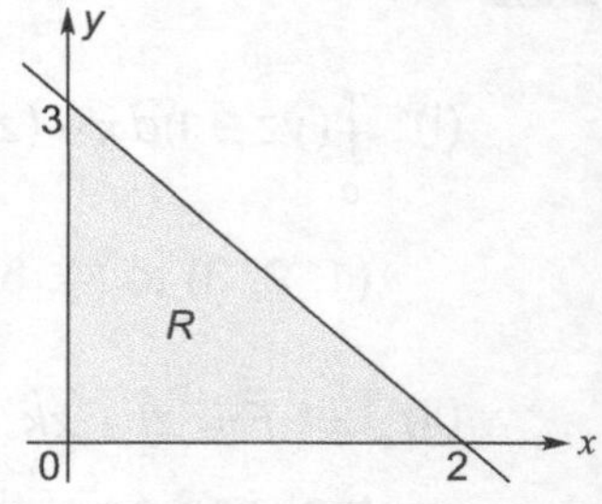

$$\Rightarrow \quad \hat{n}\cdot\hat{k} = \frac{1}{\sqrt{14}}$$

$$\vec{F}\cdot\hat{n} = (zj+zk)\cdot\frac{(3i+2j+k)}{\sqrt{14}} = \frac{3z}{\sqrt{14}}$$

$$\therefore \quad \iint_S \vec{F}\cdot\hat{n}dS = \iint_R \vec{F}\cdot\hat{n}\frac{dxdy}{\hat{n}\cdot\hat{k}}$$

$$= \iint_R \frac{3z}{\sqrt{14}}\frac{dxdy}{\frac{1}{\sqrt{14}}}$$

$$= \iint_R 3z\,dxdy$$

$$= \int_{x=0}^{2}\int_{y=0}^{\frac{6-3x}{2}} 3[6-3x-2y]dydx$$

$$= \int_0^2 3\left[(6-3x)y - y^2\right]_0^{\frac{6-3x}{2}} dx$$

$$= 3\int_0^2 \left[\frac{(6-3x)^2}{2} - \frac{(6-3x)^2}{4}\right]dx$$

$$= 3\int_0^2 \frac{1}{4}[36+9x^2-36x]dx$$

$$= \frac{3}{4}\left[36x+3x^3-18x^2\right]_0^2 = \frac{3}{4}[24]$$

So, the flux of flow $\vec{F} = 18$

2.6 **Find the volume of the solid bounded by the sphere $x^2 + y^2 + z^2 = 20$ and the paraboloid $x^2 + y^2 = z$.**

[10 marks : 2022]

Solution:

$$V = \iiint dzdydz$$

Domain bounded by $x^2 + y^2 + z^2 = 20$ and $x^2 + y^2 = z$

In cylindrical coordinates, there are

$$\rho^2 = z \text{ and } \rho^2 + z^2 = 20$$

Solving we get, $z = 4 \Rightarrow \rho^2 = 4 \Rightarrow \rho = 2$

So, $V = \iiint \rho \, d\rho \, d\phi \, dz$

$$V = \int_0^{2\pi} d\phi \int_0^2 \rho \left(\int_{\rho^2}^{\sqrt{20-\rho^2}} dz \right) d\rho$$

$$V = 2\pi \int_0^2 \rho \left(\sqrt{20-\rho^2} - \rho^2 \right) d\rho$$

On solving, $V = 28.15$ cubic unit

2.7 (i) Obtain the half-range cosine series for the funtion $f(x) = \sin x$ in $0 \le x \le \pi$ and hence, find the value of $\sum_{n=1}^{\infty} \frac{1}{4n^2 - 1}$.

(ii) Evaluate the integral $\iint_R (x-y)^4 \cos^2(x+y) \, dx \, dy$ where R is the rhombus with successive vertices at $(\pi, 0)$, $(2\pi, \pi)$, $(\pi, 2\pi)$, $(0, \pi)$.

[10 + 10 = 20 marks : 2023]

Solution:

(i) Half range cosine series is given by

$$a_o = \frac{1}{l} \int_0^l f(x) dx$$

$$a_n = \frac{2}{l} \int_0^l f(x) \cos\left(\frac{n\pi x}{l} \right) dx; \quad b_n = 0$$

Given, $f(x) = \sin x$

Range : $(0, \pi)$, i.e., $l = \pi$

$$\therefore \quad f(x) = a_o + \sum a_n \cos\left(\frac{n\pi x}{l} \right)$$

$$a_o = \frac{1}{\pi} \int_0^\pi f(x) dx$$

$$\therefore \quad a_o = \frac{1}{\pi} \int_0^\pi \sin x \, dx = \frac{1}{\pi} [-\cos x]_0^\pi = \frac{2}{\pi}$$

$$\therefore \quad a_o = \frac{2}{\pi}$$

Now, $a_n = \frac{2}{\pi} \int_0^\pi \sin x \cos nx \, dx$

$$a_n = \frac{2}{\pi} \int_0^\pi [\sin(1+n)x + \sin(1-n)x] dx$$

$$a_n = \frac{1}{\pi}\left[\frac{-\cos(1+n)x}{(1+n)} - \frac{\cos(1-n)x}{(1-n)}\right]_0^\pi$$

$$a_n = \frac{1}{\pi}\left[\frac{-\cos(\pi+\pi n)}{1+n} - \frac{\cos(\pi-\pi n)}{1-n} + \frac{1}{(1+n)} + \frac{1}{(1-n)}\right]$$

$$a_n = \frac{1}{\pi}\left[\frac{\cos\pi n}{(1+n)} - \frac{\cos\pi n}{(1-n)} + \frac{2}{1-n^2}\right]$$

$$a_n = \frac{1}{\pi}\left[\cos\pi n \times \left(\frac{2}{1-n^2}\right) + \frac{2}{1-n^2}\right]$$

$$\therefore \quad a_n = \begin{cases} 0, & \text{if } n \text{ is odd} \\ \frac{1}{\pi}\cdot\frac{4}{(1-n^2)}, & \text{if } n \text{ is even} \end{cases}$$

$$f(x) = \sin x = \frac{2}{\pi} - \frac{4}{\pi}\left(\frac{\cos 2x}{3} + \frac{\cos 4x}{15} + \frac{\cos 6x}{35} + ...\right)$$

$$\sum_{n=1}^{\infty}\frac{1}{4x^2-1} = \frac{1}{2}\sum_{n=1}^{K}\left[\frac{1}{2n-1} - \frac{1}{2n+1}\right]$$

For $K = 1, 2, ...$

$$S_1 = \frac{1}{2}\sum_{n=1}^{1}\left[\frac{1}{2n-1} - \frac{1}{2n+1}\right]$$

$$= \frac{1}{2}\left(1-\frac{1}{3}\right)$$

$$S_2 = \frac{1}{2}\sum_{n=1}^{2}\left[\frac{1}{2n-1} - \frac{1}{2n+1}\right]$$

$$= \frac{1}{2}\left[\left(1-\frac{1}{3}\right) + \left(\frac{1}{3}-\frac{1}{5}\right)\right] = \frac{1}{2}\left(1-\frac{1}{5}\right)$$

$$S_3 = \frac{1}{2}\sum_{n=1}^{3}\left[\frac{1}{2n-1} - \frac{1}{2n+1}\right]$$

$$= \frac{1}{2}\left[\left(1-\frac{1}{3}\right) + \left(\frac{1}{3}-\frac{1}{5}\right) + \left(\frac{1}{5}-\frac{1}{7}\right)\right] = \frac{1}{2}\left(1-\frac{1}{7}\right)$$

$$S_K = \frac{1}{2}\left(1-\frac{1}{2K+1}\right)$$

$$\lim_{K\to\infty} S_K = \frac{1}{2}$$

$$\sum_{n=1}^{\infty}\frac{1}{4n^2-1} = \lim_{K\to\infty} S_K = \frac{1}{2}$$

(ii) Translate rhombus to origin :

$$x' = x - \pi,\ y' = y - \pi$$

$$= \iint_R (x-y)^4\cos^2(x+y)\,dx\,dy$$

$$= \iint_{R'} (x'-y')^4\cos^2(x'+y'+2\pi)\,dx'dy'$$

Change coordinate system :

$$x' - y' = u,\ x' + y' = v$$

Now, $-\pi \le u \le \pi,\ -\pi \le v \le \pi$

$$\iint_{R'} (x' - y')^4 \cos^2(x' + y')dx'dy' = \iint_{R''} u^4 \cos^2(v)J(u,v)dudv$$

Compute Jacobian : $J = \frac{1}{2}$

$$\frac{1}{2}\int_{-\pi}^{\pi}\int_{-\pi}^{\pi} u^4 \cos^2(v)dudv = \frac{1}{2}\int_{-\pi}^{\pi} u^4 dv \int_{-\pi}^{\pi} \cos^2(v)dv$$

$$= \frac{1}{2}\cdot\frac{u^5}{5}\Bigg|_{-\pi}^{\pi} \int_{-\pi}^{\pi} \cos^2(v)dv$$

$$= \frac{1}{2}\left(\frac{\pi^5}{5} + \frac{\pi^5}{5}\right)\pi$$

$$= \frac{1}{2} \times \frac{2\pi^5}{5} \times \pi$$

$$= \frac{\pi^6}{5}$$

3. Differential Equations

3.1 **A string of flexible wire stretched on a sitar has its ends fixed at $x = 0$ and $x = 20$. Initially at $t = 0$, the string is at rest and takes the shape as defined by $h(x) = \mu(20x - x^2)$, μ being a constant, and then it is released to vibrate. Formulate this boundary value problem and solve that to find the displacement at any point x at an instant t. The solution, to be obtained, should involve definite constants not the arbitrary ones.**

[20 marks : 2019]

Solution:

The above equation is, $$\frac{\partial^2 u}{\partial x^2} = a^2 \frac{\partial^2 u}{\partial t^2} \quad \text{... (i)}$$

The boundary conditions are:

(i) $u(0, t) = 0$

(ii) $u(l, t) = 0$

(iii) $\frac{\partial u}{\partial t}(x, 0) = 0$

After applying the above boundary conditions in the solution of equation we get,

$$u(x, t) = \sum_{n=1}^{\infty} B_n \sin\left(\frac{n\pi x}{l}\right)\cos\left(\frac{n\pi at}{l}\right) \quad \text{... (ii)}$$

Now, $$u(x, 0) = \sum_{n=1}^{\infty} B_n \sin\left(\frac{n\pi x}{l}\right) \quad \text{... (iii)}$$

Given that, $$u(x, 0) = \mu(20x - x^2) \quad \text{... (iv)}$$

From (iii) and (iv), $\mu(20x - x^2) = \sum_{n=1}^{\infty} B_n \sin\left(\frac{n\pi x}{l}\right)$

which is in the form of half-range Fourier sine series.

$$B_n = \frac{2}{l}\int_0^l f(x)\cdot \sin\left(\frac{n\pi x}{l}\right)\cdot dx$$

$$= \frac{2}{l}\int_0^l \mu(20x - x^2)\cdot \sin\left(\frac{n\pi x}{l}\right)\cdot dx$$

$$B_n = \frac{2\mu}{l}\left[-\frac{l(20x-x^2)}{n\pi}\cos\left(\frac{n\pi x}{l}\right) - \frac{l^2}{n^2\pi^2}(20-2x)\sin\frac{n\pi x}{l} + \frac{2l^3}{n^3\pi^3}\cos\left(\frac{n\pi x}{l}\right)\right]_0^l$$

$$B_n = \frac{2\mu}{l}\left[-\frac{l(20l-l^2)}{n\pi}(-1)^n + \frac{2l^3}{n^3\pi^3}(-1)^n - \left(0+0+\frac{2l^3}{n^3\pi^3}\right)\right]$$

Given that, $l = 20$

$$= \frac{2\mu}{20}\left[\frac{2(20)^3}{n^3\pi^3}(-1)^n - \frac{2(20)^3}{n^3\pi^3}\right]$$

$$= \frac{2\mu}{20}\left[\frac{2(20)^3}{n^3\pi^3}\{(-1)^n - 1\}\right] = \frac{1600\mu}{n^3\pi^3}\{(-1)^n - 1\}$$

$$B_n = \begin{cases} \dfrac{-3200\mu}{n^3\pi^3}; & \text{when } n \text{ is odd} \\ 0 \quad ; & \text{when } n \text{ is even} \end{cases}$$

Substituting the value of B_n is equation (ii), we get

$$u(x,\ t) = \sum_{n=1,3,5}^{\infty} -\frac{3200\mu}{n^3\pi^3}\sin\left(\frac{n\pi x}{l}\right)\cos\left(\frac{n\pi at}{l}\right)$$

Since, $l = 20$,

$$u(x,\ t) = \sum_{n=1,3,5}^{\infty} -\frac{3200\mu}{n^3\pi^3}\sin\left(\frac{n\pi x}{20}\right)\cos\left(\frac{n\pi at}{20}\right)$$

3.2 (i) Obtain the partial differential equation governing the equations

$$f(u, v) = 0, \quad u = x + yz, \quad v = x + y + z$$

(ii) Construct a partial differential equation of all surfaces of revolution having z-axis as the axis of rotation.

[12 marks : 2020]

Solution:

(i) Given equations $\phi(u, v) = 0$

where $u = x + yz,\ v = x + y + z$

The required partial differential equation of these equation is of the form

$$P\frac{\partial z}{\partial x} + Q\frac{\partial z}{\partial y} = R$$

where P, Q, R are functions of x, y, z.

These are given by

$$P = \begin{vmatrix} \frac{\partial u}{\partial y} & \frac{\partial u}{\partial z} \\ \frac{\partial v}{\partial y} & \frac{\partial v}{\partial z} \end{vmatrix} = \begin{vmatrix} z & y \\ 1 & 1 \end{vmatrix} = z - y$$

$$Q = \begin{vmatrix} \frac{\partial u}{\partial z} & \frac{\partial u}{\partial x} \\ \frac{\partial v}{\partial z} & \frac{\partial v}{\partial x} \end{vmatrix} = \begin{vmatrix} y & 1 \\ 1 & 1 \end{vmatrix} = y - 1$$

$$R = \begin{vmatrix} \frac{\partial u}{\partial x} & \frac{\partial u}{\partial y} \\ \frac{\partial v}{\partial x} & \frac{\partial v}{\partial y} \end{vmatrix} = \begin{vmatrix} 1 & z \\ 1 & 1 \end{vmatrix} = 1 - z$$

$\therefore$ Required partial differential equation is

$$P\frac{\partial z}{\partial x} + Q\frac{\partial z}{\partial y} = R$$

$$\Rightarrow \quad (z-y)\frac{\partial z}{\partial x} + (y-1)\frac{\partial z}{\partial y} = (1-z)$$

(ii) Any surface rotated about z-axis, it forms circular planes along with z-axis.
The equation of surface can be taken as

$$z = \phi\left(\sqrt{x^2+y^2}\right) \quad \text{...(1)}$$

ϕ is an arbitrary function.
We need to form the corresponding partial differential equation.
We need to eliminate arbitrary function 'ϕ' by differentiating partially with respect to x, y separately.
Partial differentiation with respect to x

$$\frac{\partial z}{\partial x} = \phi'\left(\sqrt{x^2+y^2}\right) \times \frac{1}{2\sqrt{x^2+y^2}} \times 2x \quad \text{...(2)}$$

Differentiation with respect to y:

$$\frac{\partial z}{\partial y} = \phi'\left(\sqrt{x^2+y^2}\right) \times \frac{1}{2\sqrt{x^2+y^2}} \times 2y \quad \text{...(3)}$$

Eliminating ϕ' from (2), (3)

$$(2) \div (3) \Rightarrow \quad \frac{\frac{\partial z}{\partial x}}{\frac{\partial z}{\partial y}} = \frac{\phi'\left(\sqrt{x^2+y^2}\right) \times \frac{x}{\sqrt{x^2+y^2}}}{\phi'\left(\sqrt{x^2+y^2}\right) \times \frac{y}{\sqrt{x^2+y^2}}}$$

$$\frac{p}{q} = \frac{x}{y} \Rightarrow yp - xq = 0$$

$yp - xq = 0$ is partial differential equation which represents all the surfaces of revolutions having z-axis as the axis of rotation.

3.3 Find the general solution of the partial differential equation $\frac{\partial^3 z}{\partial x^3} - 2\frac{\partial^3 z}{\partial x^2 \partial y} = 9(e^{3x} + 10xy^2)$.

[10 marks : 2022]

Solution:

Given,
$$\frac{\partial^3 z}{\partial x^3} - \frac{2\partial^3 z}{\partial x^2 \partial y} = 9(e^{3x} + 10xy^2)$$

$$D^3 - 2D^2D' = 9(e^{3x} + 10xy^2)$$

⇒ Auxillary equation,

$$m^3 - 2m^2 = 0$$

⇒ $$m^2(m-2) = 0$$

⇒ $$m = 0, 0, 2$$

So, $$CF = F_1(y) + xF_2(y) + F_3(y+2x)$$

$$PI_1 = \frac{9e^{3x}}{D^3 - 2D^2D'}$$

Here, $$D = 3 \text{ and } D' = 0$$

$$PI_1 = \frac{1}{3}e^{3x}$$

⇒ For $$PI_2 = \frac{90xy^2}{D^3 - 2D^2D'}$$

$$PI_2 = \frac{1}{D^3\left(1 - \frac{2D'}{D}\right)}(90xy^2)$$

$$= \frac{90}{D^3}\left(1 - \frac{2D'}{D}\right)^{-1}(xy^2)$$

$$= \frac{90}{D^3}\left(1 + \frac{2D'}{D} + \frac{4D'^2}{D^2} + \frac{8D'^3}{D^3} + \dots\right)xy^2$$

$$= \frac{90}{D^3}\left(xy^2 + \frac{2}{D}\cdot D'(xy)^2 + \frac{4}{D^2}D'^2(xy^2) + \frac{8}{D^3}D'^3(xy^2) + \dots\right)$$

$$= \frac{90}{D^3}\left(xy^2 + \frac{2}{D}(2xy) + \frac{4}{D^2}(2x) + 0\right)$$

$$= 90\left(\frac{y^2}{D^3}(x) + \frac{4y}{D^4}(x) + \frac{8}{D^5}(x)\right)$$

$$= 90\left(y^2 \cdot \frac{x^4}{2.3.4} + 4y\frac{x^5}{2.3.4.5} + 8\frac{x^6}{2.3.4.5.6}\right)$$

$$= 90\left(\frac{x^4y^2}{24} + \frac{x^5y}{30} + \frac{x^6}{90}\right)$$

Hence, complete solution is

$$Z = F_1(y) + xF_2(y) + F_3(y+2x) + \frac{1}{3}e^{3x} + 90\left(\frac{x^4y^2}{24} + \frac{x^5y}{30} + \frac{x^6}{90}\right)$$

3.4 Solve the partial differential equation

$$\frac{\partial^2 z}{\partial x^2}+\frac{\partial^2 z}{\partial x \partial y}-6\frac{\partial^2 z}{\partial y^2} = e^{x+2y} + \sin(4x + 3y) + y\cos x$$

[10 marks : 2023]

Solution:

Its auxillary equation is $m^2 + m - 6 = 0$

$\Rightarrow$ $m = -3, 2$

$\therefore$ Complementary function :

$$CF = f_1(y-3x) + f_2(y+2x)$$

Find PI :

$$\text{P.I.} = \frac{e^{2y+x}}{D^2+DD'-6D'^2}+\frac{\sin(3y+4x)}{D^2+DD'-6D'^2}+\frac{y\cos x}{D^2+DD'-6D'^2}$$

$$= \text{P.I.}_1 + \text{P.I.}_2 + \text{P.I.}_3$$

P.I._1 :

$$\text{P.I.}_1 = \frac{e^{2y+x}}{D^2+DD'-6D'^2} = \frac{e^{2y+x}}{1^2+1\times 2-6\times 2^2}$$

$$= \frac{e^{2y+x}}{-21}$$

P.I._2 :

$$\text{PI}_2 = \frac{\sin(3y+4x)}{D^2+DD'-6D'^2}$$

Put $D^2 = -4^2$, $DD' = -4\times 3,\ D'^2 = -3^2$

$$\text{PI}_2 = \frac{\sin(3y+4x)}{-4^2-4\times 3-6(-3^2)} = \frac{\sin(3y+4x)}{26}$$

$$\text{PI}_3 = \frac{y\cos x}{D^2+DD'-6D'^2}$$

$$= \frac{1}{(D-2D')(D+3D')}y\cos x$$

$$= \frac{1}{D-2D'}\int (C+3x)\cos x\,dx \qquad (\because y = C - mx = C + 3x)$$

$$= \frac{1}{D-2D'}[(C+3x)\sin x + 3\cos x]_{C\to y-3x}$$

$$= \frac{1}{D-2D'}(y\sin x + 3\cos x)$$

$$= \int ((C-2x)\sin x + 3\cos x)dx$$

$$= (C-2x)(-\cos x) - (-2)(-\sin x) + 3\sin x$$

Put $y = C - 2x$

$$\text{PI}_3 = -y\cos x + \sin x$$

$\therefore$ General solution of P.D.E. is

$$Z = \text{CF} + \text{PI}$$

$$Z = f_1(y-3x)+f_2(y+2x)+\frac{e^{2y+x}}{-21}+\frac{\sin(3y+4x)}{26}+\sin x - y\cos x$$

4. Complex Analysis

4.1 Evaluate $\oint_c \dfrac{z^2+4z+8}{\left(z-\frac{1}{2}\right)^2 (z^2+2z+5)} dz$ where C is the contour $|z+1-i|=2$ in anticlockwise sense.

[20 marks : 2018]

Solution:

C is a circle of radius 2 units with centre $(-1, i)$ in z-plane.

Let us find poles of function $f(z)$,

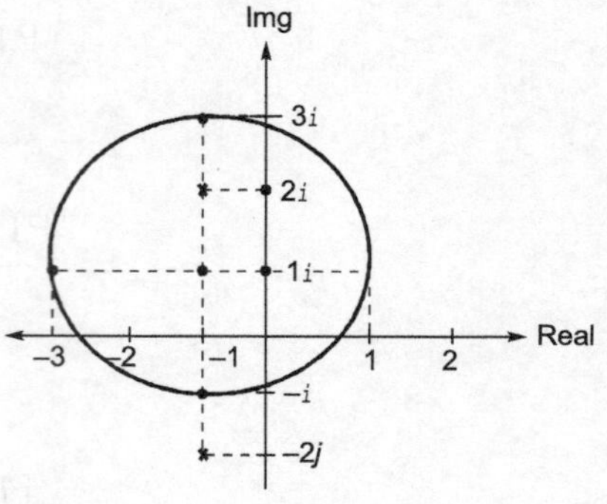

Where, $$f(z) = \frac{z^2+4z+8}{\left(z-\frac{1}{2}\right)^2 (z^2+2z+5)}$$

$$z = \frac{1}{2}, \frac{1}{2}$$

$$z = \frac{-2 \pm \sqrt{4-20}}{2} = -1 \pm 2j$$

$z = \frac{1}{2}, \frac{1}{2}$ and $(-1+2j)$ lie inside the circle

Residue at, $(z=-1+j2) = \lim_{z \to -1+j2} \dfrac{(z+1-2j)(z^2+4z+8)}{\left(z-\frac{1}{2}\right)^2 (z+1+2j)(z+1-2j)}$

$$= \frac{((-1+j2)^2 + 4(-1+j2)+8)}{\left(-1+j2-\frac{1}{2}\right)^2 (-1+j2+1+j2)}$$

$$= \frac{(1-4-4j)-4+j8+8}{\left(-4+\frac{9}{4}-6j\right)(4j)} = \frac{1+4j}{-7j+24}$$

Double pole at, $z = \frac{1}{2}$

Residue at, $\left(z=\frac{1}{2}\right) = \lim_{z \to \frac{1}{2}} \dfrac{d}{dz}\left(f(z)\left(z-\frac{1}{2}\right)^2\right)$

$$= \frac{d}{dz}\left(\frac{z^2+4z+8}{z^2+2z+5}\right)$$

$$= \frac{(2z+4)(z^2+2z+5)-(z^2+4z+8)(2z+2)}{(z^2+2z+5)^2}$$

$$= \frac{\left(2\times\frac{1}{2}+4\right)\left(\frac{1}{4}+2\times\frac{1}{2}+5\right)-\left(\frac{1}{4}+\frac{4}{2}+8\right)\left(2\times\frac{1}{2}+2\right)}{\left(\frac{1}{4}+2\times\frac{1}{2}+5\right)^2}$$

$$= \frac{0.5}{(6.25)^2}$$

$$\oint f(z)dz = 2\pi i\,[\text{sum of residues}]$$

$$= 2\pi i\left[\frac{0.5}{(6.25)^2}+\frac{1+4j}{-7j+24}\right] = -1.035 + 0.0402j$$

4.2 (i) An electrostatic field in xy-plane is given by $\phi(x, y) = 3x^2y - y^3$. Find the stream function y such that the complex potential $\omega = \phi + i\psi$ is an analytic function.

(ii) Find three Laurent's series expansions of the function $f(z) = \dfrac{1}{3z - z^2 - 2}$ and specify the regions in which those expansions are valid.

[12 marks : 2019]

Solution:

(i) $$\phi(x, y) = 3x^2y - y^3$$

Given complex function $\omega = \phi + i\psi$ is an analytic function.

C–R equations are,

$$\frac{\partial\phi}{\partial x} = \frac{\partial\psi}{\partial y}$$

$$\frac{\partial\phi}{\partial y} = -\frac{\partial\psi}{\partial x}$$

We know, that,
$$d\psi = \frac{\partial\psi}{\partial x}dx + \left(\frac{\partial\phi}{\partial x}\right)dy$$

$$= -(3x^2 - 3y^2)dx + 6xydy$$

$$= -3x^2dx + 3y^2dx + 6xydy$$

$$d\psi = -d(x^3) + 3d(xy^2)$$

$\therefore$
$$\psi = \int -d(x^3) + 3\int d(xy^2) + c$$

$$\psi = -x^3 + 3xy^2 + c$$

$\therefore$ Stream function,
$$\psi = -x^3 + 3xy^2 + c$$

(ii)
$$f(z) = \frac{1}{3z - z^2 - 2} = \frac{-1}{(z-1)(z-2)} = -\frac{1}{z-2} + \frac{1}{z-1} \quad \text{... (i)}$$

$$= \frac{1}{2}\left(1-\frac{z}{2}\right)^{-1} - (1-z)^{-1}$$

(a) For region $|z|<1$:

For $|z|<1$, both $\left|\frac{z}{2}\right|$ and $|z|$ are less than 1.

$$f(z) = \frac{1}{2}\left(1+\frac{z}{2}+\frac{z^2}{4}+\frac{z^3}{8}+....\right) - (1+z+z^2+z^3+....)$$

$$= -\left(\frac{1}{2} + \frac{3z}{4} + \frac{7z^2}{8} + \frac{15z^3}{16} + ...\right)$$

(b) For region $1<|z|<2$:

We can write equation (i) as,

$$f(z) = \frac{1}{2}\frac{1}{1-\frac{z}{2}} + \frac{1}{z\left(1-z^{-1}\right)}$$

and both $\left|\frac{z}{2}\right|$ and $\left|z^{-1}\right|$ are less than 1.

$$f(z) = \frac{1}{2}\left(1+\frac{z}{2}+\frac{z^2}{4}+\frac{z^3}{8}+....\right)+\frac{1}{z}\left(1+z^{-1}+z^{-2}+z^{-3}+....\right)$$

$$= z^{-4}+z^{-3}+z^{-2}+z^{-1}+\frac{1}{2}+\frac{1}{4}z+\frac{1}{8}z^2+\frac{1}{16}z^3+....$$

(c) For region $|z|>2$:

We write equation (i) as,

$$f(z) = \frac{-1}{z(1-2z^{-1})} + \frac{1}{z(1-z^{-1})}$$

$$= -z^{-1}\,(1 + 2z^{-1} + 4z^{-2} + 8z^{-3} + ...) + z^{-1}\,(1 + z^{-1} + z^{-2} + z^{-3} + ...)$$

$$= .....\ -7z^{-4} - 3z^{-3} - z^{-2} - .....$$

$\therefore$ For all three regions $|z|<1, 1<|z|<2, |z|>2$ expansion are valid.

4.3 **Using the contour integration, evaluate the integral $\int_{-\infty}^{\infty} \frac{dx}{1+x^4}$**

[10 marks : 2021]

Solution:

Let $$f(z) = \frac{1}{z^4+1}$$

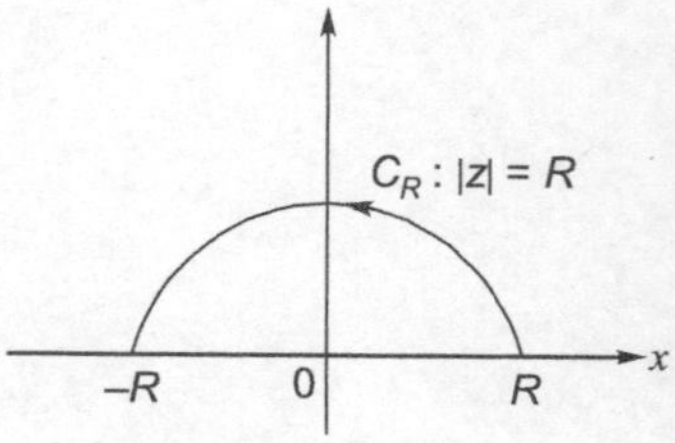

Consider integral $\int_C f(z)dz$ where 'C' is closed contour consisting of upper half C_R of a large circle $|Z| = R$ and along with real axis from $-R$ to R.

$$\oint_C f(z)dz = \int_{C_R} f(z)dz + \int_{-R}^{R} f(x)dz$$

The poles of $f(z)$ are given by $z^4 + 1 = 0$.

$$\Rightarrow \quad z^4 = -1$$

$$z^4 = \cos(2n+1)\pi + i\sin(2n+1)\pi$$

$$z^4 = e^{i(2n+1)\pi}$$

$$z = e^{i(2n+1)\pi/4},\ n = 0, 1, 2, 3$$

These poles are simple poles.

$$z = e^{i\pi/4},\ e^{i3\pi/4},\ e^{i5\pi/4},\ e^{7\pi i/4}$$

but only $z = e^{i\pi/4}, e^{3\pi i/4}$ poles lies inside upper half of plane 'C'.

∴ Residue of $f(z)$

$$\text{at } z = e^{i\pi/4} = \lim_{z \to e^{i\pi/4}} \frac{1}{4z^3}$$

$$= \lim_{z \to e^{i\pi/4}} \frac{z}{4z^4} = \frac{e^{i\pi/4}}{4(-1)}$$

Similarly residue of $f(z)$

$$\text{at } z = e^{3\pi i/4} = \lim_{z \to e^{3\pi i/4}} \frac{1}{4z^3}$$

$$= \frac{e^{i3\pi/4}}{4(-1)}$$

∴ By Cauchy's residue theorem

$$\oint_C f(z)dz = 2\pi i\left[\text{Res } e^{i\pi/4} + \text{Res } e^{3\pi i/4}\right]$$

$$= 2\pi i\left[\frac{e^{i\pi/4}}{-4} + \frac{e^{3\pi i/4}}{-4}\right]$$

$$= \frac{-\pi i}{2}\left[\cos\frac{\pi}{4} + i\sin\frac{\pi}{4} + \cos\frac{3\pi}{4} + i\sin\frac{3\pi}{4}\right]$$

$$= \frac{-\pi i}{2}\left[\frac{1}{\sqrt{2}} + i\frac{1}{\sqrt{2}} - \frac{1}{\sqrt{2}} + \frac{i}{\sqrt{2}}\right] = \frac{\pi}{\sqrt{2}}$$

$$\therefore \quad \oint_C f(z)dz = \oint_{C_R} f(z)dz + \int_{-R}^{R} f(x)dx = \frac{\pi}{\sqrt{2}}$$

$$\text{as } R \to \infty \quad \left|\int_{C_R} \frac{dz}{z^4+1}\right| \le \left|\int_0^{\pi} \frac{Rd\theta}{R^4+1}\right| \to 0 \text{ as } R \to \infty$$

$$\text{as } R \to \infty \quad \int_{-R}^{R} f(x)dx = \int_{-\infty}^{\infty} f(x)dx$$

Hence,

$$\oint_C f(z)dz = \oint_{C_R} f(z)dz + \int_{-\infty}^{\infty} f(x)dx$$

$$\Rightarrow \quad \int_{-\infty}^{\infty} \frac{1}{x^4+1}dx = \frac{\pi}{\sqrt{2}}$$

4.4 Use the residue theorem to evaluate the integral $\int_C \frac{e^z - 1}{z(z-1)^2(z-i)}dz$, where C is the circle $|z| = 3$.

[10 marks : 2022]

Solution:

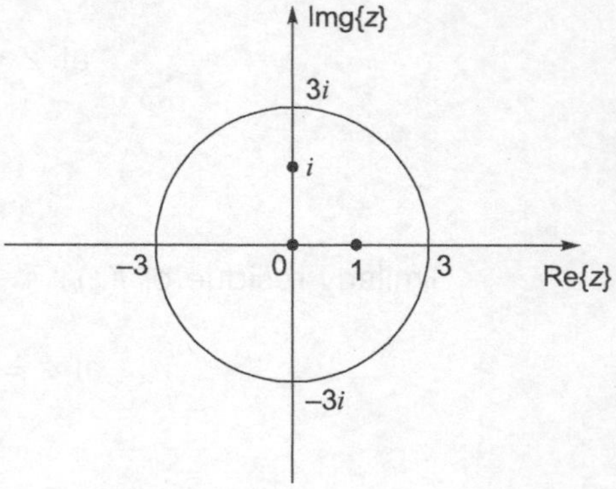

Given : Integral $$I = \int_C \frac{e^z - 1}{z(z-1)^2(z-i)}, \; |z| = 3$$

Singularity points are at $z = 0, 1, i$ and all lie inside the contour.

Residue at $z = 0$, i.e., R_1

$$R_1 = \lim_{z \to 0} z f(z)$$

where, $$f(z) = \frac{e^z - 1}{z(z-1)^2(z-i)}$$

$$\Rightarrow \quad R_1 = \lim_{z \to 0} z \cdot \frac{e^z - 1}{z(z-1)^2(z-i)}$$

$$\Rightarrow \quad R_1 = 0$$

Residue at $z = 1$, i.e., R_2

$$R_2 = \frac{1}{(m-1)!} \lim_{z \to z_0} \frac{d^{m-1}}{dz^{m-1}} (z - z_0)^m f(z)$$

where m is the order of singularity.

$$\Rightarrow \quad R_2 = \frac{1}{(2-1)!} \lim_{z \to 1} \frac{d^{2-1}}{dz^{2-1}} (z-1)^2 \frac{(e^z - 1)}{z(z-1)^2(z-i)}$$

$$\Rightarrow \quad R_2 = \lim_{z \to 1} \frac{d}{dz} \left[\frac{e^z - 1}{z^2 - iz} \right]$$

$$= \frac{e^z(z^2 - iz) - (e^z - 1)(2z - i)}{(z^2 - iz)^2}$$

$$R_2 = \frac{e(1-i) - (e-1)(2-i)}{(1-i)^2}$$

$$\Rightarrow \quad R_2 = 0.5 - 0.359i$$

Residue for $z = i$, i.e., R_3

$$R_3 = \lim_{z \to i} (z - i) f(z)$$

$$R_3 = \lim_{z \to i} (z - i) \frac{e^z - 1}{z(z-1)^2(z-i)}$$

$$\Rightarrow \quad R_3 = \frac{1}{2}[e^i - 1]$$

So, from Cauchy's Integral formula,

$$I = 2\pi i [R_1 + R_2 + R_3]$$

$$\Rightarrow \quad I = 2\pi i [0.5 - 0.359i + 0.5e^i - 0.5]$$

$$\Rightarrow \quad I = 2\pi i [0.5 \cos 1 + i\, 0.5 \sin 1 - 0.359i]$$

$$I = -0.388 + 1.6974i$$

4.5 Compute the following integral by residue theorem : $\int_0^{2\pi} \frac{\sin\theta}{3-2\sin\theta} d\theta$

[10 marks : 2023]

Solution:

Given : $\int_0^{2\pi} \frac{\sin\theta}{3-2\sin\theta} \cdot d\theta$

$$\because \quad \sin\theta = \frac{e^{i\theta}-e^{-i\theta}}{2i}$$

$$= \int_0^{2\pi} \frac{e^{i\theta}-e^{-i\theta}}{6i-2(e^{i\theta}-e^{-i\theta})} d\theta$$

$$\because \quad e^{i\theta} = z$$

$$ie^{i\theta}.d\theta = dz$$

$$d\theta = \frac{dz}{z \cdot i}$$

Integral,

$$I = \int_C \frac{\left(z-\frac{1}{z}\right)}{6i-2\left(z-\frac{1}{z}\right)} \cdot \frac{dz}{zi}$$

$$I = \frac{1}{i}\int_C \frac{(z^2-1)}{6zi-2(z^2-1)} \cdot \frac{dz}{z}$$

$$= \frac{1}{i}\int_C f(z)dz \text{ where } C \text{ is the unit circle } |z| = 1$$

$$\because \quad f(z) = \frac{-(z^2-1)}{z(2z^2-6zi-2)}$$

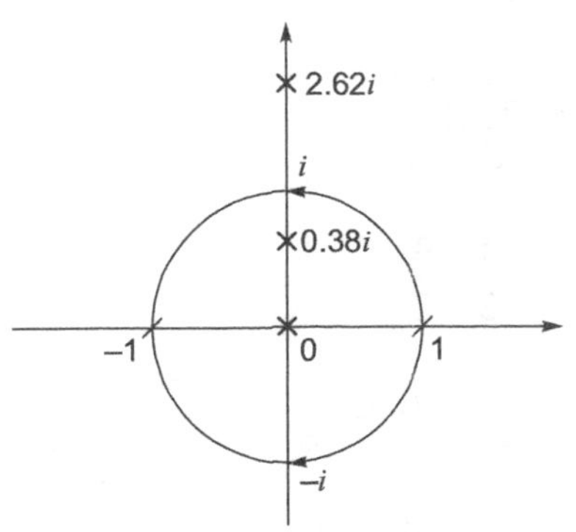

Poles of $f(z)$ are $z = 0, 0.38i, 2.6zi$. So, here only two poles lies inside the unity circle.

Residues,

$$\text{Res}(0) = \frac{-1}{2}$$

$$\text{Res}(0.38i) = \frac{-((0.38i)^2-1)}{0.38i(0.38i-2.62i)}$$

$$= \frac{1.144}{0.38i(-2.24i)}$$

$$= \frac{1.144}{0.85}$$

$$\text{Res}(0.38i) = 1.344$$

$$I = \frac{1}{i}\left(2\pi i \times \left(-\frac{1}{2} + 1.344\right)\right)$$

$$= 2\pi \times 0.844$$

$$I = 5.303$$

5. Probability and Statistics

5.1 Let the probability density function of a random variable X be given as:

$$f(x) = \frac{5}{3}e^{-5x}\,u(x) + \beta e^{2x}\,u(-x)$$

where β is a constant and $u(x)$ is the unit step function.

Calculate:

(i) The value β

(ii) Mean of X

(iii) Variance of X

[3 + 7 + 10 = 20 marks : 2018]

Solution:

$$f(x) = \frac{5}{3}e^{-5x}u(x) + \beta e^{2x}u(-x)$$

(i) Given that, $f(x)$ is a pdf.

$$\therefore \int_{-\infty}^{\infty} f(x)dx = 1$$

$$\therefore \int_0^{\infty} \frac{5}{3}e^{-5x}dx + \int_{-\infty}^{0} \beta e^{2x}dx = 1$$

$$\frac{5}{3}\left[-\frac{e^{-5x}}{5}\right]_0^{\infty} + \beta\left[\frac{e^{2x}}{2}\right]_{-\infty}^{0} = 1$$

$$\frac{1}{3} + \frac{\beta}{2} = 1$$

$$\beta = \frac{4}{3}$$

(ii) Mean of x = $E(x) = \int_{-\infty}^{\infty} xf(x)dx$

$$E(x) = \int_0^{\infty} \frac{5}{3}x.e^{-5x}dx + \int_{-\infty}^{0} \frac{4}{3}x.e^{2x}dx$$

$$E(x) = \frac{5}{3}\left[-\frac{xe^{-5x}}{5} - \frac{e^{-5x}}{25}\right]_0^{\infty} + \frac{4}{3}\left[\frac{xe^{2x}}{2} - \frac{e^{2x}}{4}\right]_{-\infty}^{0}$$

$$E(x) = \frac{5}{3}\left[-\frac{xe^{-5x}}{5} - \frac{e^{-5x}}{25}\right]_0^{\infty} + \frac{4}{3}\left[\frac{xe^{2x}}{2} - \frac{e^{2x}}{4}\right]_{-\infty}^{0} = -0.267$$

(iii) Variance of $x = \sigma^2 = E(x^2) - [E(x)]^2$

$$E(x^2) = \int_{-\infty}^{\infty} x^2 f(x)dx$$

$$E(x^2) = \int_0^{\infty} \frac{5}{3}x^2e^{-5x}dx + \int_{-\infty}^{0} \frac{4}{3}x^2e^{2x}dx$$

$$E(x^2) = \frac{5}{3}\left[\frac{-x^2e^{-5x}}{5} - \frac{2xe^{-5x}}{25} - \frac{2e^{-5x}}{125}\right]_0^{\infty} + \frac{4}{3}\left[\frac{x^2e^{2x}}{2} - \frac{2xe^{2x}}{4} + \frac{2e^{2x}}{8}\right]_{-\infty}^{0}$$

$$E(x^2) = \frac{5}{3}\left[+\frac{2}{125}\right] + \frac{4}{3}\left[\frac{2}{8}\right] = 0.36$$

$\therefore$

$$\text{Variance} = \sigma^2 = E(x^2) - (E(x))^2$$

$$\sigma^2 = 0.36 - (-0.267)^2 = 0.2887$$

$$= \text{Variance of } x$$

5.2 Prove that the most probable value is the mean value.

[6 marks : 2018]

Solution:

The deviation $\delta_i = x_i - x_m$, where x_m the most probable (best) value.

$$\delta_i^2 = x_i^2 + x_m^2 - 2x_ix_m$$

$$\sum_{i=1}^{m} \delta_i^2 = \sum_{i=1}^{n} x_i^2 + nx_m^2 - 2x_m\sum_{i=1}^{n} x_i \quad \ldots(i)$$

For the most probable value sum of the squares of deviations should be minimum, so that we gets,

$$\frac{d}{dx_m}\left[\sum \delta_i\right] = 0$$

$$2nx_m - 2\sum_{i=1}^{n} x_i = 0$$

$$x_m = \frac{\sum_{i=1}^{n} x_i}{n} = \bar{x} \quad \text{(proved)}$$

5.3 **(i) If the density function of a continuous random variable is given by**

$$f(x) = \begin{cases} 0, & x < 0 \\ ax, & 0 \le x \le 2 \\ (4-x)a, & 2 \le x \le 4 \\ 0 & x > 4 \end{cases}$$

(p) Find value of *a*.

(q) Find the cumulative distribution function (cdf).

(r) Find $P(X > 2.5)$.

(ii) Compute $\oint_C \frac{\cos z}{z^2(z-\pi)^3} dz$, where $C : |z| = 4$.

[10 + 10 = 20 marks : 2020]

Solution:

(i) Given probability density function of a continuous random variable

$$f(x) = \begin{cases} 0, & x < 0 \\ ax, & 0 < x \le 2 \\ (4-x)a, & 2 \le x \le 4 \\ 0, & x > 4 \end{cases}$$

(p) Find value of a?

Probability density function is a valid only if $\int_{-\infty}^{\infty} f(x)dx = 1$

$$\therefore \quad \int_{-\infty}^{\infty} f(x)dx = \int_{-\infty}^{0} 0.dx + \int_{0}^{2} axdx + \int_{2}^{4} a(4-x)dx + \int_{4}^{\infty} 0dx$$

$$= \left(0 + a\frac{x^2}{2}\right)_0^2 + a\left(4x - \frac{x^2}{2}\right)_0^4 + 0 = 1$$

$$\Rightarrow \quad 2a + 8a - 6a = 1$$

$$\Rightarrow \quad 4a = 1$$

$$\Rightarrow \quad a = \frac{1}{4}$$

$$f(x) = \begin{cases} 0, & x < 0 \\ \frac{x}{4}, & 0 \le x \le 2 \\ 1 - \frac{x}{4} & 2 \le x \le 4 \\ 0, & x > 4 \end{cases}$$

(q) $$F(x_K) = \int_{-\infty}^{x_K} f(x)dx$$

C.D.F. is obtained by integrating $f(x)$ within the interval.

$$\therefore \quad F(x) = \begin{cases} 0, & x < 0 \\ \frac{x^2}{8}, & 0 \le x \le 2 \\ x - \frac{x^2}{8}, & 2 \le x \le 4 \\ 0, & x > 4 \end{cases} \quad \text{is C.D.F.}$$

(r) $$P(X > 2.5) = \int_{2.5}^{\infty} f(X)dx$$

$$= \int_{2.5}^{4}\left(1-\frac{x}{4}\right)dx + \int_{4}^{\infty} 0.dx = x - \frac{x^2}{8}\Bigg]_{2.5}^{4}$$

$$= \left(4-\frac{16}{8}\right)-\left(2.5-\frac{(2.5)^2}{8}\right)$$

$$= 0.281$$

(ii) $\oint_C \frac{\cos z}{z^2(z-\pi)^3}dz,\ C: |z| = 4$

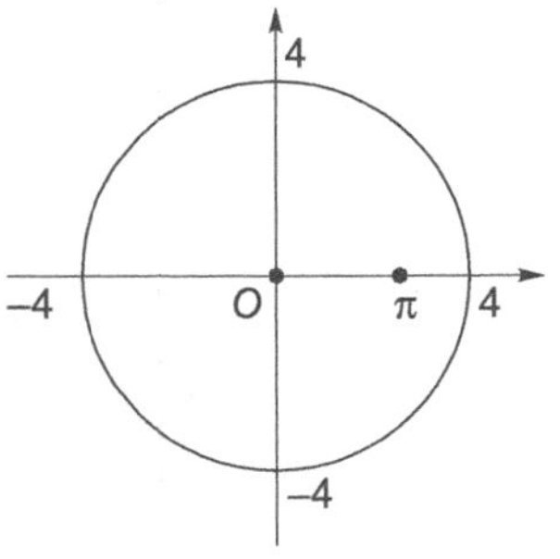

Singularities of $f(z)$ are given by

$$z^2(z-\pi)^3 = 0$$

$\Rightarrow z = 0, \pi$ both lies inside $C: |z| = 4$

Here, $z = 0$ is pole of order $m = 2$.

$z = \pi$ is pole of order $m = 3$.

By Cauchy Residue Theorem :

$$\oint_C f(z)dz = 2\pi i[\text{Residue at } 0 + \text{Residue at } \pi]$$

Residue at $z = 0$:

$$= \frac{1}{(2-1)!}\lim_{z\to 0}\frac{d}{dz}\left[(z-0)^2\frac{\cos z}{z^2(z-\pi)^3}\right] = \lim_{z\to 0}\frac{d}{dz}\left[\frac{\cos z}{(z-\pi)^3}\right]$$

$$= \lim_{z\to 0}\frac{(z-\pi)^3(-\sin z)-\cos z(3(z-\pi)^2}{(z-\pi)^6}$$

$$= \frac{-3\pi^2}{\pi^6} = \frac{-3}{\pi^4}$$

Residue at $z = \pi$:

$$= \frac{1}{(3-1)!}\lim_{z\to\pi}\frac{d^2}{dz^2}\left[(z-\pi)^3\frac{\cos z}{z^2(z-\pi)^3}\right]$$

$$= \frac{1}{2}\lim_{z\to\pi}\frac{d}{dz}\left[\frac{z^2(-\sin z)-2z\cos z}{z^4}\right]$$

$$= \frac{1}{2}\left[\frac{\pi^2-6}{\pi^4}\right]$$

$\therefore$ by Cauchy Residue Theorem

$$\oint_C f(z)dz = 2\pi i[\text{Residue at } 0 + \text{Residue at } \pi]$$

$$= 2\pi i\left[\frac{-3}{\pi^4}+\frac{1}{2}\cdot\frac{\pi^2-6}{\pi^4}\right]$$

$$= 2\pi i\left[\frac{-6}{\pi^4}+\frac{1}{2\pi^2}\right]$$

$$= 2i\left[\frac{-6}{\pi^3}+\frac{1}{2\pi}\right]$$

5.4 **The average number of mistakes commited by a typist is 3 per page. Find the probability that the typist commits**
(i) no mistake
(ii) at least two mistakes per page

[12 marks : 2021]

Solution:

Let X = Number of mistakes commited by typist per page and follows Poisson distribution with mean = λ = 3 per page (given)

(i) Probability that typist commits no mistakes,

$$P(X=0) = \frac{e^{-\lambda}\lambda^0}{\lfloor 0} = \frac{e^{-3}\cdot 3^0}{\lfloor 0} = e^{-3} = 0.049$$

(ii) Probability that at least two mistakes per page

$$= P(X \geq 2)$$
$$= 1 - P[X < 2]$$
$$= 1 - [P(X=0) + P(X=1)]$$
$$= 1 - \left[\frac{e^{-\lambda}\lambda^0}{\lfloor 0} + \frac{e^{-\lambda}\lambda^1}{\lfloor 1}\right]$$
$$= 1 - [e^{-3} + e^{-3}(3)]$$
$$= 1 - 4e^{-3} = 0.80$$

5.5 **(i) Obtain the equations of regression lines and the correlation coefficient for the following data. Justify the sign of correlation coefficient.**

x	1	2	3	4	5
y	2	5	3	8	7

(ii) The probability density function of a random variable x is

$$f(x) = \begin{cases} \alpha x(2-x), & 0 \leq x \leq 2 \\ 0, & \text{otherwise} \end{cases}$$

for some real number α. Find the value of α and the variance of x.

[10 + 10 marks : 2022]

Solution:

(i) Given data :

X	Y	$(X-\bar{X})^2$	$(Y-\bar{Y})^2$	$(X-\bar{X})(Y-\bar{Y})$
1	2	4	9	6
2	5	1	0	0
3	3	0	4	0
4	8	1	9	3
5	7	4	4	4
$\Sigma X_i = 15$	$\Sigma Y_i = 25$	$\Sigma(X-\bar{X})^2 = 10$	$\Sigma(Y-\bar{Y})^2 = 26$	13

$$\bar{X} = \frac{\Sigma x_i}{n} = \frac{15}{5} = 3$$

$$\bar{Y} = \frac{\Sigma y_i}{n} = \frac{25}{5} = 5$$

$$\sigma_X^2 = \frac{\Sigma(X-\bar{X})^2}{n} = \frac{10}{5} = 2$$

$$\sigma_Y^2 = \frac{\Sigma(Y-\bar{Y})^2}{n} = \frac{26}{5} = 5.2$$

Co-variance, $(X, Y) = \dfrac{\Sigma(X-\bar{X})(Y-\bar{Y})}{\eta} = \dfrac{13}{5} = 2.6$

$\therefore$ $\gamma = \dfrac{\text{Co-variance}(X,Y)}{\sigma_X \sigma_Y} = \dfrac{2.6}{\sqrt{2}\sqrt{5.2}} = 0.806$

Regression coefficient of Y on X

$$= b_{yx} = \frac{\gamma\sigma_y}{\sigma_x}$$

$$= \frac{0.806 \times \sqrt{5.2}}{\sqrt{2}} = 1.3$$

$\therefore$ Line of regression of Y on X is

$$(y-\bar{y}) = b_{yx}(x-\bar{x})$$

$$(y-5) = 1.3(x-3)$$

$\Rightarrow$ $1.3x - y + 1.1 = 0$ ...(1)

and line of regression of X on Y is

$$(x-\bar{x}) = b_{xy}(y-\bar{y})$$

$$(x-3) = \frac{\gamma\sigma_x}{\sigma_y}(y-5)$$

$$(x-3) = \frac{0.806 \times \sqrt{2}}{\sqrt{5.2}}(y-5)$$

$$(x-3) = 0.5(y-5)$$

$\Rightarrow$ $x - \dfrac{y}{2} - \dfrac{1}{2} = 0$ ...(2)

Both (1), (2) are of regression lines.

$\because$ Co-variance(X, Y) is positive, so that γ is also positive.

(ii) $$f(x) = \begin{cases} \alpha x(2-x), & 0 \le x \le 2 \\ 0, & \text{otherwise} \end{cases}$$

As $f(x)$ is probability density function

$$\int_{-\alpha}^{\alpha} f(x)dx = 1$$

Now, $$\int_{x=0}^{x=2} f(x)dx = 1$$

$$\int_{x=0}^{x=2} \alpha x(2-x)dx = 1$$

$$\int_{x=0}^{x=2} (2\alpha x - \alpha x^2) = 1$$

$$\left[2\alpha\frac{x^2}{2} - \alpha\frac{x^3}{3}\right]_{x=0}^{x=2} = 1$$

$$4\alpha - \frac{8\alpha}{3} = 1$$

$$\alpha = \frac{3}{4} = 0.75$$

$$f(x) = \frac{3}{4}x(2-x) = \frac{3}{2}x - \frac{3}{4}x^2$$

$$\text{Mean} = E(x) = \int_{x=0}^{x=2} xf(x)dx$$

$$= \int_{x=0}^{x=2} x\left(\frac{3}{2}x - \frac{3}{4}x^2\right)dx = \int_{x=0}^{x=2}\left(\frac{3}{2}x^2 - \frac{3}{4}x^3\right)dx$$

$$= \left[\frac{3}{2}\frac{x^3}{3} - \frac{3}{4}\times\frac{x^4}{4}\right]_0^2$$

$$= 1$$

$$E(x^2) = \int_{x=0}^{x=2} x^2 f(x)dx$$

$$= \int_{x=0}^{x=2} x^2\left(\frac{3}{2}x - \frac{3}{4}x^2\right)dx$$

$$= \int_{x=0}^{x=2}\left(\frac{3}{2}x^3 - \frac{3}{4}x^4\right)dx$$

$$= \left[\frac{3}{2}\times\frac{x^4}{4} - \frac{3}{4}\times\frac{x^5}{5}\right]_{x=0}^{x=2}$$

$$= 1.2$$

$$\text{Variance} = E(x^2) - [E(x)]^2$$

$$= 1.2 - (1)^2 = 0.2$$

5.6 (i) Suppose that X and Y are independent random variables having the common density function

$$f(x) = \begin{cases} e^{-x}, & x > 0 \\ 0, & \text{otherwise} \end{cases}$$

Find the density function of the random variable $X|Y$.

(ii) A root of the equation $xe^x - 1 = 0$ lies in the interval (0.5, 1.0). Determine this root correct to three decimal places using regula-falsi method. First find how many least decimal digits are required for three decimal places accuracy.

[8 + 12 marks : 2023]

Solution:

(i) Let $Z = \frac{X}{Y}$

Find probability distribution function of Z if X, Y are two R.V. and both are exponentially distributed with mean 1.

Assume $Y = y$ then, $Z = \frac{X}{y}$ is a scaled of X.

$$\therefore \qquad f_z\left(\frac{Z}{y}\right) = |y| f_X\left(\frac{YZ}{y}\right)$$

$\therefore$ Probability distribution function of Z is given by

$$f_Z(Z) = \int_{-\infty}^{\infty} |y| f_X\left(\frac{YZ}{y}\right) f_Y(Y) dy$$

$$= \int_{-\infty}^{\infty} |y| f_{XY}(YZ, y) dy$$

Now X, Y are independent and exponentially distributed with mean 1.

$$\therefore \qquad f_Z(Z) = \int_0^{\infty} y f_X(YZ) f_Y(Y) dy \qquad (Z > 0)$$

$$= \int_0^{\infty} y e^{-yz} e^{-y} dy = \int_0^{\infty} e^{-(1+Z)y} y dy$$

Let $1 + Z = K$

$$= \int_0^{\infty} e^{-Ky} y dy$$

Let $ky = t$, $\qquad dy = \frac{1}{k} dt$

$$= \int_0^{\infty} e^{-t} \frac{t}{k} \cdot \frac{1}{k} dt$$

$$= \frac{1}{k^2} \int_0^{\infty} e^{-t} t dt$$

$$= \frac{\lceil 2}{k^2}$$

$$f_Z(Z) = \frac{1}{(Z+1)^2} \qquad (\text{for } Z > 0)$$

(ii) Given : $\qquad xe^x - 1 = 0$

We can write $\qquad f(x) = x^{ex} - 1 = 0$

$$I_o = [0.5, 1.0]$$

$$f(0.5) = 0.5e^{0.5} - 1 = -0.1756$$

$$f(1) = 1e - 1 = 1.71828$$

By Regula-Falsi :

$$x_1 = \frac{af(b) - bf(a)}{f(b) - f(a)}$$

$$= \frac{0.5(1.71828) - 1(-0.1756)}{1.71828 - (-0.1756)}$$

$$= 0.546369$$

$$f(x_1) = x_1 e^{x_1} - 1 = -0.056436 < 0$$

$\because$ $f(1) > 0$ and $f(x_1) < 0$

Then root of $f(x) = 0$ lies in $[x_1, 1]$

2nd approximation $$x_2 = \frac{x_1 f(b) - b f(x_1)}{f(b) - f(x_1)} = 0.560794$$

$$f(x_2) = x_2 e^{x_2} - 1 = -0.01745 < 0$$

$\because$ $f(x_2) < 0$ and $f(b) > 0$

$\therefore$ Root lies in $[x_2, b]$

3rd approximation $$x_3 = \frac{x_2 f(b) - b f(x_2)}{f(b) - f(x_2)} = 0.565210$$

$\therefore$ $$f(x_3) = x_3 e^{x_3} - 1 = -0.005336 < 0$$

$\because$ $f(x_3) < 0$ and $f(b) > 0$

Root lies in $[x_3, b]$

4th approximation $$x_4 = \frac{x_3 f(b) - b f(x_3)}{f(b) - f(x_3)} = 0.566555$$

$$f(x_4) = x_4 e^{x_4} - 1 = -0.0016247 < 0$$

$\therefore$ Root lies in $[x_4, b]$

5th approximation $$x_5 = \frac{x_4 f(b) - b f(x_4)}{f(b) - f(x_4)} = 0.566964$$

$\because$ The value of x repeating upto 3 decimals.

$\therefore$ The root of $f(x) = 0$ converges to $x = 0.567$.

To get accuracy upto 3 decimals, we need to consider 5 decimals.

Hence, the minimum number of decimals requried to get accuracy upto 3 decimals is 5.

6. Numerical Method

6.1 **Find $y(0.75)$ by applying Euler's method to the initial value problem $\frac{dy}{dx} = 2\cos\left(\frac{4\pi x}{3}\right) - y$, $y(0) = 1$. Take step size $h = 0.25$.**

[12 marks : 2017]

Solution:

$$\frac{dy}{dx} = 2\cos\left(\frac{4\pi x}{3}\right) - y$$

$$y(0) = 1, \quad h = 0.25$$

$$f(x, y) = 2\cos\left(\frac{4\pi x}{3}\right) - y$$

$$y_0 = 1, \quad x_0 = 0, \quad h = 0.25$$

By Euler's method,

I. $$y_1 = y_0 + hf(x_0, y_0)$$

$$= y_0 + h\left(2\cos\left(\frac{4\pi x_0}{3}\right) - y_0\right) = 1 + 0.25(2\cos(0) - 1) = 1.25$$

$$x_1 = x_0 + h = 0 + 0.25 = 0.25$$

II. $$y_2 = y_1 + hf(x_1, y_1)$$

$$= y_1 + h\left(2\cos\left(\frac{4\pi x_1}{3}\right) - y_1\right)$$

$$= 1.25 + 0.25\left(2\cos\left(\frac{4\pi(0.25)}{3}\right) - 1.25\right) = 1.1875$$

$$x_2 = x_1 + h = 0.25 + 0.25 = 0.50$$

III. $$y_3 = y_2 + hf(x_2, y_2)$$

$$= y_2 + h\left(2\cos\left(\frac{4\pi x_2}{3}\right) - y_2\right)$$

$$= 1.1875 + 0.25\left(2\cos\left(\frac{4\pi(0.5)}{3}\right) - 1.1875\right)$$

$$= 1.1875 + 0.25(2(-0.5) - 1.1875)$$

$$= 1.1875 + 0.25(-2.1875) = 0.64062$$

$$x_3 = x_2 + h = 0.5 + 0.25 = 0.75$$

$\therefore$ $$y(0.75) = 0.64062$$

6.2 **Find the root of $xe^x - 1 = 0$ correct to four decimal places by the Newton-Raphson method starting with $x_0 = 0.6$.**

[10 marks : 2022]

Solution:

Given, Equation, $f(x) = xe^x - 1 = 0$

and $x_0 = 0.6$

For Newton-Raphson method,

$$x_{n+1} = x_n - \frac{f(x_n)}{f'(x_n)}$$

So, $$x_1 = x_0 - \frac{f(x_0)}{f'(x_0)}$$

$\Rightarrow$ $$f'(x_0) = e^x + xe^x = (x+1)e^x$$

$\Rightarrow$ $$x_1 = x_0 - \frac{(x_0 e^{x_0} 1)}{(x_0 + 1)e^{x_0}}$$

$\Rightarrow$ $$x_1 = 0.5680$$

Similarly, $$x_2 = x_1 - \frac{(x_1 e^{x_1} - 1)}{(x_1 + 1)e^{x_1}}$$

$\Rightarrow$ $$x_2 = 0.56714$$

Similarly, $$x_3 = x_2 - \frac{(x_2 e^{x_2} - 1)}{(x_2 + 1)e^{x_2}}$$

$\Rightarrow$ $$x_3 = 0.56714$$

Hence, the root of the equation correct upto 4 digit is 0.567.

7. Transform Theory

7.1 Show that in the interval (0, 1), $\cos \pi x = \frac{8}{\pi}\sum_{n=1}^{\infty}\frac{n}{4n^2-1}\cdot \sin 2n\pi x$.

[20 marks : 2018]

Solution:

Given, $f(x) = \cos \pi x$

Half range sine series of $f(x)$ in interval (0, 1)

$$f(x) = \sum_{n=1}^{\infty} b_n \sin n\pi x \text{ where } b_n = 2\int_0^1 f(x)\sin n\pi x dx$$

$$b_n = 2\int_0^1 \cos\pi x \sin n\pi\, x dx$$

$$b_n = \int_0^1 [\sin(n\pi x + \pi x) + \sin(n\pi x - \pi x)]dx$$

$$b_n = \left[\frac{-\cos(n+1)\pi x}{(n+1)\pi}\right]_0^1 + \left[\frac{-\cos(n-1)\pi x}{(n-1)\pi}\right]_0^1$$

$$= \frac{(-1)^n}{\pi}\left[\frac{2n}{n^2-1}\right] + \frac{1}{\pi}\left[\frac{2n}{n^2-1}\right] \quad \text{where } n = 1, 2, 3 \ldots$$

$$= \frac{2n}{(n^2-1)\pi}\left[(-1)^n + 1\right]$$

$$= \frac{4n}{(n^2-1)\pi} \qquad \text{(if } n = \text{even)}$$

$$= 0 \qquad \text{(if n odd)}$$

$$n = 2p \text{ where } p = 1, 2, 3 \ldots$$

Hence, $$b_n = \frac{8P}{(4P^2-1)\pi} \qquad \ldots \text{(i)}$$

Substituting value of (i) in $f(x)$

$$f(x) = \sum_{p=1}^{\infty}\frac{8p}{(4p^2-1)\pi}\sin(2p\pi x)$$

or, $$f(x) = \frac{8}{\pi}\sum_{n=1}^{\infty}\frac{n}{(4n^2-1)}\sin 2n\pi x$$

7.2 State Dirichlet's conditions for existence of Fourier series of a function. Determine the half range Fourier cosine series of

$$f(x) = \begin{cases} x, & 0 < x < \frac{\pi}{2} \\ \pi - x, & \frac{\pi}{2} \le x < \pi \end{cases}$$

[10 marks : 2020]

Solution:

Dirichlet Conditions:

If a signal $x(t)$ satisfies certain conditions, its Fourier series is guaranteed to converge point-wise at all points where $x(t)$ is continuous. These conditions are known as Dirichlet conditions.

(i) The signal $x(t)$ must be absolutely integrable over the range of time-period,

$$\int_0^T |x(t)|\,dt < \infty$$

(ii) It should have finite number of finite size discontinuities in one period.

(iii) It should have finite number of maxima and minima in one period.

Comment: Condition (i) is necessary whereas (ii) and (iii) are sufficient conditions.

Given: $C = \pi$

$$a_0 = \frac{2}{C}\int_0^C f(x)\cdot dx = \frac{2}{\pi}\int_0^C f(x)\cdot dx$$

$$= \frac{2}{\pi}\int_0^\pi f(x)\cdot dx = \frac{2}{\pi}\left[\int_0^{\pi/2} x\cdot dx + \int_{\pi/2}^{\pi} (\pi - x)dx\right]$$

$$= \frac{2}{\pi}\left[\left(\frac{x^2}{2}\right)_0^{\pi/2} + \left(\pi x - \frac{x^2}{2}\right)_{\pi/2}^{\pi}\right]$$

$$= \frac{2}{\pi}\left[\frac{\pi^2}{8} + \left(\pi^2 - \frac{\pi^2}{2}\right) - \left(\frac{\pi^2}{2} - \frac{\pi^2}{8}\right)\right]$$

$$a_0 = \frac{2}{\pi}\left[\frac{\pi^2}{8} + \pi^2 - \frac{\pi^2}{2} - \frac{\pi^2}{2} + \frac{\pi^2}{8}\right]$$

$$a_0 = \frac{2}{\pi}\left[\frac{2\pi^2}{8}\right] = \frac{\pi}{2}$$

$$a_0 = \frac{\pi}{2}$$

$$a_n = \frac{2}{C}\int_0^C f(x)\cdot \cos\frac{n\pi x}{C}\cdot dx$$

$$a_n = \frac{2}{\pi}\int_0^\pi f(x)\cdot \cos nx\cdot dx$$

$$a_n = \frac{2}{\pi}\left[\int_0^{\pi/2} x\cdot\cos nx\cdot dx + \int_{\pi/2}^{\pi} (\pi - x)\cdot\cos nx\cdot dx\right]$$

$$a_n = \frac{2}{\pi}\left[\left[\left(x\frac{\sin nx}{n} + \frac{\cos nx}{n^2}\right)\right]_0^{\pi/2} + \left[(\pi - x)\frac{\sin nx}{n} - \frac{\cos nx}{n^2}\right]_{\pi/2}^{\pi}\right]$$

$$a_n = \frac{2}{\pi}\left[\left[\frac{\pi}{2}\cdot\frac{\sin n\frac{\pi}{2}}{n}+\frac{\cos n\frac{\pi}{2}}{n^2}-\frac{1}{n^2}\right]+\left[-\frac{\cos n\pi}{n^2}-\frac{\pi}{2}\frac{\sin\frac{n\pi}{2}}{n}+\frac{\cos\frac{n\pi}{2}}{n^2}\right]\right]$$

$$a_n = \frac{2}{\pi}\left[\frac{1}{n^2}\left(2\cos\frac{n\pi}{2}-\cos n\pi-1\right)\right]$$

∴ Fourier half range cosine series is

$$= \frac{a_0}{2}+\sum_{n=1}^{\infty}a_n\cos\left(\frac{n\pi x}{c}\right)$$

$$= \frac{a_0}{2}+\sum_{n=1}^{\infty}a_n\cos(nx)$$

$$f(x) = \frac{\pi}{4}+\sum_{n=1}^{\infty}\frac{2}{\pi n^2}\left(2\cos\frac{n\pi}{2}-\cos n\pi-1\right)\cdot\cos nx$$

7.3 **(i) Explain $f(x) = e^{-x}$, $-\pi < x < \pi$ in a complex Fourier series.**

(ii) Find the frequency spectrum of the periodic square wave given by the extension of the function

$$f(x) = \begin{cases} 0, & -\frac{1}{2} < x < -\frac{1}{4} \\ 1, & -\frac{1}{4} < x < \frac{1}{4} \\ 0, & \frac{1}{4} < x < \frac{1}{2} \end{cases}$$

[10+10 = 20 marks : 2021]

Solution:

(i) Given : $f(x) = e^{-x}$, $-\pi < x < \pi$ with fundamental time period $T_o = 2\pi$

The complex exponential fourier series is given by

$$f(x) = \sum_{n=-\infty}^{\infty}C_n e^{jn\omega_o x} \quad ...(i)$$

Now,

$$C_n = \frac{1}{T_o}\int_{-T_o/2}^{T_o/2} f(x)\cdot e^{-jn\omega_o x}dx$$

$$= \frac{1}{2\pi}\int_{-\pi}^{\pi} e^{-x}\cdot e^{-jn\omega_o x}$$

$$= \frac{1}{2\pi}\int_{-\pi}^{\pi} e^{-(1+jn\omega_o)x}dx \qquad \left(\omega_o = \frac{2\pi}{T_o} = \frac{2\pi}{2\pi} = 1\right)$$

$$= \frac{1}{2\pi}\int_{-\pi}^{\pi} e^{-(1+jn)x}dx$$

$$= \frac{1}{2\pi}\frac{\left[e^{-(1+jn)x}\right]_{-\pi}^{\pi}}{-(1+jn)}$$

$$= \frac{1}{2\pi}\cdot\frac{e^{-(1+jn)\pi} - e^{(1+jn)\pi}}{-(1+jn)}$$

$$\Rightarrow \qquad C_n = \frac{1}{2\pi(1+jn)}\left[e^{(1+jn)\pi} - e^{-(1+jn)\pi}\right]$$

$$= \frac{1}{2\pi(1+jn)}\cdot 2\sinh(1+jn)\pi \qquad \left(\because \sinh\theta = \frac{e^{\theta} - e^{-\theta}}{2}\right)$$

From (i),

$$f(x) = \sum_{n=-\infty}^{\infty} \frac{2\sinh(1+jn)\pi}{2\pi(1+jn)}\cdot e^{jn\omega_o x}$$

$\Rightarrow$ Complex fourier series,

$$f(x) = \frac{1}{\pi}\sum_{n=-\infty}^{\infty} \frac{\sinh(1+jn)\pi}{(1+jn)}\cdot e^{jnx} \qquad [\because \omega_o = 1]$$

(ii)

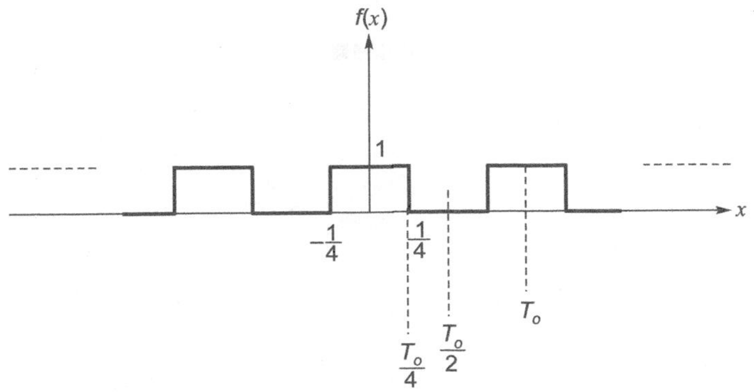

where $T_o = 1$

The frequency-spectrum of the above periodic waveform is given by

$$F(\omega) = 2\pi\sum_{n=-\infty}^{\infty} C_n\delta(\omega - n\omega_o) \qquad ...(i)$$

where,

$$C_n = \frac{1}{T_o}\int_{-T_o/2}^{T_o/2} f(x)\cdot e^{-jn\omega_o x}dx$$

$$= \frac{1}{1}\int_{-1/4}^{1/4} 1\cdot e^{-jn2\pi x}dx \qquad \left[\because \omega_o = \frac{2\pi}{T_o} = 2\pi\right]$$

$$= \left[\frac{e^{-j2\pi nx}}{-j2\pi n}\right]_{-1/4}^{1/4}$$

$$= \frac{1}{-j2\pi n}\left[e^{-j\frac{2\pi n}{4}} - e^{j\frac{2\pi n}{4}}\right]$$

$$= \frac{1}{j2\pi n}\left[e^{j\frac{\pi n}{2}} - e^{j\frac{\pi n}{2}}\right]$$

$$\Rightarrow \qquad C_n = \frac{2j\sin\left(\frac{n\pi}{2}\right)}{j2\pi n}$$

$$= \frac{\sin\left(\frac{n\pi}{2}\right)}{\pi n}$$

$$= \frac{1}{2}Sa\left(\frac{n\pi}{2}\right) \qquad \left[\because \frac{\sin\theta}{\theta} = Sa(\theta)\right]$$

From (i),

$$F(\omega) = 2\pi \sum_{n=-\infty}^{\infty} \frac{1}{2}Sa\left(\frac{n\pi}{2}\right)\cdot\delta(\omega - n\omega_o)$$

$$\Rightarrow \qquad F(\omega) = \pi \sum_{n=-\infty}^{\infty} Sa\left(\frac{n\pi}{2}\right)\cdot\delta(\omega - n\omega_o)$$

■■■■